Problems in Operator Theory

Problems in Operator Theory

Y. A. Abramovich
Indiana University-Purdue University Indianapolis

C. D. Aliprantis
Purdue University

Graduate Studies
in Mathematics
Volume 51

American Mathematical Society
Providence, Rhode Island

2000 *Mathematics Subject Classification.* Primary 46Axx, 46Bxx, 46Gxx, 47Axx, 47Bxx, 47Cxx, 47Dxx, 47Lxx, 28Axx, 28Exx, 15A48, 15A18.

Library of Congress Cataloging-in-Publication Data

Abramovich, Y. A. (Yuri A.)
Problems in operator theory / Y. A. Abramovich, C. D. Aliprantis.
p. cm. — (Graduate studies in mathematics, ISSN 1065-7339 ; v. 51)
Includes bibliographical references and index.
ISBN 0-8218-2147-4 (alk. paper)
1. Operator theory—Problems, exercises, etc. I. Aliprantis, Charalambos D. II. Abramovich, Y. A. (Yuri A.). Invitation to operator theory. III. Title. IV. Series.

QA329.A27 2002
515′.724′076—dc21 2002074421

∞ The paper used in this book is acid-free and falls within the guidelines
established to ensure permanence and durability.
Visit the AMS home page at http://www.ams.org/

10 9 8 7 6 5 4 3 2 1 07 06 05 04 03 02

To the Memory of our Parents

Contents

Foreword xi

Chapter 1. Odds and Ends 1
§1.1. Banach Spaces, Operators, and Linear Functionals 1
§1.2. Banach Lattices and Positive Operators 20
§1.3. Bases in Banach Spaces 31
§1.4. Ultrapowers of Banach Spaces 44
§1.5. Vector-valued Functions 48
§1.6. Fundamentals of Measure Theory 51

Chapter 2. Basic Operator Theory 63
§2.1. Bounded Below Operators 63
§2.2. The Ascent and Descent of an Operator 68
§2.3. Banach Lattices with Order Continuous Norms 71
§2.4. Compact and Weakly Compact Positive Operators 78

Chapter 3. Operators on AL- and AM-spaces 87
§3.1. AL- and AM-spaces 87
§3.2. Complex Banach Lattices 96
§3.3. The Center of a Banach Lattice 105
§3.4. The Predual of a Principal Ideal 111

Chapter 4. Special Classes of Operators 119
§4.1. Finite-rank Operators 119
§4.2. Multiplication Operators 125

§4.3. Lattice and Algebraic Homomorphisms 129
§4.4. Fredholm Operators 134
§4.5. Strictly Singular Operators 139

Chapter 5. Integral Operators 145
§5.1. The Basics of Integral Operators 145
§5.2. Abstract Integral Operators 154
§5.3. Conditional Expectations and Positive Projections 169
§5.4. Positive Projections and Lattice-subspaces 180

Chapter 6. Spectral Properties 189
§6.1. The Spectrum of an Operator 189
§6.2. Special Points of the Spectrum 197
§6.3. The Resolvent of a Positive Operator 201
§6.4. Functional Calculus 205

Chapter 7. Some Special Spectra 215
§7.1. The Spectrum of a Compact Operator 215
§7.2. Turning Approximate Eigenvalues into Eigenvalues 222
§7.3. The Spectrum of a Lattice Homomorphism 230
§7.4. The Order Spectrum of an Order Bounded Operator 232
§7.5. The Essential Spectrum of a Bounded Operator 237

Chapter 8. Positive Matrices 243
§8.1. The Banach Lattices $M_n(\mathbb{R})$ and $M_n(\mathbb{C})$ 243
§8.2. Operators on Finite Dimensional Spaces 251
§8.3. Matrices with Non-negative Entries 262
§8.4. Irreducible Matrices 265
§8.5. The Perron–Frobenius Theorem 268

Chapter 9. Irreducible Operators 273
§9.1. Irreducible and Expanding Operators 273
§9.2. Ideal Irreducibility and the Spectral Radius 283
§9.3. Band Irreducibility and the Spectral Radius 290
§9.4. Krein Operators and $C(\Omega)$-spaces 293

Chapter 10. Invariant Subspaces 299
§10.1. A Smorgasbord of Invariant Subspaces 299
§10.2. The Lomonosov Invariant Subspace Theorem 307

§10.3. Invariant Ideals for Positive Operators 310
§10.4. Invariant Subspaces of Families of Positive Operators 317
§10.5. Compact-friendly Operators 320
§10.6. Positive Operators on Banach Spaces with Bases 329
§10.7. Non-transitive Algebras 331

Chapter 11. The Daugavet Equation 335
§11.1. The Daugavet Equation and Uniform Convexity 335
§11.2. The Daugavet Property in AL- and AM-spaces 352
§11.3. The Daugavet Property in Banach Spaces 356
§11.4. The Daugavet Property in $C(\Omega)$-spaces 359
§11.5. Slices and the Daugavet Property 365
§11.6. Narrow Operators 369
§11.7. Some Applications of the Daugavet Equation 372

Bibliography 375

Index 379

Foreword

This book contains complete solutions to the more than six hundred exercises in the authors' book: **An Invitation to Operator Theory**, American Mathematical Society, 2002. The problems have been spread over eleven chapters following the format of that book. Each problem is identified by a triplet of numbers $x.y.z$; x designates the chapter, y the section, and z the exercise. For instance, Problem 3.4.7 indicates Exercise 7 in Section 4 of Chapter 3.

All solutions are based on the material covered in the text with frequent references to the results in the text. For example, a reference to Theorem 5.9 refers to Theorem 5.9 and a reference to Example 6.21 refers to Example 6.21 in the book **An Invitation to Operator Theory**. We have added an extra amount of material to many solutions in order to make this book as self-contained as possible.

This problem book will be beneficial to students only if they use it "properly," that is to say, if students look at a solution of a problem *only after* trying very hard to solve the problem. Students will do themselves great injustice by reading a solution without any prior attempt on the problem. It should be a real challenge to students to produce solutions which are different from the ones presented here.

Due to the extra material incorporated into the problems, the book can be used as a companion supplement to any text used for the standard functional analysis graduate courses. In addition, this solution book can be used as a reference not only for mathematical subjects but also for other disciplines that rely on functional analytic or measure theoretic techniques.

We would like to express our most sincere thanks to all people who made constructive comments and corrections regarding the text and the problems. Special thanks are due to Professors Arkady Kitover and Vladimir Troitsky who read the solutions and made numerous suggestions and corrections. A final thank you goes to Arlene O'Sean, the AMS Copy Editor, for her excellent job in editing the manuscript.

Y. A. Abramovich and C. D. Aliprantis

Indianapolis and West Lafayette, May 2002

Chapter 1

Odds and Ends

1.1. Banach Spaces, Operators, and Linear Functionals

Problem 1.1.1. *If $\|x+y\| = \|x\|+\|y\|$ for two vectors x and y in a normed space, then show that $\|\alpha x + \beta y\| = \alpha\|x\| + \beta\|y\|$ for all scalars $\alpha, \beta \geq 0$.*

Solution: If $\alpha \geq \beta$, then

$$\begin{aligned} \alpha\|x\| + \beta\|y\| &\geq \|\alpha x + \beta y\| = \|\alpha(x+y) + (\beta - \alpha)y\| \\ &\geq \alpha\|x+y\| - (\alpha - \beta)\|y\| = \alpha\|x\| + \beta\|y\| \,. \end{aligned}$$

Hence, $\|\alpha x + \beta y\| = \alpha\|x\| + \beta\|y\|$ holds for all $\alpha, \beta \geq 0$. ∎

Problem 1.1.2. *Let $T\colon V \to W$ be a surjective one-to-one operator between two vector spaces. Show that T^{-1} (the inverse of T) is a linear operator.*

Solution: Let $S\colon W \to V$ denote the inverse of T. That is, we have $TS = I_W$ and $ST = I_V$. Let $v, w \in W$ and let λ be a scalar.

For the additivity of S note that

$$T[(Sv + Sw)] = TSv + TSw = v + w = T[S(v+w)] \,.$$

Since T is one-to-one, it follows that $S(v+w) = S(v) + S(w)$. That is, S is additive. Similarly, for the homogeneity of S observe that

$$T[S(\lambda w)] = \lambda w = \lambda TSw = T[\lambda Sw] \,,$$

and so $S(\lambda w) = \lambda S(w)$. Therefore, S is a linear operator. ∎

Problem 1.1.3. *If $T\colon X \to Y$ is a bounded operator between normed spaces, then show that*

$$\|T\| = \min\{M \geq 0\colon \ \|Tx\| \leq M\|x\| \ \textit{for all } x \in X\} \,.$$

Solution: Let $M_0 = \inf\{M \geq 0\colon \ \|Tx\| \leq M\|x\| \text{ for all } x \in X\}$, and note that the infimum is, in fact, a minimum. Now if $M \geq 0$ satisfies $\|Tx\| \leq M\|x\|$ for all $x \in X$, then $\|T\| = \sup_{\|x\|=1} \|Tx\| \leq M$, and so $\|T\| \leq M_0$. On the other hand, $\|Tx\| \leq \|T\| \cdot \|x\|$ for each $x \in X$ implies $M_0 \leq \|T\|$. Thus, $\|T\| = M_0$. ∎

Problem 1.1.4. *Show that if X is a normed space and Y is a Banach space, then $\mathcal{L}(X,Y)$ equipped with the operator norm is a Banach space.*

Solution: We assume that $X \neq \{0\}$. Recall that the operator norm on $\mathcal{L}(X,Y)$ is the function $\|\cdot\|\colon \mathcal{L}(X,Y) \to \mathbb{R}$ defined by $\|T\| = \sup_{\|x\|=1} \|Tx\|$.

Clearly, $\|T\| \geq 0$ for each $T \in \mathcal{L}(X,Y)$ and $\|T\| = 0$ if and only if $T = 0$. Also, if $T \in \mathcal{L}(X,Y)$ and α is a scalar, then

$$\|\alpha T\| = \sup_{\|x\|=1} \|(\alpha T)x\| = \sup_{\|x\|=1} |\alpha| \cdot \|Tx\| = |\alpha| \sup_{\|x\|=1} \|Tx\| = |\alpha| \cdot \|T\| .$$

To establish the triangle inequality, let $S, T \in \mathcal{L}(X,Y)$. For each $x \in X$ with $\|x\| = 1$ we have $\|(S+T)x\| \leq \|Sx\| + \|Tx\| \leq \|S\| + \|T\|$, and so

$$\|S+T\| = \sup_{\|x\|=1} \|(S+T)x\| \leq \|S\| + \|T\| .$$

Next, we shall show that $\mathcal{L}(X,Y)$ is a complete normed space. To this end, let $\{T_n\}$ be a Cauchy sequence in $\mathcal{L}(X,Y)$. Fix $\epsilon > 0$ and then choose some n_0 such that $\|T_n - T_m\| < \epsilon$ for all $n, m \geq n_0$. The inequality

$$\|T_n x - T_m x\| \leq \|T_n - T_m\| \cdot \|x\| \leq \epsilon\|x\| \tag{$\star$}$$

is true for all $n, m \geq n_0$, and this shows that $\{T_n x\}$ is a Cauchy sequence in Y for each $x \in X$. Since Y is a Banach space, the sequence $\{T_n x\}$ converges in Y for each $x \in X$. Let $Tx = \lim T_n x$. This defines a linear operator $T\colon X \to Y$. Letting $m \to \infty$ in ($\star$) yields

$$\|T_n x - Tx\| \leq \epsilon\|x\| \tag{$\star\star$}$$

for all $n \geq n_0$. From $\|Tx\| \leq \|Tx - T_{n_0}x\| + \|T_{n_0}x\| \leq (\epsilon + \|T_{n_0}\|)\|x\|$, it follows that $T \in \mathcal{L}(X,Y)$. Now a glance at ($\star\star$) guarantees that $\|T_n - T\| \leq \epsilon$ for all $n \geq n_0$. That is, $T_n \to T$ holds in $\mathcal{L}(X,Y)$, and so $\mathcal{L}(X,Y)$ is a Banach space. ∎

Problem 1.1.5. *Prove the following converse of the preceding problem. If X is a non-trivial normed space, Y is another normed space and $\mathcal{L}(X,Y)$ is a Banach space, then Y is likewise a Banach space.*

Solution: Assume that $\mathcal{L}(X,Y)$ is a Banach space. Fix a non-zero continuous linear functional $f\colon X \to \mathbb{R}$ and then choose some $x_0 \in X$ such that $f(x_0) = 1$. Now let $\{y_n\}$ be a Cauchy sequence of Y.

For each n define the bounded operator $T_n\colon X \to Y$ by $T_n(x) = f(x)y_n$. From

$$\|T_n x - T_m x\| = |f(x)| \cdot \|y_n - y_m\| \leq \|y_n - y_m\| \cdot \|f\| \cdot \|x\| ,$$

we see that $\|T_n - T_m\| \leq \|f\| \cdot \|y_n - y_m\|$ for all n and m. This implies that $\{T_n\}$ is a Cauchy sequence in $\mathcal{L}(X,Y)$. Since $\mathcal{L}(X,Y)$ is a Banach space, there exists some $T \in \mathcal{L}(X,Y)$ such that $T_n \to T$. In particular, we have

$$y_n = f(x_0)y_n = T_n(x_0) \to T(x_0) .$$

This shows that Y is a Banach space. ∎

Problem 1.1.6. *Show that the complexification X_c of a real vector space X under the operations*

$$\begin{aligned}(x_1 + \imath y_1) + (x_2 + \imath y_2) &= x_1 + x_2 + \imath(y_1 + y_2), \text{ and} \\ (\alpha + \imath\beta)(x + \imath y) &= \alpha x - \beta y + \imath(\beta x + \alpha y)\end{aligned}$$

is a complex vector space.

Solution: The addition axioms can be verified as follows:

$$(x_1 + \imath y_1) + (x_2 + \imath y_2) = x_1 + x_2 + \imath(y_1 + y_2) = x_2 + x_1 + \imath(y_2 + y_1) = (x_2 + \imath y_2) + (x_1 + \imath y_1),$$

$$\begin{aligned}[(x_1 + \imath y_1) + (x_2 + \imath y_2)] + (x_3 + \imath y_3) &= [x_1 + x_2 + \imath(y_1 + y_2)] + (x_3 + \imath y_3) \\ &= (x_1 + x_2 + x_3) + \imath(y_1 + y_2 + y_3) \\ &= (x_1 + \imath y_1) + [(x_2 + \imath y_2) + (x_3 + \imath y_3)],\end{aligned}$$

$$(x_1 + \imath y_1) + (0 + \imath 0) = x_1 + \imath y_1,$$

$$(x_1 + \imath y_1) + [(-x_1) + \imath(-y_1)] = 0 + \imath 0.$$

The multiplication axioms can be verified as follows:

$$\begin{aligned}&[(\alpha + \imath\beta)(\gamma + \imath\delta)](x + \imath y) \\ &\quad = [(\alpha\gamma - \beta\delta) + \imath(\beta\gamma + \alpha\delta)](x + \imath y) \\ &\quad = [(\alpha\gamma - \beta\delta)x - (\beta\gamma + \alpha\delta)y] + \imath[(\alpha\gamma - \beta\delta)y + (\beta\gamma + \alpha\delta)x] \\ &\quad = [\alpha(\gamma x - \delta y) - \beta(\delta x + \gamma y)] + \imath[\beta(\gamma x - \delta y) + \alpha(\delta x + \gamma y)] \\ &\quad = (\alpha + \imath\beta)[(\gamma x - \delta y) + \imath(\delta x + \gamma y)] \\ &\quad = (\alpha + \imath\beta)[(\gamma + \imath\delta)(x + \imath y)],\end{aligned}$$

$$\begin{aligned}&[(\alpha + \imath\beta) + (\gamma + \imath\delta)](x + \imath y) \\ &\quad = [(\alpha + \gamma) + \imath(\beta + \delta)](x + \imath y) \\ &\quad = [(\alpha + \gamma)x - (\beta + \delta)y] + \imath[(\alpha + \gamma)y + (\beta + \delta)x] \\ &\quad = [(\alpha x - \beta y) + \imath(\alpha y + \beta x)] + [(\gamma x - \delta y) + \imath(\gamma y + \delta x)] \\ &\quad = (\alpha + \imath\beta)(x + \imath y) + (\gamma + \imath\delta)(x + \imath y),\end{aligned}$$

$$\begin{aligned}&(\alpha + \imath\beta)[(x_1 + \imath y_1) + (x_2 + \imath y_2)] \\ &\quad = (\alpha + \imath\beta)[(x_1 + x_2) + \imath(y_1 + y_2)] \\ &\quad = [\alpha(x_1 + x_2) - \beta(y_1 + y_2)] + \imath[\alpha(y_1 + y_2) + \beta(x_1 + x_2)] \\ &\quad = [(\alpha x_1 - \beta y_1) + \imath(\alpha y_1 + \beta x_1)] + [(\alpha x_2 - \beta y_2) + \imath(\alpha y_2 + \beta x_2)] \\ &\quad = (\alpha + \imath\beta)(x_1 + \imath y_1) + (\alpha + \imath\beta)(x_2 + \imath y_2),\end{aligned}$$

$$(1 + \imath 0)(x + \imath y) = x + \imath y.$$

Therefore, X_c is a complex vector space. ∎

Problem 1.1.7. *Let X be a real normed space with norm $\|\cdot\|$. Show that the function $\|\cdot\|_{\mathbf{c}}\colon X_{\mathbf{c}} \to \mathbb{R}$ defined via the formula*

$$\|z\|_{\mathbf{c}} = \sup_{\theta\in[0,2\pi]} \|x\cos\theta + y\sin\theta\|, \quad z = x + \imath y \in X_{\mathbf{c}},$$

is a norm on $X_{\mathbf{c}}$ that extends $\|\cdot\|$. Also, show that

$$\tfrac{1}{2}(\|x\| + \|y\|) \le \|z\|_{\mathbf{c}} \le \|x\| + \|y\|$$

for each $z = x + \imath y \in X_{\mathbf{c}}$. Use the preceding inequality to verify that:

(a) *A sequence $\{z_n\}$, where $z_n = x_n + \imath y_n$, satisfies $\|z_n\|_{\mathbf{c}} \to 0$ if and only if $\|x_n\| \to 0$ and $\|y_n\| \to 0$.*

(b) *If X is a Banach space, then $X_{\mathbf{c}}$ is also a Banach space.*

Solution: Let $z = x + \imath y \in X_{\mathbf{c}}$. Since for each $\phi \in \mathbb{R}$ there exists some $\theta \in [0, 2\pi]$ with $\cos\theta = \cos\phi$ and $\sin\theta = \sin\phi$, it should be clear that

$$\|z\|_{\mathbf{c}} = \sup_{\theta\in[0,2\pi]} \|x\cos\theta + y\sin\theta\| = \sup_{\theta\in\mathbb{R}} \|x\cos\theta + y\sin\theta\|.$$

From the inequality $\|x\cos\theta + y\sin\theta\| \le |\cos\theta|\cdot\|x\| + |\sin\theta|\cdot\|y\| \le \|x\| + \|y\|$, we see that $\|z\|_{\mathbf{c}} \le \|x\| + \|y\|$. In addition, from the inequalities

$$\|x\| = \|x\cos 0 + x\sin 0\| \le \|z\|_{\mathbf{c}} \quad\text{and}\quad \|y\| = \|x\cos\tfrac{\pi}{2} + y\sin\tfrac{\pi}{2}\| \le \|z\|_{\mathbf{c}},$$

we get $\|x\| + \|y\| \le 2\|z\|_{\mathbf{c}}$ or $\frac{1}{2}(\|x\| + \|y\|) \le \|z\|_{\mathbf{c}}$. Therefore,

$$\tfrac{1}{2}(\|x\| + \|y\|) \le \|z\|_{\mathbf{c}} \le \|x\| + \|y\|. \qquad (\star)$$

Clearly $\|z\|_{\mathbf{c}} \ge 0$ for each $z \in X_{\mathbf{c}}$, and from $(\star)$ we see that $\|z\|_{\mathbf{c}} = 0$ if and only if $z = 0$. For the triangle inequality note that if $z_1 = x_1 + \imath y_1, z_2 = x_2 + \imath y_2 \in X_{\mathbf{c}}$, then $z_1 + z_2 = x_1 + x_2 + \imath(y_1 + y_2)$, and so from

$$\begin{aligned}\big\|(x_1 + x_2)\cos\theta + (y_1 + y_2)\sin\theta\big\| &\le \|x_1\cos\theta + y_1\sin\theta\| + \|x_2\cos\theta + y_2\sin\theta\| \\ &\le \|z_1\|_{\mathbf{c}} + \|z_2\|_{\mathbf{c}},\end{aligned}$$

it follows that

$$\|z_1 + z_2\|_{\mathbf{c}} = \sup_{\theta\in\mathbb{R}}\big\|(x_1 + x_2)\cos\theta + (y_1 + y_2)\sin\theta\big\| \le \|z_1\|_{\mathbf{c}} + \|z_2\|_{\mathbf{c}}.$$

Next, we shall prove the homogeneity of $\|\cdot\|_{\mathbf{c}}$. To this end, fix $z = x + \imath y \in X_{\mathbf{c}}$ and a complex number $\alpha + \imath\beta \ne 0$. Then $(\alpha + \imath\beta)z = \alpha x - \beta y + \imath(\beta x + \alpha y)$. Notice that the point $\big(\frac{\beta}{\sqrt{\alpha^2+\beta^2}}, \frac{\alpha}{\sqrt{\alpha^2+\beta^2}}\big)$ lies on the unit circle. So, there exists a unique angle $0 \le \omega < 2\pi$ such that

$$\cos\omega = \frac{\beta}{\sqrt{\alpha^2+\beta^2}} \quad\text{and}\quad \sin\omega = \frac{\alpha}{\sqrt{\alpha^2+\beta^2}}.$$

Now observe that

$$\begin{aligned}\alpha\cos\theta + \beta\sin\theta &= \sqrt{\alpha^2 + \beta^2}\Big(\frac{\alpha}{\sqrt{\alpha^2+\beta^2}}\cos\theta + \frac{\beta}{\sqrt{\alpha^2+\beta^2}}\sin\theta\Big) \\ &= \sqrt{\alpha^2 + \beta^2}\,(\sin\omega\cos\theta + \cos\omega\sin\theta) \\ &= \sqrt{\alpha^2 + \beta^2}\,\sin(\omega + \theta) = |\alpha + \imath\beta|\sin(\omega + \theta),\end{aligned}$$

and similarly

$$\begin{aligned}\alpha\sin\theta-\beta\cos\theta &= \sqrt{\alpha^2+\beta^2}\,(\sin\omega\sin\theta-\cos\omega\cos\theta)\\ &= -\sqrt{\alpha^2+\beta^2}\,\cos(\omega+\theta)=-|\alpha+\imath\beta|\cos(\omega+\theta)\,.\end{aligned}$$

Consequently, we have

$$\begin{aligned}&\big\|(\alpha x-\beta y)\cos\theta+(\beta x+\alpha y)\sin\theta\big\|\\ &\qquad=\big\|(\alpha\cos\theta+\beta\sin\theta)x+(\alpha\sin\theta-\beta\cos\theta)y\big\|\\ &\qquad=\big\|\big[|\alpha+\imath\beta|\sin(\omega+\theta)\big]x-\big[|\alpha+\imath\beta|\cos(\omega+\theta)\big]y\big\|\\ &\qquad=|\alpha+\imath\beta|\cdot\|x\cos(\omega+\theta)-y\sin(\omega+\theta)\|\,,\end{aligned}$$

from which it follows that

$$\begin{aligned}\|(\alpha+\imath\beta)z\|_{\mathbf{c}} &= \sup_{\theta\in\mathbb{R}}\big\|(\alpha x-\beta y)\cos\theta+(\beta x+\alpha y)\sin\theta\big\|\\ &= \sup_{\theta\in\mathbb{R}}|\alpha+\imath\beta|\cdot\|x\cos(\omega+\theta)-y\sin(\omega+\theta)\|\\ &= |\alpha+\imath\beta|\cdot\sup_{\theta\in\mathbb{R}}\|x\cos(\omega+\theta)-y\sin(\omega+\theta)\|\\ &= |\alpha+\imath\beta|\cdot\|z\|_{\mathbf{c}}\,.\end{aligned}$$

Properties (a) and (b) follow immediately from $(\star)$. ∎

Problem 1.1.8. *Let X be an arbitrary real vector space with complexification $X_{\mathbf{C}}=X\oplus\imath X$. Establish the following properties.*

(a) *If V is a subspace of X, then $V_{\mathbf{C}}=V\oplus\imath V=\{x+\imath y\colon\ x,y\in V\}$ is a vector subspace of $X_{\mathbf{C}}$.*

(b) *If X is also a normed space and V is a vector subspace of X, then the vector subspace $V_{\mathbf{C}}=V\oplus\imath V$ of $X_{\mathbf{C}}$ is closed in $X_{\mathbf{C}}$ if and only if V is a closed subspace of X.*

(c) *Let $T\colon X\to X$ be an operator. A subspace V of X is said to be **invariant under** T (or simply T-**invariant**) whenever $T(V)\subseteq V$.*

A vector subspace V of X is T-invariant if and only if the vector subspace $V_{\mathbf{C}}=V\oplus\imath V$ of $X_{\mathbf{C}}$ is $T_{\mathbf{C}}$-invariant.

Solution: (a) Let $z_1=x_1+\imath y_1, z_2=x_2+\imath y_2\in V_{\mathbf{C}}$ and let $\alpha+\imath\beta$ be a complex number. Then we have

$$\begin{aligned}z_1+z_2 &= (x_1+\imath y_1)+(x_2+\imath y_2)=x_1+x_2+\imath(y_1+y_2)\in V_{\mathbf{c}}\,,\ \text{ and}\\ (\alpha+\imath\beta)z_1 &= (\alpha x_1-\beta y_1)+\imath(\beta x_1+\alpha y_1)\in V_{\mathbf{c}}\,.\end{aligned}$$

Therefore, $V_{\mathbf{c}}$ is a vector subspace of $X_{\mathbf{c}}$.

(b) This follows from the fact that a sequence $\{z_n\}$, where $z_n=x_n+\imath y_n$, in $X_{\mathbf{c}}$ satisfies $z_n\to x+\imath y$ in $X_{\mathbf{c}}$ if and only if $x_n\to x$ and $y_n\to y$ hold in X; see the preceding problem.

(c) If $T(V) \subseteq V$, then $T_{\mathbf{c}}(V_{\mathbf{c}}) \subseteq V_{\mathbf{c}}$ is trivially true. Now assume $T_{\mathbf{c}}(V_{\mathbf{c}}) \subseteq V_{\mathbf{c}}$ and let $x \in V$. Then $x + \imath 0 \in V_{\mathbf{c}}$, and so $T_{\mathbf{c}}(x + \imath 0) = Tx + \imath T0 = Tx + \imath 0 \in V_{\mathbf{c}} = V \oplus \imath V$. This implies $Tx \in V$, and thus $T(V) \subseteq V$. ∎

Problem 1.1.9. *Let X be a real vector space. It is very common to denote the vectors $x + \imath y$ of $X_{\mathbf{C}}$ as column vectors $\begin{bmatrix} x \\ y \end{bmatrix}$. Under this notation the vector space operations of $X_{\mathbf{C}}$ are given by*

$$\begin{bmatrix} x_1 \\ y_1 \end{bmatrix} + \begin{bmatrix} x_2 \\ y_2 \end{bmatrix} = \begin{bmatrix} x_1 + x_2 \\ y_1 + y_2 \end{bmatrix} \quad \textit{and} \quad (\alpha + \imath\beta)\begin{bmatrix} x \\ y \end{bmatrix} = \begin{bmatrix} \alpha x - \beta y \\ \beta x + \alpha y \end{bmatrix}.$$

*Now assume that X and Y are two real vector spaces. Show that a mapping $\mathcal{T}\colon X_{\mathbf{C}} \to Y_{\mathbf{C}}$ is a linear operator if and only if there exist two (uniquely determined) operators $S, T\colon X \to Y$ such that $\mathcal{T}$ has the **matrix representation** $\mathcal{T} = \begin{bmatrix} T & -S \\ S & T \end{bmatrix}$, where as usual*

$$\mathcal{T}(x + \imath y) = \begin{bmatrix} T & -S \\ S & T \end{bmatrix}\begin{bmatrix} x \\ y \end{bmatrix} = \begin{bmatrix} Tx - Sy \\ Sx + Ty \end{bmatrix}.$$

Also, show that if X and Y are normed spaces, then $\mathcal{T}$ is a bounded operator if and only if both S and T are bounded operators.

Solution: A straightforward verification shows that every mapping represented by a matrix of the form $\begin{bmatrix} T & -S \\ S & T \end{bmatrix}$, where $S, T\colon X \to Y$ are linear operators, defines a linear operator from $X_{\mathbf{C}}$ to $Y_{\mathbf{C}}$.

Now consider a linear operator $\mathcal{T}\colon X_{\mathbf{C}} \to Y_{\mathbf{C}}$. For each $x \in X$ we can write

$$\mathcal{T}(x + \imath 0) = Tx + \imath Sx\,,$$

where Tx and Sx belong to Y. A direct computation shows that T and S define linear operators from X to Y. Moreover, the linearity of $\mathcal{T}$ implies

$$\begin{aligned} \mathcal{T}(x + \imath y) &= \mathcal{T}x + \imath\mathcal{T}y = (Tx + \imath Sx) + \imath(Ty + \imath Sy) \\ &= Tx - Sy + \imath(Sx + Ty) = \begin{bmatrix} T & -S \\ S & T \end{bmatrix}\begin{bmatrix} x \\ y \end{bmatrix}. \end{aligned}$$

Now assume that $\mathcal{T} = \begin{bmatrix} T & -S \\ S & T \end{bmatrix}$ is a bounded operator. Then for each $x \in X$ we have

$$\|Tx\| = \|(Tx)\cos 0 + (Sx)\sin 0\| \leq \|\mathcal{T}(x + \imath 0)\| \leq \|\mathcal{T}\| \cdot \|x\|\,,$$

which shows that T is a bounded operator. Similarly, S is a bounded operator. Finally, suppose that S and T are bounded operators. Fix $z = x + \imath y \in X_{\mathbf{C}}$ and note that for each angle θ we have

$$\begin{aligned} &\|(Tx - Sy)\cos\theta + (Sx + Ty)\sin\theta\| \\ &\qquad = \|T(x\cos\theta + y\sin\theta) + S(x\sin\theta - y\cos\theta)\| \\ &\qquad \leq \|T\| \cdot \|x\cos\theta + y\sin\theta\| + \|S\| \cdot \|x\sin\theta - y\cos\theta\| \\ &\qquad \leq (\|T\| + \|S\|)\|z\|, \end{aligned}$$

and so

$$\|\mathcal{T}z\| = \sup_{\theta\in\mathbb{R}} \|(Tx - Sy)\cos\theta + (Sx + Ty)\sin\theta\| \leq (\|T\| + \|S\|)\|z\|.$$

This shows that $\mathcal{T}$ is norm bounded. ∎

Problem 1.1.10. *A bounded operator $T\colon X \to Y$ between normed spaces is said to be* ***invertible*** *if there exists (a uniquely determined) bounded operator $S\colon Y \to X$ satisfying $ST = I_X$ and $TS = I_Y$; the operator S is called the inverse of T and is denoted T^{-1}.*

For an operator $T\colon X \to Y$ between real vector spaces establish the following properties.

(a) *The operator T is surjective if and only if $T_{\mathbf{c}}\colon X_{\mathbf{c}} \to Y_{\mathbf{c}}$ is surjective.*

(b) *The operator T is one-to-one if and only if $T_{\mathbf{c}}$ is one-to-one.*

(c) *The operator T is invertible if and only if $T_{\mathbf{c}}$ is invertible. Also, show that if T is invertible, then $(T_{\mathbf{c}})^{-1} = (T^{-1})_{\mathbf{c}}$.*

Solution: (a) Assume that T is surjective and let $z = x + \imath y \in Y_{\mathbf{c}}$. Pick $u, v \in X$ such that $Tu = x$ and $Tv = y$ and note that $T_{\mathbf{c}}(u + \imath v) = Tu + \imath Tv = x + \imath y = z$. That is, $T_{\mathbf{c}}$ is surjective. Conversely, if $T_{\mathbf{c}}$ is surjective and $y \in Y$, then there exists some $z = u + \imath v \in X_{\mathbf{c}}$ such that $T_{\mathbf{c}}x = Tu + \imath Tv = y + \imath 0$. So, $Tu = y$ which proves that T is surjective.

(b) Assume that T is one-to-one. If $z = x + \imath y$ satisfies $Tx + \imath Ty = T_{\mathbf{c}}z = 0$, then $Tx = Ty = 0$, and so $x = y = 0$ or $z = x + \imath y = 0$. That is, $T_{\mathbf{c}}$ is one-to-one. Conversely, if $T_{\mathbf{c}}$ is one-to-one and $Tx = 0$, then $T_{\mathbf{c}}(x + \imath 0) = 0$, from which it follows that $x + \imath 0 = 0$ or $x = 0$. Thus, T is one-to-one.

(c) The first claim follows from (a) and (b) above. For the last part note that if $x + \imath y \in X_{\mathbf{c}}$ and $u + \imath v \in Y_{\mathbf{c}}$, then

$$\begin{aligned} (T^{-1})_{\mathbf{c}}T_{\mathbf{c}}(x + \imath y) &= (T^{-1})_{\mathbf{c}}(Tx + \imath Ty) = T^{-1}(Tx) + \imath T^{-1}(Ty) = x + \imath y, \text{ and} \\ T_{\mathbf{c}}(T^{-1})_{\mathbf{c}}(u + \imath v) &= T_{\mathbf{c}}(T^{-1}u + \imath T^{-1}v) = T(T^{-1}u) + \imath T(T^{-1}v) = u + \imath v. \end{aligned}$$

Therefore, $T_{\mathbf{c}}$ is invertible and $(T_{\mathbf{c}})^{-1} = (T^{-1})_{\mathbf{c}}$. ∎

Problem 1.1.11. *Let $X = X_1 \oplus \cdots \oplus X_n$ be the direct sum Banach space. Show that a mapping $A\colon X \to X$ is a bounded linear operator if and only if there exist (uniquely determined) bounded linear operators $A_{ij}\colon X_j \to X_i$ $(i, j = 1, 2, \ldots, n)$ such that A has the matrix representation $A = [A_{ij}]$,* *where for each* $x = x_1 \oplus \cdots \oplus x_n$ *we have*

$$\begin{aligned} Ax &= \begin{bmatrix} A_{11} & A_{12} & \cdots & A_{1n} \\ A_{21} & A_{22} & \cdots & A_{2n} \\ \vdots & \vdots & \ddots & \vdots \\ A_{n1} & A_{n2} & \cdots & A_{nn} \end{bmatrix} \begin{bmatrix} x_1 \\ x_2 \\ \vdots \\ x_n \end{bmatrix} = \begin{bmatrix} \sum_{j=1}^n A_{1j}x_j \\ \sum_{j=1}^n A_{2j}x_j \\ \vdots \\ \sum_{j=1}^n A_{nj}x_j \end{bmatrix} \\ &= \Big(\sum_{j=1}^n A_{1j}x_j\Big) \oplus \Big(\sum_{j=1}^n A_{2j}x_j\Big) \oplus \cdots \oplus \Big(\sum_{j=1}^n A_{nj}x_j\Big). \end{aligned}$$

Solution: To see this, let $A\colon X \to X$ be a bounded operator. For each j define the bounded operator $T_j\colon X_j \to X$ via the formula

$$T_j(x_j) = A(0 \oplus 0 \oplus \cdots \oplus 0 \oplus x_j \oplus 0 \oplus \cdots \oplus 0),$$

where the vector x_j occupies the jth coordinate. We can write

$$T_j(x_j) = A(0 \oplus 0 \oplus \cdots \oplus 0 \oplus x_j \oplus 0 \oplus \cdots \oplus 0) = A_{1j}x_j \oplus A_{2j}x_j \oplus \cdots \oplus A_{nj}x_j .$$

An easy argument shows that each $A_{ij}\colon X_j \to X_i$ is a linear operator, and A is bounded if and only if every A_{ij} is a bounded operator. Moreover, for each $x = x_1 \oplus \cdots \oplus x_n$ in X we have

$$\begin{aligned} A(x) &= A\Big(\sum_{j=1}^n (0 \oplus 0 \oplus \cdots \oplus 0 \oplus x_j \oplus 0 \oplus \cdots \oplus 0)\Big) \\ &= \sum_{j=1}^n A(0 \oplus 0 \oplus \cdots \oplus 0 \oplus x_j \oplus 0 \oplus \cdots \oplus 0) \\ &= \Big(\sum_{j=1}^n A_{1j}x_j\Big) \oplus \Big(\sum_{j=1}^n A_{2j}x_j\Big) \oplus \cdots \oplus \Big(\sum_{j=1}^n A_{nj}x_j\Big) \\ &= \begin{bmatrix} A_{11} & A_{12} & \cdots & A_{1n} \\ A_{21} & A_{22} & \cdots & A_{2n} \\ \vdots & \vdots & \ddots & \vdots \\ A_{n1} & A_{n2} & \cdots & A_{nn} \end{bmatrix} \begin{bmatrix} x_1 \\ x_2 \\ \vdots \\ x_n \end{bmatrix}. \end{aligned}$$

This establishes the desired representation of the operator A. The converse is straightforward and is left to the reader. ∎

Problem 1.1.12. *A (not necessarily surjective) operator* $T\colon X \to Y$ *between normed spaces is said to be an **isometry** (or more precisely a **linear isometry**) if* $\|Tx\| = \|x\|$ *holds for each* $x \in X$.

Show that a bounded operator $T\colon X \to Y$ *between two normed spaces is an isometry if and only if its double adjoint* $T^{**}\colon X^{**} \to Y^{**}$ *is likewise an isometry.*

Solution: Assume that $T\colon X \to Y$ is an isometry. Clearly, $\|T\| = \|T^{**}\| = 1$ and so $\|T^{**}x^{**}\| \leq \|x^{**}\|$ for all $x^{**} \in X^{**}$. Now fix $x^* \in X^*$ with $\|x^*\| \leq 1$. Then the formula $y^*(Tx) = \langle x, x^*\rangle$ defines a continuous linear functional on the range of T such that $\|y^*\| \leq 1$. Let y^* also denote a continuous linear extension of y^* to all of Y such that $\|y^*\| \le 1$; clearly $T^*y^* = x^*$. Now note that for each $x^{**} \in X^{**}$ we have

$$|\langle x^*, x^{**}\rangle| = |\langle T^*y^*, x^{**}\rangle| = |\langle y^*, T^{**}x^{**}\rangle| \le \|y^*\| \cdot \|T^{**}x^{**}\| \le \|T^{**}x^{**}\|,$$

and so $\|x^{**}\| = \sup_{\|x^*\| \le 1} |\langle x^*, x^{**}\rangle| \le \|T^{**}x^{**}\|$. Thus, $\|T^{**}x^{**}\| = \|x^{**}\|$ hold for each $x^{**} \in X^{**}$.

For the converse, assume that T^{**} is an isometry. Then $\|Tx\| = \|T^{**}x\| = \|x\|$ holds for each $x \in X$ so that T is an isometry. ∎

Problem 1.1.13. *Let $T\colon X \to X$ be a bounded operator on a normed space. The inequality $\|Tx - Ty\| \le \|T\| \cdot \|x - y\|$ implies that T is a uniformly continuous mapping. So, T has a unique (linear) continuous extension $\hat{T}\colon \hat{X} \to \hat{X}$, where $\hat{X}$ denotes the norm completion of X. Establish the following properties:*

(a) *If T is an isometry on X, then so is $\hat{T}$ on $\hat{X}$.*
(b) *If T is an invertible isometry with inverse S, then S is also a (linear) isometry.*
(c) *If T is invertible, then so is $\hat{T}$ and $(\hat{T})^{-1} = (T^{-1})\hat{\ }$.*
(d) *If $\lambda - T$ is invertible on X for some scalar λ, then $\lambda - \hat{T}$ is likewise invertible on $\hat{X}$.*
(e) *If $\hat{T}$ is an invertible isometry on $\hat{X}$, is T necessarily an invertible isometry on X?*

Solution: (a) Assume that T is an isometry and let $\hat{x} \in \hat{X}$. Pick a sequence $\{x_n\}$ in X such that $\|x_n - \hat{x}\| \to 0$. Then we have

$$\|\hat{T}\hat{x}\| = \lim_{n\to\infty} \|Tx_n\| = \lim_{n\to\infty} \|x_n\| = \|\hat{x}\|.$$

That is, $\hat{T}$ is an isometry.

(b) Assume that $T\colon X \to X$ is an isometry and that a bounded operator $S\colon X \to X$ satisfies $ST = TS = I$, i.e., $S = T^{-1}$. Now notice that for each $x \in X$ we have

$$\|x\| = \|T(Sx)\| = \|Sx\|.$$

This shows that S is an isometry.

(c) Assume that $S\colon X \to X$ is a bounded operator satisfying $ST = TS = I$ and let $\hat{S}$ denotes the unique continuous linear extension of S to all of $\hat{X}$. Fix some $\hat{x} \in \hat{X}$ and choose a sequence $\{x_n\}$ in X satisfying $\|x_n - \hat{x}\| \to 0$. Then

$$\hat{S}\hat{T}\hat{x} = \lim_{n\to\infty} \hat{S}(Tx_n) = \lim_{n\to\infty} STx_n = \lim_{n\to\infty} x_n = \hat{x},$$

and similarly $\hat{T}\hat{S}\hat{x} = \hat{x}$. These show that the operator $\hat{S}$ is the inverse of $\hat{T}$ and that $(\hat{T})^{-1} = \hat{S} = (T^{-1})\hat{\ }$.

(d) Note first that $(\lambda - T)\hat{\ } = \lambda - \hat{T}$. Now by part (c), we see that if $\lambda - T\colon X \to X$ is invertible, then $(\lambda - T)\hat{\ } = \lambda - \hat{T}\colon \hat{X} \to \hat{X}$ is likewise invertible.

(e) Let X denote the vector space of all polynomials defined on $[0,1]$ equipped with the sup norm. By the Stone–Weierstrass theorem, $\hat{X} = C[0,1]$. Consider the operator $T\colon X \to X$ defined by $Tp(t) = p(t^2)$ for all $p \in X$ and all $t \in [0,1]$. Clearly, T is a linear isometry, and T is not invertible since T is not surjective (its range consists of even polynomials). However, $\hat{T}\colon C[0,1] \to C[0,1]$ satisfies $\hat{T}x(t) = x(t^2)$ for all $x \in C[0,1]$ and each $t \in [0,1]$, which is an invertible isometry on $\hat{X}$. ∎

Problem 1.1.14. *Show that for bounded operators $X \xrightarrow{T} Y \xrightarrow{S} Z$ between normed spaces (all over the same field) we have $\|ST\| \le \|S\| \cdot \|T\|$. Use this inequality to show that:*

(a) *The map $(S,T) \mapsto ST$, from $\mathcal{L}(Y,Z) \times \mathcal{L}(X,Y)$ to $\mathcal{L}(X,Z)$, is jointly norm continuous.*

(b) *The Banach space $\mathcal{L}(U,V)$ of all bounded operators from a normed space U to a Banach space V is a unital Banach algebra.*

Solution: Notice that if $X \xrightarrow{T} Y \xrightarrow{S} Z$ are bounded operators and $x \in X$, then

$$\|STx\| \le \|S\| \cdot \|Tx\| \le \|S\| \cdot \|T\| \cdot \|x\| .$$

This implies $\|ST\| \le \|S\| \cdot \|T\|$.

For the joint continuity of the map $(S,T) \mapsto ST$ use the inequality

$$\begin{aligned} \|ST - S_0T_0\| &= \|ST - ST_0 + ST_0 - S_0T_0\| \\ &= \|S(T-T_0) + (S-S_0)T_0\| \\ &\le \|T-T_0\| \cdot \|S\| + \|S-S_0\| \cdot \|T_0\| \end{aligned}$$

and the fact that convergent sequences are norm bounded. ∎

Problem 1.1.15. *Let X and Y be two normed spaces over the same field. Show that the mapping $T \mapsto T^*$, from $\mathcal{L}(X,Y)$ into $\mathcal{L}(Y^*,X^*)$, is a linear isometry.*

Solution: We have that

$$\begin{aligned} \|T\| &= \sup_{\|x\|=1} \|Tx\| = \sup_{\|x\|=1} \Big[\sup_{\|y^*\|=1} |\langle y^*, Tx\rangle| \Big] \\ &= \sup_{\|y^*\|=1} \Big[\sup_{\|x\|=1} |\langle T^*y^*, x\rangle| \Big] = \sup_{\|y^*\|=1} \|T^*y^*\| \\ &= \|T^*\| , \end{aligned}$$

as desired. ∎

Problem 1.1.16. *Show that every continuous linear functional defined on a subspace of a complex normed vector space X has a continuous linear extension to all of X with preservation of its norm.*

Solution: Let $f\colon M \to \mathbb{C}$ be a continuous linear functional defined on a vector subspace of a complex Banach space X. Then, by the discussion preceding Theorem 1.9, we know that there exists a unique continuous linear functional $g\colon M_{\mathbb{R}} \to \mathbb{R}$

satisfying $\|g\| = \|f\|$ and

$$f(x) = g(x) - \imath g(\imath x)$$

for each $x \in M$. Notice that $M_{\mathbb{R}}$ is a vector subspace of the real normed space $X_{\mathbb{R}}$. So, by the classical Hahn–Banach theorem, there exists a continuous linear extension of g to all of $X_{\mathbb{R}}$, say $\hat{g}\colon X_{\mathbb{R}} \to \mathbb{R}$, such that $\|\hat{g}\| = \|g\|$. Now observe that the linear functional $\hat{f}\colon X \to \mathbb{C}$, defined by

$$\hat{f}(x) = \hat{g}(x) - \imath\hat{g}(\imath x)\,,\ x \in X\,,$$

is a linear extension of f to all of X satisfying $\|\hat{f}\| = \|\hat{g}\| = \|g\| = \|f\|$. ∎

Problem 1.1.17. *Let A be a non-empty subset of a normed space X and let $x_0 \in X$. We say that there is a* ***nearest*** *(or a* ***closest****)* ***point*** *to x_0 in A if there exists some $a_0 \in A$ such that $\|x_0 - a_0\| \le \|x_0 - a\|$ for all $a \in A$. Establish the following.*

(a) *The norm function $x \mapsto \|x\|$ is weakly lower semicontinuous, i.e., $x_\alpha \xrightarrow{w} x$ in X implies $\|x\| \le \liminf \|x_\alpha\|$.*

(b) *If X is a reflexive Banach space and A is a non-empty weakly closed subset of X, then for every point $x_0 \in X$ there exists a nearest point to x_0 in A.*

(c) *Assume that the norm of X is* ***uniformly convex****, that is, for each $0 < \epsilon \le 2$ there exists some $0 < \delta < 1$ such that $\|x - y\| \ge \epsilon$ with $x, y \in U_X$ implies $\|\frac{x+y}{2}\| < 1 - \delta$. A normed space is called* ***uniformly convex*** *if its norm is uniformly convex.*

If A is a non-empty, convex, and closed subset of a uniformly convex Banach space X, then for every $x_0 \in X$ there exists exactly one point in A that is nearest to x_0. In other words, there exists a mapping $\pi\colon X \to A$ defined by

$$\big\|x - \pi(x)\big\| = \min_{y \in A} \|x - y\|\,.$$

The mapping π is called the ***nearest point mapping*** *from X to A.*

(d) *Let C be a non-empty closed convex subset of a uniformly convex Banach space X. Show that the nearest point mapping π is continuous.*

Solution: (a) Assume $x_\alpha \xrightarrow{w} x$ in X, and let $x^* \in X^*$ satisfy $\|x^*\| \le 1$. Then for each α we have $|x^*(x_\alpha)| \le \|x^*\| \cdot \|x_\alpha\| \le \|x_\alpha\|$, and so

$$|x^*(x)| = \lim |x^*(x_\alpha)| \le \liminf \|x_\alpha\|\,.$$

This implies $\|x\| = \sup_{\|x^*\| \le 1} |x^*(x)| \le \liminf \|x_\alpha\|$.

(b) Assume that X is a reflexive Banach space. Also, we can assume that $x_0 = 0$. Put $d = \inf_{a \in A} \|a\|$, and then pick some sequence $\{a_n\} \subseteq A$ such that $\|a_n\| \to d$. We can suppose that $\|a_n\| \le d+1$ holds for each n. Since X is reflexive, the closed balls of X are weakly compact sets, and so (using the Eberlein–Šmulian theorem and passing to a subsequence if necessary), we can assume without loss of generality that $a_n \xrightarrow{w} a$. Since A is weakly closed, $a \in A$, and by part (a) we get $d \le \|a\| \le \lim \|a_n\| = d$. Thus, $\|a\| = d$, and so $a \in A$ is a nearest point to zero.

(c) Let X be a uniformly convex Banach space. The uniform convexity of the norm guarantees that X is a reflexive Banach space; see Problem 11.1.5. Now assume that A is a non-empty, closed, and convex subset of X, and that $x_0 = 0 \notin A$. Since A is automatically weakly closed, it follows from part (b) that there exists a nearest point u in A to zero. That is, $\|u\| = \min_{a \in A} \|a\| = d > 0$. We claim that u is uniquely determined.

To this end, assume that some $v \in A$ satisfies $\|v\| = d$. We claim that $u = v$. To this end, assume by way of contradiction that $u \ne v$, and consider the vectors $x = \frac{u}{d}$ and $y = \frac{v}{d}$. Then $\|x\| = \|y\| = 1$, and clearly $0 < \epsilon = \|x - y\| \le 2$. So, by the uniform convexity of the norm, there exists some $0 < \delta < 1$ such that $\|\frac{x+y}{2}\| < 1 - \delta$. This implies that the vector $z = \frac{u+v}{2} \in A$ satisfies $\|z\| \le (1 - \delta)d < d$, which is impossible. Consequently, $u = v$, and so there exists exactly one point in A that is closest to zero.

(d) Let $x_n \to x$ in X. We must show that $y_n = \pi(x_n) \to \pi(x) = y$. Translating and scaling appropriately, we can assume without loss of generality that $x = 0$ and that $\|y\| = 1$. Now suppose, contrary to our claim, that $\|y_n - y\| \not\to 0$. Then, by passing to a subsequence, we can also assume that there exists some $\epsilon > 0$ such that $\|y_n - y\| > 2\epsilon$ for all n. Since y_n is the nearest point in C to x_n we have

$$\|x_n - y_n\| \le \|x_n - y\| \le \|x_n\| + \|y\| = \|x_n\| + 1 \longrightarrow 1\,.$$

Similarly, since y is the nearest point in C to 0, we have

$$\|x_n - y_n\| \ge \|y_n\| - \|x_n\| \ge \|y\| - \|x_n\| = 1 - \|x_n\| \longrightarrow 1\,.$$

From these two estimates, it follows that $\|x_n - y_n\| \to 1$ and hence $\|y_n\| \to 1$. So, if we let $z_n = y_n/\|y_n\|$, then $\|y_n - z_n\| = \big(1 - \frac{1}{\|y_n\|}\big)\|y_n\| \to 0$, and consequently from $\|z_n - y\| \ge \|y_n - y\| - \|y_n - z_n\|$, we see that $\|z_n - y\| > \epsilon$ for all $n \ge n_0$.

Since X is uniformly convex, there exists a $\delta > 0$ such that $\|z + y\| < 2(1 - \delta)$ provided $\|z\| \le 1$ and $\|z - y\| \ge \epsilon$. In particular, $\|z_n + y\| < 2(1 - \delta)$ for each $n \ge n_0$. Using again that $\|y_n - z_n\| \to 0$, the last inequality yields

$$\|y_n + y\| \le \|y_n - z_n\| + \|z_n + y\| < \|y_n - z_n\| + 2(1 - \delta) \longrightarrow 2(1 - \delta) < 2\,.$$

Therefore, for all n sufficiently large we have $\|\frac{y_n + y}{2}\| < 1$. However, for any one of these sufficiently large n, the vector $\frac{y_n + y}{2}$ belongs to C and this contradicts our assumption that the unit vector $y \in C$ is the nearest point to zero. Hence, $y_n \to y$, and thus the nearest point mapping π is continuous. ∎

Problem 1.1.18. *For a non-empty closed convex subset C of a Hilbert space X establish the following properties of the nearest point map $\pi\colon X \to C$.*

(a) *If $x \in C$, then $\pi(x) = x$.*

(b) *If $x \notin C$, then $\pi(x) \in \partial C$.*

(c) *For each $x \in X$ and each $y \in C$ we have $\big(x - \pi(x), y - \pi(x)\big) \le 0$.*

(d) *For each $x, y \in X$ we have $\big(\pi(x) - \pi(y), x - y\big) \ge 0$.*

(e) *For all* $x, y \in X$ *we have* $\|\pi(x) - \pi(y)\| \leq \|x - y\|$.

(f) *If* C *is a closed vector subspace of* X, *then for each* $x \in X$ *the vector* $x - \pi(x)$ *is orthogonal to* C, *i.e.,* $\big(x - \pi(x), c\big) = 0$ *for all* $c \in C$.

Remark: Recall that a non-empty subset A *of a topological space* Ω *is called a* ***retract*** *of* Ω *if there exists a continuous mapping* $f\colon \Omega \to A$, *called a* ***retraction of*** Ω ***onto*** A, *such that* $f(a) = a$ *for each* $a \in A$. *The preceding properties show that* π *is is a retract of* X *onto* C.

Solution: (a) If $x \in C$, then clearly $\pi(x) = x$.

(b) Let $x \notin C$ and note that the arbitrary point $x_\lambda = \lambda x + (1-\lambda)\pi(x)$ on the line segment joining x and $\pi(x)$ satisfies $\|x - x_\lambda\| = (1-\lambda)\|x - \pi(x)\| < \|x - \pi(x)\|$ for each $0 \leq \lambda < 1$. So, if $\pi(x)$ is an interior point of C, then for small λ the vector x_λ belongs to C and is closer to x than $\pi(x)$, a contradiction. Hence, $\pi(x) \in \partial C$.

(c) Let $x \in X$ and $y \in C$. For each $0 < \alpha < 1$ we have $\alpha y + (1-\alpha)\pi(x) \in C$. So, from the definition of $\pi(x)$, it follows that

$$\begin{aligned}
\big\|x - \pi(x)\big\|^2 &\leq \big\|x - \big[\alpha y + (1-\alpha)\pi(x)\big]\big\|^2 \\
&= \big\|\big[x - \pi(x)\big] - \alpha\big[y - \pi(x)\big]\big\|^2 \\
&= \big\|x - \pi(x)\big\|^2 - 2\alpha\big(x - \pi(x), y - \pi(x)\big) + \alpha^2\big\|y - \pi(x)\big\|^2 .
\end{aligned}$$

This implies $0 \leq -2\alpha\big(x - \pi(x), y - \pi(x)\big) + \alpha^2\big\|y - \pi(x)\big\|^2$ or

$$-2\big(x - \pi(x), y - \pi(x)\big) + \alpha\big\|y - \pi(x)\big\|^2 \geq 0 .$$

Letting $\alpha \downarrow 0$ yields $-2\big(x - \pi(x), y - \pi(x)\big) \geq 0$ or $\big(x - \pi(x), y - \pi(x)\big) \leq 0$.

(d) Let $x, y \in X$. Replacing y in (b) with $\pi(y) \in C$, we get

$$\big(x - \pi(x), \pi(y) - \pi(x)\big) \leq 0 . \qquad (\star)$$

Also, replacing x by y and y by $\pi(x)$ in (b), we get $\big(y - \pi(y), \pi(x) - \pi(y)\big) \leq 0$ or

$$\big(\pi(y) - y, \pi(y) - \pi(x)\big) \leq 0 . \qquad (\star\star)$$

Adding $(\star)$ and $(\star\star)$ yields

$$\begin{aligned}
0 &\geq \big(x - \pi(x), \pi(y) - \pi(x)\big) + \big(\pi(y) - y, \pi(y) - \pi(x)\big) \\
&= \big(x - y, \pi(y) - \pi(x)\big) + \big\|\pi(y) - \pi(x)\big\|^2 .
\end{aligned}$$

Therefore, $0 \leq \big\|\pi(y) - \pi(x)\big\|^2 \leq \big(y - x, \pi(y) - \pi(x)\big) = \big(\pi(x) - \pi(y), x - y\big)$.

(e) From the inequality in the preceding line and the classical Cauchy–Schwarz inequality, we get

$$\big\|\pi(y) - \pi(x)\big\|^2 \leq \big(y - x, \pi(y) - \pi(x)\big) \leq \|y - x\| \cdot \big\|\pi(y) - \pi(x)\big\| .$$

This implies $\big\|\pi(x) - \pi(y)\big\| \leq \|x - y\|$ for all $x, y \in X$.

(f) Replacing x by $x - \pi(x)$, we can assume from the outset that $\pi(x) = 0$. Then for each $c \in C$ we have $x = [x - (x,c)c] + (x,c)c$ and $x - (x,c)c \perp (x,c)c$, and so

$$\|x\|^2 = \|x - (x,c)c\|^2 + \|(x,c)c\|^2 \geq \|x - (x,c)c\|^2 .$$

Therefore, $(x,c)c = 0$ or $(x,c) = 0$. This implies $x - \pi(x) \perp C$. ∎

Problem 1.1.19. (Support Points). *In this problem X will denote a real Banach space, H is a real Hilbert space, and C is a non-empty closed convex subset of H. The function $\pi\colon H \to C$ is the nearest point mapping as defined in Problem* 1.1.17. *Recall that a point a_0 in a subset A of X is called a* ***support point*** *of A if there exists some non-zero $x^* \in X^*$ such that $x^*(a_0) \leq x^*(a)$ for each $a \in A$, i.e., if x^* attains its minimum value on A at a_0. Establish the following.*

(a) *Only boundary points of a set can be support points.*

(b) *If a closed convex subset of X has a non-empty interior, then every boundary point of the set is a support point.*

(c) *A point $c_0 \in \partial C$ is a support point of C if and only if there exists some $x \notin C$ such that $\pi(x) = c_0$.*

(d) *If $H = \mathbb{R}^k$, then every boundary point of C is a support point.*

(e) (Bishop–Phelps [**19**]) *The set of support points of C is dense in the boundary ∂C of C.*[1]

(f) *Give an example of a non-empty closed convex subset of ℓ_2 for which not every boundary point is a support point.*

Solution: (a) Assume that there exist a non-zero $x^* \in X^*$ and some $a_0 \in A$ such that $x^*(a_0) \leq x^*(a)$ for all $a \in A$. Also, suppose by way of contradiction that $a_0 \in A^{\mathrm{o}}$. Then there exists some $r > 0$ such that $\|y\| < r$ implies $a_0 \pm y \in A$. Therefore, $x^*(a_0) \leq x^*(a_0 \pm y) = x^*(a_0) \pm x^*(y)$ or $\pm x^*(y) \geq 0$ for all $y \in X$ with $\|y\| < r$. This implies $x^* = 0$, which is a contradiction. This establishes that $a_0 \in \partial A$.

(b) Let A be a closed convex subset of X such that $A^{\mathrm{o}} \neq \emptyset$. We claim that A^{o} is dense in A. To see this, let $x \in A^{\mathrm{o}}$ and fix some $y \in \partial A$. Choose an open neighborhood V of zero such that $x + V \subseteq A$ and fix $0 < \alpha < 1$. Since $y - \frac{\alpha}{1-\alpha}V$ is an open neighborhood of y, there exists some $z \in A \cap (y - \frac{\alpha}{1-\alpha}V)$. This implies $(1-\alpha)(y - z) \in \alpha V$. Moreover, the convexity of A guarantees that the non-empty open set $U = (1-\alpha)z + \alpha(x + V)$ lies entirely in A. Consequently, from

$$\alpha x + (1-\alpha)y = \alpha x + (1-\alpha)(y - z) + (1-\alpha)z \in \alpha x + \alpha V + (1-\alpha)z = U \subseteq A\,,$$

we see that $\alpha x + (1-\alpha)y \in A^{\mathrm{o}}$. Letting $\alpha \downarrow 0$, we infer that $y \in \overline{A^{\mathrm{o}}}$. That is, A^{o} is dense in A.

[1] This famous result of E. Bishop and R. R. Phelps is true for any real Banach space; for a complete discussion of the Bishop–Phelps theorem see [**4**, Section 8.9]. V. I. Lomonosov [**53**] has proven recently that the Bishop–Phelps theorem fails for complex Banach spaces.

Next, we claim that A^o is a convex set. To see this, let $u, v \in A^o$ and $0 < \lambda < 1$. Put $V = \lambda A^o + (1-\lambda)A^o$ and note that V is a non-empty open set. Since A is convex, it follows that $V \subseteq C$. Clearly, $\lambda u + (1-\lambda)v \in V$. This implies that $\lambda u + (1-\lambda)v$ is an interior point of A, i.e., $\lambda u + (1-\lambda)v \in A^o$. Therefore, A^o is a convex set.

Now assume that a_0 is a boundary point of A. Since $a_0 \notin A^o$, it follows from the classical separation theorem that there exists some non-zero linear functional $x^* \in X^*$ such that $x^*(a_0) \le x^*(a)$ for all $a \in A^o$. Since A^o is dense in A, we see that $x^*(a_0) \le x^*(a)$ for all $a \in A$. That is, a_0 is a support point of A.

(c) Fix $c_0 \in \partial C$. Assume first that there exists some $x \notin C$ such that $\pi(x) = c_0$. From part (c) of Problem 1.1.18, we know that $\big(x - \pi(x), y - \pi(x)\big) \le 0$, i.e., $(x - c_0, y - c_0) \le 0$, for each $y \in C$. Since $x \ne c_0$, the vector $p = c_0 - x \ne 0$, and the last inequality can be rewritten as $(p, c_0) \le (p, y)$ for all $y \in C$. This shows that c_0 is a support point of C.

For the converse, assume that c_0 is a support point of C. So, there exists a non-zero vector $p \in H$ such that $(p, c_0) \le (p, c)$ holds for each $c \in C$. We claim that the vector $x = c_0 - p$ does not belong to C. Indeed, if $x \in C$, then from $(p, c_0) \le (p, x) = (p, c_0 - p) = (p, c_0) - \|p\|^2$, we get $0 < \|p\|^2 \le 0$, which is a contradiction. Hence, $x \notin C$. Now note that for each $c \in C$ we have $(p, c_0 - c) \le 0$, and so

$$\begin{aligned} \|x - c\|^2 &= \big\|(c_0 - p) - c\big\|^2 = \big\|(c_0 - c) - p\big\|^2 \\ &= \|c_0 - c\|^2 - 2(p, c_0 - c) + \|p\|^2 \\ &\ge \|p\|^2 = \|x - c_0\|^2 \end{aligned}$$

holds for each $c \in C$. This shows that $\pi(x) = c_0$.

(d) Let c_0 be a boundary point of C. Choose a sequence $\{x_n\} \subseteq \mathbb{R}^k$ such that $x_n \notin C$ for each n and $x_n \to c_0$. By the preceding part, for each n there exists some non-zero vector $p_n \in \mathbb{R}^k$ such that

$$p_n \cdot c \ge p_n \cdot \pi(x_n) \tag{$\star\star\star$}$$

holds for each n. Normalizing the sequence $\{p_n\}$, we can assume without loss of generality that $\|p_n\| = 1$ holds for each n. In this case, by passing to a subsequence, we can assume without loss of generality that $\{p_n\}$ also converges, say $p_n \to p$, where $p \ne 0$. Now the continuity of π implies $\pi(x_n) \to \pi(c_0) = c_0$, and so from $(\star\star\star)$ we get $p \cdot c \ge p \cdot c_0$ for all $c \in C$. That is, c_0 is a support point of C.

(e) Let $c_0 \in \partial C$ and pick a sequence $\{x_n\}$ of points such that $x_n \notin C$ for each n and $x_n \to c_0$. By part (c), each $\pi(x_n)$ is a support point of C. Now use the continuity of π to see that $\pi(x_n) \to \pi(c_0) = c_0$.

(f) We let $C = \ell_2^+$, the positive cone of ℓ_2. Clearly, C is a non-empty closed convex subset of ℓ_2. Moreover, $C^o = \emptyset$. So, every point of C is a boundary point. We claim that a point $c = (c_1, c_2, \ldots) \in C$ is a support point of C if and only if $c_i = 0$ for some i.

To see this, let $c = (c_1, c_2, \ldots) \in C$. Assume first that $c_i = 0$ for some i. It is easy to see that the basic unit vector e_i supports C at c. For the converse, assume that c is a support point of C. So, there exists a non-zero vector $p \in \ell_2$ such that $(p, x) \geq (p, c)$ for all $x \in C$. Since $x \in C$ implies $\lambda x \in C$ for all $\lambda \geq 0$, it follows that $(p, x) \geq 0$ for each $x \in \ell_2^+$. This implies that the vector $p = (p_1, p_2, \ldots)$ satisfies $p_i \geq 0$ for each i.

Since $p \neq 0$, it follows that $p_i > 0$ for some i. Therefore, from $0 \in C$, we get $0 \leq p_i c_i \leq (p, c) \leq (p, 0) = 0$, and so $c_i = 0$. Hence, the support points of C are the non-negative vectors having at least one coordinate equal to zero. In particular, the latter shows that not every boundary point of C is a support point. ∎

Problem 1.1.20. (The Finite Dimensional Separation Theorem). *Recall that a non-zero linear functional $x^* \in X^*$* ***separates*** *two non-empty subsets A and B of a Banach space X if $x^*(a) \leq x^*(b)$ for all $a \in A$ and all $b \in B$.*

Show that every pair of disjoint non-empty convex subsets of $\mathbb{R}^k$ can always be separated.

Solution: Let A and B be two arbitrary non-empty convex subsets of $\mathbb{R}^k$ such that $A \cap B = \emptyset$. Let $C = A - B$ and note that C is a non-empty convex subset of $\mathbb{R}^k$ such that $0 \notin C$. We distinguish two cases.

CASE I: $0 \notin \overline{C}$.

In this case, we apply Problem 1.1.18 with $\overline{C}$ in place of C. If we let $x = 0$ and $p = \pi(0) \neq 0$ in part (c) of that problem, then we have $\big(0 - \pi(0), y - \pi(0)\big) \leq 0$ or $-p \cdot (y - p) \leq 0$ for each $y \in \overline{C}$. This implies $0 < \|p\|^2 \leq p \cdot y$ for all $y \in \overline{C}$. In particular, letting $y = a - b$ with $a \in A$ and $b \in B$, we get $p \cdot b < p \cdot a$ for all $a \in A$ and all $b \in B$. That is, p separates A and B.

CASE II: $0 \in \overline{C}$.

In this case, we shall prove that $0 \in \partial\overline{C}$. If this is established, then from part (d) of Problem 1.1.19 we know that 0 is a support point of $\overline{C}$. So, there exists a non-zero vector $p \in \mathbb{R}^k$ such that $p \cdot y \geq p \cdot 0 = 0$ for all $y \in \overline{C}$. This implies $p \cdot a \geq p \cdot b$ for all $a \in A$ and all $b \in B$. That is, p separates A and B.

It remains to be shown that $0 \in \overline{C}$ implies $0 \in \partial\overline{C}$. In this case, we distinguish two subcases.

CASE IIa: $C^{\text{o}} = \emptyset$.

In this subcase, fix a point $c \in C$ and let $B = c - C$. Then B is a convex set, $0 \in B$ and $B^{\text{o}} = c - C^{\text{o}} = \emptyset$. We claim there do not exist k linearly independent vectors in B. To see this, assume by way of contradiction that $\{b_1, \ldots, b_k\} \subseteq B$ is a linearly independent set. Then each $x \in \mathbb{R}^k$ can be written uniquely in the form $x = \alpha_1 b_1 + \cdots + \alpha_k b_k$ with $\alpha_i \in \mathbb{R}$. So, the function $f\colon \mathbb{R}^k \to \mathbb{R}^k$, defined by

$$f(x) = f(\alpha_1 b_1 + \cdots + \alpha_k b_k) = (\alpha_1, \alpha_2, \ldots, \alpha_k),$$

is linear (and hence continuous). Since $\alpha_i \geq 0$ for each i in conjunction with the condition $\sum_{i=1}^k \alpha_i \leq 1$ and the fact $0 \in B$ implies

$$\alpha_1 b_1 + \cdots + \alpha_k b_k = \alpha_1 b_1 + \cdots + \alpha_k b_k + \big(1 - \textstyle\sum_{i=1}^k \alpha_i\big)0 \in B,$$

we see that B contains the non-empty open set $f^{-1}(U)$, where

$$U = \left\{(\alpha_1, \alpha_2, \ldots, \alpha_k) \in (0,1)^k \colon\; 0 < \textstyle\sum_{i=1}^k \alpha_i < 1\right\},$$

contradicting $B^{\mathrm{o}} = \emptyset$.

Thus, if we let $\{b_1, \ldots, b_\ell\}$ be a maximal linearly independent subset of B, then it follows that $\ell < k$ and $B \subseteq \operatorname{span}\{b_1, \ldots, b_\ell\}$, i.e., the set B is contained in a (closed) proper linear subspace. Hence, $\overline{B} \subseteq \operatorname{span}\{b_1, \ldots, b_\ell\}$. This implies $\overline{B}^{\mathrm{o}} = c - \overline{C}^{\mathrm{o}} = \emptyset$ and so $\overline{C}^{\mathrm{o}} = \emptyset$. Therefore, $\partial\overline{C} = \overline{C}$, and thus $0 \in \partial\overline{C}$.

CASE IIb: $C^{\mathrm{o}} \neq \emptyset$.

In this case, we need to invoke the following property of convex sets.

(R) *Let X be a finite dimensional vector space, let A be an open convex subset of X and let B be a convex subset of A. If B is dense in A, then $B = A$.*

The proof is by induction on the dimension of X. If $\dim X = 1$, then the result is trivial. So, for the induction step, assume that property (R) is true for any vector space of dimension n, and let A be an open convex subset of an $(n+1)$-dimensional vector space Y. Assume also that B is a convex subset of A that is dense in A. Translating A and B if necessary, we can assume that $0 \in B$.

Fix a vector $a \in A$ and then choose a non-zero vector $u \in Y$ such that $a \cdot u = 0$. Let $H = \{y \in Y \colon\; y \cdot u = 0\}$ and note that H is a vector subspace of dimension n such that $a \in H \cap A$. Also, $H \cap A$ is an open convex subset of H. If the convex subset $H \cap B$ of $H \cap A$ is dense in $H \cap A$, then it will follow from our induction hypothesis that $a \in B$ for all $a \in A$, i.e., $B = A$, which is the desired conclusion. So, we shall finish the proof by showing that $H \cap B$ is dense in $H \cap A$.

To this end, let $x \in H \cap A$ and let $\epsilon > 0$. Also, consider the open half-spaces

$$H_1 = \{y \in Y \colon\; y \cdot u > 0\} \quad \text{and} \quad H_2 = \{y \in Y \colon\; y \cdot u < 0\}.$$

Since $x \in A$ and $x + \lambda u \in H_1$ and $x - \lambda u \in H_2$ for all $\lambda > 0$, there exists some $\lambda_0 > 0$ with $\lambda_0\|u\| < \frac{\epsilon}{2}$ such that $x + \lambda_0 u \in H_1 \cap A$ and $x - \lambda_0 u \in H_2 \cap A$. Since B is dense in A, there exist $b_1 \in H_1 \cap B$ and $b_2 \in H_2 \cap B$ such that $\|x + \lambda_0 u - b_1\| < \frac{\epsilon}{2}$ and $\|x - \lambda_0 u - b_2\| < \frac{\epsilon}{2}$. From $u \cdot b_1 > 0$ and $b_2 > u \cdot c_2$, it follows that there exists some $0 < \alpha < 1$ such that the vector $b = \alpha b_1 + (1-\alpha) b_2 \in B$ satisfies $u \cdot b = 0$, i.e., $b \in H$. Now note that

$$\begin{aligned} \|x - b\| &= \big\|\alpha(x + \lambda_0 u - b_1) + (1-\alpha)(x - \lambda_0 u - b_2) + \big[(1-\alpha)\lambda_0 u - \alpha\lambda_0 u\big]\big\| \\ &\leq \alpha\|x + \lambda_0 u - b_1\| + (1-\alpha)\|x - \lambda_0 u - b_2\| + (1-\alpha)\lambda_0\|u\| + \alpha\lambda_0\|u\| \\ &< \alpha\tfrac{\epsilon}{2} + (1-\alpha)\tfrac{\epsilon}{2} + \alpha\tfrac{\epsilon}{2} + (1-\alpha)\tfrac{\epsilon}{2} = \epsilon\,. \end{aligned}$$

This shows that $H \cap B$ is dense in $H \cap A$, as desired.

An immediate conclusion of property (R) is the following.

- *If a convex C in a finite dimensional vector space has a non-empty interior, then $\overline{C}^{\mathrm{o}} = C^{\mathrm{o}}$ and $\partial\overline{C} = \partial C$.*

Indeed, if $C^{\mathrm{o}} \neq \emptyset$, then C^{o} is dense in $\overline{C}$ (see the proof of part (b) of Problem 1.1.19), and so C^{o} is also dense in $\overline{C}^{\mathrm{o}}$. This shows that $\overline{C}^{\mathrm{o}} = C^{\mathrm{o}}$.

Finally, to finish the solution of the problem, notice that $0 \notin C$ in conjunction with $C^o \neq \emptyset$ implies $0 \in \overline{C} \setminus C^o = \overline{C} \setminus \overline{C}^o = \partial \overline{C}$. ∎

Problem 1.1.21. *Let Y be a closed subspace of a normed space X, and define the function $|||\cdot|||: X/Y \to \mathbb{R}$ by*

$$|||[x]||| = \inf_{y \in [x]} \|y\| .$$

Establish the following.

(a) *The function $|||\cdot|||$ is a norm on the quotient vector space X/Y.*

(b) *The unit ball of X/Y under the quotient norm is*

$$U_{X/Y} = \{[x]: \ x \in U_X\} .$$

(c) *The quotient map $x \mapsto [x]$ is an open contraction.*

(d) *If X is a Banach space, then $(X/Y, |||\cdot|||)$ is also a Banach space.*

Solution: (a) Clearly, $|||[x]||| \geq 0$ for each $x \in X$ and $[x] = 0$ implies $|||[x]||| = 0$. Now assume that $|||[x]||| = 0$. Then there exists a sequence $\{y_n\} \subseteq [x]$ such that $y_n \to 0$. From $x - y_n \in Y$ for each n, $x - y_n \to x$, and the closedness of Y, we get $x \in Y$ or $[x] = 0$. For the homogeneity of $|||\cdot|||$, note that for each $\lambda \neq 0$ we have

$$|||\lambda[x]||| = |||[\lambda x]||| = \inf_{y \in [\lambda x]} \|y\| = |\lambda| \inf_{\frac{y}{\lambda} \in [x]} \left\| \tfrac{y}{\lambda} \right\| = |\lambda| \inf_{z \in [x]} \|z\| = |\lambda| \cdot |||[x]||| .$$

For the triangle inequality, let $u, v \in X$. Then for each $x \in [u]$ and $y \in [v]$, we have $x + y \in [u + v]$, and so

$$|||[u+v]||| \leq \|x + y\| \leq \|x\| + \|y\| .$$

Taking infima for all $x \in [u]$ and $y \in [v]$, we obtain $|||[u+v]||| \leq |||[u]||| + |||[v]|||$.

(b) If $x \in U_X$, then $|||[x]||| \leq \|x\| < 1$, and so $\{[x]: \ x \in U_X\} \subseteq U_{X/Y}$. For the converse inclusion, let $[x] \in U_{X/Y}$, i.e., $|||[x]||| = \inf_{y \in [x]} \|y\| < 1$. This implies that there exists some $y \in [x]$ satisfying $\|y\| < 1$ or $y \in U_X$, and so $[x] = [y]$. So, $U_{X/Y} \subseteq \{[x]: \ x \in U_X\}$ is also true, and thus $U_{X/Y} = \{[x]: \ x \in U_X\}$.

(c) It should be clear that the quotient map $x \mapsto \pi(x) = [x]$ is linear. Moreover, from part (b), we have $\pi(U_X) = U_{X/Y}$. From this it easily follows that π is an open mapping. In addition, the inequality $|||\pi(x)||| \leq \|x\|$ for each $x \in X$ shows that the quotient map π is also a contraction.

(d) Assume now that X is a Banach space and let $\{[x_n]\}$ be a Cauchy sequence in X/Y. By passing to a subsequence if necessary, we can assume without loss of generality that $|||[x_{n+1}] - [x_n]||| = |||[x_{n+1} - x_n]||| < \frac{1}{2^n}$ holds for each n. Now for each n pick some $y_n \in [x_{n+1} - x_n]$ satisfying $\|y_n\| < \frac{1}{2^n}$. Since X is a Banach space, the series $\sum_{n=1}^{\infty} y_n$ is norm convergent in X; let $y = \sum_{n=1}^{\infty} y_n$. Then the continuity of the quotient map implies

$$[y] = \sum_{n=1}^{\infty} [y_n] = \sum_{n=1}^{\infty} [x_{n+1} - x_n] = \lim_{n \to \infty} \sum_{k=1}^{n-1} \big([x_{k+1}] - [x_k]\big) = \lim_{n \to \infty} \big([x_n] - [x_1]\big) .$$

Therefore, the sequence $\{[x_n]\}$ is norm convergent in X/Y and $[x_n] \to [y] - [x_1]$. This shows that X/Y is a Banach space. ∎

Problem 1.1.22. *If V is a vector subspace of a normed vector space X, then its **annihilator** is the norm closed subspace of X^* defined by*

$$V^\perp = \{x^* \in X^*\colon \ x^*(v) = 0 \ \text{for all} \ v \in V\}.$$

For a closed subspace Y of a normed space X establish that

$$(X/Y)^* = Y^\perp \qquad \text{and} \qquad Y^* = X^*/Y^\perp$$

by proving the following statements.

(1) *The mapping $\phi\colon Y^\perp \to (X/Y)^*$, defined via the formula*

$$\phi(x^*)([x]) = x^*(x)\,, \quad x^* \in Y^\perp, \ \ [x] \in X/Y\,,$$

is a well-defined surjective linear isometry.

(2) *The mapping $\psi\colon X^*/Y^\perp \to Y^*$, defined via the formula*

$$\psi([x^*]) = x^*|_Y \ \ (\text{the restriction of } x^* \text{ to } Y)$$

for each $[x^*] \in X^*/Y^\perp$, is a well-defined surjective linear isometry.

Solution: (1) We check first that ϕ is well defined. To see this, let $x^* \in Y^\perp$ and $[x] = [y]$, i.e., $x - y \in Y$. Then $x^*(x) - x^*(y) = x^*(x - y) = 0$ or $x^*(x) = x^*(y)$, and so ϕ is well defined. Clearly, ϕ is linear.

To see that ϕ is a surjective mapping, let $\Psi \in (X/Y)^*$. Since the quotient map $\pi\colon X \to X/Y$ is continuous, it follows that the linear functional $x^* = \Psi \circ \pi$ is also continuous on X, i.e., $x^* \in X^*$. From $x^*(x) = \Psi(\pi(x))$ and $\pi(x) = 0$ for each $x \in Y$, we see that $x^*(x) = 0$ for each $x \in Y$, i.e., $x^* \in Y^\perp$. Moreover,

$$\phi(x^*)([x]) = x^*(x) = \Psi(\pi(x)) = \Psi([x])$$

holds for all $x \in X$. That is, $\phi(x^*) = \Psi$, and so ϕ is also surjective.

It remains to be shown that ϕ is an isometry. To see this, let $x^* \in Y^\perp$. Then for each $[x] \in U_{X/Y}$ and each $y \in [x]$ we have

$$\big|\phi(x^*)([x])\big| = \big|x^*(x)\big| = \big|x^*(y)\big| \le \|x^*\| \cdot \|y\|\,.$$

This implies $\big|\phi(x^*)([x])\big| \le \|x^*\| \cdot \big\|[x]\big\|$ for each $[x] \in U_{X/Y}$. So, $\big\|\phi(x^*)\big\| \le \|x^*\|$. On the other hand, for each $x \in U_X$ we have $[x] \in U_{X/Y}$, and thus

$$\big\|\phi(x^*)\big\| \ge \big|\phi(x^*)([x])\big| = \big|x^*(x)\big|\,.$$

This implies that $\big\|\phi(x^*)\big\| \ge \sup_{x \in U_X} \big|x^*(x)\big| = \|x^*\|$. Therefore, $\big\|\phi(x^*)\big\| = \|x^*\|$ for each $x^* \in Y^\perp$. Consequently, ϕ is a surjective linear isometry.

(2) We shall verify first that ψ is well defined. To this end, let $[z^*] = [x^*]$ in $X^*/Y^\perp$, that is, $z^* - x^* \in Y^\perp$. Consequently, for each vector $y \in Y$ we have $z^*(y) - x^*(y) = (z^* - x^*)(y) = 0$, i.e., $z^*(y) = x^*(y)$. This shows that the mapping ψ is well defined. Clearly, ψ is linear.

To see that ψ is surjective, let $y^* \in Y^*$. According to the Hahn–Banach theorem, there exists some $z^* \in X^*$ such that $z^*(y) = y^*(y)$ for each $y \in Y$. Any such $z^* \in X^*$ satisfies $\psi([z^*]) = y^*$. So, ψ is surjective.

To finish the solution, it remains to be shown that ψ is an isometry. To see this, fix $[x^*] \in X^*/Y^\perp$. Pick some $z^* \in X^*$ such that $z^*(y) = x^*(y)$ for each $y \in Y$ (and so $z^* \in [x^*]$) and $\|z^*\| = \|x^*|_Y\|$, where $\|z^*\|$ is the norm of z^* in X^* and $\|x^*|_Y\|$ is the norm of $x^*|_Y$ in Y^*. Then for each vector $y \in Y$ with $\|y\| = 1$ we have $\|\psi([x^*])\| \geq |\psi([x^*])(y)| = |x^*(y)|$, and so

$$\|\psi([x^*])\| \geq \sup\{|x^*(y)|:\ y \in Y \text{ and } \|y\| = 1\} = \|x^*|_Y\| = \|z^*\| \geq \|[x^*]\|.$$

On the other hand, for each $u^* \in [x^*]$ and each $y \in Y$ we have

$$\|\psi([x^*])(y)\| = |x^*(y)| = |u^*(y)| \leq \|u^*\| \cdot \|y\|.$$

This implies $\|\psi([x^*])(y)\| \leq \|[x^*]\| \cdot \|y\|$ for all $y \in Y$, from which it follows that $\|\psi([x^*])\| \leq \|[x^*]\|$. Hence $\|\psi([x^*])\| = \|[x^*]\|$, and the solution is finished. ∎

1.2. Banach Lattices and Positive Operators

Problem 1.2.1. *Let C be a cone in a real vector space X. Show that the relation $\geq$ on X defined by $x \geq y$ whenever $x - y \in C$ is a vector space order such that $X^+ = C$.*

Solution: Recall that the cone C is characterized by the properties:

(a) $C + C \subseteq C$,
(b) $\alpha C \subseteq C$ for all $\alpha \geq 0$, and
(c) $C \cap (-C) = \{0\}$.

Now define the binary relation $\geq$ on X by letting $x \geq y$ whenever $x - y \in C$. Next, we shall verify that $\geq$ is a vector space order.

(1) (*Reflexivity*) We have $x \geq x$ for each $x \in X$. This follows immediately from $x - x = 0 \in C$.

(2) (*Antisymmetry*) Let $x \geq y$ and $y \geq z$. That is, $x - y \in C$ and similarly $-(x - y) = y - x \in C$. By property (c), we get $x - y = 0$ or $x = y$.

(3) (*Transitivity*) Assume $x \geq y$ and $y \geq z$. That is, $x - y \in C$ and $y - z \in C$. From property (a), we see that $x - z = (x - y) + (y - z) \in C$. So, $x \geq z$.

(4) Let $x \geq y$, i.e., $x - y \in C$. Then for each $z \in X$ we have $(x + z) - (y + z) = x - y \in C$, that is, $x + z \geq y + z$.

(5) Suppose $x \geq y$ and let $\alpha \geq 0$. Then $x - y \in C$ and using property (b), we get $\alpha x - \alpha y = \alpha(x - y) \in C$, so that $\alpha x \geq \alpha y$.

Therefore, $\geq$ is a vector space order.

Finally, notice that $X^+ = \{x \in X\colon\ x \geq 0\} = \{x \in X\colon\ x - 0 = x \in C\} = C$. ∎

Problem 1.2.2. *Let X and Y be two ordered vector spaces such that Y is Archimedean and the cone of X is generating. If $T\colon X^+ \to Y^+$ is an additive function (i.e., $T(x+y) = Tx + Ty$ for all $x, y \in X^+$), then show that T has a unique linear positive extension from X to Y.*

Moreover, if for each $x \in X$ we pick two vectors $x_1, x_2 \in X^+$ such that $x = x_1 - x_2$, then the unique extension $\hat{T}\colon X \to Y$ of T satisfies

$$\hat{T}x = Tx_1 - Tx_2 \,. \tag{$\star$}$$

Solution: First notice that if $x = x_1 - x_2 = y_1 - y_2$ with $x_i, y_i \in X^+$ for each i, then $x_1 - x_2 = y_1 - y_2$ or $x_1 + y_2 = y_1 + x_2$, and so by the additivity of T, we get

$$Tx_1 + Ty_2 = T(x_1 + y_2) = T(y_1 + x_2) = Ty_1 + Tx_2 \,.$$

This implies $Tx_1 - Tx_2 = Ty_1 - Ty_2$ and so $\hat{T}$ as given by $(\star)$ is well defined. From $(\star)$ it should be obvious that $\hat{T}x = Tx \geq 0$ for each $x \in X^+$ and $\hat{T}(-x) = -\hat{T}x$ for all $x \in X$. Next, we shall verify that $\hat{T}$ is linear.

For the additivity of $\hat{T}$ notice that if $x = x_1 - x_2$ and $y = y_1 - y_2$ with the x_i and y_i all positive, then

$$\begin{aligned}
\hat{T}(x+y) &= \hat{T}\big(x_1 + y_1 - (x_2 + y_2)\big) = T(x_1 + y_1) - T(x_2 + y_2) \\
&= Tx_1 + Ty_1 - Tx_2 - Ty_2 = (Tx_1 - Tx_2) + (Ty_1 - Ty_2) \\
&= \hat{T}x + \hat{T}y \,.
\end{aligned}$$

In particular, the additivity implies $T(rx) = rT(x)$ for each rational number r and all $x \in X$. For the homogeneity of $\hat{T}$ we need two properties.

(i) *T is monotone on X^+. That is, $0 \leq x \leq y$ implies $Tx \leq Ty$.*

Indeed, if $0 \leq x \leq y$, then $Ty = T(x + (y - x)) = Tx + T(y - x) \geq Tx$.

(ii) *Let Z be an Archimedean ordered vector space, let $x \in Z^+$, and let $y \in Z$. Assume that a sequence of real numbers $\{\alpha_n\}$ converges to α, i.e., $\alpha_n \to \alpha$. If $\alpha_n x \leq y$ (resp. $y \leq \alpha_n x$) for each n, then $\alpha x \leq y$ (resp. $y \leq \alpha x$).*

To see this, fix k. Then for all sufficiently large n we have $\alpha - \alpha_n \leq \frac{1}{k}$, and so $(\alpha - \alpha_n)x \leq \frac{1}{k}x$ for all large n. Hence, $\alpha x - y \leq \alpha x - \alpha_n x = (\alpha - \alpha_n)x \leq \frac{1}{k}x$ or $k(\alpha x - y) \leq x$ for each k. Since X is Archimedean, we get $\alpha x - y \leq 0$ or $\alpha x \leq y$.

Now let $x \in X^+$ and $\alpha > 0$. Pick two sequences of rational numbers $\{r_n\}$ and $\{t_n\}$ satisfying $r_n \uparrow \alpha$ and $t_n \downarrow \alpha$. Using (i) and the homogeneity of T with respect to the rational numbers, we get

$$r_n Tx = T(r_n x) \leq T(\alpha x) \leq T(s_n x) = s_n Tx \,.$$

From (ii), we get immediately that $\alpha Tx \leq T(\alpha x) \leq \alpha Tx$, or $\alpha Tx = T(\alpha x)$.

Finally, fix $\alpha \geq 0$ and $x = x_1 - x_2 \in X$ with $x_1, x_2 \in X^+$. Then we have

$$\begin{aligned}
\hat{T}(\alpha x) &= \hat{T}(\alpha x_1 - \alpha x_2) = T(\alpha x_1) - T(\alpha x_2) = \alpha\big(Tx_1 - Tx_2\big) = \alpha \hat{T}x \,, \text{ and} \\
\hat{T}(-\alpha x) &= \hat{T}(\alpha x_2 - \alpha x_1) = \alpha(Tx_2 - Tx_1) = (-\alpha)\hat{T}(x) \,.
\end{aligned}$$

Thus, $\hat{T}$ is also homogeneous, and hence linear. ∎

Problem 1.2.3. *Show that order complete Riesz spaces and normed Riesz spaces are Archimedean.*

Solution: Let E be an order complete Riesz space and assume that $nx \le y$ holds for all n and some $x, y \in E$. Since E is order complete, $u = \sup\{nx\colon\ n = 1, 2, \ldots\}$ exists in E. Now from $nx = (n+1)x - x \le u - x$, it follows that $u \le u - x$ or $x \le 0$. Thus, E is Archimedean.

Now assume that E is a normed Riesz space and let $nx \le y$ for all n and some $x, y \in E$. Notice that $nx \le y \le y^+$ holds for all n, and from this we infer that $0 \le nx^+ \le y^+$ for all n. Therefore, $n\|x^+\| = \|nx^+\| \le \|y^+\| < \infty$ also holds for each n. This implies $\|x^+\| = 0$ or $x^+ = 0$. Consequently $x \le x^+ = 0$, proving that E is Archimedean in this case too. ∎

Problem 1.2.4. *Establish the following properties of order convergence.*

(a) *A net in a partially ordered set can have at most one order limit.*

(b) *If $x_\alpha \uparrow x$ or $x_\alpha \downarrow x$ in a partially ordered set, then $x_\alpha \xrightarrow{o} x$.*

(c) *If a net $\{x_\alpha\}_{\alpha\in A}$ in a partially ordered set satisfies $x_\alpha \uparrow$ (resp. $x_\alpha \downarrow$) and $x_\alpha \xrightarrow{o} x$, then $x_\alpha \uparrow x$ (resp. $x_\alpha \downarrow x$).*

(d) *A positive operator $T\colon X \to Y$ between two ordered vector spaces is order continuous if $x_\alpha \downarrow 0$ in X implies $Tx_\alpha \downarrow 0$ in Y.*

(e) *A net $\{x_\alpha\}_{\alpha\in A}$ in a Riesz space satisfies $x_\alpha \xrightarrow{o} x$ if and only if there exists another net $\{u_\lambda\}_{\lambda\in\Lambda}$ such that $u_\lambda \downarrow 0$ and for each $\lambda \in \Lambda$ there exists some α_0 (depending on λ) such that $|x_\alpha - x| \le u_\lambda$ holds for each $\alpha \ge \alpha_0$.*

(f) *The lattice operations in any Riesz space are order continuous.*

Solution: (a) Let $\{x_\alpha\}_{\alpha\in A}$ be a net in a partially ordered set, and assume that $x_\alpha \xrightarrow{o} x$ and $x_\alpha \xrightarrow{o} y$. By the definition of order convergence there exist two nets $\{y_\beta\}_{\beta\in B}$ and $\{z_\gamma\}_{\gamma\in\Gamma}$ such that $y_\beta \uparrow x$, $z_\gamma \downarrow x$, and for each $\beta \in B$ and $\gamma \in \Gamma$ there exists some $\alpha(\beta,\gamma) \in A$ such that $y_\beta \le x_\alpha \le z_\gamma$ holds for each $\alpha \ge \alpha(\beta,\gamma)$. Similarly, there exist two nets $\{u_\lambda\}_{\lambda\in\Lambda}$ and $\{v_\mu\}_{\mu\in M}$ such that $u_\lambda \uparrow y$, $v_\mu \downarrow y$, and for each $\lambda \in \Lambda$ and $\mu \in M$ there is some $\alpha'(\lambda,\mu) \in A$ such that $u_\lambda \le x_\alpha \le v_\mu$ for each $\alpha \ge \alpha'(\lambda,\mu)$.

Fix $\beta \in B$, $\gamma \in \Gamma$, $\lambda \in \Lambda$, and $\mu \in M$. Pick some index $\alpha \in A$ with $\alpha \ge \alpha(\beta,\gamma)$ and $\alpha \ge \alpha'(\lambda,\mu)$. Then we have $y_\beta \le x_\alpha \le v_\mu$ and $u_\lambda \le x_\alpha \le z_\gamma$. This implies

$$y_\beta \le v_\mu \qquad \text{and} \qquad u_\lambda \le z_\gamma$$

for all indices $\beta \in B$, $\gamma \in \Gamma$, $\lambda \in \Lambda$, and $\mu \in M$. The latter inequalities yield $x \le y$ and $y \le x$. That is, $x = y$, and the uniqueness of the order limits has been established.

(b) Assume that $x_\alpha \uparrow x$ in a partially ordered set. Consider the two nets $\{y_\alpha\}_{\alpha\in A}$ and $\{z_\alpha\}_{\alpha\in A}$ defined by $y_\alpha = x_\alpha$ and $z_\alpha = x$ for each $\alpha \in A$. Since $y_\alpha \le x_\alpha \le z_\alpha$ for each $\alpha \in A$, it follows immediately that $x_\alpha \xrightarrow{o} x$.

(c) Assume that a net $\{x_\alpha\}_{\alpha\in A}$ in a partially ordered set satisfies $x_\alpha \uparrow$ and $x_\alpha \xrightarrow{o} x$. Pick two nets $\{y_\beta\}_{\beta\in B}$ and $\{z_\gamma\}_{\gamma\in\Gamma}$ such that $y_\beta \uparrow x$, $z_\gamma \downarrow x$ and for each $\beta \in B$ and $\gamma \in \Gamma$ there exists some $\alpha(\beta, \gamma) \in A$ such that $y_\beta \le x_\alpha \le z_\gamma$ for each $\alpha \ge \alpha(\beta, \gamma)$.

Now fix any index $\alpha \in A$ and note that for each $\beta \in B$ and $\gamma \in \Gamma$ there exists some index $\alpha' \in A$ such that $\alpha' \ge \alpha$ and $\alpha' \ge \alpha(\beta, \gamma)$. This implies $x_\alpha \le x_{\alpha'} \le z_\gamma$, and so $x_\alpha \le z_\gamma$ for all $\alpha \in A$ and all $\gamma \in \Gamma$. From $z_\gamma \downarrow x$, we see that $x_\alpha \le x$ for each $\alpha \in A$, i.e., the element x is an upper bound of the net $\{x_\alpha\}_{\alpha \in A}$.

If $x_\alpha \le y$ holds for each $\alpha \in A$, then as above we see that $y_\beta \le x_{\alpha'} \le y$, and so $y_\beta \le y$ for each $\beta \in B$. Taking into account that $y_\beta \uparrow x$, the latter shows that $x \le y$. Therefore, x is the least upper bound of the net $\{x_\alpha\}_{\alpha \in A}$, i.e., $x_\alpha \uparrow x$. The decreasing case can be established in a similar manner.

(d) Let $T\colon X \to Y$ be a positive operator between two ordered vector spaces. Assume first that T is order continuous, and let $x_\alpha \downarrow 0$ in X. Then $Tx_\alpha \downarrow$ in Y. By part (b), we know that $x_\alpha \xrightarrow{o} 0$, and so, by the order continuity of T, we have $Tx_\alpha \xrightarrow{o} 0$. By part (c) we get $Tx_\alpha \downarrow 0$ in Y.

For the converse, assume that $u_\alpha \downarrow 0$ in X implies $Tu_\alpha \downarrow 0$ in Y. Let $x_\alpha \xrightarrow{o} x$ in X. Pick two nets $\{y_\beta\}_{\beta \in B}$ and $\{z_\gamma\}_{\gamma \in \Gamma}$ such that $y_\beta \uparrow x$, $z_\gamma \downarrow x$ and for each $\beta \in B$ and $\gamma \in \Gamma$ there exists some $\alpha(\beta, \gamma) \in A$ such that $y_\beta \le x_\alpha \le z_\gamma$ holds for each $\alpha \ge \alpha(\beta, \gamma)$. Then the positivity of T implies that for all $\alpha \ge \alpha(\beta, \gamma)$ we have $Ty_\beta \le Tx_\alpha \le Tz_\gamma$. Moreover, our condition implies $Ty_\beta \uparrow Tx$ and $Tz_\gamma \downarrow Tx$. These show that $Tx_\alpha \xrightarrow{o} x$ in Y.

(e) Let $\{x_\alpha\}_{\alpha \in A}$ be a net in a Riesz space. Assume first that $x_\alpha \xrightarrow{o} x$. Pick two nets $\{y_\beta\}_{\beta \in B}$ and $\{z_\gamma\}_{\gamma \in \Gamma}$ such that $y_\beta \uparrow x$, $z_\gamma \downarrow x$ and for each $\beta \in B$ and $\gamma \in \Gamma$ there exists some $\alpha(\beta, \gamma) \in A$ such that $y_\beta \le x_\alpha \le z_\gamma$ holds for each $\alpha \ge \alpha(\beta, \gamma)$. Now note that for each $\alpha \ge \alpha(\beta, \gamma)$ we have $x_\alpha - x \le z_\gamma - x \le z_\gamma - y_\beta$ and $-(x_\alpha - x) = x - x_\alpha \le x - y_\beta \le z_\gamma - y_\beta$, and so $|x_\alpha - x| \le z_\gamma - y_\beta$. Therefore, if we let $\Lambda = B \times \Gamma$ (the product index set) and define the net $\{u_\lambda\}_{\lambda \in \Lambda}$ by $u_\lambda = z_\gamma - y_\beta$ for each $\lambda = (\beta, \gamma)$, then $|x_\alpha - x| \le u_\lambda$ for all $\alpha \ge \alpha(\beta, \gamma) = \alpha(\lambda)$.

For the converse, assume that there exists a net $\{u_\lambda\}_{\lambda \in \Lambda}$ such that $u_\lambda \downarrow 0$ and for each $\lambda \in \Lambda$ there exists some α_0 (depending on λ) such that $|x_\alpha - x| \le u_\lambda$ for each $\alpha \ge \alpha_0$. The latter inequality is equivalent to $x - u_\lambda \le x_\alpha \le x + u_\lambda$ for each $\alpha \ge \alpha_0$. So, if we put $y_\lambda = x - u_\lambda$ and $z_\lambda = x + u_\lambda$, then we easily infer that $x_\alpha \xrightarrow{o} x$.

(f) Assume that $x_\alpha \xrightarrow{o} x$ in a Riesz space. By (e) there exists some net $\{u_\lambda\}_{\lambda \in \Lambda}$ such that $u_\lambda \downarrow 0$ and for each $\lambda \in \Lambda$ there exists some α_0 (depending on λ) such that $|x_\alpha - x| \le u_\lambda$ for each $\alpha \ge \alpha_0$. Now use the inequalities

$$\left|x_\alpha^+ - x^+\right| \le |x_\alpha - x|\,, \quad \left|x_\alpha^- - x^-\right| \le |x_\alpha - x|\,, \quad \text{and} \quad \big||x_\alpha| - |x|\big| \le |x_\alpha - x|$$

and (e) to infer that

$$x_\alpha^+ \xrightarrow{o} x^+\,, \quad x_\alpha^- \xrightarrow{o} x^-\,, \quad \text{and} \quad |x_\alpha| \xrightarrow{o} |x|\,.$$

Thus, the lattice operations $x \mapsto x^+$, $x \mapsto x^-$, and $x \mapsto |x|$ are order continuous. ∎

Problem 1.2.5. *Show that a solid subset D of a Riesz space E is order closed if and only if $\{x_\alpha\} \subseteq D \cap E^+$ and $x_\alpha \uparrow x$ in E imply $x \in D$.*

Solution: Let D be a solid subset in a Riesz space. If D is order closed and a net $\{x_\alpha\}$ in $D \cap E^+$ satisfies $x_\alpha \uparrow x$, then clearly $x \in D$.

Now assume that whenever a net $\{y_\lambda\} \subseteq D \cap E^+$ satisfies $y_\lambda \uparrow y$, then y belongs to D. Let $\{x_\alpha\}$ be a net in D satisfying $x_\alpha \xrightarrow{o} x$. So, according to Problem 1.2.4(c), there exists another net $\{u_\lambda\}_{\lambda \in \Lambda}$ such that $u_\lambda \downarrow 0$ and for each λ there exists some α_0 such that $|x_\alpha - x| \leq u_\lambda$ for each $\alpha \geq \alpha_0$. From $x - u_\lambda \leq x_\alpha \leq |x_\alpha|$, we see that $(x - u_\lambda)^+ \leq |x_\alpha|$ for each $\alpha \geq \alpha_0$. Since D is a solid set, it follows that $\{(x - u_\lambda)^+\} \subseteq D$. Now from $(x - u_\lambda)^+ \uparrow x$ and our hypothesis, we infer that $x \in D$. Hence, D is order closed. ∎

Problem 1.2.6. *If J is an ideal in a Riesz space E, then show that the band B_J generated by J is given by*

$$B_J = \big\{x \in E\colon\ \exists\ \{x_\alpha\} \subseteq J \ \textit{ such that } \ 0 \leq x_\alpha \uparrow |x|\,\big\}.$$

Moreover, show that for each $x \in E$ the principal band B_x generated by x is given by

$$B_x = \big\{y \in E\colon\ |y| \wedge n|x| \uparrow |y|\,\big\}.$$

Solution: Fix an ideal J in a Riesz space E and let

$$B = \big\{x \in E\colon\ \exists\ \{x_\alpha\} \subseteq J \ \text{ with } \ 0 \leq x_\alpha \uparrow |x|\,\big\}.$$

Clearly, every band that includes J must include B. Thus, in order to establish that $B_J = B$, it suffices to show that B is a band.

To this end, let $x, y \in B$. Pick two nets $\{x_\alpha\}$ and $\{y_\lambda\}$ satisfying $0 \leq x_\alpha \uparrow |x|$ and $0 \leq y_\lambda \uparrow |y|$. From

$$|x+y| \wedge (x_\alpha + y_\lambda) \uparrow |x+y| \wedge (|x| + |y|) = |x+y|$$

and $|\mu| x_\alpha \uparrow |\mu||x| = |\mu x|$ for each scalar μ, it follows that B is a vector subspace. Also, if $|z| \leq |x|$, then from $\{|z| \wedge x_\alpha\} \subseteq J$ (for this use the fact that J is an ideal) and $|z| \wedge x_\alpha \uparrow |z| \wedge |x| = |z|$, we see that $z \in B$, and so B is an ideal.

Now assume that a net $\{u_\beta\}$ in B satisfies $0 \leq u_\beta \uparrow u$ in E. If we let

$$D = \{v \in J\colon\ v \leq u_\beta \ \text{ for some } \beta\},$$

then an easy argument shows that $D \uparrow u$. Therefore, considering D as a net in J, we infer that $u \in B$. Now a glance at Problem 1.2.5 guarantees that B is a band, as desired.

To establish the formula for the principal band B_x, note first that $B_x = B_{E_x}$, and then let $y \in B_x$. By the preceding case, there exists a net $\{x_\alpha\}$ of E_x satisfying $0 \leq x_\alpha \uparrow |y|$. For each index α, there exists some n such that $x_\alpha \leq |y| \wedge n|x| \leq |y|$. This implies $|y| \wedge n|x| \uparrow |y|$, and from this the desired conclusion follows. ∎

Problem 1.2.7. *Show that in an Archimedean Riesz space the band generated by a non-empty set A coincides with $A^{\mathrm{dd}} = (A^{\mathrm{d}})^{\mathrm{d}}$, i.e., $B_A = A^{\mathrm{dd}}$. In particular, conclude from this that every band B in an Archimedean Riesz space satisfies $B = B^{\mathrm{dd}}$.*

Solution: It can be easily verified that S^{d} is a band for every non-empty subset S of a Riesz space. To establish this, one must use just the order continuity of the lattice operations.

Now assume that E is an Archimedean Riesz space, and let A be a non-empty subset of E. Clearly, $A \subseteq A^{\mathrm{dd}}$, and this implies that $B_A \subseteq A^{\mathrm{dd}}$. To finish the proof, we must verify that $A^{\mathrm{dd}} \subseteq B_A$. To this end, let $0 < x \in A^{\mathrm{dd}}$ and consider the set

$$D = \{z \in B_A\colon\ 0 \le z \le x\}\,.$$

Clearly, $D\uparrow$, and we claim that $D \uparrow x$.

To see this, suppose that a vector $z \in E$ satisfies $a \le z$ for all $a \in D$. Then $a \le x \wedge z$ also holds for all $a \in D$. We must establish that $x \le z$ or, equivalently, that $x = x \wedge z$. Assume by way of contradiction that $u = x - x \wedge z > 0$. Clearly, $u \in A^{\mathrm{dd}}$. Since $A^{\mathrm{d}} \cap A^{\mathrm{dd}} = \{0\}$, we see that $u \notin A^{\mathrm{d}}$. This means that there exists some $y \in A$ such that $v = |y| \wedge u > 0$, and clearly $v \in B_A$. Then $v \in D$ and $2v = v + v \le x \wedge z + (x - x \wedge z) = x$. Now an easy inductive argument shows that $nv \le x$ for each n. Since E is Archimedean, the latter shows that $v = 0$, which is a contradiction. Hence, $u = 0$ and so $x = x \wedge z \le z$.

Thus, we have shown that $D \uparrow x$. Since $D \subseteq B_A$ and B_A is a band, we get $x \in B_A$. Therefore, $B_A = A^{\mathrm{dd}}$. ∎

Problem 1.2.8. *Show that a vector $e > 0$ in an Archimedean Riesz space is a weak unit if and only if $|x| \wedge e = 0$ implies $x = 0$.*

Solution: Let $e > 0$ be a vector in an Archimedean Riesz space E. Assume first that e is a weak unit (i.e., $B_e = E$), and let $|x| \wedge e = 0$. This implies $|x| \wedge ne = 0$ for all n. From Problem 1.2.6 we know that $0 = |x| \wedge ne \uparrow |x|$. Since the order limits are uniquely determined, it follows that $|x| = 0$ or $x = 0$. (Note that this implication does require E to be Archimedean.)

For the converse, assume that $|x| \wedge e = 0$ implies $x = 0$, i.e., $B_e^{\mathrm{d}} = \{0\}$. Since E is Archimedean, it follows from Problem 1.2.7 that $B_e = B_e^{\mathrm{dd}} = \{0\}^{\mathrm{d}} = E$, i.e., e is a weak unit. ∎

Problem 1.2.9. *Show that a positive operator $T\colon X \to Y$ between two partially ordered vector spaces is order continuous if and only if $x_\alpha \downarrow 0$ in X implies $Tx_\alpha \downarrow 0$ in Y.*

Solution: Let $T\colon X \to Y$ be a positive operator between ordered spaces. Assume first that T is order continuous and let $x_\alpha \downarrow 0$ in X. Since $x_\alpha \xrightarrow{o} 0$ is trivially true in X, we have $Tx_\alpha \xrightarrow{o} 0$ in Y. Also, since T is positive, we have $0 \le Tx_\alpha \downarrow$ in Y. To see that $Tx_\alpha \downarrow 0$, assume that $Tx_\alpha \ge y$ for all α and some $y \in Y$. Pick two nets $\{u_\alpha\}$ and $\{v_\alpha\}$ in Y satisfying $u_\alpha \le Tx_\alpha \le v_\alpha$ for each α, $u_\alpha \uparrow 0$, and $v_\alpha \downarrow 0$. Now notice that $y \le Tx_\alpha \le v_\alpha$ holds for all α, and from this we easily get that $y \le 0$. Hence, $Tx_\alpha \downarrow 0$.

Now assume that $x_\alpha \downarrow 0$ in X implies $Tx_\alpha \downarrow 0$ in Y, and suppose $z_\lambda \xrightarrow{o} z$ in X. Pick two nets $\{a_\lambda\}$ and $\{b_\lambda\}$ in X satisfying $a_\lambda \le z_\lambda \le b_\lambda$ for each λ, $a_\lambda \uparrow z$ and $b_\lambda \downarrow z$. Now the positivity of T implies

$$Ta_\lambda \le Tz_\lambda \le Tb_\lambda \qquad \text{and} \qquad Ta_\lambda \le Tz \le Tb_\lambda\,, \tag{$\star$}$$

as well as $Ta_\lambda \uparrow$ and $Tb_\lambda \downarrow$. Next, notice that $b_\lambda - a_\lambda \downarrow 0$ holds in X, and so (according to our hypothesis) $Tb_\lambda - Ta_\lambda = T(b_\lambda - a_\lambda) \downarrow 0$ in Y. Now a glance at $(\star)$ guarantees that $Ta_\lambda \uparrow Tz$ and $Tb_\lambda \downarrow Tz$ in Y, proving that $Tz_\lambda \xrightarrow{o} Tz$ in Y. That is, T is order continuous. ∎

Problem 1.2.10. *Show the following:*

(a) *If Ω is a compact topological space, then in $C(\Omega)$ each component of the constant function $\mathbf{1}$ is of the form χ_V for some clopen set V.*

(b) *If μ is a finite measure, then in $L_1(\mu)$ each component of the constant function $\mathbf{1}$ is of the form χ_A for some measurable set A.*

Solution: (a) If V is a clopen subset of Ω, then χ_V is clearly a component of $\mathbf{1}$ (the constant function one on Ω). Now let $f \in C(\Omega)$ be a component of $\mathbf{1}$, i.e., $\mathbf{1} \wedge (\mathbf{1} - f) = 0$. Therefore, $0 \le f \le \mathbf{1}$ and $\min\{f(\omega), 1 - f(\omega)\} = 0$ for each $\omega \in \Omega$. Hence, $f = \chi_V$ for some subset V of Ω. Since f is a continuous function, it is easy to see that V must be a clopen subset of Ω.

(b) Let (X, Σ, μ) be a finite measure space. As before, if $f \in L_1(\mu)$ is a component of $\mathbf{1}$, then $f = \chi_A$ for some subset A of X. Since f is measurable, it follows that A must be a measurable subset of X. ∎

Problem 1.2.11. *If E and F are Riesz spaces with F Dedekind complete, then show that the lattice operations of any two operators $S, T \in \mathcal{L}_b(E, F)$ are given by the Riesz–Kantorovich formulas*

$$\begin{aligned}(S \vee T)(x) &= \sup\{Sy + Tz\colon\ y, z \in E^+ \ \text{ and } \ y + z = x\}, \text{ and}\\ (S \wedge T)(x) &= \inf\{Sy + Tz\colon\ y, z \in E^+ \ \text{ and } \ y + z = x\}\end{aligned}$$

for each $x \in E^+$.

Solution: We prove the first formula only. From the well-known lattice identity $S \vee T = T + (S - T) \vee 0$, it suffices to establish the formula when $T = 0$. So, assume $T = 0$ and $S \in \mathcal{L}_b(E, F)$. Define the mapping $R\colon E^+ \to F^+$ via

$$Rx = \sup_{0 \le y \le x} Sy\,.$$

Since F is Dedekind complete and S is order bounded, it follows that R is indeed a well-defined mapping from E^+ to F^+. We claim that R is additive.

For this, let $u, v \in E^+$. Since $0 \le y \le u$ and $0 \le z \le v$ imply $0 \le y + z \le u + v$, we get $Sy + Sz = S(y + z) \le R(u + v)$. This implies $Ru + Rv \le R(u + v)$. For the reverse inequality, assume that $0 \le y \le u + v$. Then, by the Riesz Decomposition Property, there exist $0 \le y_1 \le u$ and $0 \le y_2 \le v$ satisfying $y_1 + y_2 = y$, and so $Sy = Sy_1 + Sy_2 \le Ru + Rv$. This implies $R(u + v) \le Ru + Rv$, and thus $R(u + v) = Ru + Rv$. That is, R is additive.

Now by Problem 1.2.2, R has a unique linear positive extension (which we shall denote by R again) from E to F. To finish the proof, we shall show that $R = S \vee 0$ holds in $\mathcal{L}_b(E, F)$.

To this end, note first that $Sx \le Rx$ for each $x \in E^+$, and hence $S \le R$. Now assume that another operator $Q \in \mathcal{L}_b(E, F)$ satisfies $Q \ge S$ and $Q \ge 0$. If $x \in E^+$, then for each $0 \le y \le x$ we have $Sy \le Qy \le Qx$, and so $Rx = \sup_{0 \le y \le x} Sy \le Qx$. Thus, $Q \ge R$ in $\mathcal{L}_b(E, F)$. That is, R is the least upper bound of S and 0 in $\mathcal{L}_b(E, F)$ or $R = S \vee 0$, as desired. ∎

Problem 1.2.12. *For a normed Riesz space E establish the following properties.*

(a) *Its cone E^+ is norm (and hence weakly) closed.*

(b) *The order intervals of E are norm (and hence weakly) closed.*

(c) *If a net $\{x_\alpha\}$ in E satisfies $x_\alpha \uparrow$ and $\|x_n - x\| \to 0$, then $x_\alpha \uparrow x$.*

Solution: (a) To see that the cone E^+ is closed, assume that a net $\{y_\lambda\}$ in E^+ satisfies $y_\lambda \to y$ in E. From $|y_\lambda - y^+| \le |y_\lambda - y|$, it follows that $\|y_\lambda - y^+\| \le \|y_\lambda - y\|$, and so $y_\lambda \to y^+$. Thus, $y = y^+ \ge 0$, proving that E^+ is closed. Since, E^+ is a convex set, it follows that E^+ is also weakly closed.

(b) If $x \le y$, then $[x, y] = (x + E^+) \cap (y - E^+)$, and so the order interval $[x, y]$ is norm closed. Since $[x, y]$ is also convex, it must also be weakly closed.

(c) Assume that a net $\{x_\alpha\}$ in E satisfies $x_\alpha \uparrow$ and $x_\alpha \to x$. If α is fixed, then note that the net $\{x_\beta - x_\alpha\}_{\beta \ge \alpha}$ in E^+ satisfies $x_\beta - x_\alpha \xrightarrow[\beta]{} x - x_\alpha$. So, $x - x_\alpha \in E^+$ and thus $x \ge x_\alpha$ for each α. This implies that x is an upper bound of the net $\{x_\alpha\}$. Now assume that $y \ge x_\alpha$ for each α. From $y - x_\alpha \in E^+$ for each α and $y - x_\alpha \to y - x$, it follows that $y - x \in E^+$ or $y \ge x$. Thus, $x = \sup\{x_\alpha\}$. ∎

Problem 1.2.13. *Show that if $x_n \to x$ holds in a Banach lattice, then there exist a subsequence $\{y_n\}$ of $\{x_n\}$ and some $u \ge 0$ satisfying $|y_n - x| \le \frac{1}{n}u$ for each n.*

Solution: Assume that $x_n \to x$ holds in a Banach lattice E. Pick a subsequence $\{y_n\}$ of $\{x_n\}$ satisfying $\|y_n - x\| \le \frac{1}{n2^n}$ for each n. Then $\|n|y_n - x|\| \le \frac{1}{2^n}$ for each n and since E is a Banach space, it follows that the series $\sum_{n=1}^{\infty} n|y_n - x|$ is norm convergent in E. Let $u = \sum_{n=1}^{\infty} n|y_n - x| \in E$. Now a glance at the preceding Problem 1.2.12 guarantees that $n|y_n - x| \le u$, that is, $|y_n - x| \le \frac{1}{n}u$ for each n. ∎

Problem 1.2.14. *Show that every order bounded operator from a Banach lattice to a normed Riesz space is continuous.*

Solution: Let $T\colon E \to F$ be an order bounded operator from a Banach lattice to a normed Riesz space. Assume by way of contradiction that T is not continuous. So, there exists a sequence $\{x_n\}$ in E satisfying $x_n \to 0$ in E and $Tx_n \not\to 0$ in F. This means that there exist a subsequence $\{y_n\}$ of $\{x_n\}$ and some $\epsilon > 0$ such that $\|Ty_n\| > \epsilon$ for each n (and, of course, $y_n \to 0$ in E). Now, by the preceding Problem 1.2.13, there exist a subsequence $\{z_n\}$ of $\{y_n\}$ and some $u \in E^+$ satisfying $|z_n| \le \frac{1}{n}u$ for each n. Since T is order bounded, there exists some $w \in F$ such that $T\big([-u, u]\big) \subseteq [-w, w]$. From $|nz_n| \le u$, it follows that $n|T(z_n)| = \big|T(nz_n)\big| \le w$ or $|Tz_n| \le \frac{1}{n}w$ for each n. But then we have

$$0 < \epsilon \le \|Tz_n\| \le \tfrac{1}{n}\|w\| \to 0\,,$$

which is impossible. Hence, T is a continuous operator. ∎

Problem 1.2.15. *Show that for a given Riesz space E there is, up to an equivalence, at most one lattice norm which makes E a Banach lattice.*

Solution: Assume that a Riesz space E is a Banach lattice under two lattice norms $\|\cdot\|_1$ and $\|\cdot\|_2$. Then the identity operator $I\colon (E, \|\cdot\|_1) \to (E, \|\cdot\|_2)$ is positive and hence (by the preceding Problem 1.2.14) it must be a homeomorphism. ∎

Problem 1.2.16. *Show that a normed Riesz space E is a Banach lattice if and only if every increasing norm Cauchy sequence of E^+ is norm convergent.*

Solution: The "only if" part is trivial. To establish the "if" part let $\{x_n\}$ be a norm Cauchy sequence in E. It suffices to establish that $\{x_n\}$ has a convergent subsequence.

Pick a subsequence $\{u_n\}$ of $\{x_n\}$ satisfying $\|u_{n+1} - u_n\| < \frac{1}{2^n}$ for each n, and then consider the sequences $\{y_n\}$ and $\{z_n\}$ in E^+ defined by

$$y_n = \sum_{i=1}^{n} (u_{i+1} - u_i)^+ \quad \text{and} \quad z_n = \sum_{i=1}^{n} (u_{i+1} - u_i)^- .$$

Clearly, $0 \le y_n \uparrow$ and from

$$\begin{aligned} \|y_{n+p} - y_n\| &= \Big\| \sum_{i=n+1}^{n+p} (u_{i+1} - u_i)^+ \Big\| \le \sum_{i=n+1}^{n+p} \|(u_{i+1} - u_i)^+\| \\ &\le \sum_{i=n+1}^{n+p} \|u_{i+1} - u_i\| \le \sum_{i=n+1}^{n+p} \tfrac{1}{2^i} < \tfrac{1}{2^n} , \end{aligned}$$

we see that $\{y_n\}$ is a norm Cauchy sequence. Similarly, the increasing sequence $\{z_n\}$ in E^+ is norm Cauchy. By our hypothesis both sequences are norm convergent. Let $y_n \to y$ and $z_n \to z$.

Next, notice that the lattice identity $x = x^+ - x^-$ implies $y_n - z_n = u_{n+1} - u_1$ for each n. In particular, it follows that $u_n = y_{n-1} - z_{n-1} + u_1 \to y - z + u_1$. This guarantees that E is norm complete. ∎

Problem 1.2.17. *Show that if E and F are two Banach lattices with F Dedekind complete, then the Dedekind complete Riesz space $\mathcal{L}_r(E, F)$ of all regular operators from E to F equipped with the r-norm is a Banach lattice.*

Solution: The verification of the fact that $\|\cdot\|_r$ is a norm on $\mathcal{L}_r(E, F)$ is straightforward. Next, we shall verify that $\|\cdot\|_r$ is a lattice norm.

Notice first that if T is a positive operator, then $\|T\|_r = \big\||T|\big\| = \|T\|$. Moreover, if two operators $S, T \in \mathcal{L}_r(E, F)$ satisfy $0 \le S \le T$ and $\|x\| = \big\||x|\big\| \le 1$, then the inequalities $|Sx| \le S|x| \le T|x|$ imply

$$\|Sx\| = \big\||Sx|\big\| \le \big\|S|x|\big\| \le \big\|T|x|\big\| \le \|T\| .$$

So, $\|S\| = \sup_{\|x\| \le 1} \|Sx\| \le \|T\|$. In particular, if $S, T \in \mathcal{L}_r(E, F)$ satisfy $|S| \le |T|$, then

$$\|S\|_r = \big\||S|\big\| \le \big\||T|\big\| = \|T\|_r .$$

This proves that $\|\cdot\|_r$ is a lattice norm on $\mathcal{L}_r(E,F)$.

To establish that $\|\cdot\|_r$ is a complete norm, it suffices (according to Problem 1.2.16) to show that every increasing $\|\cdot\|_r$-Cauchy sequence of positive operators is $\|\cdot\|_r$-convergent in $\mathcal{L}_r(E,F)$. To this end, let $\{T_n\}$ be an increasing $\|\cdot\|_r$-Cauchy sequence of positive operators. From $\|T_n - T_m\| \le \|T_n - T_m\|_r$, it follows that $\{T_n\}$ is a $\|\cdot\|$-norm Cauchy sequence. So, by Problem 1.1.4, there exists an operator $T \in \mathcal{L}(E,F)$ satisfying $\|T_n - T\| \to 0$.

For each $x \in E^+$ we have $0 \le T_n x \uparrow$ and $\|T_n x - Tx\| \to 0$, and hence Problem 1.2.12(c) guarantees that $T_n x \uparrow Tx$ for each $x \in E^+$. In particular, T is a positive operator (and so $T \in \mathcal{L}_r(E,F)$) and $0 \le T_n \le T$ holds for each n. This implies

$$\|T_n - T\|_r = \big\||T_n - T|\big\| = \|T - T_n\| \to 0\,,$$

proving that $\mathcal{L}_r(E,F)$ under the r-norm is a Banach lattice. ∎

Problem 1.2.18. *Give an example of a regular operator $T\colon E \to F$ between two Banach lattices with F Dedekind complete satisfying $\|T\| < \|T\|_r$.*

Solution: Let $E = F = \mathbb{R}^2$ be equipped with its Euclidean norm $\|(x,y)\| = (x^2+y^2)^{\frac{1}{2}}$. The matrix $A = \begin{bmatrix} 1 & -1 \\ 1 & 1 \end{bmatrix}$ defines an order bounded operator on E whose modulus is given by the matrix $|A| = \begin{bmatrix} 1 & 1 \\ 1 & 1 \end{bmatrix}$. (For the computation of the modulus of a matrix see also Problem 8.1.11.) Now let us compute the norms of these two matrices. To this end, fix some vector $(x,y) \in \mathbb{R}^2$ with $x^2+y^2 = 1$. Then we have

$$\left\|A\begin{bmatrix} x \\ y \end{bmatrix}\right\|^2 = \left\|\begin{bmatrix} 1 & -1 \\ 1 & 1 \end{bmatrix}\begin{bmatrix} x \\ y \end{bmatrix}\right\|^2 = \left\|\begin{bmatrix} x-y \\ x+y \end{bmatrix}\right\|^2 = (x-y)^2 + (x+y)^2 = 2(x^2+y^2) = 2\,.$$

This implies $\|A\| = \sqrt{2}$.

For the modulus $|A|$ of A note that:

$$\begin{aligned} \left\||A|\begin{bmatrix} x \\ y \end{bmatrix}\right\|^2 &= \left\|\begin{bmatrix} 1 & 1 \\ 1 & 1 \end{bmatrix}\begin{bmatrix} x \\ y \end{bmatrix}\right\|^2 = \left\|\begin{bmatrix} x+y \\ x+y \end{bmatrix}\right\|^2 \\ &= 2(x+y)^2 = 2(1+2xy)\,. \end{aligned}$$

Now an easy computation shows that $\big\||A|\big\| = \sqrt{2(1+1)} = 2$. In particular, we have

$$\|A\| = \sqrt{2} < 2 = \big\||A|\big\| = \|A\|_r\,.$$

So, A is a desired example. ∎

Problem 1.2.19. *Show that a band projection P on a normed Riesz space is a contraction.*

Solution: Let $P\colon E \to E$ be a band projection on a normed Riesz space. That is, the range $B = P(E)$ of P is a projection band. Hence, each $x \in E$ can be written in the form $x = y + z$ with $y \in B$ and $z \in B^{\mathrm{d}}$. From $y \perp z$, it follows that

$$|Px| = |y| \le |y| + |z| = |y+z| = |x|\,,$$

and so $\|Px\| = \||Px|\| \le \||x|\| = \|x\|$ for all $x \in E$. Hence, $\|P\| \le 1$, that is, P is a contraction. (If $P \ne 0$, then of course $\|P\| = 1$.) ∎

Problem 1.2.20. *If E is a normed Riesz space and $x^* \in E^*$, then show that $\|x^*\| = \|x^*\|_r = \|\,|x^*|\,\|$.*

Solution: From the inequalities

$$|x^*(x)| \le |x^*|(|x|) \le \|\,|x^*|\,\| \cdot \|x\|\,,$$

we infer that $\|x^*\| \le \|\,|x^*|\,\|$. On the other hand, if $\|x\| \le 1$ and $\epsilon > 0$ are fixed, then from $|x^*|(|x|) = \sup\{|x^*(y)|\colon\ |y| \le |x|\,\}$, we see that there exists some $y \in E$ with $|y| \le |x|$ such that $|x^*|(|x|) - \epsilon < |x^*(y)| \le \|x^*\|$. Therefore,

$$\|\,|x^*|\,\| = \sup\{|x^*|(|x|)\colon\ \|x\| \le 1\} \le \|x^*\| + \epsilon$$

for each $\epsilon > 0$. This implies $\|\,|x^*|\,\| \le \|x^*\|$ and so $\|\,|x^*|\,\| = \|x^*\|$. ∎

Problem 1.2.21. *A Dedekind complete Riesz space $\hat{E}$ is said to be a **Dedekind completion** of a Riesz space E if there exists a lattice isomorphism $\pi\colon E \to \hat{E}$ such that*

$$\hat{x} = \sup\bigl\{\pi(v)\colon\ v \in E \text{ and } \pi(v) \le \hat{x}\bigr\} = \inf\bigl\{\pi(w)\colon\ w \in E \text{ and } \hat{x} \le \pi(w)\bigr\}$$

for each $\hat{x} \in \hat{E}$. We can identify E with $\pi(E)$ so that E can be viewed as a Riesz subspace of $\hat{E}$. It should be clear that only Archimedean Riesz spaces can have Dedekind completions. As a matter of fact, every Archimedean Riesz space E has a unique (up to a lattice isomorphism) Dedekind completion E^δ.

Now assume that E is a Banach lattice (so E is an Archimedean Riesz space) and let E^δ be its Dedekind completion. On E^δ define the norm

$$|||\hat{x}||| = \inf\bigl\{\|x\|\colon\ x \in E^+ \text{ and } |\hat{x}| \le x\bigr\}\,.$$

Show that $|||x||| = \|x\|$ for each $x \in E$ and that $(E^\delta, |||\cdot|||)$ is a Banach lattice.

Solution: The verification of the norm properties by $|||\cdot|||$ (including its lattice property) is straightforward and is omitted. Also, the equality $|||x||| = \|x\|$ for each $x \in E$ should be obvious. For simplicity, we shall denote $|||\cdot|||$ by $\|\cdot\|$ again.

To check the norm completeness of E^δ, we use Problem 1.2.16. Assume that a sequence $\{\hat{x}_n\}$ in E^δ satisfies $0 \le \hat{x}_n \uparrow$ and $\|\hat{x}_{n+1} - \hat{x}_n\| < \frac{1}{2^n}$ for all n. For each n pick some $y_n \in E^+$ such that $0 \le \hat{x}_{n+1} - \hat{x}_n \le y_n$ and $\|y_n\| < \frac{1}{2^n}$. If we let $x_n = \sum_{i=1}^n y_i$, then $\{x_n\}$ is a norm Cauchy sequence in E and so $\{x_n\}$ is norm convergent in E, say to $x = \sum_{n=1}^\infty y_n \in E$. From Problem 1.2.12(c), we know that $x_n \uparrow x$ in E. In particular, if we let $z_n = \sum_{i=n}^\infty y_i \in E$, then $\|z_n\| \to 0$. Letting $\hat{x}_0 = 0$ and choosing some $y_0 \in E$ such that $y_0 \ge \hat{x}_1$, we see that

$$\hat{x}_n = \sum_{i=1}^n (\hat{x}_i - \hat{x}_{i-1}) \le \sum_{i=1}^n y_{i-1} = x_{n-1} + y_0\,.$$

This implies $\hat{x}_n \le x + y_0$ for each n. Since E^δ is Dedekind complete, there exists some $\hat{x} \in E^\delta$ such that $\hat{x}_n \uparrow \hat{x}$.

Next, note that for each n and p we have

$$0 \le \hat{x}_{n+p} - \hat{x}_n = \sum_{i=n}^{n+p-1} (\hat{x}_{i+1} - \hat{x}_i) \le \sum_{i=n}^{n+p-1} y_i \le \sum_{i=n}^{\infty} y_i = z_n ,$$

where the last inequality holds true by virtue of Problem 1.2.12(c). So, by letting $p \to \infty$, we get $0 \le \hat{x} - \hat{x}_n \le z_n$ for each n. This implies $\|\hat{x} - \hat{x}_n\| \le \|z_n\|$, and so $\|\hat{x}_n - \hat{x}\| \to 0$. Consequently, $(E^\delta, |||\cdot|||)$ is a Banach lattice. ∎

Problem 1.2.22 (Nakano [**60**]). *Let E be a Riesz space and let $f \in E^\sim$. The **null ideal** of the functional f is the ideal $N_f = \{x \in E\colon\ |f|(|x|) = 0\}$. The band $C_f = N_f^{\mathrm{d}}$ is called the **carrier** of f.*

Assume that E is Archimedean and $f, g \in E_{\mathrm{n}}^\sim$. Show that $|f| \wedge |g| = 0$ if and only if $C_f \wedge C_g = \{0\}$.

Proof. We can assume $0 \le f, g \in E_{\mathrm{n}}^\sim$. Suppose first that $f \wedge g = 0$. Fix a vector $0 \le x \in C_f = N_f^{\mathrm{d}}$ and let $\epsilon > 0$. From $[f \wedge g](x) = \inf\{f(y) + g(x-y)\colon\ y \in [0,x]\} = 0$, we see that there is a sequence $\{y_n\} \subseteq [0,x]$ such that $f(y_n) + g(x - y_n) < 2^{-n}\epsilon$ for each n. If $x_n = \bigwedge_{i=1}^n y_i$, then note that $x_n \downarrow 0$. Indeed, if $0 \le z \le x_n$ for each n, then $0 \le f(z) \le f(x_n) \le f(y_n) < 2^{-n}\epsilon$, and hence $f(z) = 0$. This shows that $z \in N_f \cap C_f = \{0\}$ or $z = 0$.

From the inequality $x - x_n = x - \bigwedge_{i=1}^n y_i = \bigvee_{i=1}^n (x - y_i) \le \sum_{i=1}^n (x - y_i)$, it follows that $0 \le g(x - x_n) \le \sum_{i=1}^n g(x - y_i) < \epsilon$ holds for each n. Since g is order continuous and $x - x_n \uparrow x$, we get that $g(x) = \lim_{n\to\infty} g(x - x_n) \le \epsilon$ for all $\epsilon > 0$. Thus, $g(x) = 0$. This implies $C_f \subseteq N_g$, and consequently $C_f \cap C_g = \{0\}$.

For the converse, assume that $C_f \cap C_g = \{0\}$. Since f is order continuous, it follows that N_f is a band. This implies $C_g \subseteq C_f^{\mathrm{d}} = N_f^{\mathrm{dd}} = N_f$. Consequently, if $0 \le x = y + z \in N_g \oplus C_g$, then

$$0 \le (f \wedge g)(x) \le (f \wedge g)(y) + (f \wedge g)(z) \le g(y) + f(z) = 0 .$$

That is, $f \wedge g = 0$ on the order dense ideal $N_g \oplus C_g$. Since E is Archimedean, this guarantees that $f \wedge g = 0$. ∎

1.3. Bases in Banach Spaces

Problem 1.3.1. *Let X be a Banach space with a basis $\{x_n\}$, and for each vector $x \in X$ let $x = \sum_{n=0}^\infty \alpha_n x_n$ be its series representation with respect to the basis $\{x_n\}$. Show that the function $|||\cdot|||\colon X \to \mathbb{R}$, defined by*

$$|||x||| = \sup_{n\in\mathbb{N}} \Big\| \sum_{i=1}^n \alpha_i x_i \Big\| ,$$

is a norm that is equivalent to the norm $\|\cdot\|$ *of* X.

Solution: We shall verify first that the function $|||\cdot|||$ is a norm. Clearly, $|||x||| \geq 0$ for all $x \in X$, and $|||0||| = 0$. Conversely, $|||x||| = 0$ is equivalent to $\left\|\sum_{i=1}^{n} \alpha_i x_i\right\| = 0$ for each n, which in turn (by the definition of the basis) is equivalent to $\alpha_i = 0$ for each $i \in \mathbb{N}$, that is, $x = 0$.

For the homogeneity of $|||\cdot|||$ note that if λ is a scalar, then

$$|||\lambda x||| = \sup_{n\in\mathbb{N}} \Big\|\sum_{i=1}^{n} \lambda\alpha_i x_i\Big\| = \sup_{n\in\mathbb{N}} |\lambda| \Big\|\sum_{i=1}^{n} \alpha_i x_i\Big\| = |\lambda| \sup_{n\in\mathbb{N}} \Big\|\sum_{i=1}^{n} \alpha_i x_i\Big\| = |\lambda| \cdot |||x|||\,.$$

For the triangle inequality note that if $y = \sum_{n=1}^{\infty} \beta_n x_n$, then from

$$\Big\|\sum_{i=1}^{n} (\alpha_i + \beta_i) x_i\Big\| \leq \Big\|\sum_{i=1}^{n} \alpha_i x_i\Big\| + \Big\|\sum_{i=1}^{n} \beta_i x_i\Big\| \leq |||x||| + |||y|||\,,$$

it follows that $|||x + y||| \leq |||x||| + |||y|||$.

The non-trivial part is to prove that $|||\cdot|||$ is a complete norm, i.e., that $(X, |||\cdot|||)$ is a Banach space. To this end, let $\{z_n\}$ be a $|||\cdot|||$-Cauchy sequence in X, where $z_n = \sum_{i=1}^{\infty} \alpha_i^n x_i$. From $|||z_n - z_m||| = \sup_k \left\|\sum_{i=1}^{k} (\alpha_i^n - \alpha_i^m) x_i\right\|$, we see that

$$\begin{aligned} |\alpha_j^n - \alpha_j^m| \cdot \|x_j\| &= \Big\|\sum_{i=1}^{j} (\alpha_i^n - \alpha_i^m) x_i - \sum_{i=1}^{j-1} (\alpha_i^n - \alpha_i^m) x_i\Big\| \\ &\leq \Big\|\sum_{i=1}^{j} (\alpha_i^n - \alpha_i^m) x_i\Big\| + \Big\|\sum_{i=1}^{j-1} (\alpha_i^n - \alpha_i^m) x_i\Big\| \\ &\leq 2|||z_n - z_m|||\,. \end{aligned}$$

This implies that the sequence $\{\alpha_j^n\}_{n\in\mathbb{N}}$ is a Cauchy sequence of scalars for each j and so $\alpha_j = \lim_{n\to\infty} \alpha_j^n$ exists. To complete the proof of this part, we shall show that the series $z = \sum_{i=1}^{\infty} \alpha_i x_i$ is norm convergent in X and that $|||z_n - z||| \to 0$.

To this end, fix $\epsilon > 0$, and then pick a natural number n_0 such that

$$\Big\|\sum_{i=1}^{k} (\alpha_i^n - \alpha_i^m) x_i\Big\| \leq \sup_{r\in\mathbb{N}} \Big\|\sum_{i=1}^{r} (\alpha_i^n - \alpha_i^m) x_i\Big\| = |||z_n - z_m||| < \epsilon \qquad (\star)$$

for all $n, m \geq n_0$ and all $k \in \mathbb{N}$. Letting $n \to \infty$, we get $\left\|\sum_{i=1}^{k} (\alpha_i - \alpha_i^m) x_i\right\| \leq \epsilon$ for all $m \geq n_0$ and each $k \in \mathbb{N}$. To verify the norm convergence of the series $\sum_{i=1}^{\infty} \alpha_i x_i$, let $\sigma_n = \sum_{i=1}^{n} \alpha_i x_i$. Then for $m > n \geq n_0$ we have

$$\begin{aligned} \|\sigma_m - \sigma_n\| &= \Big\|\sum_{i=n+1}^{m} \alpha_i x_i\Big\| \leq \Big\|\sum_{i=n+1}^{m} (\alpha_i - \alpha_i^{n_0}) x_i\Big\| + \Big\|\sum_{i=n+1}^{m} \alpha_i^{n_0} x_i\Big\| \\ &\leq \Big\|\sum_{i=1}^{m} (\alpha_i - \alpha_i^{n_0}) x_i\Big\| + \Big\|\sum_{i=1}^{n} (\alpha_i - \alpha_i^{n_0}) x_i\Big\| + \Big\|\sum_{i=n+1}^{m} \alpha_i^{n_0} x_i\Big\| \\ &\leq 2\epsilon + \Big\|\sum_{i=n+1}^{m} \alpha_i^{n_0} x_i\Big\|\,. \end{aligned}$$

Now use the fact that the series $\sum_{i=1}^{\infty} \alpha_i^{n_0} x_i$ is norm convergent in X to select some $k_0 \geq n_0$ such that $\big\|\sum_{i=n+1}^{m} \alpha_i^{n_0} x_i\big\| < \epsilon$ holds for all $m \geq k_0$. So, for all $m > n \geq k_0$ we have $\|\sigma_m - \sigma_n\| \leq 3\epsilon$. This shows that $\{\sigma_n\}$ is a $\|\cdot\|$-Cauchy sequence, and hence $z = \lim_{n\to\infty} \sigma_n = \sum_{i=1}^{\infty} \alpha_i x_i$ exists in X.

Next, notice that by letting $m \to \infty$ in ($\star$), we get $\big\|\sum_{i=1}^{k}(\alpha_i^n - \alpha_i)x_i\big\| \leq \epsilon$ for all $k \in \mathbb{N}$ and all $n \geq n_0$. Therefore, for each $n \geq n_0$ we have

$$|||z_n - z||| = \sup_{k\in\mathbb{N}} \Big\|\sum_{i=1}^{k}(\alpha_i^n - \alpha_i)x_i\Big\| \leq \epsilon\,,$$

so that $|||z_n - z||| \to 0$.

Since $\|x\| \leq |||x|||$ holds for each $x \in X$, it follows from the Open Mapping Theorem that $\|\cdot\|$ and $|||\cdot|||$ are equivalent norms. So, there exists some constant $c > 0$ such that $c|||x||| \leq \|x\| \leq |||x|||$ for each $x \in X$. ∎

Problem 1.3.2. *Show that a sequence $\{x_n\}$ of non-zero vectors in a Banach space X is a basis if and only if it satisfies the following properties:*

(a) *The linear span of the set $\{x_1, x_2, \ldots\}$ is dense in X.*

(b) *There exists some $K > 0$ such that if $m > n$ and $\lambda_1, \ldots, \lambda_m$ are arbitrary scalars, then*

$$\Big\|\sum_{i=1}^{n} \lambda_i x_i\Big\| \leq K\Big\|\sum_{i=1}^{m} \lambda_i x_i\Big\|\,.$$

Solution: Assume first that $\{x_n\}$ is a basis. Then (a) is obvious. For (b) note that if $P_n\colon X \to X$ denotes the natural projection defined by $P_n\big(\sum_{i=1}^{\infty}\alpha_i x_i\big) = \sum_{i=1}^{n}\alpha_i x_i$, then according to Theorem 1.37(b) there exists some constant $K > 0$ such that $\|P_n\| \leq K$ holds for each n. Therefore, if $m > n$ and $\lambda_1, \ldots, \lambda_m$ are arbitrary scalars, then

$$\Big\|\sum_{i=1}^{n} \lambda_i x_i\Big\| = \Big\|P_n\Big(\sum_{i=1}^{m} \lambda_i x_i\Big)\Big\| \leq \|P_n\|\cdot\Big\|\sum_{i=1}^{m} \lambda_i x_i\Big\| \leq K\Big\|\sum_{i=1}^{m} \lambda_i x_i\Big\|$$

holds.

For the converse, assume that conditions (a) and (b) are satisfied. Notice that if $x = \sum_{i=1}^{\infty}\alpha_i x_i = \sum_{i=1}^{\infty}\beta_i x_i$, then from (b) it follows that for each n we have

$$\Big\|\sum_{i=1}^{n}(\alpha_i - \beta_i)x_i\Big\| \leq K\Big\|\sum_{i=1}^{m}(\alpha_i - \beta_i)x_i\Big\| = \Big\|\sum_{i=1}^{m}\alpha_i x_i - \sum_{i=1}^{m}\beta_i x_i\Big\| \xrightarrow[m\to\infty]{} 0\,.$$

That is, $\sum_{i=1}^{n}\alpha_i x_i = \sum_{i=1}^{n}\beta_i x_i$ for each $n \geq 1$. This easily implies $\alpha_i = \beta_i$ for each i. Therefore, if $x = \sum_{i=1}^{\infty}\alpha_i x_i$, then the sequence $\{\alpha_n\}$ is uniquely determined. Now let

$$Y = \Big\{x \in X\colon\ \exists \text{ a sequence } \{\alpha_n\} \text{ of scalars such that } x = \sum_{n=1}^{\infty}\alpha_n x_n\Big\}\,.$$

Clearly, Y is a vector subspace of X. To complete the solution, we need to show that Y is closed. If this is done, then by (a) we have $Y = X$ and this will establish that $\{x_n\}$ is a basis of X.

To see that Y is closed, assume that a sequence $\{y_n\} \subseteq Y$ satisfies $y_n \to y$ in X. Let $y_n = \sum_{i=1}^{\infty} \alpha_i^n x_i$ and fix $\epsilon > 0$. Fix some $k \in \mathbb{N}$ such that $\|y_n - y\| < \epsilon$ for each $n \geq k$, and so $\|y_n - y_m\| < 2\epsilon$ for all $n, m \geq k$. From $\|y - y_k\| < \epsilon$, we see that there exists some $r_0 > k$ such that for all $r \geq r_0$ we have

$$\Big\| y - \sum_{i=1}^{r} \alpha_i^k x_i \Big\| < \epsilon . \tag{$\star$}$$

Since for any $\ell > r$ we have $\| \sum_{i=1}^{r} (\alpha_i^n - \alpha_i^m) x_i \| \leq K \| \sum_{i=1}^{\ell} (\alpha_i^n - \alpha_i^m) x_i \|$, by letting $\ell \to \infty$ we get

$$\Big\| \sum_{i=1}^{r} (\alpha_i^n - \alpha_i^m) x_i \Big\| \leq K \|y_n - y_m\| \tag{$\star\star$}$$

for all n, m, and r. Therefore, if $j > 1$ is fixed and $m > n \geq k$, then

$$\begin{aligned} |\alpha_j^n - \alpha_j^m| \cdot \|x_j\| &= \Big\| \sum_{i=1}^{j} (\alpha_i^n - \alpha_i^m) x_i - \sum_{i=1}^{j-1} (\alpha_i^n - \alpha_i^m) x_i \Big\| \\ &\leq \Big\| \sum_{i=1}^{j} (\alpha_i^n - \alpha_i^m) x_i \Big\| + \Big\| \sum_{i=1}^{j-1} (\alpha_i^n - \alpha_i^m) x_i \Big\| \\ &\leq 2K \|y_n - y_m\| < 4K\epsilon . \end{aligned}$$

This guarantees that for each j (including $j = 1$) the sequence of real numbers $\{a_j^n\}$ is a Cauchy sequence. Let $\alpha_j^n \xrightarrow[n\to\infty]{} \alpha_j$ for each j. Letting $n = k$ and $m \to \infty$ in ($\star\star$), we get $\| \sum_{i=1}^{r} (\alpha_i^k - \alpha_i) x_i \| \leq K\epsilon$ for each r. Therefore, it follows from ($\star$) that for $r \geq r_0$ we have

$$\Big\| y - \sum_{i=1}^{r} \alpha_i x_i \Big\| \leq \Big\| y - \sum_{i=1}^{r} \alpha_i^k x_i \Big\| + \Big\| \sum_{i=1}^{r} (\alpha_i^k - \alpha_i) x_i \Big\| < \epsilon + 4K\epsilon = (1 + 4K)\epsilon .$$

This shows that $\lim_{r\to\infty} \sum_{i=1}^{r} \alpha_i x_i = y$ or $y = \sum_{i=1}^{\infty} \alpha_i x_i$, as desired. ∎

Problem 1.3.3. *This problem describes a basis for the Banach spaces $L_p[0,1]$ with $1 \leq p < \infty$. Consider the sequence of functions $\{h_n\}$ defined by $h_1(t) = 1$ for each $t \in [0,1]$ and*

$$h_{2^k+\ell}(t) = \begin{cases} 1 & \text{if } \frac{2\ell-2}{2^{k+1}} \leq t < \frac{2\ell-1}{2^{k+1}} \\ -1 & \text{if } \frac{2\ell-1}{2^{k+1}} \leq t \leq \frac{2\ell}{2^{k+1}} \\ 0 & \text{otherwise}, \end{cases}$$

*for $k = 0, 1, 2, \ldots$ and $1 \leq \ell \leq 2^k$. The sequence $\{h_n\}$ is known as the **Haar system**. Establish the following.*

(a) *$h_n \neq 0$ for each n.*

(b) *The linear span of the set $\mathcal{H} = \{h_1, h_2, \ldots\}$ is norm dense in $L_p[0,1]$.*

(c) *For any collection of scalars* $\lambda_1, \ldots, \lambda_n, \lambda_{n+1}$ *we have*

$$\Big\|\sum_{i=1}^{n} \lambda_i h_i\Big\|_p \le \Big\|\sum_{i=1}^{n+1} \lambda_i h_i\Big\|_p .$$

(d) *The Haar system* $\{h_n\}$ *is a monotone basis in each* $L_p[0,1]$.

Solution: (a) This is obvious. The graphs of the three functions h_2, h_3, and h_4 are shown in Figure 1.

(b) Let Y denote the linear span of the set of functions in the Haar system $\mathcal{H} = \{h_1, h_2, \ldots\}$. We claim that the characteristic functions of the intervals of the form

$$\big[\tfrac{\ell}{2^k}, \tfrac{\ell+1}{2^k}\big],$$

where $0 \le \ell \le 2^k - 1$ and $k = 1, 2, \ldots$, belong to Y.

To see this, let $[a,b] = \big[\frac{\ell}{2^k}, \frac{\ell+1}{2^k}\big]$ for some $0 \le \ell \le 2^k - 1$. Also, let $[c,d] = \big[\frac{\ell+1}{2^k}, \frac{\ell+2}{2^k}\big]$ if ℓ is even, and let $[c,d] = \big[\frac{\ell-1}{2^k}, \frac{\ell}{2^k}\big]$ if ℓ is odd. We consider the first case; the second case can be established similarly. Define the function $f_0 = \chi_{[a,b]} - \chi_{[c,d]}$. A look at the definition of the Haar system reveals that $f_0 \in \mathcal{H}$.

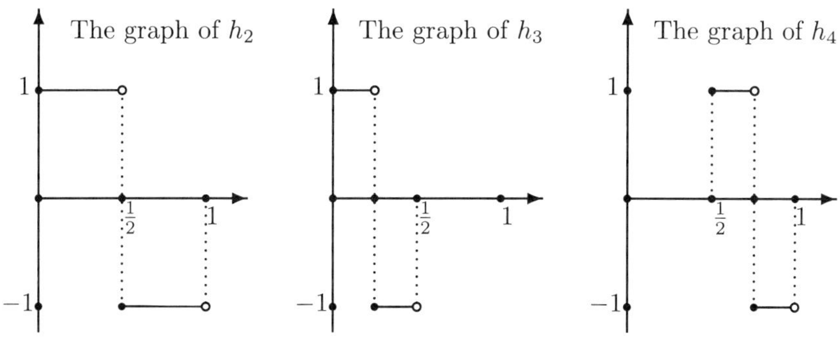

Figure 1

If c_1 is the midpoint of the interval $[c,d]$, then $g_1 = \chi_{[c,c_1)} - \chi_{[c_1,d]} \in \mathcal{H}$. In turn, this implies that the function

$$f_1 = f_0 + g_1 = \chi_{[a,b]} - \chi_{[c_1,d]}$$

belongs to Y. Now if c_2 is the midpoint of the interval $[c_1,d]$, then the function $g_2 = \chi_{[c_1,c_2)} - \chi_{[c_2,d]}$ belongs to $\mathcal{H}$, and so the function

$$f_2 = f_1 + g_2 = \chi_{[a,b]} - 2\chi_{[c_2,d]}$$

belongs to Y. Continuing in an inductive manner, we can construct a sequence $\{f_n\}$ in Y and a sequence $\{g_n\}$ in $\mathcal{H}$ such that

$$f_{n+1} = f_n - g_{n+1} = \chi_{[a,b]} - (n+1)\chi_{[c_{n+1},d]} \qquad (\dagger)$$

holds for each n. It is easy to see now that

$$\big\|f_{n+1} - \chi_{[a,b]}\big\|_p = \Big[\int_0^1 (n+1)^p \chi_{[c_{n+1},d]}(t)\,dt\Big]^{\frac{1}{p}} = \tfrac{1}{2^{pk}} \cdot \tfrac{n+1}{2^{pn}} \xrightarrow[n\to\infty]{} 0 .$$

This implies $f_n \to \chi_{[a,b]}$ in $L_p[0,1]$ and so $\chi_{[a,b]} \in \overline{Y}$. Therefore, the characteristic function of any interval of the form $[\frac{\ell}{2^k}, \frac{m}{2^k}]$, $0 \le \ell < m \le 2^k$, belongs to $\overline{Y}$.

Next, let $[u,v]$ be any subinterval of $[0,1]$ and let $\epsilon > 0$. Fix some natural number k such that $\frac{1}{2^k} < \epsilon$, and then pick two integers $0 \le \ell < m \le 2^k - 1$ such that $\frac{\ell}{2^k} \le u < \frac{\ell+1}{2^k} \le \frac{m-1}{2^k} < b \le \frac{m}{2^k}$. This yields $\int_0^1 |\chi_{[u,v]}(t) - \chi_{[\frac{\ell}{2^k},\frac{m}{2^k}]}(t)|^p\,dt \le 2\epsilon$, and so $\chi_{[u,v]} \in \overline{Y}$. The latter easily implies that $\overline{Y} = L_p[0,1]$.

(c) Assume $1 \le p < \infty$. The function $\theta(t) = t^p$, $t \ge 0$, is increasing and convex. (This follows from $\theta'(t) = pt^{p-1} \ge 0$ and $\theta''(t) = p(p-1)t^{p-2} \ge 0$ for all $t \ge 0$.) This implies that $\psi(t) = |t|^p$, $t \in \mathbb{R}$, is also a convex function. For if $x, y \in \mathbb{R}$ and $\alpha, \beta \in [0,1]$ satisfy $\alpha + \beta = 1$, then

$$\psi(\alpha x + \beta y) = |\alpha x + \beta y|^p \le (\alpha|x| + \beta|y|)^p \le \alpha|x|^p + \beta|y|^p = \alpha\psi(x) + \beta\psi(y).$$

In particular, from $x = \frac{1}{2}(x+y) + \frac{1}{2}(x-y)$, we get $|x|^p \le \frac{1}{2}|x+y|^p + \frac{1}{2}|x-y|^p$ or

$$2|x|^p \le |x+y|^p + |x-y|^p \qquad (\star)$$

for all $x, y \in \mathbb{R}$.

Now let $\lambda_1, \ldots, \lambda_n, \lambda_{n+1}$ be arbitrary scalars and consider the two functions $f = \sum_{i=1}^n \lambda_i h_i$ and $g = \sum_{i=1}^{n+1} \lambda_i h_i = f + \lambda_{n+1} h_{n+1}$. Also, let $[a,b]$ be the support-interval of h_{n+1}. Since the functions $h_1, \ldots, h_n$ are all constant on $[a,b]$, it follows that there exists some scalar c such that $f(t) = c$ for all $t \in [a,b]$. This implies that the functions f and g agree outside $[a,b]$, $f = c$ on $[a,b]$, and g takes the value $c + \lambda_{n+1}$ on the interval $[a, \frac{a+b}{2})$ and the value $c - \lambda_{n+1}$ on $[\frac{a+b}{2}, b]$. Now a glance at $(\star)$ yields

$$\begin{aligned} \int_a^b |f(t)|^p\,dt &= |c|^p(b-a) = 2|c|^p \tfrac{b-a}{2} \\ &\le |c+\lambda_{n+1}|^p \tfrac{b-a}{2} + |c-\lambda_{n+1}|^p \tfrac{b-a}{2} = \int_a^b |g(t)|^p\,dt. \end{aligned}$$

Consequently, $\int_0^1 |f(t)|^p\,dt \le \int_0^1 |g(t)|^p\,dt$ or

$$\Big\|\sum_{i=1}^n \lambda_i h_i\Big\|_p \le \Big\|\sum_{i=1}^{n+1} \lambda_i h_i\Big\|_p.$$

This shows that $\{h_n\}$ is a monotone basis of $L_p[0,1]$.

(d) The properties (a), (b), and (c) guarantee (according to Problem 1.3.2) that $\{h_n\}$ is a basis in $L_p[0,1]$ for each $1 \le p < \infty$.

Remark. It is considerably more difficult to show that the Haar system is also an unconditional basis in each $L_p[0,1]$ with $1 < p < \infty$; for details see [**72**, p. 407]. In the Banach space $L_1[0,1]$ the Haar system fails to be an unconditional basis. See Corollary 11.61 in the text where we establish that $L_1[0,1]$ does not have any unconditional basis. Moreover, A. Pelczynski [**63**] has established the following general result.

- *The Banach space $L_1[0,1]$ is not embeddable in any Banach space with an unconditional basis.*

For an alternative proof of the above result see [**52**, Proposition 1.d.1]. ∎

Problem 1.3.4. *This problem describes a basis in the Banach space $C[0,1]$. Let $\{h_n\}$ be the Haar system that was introduced in Problem* 1.3.3. *Consider the sequence of functions $\{s_n\}$ in $C[0,1]$ defined by*

$$s_1 = \mathbf{1} \quad \textit{and} \quad s_n(t) = \int_0^t h_{n-1}(\tau)\,d\tau \quad \textit{for } n = 2,3,4,\dots\,.$$

The sequence $\{s_n\}$ is called the ***Schauder system****. Show that $\{s_n\}$ is a monotone basis for $C[0,1]$.*

Solution: Clearly, each s_n is a continuous non-zero function. The graphs of the functions s_1, s_2, s_3, s_4, and s_5 are shown in Figures 2 and 3. We need to verify that the sequence $\{s_n\}$ satisfies properties (a) and (b) of Problem 1.3.2.

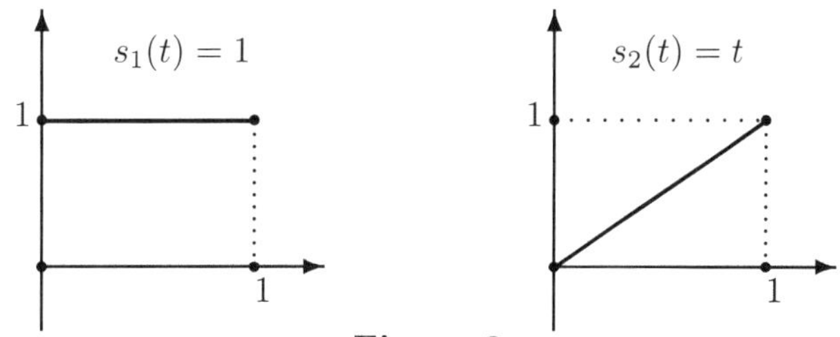

Figure 2

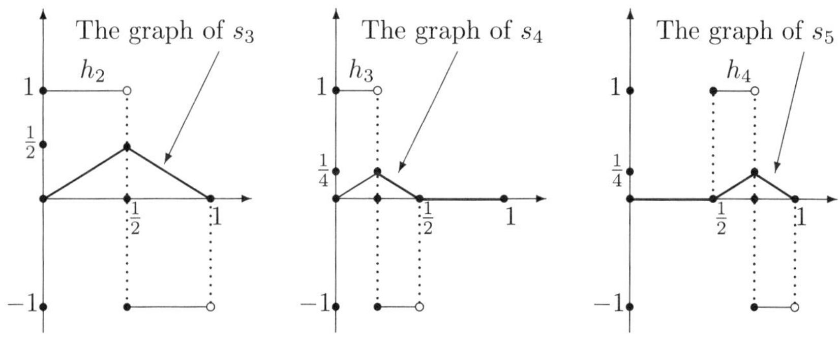

Figure 3

(a) Let Z denote the linear span of $\{s_1, s_2, \dots\}$. We must show that the uniform closure $\overline{Z}$ of Z coincides with $C[0,1]$. According to the Hahn–Banach theorem, it suffices to show that every finite Borel signed measure on $[0,1]$ that vanishes on Z also vanishes on $C[0,1]$. To this end, fix a finite Borel signed measure μ on $[0,1]$ that vanishes on Z. That is, assume that $\int_0^1 z(\tau)\,d\mu(\tau) = 0$ for each $z \in Z$.

Before proving that $\mu = 0$, we need to establish a few facts. Consider a closed interval $[a,b]$ of the form $\left[\frac{\ell}{2^k}, \frac{\ell+1}{2^k}\right]$, where $0 \le \ell < 2^k - 1$. From property (†) in the solution of part (b) of Problem 1.3.3, it follows that there exists a sequence $\{f_n\}$

in the linear span of the Haar system satisfying

$$\left| \int_0^t \chi_{[a,b]}(\tau)\, d\tau - \int_0^t f_{n+1}(\tau)\, d\tau \right| \le \tfrac{n+1}{2^n} \qquad (\dagger\dagger)$$

for each $t \in [0,1]$. Since the function $\theta_n(t) = \int_0^t f_n(\tau)\, d\tau$ belongs to Z, it follows from ($\dagger\dagger$) that the function $\theta(t) = \int_0^t \chi_{[a,b]}(\tau)\, d\tau$ belongs to $\overline{Z}$. The latter implies that if $[a,b]$ is any closed subinterval of $[0,1]$ with "dyadic" endpoints (i.e., a and b are of the form $a = \frac{\ell}{2^k}$ and $b = \frac{r}{2^m}$), then the function $\phi(t) = \frac{1}{b-a} \int_0^t \chi_{[a,b]}(\tau)\, d\tau$ also belongs to $\overline{Z}$. The graph of ϕ is shown in Figure 4(a).

In turn, the last conclusion implies that if $0 \le a < b < c < d \le 1$ are dyadic points, then the function $\psi \in C[0,1]$ defined by

$$\psi(t) = \tfrac{1}{b-a} \int_0^t \chi_{[a,b]}(\tau)\, d\tau - \tfrac{1}{c-d} \int_0^t \chi_{[c,d]}(\tau)\, d\tau$$

also belongs to $\overline{Z}$. Its graph is shown in Figure 4(b).

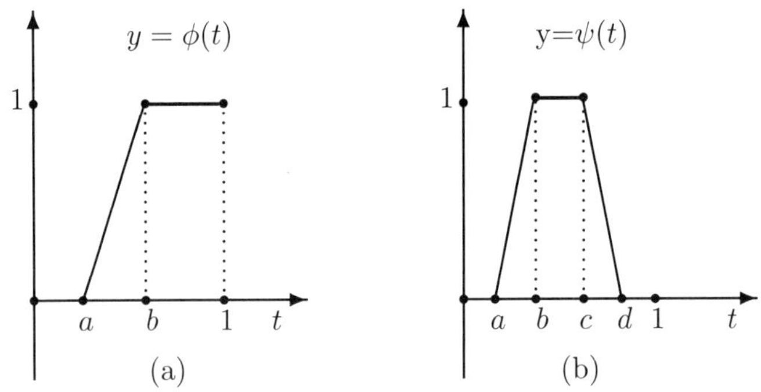

Figure 4

We are now ready to establish that $\mu = 0$. Let (a,b) be an arbitrary open subinterval of $[0,1]$. Pick four sequences $\{a_n\}$, $\{b_n\}$, $\{c_n\}$, and $\{d_n\}$ of dyadic points such that $a < a_n < b_n < c_n < d_n < b$ for each n, $b_n \downarrow a$, and $c_n \uparrow b$, and consider the sequence of functions $\{\psi_n\} \subseteq \overline{Z}$ defined by

$$\psi_n(t) = \tfrac{1}{b_n - a_n} \int_0^t \chi_{[a_n,b_n]}(\tau)\, d\tau - \tfrac{1}{d_n - c_n} \int_0^t \chi_{[c_n,d_n]}(\tau)\, d\tau\,.$$

Clearly, $\psi_n(t) \to \chi_{(a,b)}(t)$ for each $t \in [0,1]$. So, by the Lebesgue Dominated Convergence Theorem, we see that

$$\mu\big((a,b)\big) = \int_0^1 \chi_{(a,b)}(\tau)\, d\mu(\tau) = \lim_{n\to\infty} \int_0^1 \psi_n(\tau)\, d\mu(\tau) = 0\,.$$

If $\theta_n(t) = 1 - \int_0^t \chi_{[0,\frac{1}{2^k}]}(\tau)\, d\tau$, then $\{\theta_n\} \subseteq \overline{Z}$ and $\theta_n(t) \to \chi_{\{0\}}(t)$ for each $t \in [0,1]$. As above, this implies $\mu(\{0\}) = \lim_{n\to\infty} \int_0^1 \theta_n(\tau)\, d\mu(\tau) = 0$. Similarly, $\mu(\{1\}) = 0$.

The above conclusions easily imply that $\mu(J) = 0$ for each open subset of $[0, 1]$. Since μ (as a finite Borel signed measure on $[0, 1]$) is regular, we conclude that $\mu(B) = 0$ for each Borel subset B of $[0, 1]$. That is, $\mu = 0$, as desired.

(b) Assume that $\lambda_1, \lambda_2 \ldots, \lambda_n, \lambda_{n+1}$ are arbitrary scalars, and consider the two functions $f = \sum_{i=1}^{n} \lambda_i s_i$ and $g = \sum_{i=1}^{n+1} \lambda_i s_i = f + \lambda_{n+1} s_{n+1}$. It suffices to show that

$$\Big\| \sum_{i=1}^{n} \lambda_i s_i \Big\|_\infty = \|f\|_\infty \le \|g\|_\infty = \Big\| \sum_{i=1}^{n+1} \lambda_i s_i \Big\|_\infty .$$

Let $[a, b]$ be the support-interval of s_{n+1}. That is, $s_{n+1}(t) = 0$ holds for all $t \notin [a, b]$ and $s_{n+1}(t) > 0$ for each $a < t < b$. This implies $f(t) = g(t)$ for all $t \notin [a, b]$, and so in order to establish that $\|f\|_\infty \le \|g\|_\infty$, it is enough to show that $\max_{t\in[a,b]} |f(t)| \le \max_{t\in[a,b]} |g(t)|$.

To this end, notice that for each $1 \le i \le n$ the function s_i is linear over the interval $[a, b]$. This implies that there exist two constants m and β such that $f(t) = mt+\beta$ for all $t \in [a, b]$. Now note that $\max_{t\in[a,b]} |f(t)| = \max\{|f(a)|, |f(b)|\}$, i.e., the maximum of the function $|f|$ over $[a, b]$ is achieved at the endpoints of the interval $[a, b]$. Since $f(a) = g(a)$ and $f(b) = g(b)$, we infer that

$$\max_{t\in[a,b]} |f(t)| = \max\{|f(a)|, |f(b)|\} = \max\{|g(a)|, |g(b)|\} \le \max_{t\in[a,b]} |g(t)| .$$

Thus, $\|f\|_\infty \le \|g\|_\infty$, and the solution is finished. ∎

Problem 1.3.5 (Bessaga–Pelczynski [**18**]). *Let $\{x_n\}$ be a sequence in an infinite dimensional Banach space satisfying $x_n \xrightarrow{w} 0$ and $\liminf \|x_n\| > 0$. Show that $\{x_n\}$ has a subsequence that is a basic sequence.*

Solution: Assume that a sequence $\{x_n\}$ in an infinite dimensional Banach space X satisfies $\liminf \|x_n\| > 0$ and $x_n \xrightarrow{w} 0$. By passing to a subsequence, we can assume that $\|x_n\| \ge c > 0$ holds for each n and some constant $c > 0$. Obviously, $\frac{x_n}{\|x_n\|} \xrightarrow{w} 0$. So, replacing x_n by $\frac{x_n}{\|x_n\|}$, we can suppose without loss of generality that $\|x_n\| = 1$ for each n. We claim that the following property is true:

(P) *If Y is a finite dimensional vector subspace of X and $\epsilon > 0$, then for each k there exists some $n > k$ such that $\|y\| \le (1+\epsilon)\|y + \alpha x_n\|$ for all $y \in Y$ and all scalars α.*

To see this, assume that Y is finite dimensional, let $\epsilon > 0$, and fix k. Also, pick some $0 < \delta < \frac{1}{2}$ such that $2\delta < \epsilon$. Since the closed unit ball of Y is norm compact, there exist unit vectors $y_1, \ldots, y_m$ in Y such that for each $y \in Y$ with $\|y\| = 1$ there exists some i with $\|y - y_i\| < \frac{\delta}{2}$. For each $1 \le i \le m$ choose some $x_i^* \in X^*$ satisfying $\|x_i^*\| = 1$ and $x_i^*(y_i) = 1$. From $x_n \xrightarrow{w} 0$, it follows that there exists some $n > k$ satisfying $\big|x_i^*(x_n)\big| < \frac{\delta}{4}$ for $i = 1, \ldots, m$.

Now let α be an arbitrary scalar and fix some $y \in Y$ with $\|y\| = 1$. If $|\alpha| \ge 2$, then we have

$$(1+\epsilon)\|y + \alpha x_n\| \ge \|y + \alpha x_n\| \ge |\alpha| - \|y\| = |\alpha| - 1 \ge 1 .$$

On the other hand, if $|\alpha| < 2$, then choose some $y_i \in Y$ with $\|y - y_i\| < \frac{\delta}{2}$ and note that $|\alpha x_i^*(x_n)| < \frac{\delta}{2}$ implies

$$\begin{aligned}\|y + \alpha x_n\| &\geq \|y_i + \alpha x_n\| - \tfrac{\delta}{2} \geq x_i^*(y_i + \alpha x_n) - \tfrac{\delta}{2}\\ &= 1 + \alpha x_i^*(x_n) - \tfrac{\delta}{2} \geq 1 - \tfrac{\delta}{2} - \tfrac{\delta}{2} = 1 - \delta\,.\end{aligned}$$

This implies

$$1 \leq \tfrac{1}{1-\delta}\|y + \alpha x_n\| \leq (1 + 2\delta)\|y + \alpha x_n\| \leq (1 + \epsilon)\|y + \alpha x_n\|$$

in this case too. Thus, we have established the validity of (P) for each unit vector y. The general case for an arbitrary y follows from the homogeneity.

Fix some real number $\epsilon > 0$ and then choose a sequence $\{\epsilon_n\}$ of positive real numbers such that $K = \prod_{n=1}^{\infty}(1 + \epsilon_n) \leq 1 + \epsilon$. Using property (P) and an easy inductive argument, we see that there exists a subsequence $\{z_n\}$ of $\{x_n\}$ satisfying

$$\|z\| \leq (1 + \epsilon_n)\|z + \alpha x_{n+1}\|$$

for all scalars α and all z in the vector subspace generated by $\{z_1, \ldots, z_n\}$. Now note that if $m > n$, then for any choice of scalars $\alpha_1, \ldots, \alpha_m$ we have

$$\Big\|\sum_{i=1}^{n} \alpha_i z_i\Big\| \leq K\Big\|\sum_{i=1}^{m} \alpha_i z_i\Big\|\,.$$

By Theorem 1.41, the subsequence $\{z_n\}$ of $\{x_n\}$ is a basic sequence in X. ∎

Problem 1.3.6. *Assume that X is a reflexive Banach space with a basis $\{x_n\}$. Show that its biorthogonal sequence $\{c_n^*\}$ is a basis in X^*.*

Solution: We already know that $\{c_n^*\}$ is a basic sequence in X^*; see the discussion at end of Section 1.3. What needs verification is the fact that the closed linear span Y of $\{c_n^*\}$ is all of X^*.

If $Y \neq X^*$, then there exists (by virtue of the reflexivity of X) some non-zero $x = \sum_{n=1}^{\infty} \alpha_n x_n \in X$ such that $x(c_n^*) = c_n^*(x) = \alpha_n = 0$ holds for each n. But then $x = \sum_{n=1}^{\infty} \alpha_n x_n = 0$, a contradiction. This contradiction establishes that $Y = X^*$, and so the biorthogonal sequence $\{c_n^*\}$ is a basis in X^*. ∎

Problem 1.3.7. *Let $\sum_{n=1}^{\infty} x_n$ be a series in a Banach space X.*

(a) *If X is finite dimensional, show that $\sum_{n=1}^{\infty} x_n$ converges unconditionally if and only if $\sum_{n=1}^{\infty} \|x_n\| < \infty$.*

(b) *Give an example of an unconditionally convergent series $\sum_{n=1}^{\infty} x_n$ such that $\sum_{n=1}^{\infty} \|x_n\| = \infty$.*

Solution: (a) We can assume that $X = \mathbb{R}^k$ for some $k \in \mathbb{N}$. Let $x_n = (x_1^n, \ldots, x_k^n)$ and note that for each choice $s = (s_1, s_2, \ldots)$ of signs we have

$$\sum_{n=1}^{\infty} s_n x_n = \Big(\sum_{n=1}^{\infty} s_n x_1^n, \sum_{n=1}^{\infty} s_n x_2^n, \ldots, \sum_{n=1}^{\infty} s_n x_k^n\Big)\,.$$

This implies that the series $\sum_{n=1}^{\infty} s_n x_n$ is norm convergent in X if and only if the series $\sum_{n=1}^{\infty} s_n x_i^n$ converges in $\mathbb{R}$ for each $1 \leq i \leq k$. Now notice that the latter is equivalent to saying that $\sum_{n=1}^{\infty} |x_i^n| < \infty$ holds for each $1 \leq i \leq k$.

If we consider X equipped with the L_1-norm $\|x\|_1 = \sum_{i=1}^k |x_i|$, then it is easy to see that

$$\Big\|\sum_{n=1}^{\infty} x_n\Big\|_1 = \sum_{i=1}^{k}\Big|\sum_{n=1}^{\infty} x_i^n\Big| \le \sum_{i=1}^{k}\sum_{n=1}^{\infty}|x_i^n| = \sum_{n=1}^{\infty}\sum_{i=1}^{k}|x_i^n| = \sum_{n=1}^{\infty}\|x_n\|_1 .$$

This inequality in connection with the preceding discussion shows that the series $\sum_{n=1}^\infty x_i^n$ is unconditionally convergent if and only if $\sum_{n=1}^\infty \|x_n\| < \infty$.

(b) Let $X = \ell_2$ and let e_n denote the standard unit coordinate vector. If $x_n = \frac{1}{n}e_n$, then $\|x_n\| = \frac{1}{n}$, and so $\sum_{n=1}^\infty \|x_n\| = \sum_{n=1}^\infty \frac{1}{n} = \infty$. On the other hand, it is easy to see that the series $\sum_{n=1}^\infty x_n$ converges unconditionally in ℓ_2. ∎

Problem 1.3.8. *Show that for a series $\sum_{n=1}^\infty x_n$ in a Banach space the following statements are equivalent.*

(1) *The series $\sum_{n=1}^\infty x_n$ converges unconditionally.*

(2) *For any sequence $\{s_n\}$ of signs (i.e., $s_n = \pm 1$ for each n) the series $\sum_{n=1}^\infty s_n x_n$ is norm convergent.*

(3) *For every strictly increasing sequence $\{k_n\}$ of natural numbers the series $\sum_{n=1}^\infty x_{k_n}$ is norm convergent.*

(4) *For each $\epsilon > 0$ there exists a natural number k such that for each finite subset A of $\mathbb{N}$ with $\min A \ge k$ we have $\big\|\sum_{n\in A} x_n\big\| < \epsilon$.*

Solution: (1) $\Longrightarrow$ (2) Let $\{s_n\}$ be a sequence of signs, and assume by way of contradiction that the series $\sum_{n=1}^\infty s_n x_n$ is not norm convergent. This means that there exist some $\epsilon > 0$ and a sequence $\{k_n\}$ of strictly increasing natural numbers such that $\big\|\sum_{i=k_n+1}^{k_{n+1}} s_i x_i\big\| > 2\epsilon$ holds for each $n \in \mathbb{N}$. We shall obtain a contradiction by demonstrating the existence of a permutation π of $\mathbb{N}$ and two sequences $\{m_n\}$ and $\{r_n\}$ of natural numbers satisfying $k_n < m_n \le r_n$ and $\big\|\sum_{i=m_n}^{r_n} x_{\pi(i)}\big\| > \epsilon$ for each n.

Let $I_n = \{k \in \mathbb{N}\colon\ k_n < i \le k_{n+1}\}$. Now fix $n \in \mathbb{N}$. If $s_i = 1$ for each $i \in I_n$, let $\pi_n(i) = i$. Similarly, if $s_n = -1$ for each $i \in I_n$, we let $\pi_n(i) = i$. In these cases, we let $m_n = k_n + 1$ and $r_n = k_{n+1}$. We also let $I_0 = \{1, 2, \ldots, k_1\}$ and $\pi_0(i) = i$ for each $i \in I_0$.

Now assume that for some $i \in I_n$ we have $s_i = 1$ and for some $j \in I_n$ we have $s_j = -1$. In this case, there exists some $k_n + 1 \le p_n < k_{n+1}$ and some permutation π_n of I_n such that $s_{\pi_n(i)} = 1$ if $i \le p_n$ and $s_{\pi_n(i)} = -1$ if $i > p_n$. Let $s_n = 1$ for all $k_n + 1 \le i \le p_n$ and $s_n = -1$ for all $p_n < i \le k_{n+1}$. In view of

$$\Big\|\sum_{i=k_n+1}^{k_{n+1}} s_i x_i\Big\| = \Big\|\sum_{i=k_n+1}^{p_n} x_{\pi_n(i)} - \sum_{i=p_n+1}^{k_{n+1}} x_{\pi_n(i)}\Big\| \ge 2\epsilon\,,$$

it follows that either $\big\|\sum_{i=k_n+1}^{p_n} x_{\pi_n(i)}\big\| > \epsilon$ or $\big\|\sum_{i=p_n+1}^{k_{n+1}} x_{\pi_n(i)}\big\| > \epsilon$. In the first case, we let $m_n = k_n + 1$ and $r_n = p_n$, and in the second case $m_n = p_n + 1$ and $r_n = k_{n+1}$. Now if π is the permutation of $\mathbb{N}$ that coincides on each I_n with π_n, then we have

$$\Big\|\sum_{i=m_n}^{r_n} x_{\pi(i)}\Big\| > \epsilon$$

for each n. This shows that the series $\sum_{n=1}^{\infty} x_{\pi(i)}$ is not norm convergent, which contradicts (1).

(2) $\Longrightarrow$ (3) Letting $s_n = 1$ for each n, we get that $\sum_{n=1}^{\infty} x_n$ is norm convergent. Now assume that $\{k_n\}$ is a strictly increasing sequence of natural numbers. Put $t_{k_n} = 1$ for each n and $t_i = -1$ if $i \neq k_n$ for each n, and note that

$$\sum_{n=1}^{\infty} x_{k_n} = \tfrac{1}{2}\Big[\sum_{n=1}^{\infty} x_n + \sum_{n=1}^{\infty} t_n x_n\Big].$$

This implies that $\sum_{n=1}^{\infty} x_{k_n}$ is norm convergent.

(3) $\Longrightarrow$ (4) If (4) is not true, then there exist some $\epsilon > 0$ and a sequence $\{A_n\}$ of non-empty subsets of $\mathbb{N}$ such that for each n we have

$$\max A_n < \min A_{n+1} \quad \text{and} \quad \Big\|\sum_{i\in A_n} x_i\Big\| \geq \epsilon.$$

Let $A = \bigcup_{n=1}^{\infty} A_n$. Clearly, this is a subset of $\mathbb{N}$, and so $A = \{k_1 < k_2 < \cdots\}$. It is easy to see that the sequence of partial sums of the series $\sum_{n=1}^{\infty} x_{k_n}$ is not a norm Cauchy sequence. This implies that the series $\sum_{n=1}^{\infty} x_{k_n}$ is not norm convergent, a contradiction.

(4) $\Longrightarrow$ (1) Assume by way of contradiction that the series $\sum_{n=1}^{\infty} x_n$ is not unconditionally convergent. This means that there exists a permutation π of $\mathbb{N}$ such that the series $\sum_{n=1}^{\infty} x_{\pi(n)}$ is not norm convergent. So, there exist some $\epsilon > 0$ and a strictly increasing sequence $\{k_n\}$ of natural numbers such that $\big\|\sum_{i=k_n+1}^{k_{n+1}} x_{\pi(i)}\big\| > \epsilon$ for each n.

Now fix $k \in \mathbb{N}$. Since π is a permutation of $\mathbb{N}$, there exists some r such that $\{1, 2, \ldots, k\} \subseteq \{\pi(1), \pi(2), \ldots, \pi(k_r)\}$. So, if $A = \{\pi(i)\colon\ k_r < i \leq k_{r+1}\}$, then

$$\min A > k \quad \text{and} \quad \Big\|\sum_{i\in A} x_{\pi(i)}\Big\| = \Big\|\sum_{i=k_r+1}^{k_{r+1}} x_{\pi(i)}\Big\| > \epsilon.$$

Thus, for each $k \in \mathbb{N}$ there exists a non-empty finite subset A of $\mathbb{N}$ with $\min A > k$ and $\big\|\sum_{i\in A} x_{\pi(i)}\big\| > \epsilon$. This contradicts (4), and so (4) implies (1). ∎

Problem 1.3.9. *If a series $\sum_{n=1}^{\infty} x_n$ in a Banach space converges unconditionally, then show that $\sum_{n=1}^{\infty} x_n = \sum_{n=1}^{\infty} x_{\pi(n)}$ holds for each permutation π of $\mathbb{N}$.*

Solution: Assume that a series $\sum_{n=1}^{\infty} x_n$ in a Banach space converges unconditionally, and let π be a permutation of $\mathbb{N}$. Put $x = \sum_{n=1}^{\infty} x_n$ and $y = \sum_{n=1}^{\infty} x_{\pi(n)}$, and fix $\epsilon > 0$.

By statement (4) of Problem 1.3.8 there exists some k such that $\|\sum_{i\in A} x_i\| < \epsilon$ for each finite subset A of $\mathbb{N}$ with $\min A \geq k$. Next, choose some $r > k$ such that $\big\|x - \sum_{i=1}^{r} x_i\big\| < \epsilon$, and then select some $m > r$ such that:

(a) $\{1, 2, \ldots, r\} \subseteq \{\pi(1), \pi(2), \ldots, \pi(m)\}$.

(b) The finite subset $A = \{\pi(1), \pi(2), \ldots, \pi(m)\} \setminus \{1, 2, \ldots, r\}$ of $\mathbb{N}$ is non-empty.

(c) $\left\| y - \sum_{i=1}^m x_{\pi(i)} \right\| < \epsilon$.

(The existence of such a number m follows from the fact that π is a permutation of $\mathbb{N}$.) Clearly, $\min A > k$. Now note that

$$\begin{aligned} \|x - y\| &\leq \left\| x - \sum_{i=1}^r x_i \right\| + \left\| \sum_{i=1}^r x_i - \sum_{i=1}^m x_{\pi(i)} \right\| + \left\| \sum_{i=1}^m x_{\pi(i)} - y \right\| \\ &< \epsilon + \left\| \sum_{i \in A} x_i \right\| + \epsilon < \epsilon + \epsilon + \epsilon = 3\epsilon \,. \end{aligned}$$

Since $\epsilon > 0$ is arbitrary, it follows that $\|x - y\| = 0$, that is, $x = y$, as desired. ∎

Problem 1.3.10. *Let $\sum_{n=1}^\infty x_n$ be an unconditionally convergent series in a Banach space X. If $(\alpha_1, \alpha_2, \ldots) \in \ell_\infty$, then show that the series $\sum_{n=1}^\infty \alpha_n x_n$ is norm convergent. Moreover, show that the mapping $T\colon \ell_\infty \to X$, defined by*

$$T(\alpha_1, \alpha_2, \ldots) = \sum_{n=1}^\infty \alpha_n x_n \,,$$

is a bounded linear operator.

Solution: We will consider the case of a real Banach space only. The complex case can be dealt with similarly. Let $(\alpha_1, \alpha_2, \ldots) \in \ell_\infty$. We can suppose that $|\alpha_n| \leq 1$ holds for each n. Fix $\epsilon > 0$. Since the series $\sum_{n=1}^\infty x_n$ is unconditionally convergent, statement 4 in Problem 1.3.8 implies immediately that there exists some m such that for any sequence $\{s_n\}$ of signs and any non-empty finite subset F of $\mathbb{N}$ with $\min F > m$ we have $\left\| \sum_{n \in F} s_n x_n \right\| < \epsilon$. Fix for a moment any such set F.

Pick $x^* \in X^*$ of norm one satisfying $x^*\left(\sum_{n \in F} \alpha_n x_n\right) = \left\| \sum_{n \in F} \alpha_n x_n \right\|$. Next, consider the sequence of signs $s = (s_1, s_2, \ldots)$ defined by $s_n = 1$ if $\alpha_n x^*(x_n) \geq 0$ and $s_n = -1$ if $\alpha_n x^*(x_n) < 0$. This implies

$$\begin{aligned} \left\| \sum_{n \in F} \alpha_n x_n \right\| &= x^*\Big(\sum_{n \in F} \alpha_n x_n \Big) \leq \sum_{n \in F} |\alpha_n| |x^*(x_n)| = \sum_{n \in F} s_n x^*(x_n) \\ &= x^*\Big(\sum_{n \in F} s_n x_n \Big) \leq \left\| \sum_{n \in F} s_n x_n \right\| < \epsilon \,. \end{aligned}$$

Therefore, the sequence of partial sums of the series $\sum_{n=1}^\infty \alpha_n x_n$ is a norm Cauchy sequence. Hence, the series $\sum_{n=1}^\infty \alpha_n x_n$ is norm convergent. Now it is a routine matter to verify that the mapping $T\colon \ell_\infty \to X$, defined by

$$T(\alpha_1, \alpha_2, \ldots) = \sum_{n=1}^\infty \alpha_n x_n \,,$$

is a linear operator.

Finally, notice that $\left\| \sum_{n=m+1}^\infty \alpha_n x_n \right\| \leq \epsilon$ holds for all $(\alpha_1, \alpha_2, \ldots) \in \ell_\infty$ satisfying $|\alpha_n| \leq 1$ for each n. This implies that the linear operator carries the closed unit ball of ℓ_∞ to a norm bounded subset of X, and so $T\colon \ell_\infty \to X$ is also a bounded operator. ∎

1.4. Ultrapowers of Banach Spaces

Problem 1.4.1. *For a filter $\mathcal{F}$ of subsets of an infinite set Δ show that the following statements are equivalent.*

(1) *$\mathcal{F}$ is an ultrafilter.*

(2) *If $A \cup B \in \mathcal{F}$, then either $A \in \mathcal{F}$ or $B \in \mathcal{F}$.*

(3) *If $A \cap B \neq \emptyset$ for all $B \in \mathcal{F}$, then $A \in \mathcal{F}$.*

Solution: (1) $\Longrightarrow$ (2) Assume that $\mathcal{F}$ is an ultrafilter and that $A \cup B \in \mathcal{F}$. If $A \in \mathcal{F}$, then (2) is true. So, we can suppose that $A \notin \mathcal{F}$. Consider the collection $\mathcal{G} = \{C \subseteq \Delta\colon\ A \cup C \in \mathcal{F}\}$. A direct verification shows that $\mathcal{G}$ is a filter, $\mathcal{F} \subseteq \mathcal{G}$, and that $B \in \mathcal{F}$. Since $\mathcal{F}$ is an ultrafilter, $\mathcal{G} = \mathcal{F}$. Therefore, $B \in \mathcal{G} = \mathcal{F}$.

(2) $\Longrightarrow$ (3) Assume that a subset A of Δ satisfies $A \cap B \neq \emptyset$ for each $B \in \mathcal{F}$ and suppose by way of contradiction that $A \notin \mathcal{F}$. From $A \cup A^{c} = X \in \mathcal{F}$ and (2), it follows that $A^{c} \in \mathcal{F}$. In particular, we have $\emptyset = A \cap A^{c} \in \mathcal{F}$, a contradiction. Hence, $A \in \mathcal{F}$.

(3) $\Longrightarrow$ (1) Assume that a filter $\mathcal{G}$ satisfies $\mathcal{F} \subseteq \mathcal{G}$. Fix $A \in \mathcal{G}$. If $B \in \mathcal{F}$, then $B \in \mathcal{G}$, and so $A \cap B \neq \emptyset$ for each $B \in \mathcal{F}$. Now (3) implies $A \in \mathcal{F}$, i.e., $\mathcal{G} \subseteq \mathcal{F}$. Hence, $\mathcal{G} = \mathcal{F}$, and therefore $\mathcal{F}$ is an ultrafilter. ∎

Problem 1.4.2. *If an ultrafilter $\mathcal{U}$ on a set Δ satisfies $\bigcap_{A \in \mathcal{U}} A \neq \emptyset$, then show that there exists a unique $\delta \in \Delta$ such that $\mathcal{U} = \{A \subseteq \Delta\colon\ \delta \in A\}$.*

Solution: Let $\mathcal{U}$ be an ultrafilter on a set Δ such that $\bigcap_{A \in \mathcal{U}} A \neq \emptyset$. Fix some $\delta \in \bigcap_{A \in \mathcal{U}} A$ and note that the set $\mathcal{F} = \{A \subseteq \Delta\colon\ \delta \in A\}$ is a filter of subsets of Δ satisfying $\mathcal{U} \subseteq \mathcal{F}$. Since $\mathcal{U}$ is an ultrafilter, it follows that $\mathcal{U} = \mathcal{F}$.

If another point $\eta \in \Delta$ satisfies $\mathcal{U} = \{A \subseteq \Delta\colon\ \eta \in A\}$, then $\{\delta\}, \{\eta\} \in \mathcal{U}$. This implies $\emptyset = \{\delta\} \cap \{\eta\} \in \mathcal{U}$, which is impossible. Hence, there exists exactly one $\delta \in \Delta$ such that $\mathcal{U} = \{A \subseteq \Delta\colon\ \delta \in A\}$. ∎

Problem 1.4.3. *Show that every filter coincides with the intersection of all ultrafilters that include it.*

Solution: Let $\mathcal{F}$ be a filter on an infinite set Δ and let $A \notin \mathcal{F}$. It suffices to show that there exists an ultrafilter $\mathcal{U}$ such that $\mathcal{F} \subseteq \mathcal{U}$ and $A \notin \mathcal{U}$.

To see this, start by observing that $F \cap A^{c} \neq \emptyset$ holds for each $F \in \mathcal{F}$. Indeed, if for some $F \in \mathcal{F}$ we have $F \cap A^{c} = \emptyset$, then from

$$F = F \cap X = \big(F \cap A\big) \cup \big(F \cap A^{c}\big) = F \cap A \subseteq A$$

it follows that $A \in \mathcal{F}$, a contradiction. Hence, $F \cap A^{c} \neq \emptyset$ for each $F \in \mathcal{F}$.

Now consider the collection of sets

$$\mathcal{G} = \{B \subseteq \Delta\colon \ F \cap A^c \subseteq B \ \text{ for some } \ F \in \mathcal{F}\}.$$

A direct verification shows that $\mathcal{G}$ is a filter satisfying $\mathcal{F} \subseteq \mathcal{G}$ and $A^c \in \mathcal{G}$. If $\mathcal{U}$ is any ultrafilter containing $\mathcal{G}$, then $\mathcal{F} \subseteq \mathcal{U}$ and $A = (A^c)^c \notin \mathcal{U}$, as desired. ∎

Problem 1.4.4. *If in a topological vector space $x_n \xrightarrow{\mathcal{U}} x$ and $y_n \xrightarrow{\mathcal{U}} y$, then show that $x_n + y_n \xrightarrow{\mathcal{U}} x + y$ and $\lambda x_n \xrightarrow{\mathcal{U}} \lambda x$ for each scalar λ.*

Solution: Assume that $x_n \xrightarrow{\mathcal{U}} x$ and $y_n \xrightarrow{\mathcal{U}} y$ in a topological vector space. Let V be a neighborhood of zero, and then select another neighborhood W of zero with $W + W \subseteq V$. Now notice that

$$\{n \in \mathbb{N}\colon \ x_n \in x + W\} \cup \{n \in \mathbb{N}\colon \ y_n \in x + W\} \subseteq \{n \in \mathbb{N}\colon \ x_n + y_n \in x + y + V\}.$$

Since the sets $\{n \in \mathbb{N}\colon \ x_n \in x + W\}$ and $\{n \in \mathbb{N}\colon \ y_n \in x + W\}$ belong to $\mathcal{U}$, it follows that $\{n \in \mathbb{N}\colon \ x_n + y_n \in x + y + V\} \in \mathcal{U}$. This shows that $x_n + y_n \xrightarrow{\mathcal{U}} x + y$.

Next, suppose that $\lambda \neq 0$. Clearly,

$$\{n \in \mathbb{N}\colon \ \lambda x_n \in \lambda x + V\} = \{n \in \mathbb{N}\colon \ x_n \in x + \tfrac{1}{\lambda} V\} \in \mathcal{U}.$$

This shows that $\lambda x_n \xrightarrow{\mathcal{U}} \lambda x$. ∎

Problem 1.4.5. *Let Ω be a topological space, let A be a subset of Ω, and let $\{x_n\}$ be a sequence in Ω such that for some $N \in \mathcal{U}$ we have $x_n \in A$ for each $n \in N$. Show that if $x_n \xrightarrow{\mathcal{U}} x$, then $x \in \overline{A}$.*

Use this conclusion to show that the $\mathcal{U}$-limits preserve inequalities in the following sense: If in a Banach lattice $x_n \xrightarrow{\mathcal{U}} x$, $y_n \xrightarrow{\mathcal{U}} y$, and for some $N \in \mathcal{U}$ we have $x_n \geq y_n$ for each $n \in N$, then $x \geq y$.

Solution: Let A be a subset of a topological space Ω and let a sequence $\{x_n\}$ in Ω be such that for some $N \in \mathcal{U}$ we have $x_n \in A$ for each $n \in N$ and $x_n \xrightarrow{\mathcal{U}} x$. If V is a neighborhood of x, then the set $J = \{n \in \mathbb{N}\colon \ x_n \in V\}$ belongs to $\mathcal{U}$. In particular, we have $N \cap J \neq \emptyset$. Now if $k \in N \cap J$, then $x_k \in V \cap A$. This shows that $V \cap A \neq \emptyset$ for each neighborhood V of x, i.e., $x \in \overline{A}$.

For the last part, notice that $w \geq v$ in a Banach lattice E is equivalent to $w - v \in E_+$. Since E_+ is norm closed, it follows from the preceding conclusion that $x - y = \lim_{\mathcal{U}} x_n - \lim_{\mathcal{U}} y_n = \lim_{\mathcal{U}}(x_n - y_n) \in E_+$. This shows that $x \geq y$. ∎

Problem 1.4.6. *For a non-zero bounded operator $T\colon X \to Y$ between two Banach spaces and a sequence $\{x_n\}$ in X establish the following.*

(a) *If $x_n \xrightarrow{\mathcal{U}} 0$ in X, then $Tx_n \xrightarrow{\mathcal{U}} 0$ in Y.*

(b) *If $Tx_n \xrightarrow{\mathcal{U}} 0$ in Y and $\lim_{\mathcal{U}} x_n \neq 0$ in X, then there exist some $\eta > 0$ and a subsequence $\{x_{k_n}\}$ of $\{x_n\}$ satisfying $\|x_{k_n}\| \geq \eta$ for each n and $\|Tx_{k_n}\| \to 0$.*

Solution: (a) Assume that $x_n \xrightarrow{\mathcal{U}} 0$ holds in X and let $\epsilon > 0$. Then the set $A = \{n \in \mathbb{N}\colon \ \|x_n\| < \frac{\epsilon}{\|T\|}\}$ belongs to $\mathcal{U}$. Clearly, $A \subseteq \{n \in \mathbb{N}\colon \ \|Tx_n\| < \epsilon\}$, and therefore $\{n \in \mathbb{N}\colon \ \|Tx_n\| < \epsilon\} \in \mathcal{U}$. This shows that $Tx_n \xrightarrow{\mathcal{U}} 0$ in Y.

(b) From $\lim_{\mathcal{U}} x_n \neq 0$, it follows that there exists some $\eta > 0$ such that the set $A = \{n \in \mathbb{N}\colon \ \|x_n\| < \eta\}$ does not belong to $\mathcal{U}$. So, $A^c = \{n \in \mathbb{N}\colon \ \|x_n\| \geq \eta\} \in \mathcal{U}$.

Next, for each k the set $B_k = \{n \in \mathbb{N}\colon \ \|Tx_n\| < \frac{1}{k}\}$ belongs to the free ultrafilter $\mathcal{U}$. Put $A_k = B_k \cap A^c \in \mathcal{U}$; and so each A_k is an infinite set. Now if we select a sequence $\{m_k\}$ of strictly increasing natural numbers such that $m_k \in A_k$ for each k, then $\|x_{m_k}\| \geq \eta$ and $\|Tx_{m_k}\| < \frac{1}{k}$ hold for each k, and the desired conclusion follows. ∎

Problem 1.4.7. *Let Z be a normed space and let $\mathcal{U}$ be a free ultrafilter on $\mathbb{N}$. If, as usual, we let $\mathcal{N}_\mathcal{U} = \{(z_1, z_2, \ldots) \in \ell_\infty(Z)\colon \ z_n \xrightarrow{\mathcal{U}} 0\}$, then $\mathcal{N}_\mathcal{U}$ is a closed vector subspace of Z, and so $Z_\mathcal{U} = \ell_\infty(Z)/\mathcal{N}_\mathcal{U}$ is a normed space. That is, in the definition of the ultrapowers we do not need Z to be complete.*

Suppose that X is a normed space, Y is a subspace of X, and $\mathcal{U}$ is an ultrafilter on $\mathbb{N}$. Also, let $\mathcal{N}_\mathcal{U}^Y = \{(y_1, y_2, \ldots) \in \ell_\infty(Y)\colon \ y_n \xrightarrow{\mathcal{U}} 0\}$ and $\mathcal{N}_\mathcal{U}^X = \{(x_1, x_2, \ldots) \in \ell_\infty(X)\colon \ x_n \xrightarrow{\mathcal{U}} 0\}$, and define $J\colon Y_\mathcal{U} \to X_\mathcal{U}$ by

$$J\big(y + \mathcal{N}_\mathcal{U}^Y\big) = y + \mathcal{N}_\mathcal{U}^X$$

for each $y = (y_1, y_2, \ldots) \in \ell_\infty(Y)$. Show that J is a linear isometry—and so, under this isometry, $Y_\mathcal{U}$ can be considered as a vector subspace (which is norm closed if Y is also norm closed) of $X_\mathcal{U}$.

Solution: We first check that J is a well-defined mapping. To this end, assume that $y, z \in \ell_\infty(Y)$ satisfy $y + \mathcal{N}_\mathcal{U}^Y = z + \mathcal{N}_\mathcal{U}^Y$, i.e., $y - z \in \mathcal{N}_\mathcal{U}^Y$. This implies $y_n - z_n \xrightarrow{\mathcal{U}} 0$. Since $\ell_\infty(Y) \subseteq \ell_\infty(X)$, it follows that $y + \mathcal{N}_\mathcal{U}^X = z + \mathcal{N}_\mathcal{U}^X$. This shows that J is well defined.

Clearly, J is a linear mapping, and it follows from Lemma 1.62 that for each $y = (y_1, y_2, \ldots) \in \ell_\infty(Y)$ we have

$$\big\|J\big(y + \mathcal{N}_\mathcal{U}^Y\big)\big\| = \big\|y + \mathcal{N}_\mathcal{U}^X\big\| = \lim_{\mathcal{U}} \|y_n\| = \big\|y + \mathcal{N}_\mathcal{U}^Y\big\| .$$

Therefore, J is a linear isometry, as claimed. ∎

Problem 1.4.8. *For a bounded operator $T\colon X \to Y$ between two Banach lattices the following statements are equivalent:*

(1) *T is a positive operator if and only if $T_\mathcal{U}$ is a positive operator.*

(2) *T is a lattice homomorphism if and only if $T_\mathcal{U}$ is a lattice homomorphism.*

(3) *If T is interval preserving, then so is $T_\mathcal{U}$.*

Solution: (1) If $T_\mathcal{U}$ is positive, then T is clearly a positive operator. Now assume that T is positive, and let $[x] \geq 0$ in $X_\mathcal{U}$. There exists some $y \in \ell_\infty^+(X)$ such that $y \in [x]$. The positivity of T implies $Ty_n \geq 0$ for each n. Therefore,

$$T_\mathcal{U}[x] = T_\mathcal{U}[y] = [(Ty_1, Ty_2, \ldots)] \in X_\mathcal{U}^+ .$$

This shows that $T_\mathcal{U}$ is a positive operator.

(2) If $T_{\mathcal{U}}$ is a lattice homomorphism, then T is clearly a lattice homomorphism. For the converse suppose that the operator T is a lattice homomorphism. Then for each $x, y \in \ell_\infty(X)$ we have

$$\begin{aligned} T_{\mathcal{U}}([x] \vee [y]) &= T_{\mathcal{U}}([x \vee y]) = T_{\mathcal{U}}([(x_1 \vee y_1, x_2 \vee y_2, \ldots)]) \\ &= [(T(x_1 \vee y_1), T(x_2 \vee y_2), \ldots)] = [(Tx_1 \vee Ty_1, Tx_2 \vee Ty_2, \ldots)] \\ &= [(Tx_1, Tx_2, \ldots) \vee (Ty_1, Ty_2, \ldots)] \\ &= [(Tx_1, Tx_2, \ldots)] \vee [(Ty_1, Ty_2, \ldots)] \\ &= T_{\mathcal{U}}([x]) \vee T_{\mathcal{U}}([y]) \,. \end{aligned}$$

This shows that $T_{\mathcal{U}}$ is a lattice homomorphism.

(3) Assume that the operator T is interval preserving and let $x \in \ell_\infty^+(X)$ and $y \in \ell_\infty^+(Y)$ satisfy

$$0 \le [y] = [(y_1, y_2, \ldots)] \le T_{\mathcal{U}}[x] = [(Tx_1, Tx_2, \ldots)]$$

in $Y_{\mathcal{U}}$. Without loss of generality, we can suppose that $0 \le y_n \le Tx_n$ holds for each n. Since the operator T is interval preserving, for each n there exists some $0 \le z_n \le x_n$ such that $Tz_n = y_n$. To complete the solution notice that the vector $z = (z_1, z_2, \ldots) \in \ell_\infty(X)$ satisfies $0 \le [z] \le [x]$ and $T_{\mathcal{U}}[z] = [y]$. This shows that $T_{\mathcal{U}}$ is interval preserving. ∎

Problem 1.4.9. *For a bounded operator $T\colon X \to Y$ between two Banach spaces establish the following properties.*

(a) *If T is surjective, then $T_{\mathcal{U}}$ is surjective.*

(b) *If T has a closed range and $T_{\mathcal{U}}$ is surjective, then T is surjective.*

(c) *The operator T is invertible if and only if $T_{\mathcal{U}}$ is invertible—in which case we have $(T_{\mathcal{U}})^{-1} = (T^{-1})_{\mathcal{U}}$.*

Solution: (a) Assume first that T is surjective and let $y = (y_1, y_2, \ldots) \in \ell_\infty(Y)$. By Corollary 2.15 there exists some constant $c > 0$ such that for each n there exists some vector $x_n \in X$ satisfying $\|x_n\| \le c\|y_n\|$ and $Tx_n = y_n$. Clearly, $x = (x_1, x_2, \ldots) \in \ell_\infty(X)$ and $T_{\mathcal{U}}[x] = [y]$. Therefore, $T_{\mathcal{U}}$ is surjective.

(b) Suppose that T has a closed range and that the operator $T_{\mathcal{U}}$ is surjective. Fix some $y \in Y$ and let $\hat{y} = (y, y, y, \ldots) \in \ell_\infty(Y)$. Since $T_{\mathcal{U}}$ is surjective, there exists some $x \in \ell_\infty(X)$ such that $T_{\mathcal{U}}[x] = [\hat{y}]$. That is, there exists some $z = (z_1, z_2, \ldots)$ in $\mathcal{N}_{\mathcal{U}}$ such that $Tx_n = y + z_n$ holds for each n. From $\lim_{\mathcal{U}} z_n = 0$ (and passing to a subsequence if necessary), we can assume that $z_n \to 0$ in Y. This implies $Tx_n \to y$. Thus, y belongs to the closure of the range of T. Since T has a closed range, it follows that $y \in R(T)$. Therefore, $R(T) = Y$ and so T is surjective.

(c) Assume first that T is invertible. That is, there exists a bounded operator $T^{-1}\colon Y \to X$ satisfying $T^{-1}T = I_X$ and $TT^{-1} = I_Y$. Consequently,

$$(T^{-1})_{\mathcal{U}} T_{\mathcal{U}} = (I_X)_{\mathcal{U}} = I_{X_{\mathcal{U}}} \quad \text{and} \quad T_{\mathcal{U}}(T^{-1})_{\mathcal{U}} = (I_Y)_{\mathcal{U}} = I_{Y_{\mathcal{U}}} \,.$$

This shows that $T_{\mathcal{U}}\colon X_{\mathcal{U}} \to Y_{\mathcal{U}}$ is invertible and that $(T_{\mathcal{U}})^{-1} = (T^{-1})_{\mathcal{U}}$.

For the converse, assume that $T_{\mathcal{U}}$ is invertible. This implies immediately that T is one-to-one. Moreover, we claim that T has closed range. To see this, assume that $Tx_n \to y$ in Y. Then we have

$$x_n = (T_{\mathcal{U}})^{-1}[(Tx_n, Tx_n, \ldots)] \to (T_{\mathcal{U}})^{-1}y$$

in $X_{\mathcal{U}}$. In particular, $\{x_n\}$ is a Cauchy sequence in X. If $x_n \to x$ holds in X, then $Tx_n \to Tx$ in Y, and so $y = Tx$. Thus, T has a closed range. By part (c), T is surjective, and consequently invertible. ∎

1.5. Vector-valued Functions

Problem 1.5.1. *Establish the following "product rule" for derivatives: Consider two functions $g\colon \mathcal{O} \to \mathbb{C}$ and $f\colon \mathcal{O} \to X$ (where $\mathcal{O}$ is an open subset of $\mathbb{C}$ and X is a complex Banach space). If both functions f and g are differentiable at some $\lambda_0 \in \mathcal{O}$, then show that the X-valued function $\lambda \mapsto [gf](\lambda) = g(\lambda)f(\lambda)$ is also differentiable at λ_0 and*

$$(gf)'(\lambda_0) = g'(\lambda_0)f(\lambda_0) + g(\lambda_0)f'(\lambda_0)\,.$$

Solution: From $f(\lambda) - f(\lambda_0) = (\lambda - \lambda_0)\frac{f(\lambda)-f(\lambda_0)}{\lambda-\lambda_0} \xrightarrow[\lambda\to\lambda_0]{} 0 \cdot f'(\lambda_0) = 0$, it follows that $\lim_{\lambda\to\lambda_0} f(\lambda) = f(\lambda_0)$. That is, f is continuous at λ_0. Now note that

$$\tfrac{(gf)(\lambda)-(gf)(\lambda_0)}{\lambda-\lambda_0} = \tfrac{g(\lambda)f(\lambda)-g(\lambda_0)f(\lambda_0)}{\lambda-\lambda_0} = \tfrac{g(\lambda)-g(\lambda_0)}{\lambda-\lambda_0}f(\lambda) + g(\lambda_0)\tfrac{f(\lambda)-f(\lambda_0)}{\lambda-\lambda_0}\,.$$

This easily implies

$$\lim_{\lambda\to\lambda_0} \tfrac{(gf)(\lambda)-(gf)(\lambda_0)}{\lambda-\lambda_0} = g'(\lambda_0)f(\lambda_0) + g(\lambda_0)f'(\lambda_0)\,.$$

So, gf is differentiable at λ_0 and $(gf)'(\lambda_0) = g'(\lambda_0)f(\lambda_0) + g(\lambda_0)f'(\lambda_0)$. ∎

Problem 1.5.2. ***Morera's theorem*** *is a converse of Cauchy's theorem and it states that:* If a continuous function $f\colon \mathcal{O} \to \mathbb{C}$ (where $\mathcal{O}$ is an open subset of $\mathbb{C}$) satisfies $\int_C f(\lambda)\,d\lambda = 0$ for each simple closed rectifiable curve C which lies with its interior in $\mathcal{O}$, then f is an analytic function.

Show that Morera's theorem is valid for Banach-valued functions.

Solution: Assume that a continuous function $f\colon \mathcal{O} \to X$ (where $\mathcal{O}$ is an open subset of $\mathbb{C}$ and X is a complex Banach space) satisfies $\int_C f(\lambda)\,d\lambda = 0$ for each simple closed rectifiable curve C which lies with its interior in $\mathcal{O}$.

Fix an arbitrary continuous linear functional $x^* \in X^*$ and consider the continuous function $x^* \circ f\colon \mathcal{O} \to \mathbb{C}$. Now if C is a simple closed rectifiable curve which lies with its interior in $\mathcal{O}$, then it follows from Theorem 1.69 that

$$\int_C (x^* \circ f)(\lambda)\,d\lambda = x^*\Big(\int_C f(\lambda)\,d\lambda\Big) = x^*(0) = 0\,.$$

This, in connection with the standard Morera theorem, yields that $x^* \circ f$ is an analytic function for each $x^* \in X^*$. That is, the function $f\colon \mathcal{O} \to X$ is weakly analytic. By Theorem 1.76, the function f is also analytic. ∎

Problem 1.5.3. *Let C be a rectifiable curve lying in $\mathbb{C}$, and let X and Y be two complex Banach spaces. Also, let $A\colon C \to \mathcal{L}(X,Y)$ be a continuous operator-valued function. Show that for each $x \in X$ we have*

$$\Big[\int_C A(\lambda)\,d\lambda\Big](x) = \int_C A(\lambda)x\,d\lambda\,.$$

Solution: Since the vectors on both sides belong to Y, it suffices to verify that for each $x \in X$ and each $y^* \in Y^*$ we have

$$\Big\langle y^*, \Big[\int_C A(\lambda)\,d\lambda\Big](x)\Big\rangle = \Big\langle y^*, \int_C A(\lambda)x\,d\lambda\Big\rangle\,.$$

In view of Theorem 1.69, the right-hand side equals $\int_C y^*(A(\lambda)x)\,d\lambda$. Now consider the functional $y^* \otimes x$ on $\mathcal{L}(X,Y)$ defined by $\langle y^* \otimes x, T\rangle = \langle y^*, Tx\rangle$ for each T in $\mathcal{L}(X,Y)$. So, on one hand, we have

$$\Big\langle y^* \otimes x, \int_C A(\lambda)\,d\lambda\Big\rangle = \Big\langle y^*, \Big[\int_C A(\lambda)\,d\lambda\Big](x)\Big\rangle\,,$$

and on the other hand, applying $y^* \otimes x$ to the operator $\int_C A(\lambda)\,d\lambda$ and again using Theorem 1.69, we get

$$\Big\langle y^* \otimes x, \int_C A(\lambda)\,d\lambda\Big\rangle = \int_C \langle y^* \otimes x, A(\lambda)\rangle\,d\lambda = \int_C y^*(A(\lambda)x)\,d\lambda\,.$$

Therefore,

$$\Big\langle y^* \otimes x, \int_C A(\lambda)\,d\lambda\Big\rangle = \Big\langle y^*, \Big[\int_C A(\lambda)\,d\lambda\Big](x)\Big\rangle = \Big\langle y^*, \int_C A(\lambda)x\,d\lambda\Big\rangle$$

for each $x \in X$ and all $y^* \in Y^*$, as desired. ∎

Problem 1.5.4. *This is a generalization of Theorem 1.76. Let X be a Banach space and let G be a (not necessarily closed)* ***norming subspace*** *of X^*, that is, there exists a constant $\gamma > 0$ such that*

$$\sup\{|x^*(x)|\colon\ x^* \in G \ \text{ and } \ \|x^*\| \le 1\} \ge \gamma\|x\|$$

for each $x \in X$.

Show that a continuous function $f\colon \mathcal{O} \to X$ defined on an open subset $\mathcal{O}$ of the complex plane $\mathbb{C}$ is analytic if and only if the scalar-valued function $x^* \circ f$ is analytic for each $x^* \in G$.

Solution: The proof mimics that of Theorem 1.76. The only difference is that instead of the inequality $\|x_n - x_m\| \le \frac{M}{r^2}|\lambda_n - \lambda_m|$ we will obtain (using the analyticity of the functions $x^* \circ f$ for $x^* \in G$) that $\gamma\|x_n - x_m\| \le \frac{M}{r^2}|\lambda_n - \lambda_m|$. This is enough to guarantee that $\{x_n\}$ is a Cauchy sequence. The other arguments in the proof remain the same. ∎

Problem 1.5.5. *For a continuous function $A\colon \mathcal{O} \to \mathcal{L}(X,Y)$, where $\mathcal{O}$ is an open subset of $\mathbb{C}$, show that the following statements are equivalent.*

(a) *The function $\lambda \mapsto A(\lambda)$ is analytic on $\mathcal{O}$.*

(b) *For each $x \in X$ the Y-valued function $\phi_x(\lambda) = A(\lambda)x$ is analytic on $\mathcal{O}$.*

(c) *For every $x \in X$ and $y^* \in Y^*$ the standard complex-valued function*

$$\phi_{x,y^*}(\lambda) = \langle A(\lambda)x, y^* \rangle = y^*\big(A(\lambda)x\big)$$

is analytic on $\mathcal{O}$.

Solution: To prove that the continuous function $A(\lambda)$ is analytic, it is sufficient to show (according to Problem 1.5.4) that there exists a norming subspace G of $\mathcal{L}(X,Y)^*$ such that $g^* \circ A(\lambda)$ is analytic for each $g^* \in G$. Let $G = Y^* \otimes X$. Recall that each expression $y^* \otimes x$ can be considered as an element of $\mathcal{L}(X,Y)^*$ by letting $T \mapsto y^*(Tx)$. Obviously, G is norming and this proves the equivalence of (a) and (c). The equivalence of (c) and (b) is a direct consequence of Theorem 1.76. ∎

Problem 1.5.6. *For each $r > 0$ let $A_r = \{\lambda \in \mathbb{C}\colon\ |\lambda| > r\}$. Assume that X is a complex Banach space and for some $r > 0$ a function $f\colon A_r \to X$ is analytic having the Laurent series expansion $f(\lambda) = \sum_{n=-\infty}^{\infty} \lambda^n c_n$ on the open annulus A_r. Show that $\lim_{\lambda\to\infty} f(\lambda) = 0$ if and only if $c_n = 0$ for each $n = 0, 1, 2, \ldots$.*

Solution: Assume first that $\lim_{\lambda\to\infty} f(\lambda) = 0$. Fix any $\epsilon > 0$ and then choose some $r_1 > \max\{1, r\}$ such that $\|f(\lambda)\| < \epsilon$ for each $|\lambda| > r_1$. If C_R denotes the positively oriented circle with radius R and center zero, then according to Laurent's Theorem 1.78, for each $R > r_1$ and each $n \geq 0$ we have

$$\|c_n\| = \Big\| \tfrac{1}{2\pi\imath} \int_{C_R} \tfrac{f(\lambda)}{\lambda^{n+1}}\, d\lambda \Big\| \leq \tfrac{1}{2\pi} \cdot \tfrac{\epsilon}{R^{n+1}} \cdot 2\pi R = \tfrac{\epsilon}{R^n} \leq \epsilon\,.$$

Since $\epsilon > 0$ is arbitrary, it follows that $c_n = 0$ for all $n \geq 0$.

Now assume that $c_n = 0$ for each $n \geq 0$. From Theorem 1.77, we know that $\limsup \sqrt[n]{\|c_{-n}\|} \leq r$. So, there exists some $r_2 > r$ such that $\sqrt[n]{\|c_{-n}\|} < r_2$ for each $n \geq 1$. For each $\lambda \in \mathbb{C}$ with $|\lambda| > r_2$, let $\delta_\lambda = \frac{r_2}{|\lambda|}$. Clearly, $0 < \delta_\lambda < 1$ and $\lim_{\lambda\to\infty} \delta_\lambda = 0$. From $\big\|\frac{c_{-n}}{\lambda^n}\big\| = \frac{\|c_{-n}\|}{|\lambda|^n} \leq \frac{r_2^n}{|\lambda|^n} = \delta_\lambda^n$, it follows that for $\lambda \in \mathbb{C}$ with $|\lambda| > r_2$ we have

$$\|f(\lambda)\| = \Big\|\sum_{n=1}^{\infty} \tfrac{c_{-n}}{\lambda^n}\Big\| \leq \sum_{n=1}^{\infty} \Big\|\tfrac{c_{-n}}{\lambda^n}\Big\| \leq \sum_{n=1}^{\infty} \delta_\lambda^n = \tfrac{\delta_\lambda}{1-\delta_\lambda}\,.$$

Since $\lim_{\lambda\to\infty} \frac{\delta_\lambda}{1-\delta_\lambda} = 0$, the preceding inequality shows that $\lim_{\lambda\to\infty} f(\lambda) = 0$. ∎

Problem 1.5.7. *Assume that two points a and b in a topological space X are joined by a curve C. If C is covered by a collection $\mathcal{V}$ of open sets, then show that there exist (not necessarily distinct) sets $V_1, V_2, \ldots, V_n$ in $\mathcal{V}$ such that:*

(a) $a \in V_1$ *and* $b \in V_n$.

(b) $C \subseteq V_1 \cup V_2 \cup \cdots \cup V_n$.

(c) $V_i \cap V_{i+1} \neq \emptyset$ *for each* $i = 1, 2, \ldots, n-1$.

(*Such a collection* $V_1, V_2, \ldots, V_n$ *of open sets is called a* ***chain*** *of open sets covering the curve* C.)

Solution: Fix a continuous function $x\colon [\alpha, \beta] \to X$ such that $x(\alpha) = a$, $x(\beta) = b$, and $x([\alpha, \beta]) = C$. Since x is continuous, it follows that $\{x^{-1}(V)\colon\ V \in \mathcal{V}\}$ is an open cover of the closed interval $[\alpha, \beta]$. Now notice that for each $V \in \mathcal{V}$ the open subset $x^{-1}(V)$ of $[\alpha, \beta]$ is a union of open subintervals of $[\alpha, \beta]$. So, there exists a family of open subintervals $\mathcal{I}$ of $[\alpha, \beta]$ such that $\mathcal{I}$ covers $[\alpha, \beta]$ and for each $I \in \mathcal{I}$ there exists some $V \in \mathcal{V}$ with $I \subseteq x^{-1}(V)$. We claim that there exists a chain $I_1, I_2, \ldots, I_k$ in $\mathcal{I}$ that covers $[\alpha, \beta]$. If this is the case, then for each $1 \leq r \leq k$ choose some $V_r \in \mathcal{V}$ with $I_r \subseteq x^{-1}(V_r)$ and note that $V_1, V_2, \ldots, V_r$ is a desired chain for the curve C.

In view of the compactness of $[\alpha, \beta]$, in order to establish our claim, it suffices to prove that whenever a closed interval is a finite union of open subintervals, then the open cover can be reduced to a chain. The proof is by induction on the number n of the subintervals of the open cover. For $n = 1$ the conclusion should be obvious. For the induction step, assume that every open cover of any closed interval consisting of at most n open subintervals can be reduced to a chain.

Suppose that a closed interval $[\alpha, \beta]$ is the union of a family $\mathcal{I}$ of $n+1$ open subintervals of $[\alpha, \beta]$. Pick some $I_0 \in \mathcal{I}$ such that $\beta \in I_0$, i.e., $I_0 = (\gamma, \beta]$ for some $\alpha < \gamma < \beta$. Clearly, the family $\{I \cap [\alpha, \gamma]\colon\ I \in \mathcal{I} \text{ and } I \cap [\alpha, \gamma] \neq \emptyset\}$ of open subintervals of $[\alpha, \gamma]$ covers $[\alpha, \gamma]$ and has at most n open subintervals. By our induction hypothesis, this open cover of $[\alpha, \gamma]$ can be reduced to a chain, say $I_1 \cap [\alpha, \gamma], I_2 \cap [\alpha, \gamma], \ldots, I_k \cap [\alpha, \gamma]$. To finish the induction argument, it remains to observe that $I_1, I_2, \ldots, I_k, I_0$ is a chain for $\mathcal{I}$. ∎

1.6. Fundamentals of Measure Theory

Problem 1.6.1. *Let E be a Riesz space with the countable sup property, and let D be a subset of E such that the supremum $u = \sup D$ exists. Does there exist a countable subset D_1 of D such that for each $x \in D$ there is an element $a \in D_1$ such that $a \geq x$? (Any subset D_1 of D with this property is called* ***cofinal*** *in D.)*

Solution: The answer to this question is negative. That is, in general, cofinal countable subsets need not exist. Here is an example.

Let $E = C[0, 1]$. Since the Riemann integral is a strictly positive linear function on $C[0, 1]$, it is easy to verify that E has the countable sup property. For each $t \in [0, 1]$ choose a function $f_t \in E$ such that $f_t(t) = 1$ and $0 \leq f_t(s) < 1$ for each $s \neq t$. Consider the set $D = \{f_t\colon\ t \in [0, 1]\}$. Obviously, $\sup D = \mathbf{1} = \chi_{[0,1]}$.

Now let $D_1 = \{f_{t_n}\colon n = 1, 2, \ldots\}$ be an arbitrary subset of D. Pick any $t \in [0,1]$ that is distinct from each t_n. Since $f_t(t) = 1$ and $f_{t_n}(t) < 1$ for each n, it should be clear that f_t is not dominated by any function from D_1. ∎

Problem 1.6.2. *Let E be an Archimedean Riesz space satisfying the countable sup property and having a weak unit. Show that for each non-empty subset D of E there exists a countable subset D_0 of D such that D_0 has the same set of upper bounds as D. That is, whenever $u \in E$ satisfies $u \geq x$ for each $x \in D_0$, then $u \geq x$ for each $x \in D$.*

Solution: Let e be a weak unit in E. Assume first that E is Dedekind complete. For each n consider the set $D_n = D \wedge ne = \{x \wedge ne\colon x \in D\}$. This set is bounded from above and so $u_n = \sup D_n$ exists. Since E has the countable sup property, there is a countable subset A_n of D such that $\sup(A_n \wedge ne) = u_n$. We claim that the set $D_0 = \bigcup_{n=1}^{\infty} A_n$ satisfies the desired property. To see this, let $u \in E$ satisfy $u \geq a$ for all $a \in D_0$ and fix some $d \in D$. Fix for a moment some n. Then, of course, $u \geq a \wedge ne$ for each $a \in A_n$, and so $u \geq u_n$ holds for each n. This implies $u \geq d \wedge ne$ for each $d \in D$ and each n. From $d \wedge ne \uparrow d$, we get $u \geq d$.

Now let E be an arbitrary Riesz space, then consider $\hat{E}$, the Dedekind completion of E. Clearly the weak unit e in E remains a weak unit in $\hat{E}$. Also, it is easy to see that $\hat{E}$ has the countable sup property since E has it. Therefore, the general case is reduced to the previous one. The proof is finished. Also notice that the assumption that E has a weak unit is essential. ∎

Problem 1.6.3. *Let (Ω, Σ, π) be an arbitrary measure space, and let $\widehat{\Sigma}$ be the collection consisting of all subsets of Ω that can be written in the form $A\Delta M$, where $A \in \Sigma$ and $M \subseteq N$ for some $N \in \Sigma$ with $\pi(N) = 0$. (As always, $A\Delta M = (A \setminus M) \cup (M \setminus A)$ is the symmetric difference of the sets A and M.) Verify that:*

(a) *$\widehat{\Sigma}$ is a σ-algebra.*

(b) *The function $\widehat{\pi}\colon \widehat{\Sigma} \to [0, \infty]$, defined by $\widehat{\pi}(A\Delta M) = \pi(A)$, is well defined and extends π from Σ to a measure on $\widehat{\Sigma}$.*

(c) *If some $A \in \Sigma_\pi$ satisfies $\pi^*(A) < \infty$, then $A \in \widehat{\Sigma}$. In particular, $\widehat{\pi}$ is a **complete measure**, that is, if some subset A of Ω satisfies $\pi^*(A) = 0$, then $A \in \widehat{\Sigma}$.*

(d) *If π is σ-finite, then $\widehat{\Sigma} = \Sigma_\pi$.*

Solution: (a) Clearly, $\Sigma \subseteq \widehat{\Sigma}$. Now take an arbitrary set $A\Delta M$ in $\widehat{\Sigma}$, where $A \in \Sigma$ and $M \subseteq N$ for some $N \in \Sigma$ with $\pi(N) = 0$. An easy verification shows that

$$(A\Delta M)^{\mathrm{c}} = A^{\mathrm{c}}\Delta M = (A \cap M) \cup (A \cup M)^{\mathrm{c}}\,.$$

This implies that $\widehat{\Sigma}$ is closed under complementation.

To establish that $\widehat{\Sigma}$ is closed under countable unions, we need to verify that

$$\widehat{\Sigma} = \big\{D \in 2^{\Omega}\colon\ \exists\, A, N \in \Sigma \text{ such that } \pi(N) = 0 \text{ and } A \setminus N \subseteq D \subseteq A \cup N\big\}\,.$$

To see this, note that if $A \setminus N \subseteq D \subseteq A \cup N$ holds with $\pi(N) = 0$, then letting $M = D \cap N$, we have $D = (A \setminus N)\Delta M$. On the other hand, if $D = A\Delta M$, where $A \in \Sigma$ and $M \subseteq N$ for some $N \in \Sigma$ with $\pi(N) = 0$, then we have

$$A \setminus N \subseteq A \setminus M \subseteq D = (A \setminus M) \cup (M \setminus A) \subseteq A \cup N\,.$$

We are now ready to show that $\widehat{\Sigma}$ is closed under countable unions. To see this, let $\{D_n\} \subseteq \widehat{\Sigma}$ be given, and put $D = \bigcup_{n=1}^{\infty} D_n$. Pick two sequences $\{A_n\}$ and $\{N_n\}$ in Σ such that $\pi(N_n) = 0$ and $A_n \setminus N_n \subseteq D_n \subseteq A_n \cup N_n$ holds for each n. If $A = \bigcup_{n=1}^{\infty} A_n \in \Sigma$ and $N = \bigcup_{n=1}^{\infty} N_n \in \Sigma$, then $\pi(N) = 0$ and

$$A \setminus N \subseteq \bigcup_{n=1}^{\infty} (A_n \setminus N_n) \subseteq \bigcup_{n=1}^{\infty} D_n = D \subseteq \bigcup_{n=1}^{\infty} (A_n \cup N_n) = A \cup N\,.$$

This shows that $D \in \widehat{\Sigma}$. Therefore, $\widehat{\Sigma}$ is a σ-algebra.

(b) Clearly, we have $\Sigma \subseteq \widehat{\Sigma} \subseteq \Sigma_\pi$. Now let $D \in \widehat{\Sigma}$. Pick $A \in \Sigma$, $N \in \Sigma$ with $\pi(N) = 0$ and a subset M of N such that $D = A\Delta M$. This implies that $\pi(D) = \pi(A) = \pi^*(A)$. Since π^* is a measure on Σ_π, it follows that π^* (and hence $\widehat{\pi}$) is also a measure on $\widehat{\Sigma}$.

(c) Assume that a set $A \in \Sigma_\pi$ satisfies $\pi^*(A) < \infty$. Then for each n there exists some $B_n \in \Sigma$ such that $\pi(B_n) < \pi^*(A) + \frac{1}{n}$. Let $B = \bigcap_{n=1}^{\infty} B_n \in \Sigma$. From $\pi(B) \le \pi(B_n) < \pi^*(A) + \frac{1}{n} \le \pi(B) + \frac{1}{n}$, it follows that $\pi(B) = \pi^*(A)$.

Since $\pi^*(A) < \infty$, the set $M = B \setminus A$ satisfies $\pi^*(M) = \pi(B) - \pi^*(A) = 0$. Now choose some $N \in \Sigma$ with $M \subseteq N$ and $\pi(N) = 0$ and note that $A = B\Delta M$. Therefore, $A \in \widehat{\Sigma}$.

(d) We already know that $\widehat{\Sigma} \subseteq \Sigma_\pi$. Now assume that π is σ-finite. Pick a sequence $\{\Omega_n\} \subseteq \Sigma$ with $\pi(\Omega_n) < \infty$ for each n and $\Omega = \bigcup_{n=1}^{\infty} \Omega_n$.

If $A \in \Sigma_\pi$, then by part (c) we have $A \cap \Omega_n \in \widehat{\Sigma}$ for each n. This implies $A = \bigcup_{n=1}^{\infty} A \cap \Omega_n \in \widehat{\Sigma}$. Therefore, in this case, $\widehat{\Sigma} = \Sigma_\pi$ holds true. ∎

Problem 1.6.4. *Let J be an ideal in some $L_0(\pi)$-space and let $\{x_n\}$ be an order bounded sequence in J. Show that $x_n \xrightarrow{o} x$ holds in J if and only if $x_n(\omega) \to x(\omega)$ for π-almost all $\omega \in \Omega$.*

Solution: Let $\{x_n\}$ be an order bounded sequence in some ideal J of $L_0(\pi)$. Fix some vector $u \in J$ such that $|x_n| \le u$ holds for each $n \in \mathbb{N}$. Assume first that $x_n \xrightarrow{o} x$. This means that there exists some sequence $\{y_n\} \subseteq J$ satisfying $y_n \downarrow 0$ and $|x_n - x| \le y_n$ for each n. If we pick a π-null set A such that $|x_n(\omega) - x(\omega)| \le y_n(\omega)$ and $y_n(\omega) \downarrow 0$ hold for each $\omega \notin A$, then $x_n(\omega) \to x(\omega)$ for each $\omega \notin A$. This shows that $x_n \xrightarrow{p} x$.

For the converse, suppose there is a π-null set A such that $|x_n(\omega)| \le u(\omega) < \infty$ and $x_n(\omega) \to x(\omega)$ for all $\omega \notin A$. Now define the two sequences $\{y_n\}$ and $\{z_n\}$ on $\Omega \setminus A$ via the formulas

$$y_n(\omega) = \sup\{x_i(\omega)\colon\ i \ge n\} \quad \text{and} \quad z_n(\omega) = \inf\{x_i(\omega)\colon\ i \ge n\}\,.$$

Clearly, y_n and z_n are functions in $L_0(\pi)$ and in view of $|y_n| \le u$ and $|z_n| \le u$, we see that $\{y_n\}$ and $\{z_n\}$ are sequences in J. Moreover, notice that $z_n(\omega) \uparrow x(\omega)$ and

$y_n(\omega) \downarrow x(\omega)$ for each $\omega \notin A$. Now if $v_n = y_n - z_n$, then $v_n(\omega) \downarrow 0$ for each $\omega \notin A$ (and so $v_n \downarrow 0$ in J) and $|x_n(\omega) - x(\omega)| \le v_n(\omega)$ for all $\omega \notin A$. Thus, $|x_n - x| \le v_n$ in J for each n, and therefore $x_n \xrightarrow{o} x$ in J. ∎

Problem 1.6.5. *If π is σ-finite, then show that $L_0(\pi)$ is a Dedekind complete Riesz space with the countable sup property.*

Solution: Fix a sequence $\{\Omega_n\} \subseteq \Sigma$ satisfying $0 < \pi(\Omega_n) < \infty$ for each n and $\Omega = \bigcup_{n=1}^{\infty} \Omega_n$. Also, define $\rho\colon L_0^+(\pi) \to [0, \infty)$ by

$$\rho(x) = \sum_{n=1}^{\infty} \tfrac{1}{2^n} \cdot \tfrac{1}{\pi(\Omega_n)} \int_{\Omega_n} \tfrac{x}{1+x}\, d\pi\,.$$

It is easy to see that ρ has the following properties:

(1) A function $x \in L_0^+(\pi)$ satisfies $\rho(x) = 0$ if and only if $x = 0$.

(2) If $x, y \in L_0^+(\pi)$ satisfy $0 \le x < y$, then $\rho(x) < \rho(y)$, i.e., ρ is strictly monotone.

(3) If $0 \le x_n \uparrow x$ holds in $L_0^+(\pi)$, then $\rho(x_n) \uparrow \rho(x)$.

Now let D be a non-empty subset of $L_0(\pi)$ that is bounded from above in $L_0(\pi)$. Since the collection of all finite suprema of D has the same upper bounds as D, we can assume that D is directed upward. Now fix some $x_0 \in D$ and consider the set $D_1 = \{x \in D\colon\ x \ge x_0\} - x_0$. Notice that $\sup D$ exists if and only $\sup D_1$ exists and $\sup D_1 = (\sup D) - x_0$. So, replacing D with the set consisting of all finite suprema of D_1, we can assume without loss of generality that D is an upward directed subset of $L_0^+(\pi)$.

Put $m = \sup_{x \in D} \rho(x) < \infty$, and then choose a sequence $\{x_n\} \subseteq D$ such that $x_n \uparrow$ and $\rho(x_n) \uparrow m$. Since D is bounded from above, it follows that there exists some $x \in L_0^+(\pi)$ such that $x_n \uparrow x$. This implies $\rho(x_n) \uparrow \rho(x)$. We claim that $x = \sup D$.

First, we shall show that x is an upper bound of D. To see this, let $y \in D$. Then $y \vee x_n \in D$ and $y \vee x_n \uparrow y \vee x$. From $x_n \le y \vee x_n$ and $\rho(y \vee x_n) \le m$, it follows that $\rho(y \vee x) = m$. Now from $0 \le x \le y \vee x$, $\rho(y \vee x) = \rho(x) = m$, and the strict monotonicity of ρ, we get $x = y \vee x$. This implies $y \le x$, i.e., x is an upper bound of D. To see that x is the least upper bound of D, let $z \in L_0(\pi)$ be an upper bound of D, i.e., $x \le z$ for each $x \in D$. Then $x_n \le z$ also holds in $L_0(\pi)$ for each n. Since x is the π-a.e. pointwise limit of the sequence $\{x_n\}$, we see that $x \le z$. Thus, $x = \sup D$.

The above arguments show that $L_0(\pi)$ is a Dedekind complete Riesz space with the countable sup property. ∎

Problem 1.6.6. *Recall that a Riesz subspace G of a Riesz space L is said to be* ***order dense*** *if for each $0 < x \in L$ there exists some $0 < y \in G$ satisfying $0 < y \le x$.*

Show that a Riesz subspace G of an Archimedean Riesz space L is order dense if and only if for each $x \in L^+$ we have $\{y \in G^+ : y \le x\} \uparrow x$ in L.

Solution: Let G be an order dense Riesz subspace of an Archimedean Riesz space L, and let $x \in L^+$. Assume by way of contradiction that $\sup\{y \in G\colon\ y \leq x\} \neq x$. This means that there exists some $z < x$ such that $y \in G$ and $y \leq x$ imply $y \leq z$. Put $u = x - z > 0$, and then select some $v \in G$ with $0 < v \leq u < x$. This implies $v \leq z$ and so $2v = v + v \leq u + z = x$. Now assuming that $nv \leq x$, we get $nv \leq z$, and so $(n+1)v = nv + v \leq z + u = x$. Therefore, we have shown by induction that $nv \leq x$ holds for each n. Since L is Archimedean, the latter yields $v = 0$, which is a contradiction. Hence, $\sup\{y \in G\colon\ y \leq x\} = x$. The converse is obvious. ∎

Problem 1.6.7. *Consider the following property* (P) *for subsets D of Riesz spaces:*

(P) If $\{x_n\} \subseteq D$ and $\lambda_n \downarrow 0$ in $\mathbb{R}$, then $\lambda_n x_n \xrightarrow{o} 0$.

Assume that a subset D of a Riesz space E satisfies property (P). *Prove that the set consisting of all finite suprema of the set $|D| = \{|d|\colon\ d \in D\}$ also satisfies* (P).

Solution: Let D be a non-empty subset in a Riesz space E that satisfies property (P), and let $\hat{D}$ denote the set of all finite suprema of $|D|$. Assume that $\{\lambda_n\}$ is a sequence of scalars such that $\lambda_n \downarrow 0$, and let $\{y_n\}$ be a sequence in $\hat{D}$. By definition of $\hat{D}$ for each n there is a finite subset $\{d_{n,i}\colon\ 1 \leq i \leq k_n\}$ of D such that $y_n = |d_{n,1}| \vee \cdots \vee |d_{n,k_n}|$. Let $\{x_n\}$ denote the sequence in D defined by:

$$d_{1,1}, d_{1,2}, \ldots, d_{1,k_1}, d_{2,1}, d_{2,2}, \ldots, d_{2,k_2}, \ldots .$$

For each $n \in \mathbb{N}$ let $r_n = k_1 + \cdots + k_n$, and then for each n satisfying $r_{m-1} < n \leq r_m$ (we assume that $r_0 = 0$) let $\beta_n = \lambda_m$. Clearly $\beta_n \downarrow 0$ and therefore, by our hypothesis $\beta_n x_n \xrightarrow{o} 0$. Let $\{z_n\}$ be a sequence such that $z_n \downarrow 0$ and $|\beta_n x_n| \leq z_n$ for each n. Put $u_m = z_{r_{m-1}}$ and note that $u_m \downarrow 0$ in E. Moreover, we have

$$|\lambda_m y_m| = \lambda_m |y_m| = \bigvee_{n=r_{m-1}+1}^{r_m} |\lambda_m x_n| = \bigvee_{n=r_{m-1}+1}^{r_m} |\beta_n x_n| \leq z_{r_{m-1}} = u_m .$$

This shows that $\lambda_n y_n \xrightarrow{o} 0$, as desired. ∎

Problem 1.6.8. *For a sequence $\{x_n\}$ in some $L_0(\pi)$-space establish the following.*

(a) *If $x_n \xrightarrow{\pi} x$ and $x_n \xrightarrow{\pi} y$ hold in $L_0(\pi)$, then $x = y$.*

(b) *If $x_n \xrightarrow{\pi} x$ in $L_0(\pi)$ and $\{x_n\}$ is a positive sequence, then $x \geq 0$.*

Solution: (a) From Theorem 1.82 it follows that there exists a subsequence $\{z_n\}$ of $\{x_n\}$ such that $z_n \xrightarrow{p} x$ and $z_n \xrightarrow{p} y$. Therefore, $x(\omega) = y(\omega)$ for π-almost all $\omega \in \Omega$.

(b) Again, there exists a subsequence $\{y_n\}$ of $\{x_n\}$ such that $y_n \xrightarrow{p} x$. This implies that $x(\omega) \geq 0$ for π-almost all $\omega \in \Omega$. ∎

Problem 1.6.9. *We say that a sequence $\{x_n\}$ in some $L_0(\pi)$ is* **$\star$-convergent** *to x (in symbols, $x_n \xrightarrow{\star} x$) if every subsequence of $\{x_n\}$ has a subsequence that converges pointwise π-almost everywhere to x.*

Let (Ω, Σ, π) be a σ-finite measure space and let A be a measurable set of finite measure. Also let $\phi\colon A \to \mathbb{R}$ be a π-integrable function such that $\phi(\omega) > 0$ for π-almost all $\omega \in A$. Show that for each sequence $\{B_n\}$ of measurable subsets of A the following statements are equivalent:

(a) $\chi_{B_n} \xrightarrow{\pi} 0$.

(b) $\chi_{B_n} \xrightarrow{\star} 0$.

(c) $\int_{B_n} \phi\, d\pi \longrightarrow 0$.

(d) $\pi(B_n) \longrightarrow 0$.

Solution: (a) $\Longleftrightarrow$ (b) This follows immediately from Theorem 1.82.

(b) $\Longrightarrow$ (c) If $\int_{B_n} \phi\, d\pi \not\to 0$, then there exists a subsequence $\{C_n\}$ of $\{B_n\}$ and some $\alpha > 0$ such that $\int_{C_n} \phi\, d\pi \geq \alpha$ holds for each n. Since $\chi_{B_n} \xrightarrow{\star} 0$, there exists a subsequence $\{D_n\}$ of $\{C_n\}$ such that $\chi_{D_n} \xrightarrow{p} 0$. But then, by the Lebesgue Dominated Convergence Theorem, we have $\alpha \leq \int_{D_n} \phi\, d\pi = \int_A \phi \chi_{D_n}\, d\pi \longrightarrow 0$, which is impossible. Hence, $\int_{B_n} \phi\, d\pi \longrightarrow 0$.

(c) $\Longrightarrow$ (d) If $\pi(B_n) \not\to 0$, then we can assume (by passing to a subsequence if necessary) that there exists some $\eta > 0$ such that $\pi(B_n) \geq \eta$ for each n. We claim that there is some constant $\alpha > 0$ such that $\int_{B_n} \phi\, d\pi \geq \alpha$ for each n.

To see this, assume that there is no such constant α. This means that there exists a subsequence $\{C_n\}$ of $\{B_n\}$ satisfying $\int_{C_n} \phi\, d\pi < \frac{1}{2^n}$ for every $n \in \mathbb{N}$. Put $D_n = \bigcup_{i=n}^{\infty} C_i$ and note that $\int_{D_n} \phi\, d\pi \leq \sum_{i=n}^{\infty} \int_{C_i} \phi\, d\pi \leq 2^{1-n}$ for each n. If $D_n \downarrow D$, then from $0 \leq \int_D \phi\, d\pi = \lim_{n\to\infty} \int_{D_n} \phi\, d\pi = 0$, we get $\int_D \phi\, d\pi = 0$. Since $\phi(\omega) > 0$ for π-almost all $\omega \in A$, we obtain $\pi(D) = 0$. On the other hand, $D_1 \subseteq A$ implies $\pi(D_1) < \infty$, and from $C_n \subseteq D_n$ and the continuity of the measure π we get

$$0 = \pi(D) = \lim_{n\to\infty} \pi(D_n) \geq \limsup_{n\to\infty} \pi(C_n) \geq \eta > 0\,,$$

which is a contradiction. This establishes that $\pi(B_n) \to 0$.

(d) $\Longrightarrow$ (b) Let $\{C_n\}$ be an arbitrary subsequence of $\{B_n\}$. Then we have $\int \chi_{C_n}\, d\pi = \pi(C_n) \to 0$, and from this we infer that the sequence $\{\chi_{C_n}\}$ has a π-a.e. pointwise convergent subsequence. So, $\chi_{B_n} \xrightarrow{\star} 0$. ∎

Problem 1.6.10. *Assume that (Ω, Σ, π) is a σ-finite measure space and that a sequence $\{\Omega_n\} \subseteq \Sigma$ satisfies $\pi(\Omega_n) < \infty$ for each n and $\bigcup_{n=1}^{\infty} \Omega_n = \Omega$. Suppose also that $\phi\colon [0,\infty) \to [0,\infty)$ is a continuous strictly increasing bounded function with $\phi(0) = 0$; for instance, we can consider the functions $\phi(t) = \frac{t}{1+t}$ and $\phi(t) = 1 - e^{-t}$.*

Show that a sequence $\{x_n\} \subseteq L_0(\pi)$ satisfies $x_n \xrightarrow{\pi} 0$ if and only if $\lim_{n\to\infty} \int_{\Omega_k} \phi(|x_n|)\, d\pi = 0$ for each k.

Solution: Let $\phi\colon [0,\infty) \to [0,\infty)$ be a continuous strictly increasing bounded function with $\phi(0) = 0$. Fix some $M > 0$ such that $0 \leq \phi(t) \leq M$ for each $t \geq 0$.

Suppose first that $\lim_{n\to\infty}\int_{\Omega_k}\phi(|x_n|)\,d\pi=0$ for each k. Fix some $A\in\Sigma$ with $\pi(A)<\infty$, and let $\epsilon>0$. Put $\Theta_n=\bigcup_{k=1}^n\Omega_k$, and note that $A\cap\Theta_n\uparrow A$. So, by the continuity of the measure, there exists some m such that $\pi(A\setminus\Theta_m)<[\phi(\epsilon)]^2$. Now define $A_n=\{\omega\in A\colon\ |x_n(\omega)|\geq\epsilon\}$ for each n, and observe that since ϕ is strictly increasing, we have $A_n=\{\omega\in A\colon\ \phi(|x_n(\omega)|)\geq\phi(\epsilon)\}$. This implies $\phi(\epsilon)\chi_{A_n}\leq\chi_A\phi(|x|)$. Integrating yields

$$\begin{aligned}\phi(\epsilon)\pi(A_n) &\leq \int_A\phi(|x_n|)\,d\pi\leq\int_{A\setminus\Theta_m}\phi(|x_n|)\,d\pi+\int_{\Theta_m}\phi(|x_n|)\,d\pi\\ &\leq M\pi(A\setminus\Theta_m)+\sum_{k=1}^m\int_{\Omega_k}\phi(|x_n|)\,d\pi\\ &\leq M[\phi(\epsilon)]^2+\sum_{k=1}^m\int_{\Omega_k}\phi(|x_n|)\,d\pi\,.\end{aligned}$$

So, $\limsup_{n\to\infty}\phi(\epsilon)\pi(A_n)\leq M[\phi(\epsilon)]^2$ or $\limsup_{n\to\infty}\pi(A_n)\leq M\phi(\epsilon)$ for each $\epsilon>0$. Since $\epsilon>0$ is arbitrary, $\phi(0)=0$ and ϕ is continuous at zero, we get $\limsup_{n\to\infty}\pi(A_n)=0$. Consequently, $\lim_{n\to\infty}\pi(A_n)=0$, and thus $x_n\xrightarrow{\pi}0$.

For the converse, assume that $x_n\xrightarrow{\pi}0$. Fix some k and let $\epsilon>0$. Also, for each n let $A_n=\{\omega\in\Omega_k\colon\ |x_n(\omega)|\geq\epsilon\}$. Clearly, $A_n=\{\omega\in\Omega_k\colon\ \phi(|x_n(\omega)|)\geq\phi(\epsilon)\}$. According to our assumption we have $\lim_{n\to\infty}\pi(A_n)=0$. Now note that

$$\int_{\Omega_k}\phi(|x_n|)d\pi=\int_{A_n}\phi(|x_n|)\,d\pi+\int_{\Omega_k\setminus A_n}\phi(|x_n|)\,d\pi\leq M\pi(A_n)+\phi(\epsilon)\pi(\Omega_k)\,.$$

This implies $\limsup_{n\to\infty}\int_{\Omega_k}\phi(|x_n|)\,d\pi\leq\phi(\epsilon)\pi(\Omega_k)$. Since $\epsilon>0$ is arbitrary, it follows from the properties of the function ϕ that $\limsup_{n\to\infty}\int_{\Omega_k}\phi(|x_n|)\,d\pi=0$. Consequently, $\lim_{n\to\infty}\int_{\Omega_k}\phi(|x_n|)\,d\pi=0$, and the solution is finished. ∎

Problem 1.6.11. *Let (Ω,Σ,π) be a σ-finite measure. Fix a sequence $\{\Omega_k\}$ in Σ such that $0<\pi(\Omega_k)<\infty$ for each k and $\Omega=\bigcup_{k=1}^\infty\Omega_k$. Also, define the function $\rho\colon L_0(\pi)\to[0,\infty)$ by*

$$\rho(x)=\sum_{k=1}^\infty\tfrac{1}{2^k}\cdot\tfrac{1}{\pi(\Omega_k)}\int_{\Omega_k}\tfrac{|x|}{1+|x|}\,d\pi\,.$$

Show that for a sequence $\{x_n\}\subseteq L_0(\pi)$ and a function $x\in L_0(\pi)$ the following statements are equivalent.

(a) $x_n\xrightarrow{\pi}x$.

(b) *For each k we have* $\lim\limits_{n\to\infty}\int_{\Omega_k}\frac{|x_n-x|}{1+|x_n-x|}\,d\pi=0\,.$

(c) $\lim\limits_{n\to\infty}\rho(x_n-x)=0$.

Solution: (a) $\Longleftrightarrow$ (b) This follows immediately from Problem 1.6.10.

(b) $\Longrightarrow$ (c) Fix $\epsilon>0$, and then pick some m such that $\sum_{k=m+1}^\infty\frac{1}{2^k}<\epsilon$. Also, pick some n_0 such that $n\geq n_0$ implies $\sum_{k=1}^m\int_{\Omega_k}\frac{1}{2^k}\cdot\frac{1}{\pi(\Omega_k)}\frac{|x_n-x|}{1+|x_n-x|}\,d\pi<\epsilon$ for all

$n \geq n_0$. So, if $n \geq n_0$, then

$$\begin{aligned}\rho(x_n - x) &= \sum_{k=1}^{m} \int_{\Omega_k} \tfrac{1}{2^k} \cdot \tfrac{1}{\pi(\Omega_k)} \tfrac{|x_n - x|}{1+|x_n - x|}\, d\pi + \sum_{k=m+1}^{\infty} \int_{\Omega_k} \tfrac{1}{2^k} \cdot \tfrac{1}{\pi(\Omega_k)} \tfrac{|x_n - x|}{1+|x_n - x|}\, d\pi \\ &< \epsilon + \epsilon = 2\epsilon\,.\end{aligned}$$

This shows that $\rho(x_n - x) \to 0$.

(c) $\Longrightarrow$ (b) This implication is trivial. ∎

Problem 1.6.12. *Let $(\mathcal{S}, \Sigma_1, \mu)$ and $(\mathcal{T}, \Sigma_2, \nu)$ be two measure spaces. If a function $f\colon \mathcal{S} \times \mathcal{T} \to \mathbb{R}$ is jointly measurable, then $f(\cdot, t)$ is Σ_1-measurable for* each *$t \in \mathcal{T}$ and $f(s, \cdot)$ is ν-measurable for* each *$s \in \mathcal{S}$.*

Solution: For each subset D of $\mathcal{S} \times T$, each $s \in \mathcal{S}$, and each $t \in \mathcal{T}$ we put:

$$D_s = \{t \in \mathcal{T}\colon\ (s,t) \in D\} \quad \text{and} \quad D^t = \{s \in \mathcal{S}\colon\ (s,t) \in D\}\,.$$

Consider the collection of subsets of $\mathcal{S} \times \mathcal{T}$ given by

$$\mathcal{A} = \big\{D \subseteq \mathcal{S} \times \mathcal{T}\colon\ D_s \in \Sigma_2 \text{ and } D^t \in \Sigma_1 \text{ for all } s \in \mathcal{S} \text{ and } t \in \mathcal{T}\big\}\,.$$

A straightforward verification shows that $\mathcal{A}$ is a σ-algebra. Moreover, from

$$(A \times B)_s = \begin{cases} B & \text{if } s \in A \\ \emptyset & \text{if } s \notin A \end{cases} \quad \text{and} \quad (A \times B)^t = \begin{cases} A & \text{if } t \in B \\ \emptyset & \text{if } t \notin B\,, \end{cases}$$

we see that $\mathcal{A}$ contains all measurable rectangles. Therefore, $\Sigma_1 \otimes \Sigma_2 \subseteq \mathcal{A}$.

Now assume that a function $f\colon \mathcal{S} \times \mathcal{T} \to \mathbb{R}$ is jointly measurable. Let B be a Borel subset of $\mathbb{R}$, and let $s \in \mathcal{S}$. Then $f^{-1}(B) \in \Sigma_1 \otimes \Sigma_2$, and so

$$\big[f(s,\cdot)\big]^{-1}(B) = \big\{t \in \mathcal{T}\colon\ f(s,t) \in B\big\} = \big[f^{-1}(B)\big]_s \in \Sigma_2\,.$$

This shows that the function $f(s, \cdot)$ is Σ_2-measurable for each $s \in \mathcal{S}$. Similarly, $f(\cdot, t)$ is Σ_1-measurable for each $t \in \mathcal{T}$. ∎

Problem 1.6.13. *If (Ω, Σ, π) is a σ-finite measure space and E is an order dense ideal of $L_0(\pi)$, then show that for each $f \in L_0(\pi)$ there exists a sequence $\{f_n\} \subseteq E \cap L_1(\pi)$ satisfying $|f_n| \leq |f|$ for each n and $f_n(\omega) \to f(\omega)$ for π-almost all ω.*

If, in addition, $f \in L_0^+(\pi)$, then show that the sequence $\{f_n\}$ can be chosen in $E \cap L_1^+(\pi)$ such that $f_n \uparrow f$.

Solution: We claim that $L_1(\pi)$ is an order dense ideal of $L_0(\pi)$. To see this, let $0 < f \in L_0(\pi)$. Then for some $\epsilon > 0$ the π-measurable set $A = \{\omega \in \Omega\colon\ f(\omega) \geq \epsilon\}$ has positive measure. By the σ-finiteness of π, there exists some $B \in \Sigma$ satisfying $B \subseteq A$ and $0 < \pi(B) < \infty$. Now note that function $g = \epsilon\chi_B \in L_1(\pi)$ satisfies $0 < g \leq f$. This shows that $L_1(\pi)$ is order dense in $L_0(\pi)$.

Since E is order dense, it follows that $E \cap L_1(\pi)$ is also an order dense ideal (why?). Now let $0 \leq f \in L_0(\pi)$. Pick a net $\{f_\alpha\} \subseteq E \cap L_1(\pi)$ with $0 \leq f_\alpha \uparrow f$ in $L_0(\pi)$. Since $L_0(\pi)$ has the countable sup property (see Problem 1.6.5), there exists a sequence of indices $\{\alpha_n\}$ such that $0 \leq f_{\alpha_n} \uparrow f$.

Next, let $f \in L_0(\pi)$ be an arbitrary function. Pick two sequences $\{g_n\}$ and $\{h_n\}$ in $E \cap L_1(\pi)$ such that $0 \le g_n \uparrow f^+$ and $0 \le h_n \uparrow f^-$. Now if $\phi_n = g_n - h_n$, then $\phi_n \in E \cap L_1(\pi)$ and $|f_n| \le |f|$ hold for each n. To finish the solution, note that $\phi_n(\omega) \to f(\omega)$ for π-almost all $\omega \in \Omega$. ∎

Problem 1.6.14. *Let $m\colon \Gamma \to [0,\infty]$ be a measure on a semiring. Then for a function $f \in L_1(m)$ establish the following:*

(1) *$f \ge 0$ m-a.e. if and only if $\int_A f\,dm \ge 0$ holds for all $A \in \Gamma$ with $m(A) < \infty$.*

(2) *$f = 0$ m-a.e. if and only if $\int_A f\,dm = 0$ holds for all $A \in \Gamma$ with $m(A) < \infty$.*

Solution: (a) Let Γ be a semiring of subsets of a set Ω and assume that a function $f \in L_1(m)$ satisfies $\int_A f\,dm \ge 0$ for all $A \in \Gamma$ with $m(A) < \infty$. Consider the set $C = \{\omega \in \Omega\colon\ f(\omega) < 0\}$. We must prove that $m^*(C) = 0$. For this, it suffices to show that for each $\epsilon > 0$ the measurable set $C_\epsilon = \{\omega \in \Omega\colon\ f(\omega) < -\epsilon\}$ has outer measure zero. Since $|f| \ge \epsilon\chi_{C_\epsilon}$, it follows that $m^*(C_\epsilon) < \infty$. We shall verify that $m^*(C_\epsilon) = 0$ for each $\epsilon > 0$.

To this end, fix $\epsilon > 0$ and let $\delta > 0$. Since f has an absolutely continuous integral, there exists some $\eta > 0$ such that if $A \in \Sigma_m$ satisfies $m^*(A) < \eta$, then $\int_A |f| dm < \delta$. Since $m^*(C_\epsilon) < \infty$, there exists a sequence $\{A_n\}$ of pairwise disjoint sets in Γ satisfying $C_\epsilon \subseteq A = \bigcup_{n=1}^\infty A_n$ and $m^*(A) = \sum_{n=1}^\infty m(A_n) < m^*(C_\epsilon) + \eta$. (Here we use the obvious fact that Γ is a semiring; see [**8**, Problem 15.2].) Clearly, $m^*(A \setminus C_\epsilon) = m^*(A) - m^*(C_\epsilon) < \eta$. This implies

$$\begin{aligned} 0 \le -\int_{C_\epsilon} f\,dm &= \Big[\int_A f\,dm - \int_{C_\epsilon} f\,dm\Big] - \int_A f\,dm \\ &= \int_{A\setminus C_\epsilon} f\,dm - \int_A f\,dm \\ &\le \int_{A\setminus C_\epsilon} |f|\,dm - \sum_{n=1}^\infty \int_{A_n} f\,dm \\ &\le \int_{A\setminus C_\epsilon} |f|\,dm < \delta \end{aligned}$$

for each $\delta > 0$. Therefore, $\int_{C_\epsilon} f\,dm = 0$, and from this (since $f(\omega) < -\epsilon$ for each $\omega \in C_\epsilon$) we obtain $m^*(C_\epsilon) = 0$. Thus $f \ge 0$ m-a.e. The converse should be obvious.

(b) Assume that a function $f \in L_1(m)$ satisfies $\int_A f\,dm = 0$ for all $A \in \Gamma$ with $m(A) < \infty$. By part (a), $f \ge 0$ and $f \le 0$ m-a.e. are both true. Hence, $f = 0$ m-a.e. The converse is trivial. ∎

Problem 1.6.15. *Let Γ be a semiring of subsets of set Ω, and let two set functions $\theta, \xi\colon \Gamma \to [0,\infty)$ satisfy $\theta(A) \le \xi(A)$ for each $A \in \Gamma$. If θ is finitely additive and ξ is a measure (i.e., countably additive), then show that θ is also a measure.*

Solution: Assume that θ and ξ satisfy the stated properties. Since ξ is a measure, the outer measure ξ^* of ξ is also a measure on the σ-algebra of all ξ-measurable sets that is an extension of ξ. Now suppose that the sets $A, A_1, A_2, \dots$ in Γ satisfy $A = \bigcup_{n=1}^\infty A_n$ and $A_i \cap A_j = \emptyset$ for $i \neq j$. Fix some $\epsilon > 0$. From the equality $\sum_{i=1}^\infty \xi(A_i) = \xi(A) < \infty$, it follows that there exists some k such that

$$\xi^*\Big(A \setminus \bigcup_{i=1}^n A_i\Big) = \xi^*(A) - \sum_{i=1}^n \xi^*(A_i) = \xi(A) - \sum_{i=1}^n \xi(A_i) < \epsilon$$

holds for all $n \geq k$. Now fix some integer $n \geq k$. Since Γ is a semiring, there exist pairwise disjoint sets $C_1, \dots, C_r$ in Γ such that $A \setminus \bigcup_{i=1}^n A_i = \bigcup_{j=1}^r C_j$ or $A = \big(\bigcup_{i=1}^n A_i\big) \cup \big(\bigcup_{j=1}^r C_j\big)$. This, coupled with the finite additivity of θ, yields $\theta(A) = \sum_{i=1}^n \theta(A_i) + \sum_{j=1}^r \theta(C_j)$. Therefore,

$$0 \leq \theta(A) - \sum_{i=1}^n \theta(A_i) = \sum_{j=1}^r \theta(C_j) \leq \sum_{j=1}^r \xi(C_j) = \xi^*\Big(\bigcup_{j=1}^r C_j\Big) = \xi^*\Big(A \setminus \bigcup_{i=1}^n A_i\Big) < \epsilon$$

for all $n \geq k$. This shows that $\theta(A) = \bigcup_{i=1}^\infty \theta(A_i)$, and so θ is a measure too. ∎

Problem 1.6.16. *Establish the following property of the product measure: If the measures μ and ν are σ-finite, then for each set $C \in \Sigma_{\mu\times\nu}$ and for each $\epsilon > 0$ there exists a set of the form $D = \bigcup_{n=1}^k A_n \times B_n$, where $A_n \in \Sigma_1$ and $B_n \in \Sigma_2$ for each n, such that $(\mu \times \nu)(C \Delta D) < \epsilon$.*

Solution: Without loss of generality we can assume that both measures are finite. By the definition of the outer measure we can find a sequence $\{A_n \times B_n\}$ of rectangles such that $C \subseteq \bigcup_{n=1}^\infty A_n \times B_n$ and $(\mu\times\nu)\big(\big(\bigcup_{n=1}^\infty A_n \times B_n\big) \setminus C\big) < \frac{\epsilon}{2}$. Pick k large enough so that $(\mu \times \nu)\big(\bigcup_{n=k+1}^\infty A_n \times B_n\big) < \frac{\epsilon}{2}$. Then the set $D = \bigcup_{n=1}^k (A_n \times B_n)$ satisfies the desired property. ∎

Problem 1.6.17. *If $f \in L_0(\pi)$, then show that*

$$\max\big\{\operatorname{ess\,sup} f, \operatorname{ess\,sup}(-f)\big\} = \operatorname{ess\,sup}|f| = \|f\|_\infty .$$

Solution: Let $a = \operatorname{ess\,sup} f$, $b = \operatorname{ess\,sup}(-f)$, and $c = \operatorname{ess\,sup}|f|$. From $f \leq |f|$ and $-f \leq |f|$, it easily follows that $a \leq c$ and $b \leq c$, and so $a \vee b \leq c$. If $c = 0$, then $f = 0$ π-a.e., and so $a = b = c = 0$ is trivially true. So, we can assume that $0 < c \leq \infty$.

Fix $0 < \delta < c$, and note that the measurable set $D = \{\omega \in \Omega\colon\ |f(\omega)| > \delta\}$ has positive measure. It follows that at least one of the measurable sets

$$D_1 = \{\omega \in \Omega\colon\ f(\omega) > \delta\} \quad\text{and}\quad D_2 = \{\omega \in \Omega\colon\ -f(\omega) > \delta\}$$

must have positive measure. This implies that either $a \geq \delta$ or $b \geq \delta$ must be true. Therefore, $a \vee b \geq \delta$ for each $0 < \delta < c$, and so $a \vee b \geq c$. Consequently, $c = a \vee b$, as desired. ∎

Problem 1.6.18. *Assume that (Ω, Σ, π) is a non-atomic measure space and let $\{x_n\} \subseteq L_1^+(\pi)$. Also let $x = \sup_n x_n$ be the π-a.e. pointwise supremum of the sequence $\{x_n\}$. Show that $x \notin L_1(\pi)$ if and only if there exists a pairwise disjoint sequence $\{A_n\} \subseteq \Sigma$ such that $\sum_{n=1}^\infty \int_{A_n} x_n \, d\pi = \infty$.*

Solution: Let $\{x_n\}$ be a sequence in $L_1^+(\pi)$. We can suppose that each function x_n is real-valued, i.e, $x_n : \Omega \to \mathbb{R}$.

Assume first that there exists a pairwise disjoint sequence $\{A_n\} \subseteq \Sigma$ such that $\sum_{n=1}^{\infty} \int_{A_n} x_n \, d\pi = \infty$. Since $\sum_{n=1}^{k} x_n \chi_{A_n} \leq x$ holds for each k, it easily follows that $\sum_{n=1}^{k} \int_{A_n} x_n \, d\pi \leq \int_\Omega x \, d\pi$ for each k. This implies $\int_\Omega x \, d\pi = \infty$, and therefore $x \notin L_1(\pi)$.

For the converse, assume that $x \notin L_1(\pi)$. Let $E = \{\omega \in \Omega \colon x(\omega) = \infty\}$ and note that $E \in \Sigma$. Suppose first that the function x is π-a.e. finite, that is, $\pi(E) = 0$. Put $A_1 = \{t \in \Omega \colon\ x_1(t) \geq \frac{1}{2} x(t)\}$ and for each $n > 1$ let

$$A_n = \{t \in \Omega \setminus \textstyle\bigcup_{k=1}^{n-1} A_k \colon\ x_n(t) \geq \frac{1}{2} x(t)\}.$$

Clearly, $\{A_n\}$ is a disjoint sequence in Σ and $\bigcup_{k=1}^{\infty} A_k = \Omega$. Consequently,

$$\sum_{n=1}^{\infty} \int_{A_n} x_n \, d\pi \geq \sum_{n=1}^{\infty} \tfrac{1}{2} \int_{A_n} x \, d\pi = \tfrac{1}{2} \int_\Omega x \, d\pi = \infty.$$

We now consider the case $\pi(E) > 0$. Using this and the non-atomicity of the measure, it is easy to construct a measurable function $u \colon \Omega \to \mathbb{R}$ such that $0 \leq u \leq x$ and $u \notin L_1(\pi)$. For instance, the non-atomicity of the measure guarantees the existence of a pairwise disjoint sequence $\{E_n\}$ in Σ satisfying $E_n \subseteq E$ and $\pi(E_n) > 0$ for each n. If we consider the "infinite step function" $u = \sum_{n=1}^{\infty} \frac{1}{\pi(E_n)} \chi_{E_n}$, then u is a real-valued measurable function such that $0 \leq u \leq x$ and $u \notin L_1(\pi)$.

Finally, for each n let $y_n = u \wedge x_n$ and note that the sequence $\{y_n\} \subseteq L_1^+(\pi)$ satisfies $\sup_n y_n = u$. By the previous case, there exists a disjoint sequence $\{A_n\}$ in Σ such that $\sum_{n=1}^{\infty} \int_{A_n} y_n \, d\pi = \infty$. From $x_n \geq y_n$, we get $\sum_{n=1}^{\infty} \int_{A_n} x_n \, d\pi = \infty$, and the solution is finished. ∎

Chapter 2

Basic Operator Theory

2.1. Bounded Below Operators

Problem 2.1.1. *Let $T: X \to Y$ be a bounded below continuous operator between normed spaces. Then show the following:*

(a) *If X is not a Banach space, then T need not have a closed range.*

(b) *The unique continuous linear extension of T to the norm completion of X is also bounded below.*

Solution: (a) Let X be a non-closed subspace of a Banach space Y and consider the operator $T: X \to Y$ defined by $Tx = x$ for each $x \in X$. Clearly, $\|Tx\| = \|x\|$ for each $x \in X$ and so T is bounded below. However, the range of T is X which is not closed in Y.

(b) Assume $\|Tx\| \geq \gamma\|x\|$ holds for all $x \in X$ and some $\gamma > 0$. Fix $\hat{x}$ in $\hat{X}$ (the norm completion of X), and then pick a sequence $\{x_n\}$ of X satisfying $x_n \to \hat{x}$. Now note that

$$\|\hat{T}\hat{x}\| = \lim_{n\to\infty} \|Tx_n\| \geq \lim_{n\to\infty} \gamma\|x_n\| = \gamma\|\hat{x}\|.$$

So, $\hat{T}$ is bounded below. ∎

Problem 2.1.2. *Prove that an operator $T: X \to Y$ between two real normed spaces is bounded below if and only if $T_{\mathbf{c}}: X_{\mathbf{c}} \to Y_{\mathbf{c}}$ is likewise bounded below.*

Solution: Let $T: X \to Y$ be an operator between two real normed spaces. Assume first that T is bounded below. So, there exists some $\gamma > 0$ such that

$\|Tx\| \geq \gamma\|x\|$ for all $x \in X$. Now if $z = x + \imath y \in X_{\mathbf{c}}$, then we have

$$\begin{aligned} \|T_{\mathbf{c}}z\| &= \|Tx + \imath Ty\| \geq \tfrac{1}{2}\big(\|Tx\| + \|Ty\|\big) \geq \tfrac{1}{2}\big(\gamma\|x\| + \gamma\|y\|\big) \\ &= \tfrac{\gamma}{2}\big(\|x\| + \|y\|\big) \geq \tfrac{\gamma}{2}\|z\| \,. \end{aligned}$$

This shows that the operator $T_{\mathbf{c}}\colon X_{\mathbf{c}} \to Y_{\mathbf{c}}$ is bounded below.

For the converse, assume that $T_{\mathbf{c}}\colon X_{\mathbf{c}} \to Y_{\mathbf{c}}$ is bounded below. Pick some constant $\gamma > 0$ satisfying $\|T_{\mathbf{c}}z\| \geq \gamma\|z\|$ for all $z = x + \imath y \in X_{\mathbf{c}}$. In particular if $x \in X$, then

$$\|Tx\| = \|T_{\mathbf{c}}(x + \imath 0)\| \geq \gamma\|x + \imath 0\| = \gamma\|x\| \,,$$

and so the operator $T\colon X \to X$ is also bounded below. ∎

Problem 2.1.3. *Let $T\colon X \to Y$ be a bounded, one-to-one, and surjective operator between two Banach spaces. Show that T is an isometry if and only if T and T^{-1} are both contractions. (Recall that a bounded operator between two normed spaces is called a **contraction** if its norm is no more than one.)*

Solution: If T is an isometry, then clearly $\|T\| = \|T^{-1}\| = 1$ holds. For the converse, assume that T and T^{-1} are both contractions, i.e., that $\|T\| \leq 1$ and $\|T^{-1}\| \leq 1$ both hold.

Now let $x \in X$. Clearly, $\|Tx\| \leq \|T\| \cdot \|x\| \leq \|x\|$. On the other hand, note that $\|x\| = \big\|T^{-1}(Tx)\big\| \leq \big\|T^{-1}\big\| \cdot \|Tx\| \leq \|Tx\|$. This implies $\|Tx\| = \|x\|$ for each $x \in X$. That is, T is an isometry. ∎

Problem 2.1.4. *Let $T \in \mathcal{L}(X)$ be bounded below but not invertible. Show that there exists some $\epsilon > 0$ such that the operator $\lambda - T$ is not surjective whenever $|\lambda| < \epsilon$.*

Solution: Assume that the claim is false. Then there exists a sequence $\{\lambda_n\}$ of scalars satisfying $\lambda_n \to 0$ and such that $\lambda_n - T$ is surjective for each n. Now notice that $\lambda_n - T \to -T$ in $\mathcal{L}(X)$. Since $-T$ is bounded below, it follows from Theorem 2.9 that $-T$ (and hence T) is surjective. From the fact that T is bounded below and surjective, it follows that T is invertible, a contradiction. ∎

Problem 2.1.5. *This is a generalization of the previous problem. Assume that X and Y are Banach spaces and let $T \in \mathcal{L}(X,Y)$ be bounded below but not invertible. Then show that there exists some $\epsilon > 0$ such that the operator $T - S$ is not surjective whenever $S \in \mathcal{L}(X,Y)$ satisfies $\|S\| < \epsilon$. That is, show that the set of all bounded below and non-surjective operators from X to Y is an open subset of $\mathcal{L}(X,Y)$.*

Solution: If the claim is false, then there exists a sequence $\{S_n\}$ of surjective operators satisfying $S_n \to 0$ in $\mathcal{L}(X,Y)$ and such that $T - S_n$ is surjective for each n. Since $T - S_n \to T$ in $\mathcal{L}(X,Y)$, it follows from Theorem 2.9 that T is also surjective. Since T is bounded below, T must be invertible, which is impossible. ∎

Problem 2.1.6. *Establish the following property that was used in the proof of Corollary* 2.6. Assume that X is a vector space such that $X = V \oplus W$ with

V finite dimensional. If Y is an arbitrary vector space, then any operator $T\colon X \to Y$ that is one-to-one on W satisfies $\dim N(T) \le \dim V$.

Solution: Assume that $k = \dim N(T) < \infty$. To show that $N(T)$ has dimension at most k, it suffices to show that every collection of $k+1$ vectors in $N(T)$ is linearly dependent. So, let $x_1, x_2, \ldots, x_{k+1}$ be $k+1$ vectors in $N(T)$. For each $1 \le i \le k+1$ choose vectors $v_i \in V$ and $w_i \in W$ such that $x_i = v_i + w_i$. Since $v_1, v_2, \ldots, v_{k+1}$ are linearly dependent, there exist scalars $c_1, c_2, \ldots, c_{k+1}$ not all zero such that $\sum_{i=1}^{k+1} c_i v_i = 0$. It follows that $\sum_{i=1}^{k+1} c_i x_i = \sum_{i=1}^{k+1} c_i w_i$, and so $T\big(\sum_{i=1}^{k+1} c_i w_i\big) = T\big(\sum_{i=1}^{k+1} c_i x_i\big) = 0$. Since T is one-to-one on W, it follows that $\sum_{i=1}^{k+1} c_i w_i = 0$. This implies $\sum_{i=1}^{k+1} c_i x_i = 0$. Thus, the vectors $x_1, x_2, \ldots, x_{k+1}$ are linearly dependent, and consequently $\dim N(T) \le \dim V$. ∎

Problem 2.1.7. *Show that the assumption* $\dim N(T) < \infty$ *in the last conclusion of Corollary* 2.6 *cannot be replaced by the assumption that* $N(T)$ *is complemented.*

Solution: We must show that there exists a Banach space X such that the collection $\mathcal{B}$ of all bounded operators $T \in \mathcal{L}(X)$ having a closed range $R(T)$ and a complemented null space $N(T)$ is not open in $\mathcal{L}(X)$.

Let $X = \ell_2$, fix a number $\alpha \in (0,1)$, and consider the operators T and T_n on X defined as follows:

$$T(x_1, x_2, x_3, x_4, x_5, \ldots) = (x_1, 0, x_3, 0, x_5, \ldots)$$

and

$$T_n(x_1, x_2, x_3, x_4, x_5, \ldots) = (x_1, \alpha^n x_2, x_3, \alpha^{n+1} x_4, x_5, \alpha^{n+2} x_6, x_7 \ldots)\,.$$

It is obvious that $T \in \mathcal{B}$ and that $T_n \to T$. It remains to notice that none of the operators T_n has a closed range. Indeed, each operator T_n is clearly one-to-one. Therefore if some T_n were to have a closed range, then this T_n would be bounded from below. However, a direct inspection shows that this is not true. ∎

Problem 2.1.8. *Let* $T\colon X \to Y$ *be a bounded operator between two Banach spaces. If the quotient vector space* $Y/R(T)$ *is finite dimensional, then show that the range of the operator* T *is closed.*

Solution: Let $T\colon X \to Y$ be a bounded operator between two Banach spaces such that the quotient vector space $Y/R(T)$ is finite dimensional, say of dimension n. Pick a basis $\{[y_1], [y_2], \ldots, [y_n]\}$ in $Y/R(T)$ and note that the vectors $y_1, y_2, \ldots, y_n$ are linearly independent in Y. Indeed, if $\lambda_1 y_1 + \lambda_2 y_2 + \cdots + \lambda_n y_n = 0$, then $\lambda_1[y_1] + \lambda_2[y_2] + \cdots + \lambda_n[y_n] = 0$ holds in $Y/R(T)$, and so $\lambda_i = 0$ for each $i = 1, \ldots, n$. Let V be the n-dimensional vector subspace generated by $\{y_1, y_2, \ldots, y_n\}$ in Y.

We claim first that $R(T) \cap V = \{0\}$. To see this, let $y \in R(T) \cap V$ and write $y = \sum_{i=1}^n \alpha_i y_i$. This implies $[y] = \sum_{i=1}^n \alpha_i [y_i] = 0$ in $Y/R(T)$, and so $\alpha_i = 0$ for each $i = 1, \ldots, n$. Therefore, $y = 0$, and thus $R(T) \cap V = \{0\}$. Next, we assert that $R(T) \oplus V = Y$. To this end, let $y \in Y$. Pick scalars $\beta_1, \beta_2, \ldots, \beta_n$ such that $[y] = \sum_{i=1}^n \beta_i [y_i]$ or, equivalently, $[y - \sum_{i=1}^n \beta_i y_i] = 0$. This implies that the vector $z = y - \sum_{i=1}^n \beta_i y_i$ belongs to $R(T)$. That is, $y = z + \sum_{i=1}^n \beta_i y_i \in R(T) \oplus V$. Hence $R(T) \oplus V = Y$ holds true.

Now Theorem 2.16 guarantees that $R(T)$ is closed. That is, the operator T has a closed range. ∎

Problem 2.1.9. *Let $T\colon X \to Y$ be a bounded operator between Banach spaces such that its range $R(T)$ is not closed. If $X = V \oplus W$, where V is closed and W is finite dimensional, then show that $T\colon V \to Y$ is not bounded below.*

Solution: Assume that $T\colon V \to Y$ is bounded below, i.e., there exists some $c > 0$ such that $c\|v\| \le \|Tv\|$ holds for each $v \in V$. This implies that $T(V)$ is a closed subspace of Y. Also $T(X) = T(V) + T(W)$, and clearly $T(W)$ is finite dimensional. Since the sum of a closed subspace and a finite dimensional subspace is closed (see Problem 3.4.1), we infer that $T(X)$ is closed, which is impossible. This contradiction establishes that T cannot be bounded below on V. ∎

Problem 2.1.10. *For a bounded operator $T\colon X \to Y$ between Banach spaces establish the following properties regarding the ranges of T and T^*.*

(a) *The range of the operator T is dense if and only if the adjoint operator $T^*\colon Y^* \to X^*$ (defined as usual via the duality identity $\langle Tx, y^*\rangle = \langle x, T^*y^*\rangle$) is one-to-one.*

(b) *If the adjoint operator T^* is one-to-one and has closed range, then show that T is surjective.*

(c) *If T carries norm bounded closed subsets of X to closed subsets of Y, then T has closed range. Moreover, in this case, all powers of T also have closed ranges.*

Solution: (a) Assume that $R(T)$ is dense and let $T^*y^* = 0$ for some $y^* \in Y^*$. Then for each $x \in X$ we have $\langle Tx, y^*\rangle = \langle x, T^*y^*\rangle = \langle x, 0\rangle = 0$, i.e., y^* vanishes on $R(T)$. Since $R(T)$ is dense, this implies $y^* = 0$. Thus, T^* is one-to-one.

For the converse, assume that T^* is one-to-one. If $R(T)$ is not dense in Y^*, then by the Separation Theorem there exists some non-zero $y^* \in Y^*$ satisfying $\langle y^*, Tx\rangle = \langle T^*y^*, x\rangle = 0$ for all $x \in X$. This implies $T^*y^* = 0$, that is, T^* is not one-to-one, a contradiction. Hence, $R(T)$ is dense in Y.

(b) According to Theorem 2.18 the range $R(T)$ of the operator T is also closed and by (a) the range $R(T)$ is dense. Hence, $R(T) = Y$ and so T is surjective.

(c) Assume T is also one-to-one and let $Tx_n \to y$. If $\{x_n\}$ has a norm bounded subsequence, say $\{z_n\}$, then y belongs to the closed set $\overline{T(\{z_1, z_2, \ldots\})} \subseteq R(T)$. The case $\|x_n\| \to \infty$ cannot occur. For if $\|x_n\| \to \infty$, then $T(\frac{x_n}{\|x_n\|}) = \frac{Tx_n}{\|x_n\|} \to 0$, and so zero belongs to the closed set $T(\{x \in X\colon \|x\| = 1\})$, contrary to the fact that T is one-to-one.

To see that we can assume that T is one-to-one, we consider the closed subspace $M = \operatorname{Ker} T$ and the quotient Banach space X/M. Let $\pi\colon X \to X/M$ be the quotient map. Note that the function $\hat{T}\colon X/M \to Y$ defined by $T(\pi(x)) = Tx$ is a well-defined one-to-one bounded linear operator that satisfies $R(T) = R(\hat{T})$. Moreover, we claim that $\hat{T}$ carries norm bounded closed subsets of X/M to closed subsets of

Y. To see this, let C be a bounded closed subset of X/M and pick an open ball $rU_{X/M}$ satisfying $C \subseteq rU_{X/M} = r\pi(U_X)$. Since π is continuous, the set $\pi^{-1}(C)$ is a closed subset of X. Now notice that if we consider the closed bounded subset $K = (rU_X) \cap \pi^{-1}(C)$ of X, then $\hat{T}(C) = T(K)$, which (in view of our hypothesis) shows that $\hat{T}(C)$ is a closed subset of Y.

Thus, we have shown that the one-to-one bounded operator $\hat{T}\colon X/M \to Y$ carries norm bounded closed subsets of X/M to closed subsets of Y. By the one-to-one case considered above, $\hat{T}$ has a closed range. Since $R(T) = R(\hat{T})$, the operator T likewise has a closed range, as desired. ∎

Problem 2.1.11. *Let X and Y be two Banach spaces, and let $\mathcal{I}(X,Y)$ denote the open subset of $\mathcal{L}(X,Y)$ consisting of all isomorphisms from X to Y. Show that the mapping $T \mapsto T^{-1}$, from $\mathcal{I}(X,Y)$ to $\mathcal{I}(Y,X)$, is (norm) continuous.*

Solution: Fix some $R \in \mathcal{I}(X,Y)$ and assume that an operator $S \in \mathcal{I}(X,Y)$ satisfies $\|S-R\| < \frac{1}{\|R^{-1}\|}$ or $1 - \|R^{-1}\| \cdot \|S-R\| > 0$. From

$$\begin{aligned} S^{-1} - R^{-1} &= R^{-1}(R-S)S^{-1} = R^{-1}(R-S)\big[(S^{-1}-R^{-1}) + R^{-1}\big] \\ &= R^{-1}(R-S)(S^{-1}-R^{-1}) + R^{-1}(R-S)R^{-1}, \end{aligned}$$

we get $\|S^{-1}-R^{-1}\| \le \|R^{-1}\| \cdot \|R-S\| \cdot \|R^{-1}-S^{-1}\| + \|R^{-1}\|^2 \cdot \|R-S\|$. It follows that

$$\|S^{-1}-R^{-1}\| \le \frac{\|R^{-1}\|^2}{1-\|R^{-1}\|\cdot\|S-R\|}\|S-R\|,$$

holds for all $S \in \mathcal{I}(X,Y)$ sufficiently close to R. This implies that $T \mapsto T^{-1}$ is norm continuous at R. ∎

Problem 2.1.12. *For a bounded operator $T\colon X \to Y$ between Banach spaces establish the following.*

(a) *The operator T has a closed range if and only if $R(T^*) = N(T)^\perp$.*

(b) *If T has a closed range, then $X^*/R(T^*) = [N(T)]^*$.*

Solution: (a) Assume first that $R(T^*) = N(T)^\perp$. Then $R(T^*)$ is closed. By Theorem 2.18, it follows immediately that $R(T)$ is also closed.

For the converse, assume that $R(T)$ is closed. From Corollary 2.15 we know that there exists some constant $c > 0$ such that for each $y \in R(T)$ there exists some $x \in X$ with $y = Tx$ and $\|x\| \le c\|y\|$.

The inclusion $R(T^*) \subseteq N(T)^\perp$ is obvious. Indeed, for each $y^* \in Y^*$ and each $x \in N(T)$ we have $T^*y^*(x) = y^*(Tx) = y^*(0) = 0$, and hence $T^*y^* \in N(T)^\perp$. For the reverse inclusion let $x^* \in N(T)^\perp$. Define a scalar-valued mapping y^* on $R(T)$ via the formula

$$y^*(Tx) = x^*(x).$$

Notice that if $Tx = Tz$, then $z - x \in N(T)$ and $x^* \in N(T)^\perp$ implies $x^*(z) = x^*(x)$. This shows that y^* is well defined. Clearly, y^* is also linear. Now for each $y \in R(T)$ pick some $x \in X$ with $y = Tx$ and $\|x\| \le c\|y\|$, and note that

$$\|y^*(y)\| = \|y^*(Tx)\| = |x^*(x)| \le \|x^*\| \cdot \|x\| \le c\|x^*\| \cdot \|y\|.$$

This shows that y^* is a continuous linear functional defined on $R(T)$. If z^* is any continuous linear extension of y^* to all of Y^*, then for each $x \in X$ we have

$$T^*z^*(x) = z^*(Tx) = y^*(Tx) = x^*(x).$$

Thus, $T^*z^* = x^*$, and so $x^* \in R(T^*)$. This shows that $N(T)^\perp \subseteq R(T^*)$ is also true, and therefore $R(T^*) \subseteq N(T)^\perp$.

(b) By Problem 1.1.22 we know that $[N(T)]^* = X^*/N(T)^\perp$. Moreover, from part (a) it follows that $R(T^*) = N(T)^\perp$. Therefore, $[N(T)]^* = X^*/R(T^*)$. ∎

2.2. The Ascent and Descent of an Operator

Problem 2.2.1. *Let $T: X \to X$ be an operator on a vector space. If $R(T^k) = R(T^{k+1})$ holds for some k, then show that $R(T^m) = R(T^k)$ for all $m \geq k$.*

Solution: The verification of the claim is by induction. Assume that $R(T^k) = R(T^{k+n})$ for some $n \geq 1$. We must show that $R(T^k) = R(T^{k+n+1})$. Clearly, $R(T^{k+n+1}) \subseteq R(T^{k+n})$. For the reverse inclusion, let $y \in R(T^{k+n})$. So, there exists some $x \in X$ such that $y = T^{k+n}x$. Since $T^kx \in R(T^k) = R(T^{k+n})$, there exists some $z \in X$ such that $T^kx = T^{k+n}z$. This implies

$$y = T^{k+n}x = T^n(T^kx) = T^k(T^{k+n}z) = T^{k+n+1}(T^{k-1}z) \in R(T^{k+n+1}),$$

and so $R(T^{k+n+1}) = R(T^{k+n}) = R(T^k)$. Therefore, $R(T^m) = R(T^k)$ holds for all $m \geq k$. ∎

Problem 2.2.2. *Show that every operator on a finite dimensional vector space has finite ascent and descent.*

Solution: This follows easily from the fact that in a finite dimensional vector space there is neither a strictly increasing nor a strictly decreasing sequence of vector subspaces. ∎

Problem 2.2.3. *Consider the operator $T: \mathbb{R}^3 \to \mathbb{R}^3$ defined by the matrix*

$$T = \begin{bmatrix} 0 & 0 & 1 \\ 0 & 1 & 0 \\ 0 & 0 & 0 \end{bmatrix}.$$

Find the ascent and descent of T.

Solution: Note that

$$T^2 = \begin{bmatrix} 0 & 0 & 0 \\ 0 & 1 & 0 \\ 0 & 0 & 0 \end{bmatrix} \quad \text{and} \quad T^3 = T^2 = \begin{bmatrix} 0 & 0 & 0 \\ 0 & 1 & 0 \\ 0 & 0 & 0 \end{bmatrix}.$$

From $N(T) = \{x = (x_1, x_2, x_3) \in \mathbb{R}^3 \colon Tx = (x_3, x_2, 0) = 0\} = \{(x_1, 0, 0) \colon x_1 \in \mathbb{R}\}$ and

$$N(T^2) = \{x = (x_1, x_2, x_3) \in \mathbb{R}^3 \colon T^2x = (0, x_2, 0) = 0\} = \{(x_1, 0, x_3) \colon x_1, x_2 \in \mathbb{R}\},$$

it follows that $N(T)$ is a proper subspace of $N(T^2)$. The identity $T^2 = T^3$ implies $N(T^2) = N(T^3)$. Therefore, $\alpha(T) = 2$, and consequently $\delta(T) = 2$. ∎

Problem 2.2.4. *Give examples of bounded operators $T \colon X \to X$ on a Banach space such that:*

(i) $\alpha(T) < \infty$ *and* $\delta(T) = \infty$.

(ii) $\alpha(T) = \infty$ *and* $\delta(T) < \infty$.

(iii) $\alpha(T) = \infty$ *and* $\delta(T) = \infty$.

Solution: Let $B, S \colon \ell_2 \to \ell_2$ be the **backward** and **forward shift**, respectively. That is,

$$B(x_1, x_2, x_3, \dots) = (x_2, x_3, \dots) \quad \text{and} \quad S(x_1, x_2, x_3, \dots) = (0, x_1, x_2, \dots).$$

(i) Clearly, S is one-to-one, and so $N(S) = N(S^2)$. This implies $\alpha(S) = 1$. Since

$$R(S^k) = \{y = (y_1, y_2, y_3, \dots) \in \ell_2 \colon y_i = 0 \text{ for each } 1 \le i \le k\},$$

it follows that $R(S^{k+1})$ is a proper subspace of $R(S^k)$ for each k. This implies $\delta(S) = \infty$.

(ii) For the backward shift B we have

$$N(B^k) = \{x = (x_1, x_2, x_3, \dots) \in \ell_2 \colon x_i = 0 \text{ for each } i > k\}.$$

This implies that $N(B^k)$ is a proper subspace of $N(B^{k+1})$ for each k. Hence, $\alpha(B) = \infty$. Since $R(B) = R(B^2) = \ell_2$, it follows that $\delta(B) = 1$.

(iii) Consider the operator $T \colon \ell_2 \otimes \ell_2 \to \ell_2 \otimes \ell_2$ defined by

$$\begin{aligned} T\big((x_1, x_2, \dots) \otimes (z_1, z_2, \dots)\big) &= S(x_1, x_2, \dots) \otimes B(z_1, z_2, \dots) \\ &= (0, x_1, x_2, \dots) \otimes (z_2, z_3, \dots). \end{aligned}$$

That is, $T = S \otimes B$. Then it is easy to see that $\alpha(T) = \delta(T) = \infty$. ∎

Problem 2.2.5. *Let $T \colon X \to X$ be a bounded operator on a Banach space with $p = \alpha(T) = \delta(T) < \infty$. Show that the adjoint operator $T^* \colon X^* \to X^*$ also has finite ascent and descent and that $\alpha(T^*) = \delta(T^*) = p$.*

Solution: Let $V = N(T^p)$ and $W = R(T^p)$. By Theorem 2.23 we know that T is nilpotent on V and invertible on W. Then $X^* = V^* \oplus W^*$, T^* is nilpotent on V^* and invertible on W^*, and thus the pair (V^*, W^*) reduces T^*. So, by Theorem 2.23, the operator T^* has finite ascent and descent. Let $q = \alpha(T^*) = \delta(T^*)$. Since $(T^*)^p = (T^p)^* = 0$ on V^*, it follows that $q \le p$.

Similarly, T^{**} has finite ascent and descent. Let $r = \alpha(T^{**}) = \delta(T^{**})$. By the above $r \le q$. Now note that $T^r = 0$ on V. This implies $p \le r \le q \le p$. Consequently, $p = q$, and the solution is finished. ∎

Problem 2.2.6. *Show that a bounded operator $T\colon X \to X$ on a Banach space is nilpotent if and only if for each $x \in X$ there exists some $n \in \mathbb{N}$ such that $T^n x = 0$.*

Solution: If T is nilpotent, then the condition is satisfied automatically. For the converse, assume for each $x \in X$ there exists some $n \in \mathbb{N}$ (depending on x) such that $T^n x = 0$. This implies that

$$X = \bigcup_{n=1}^{\infty} N(T^n).$$

Since each $N(T^n)$ is a closed vector subspace of X and X is complete, it follows from the Baire Category Theorem that for some k the closed vector subspace $N(T^k)$ has an interior point. This implies $N(T^k) = X$ or $T^k = 0$ for some k. That is, T is a nilpotent operator. ∎

Problem 2.2.7. *If a nilpotent operator $T\colon V \to V$ on a vector space has index p, then show that $\alpha(T) = \delta(T) = p$.*

Solution: Let V be a vector space and let $T\colon V \to V$ be a nilpotent operator of index p. Clearly, $N(T^p) = N(T^{p+1}) = V$ and $N(T^{p-1}) \neq N(T^p)$. This implies that T has finite ascent and that $\alpha(T) = p$. From $R(T^p) = R(T^{p+1}) = \{0\}$, it follows that $\delta(T) = p$ is also true. ∎

Problem 2.2.8. *For an operator $T\colon V \to V$ on a vector space establish the following algebraic properties.*

(a) *For each $n, m \in \mathbb{N}$ we have $T^n\big(N(T^{n+m})\big) \subseteq R(T^n) \cap N(T^m)$.*

(b) *For each $n, m \in \mathbb{N}$ the operator $T^n\colon N(T^{n+m}) \to R(T^n) \cap N(T^m)$ is surjective—and consequently the vector spaces $N(T^{n+m})/N(T^m)$ and $R(T^n) \cap N(T^m)$ are linearly isomorphic.*

(c) *If the ascent of T is finite, then $R(T^p) \cap N(T^n) = \{0\}$ for each $n \in \mathbb{N}$ and all $p \geq \alpha(T)$.*

(d) *If for some $p, n \in \mathbb{N}$ we have $R(T^p) \cap N(T^n) = \{0\}$, then T has finite ascent and $\alpha(T) \leq p$ holds.*

Solution: (a) Let $v \in N(T^{n+m})$. Then $T^m(T^n v) = T^{n+m} v = 0$. This implies $T^n v \in N(T^m)$, and therefore $T^n v \in R(T^n) \cap N(T^m)$. Consequently, we have $T^n(N(T^{n+m})) \subseteq R(T^n) \cap N(T^m)$.

(b) To see that the operator $T^n\colon N(T^{n+m}) \to R(T^n) \cap N(T^m)$ is surjective, let $u \in R(T^n) \cap N(T^m)$. This means $T^m u = 0$ and $u = T^n v$ for some $v \in V$. Hence, $T^{n+m} v = T^m(T^n v) = T^m u = 0$, and so $v \in N(T^{n+m})$. The latter conclusion shows that the operator $T^n : N(T^{n+m}) \to R(T^n) \cap N(T^m)$ is surjective.

Now if we define the operator $S\colon N(T^{n+m})/N(T^m) \to R(T^n) \cap N(T^m)$ via the formula $S[v] = T^n v$, then it is easy to see that S is a surjective linear isomorphism. Therefore, $N(T^{n+m})/N(T^m)$ and $R(T^n) \cap N(T^m)$ are linearly isomorphic vector spaces.

(c) Assume that $u \in R(T^p) \cap N(T^n)$ for some $n \in \mathbb{N}$ and some $p \geq \alpha(T)$. By part (b), there exists some vector $v \in N(T^{p+n})$ such that $u = T^p v$. Since $N(T^{p+n}) = N(T^p) = N(T^{\alpha(T)})$, it follows that $v \in N(T^p)$. So, $u = T^p v = 0$ and hence $R(T^p) \cap N(T^n) = \{0\}$.

(d) Now suppose that for some $p, n \in \mathbb{N}$ we have $R(T^p) \cap N(T^n) = \{0\}$. From part (b), we see that $N(T^{p+n})/N(T^p) = \{0\}$ or $N(T^p) = N(T^{p+n})$. This implies $N(T^p) = N(T^{p+1})$. Consequently, T has finite ascent and $\alpha(T) \leq p$ holds. ∎

Problem 2.2.9. *Let $T\colon V \to V$ be an operator on a vector space. Show that for each $n, m \in \mathbb{N}$ the mapping $J\colon V/[N(T^n)+R(T^m)] \to R(T^n)/R(T^{n+m})$, defined by*

$$J[v] = T^n v + R(T^{n+m}),$$

is a well-defined surjective linear isomorphism.

Solution: We shall verify first that J is a well-defined mapping. To this end, let $[u] = [v] \in V/[N(T^n) + R(T^m)]$, i.e, let $u - v \in N(T^n) + R(T^m)$. Pick $x \in N(T^n)$ and $y \in V$ such that $u - v = x + T^m y$, and note that

$$T^n u - T^n v = T^n x + T^{n+m} y = T^{n+m} y \in R(T^{n+m}).$$

This shows that $J[u] = J[v]$, and thus J is well defined. Clearly, J is a surjective linear operator.

To finish the solution, we must verify that J is one-to-one. To see this, assume that $J[v] = T^n v + R(T^{n+m}) = 0$, i.e, $T^n v \in R(T^{n+m})$. This implies the existence of some $u \in V$ such that $T^n v = T^{n+m} u$ or $T^n(v - T^m u) = 0$. So, the vector $w = v - T^m u$ belongs to $N(T^n)$, and consequently $v = w + T^m u \in N(T^n) + R(T^m)$. That is, $[v] = 0$, and so J is one-to-one. ∎

2.3. Banach Lattices with Order Continuous Norms

Problem 2.3.1. *Let Ω be a compact Hausdorff space. Show that the Banach lattice $C(\Omega)$ has order continuous norm if and only if Ω is a finite set.*

Solution: If Ω is a finite set, say $\Omega = \{\omega_1, \ldots, \omega_n\}$, then $C(\Omega) = \mathbb{R}^n$, and so $C(\Omega)$ has order continuous norm. For the converse, assume that $C(\Omega)$ has order continuous norm. To finish the proof, it suffices to show that every point in Ω is an isolated point.

To this end, fix some $\omega \in \Omega$ and suppose by way of contradiction that ω is an accumulation point. Consider the non-empty collection of functions

$$D_\omega = \big\{f \in C(\Omega)\colon\ 0 \leq f \leq \mathbf{1},\ f(\omega) = 1 \text{ and } f \text{ vanishes outside of an open neighborhood of } \omega\big\}.$$

Clearly, $D_\omega \downarrow$ and $\inf_{f \in D_\omega} f(\omega') = 0$ for each $\omega' \neq \omega$. Since ω is an accumulation point, it follows that $D_\omega \downarrow 0$. But then $\|f\|_\infty = 1$ for each $f \in D_\omega$ contradicts the order continuity of the norm $\|\cdot\|_\infty$. This contradiction establishes that each point $\omega \in \Omega$ is an isolated point, and so Ω is a finite set. ∎

Problem 2.3.2. *Show that in an Archimedean Riesz space E a positive vector is an atom if and only if it is a discrete vector. Also show that if u is an atom in E, then $\{\lambda u\colon\ \lambda \in \mathbb{R}\}$ (the vector space generated by u) is a projection band.*

Solution: To begin with, we need two properties of Archimedean Riesz spaces. So, let E be a Riesz space.

- *E is Archimedean if and only if $\epsilon_n \downarrow 0$ in $\mathbb{R}$ implies $\epsilon_n x \downarrow 0$ in E for each $x \in E^+$.*

To see this, suppose first that E is Archimedean, $\epsilon_n \downarrow 0$ in $\mathbb{R}$, and $x \in E^+$. Assume that $y \leq \epsilon_n x$ holds for each n and some $y \in E$. If k is a natural number, then $\epsilon_n < \frac{1}{k}$ for all n sufficiently large, and therefore if we choose any such n, then $y \leq \epsilon_n x \leq \frac{1}{k}x$ and so $ky \leq x$ for all k. This implies $y \leq 0$ and so $\epsilon_n x \downarrow 0$.

Now suppose that $\epsilon_n \downarrow 0$ in $\mathbb{R}$ implies $\epsilon_n x \downarrow 0$ in E for each $x \in E^+$. Assume $nx \leq y$ in E for all n and some $x, y \in E$. This yields $nx^+ \leq y^+$ or $0 \leq x^+ \leq \frac{1}{n}y^+$ for each n. Since $\frac{1}{n}y^+ \downarrow 0$, it follows that $x^+ = 0$. Hence, $x \leq x^+ = 0$ so that E is Archimedean.

- *If E is Archimedean and $\alpha_n \to \alpha$ holds in $\mathbb{R}$, then $\alpha_n x \xrightarrow{o} \alpha x$ in E for each $x \in E$.*

To see this, for each n let $\epsilon_n = \sup_{k \geq n} |\alpha_k - \alpha|$ and note that $\epsilon_n \downarrow 0$. Hence, for each $x \in E$ we have $|\alpha_n x - \alpha x| = |\alpha_n - \alpha||x| \leq \epsilon_n |x| \downarrow 0$.

We now return to our problem. We must keep in mind that the lattice operations of a Riesz space are order continuous. Assume that $u > 0$ is an atom in an Archimedean Riesz space E and let $0 < x \leq u$. Put $\alpha = \sup\{\beta \geq 0\colon\ \beta x \leq u\}$. Clearly, $\alpha \geq 1$, and (since E is Archimedean) we see that $\alpha < \infty$. Moreover, $\alpha x \leq u$, and we claim that $\alpha x = u$ or $x = \frac{1}{\alpha}u$.

To see this, assume by way of contradiction that $v = u - \alpha x > 0$. Since $(v - \frac{1}{n}x)^+ \uparrow v$ (here we use the above properties in connection with the order continuity of the lattice operations), it follows that $(v - \frac{1}{k}x)^+ > 0$ for some $k > 1$. Clearly,

$$(v - \tfrac{1}{k}x)^+ = [u - (\alpha + \tfrac{1}{k})x]^+ \leq u^+ = u \leq 2\alpha u\,.$$

On the other hand, $[u - (\alpha + \frac{1}{k})x]^- > 0$. (Otherwise, $[u - (\alpha + \frac{1}{k})x]^- = 0$ implies $(\alpha + \frac{1}{k})x \leq u$, contrary to the definition of α.) Next, observe that

$$[u - (\alpha + \tfrac{1}{k})x]^- = [(\alpha + \tfrac{1}{k})x - u]^+ \leq 2\alpha x \leq 2\alpha u\,.$$

So, if $y = \frac{1}{2\alpha}[u - (\alpha + \frac{1}{k})x]^+$ and $z = \frac{1}{2\alpha}[u - (\alpha + \frac{1}{k})x]^-$, then we have $0 < y \leq u$, $0 < z \leq u$ and $y \wedge z = 0$, contradicting the fact that u is an atom. Hence, $\alpha x = u$, and so u is indeed a discrete element. It should be obvious that a discrete element is automatically an atom regardless of whether or not E is Archimedean.

Now let $A = \{\lambda u\colon\ \lambda \in \mathbb{R}\}$. By the above, A is an ideal. To see that A is a band, assume $0 \leq \lambda_n u \uparrow x$ in E. Clearly, $0 \leq \lambda_n \uparrow$ in $\mathbb{R}$ and, since E is Archimedean, $\lambda_n \uparrow \lambda$ in $\mathbb{R}$. So, $\lambda_n u \uparrow \lambda u$, and hence $x = \lambda u$, i.e., $x \in A$. Thus, A is a band of E.

To see that A is a projection band, let $0 < x \in E$ and put

$$\alpha = \sup\{\beta \geq 0\colon\ \beta u \leq x\}.$$

Since E is Archimedean, we have $0 \leq \alpha < \infty$ and $\alpha u \leq x$. In view of the identity $x = \alpha u + (x - \alpha u)$, it suffices to show that $x - \alpha u \perp \alpha u$ or $x - \alpha u \perp u$.

To this end, observe first that the definition of α implies

$$0 < \big[x - \big(\alpha + \tfrac{1}{n}\big)u\big]^- = \big[\big(\alpha + \tfrac{1}{n}\big)u - x\big]^+ \leq (1 + \alpha)u.$$

Now from the fact that $(1+\alpha)u$ is an atom, $0 \leq \big[x - \big(\alpha + \frac{1}{n}\big)u\big]^+ \wedge u \leq (1+\alpha)u$, and $\big(\big[x - \big(\alpha + \frac{1}{n}\big)u\big]^+ \wedge u\big) \wedge \big(\big[x - \big(\alpha + \frac{1}{n}\big)u\big]^-\big) = 0$, it follows that $\big[x - \big(\alpha + \frac{1}{n}\big)u\big]^+ \wedge u = 0$ for each n. Since $\alpha + \frac{1}{n} \to \alpha$ and the lattice operations are order continuous, the latter implies $(x - \alpha u) \wedge u = (x - \alpha u)^+ \wedge u = 0$. Thus, $x - \alpha u \in A^{\mathrm{d}}$, and so A is a projection band. ∎

Problem 2.3.3. *Show that $C[0,1]$ and $L_1[0,1]$ are atomless Banach lattices.*

Solution: Let $0 < u \in C[0,1]$. Pick $0 < t_0 < 1$ and $\epsilon > 0$ with $0 < t_0 - \epsilon < t_0 + \epsilon < 1$ such that $u(t) > 0$ holds for each $t_0 - \epsilon \leq t \leq t_0 + \epsilon$. Now choose two non-zero functions $f, g \in C[0,1]$ such that $0 \leq f, g \leq \mathbf{1}$, f is supported by the interval $[t_0 - \epsilon, t_0]$ and g is supported by the interval $[t_0, t_0 + \epsilon]$. Then $0 < uf \leq u$, $0 < ug \leq u$ and $(uf) \wedge (ug) = 0$ in $C[0,1]$. This shows that u cannot be an atom, and so $C[0,1]$ is an atomless Riesz space.

Next, let $0 < f \in L_1[0,1]$. Then the measurable set $A = \{x \in [0,1]\colon\ f(x) > 0\}$ has positive measure. Pick a measurable subset B of A such that B and $C = A \setminus B$ both have positive measure, and note that the functions $f\chi_B, f\chi_C \in L_1[0,1]$ satisfy $0 < f\chi_B \leq f$, $0 < f\chi_C \leq f$, and $(f\chi_B) \wedge (f\chi_C) = 0$ in $L_1[0,1]$. Therefore, f cannot be an atom, and so $L_1[0,1]$ is also an atomless Banach lattice. ∎

Problem 2.3.4. *Show that in general the norm dual of an atomless Banach lattice need not be atomless by proving that the norm dual of the atomless Banach lattice $E = C[0,1]$ has atoms.*

Solution: Fix an arbitrary point $t \in [0,1]$ and consider the continuous linear functional $\delta_t\colon C[0,1] \to \mathbb{R}$ defined by $\delta_t(f) = f(t)$. We claim that δ_t is a discrete element, and so Problem 2.3.2 will guarantee that δ_t is an atom in $C[0,1]^*$.

To see that this is the case, pick some $x^* \in C[0,1]^*$ satisfying $0 < x^* \leq \delta_t$. By the Riesz Representation Theorem, there exists a unique regular Borel measure μ on $[0,1]$ satisfying $x^*(f) = \int f\, d\mu$ for each $f \in C[0,1]$. We claim that $\operatorname{Supp}\mu = \{t\}$. Since $x^* \neq 0$, we know that $\operatorname{Supp}\mu \neq \emptyset$. We verify next that $\operatorname{Supp}\mu = \{t\}$. Indeed, if $s \in \operatorname{Supp}\mu$ and $s \neq t$, then by choosing a function $0 < f \in C[0,1]$ satisfying $f(s) = 1$ and $f(t) = 0$, we see that

$$0 < \int f\, d\mu = x^*(f) \leq \delta_t(f) = f(t) = 0,$$

which is impossible. Hence, $\operatorname{Supp}\mu = \{t\}$. Thus, for each $g \in C[0,1]$ we have

$$x^*(g) = \int g\, d\mu = \int_{\{t\}} g\, d\mu = g(t)\mu(\{t\}) = \mu(\{t\})\delta_t(g).$$

Therefore, $x^* = \mu(\{t\})\delta_t$. Thus, δ_t is a discrete element (and hence an atom) in the Banach lattice $C[0,1]^*$. ∎

Problem 2.3.5. *If $g, h\colon E \to \mathbb{R}$ are two disjoint non-zero positive linear functionals on a Riesz space, then show that there exist disjoint vectors u, v in E^+ such that $g(u) > 0$ and $h(v) > 0$.*

Solution: Since g and h are non-zero and positive, we can find some $0 < x \in E$ such that $g(x) > 0$ and $h(x) > 0$. By scaling, we can assume that $g(x) > 2$ and $h(x) > 2$. Since g and h are disjoint, we have

$$0 = (g \wedge h)(x) = \inf\{g(y) + h(x-y)\colon\ 0 \le y \le x\}\,.$$

Pick any $y \in [0, x]$ such that $g(y) + h(x-y) < 1$. Then $g(x-y) > 1$ and $h(y) > 1$. Let $u = (x-y) - y \wedge (x-y)$ and $v = y - y \wedge (x-y)$. These two elements are disjoint, and

$$g(u) = g(x-y) - g(y \wedge (x-y)) \ge g(x-y) - g(y) > 0\,.$$

Similarly,

$$h(v) = h(y) - h(y \wedge (x-y)) \ge h(y) - h((x-y)) > 0\,.$$

Thus, the elements u and v satisfy the desired properties. ∎

Problem 2.3.6. *Let E be a Riesz space and let $0 < f \in E^\sim$. Show that f is an atom in $E^\sim$ if and only if f is a lattice homomorphism, that is, if and only if $f(x \vee y) = \max\{\, f(x), f(y)\,\}$ for all $x, y \in E$.*

Solution: Assume first that f is an atom in the order dual $E^\sim$. To show that f is a lattice homomorphism, it is enough to show (according to Theorem 1.34) that $f(u)f(v) = 0$ whenever $u \wedge v = 0$. Assume, contrary to our claim, that there are two disjoint positive elements $u, v \in E$ such that $f(u) > 0$ and $f(v) > 0$.

For each $x \in E^+$ let $g(x) = \lim f(x \wedge nu)$. A straightforward verification shows that g is additive on E^+ and hence the formula $g(x) = g(x^+) - g(x^-)$ defines a linear functional on E. Clearly, $0 \le g \le f$ and $g(u) = f(u) > 0$.

The Riesz space $E^\sim$ is necessarily Archimedean and so, in view of Lemma 2.30, the functional f is a discrete element in $E^\sim$. Therefore, there exists a constant $\lambda \ge 0$ such that $g = \lambda f$. Clearly $\lambda = 1$ because $g(u) = f(u) > 0$, and hence we also have $g(v) = f(v) > 0$. On the other hand, by the definition of the functional g, we obviously have $g(x) = 0$ for each x that is disjoint to u. In particular, $g(v) = 0$, a contradiction.

For the converse, assume that f is a lattice homomorphism, and assume that g, h are two disjoint functionals satisfying $0 \le g, h \le f$. We must show that at least one of them is zero. If not, then we can find two disjoint elements $u, v \in E^+$ such that $g(u) > 0$ and $h(v) > 0$; see Problem 2.3.5. Hence, $f(u)f(v) \ge g(u)h(v) > 0$. On the other hand, since f is a lattice homomorphism, we have $0 = f(u \wedge v) = f(u)f(v)$, which is impossible. ∎

Problem 2.3.7. *Let $S, T\colon E \to F$ be two regular operators between Banach lattices with F Dedekind complete. Show that:*

(a) *$S \le T$ in $\mathcal{L}_r(E, F)$ implies $S^* \le T^*$ in $\mathcal{L}_r(F^*, E^*)$.*

(b) $|T^*| \leq |T|^*$.

Give an example where $|T^*| < |T|^*$ holds true.

Solution: (a) Since E is a Banach lattice, we have $\mathcal{L}_r(E,F) \subseteq \mathcal{L}(E,F)$. Assume that two operators $S, T \in \mathcal{L}_r(E,F)$ satisfy $S \leq T$, i.e., $Sx \leq Tx$ for each $x \in E^+$. Let $0 \leq y^* \in F^*$. If $x \in E^+$, then note that

$$S^*y^*(x) = y^*(Sx) \leq y^*(Tx) = T^*y^*(x).$$

This shows that $S^*y^* \leq T^*y^*$ for each $0 \leq y^* \in F^*$, i.e., $S^* \leq T^*$ in $\mathcal{L}_r(F^*, E^*)$. In other words, the linear isometry $T \mapsto T^*$ from $\mathcal{L}_r(E,F)$ to $\mathcal{L}_r(F^*, E^*)$ is a positive operator.

(b) From $\pm T \leq |T|$ and part (a), it follows that $\pm T^* \leq |T|^*$. Consequently, we have $|T^*| = T^* \vee (-T^*) \leq |T|^*$.

For an example of a regular operator T satisfying $|T^*| < |T|^*$, consider the operator $T\colon \ell_1 \to \ell_\infty$ defined by

$$T(x) = (x_1 - x_2, x_2 - x_3, x_3 - x_4, \ldots) = (x_1, x_2, x_3, \ldots) - (x_2, x_3, x_4, \ldots)$$

for each $x = (x_1, x_2, \ldots) \in \ell_1$. Clearly, T is a regular operator, and an easy argument shows that

$$|T|x = \sup_{|y| \leq x} |Ty| = (x_1 + x_2, x_2 + x_3, x_3 + x_4, \ldots)$$

for each $x \in \ell_1^+$. We shall establish that $|T^*| \neq |T|^*$ by proving that $|T^*|\phi \neq |T|^*\phi$ for each Banach–Mazur limit ϕ on ℓ_∞.

To see this, let $\phi\colon \ell_\infty \to \mathbb{R}$ be a Banach–Mazur limit and let $e = (1,1,1,\ldots)$, the constant sequence one. Clearly, $\langle |T|^*\phi, e\rangle = \langle \phi, |T|e\rangle = \phi(2,2,2,\ldots) = 2$, and so $|T^*|\phi \neq 0$. On the other hand, if $\psi \in \ell_\infty^*$ satisfies $|\psi| \leq \phi$ and $y = (y_1, y_2, \ldots)$ is a null sequence, then from $|\psi(y)| \leq |\psi|(|y|) \leq \phi(|y|) = \lim |y_n| = 0$, we see that $\psi(y) = 0$. So, since $y = Tx$ is a null sequence in ℓ_∞, we have $T^*\psi(x) = \psi(Tx) = 0$ for each $x \in \ell_1$. In other words, $T^*\psi = 0$ for all $\psi \in \ell_\infty^*$ with $|\psi| \leq \phi$. This implies

$$|T^*|\phi = \sup_{|\psi| \leq \phi} |T^*\psi| = 0 \neq |T|^*\phi,$$

and so $|T^*| \neq |T|^*$. Therefore, $|T^*| < |T|^*$. ∎

Problem 2.3.8. *Show that a Banach lattice E is a KB-space if and only if E is a band in its double dual E^{**}.*

Solution: Assume first that E is a KB-space. Clearly, E has order continuous norm, and so E is (by Theorem 2.26) an ideal in E^{**}. To see that E is a band in E^{**}, suppose that a net $\{x_\alpha\} \subseteq E$ satisfies $0 \leq x_\alpha \uparrow x^{**}$ in E^{**}. Then $\|x_\alpha\| \leq \|x^{**}\| < \infty$ for each α, and we claim that $\{x_\alpha\}$ is a Cauchy net.

To see this, assume by way of contradiction that $\{x_\alpha\}$ is not a Cauchy net. This means that there exist a sequence of indices $\{\alpha_n\}$ satisfying $\alpha_n \uparrow$ and some $\epsilon > 0$ such that $\|x_{\alpha_{n+1}} - x_{\alpha_n}\| \geq \epsilon$ for each n. Since the sequence $\{x_{\alpha_n}\}$ is norm bounded and increasing, and E is a KB-space, we obtain that $\{x_{\alpha_n}\}$ is a Cauchy sequence. However, the latter conclusion contradicts our assumption that $\|x_{\alpha_{n+1}} - x_{\alpha_n}\| \geq \epsilon$ for each n. So, $\{x_\alpha\}$ is a Cauchy net. If $x \in E$ is its norm limit, then $x_\alpha \uparrow x$ in E;

see Problem 1.2.12(c). Since E is an ideal in E^{**}, it follows that $x_\alpha \uparrow x$ also holds in E^{**}. This implies $x^{**} = x \in E$, and so E is a band in E^{**}.

For the converse, suppose that E is a band in E^{**}, and let a norm bounded sequence $\{x_n\}$ in E satisfy $0 \le x_n \uparrow$. We must show that the sequence $\{x_n\}$ is norm convergent. The formula

$$x^{**}(x^*) = \lim_{n\to\infty} x^*(x_n)\,, \quad x^* \in E^*\,,$$

defines a positive linear functional on E^*, i.e., $x^{**} \in E^{**}_+$. An easy argument shows that $x_n \uparrow x^{**}$ in E^{**}. Since, E is a band in E^{**}, it follows that $x^{**} \in E$. Now from $x_n(x^*) = x^*(x_n) \uparrow x^{**}(x^*)$ for each $x^* \in U^+_{E^*} = U_{E^*} \cap E^*_+$ and the fact that $U^+_{E^*}$ is a w^*-compact subset of E^*, it follows from Dini's classical theorem (see for instance [**7**, Theorem 9.4, p. 70]) that

$$\|x^{**} - x_n\| = \sup_{x^* \in U^+_{E^*}} \big|x^{**}(x^*) - x_n(x^*)\big| \downarrow 0\,.$$

This shows that E is a KB-space, and the solution is finished. ■

Problem 2.3.9 (Dodds–Fremlin [**25**]). *Let E and F be two Banach lattices with F having order continuous norm. If $x \in E^+$, then show that the set*

$$\mathcal{B} = \big\{T \in \mathcal{L}_r(E,F)\colon\ T[0,x] \ \textit{is norm totally bounded}\big\}$$

is a band in $\mathcal{L}_r(E,F)$.

Solution: Clearly, $\mathcal{B}$ is a vector subspace of the Banach lattice $\mathcal{L}_r(E,F)$. The verification that $\mathcal{B}$ is a band will be done by steps.

STEP 1: *If $T \in \mathcal{B}$ and $R, S \in \mathcal{L}_r(E,F)$ satisfy $T = R + S$ and $|R| \wedge |S| = 0$, then $R, S \in \mathcal{L}_r(E,F)$.*

To establish this claim, fix $x \in E^+$ and let $\epsilon > 0$. From

$$\Big\{\sum_{i=1}^n |R|x_i \wedge |S|x_i\colon\ x_i \in E^+ \text{ for each } i \text{ and } \sum_{i=1}^n x_i = x\Big\} \downarrow \big[|R| \wedge |S|\big]x = 0\,,$$

(see [**6**, Theorem 1.16, p. 15]) and the order continuity of the norm in F, there exist vectors $x_1, \dots, x_n \in E^+$ with $\sum_{i=1}^n x_i = x$ and $\big\|\sum_{i=1}^n |R|x_i \wedge |S|x_i\big\| < \frac{\epsilon}{2}$. From $T[0,x_i] \subseteq T[0,x]$ and the norm total boundedness of $T[0,x]$, we see that all sets $T[0,x_i]$ are norm totally bounded. So, for each $1 \le i \le n$ there exists a finite subset Φ_i of $[0,x_i]$ such that $T[0,x_i] \subseteq T(\Phi_i) + \frac{\epsilon}{2n}U_F$ for each i, where U_F is the closed unit ball of F. Put $\Phi = \big\{\sum_{i=1}^n y_i\colon\ y_i \in \Phi_i \text{ for each } i = 1, \dots, n\big\}$, and note that Φ is a finite set.

Now if $z \in [0,x]$, then using the Riesz Decomposition Property we can write $z = \sum_{i=1}^n z_i$, where $0 \le z_i \le x_i$ for each i. For each $1 \le i \le n$ pick some $y_i \in \Phi_i$ such that $Tz_i \in T(y_i) + \frac{\epsilon}{2n}U_F$. Let $y = \sum_{i=1}^n y_i \in \Phi$ and for simplicity put $w_i = z_i - y_i$. Clearly, $|w_i| \le x_i$ and $\|R(w_i)\| < \frac{\epsilon}{2n}$ for each i. Next, taking into account the lattice inequality $0 \le |u| - |u| \wedge |v| \le |u + v|$ (valid in Riesz spaces), we get

$$\begin{aligned}
\|R(z-y)\| &= \Big\|\sum_{i=1}^{n} R(w_i)\Big\| \le \sum_{i=1}^{n} \|R(w_i)\| \\
&= \sum_{i=1}^{n} \big\| |R(w_i)| - |R(w_i)| \wedge |S(w_i)| \big\| + \Big\|\sum_{i=1}^{n} |R(w_i)| \wedge |S(w_i)|\Big\| \\
&\le \sum_{i=1}^{n} \|(R+S)(w_i)\| + \Big\|\sum_{i=1}^{n} |R|x_i \wedge |S|x_i\Big\| \\
&< \sum_{i=1}^{n} \|T(w_i)\| + \Big\|\sum_{i=1}^{n} |R|x_i \wedge |S|x_i\Big\| \\
&< \sum_{i=1}^{n} \tfrac{\epsilon}{2n} + \tfrac{\epsilon}{2} = \epsilon\,.
\end{aligned}$$

Thus, $R(z) \in R(\Phi) + \epsilon U_F$, and so $R[0,x] \subseteq R(\Phi) + \epsilon U_F$. Therefore, $R[0,x]$ is norm totally bounded. That is, $R \in \mathcal{B}$. By symmetry, $S \in \mathcal{B}$ is also true.

STEP 2: *If $T \in \mathcal{B}$, then $T^+ \in \mathcal{B}$, i.e., $\mathcal{B}$ is a Riesz subspace in $\mathcal{L}_r(E,F)$.*

This follows from the preceding part by observing that every $T \in \mathcal{L}_r(E,F)$ satisfies $T = T^+ + (-T^-)$ and $|T^+| \wedge |-T^-| = 0$.

STEP 3: *If $\{T_\alpha\} \subseteq \mathcal{B}$ satisfies $T_\alpha \uparrow T$ in $\mathcal{L}_r(E,F)$, then $T \in \mathcal{B}$.*

Fix $x \in E^+$ and let $\epsilon > 0$. Since F has order continuous norm, there exists some β with $\|Tx - T_\beta x\| < \frac{\epsilon}{2}$. Next, pick an arbitrary finite subset A of F such that $T_\beta[0,x] \subseteq A + \frac{\epsilon}{2}U_F$. Now if $0 \le y \le x$, choose some $a \in A$ with $\|T_\beta y - a\| < \frac{\epsilon}{2}$ and note that

$$\|a - Ty\| \le \|a - T_\beta y\| + \|Ty - T_\beta y\| \le \|a - T_\beta y\| + \|Tx - T_\beta x\| < \tfrac{\epsilon}{2} + \tfrac{\epsilon}{2} = \epsilon\,.$$

That is, $T[0,x] \subseteq A + \epsilon V_F$, and so $T[0,x]$ is norm totally bounded, i.e., $T \in \mathcal{B}$.

STEP 4: *If $0 \le S \le T$ holds in $\mathcal{L}_r(E,F)$ with $T \in \mathcal{B}$, then $S \in \mathcal{B}$.*

This conclusion can be obtained from a result of H. H. Schaefer [**70**, p. 142] (see also [**6**, Theorem 11.17, p. 172]), a special case of which is stated below.

- *If E is a Dedekind complete Banach lattice and $x \in E^+$, then the order interval $[0,x]$ is the closed convex hull of the set of all components of x.*

A short proof of this result goes as follows. Let $\mathcal{C} \subseteq [0,x]$ be the closed convex hull of the set C of all components of x. If $\mathcal{C} \ne [0,x]$, then there exists some $z \in [0,x]$ that does not belong to $\mathcal{C}$. Now, by the Separation Theorem, there exists some non-zero linear functional $f \in E^*$ and some scalar α such that $f(z) > \alpha$ and $f(y) < \alpha$ for all $y \in \mathcal{C}$. But then, by Theorem 1.17, we have

$$f^+(x) = \sup\big\{f(y)\colon\ y \in [0,x] \text{ and } y \wedge (x-y) = 0\big\} = \sup_{y \in C} f(y) \le \alpha\,.$$

So, we have $\alpha < f(z) \le f^+(z) \le f^+(x) \le \alpha$, a contradiction. Thus, $\mathcal{C} = [0,x]$.

We now establish the claim in Step 4. Let $\mathcal{A}$ denote the set of all convex combinations of components of T, and note that according to Step 1 we have

$\mathcal{A} \subseteq \mathcal{B}$. By the preceding property, we conclude that $\mathcal{A}$ is r-norm dense in $[0, T]$. Since $\|R\| \leq \|R\|_r$ holds for each $R \in \mathcal{L}_r(E, F)$, it follows that $\mathcal{A}$ is also dense in $[0, T]$ with respect to the operator norm. Now it should be straightforward to verify that S (as being the norm limit of a sequence of $\mathcal{A}$) belongs to $\mathcal{B}$. This complete the solution. ∎

2.4. Compact and Weakly Compact Positive Operators

Problem 2.4.1. *Show that a positive operator $T\colon E \to F$ between two Riesz spaces dominates an operator $S\colon E \to F$ if and only if $-T \leq S \leq T$.*

Solution: Assume first that T dominates S. Then for each $x \in E^+$ we have

$$\pm Sx \leq |Sx| \leq T|x| = Tx\,.$$

Thus, $\pm S \leq T$ or, equivalently, $-T \leq S \leq T$.

For the converse, assume that $-T \leq S \leq T$. This implies that for each $y \in E^+$ we have $-Ty \leq Sy \leq Ty$ or $\pm Sy \leq Ty$, and so $|Sy| = (Sy) \vee (-Sy) \leq Ty$. Consequently, if $x \in E$, then

$$|Sx| = |Sx^+ - Sx^-| \leq |Sx^+| + |Sx^-| \leq Tx^+ + Tx^- = T(x^+ + x^-) = T|x|\,.$$

This shows that T dominates S. ∎

Problem 2.4.2. *Let $T\colon E \to F$ be a positive operator between two Banach lattices. Show that for an operator $S\colon E \to F$ the following statements are equivalent.*

(a) *T dominates S.*

(b) *T^* dominates S^**

(c) *T^{**} dominates S^{**}.*

Solution: (a) $\Longrightarrow$ (b) Assume that T dominates S. By Problem 2.4.1 this is equivalent to $-T \leq S \leq T$. Since the adjoint mapping $R \mapsto R^*$ is a positive operator, it follows that $-T^* \leq S^* \leq T^*$, which (by Problem 2.4.1 again) is equivalent to saying that the positive operator T^* dominates S^*.

(b) $\Longrightarrow$ (c) The validity of this implication follows immediately from the previous implication (a) $\Longrightarrow$ (b).

(c) $\Longrightarrow$ (a) Assume that the positive operator $T^{**}\colon E^{**} \to F^{**}$ dominates S^{**}. By Problem 2.4.1 this is equivalent to $-T^{**} \leq S^{**} \leq T^{**}$. Since E is a Riesz subspace of E^{**} and T agrees with T^{**}, it follows that $-Tx \leq Sx \leq Tx$ holds for each $x \in E^+$. Using Problem 2.4.1 once more, we see that T dominates S. ∎

Problem 2.4.3. *Prove the following basic properties of compact operators.*

(a) *Sums and scalar multiples of compact operators are compact operators.*

(b) *The composition of two (or more) bounded operators at least one of which is compact is likewise a compact operator.*

(c) *Bounded finite-rank operators are compact. (Recall that an operator between two vector spaces is said to be* ***finite-rank*** *if its range is finite dimensional.)*

(d) *A bounded operator is compact if and only if its adjoint is likewise a compact operator.*

(e) *The norm limit of a sequence of compact operators is a compact operator.*

Solution: (a) That the sum of two compact operators is compact follows immediately from the fact that the sum of two totally bounded sets of a normed space is always totally bounded. (If $C \subseteq \bigcup_{i=1}^{k} B(x_i, \epsilon)$ and $D \subseteq \bigcup_{j=1}^{m} B(y_j, \epsilon)$, then $C + D \subseteq \bigcup_{i=1}^{k} \bigcup_{j=1}^{m} B(x_i + y_j, 2\epsilon)$.) The fact that scalar multiples of compact operators are compact is obvious.

(b) Let $X \xrightarrow{S} Y \xrightarrow{T} Z$ be a scheme of bounded operators between Banach spaces. If either S or T is compact, then $TS(U_X) = T\big(S(U_X)\big)$ must be a totally bounded subset of Z. (Keep in mind that every bounded linear operator carries norm totally bounded sets to norm totally bounded sets.) That is, TS is a compact operator.

(c) Let $T\colon X \to Y$ be a bounded finite-rank operator between Banach spaces. This means that there exist $x_1^*, x_2^*, \ldots, x_k^* \in X^*$ and $y_1, y_2, \ldots, y_k \in Y$ such that $Tx = \sum_{i=1}^{k} x_i^*(x) y_i$ holds for all $x \in X$. Notice that each $x_i^*(U_X)$ is a bounded subset of $\mathbb{R}$. So, if $\{x_n\}$ is an arbitrary sequence of U_X, then there exists a subsequence of $\{x_n\}$, say $\{z_n\}$, such that $x_i^*(z_n) \to \alpha_i$ for each i. Now note that

$$Tz_n = \sum_{i=1}^{k} x_i^*(z_n) y_i \longrightarrow \sum_{i=1}^{k} \alpha_i y_i$$

in Y, proving that T is a compact operator.

(d) Let $T\colon X \to Y$ be compact and consider Y as a closed vector subspace of $C(U_Y^*)$, where U_Y^* is the closed unit ball of Y^* equipped with its w^*-topology. Now fix $y^* \in U_Y^*$ and let $\epsilon > 0$. Since $T(U_X)$ is a norm totally bounded subset of Y, it follows from the classical Ascoli–Arzelà theorem that $T(U_X)$ is an equicontinuous bounded subset of $C(U_Y^*)$. So, there exists a w^*-neighborhood V_{y^*} of y^* satisfying

$$|Tx(y^*) - Tx(z^*)| = |x(T^*y^* - T^*z^*)| < \tfrac{\epsilon}{2}$$

for all $z^* \in V_{y^*}$ and $x \in U_X$. So, $\|T^*y^* - T^*z^*\| = \sup_{x \in U_X} |x(T^*y^* - T^*z^*)| \leq \frac{\epsilon}{2} < \epsilon$ for all $z^* \in V_{y^*}$. If the linear functionals $y_1^*, \ldots, y_m^* \in U_Y^*$ are chosen so that $U_Y^* \subseteq \bigcup_{i=1}^{m} V_{y_i^*}$, then $T^*(U_Y^*) \subseteq \bigcup_{i=1}^{m} B(T^*y_i^*, \epsilon)$, proving that $T^*(U_Y^*)$ is a norm totally bounded set.

Now assume that $T^*\colon Y^* \to X^*$ is a compact operator. By the preceding case, we know that $T^{**}\colon X^{**} \to Y^{**}$ is also a compact operator. In particular, the set

$$T(U_X) = T^{**}(U_X) \subseteq T^{**}(U_X^{**})$$

is a norm totally bounded subset of X^{**}. Since X is a subspace of X^{**}, it follows that $T(U_X)$ is a norm totally bounded subset of X. Thus, T is a compact operator.

(e) Let $\{T_n\} \subseteq \mathcal{L}(X,Y)$ be an arbitrary sequence of compact operators such that $\|T_n - T\| \to 0$ holds for some operator $T \in \mathcal{L}(X,Y)$. Fix $\epsilon > 0$, and then choose some k such that $\|T_k - T\| < \epsilon$. Since T_k is a compact operator, there exist $y_1, \ldots, y_m \in Y$ such that $T_k(U_X) \subseteq \bigcup_{i=1}^m B(y_i, \epsilon)$. Next, fix $x \in U_X$ and then pick some $1 \leq i \leq m$ such that $\|T_k x - y_i\| < \epsilon$. Now notice that

$$\|Tx - y_i\| \leq \|(T - T_k)x\| + \|T_k x - y_i\| \leq \|T - T_k\| + \|T_k x - y_i\| < \epsilon + \epsilon = 2\epsilon\,.$$

Therefore, $T(U_X) \subseteq \bigcup_{i=1}^m B(y_i, 2\epsilon)$ so that $T(U_X)$ is a norm totally bounded subset of Y. This shows that T is a compact operator. ∎

Problem 2.4.4 (Gantmacher [**30**]). *For a bounded operator $T\colon X \to Y$ between Banach spaces show that the following statements are equivalent.*

(a) *T is weakly compact.*

(b) *$T^{**}(X^{**}) \subseteq Y$.*

(c) *$T^*\colon \big(Y^*, \sigma(Y^*, Y)\big) \to \big(X^*, \sigma(X^*, X^{**})\big)$ is continuous.*

(d) *$T^*\colon Y^* \to X^*$ is weakly compact.*

Solution: (a) $\Longrightarrow$ (b) We know that U_X is $\sigma(X^{**}, X^*)$-dense in U_X^{**}. That is, if bar denotes the $\sigma(X^{**}, X^*)$-closure operation in X^{**}, we have $\overline{U_X} = U_X^{**}$. Taking into account that $T^{**}\colon \big(X^{**}, \sigma(X^{**}, X^*)\big) \to \big(Y^{**}, \sigma(Y^{**}, Y^*)\big)$ is continuous and that (by (a)) the $\sigma(Y^{**}, Y^*)$-closure of $T(U)$ in Y^{**} lies in Y, we get

$$T^{**}(U_X^{**}) = T^{**}\big(\overline{U_X}\big) \subseteq \overline{T^{**}(U_X)} = \overline{T(U_X)} \subseteq Y\,.$$

Hence, $T^{**}(X^{**}) \subseteq Y$.

(b) $\Longrightarrow$ (c) Let $y_\alpha^* \xrightarrow{\sigma(Y^*,Y)} 0$ in Y^*, and let $x^{**} \in X^{**}$. From $T^{**}x^{**} \in Y$, we see that

$$\langle T^* y_\alpha^*, x^{**}\rangle = \langle y_\alpha^*, T^{**}x^{**}\rangle \longrightarrow 0\,.$$

This shows that $T^* y_\alpha^* \xrightarrow{\sigma(X^*,X^{**})} 0$ in X^*.

(c) $\Longrightarrow$ (d) By Alaoglu's theorem, we know that the closed unit ball of U_Y^* is $\sigma(Y^*, Y)$-compact. Therefore, $T^*(U_Y^*)$ is a $\sigma(X^*, X^{**})$-compact subset of X^*. Thus, T^* is a weakly compact operator.

(d) $\Longrightarrow$ (a) From the above established implications, it follows that the operator

$$T^{**}\colon \big(X^{**}, \sigma(X^{**}, X^*)\big) \to \big(Y^{**}, \sigma(Y^{**}, Y^{***})\big)$$

is continuous. Thus, $T^{**}(U_X^{**})$ is a weakly compact subset of Y^{**}. Since Y is norm closed in Y^{**}, it follows that Y is also weakly closed, i.e., $\big(Y^{**}, \sigma(Y^{**}, Y^{***})\big)$-closed, in Y^{**}. From $\sigma(Y^{**}, Y^*) \subseteq \sigma(Y^{**}, Y^{***})$, it follows that $T^{**}(U_Y^{**}) \cap Y$ is also $\sigma(Y^{**}, Y^*)$-compact. Now use that $T(U_X) = T^{**}(U_X) \subseteq T^{**}(U_Y^{**}) \cap Y$ and the fact that $\sigma(Y^{**}, Y^*)$ and $\sigma(Y, Y^*)$ agree on Y to conclude that $T(U_X)$ is a relatively weakly compact subset of Y. That is, T is weakly compact, and we are done. ∎

Problem 2.4.5. *Prove properties similar to the ones listed in Problem* 2.4.3 *for weakly compact operators.*

Solution: (a) Since the sum of two relatively weakly compact sets is relatively weakly compact, it follows that the sum of two weakly compact operators is weakly compact. Also, it should be clear that scalar multiples of weakly compact operators are weakly compact.

(b) The composition of two operators one of which is weakly compact is also weakly compact. This follows from the fact that an operator between two Banach spaces is bounded if and only if it is weakly continuous.

(c) From Problem 2.4.4, we know that a bounded operator between Banach spaces is weakly compact if and only if its adjoint is weakly compact.

(d) Consider a sequence $\{T_n\} \subseteq \mathcal{L}(X, Y)$ of weakly compact operators satisfying $\|T_n - T\| \to 0$ for some $T \in \mathcal{L}(X, Y)$, where X and Y are Banach spaces. Fix $x^{**} \in X^{**}$. By Problem 2.4.4, we know that $\{T_n^{**}x^{**}\} \subseteq Y$, and from the relation $\|T_n^{**}x^{**} - T^{**}x^{**}\| \to 0$, we see that $T^{**}x^{**} \in Y$. That is, $T^{**}(X^{**}) \subseteq Y$ holds, and so (by Problem 2.4.4) the operator T is weakly compact. ∎

Problem 2.4.6. *Show that the range of an arbitrary compact operator is separable.*

Solution: Let $T\colon X \to Y$ be a compact operator between Banach spaces. Then $T(U_X)$ is a norm totally bounded subset of Y and hence a separable metric space. Let A be a countable subset of $T(U_X)$ that is dense in $T(U_X)$, and let $B = \bigcup_{n=1}^{\infty} nA$. We claim that the countable set B is dense in $R(T)$.

To see this, let $y = Tx \in R(T)$. Choose some natural number k such $x \in kU_X$. This implies $y = Tx \in kT(U_X) \subseteq k\overline{A} = \overline{kA} \subseteq \overline{B}$. Hence, B is a countable dense subset of $R(T)$, proving that the range of T is separable. ∎

Problem 2.4.7. *Show that a subset of a normed space is norm totally bounded if and only if it is contained in the closed convex hull of a sequence that converges to zero.*

Solution: Let A be a non-empty subset of a normed space X. Assume first that there exists a sequence $\{x_n\}$ such that $\|x_n\| \to 0$ and $A \subseteq \overline{\text{co}}\,\{x_n\}$. Since the set $\{0, x_1, x_2, \ldots\}$ is a compact subset of X, it follows (from Mazur's theorem) that $\overline{\text{co}}\,\{x_n\}$ is a norm totally bounded set. This implies that A is also norm totally bounded.

For the converse assume that A is norm totally bounded. Let us denote the closed unit ball of X by U. The construction of the desired sequence will be done by induction. Start by letting $A_0 = A$ and $k_0 = 0$. Pick a finite subset $F_1 = \{x_1, x_2, \ldots, x_{k_1}\}$ of $2A_0$ such that $2A_0 \subseteq F_1 + 2^{-1}U$. (Clearly, such a set exists since $2A_0$ is norm totally bounded.) Put $A_1 = (2A_0 - F_1) \cap 2^{-1}U$, and note that A_1 is a norm totally bounded set.

Now for the induction hypothesis, assume that a pair (A_{n-1}, F_n) of subsets of X has been constructed such that A_{n-1} is a norm totally bounded subset of $2^{-n}U$ and the finite set $F_n = \{x_{k_{n-1}+1}, x_{k_{n-1}+2}, \ldots, x_{k_n}\}$ is a subset of the totally bounded set $2A_{n-1}$ satisfying $2A_{n-1} \subseteq F_n + 2^{-n}U$. Let $A_n = (2A_{n-1} - F_n) \cap 2^{-n}U$, and note that each A_n is a norm totally bounded set. Now choose a finite subset $F_{n+1} = \{x_{k_n+1}, \ldots, x_{k_{n+1}}\}$ of $2A_n$ satisfying $2A_n \subseteq F_{n+1} + 2^{-n-1}U$. Thus, the induction step produces the pair (A_n, F_{n+1}) with the same properties. Notice also that $A_n \subseteq 2^{-n}U$ for each $n \geq 1$. We claim that the sequence $\{x_n\}$ produced by the above induction process satisfies $A \subseteq \overline{\text{co}}\,\{x_n\}$.

To see this, let $n > 1$ and choose some k such that $k_{n-1}+1 \leq k \leq k_n$. It follows that $x_k \in F_n \subseteq 2A_n \subseteq 2 \cdot 2^{-n}U$ or $\|x_k\| \leq 2^{1-n}$. This implies $\|x_n\| \to 0$. Now let $x \in A$. Then an easy inductive argument shows that for each $n > 1$ there exist integers $m_1, \ldots, m_n$ with $k_{i-1}+1 \leq m_i \leq k_i$ for each $1 \leq i \leq n$ such that the vector $\frac{x_{m_1}}{2} + \cdots + \frac{x_{m_n}}{2^n} \in \overline{\text{co}}\,\{x_n\}$ satisfies

$$\left\|x - \left(\tfrac{x_{m_1}}{2} + \tfrac{x_{m_2}}{2^2} + \cdots + \tfrac{x_{m_n}}{2^n}\right)\right\| \leq 4^{-n}.$$

Thus, $x \in \overline{\text{co}}\,\{x_n\}$ and so $A \subseteq \overline{\text{co}}\,\{x_n\}$, as desired. ■

Problem 2.4.8 (Terzioglu [**78**]). *Show that an operator $T\colon X \to Y$ between two Banach spaces is compact if and only if there exists a sequence $\{x_n^*\}$ in X^* satisfying $\|x_n^*\| \to 0$ and $\|Tx\| \leq \sup_n |x_n^*(x)|$ for each $x \in X$.*

Solution: Assume first that T is a compact operator. Then T^* is also a compact operator, and so $T^*(U_Y^*)$ is a norm totally bounded subset of X^*. By Problem 2.4.7 there exists a null sequence $\{x_n^*\}$ in Y^* satisfying $T^*(U_Y^*) \subseteq \overline{\text{co}}\,\{x_n^*\}$. Now notice that for each $x \in X$ we have

$$\|Tx\| = \sup_{y^* \in U_Y^*} \left|y^*(Tx)\right| = \sup_{y^* \in U_Y^*} \left|T^*y^*(x)\right| \leq \sup_n |x_n^*(x)|.$$

For the converse, assume that there exists a sequence $\{x_n^*\}$ of X^* satisfying $\|x_n^*\| \to 0$ and $\|Tx\| \leq \sup_n |x_n^*(x)|$ for all $x \in X$. Now define the operator $S\colon X \to c_0$ by

$$S(x) = \left(x_1^*(x), x_2^*(x), \ldots\right).$$

Let $u = \left(\|x_1^*\|, \|x_2^*\|, \ldots\right)$ and note that $u \in c_0$. Moreover, $S(U_X) \subseteq [-u, u]$ holds. Since the order interval $[-u, u]$ is a compact subset of c_0, it follows that $S(U_X)$ is a norm totally bounded subset of c_0, and, of course, a norm totally bounded subset of the vector subspace $Z = S(X)$ of c_0. Next, define the operator $R\colon Z \to Y$ by

$$R(Sx) = Tx.$$

To see that R is well defined, notice that if $Sx = Sy$, then $x_n^*(x - y) = 0$ holds for each n, and so from $\|Tx - Ty\| = \|T(x - y)\| \leq \sup_n |x_n^*(x - y)| = 0$ we get $Tx = Ty$. Observe that the inequality

$$\left\|R(Sx)\right\| = \|Tx\| \leq \sup_n |x_n^*(x)| = \|Sx\|_\infty$$

implies that R is a continuous operator. Now from $T(U_X) = R\big(S(U_X)\big)$, it follows that $T(U_X)$ is a norm totally bounded subset of Y, and so T is compact. ■

Problem 2.4.9. *Show that an operator $T\colon X \to Y$ between Banach spaces is compact if and only if T factors with compact factors through a closed subspace of c_0.*

Solution: The "if" part is trivial. For the "only if" part assume that $T: X \to Y$ is a compact operator. By Problem 2.4.8, there exists a null sequence $\{x_n^*\}$ of X^* satisfying

$$\|Tx\| \leq \sup_n |x_n^*(x)| \tag{$\star$}$$

for each $x \in X$. We can assume that $x_n^* \neq 0$ for each n. Put $\alpha_n = \|x_n^*\|^{\frac{1}{2}}$ and $y_n^* = \|x_n^*\|^{-\frac{1}{2}} x_n^*$. Clearly, $\|y_n^*\| = \alpha_n \to 0$. Now define the operator $S: X \to c_0$ by

$$S(x) = \big(y_1^*(x), y_2^*(x), \ldots\big),$$

and note that $\|Sx\|_\infty = \sup_n |y_n^*(x)|$ for each $x \in X$. By Problem 2.4.8, S is a compact operator. Let $Z = \overline{S(X)}$ be the norm closure of $S(X)$ in c_0. Clearly, $S: X \to Z$ is a compact operator.

Next consider the operator $R: S(X) \to Y$ defined by $R(Sx) = Tx$. If $Sx = Sy$, then $x_n^*(x - y) = 0$ for each n, and so from $(\star)$, we get $Tx = Ty$. This shows that R is well defined. Moreover, if $M = \sup_n |\alpha_n|$, then it follows from the relation $|x_n^*(x)| = |\alpha_n y_n^*(x)| \leq M|y_n^*(x)|$ that

$$\big\|R(Sx)\big\| = \|Tx\| \leq \sup_n |x_n^*(x)| \leq M \sup_n |y_n^*(x)| = M\|Sx\|_\infty .$$

This shows that $R: S(X) \to Y$ is a continuous operator, and so R has a continuous extension to $Z = \overline{S(X)}$ (that will be denoted by R again). Clearly, $T = RS$. To complete the proof, we shall show that R is also a compact operator.

To this end, define $f_n \in c_0^*$ by $f_n(\lambda_1, \lambda_2, \ldots) = \alpha_n \lambda_n$. Then $\|f_n\| = \alpha_n \to 0$ and also $f_n(Sx) = \alpha_n y_n^*(x) = x_n^*(x)$. Since $\|R(Sx)\| = \|Tx\| \leq \sup_n |f_n(Sx)|$ for each $x \in X$, it follows that $\|R(z)\| \leq \sup_n |f_n(z)|$ for all z. Now Problem 2.4.8 guarantees that R is indeed a compact operator. ∎

Problem 2.4.10. *Let $\tau: \Omega \to \Omega$ be a continuous mapping on a compact Hausdorff space Ω, and let $T: C(\Omega) \to C(\Omega)$ be the composition operator defined by $Tf = f \circ \tau$. When is T a compact operator?*

Solution: Assume that the composition operator is compact. This means that the set $T([-\mathbf{1}, \mathbf{1}])$ is a totally bounded subset of $C(\Omega)$. By the classical Ascoli–Arzelà theorem, the set $T([-\mathbf{1}, \mathbf{1}])$ is an equicontinuous subset of $C(\Omega)$. In particular, for each $\omega_0 \in \Omega$, there exists an open neighborhood V_{ω_0} of ω_0 such that

$$\big|f(\tau(\omega)) - f(\tau(\omega_0))\big| < \tfrac{1}{2} \tag{$\star$}$$

for all $\omega \in V_{\omega_0}$ and all $f \in C(\Omega)$ with $-\mathbf{1} \leq f \leq \mathbf{1}$.

We claim that $\tau(\omega) = \tau(\omega_0)$ for all $\omega \in V_{\omega_0}$. To see this, assume by way of contradiction that for some $\omega_1 \in V_{\omega_0}$ we have $\tau(\omega_1) \neq \tau(\omega_0)$. Pick a continuous function satisfying $0 \leq f \leq \mathbf{1}$, $f\big(\tau(\omega_1)\big) = 1$, and $f\big(\tau(\omega_0)\big) = 0$, and note that $\big|f\big(\tau(\omega_1)\big) - f\big(\tau(\omega_0)\big)\big| = 1$ contradicts $(\star)$. Thus, we have shown that for each $\omega \in \Omega$ there exists an open neighborhood V_ω of ω on which τ is constant.

From $\Omega = \bigcup_{\omega \in \Omega} V_\omega$ and the compactness of Ω, it follows that there exist $\omega_1, \ldots, \omega_k \in \Omega$ such that $\Omega = \bigcup_{i=1}^k V_{\omega_i}$. In particular, τ has a finite range. Let $t_1, \ldots, t_n \in \Omega$ be the distinct values that τ attains. From

$$W_i = \big\{\omega \in \Omega\colon\ \tau(\omega) = t_i\big\} = \tau^{-1}\big(\{t_i\}\big) = \tau^{-1}\big(\Omega \setminus \{t_1, \ldots, t_{i-1}, t_{i+1}, \ldots, t_n\}\big),$$

we see that each W_i is a clopen subset of Ω. Moreover, $\Omega = \bigcup_{i=1}^n W_i$ is a finite partition of Ω and τ is constant on each W_i.

It is easy to see that these are the only types of continuous functions $\tau\colon \Omega \to \Omega$ whose composition operators are compact. Also, observe that if Ω is a connected topological space, then for a composition operator to be compact, it is necessary and sufficient that the continuous function τ be constant. ∎

Problem 2.4.11. *Let $T\colon \ell_1 \to Y$ be a bounded operator, where Y is a Banach space. Show that the operator T is weakly compact if and only if the set $\{T(e_n)\colon\ n = 1, 2, \ldots\}$ is a relatively weakly compact subset of Y (where e_n denotes the sequence whose n^{th} entry is one and every other is zero).*

Solution: Let $T\colon \ell_1 \to Y$ be a bounded operator. Put $A = \{T(e_n)\colon\ n = 1, 2, \ldots\}$ and $U = U_{\ell_1}$. Clearly, $A \subseteq T(U)$. So, if T is weakly compact, then obviously A is a relatively weakly compact subset of Y.

For the converse, assume that A is a relatively weakly compact subset of Y. By the Krein–Šmulian theorem (see, for instance, [**6**, Theorem 10.15, p. 158]) the closed convex circled hull C of A is also a weakly compact subset of Y. Now assume that $x = (x_1, x_2, \ldots) \in \ell_1$ satisfies $\sum_{i=1}^\infty |x_i| \le 1$, i.e., $x \in U$. For each n let $u_n = \sum_{i=1}^n x_i T(e_i) = T\big(\sum_{i=1}^n x_i e_i\big)$. Clearly, $u_n \in C$ for each n and from $u_n \to Tx$, we see that $Tx \in C$. Thus, $T(U) \subseteq C$, proving that T is a weakly compact operator. ∎

Problem 2.4.12. *For a Banach space X prove the following.*

(a) *An operator $T\colon X \to \ell_1$ is weakly compact if and only if T is compact.*

(b) *An operator $T\colon c_0 \to Y$ is weakly compact if and only if T is compact.*

Solution: For this problem we must use the fact that ℓ_1 has the **Schur property**. That is: *A subset of ℓ_1 is relatively weakly compact if and only if it is norm totally bounded*; see, for instance, [**6**, Theorem 13.1, p. 200].

(a) This follows immediately from the fact that ℓ_1 has the Schur property.

(b) Notice that:

$$\begin{aligned} T\colon c_0 \to Y \text{ is weakly compact} &\iff T^*\colon Y^* \to c_0^* = \ell_1 \text{ is weakly compact}\\ &\iff T^*\colon Y^* \to \ell_1 \text{ is compact}\\ &\iff T\colon c_0 \to Y \text{ is compact.}\end{aligned}$$

So, T is weakly compact if and only if it is compact. ∎

Problem 2.4.13. *Let $\{y_1, y_2, \ldots\}$ be a countable relatively weakly compact subset of a Banach space Y; in particular, this is so if $\{y_n\}$ is a weakly convergent sequence in Y. Show that the operator $T\colon \ell_1 \to Y$, defined by*

$$T(\lambda_1, \lambda_2, \ldots) = \sum_{n=1}^\infty \lambda_n y_n\,,$$

is weakly compact.

Solution: Notice first that the set $\{y_1, y_2, \ldots\}$ (as being relatively weakly compact) is norm bounded. This guarantees that the series $\sum_{n=1}^{\infty} \lambda_n y_n$ is norm convergent in the Banach space X for each sequence $\lambda = (\lambda_1, \lambda_2, \ldots) \in \ell_1$. So, T is a well-defined bounded linear operator. Now invoke Problem 2.4.11 to conclude that T is a weakly compact operator. ∎

Chapter 3

Operators on AL- and AM-spaces

3.1. AL- and AM-spaces

Problem 3.1.1. *A vector u in a subset A of a vector space X is called an* ***internal point*** *of A if for each $x \in X$ there exists some $\lambda_0 > 0$ such that $u + \lambda x \in A$ for all $|\lambda| \le \lambda_0$.*

Show that in an ordered vector space X the strong units are precisely the internal points of the positive cone X^+.

Solution: Assume first that e is an internal point of X^+ and let $x \in X$. Then there exists some $\alpha > 0$ such that $e + \alpha(-x) \ge 0$. This implies $x \le \frac{1}{\alpha}e$, so that e is an order unit.

Conversely, suppose that e is an order unit and let $x \in X$. Fix some $\lambda_0 > 0$ such that $\lambda_0 x \le e$. Now notice that for each $0 \le \lambda \le \lambda_0$ we have

$$\lambda x = \big(\tfrac{\lambda}{\lambda_0}\big)\lambda_0 x \le \tfrac{\lambda}{\lambda_0} e \le e\,.$$

So, $e - \lambda x \ge 0$ for all $0 \le \lambda \le \lambda_0$. Similarly, replacing x by $-x$, there exists some $\lambda_1 > 0$ such that $e + \lambda x \ge 0$ for all $0 \le \lambda \le \lambda_1$. These show that e is an internal point of X^+. ∎

Problem 3.1.2. *If E is an AM-space with a unit e, then show that the formula*

$$\|x\|_\infty = \inf\{\lambda > 0\colon\ |x| \le \lambda e\} = \min\{\lambda \ge 0\colon\ |x| \le \lambda e\} \qquad (\star)$$

defines an equivalent lattice norm on E. Under this norm E remains an AM-space and its closed unit ball coincides with the order interval $[-e, e]$.

Solution: Let us verify that $\|\cdot\|_\infty$ is a norm. Since e is a unit, the set $\{\lambda > 0\colon |x| \le \lambda e\}$ is non-empty for each $x \in E$ and so the function $\|\cdot\|_\infty$ is well defined. Clearly, $\|x\|_\infty \ge 0$ and $\|0\|_\infty = 0$. If $\|x\|_\infty = 0$, then $|x| \le \frac{1}{n}e$ for each n. This implies that $x = 0$. For the triangle inequality note that if $|x| \le \lambda e$ and $|y| \le \mu e$, then

$$|x+y| \le |x| + |y| \le (\lambda+\mu)e\,,$$

and so $\|x+y\|_\infty \le \lambda+\mu$. This implies $\|x+y\|_\infty \le \|x\|_\infty + \|y\|_\infty$. The homogeneity $\|\lambda x\|_\infty = |\lambda|\cdot\|x\|_\infty$ is obvious.

Since $x \in [-e,e]$ is equivalent to $|x| \le e$, it follows that every x in $[-e,e]$ satisfies $\|x\|_\infty \le 1$, that is, $[-e,e] \subseteq B = \{x \in E\colon\ \|x\|_\infty \le 1\}$. Now assume that $x \in B$. This implies that for each n we have $|x| \le (1+\frac{1}{n})e$, whence $|x| \le e$. That is, $x \in [-e,e]$, and consequently $B = [-e,e]$. This equality implies immediately that for arbitrary positive elements $x, y \in B$ their supremum $x \vee y$ also belongs to B. In particular, it follows that $\|\cdot\|_\infty$ is an M-norm.

It remains to show that the (original) norm $\|\cdot\|$ of E is equivalent to $\|\cdot\|_\infty$. From $|x| \le \|x\|_\infty e$, we see that

$$\|x\| \le \|e\|\cdot\|x\|_\infty$$

for each $x \in E$. Since $(E, \|\cdot\|)$ is a Banach lattice, Theorem 1.31 guarantees that the identity operator $I\colon (E,\|\cdot\|) \to (E,\|\cdot\|_\infty)$ is continuous. That is, there exists some constant $c > 0$ satisfying $\|x\| \ge c\|x\|_\infty$ for each $x \in E$. Consequently, $c\|x\|_\infty \le \|x\| \le \|e\|\cdot\|x\|_\infty$ holds for each $x \in E$, and so the norms $\|\cdot\|$ and $\|\cdot\|_\infty$ are equivalent. ∎

Problem 3.1.3. *If E is an AM-space with a unit e, F is an arbitrary Banach lattice and $T\colon E \to F$ is a positive operator, then $\|T\| = \|Te\|$.*

Solution: The positivity of T implies $|Tx| \le T|x|$ for each $x \in E$. Now notice that $U_E = [-e,e]$, i.e., $x \in U_E$ if and only if $|x| \le e$. This implies $\|Tx\| = \big\||Tx|\big\| \le \|Te\|$ for each $x \in U_E$. Therefore, $\|T\| = \sup_{x\in U_E}\|Tx\| = \|Te\|$. ∎

Problem 3.1.4. *Show that a Banach lattice E is an AL-space (resp. an AM-space) if and only if its norm dual E^* is an AM-space (resp. an AL-space). Moreover, establish the following:*

(a) *If E is an AL space, then E^* is a Dedekind complete AM-space with unit e^*, where $e^*(x) = \|x^+\| - \|x^-\|$ for each $x \in E$.*

(b) *If E is an AM-space with unit e, then E^{**} is also an AM-space with unit e.*

Solution: We show first that if E is an AL-space, then E^* is an AM-space. To this end, assume that $x^* \wedge y^* = 0$ in E^*. Put $m = \max\{\|x^*\|, \|y^*\|\}$, and note that $m \le \|x^* + y^*\|$. Fix $\epsilon > 0$ and then select some $x \in E^+$ with $\|x\| = 1$ and $\|x^*+y^*\| \le (x^*+y^*)(x)+\epsilon$. From $[x^*\wedge y^*](x) = 0$, it follows that there exist u, v in E^+ satisfying $u+v = x$ and $x^*(u)+y^*(v) < \epsilon$. Next, from $(u-u\wedge v)\wedge(v-u\wedge v) = 0$, $0 \le u+v-2(u\wedge v) \le x$, and the fact that E is an AL-space we get

$$\|u - v\wedge u\| + \|v - v\wedge u\| = \|u+v-2(u\wedge v)\| \le \|x\| = 1\,.$$

Therefore,

$$\begin{aligned}\|x^*+y^*\| &\le x^*(x)+y^*(x)+\epsilon = x^*(v)+y^*(u)+x^*(u)+y^*(v)+\epsilon\\ &\le x^*(v)+y^*(u)+2\epsilon \le x^*(v-v\wedge u)+y^*(u-v\wedge u)+3\epsilon\\ &\le m\big(\|v-v\wedge u\|+\|u-v\wedge u\|\big)+3\epsilon\\ &\le m+3\epsilon\,.\end{aligned}$$

Since $\epsilon>0$ is arbitrary, it follows that $\|x^*+y^*\|\le m$, and hence

$$\|x^*\vee y^*\| = \|x^*+y^*\| = \max\{\|x^*\|,\|y^*\|\}\,,$$

proving that E^* is an AM-space.

In order to establish that the norm dual of an AM-space is an AL-space, we need the following property.

(P) *If E is a Banach lattice and $x^*\wedge y^*=0$ holds in E^*, then for each $\epsilon>0$ there exist $x,y\in E$ such that $\|x\|\le 1$, $\|y\|\le 1$, $x\wedge y=0$, and*

$$\|x^*\|\le x^*(x)+\epsilon \quad \textit{and} \quad \|y^*\|\le y^*(y)+\epsilon\,.$$

To see this, fix $\epsilon>0$ put $\delta=\frac{\epsilon}{3}$, and then pick two vectors $u,v\in U_E$ satisfying $\|x^*\|\le x^*(u)+\delta$ and $\|y^*\|\le y^*(v)+\delta$. From $[x^*\wedge y^*](u)=0$, we see that there exist $u_1,u_2\in E^+$ with $u_1+u_2=u$ and $x^*(u_1)+y^*(u_2)<\delta$. Similarly, there exist $v_1,v_2\in U_E$ satisfying $v_1+v_2=v$ and $x^*(v_1)+y^*(v_2)<\delta$. Now consider the vectors $x=u_2-v_1\wedge u_2$ and $y=v_1-v_1\wedge u_2$. Clearly, $x,y\in U_E$, $x\wedge y=0$, and

$$\begin{aligned}x^*(x) &= x^*(u_2)-x^*(v_1\wedge u_2)\ge x^*(u_2)-\delta = x^*(u)-x^*(u_1)-\delta\\ &\ge \big(\|x^*\|-\delta\big)-\delta-\delta = \|x^*\|-3\delta\\ &= \|x^*\|-\epsilon\,.\end{aligned}$$

Similarly, $y^*(y)\ge\|y^*\|-\epsilon$.

Next, we shall show that the norm dual of an AM-space is an AL-space. To this end, let E be an AM-space and let $x^*\wedge y^*=0$ hold in E^*. Fix $\epsilon>0$ and then use property (P) to choose two vectors $x,y\in U_E$ satisfying

$$x\wedge y=0,\quad \|x^*\|\le x^*(x)+\epsilon,\quad \text{and}\quad \|y^*\|\le y^*(y)+\epsilon\,.$$

Now note that

$$\begin{aligned}\|x^*+y^*\| &\le \|x^*\|+\|y^*\|\\ &\le x^*(x+y)+y^*(x+y)+2\epsilon \le \|x^*+y^*\|\cdot\|x+y\|+2\epsilon\\ &= \|x^*+y^*\|\cdot\|x\vee y\|+2\epsilon \le \|x^*+y^*\|+2\epsilon\\ &\le \|x^*\|+\|y^*\|+2\epsilon\,.\end{aligned}$$

Since $\epsilon>0$ is arbitrary, we see that $\|x^*+y^*\|=\|x^*\|+\|y^*\|$. This shows that E^* is an AL-space.

Next, note that if E^* is an AL-space, then by the above conclusions E^{**} is an AM-space, and hence E (as a closed subspace of E^{**}) is also an AM-space. Similarly, if E^* is an AM-space, then E is an AL-space.

Now we shall verify the claims (a) and (b). Assume first that E is an AL-space. So, the formula $e^*(x)=\|x^+\|-\|x^-\|$ defines a positive linear functional $e^*\colon E\to\mathbb{R}$. For this you must use Problems 3.1.7 (listed below) and 1.2.2. Clearly,

$[-e^*, e^*] \subseteq U^*$. Now let $x^* \in U^*$. Then $|x^*| \in U^*$, and so for each $x \in E^+$ we have $|x^*|(x) \leq \||x^*|\| \cdot \|x\| \leq \|x\| = e^*(x)$ or $|x^*| \leq e^*$. Thus, $-e^* \leq x^* \leq e^*$, and so $U^* \subseteq [-e^*, e^*]$. Consequently, $U^* = [-e^*, e^*]$ proving that E^* is an AM-space with unit e^*.

Finally, let E be an AM-space with unit e. Put

$$[\![-e, e]\!] = \{x^{**} \in E^{**}:\ -e \leq x^{**} \leq e\}$$

and note that $[\![-e, e]\!] \subseteq U^{**}$. Now assume $x^{**} \in U^{**}$. Then $|x^{**}| \in U^{**}$, and for each $0 \leq x^* \in E^*$ we have

$$|x^{**}|(x^*) \leq \|x^{**}\| \cdot \|x^*\| \leq \|x^*\| = x^*(e) = e(x^*).$$

Therefore, $|x^{**}| \leq e$ or $x^{**} \in [\![-e, e]\!]$. This shows that $U^{**} \subseteq [\![-e, e]\!]$ is also true, and so $U^{**} = [\![-e, e]\!]$. Consequently, E^{**} is also an AM-space with unit e. ∎

Problem 3.1.5. *Let E be a Banach lattice and let E_x be the principal ideal generated by a vector $x \in E$. Show that E_x under the norm*

$$\|y\|_\infty = \inf\{\lambda \geq 0:\ |y| \leq \lambda|x|\}$$

is an AM-space with unit $|x|$.

Solution: Let $x \in E$. The verification of the fact that $\|\cdot\|_\infty$ is an M-norm on E_x is straightforward and is left to the reader. We shall show that $\|\cdot\|_\infty$ is a complete norm. To this end, notice first that $\|u - v\|_\infty \leq \epsilon$ in E_x is equivalent to $|u - v| \leq \epsilon|x|$. In particular, we have

$$U_{E_x} = \{y \in E_x:\ \|y\|_\infty \leq 1\} = \{y \in E_x:\ |y| \leq |x|\} = [-|x|, |x|].$$

To establish that the lattice norm $\|\cdot\|_\infty$ is complete, it suffices (according to Problem 1.2.16) to show that every monotone $\|\cdot\|_\infty$-Cauchy sequence in E_x^+ is $\|\cdot\|_\infty$-convergent in E_x. So, let $\{u_n\}$ be a $\|\cdot\|_\infty$-Cauchy sequence in E_x satisfying $0 \leq u_n \uparrow$. Fix $\epsilon > 0$, and then select some n_0 such that $\|u_n - u_m\|_\infty \leq \epsilon$ or $|u_n - u_m| \leq \epsilon|x|$ for all $n, m \geq n_0$. This implies $\|u_n - u_m\| \leq \epsilon\|x\|$ for all $n, m \geq n_0$. In particular, $\{u_n\}$ is a Cauchy sequence with respect to the original norm $\|\cdot\|$ on E. Since E is $\|\cdot\|$-complete, there exists some $u \in E$ such that $\|u_n - u\| \to 0$. Now a glance at Problem 1.2.12 guarantees that $u_n \uparrow u$ in E. Since $0 \leq u_m - u_n \leq \epsilon|x|$ holds for all $m \geq n \geq n_0$, it follows that $0 \leq u - u_n \leq \epsilon|x|$ for all $n \geq n_0$. From, $0 \leq u \leq u_{n_0} + \epsilon|x|$, it follows that $u \in E_x$. Now notice that $\|u_n - u\|_\infty \leq \epsilon$ holds for all $n \geq n_0$. This shows that E_x is $\|\cdot\|_\infty$-complete, and so E_x under the norm $\|\cdot\|_\infty$ is an AM-space with unit $|x|$. ∎

Problem 3.1.6. *Show that every AL-space has order continuous norm.*

Solution: We know that every $L_1(\mu)$-space has order continuous norm. Now if E is an AL-space, then (by Theorem 3.5) E is lattice isometric to some $L_1(\mu)$-space, and hence it must have order continuous norm. ∎

Problem 3.1.7. *Show that a Banach lattice E is:*

(a) *an AL-space if and only if $\|x + y\| = \|x\| + \|y\|$ for $x, y \in E^+$ and*

(b) *an AM-space if and only if $\|x \vee y\| = \max\{\|x\|, \|y\|\}$ for $x, y \in E^+$.*

Solution: (a) If $\|x+y\| = \|x\| + \|y\|$ holds for all $x, y \in E^+$, then E is clearly an AL-space. Now assume that E is an AL-space. Then (by Theorem 3.5) E is lattice isometric to some $L_1(\mu)$-space. Now notice that if $0 \le x, y \in L_1(\mu)$, then

$$\|x+y\|_1 = \int (x+y)\,d\mu = \int x\,d\mu + \int y\,d\mu = \|x\|_1 + \|y\|_1 \,.$$

(b) If $\|x \vee y\| = \|x\| + \|y\|$ for all $x, y \in E^+$, then E is clearly an AM-space. For the converse, assume that E is an AM-space. Then (by Theorem 3.6) E is lattice isometric to a closed Riesz subspace of some $C(\Omega)$-space. Now notice that if $0 \le x, y \in C(\Omega)$, then $[x \vee y](\omega) = x(\omega) \vee y(\omega) \le \max\{\|x\|_\infty, \|y\|_\infty\}$ for each $\omega \in \Omega$. Therefore,

$$\|x \vee y\|_\infty = \sup_{\omega\in\Omega}\,[x \vee y](\omega) \le \max\{\|x\|_\infty, \|y\|_\infty\} \le \|x \vee y\|_\infty \,,$$

and so $\|x \vee y\|_\infty = \max\{\|x\|_\infty, \|y\|_\infty\}$ for all $0 \le x, y \in C(\Omega)$. ∎

Problem 3.1.8. *Show that the lattice operations in any AM-space are weakly sequentially continuous.*

Solution: Let E be an AM-space. By Theorem 3.6, we can assume that E is a closed subspace of a $C(\Omega)$-space for some compact Hausdorff space Ω. By the Riesz Representation Theorem, we know that a sequence $\{f_n\}$ in $C(\Omega)$ satisfies $f_n \xrightarrow{w} f$ if and only if $\{f_n\}$ is norm bounded and $f_n(\omega) \to f(\omega)$ for each $\omega \in \Omega$. Therefore, if $f_n \xrightarrow{w} f$ in $C(\Omega)$, then $|f_n| \xrightarrow{w} |f|$ also holds true in $C(\Omega)$. Since $f_n \xrightarrow{w} f$ in E if and only if $f_n \xrightarrow{w} f$ in $C(\Omega)$, it follows that E has weakly sequentially continuous lattice operations. ∎

Problem 3.1.9. *Show that the lattice operations in an AL-space need not be weakly sequentially continuous.*

Solution: We shall show that the AL-space $E = L_1[0,1]$ does not have weakly sequentially continuous lattice operations. Keep in mind that $E^* = L_\infty[0,1]$.

For each n consider the partition

$$\{0, 2^{-n}, 2\cdot 2^{-n}, 3\cdot 2^{-n}, \ldots, (2^n-1)\cdot 2^{-n}, 1\}$$

of the interval $[0,1]$ and define the nth- **Rademacher** function $r_n\colon [0,1) \to \mathbb{R}$ by

$$r_n(t) = (-1)^{k-1} \quad \text{for} \quad (k-1)2^{-n} \le t < k2^{-n} \quad (k = 1, 2, \ldots, 2^n)\,.$$

The graphs of r_1 and r_2 are shown in Figure 1. We claim that if $f\colon [0,1] \to \mathbb{R}$ is a Lebesgue integrable function, then

$$\lim_{n\to\infty} \int_0^1 r_n(t) f(t)\, d\lambda(t) = 0\,.$$

That is, $r_n \xrightarrow{w} 0$ in E.[1] Since $|r_n| = \mathbf{1} \xrightarrow{w} \mathbf{1}$ also holds in E, this will establish that E does not have weakly sequentially continuous lattice operations.

To show that $\lim_{n\to\infty} \int_0^1 r_n(t) f(t)\, d\lambda(t) = 0$ for each Lebesgue integrable function f, it suffices to establish the claim for the case $f = \chi_{[a,b)}$, where $[a,b)$ is a

[1] This is the famous **Riemann–Lebesgue lemma** from the theory of integration. For completeness we present its proof.

subinterval of $[0,1]$. (Why?) Clearly, $\int_0^1 r_n(t)\chi_{[a,b)}\,d\lambda(t) = \int_a^b r_n(t)\,dt$. Therefore, it is enough to show that $\lim \int_a^b r_n(t)\,dt = 0$ for each $0 \le a < b \le 1$.

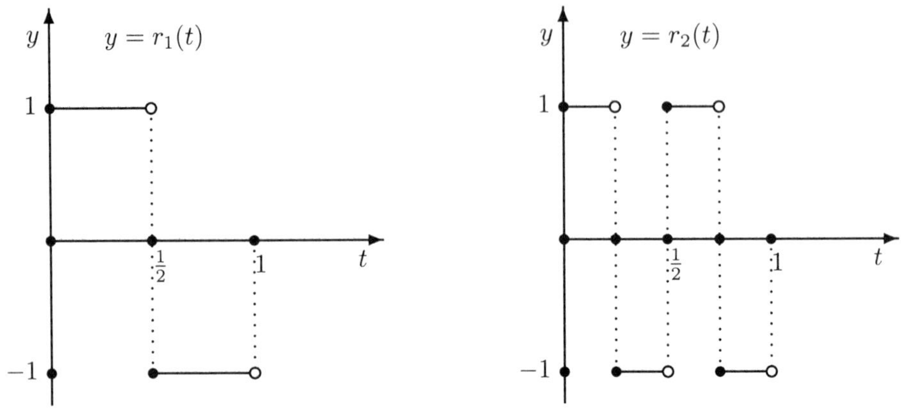

Figure 1. The graphs of r_1 and r_2

To this end, fix $0 \le a < b \le 1$ and $\epsilon > 0$. Fix n_0 such that $2^{-n_0} < \min\{\epsilon, \frac{b-a}{4}\}$. Next, choose any $n \ge n_0$ and consider the partition $\{0, \frac{1}{2^n}, \frac{2}{2^n}, \dots, \frac{2^n-1}{2^n}, 1\}$; for simplicity, let $t_i = \frac{i}{2^n}$ and note that the points a and b are related to the t_i as shown in Figure 2. Since for any three consecutive points t_{i-1}, t_i, t_{i+1} we have $\int_{t_{i-1}}^{t_{i+1}} r_n(t)\,dt = 0$, we conclude that $\int_a^b r_n(t)\,dt = \int_a^{t_k} r_n(t)\,dt + \int_c^b r_n(t)\,dt$, where $c = t_{m-1}$ or $c = t_m$; see Figure 2.

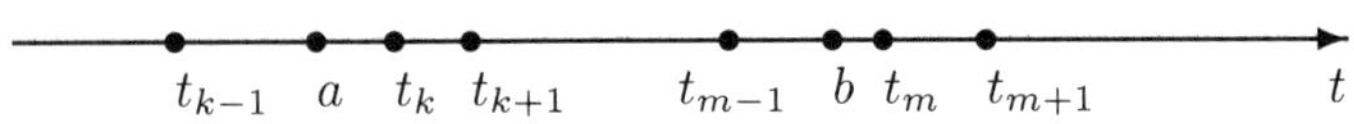

Figure 2

Consequently, for each $n \ge n_0$ we have

$$
\begin{aligned}
\left|\int_a^b r_n(t)\,dt\right| &\le \int_a^{t_k} |r_n(t)|\,dt + \int_c^b |r_n(t)|\,dt \\
&= (t_k - a) + (b - c) < \epsilon + 2\epsilon = 3\epsilon\,.
\end{aligned}
$$

This means that $\lim \int_a^b r_n(t)\,dt = 0$, as desired. (For a different solution see the proof of Lemma 11.56.) ∎

Problem 3.1.10. *Let E be a Banach lattice. For $x \in E^+$ put*

$$
\begin{aligned}
[-x,x] &= \{y \in E\colon\ -x \le y \le x\} \ \textit{and} \\
[\![-x,x]\!] &= \{y^{**} \in E^{**}\colon\ -x \le y^{**} \le x\}\,.
\end{aligned}
$$

Establish the following:

(a) $[-x,x]$ *and* $[\![-x,x]\!]$ *are both convex sets.*

(b) $[-x,x]$ *is* $\sigma(E,E^*)$*-closed and* $[\![-x,x]\!]$ *is* $\sigma(E^{**},E^*)$*-compact.*

(c) $[-x,x]$ *is* $\sigma(E^{**},E^*)$*-dense in* $[\![-x,x]\!]$.

Solution: (a) Obvious.

(b) This follows from the fact that the cones of E and E^{**} are both weakly closed; see Problem 1.2.12.

(c) Assume by way of contradiction that $[-x,x]$ is not $\sigma(E^{**},E^*)$-dense in $[\![-x,x]\!]$. By the Separation Theorem, there exist some $x^* \in E^*$ and some y^{**} in $[\![-x,x]\!]$ such that $x^*(y)=0$ for each $y \in [-x,x]$ and $x^*(y^{**}) \neq 0$. It follows that

$$|x^*|(x) = \sup\{x^*(y)\colon\ y \in E \text{ and } -x \le y \le x\} = 0\,.$$

But then we have $0 \le |x^*(y^{**})| \le |x^*|(y^{**}) \le |x^*|(x) = 0$, and so $x^*(y^{**}) = 0$, which is impossible. Hence, $[-x,x]$ is $\sigma(E^{**},E^*)$-dense in $[\![-x,x]\!]$. ∎

Problem 3.1.11. *Show that every regular operator from an AM-space with unit to any* ℓ_p*-space, where* $1 \le p < \infty$*, is compact.*

Solution: Let $T\colon E \to \ell_p$ be a positive operator, where E is an AM-space with unit e and $1 \le p < \infty$. If $U = [-e,e]$ is the unit ball of E, then $T(U) \subseteq \big[-Te, Te\big]$. To see that T is compact, notice that the order intervals in ℓ_p (for $1 \le p < \infty$) are norm compact. Hence, every regular operator (which is the difference of two positive operators) from E to ℓ_p with $1 \le p < \infty$ is compact. ∎

Problem 3.1.12. *Let E be an AM-space. Extend the norm of E to its Dedekind completion* E^δ *as in Problem* 1.2.21*. Show that the Banach lattice* E^δ *is also an AM-space. If E is an AM-space with unit e, then* E^δ *is also an AM-space with unit e.*

Solution: Let E be an AM-space. We shall show that the standard extension of the norm $\|\cdot\|$ of E to the Dedekind completion E^δ is also an M-norm.

To this end, let $0 \le \hat{x}, \hat{y} \in E^\delta$. Choose two sequences of $\{x_n\}$ and $\{y_n\}$ of E satisfying $\hat{x} \le x_n$ and $\hat{y} \le y_n$ for each n, $\|x_n\| \to \|\hat{x}\|$, and $\|y_n\| \to \|\hat{y}\|$. Since E is an AM-space, it follows from Problem 3.1.7 that

$$\max\big\{\|\hat{x}\|, \|\hat{y}\|\big\} \le \|\hat{x} \vee \hat{y}\| \le \|x_n \vee y_n\| = \max\big\{\|x_n\|, \|y_n\|\big\} \longrightarrow \max\big\{\|\hat{x}\|, \|\hat{y}\|\big\}\,.$$

Hence, $\|\hat{x} \vee \hat{y}\| = \max\big\{\|\hat{x}\|, \|\hat{y}\|\big\}$, and so E^δ is an AM-space.

Now assume that E is an AM-space with unit e. Let us denote the closed unit balls of E and E^δ by U and U^δ, respectively. Also let

$$[-e,e] = \{y \in E\colon\ -e \le y \le e\} \quad\text{and}\quad [\![-e,e]\!] = \{\hat{x} \in E^\delta\colon\ -e \le \hat{x} \le e\}\,.$$

Since E is an AM-space with unit e, we have $U = [-e,e]$. To finish the proof, we must show that $U^\delta = [\![-e,e]\!]$.

The inclusion $[\![-e,e]\!] \subseteq U^\delta$ is trivially true. For the reverse inclusion, let $\hat{x} \in U^\delta$, i.e., $\|\hat{x}\| \le 1$. Now fix some $0 < \epsilon < 1$ and note that $\|\epsilon\hat{x}\| < 1$. So, there exists some $x \in E^+$ satisfying $\epsilon|\hat{x}| \le x$ and $\|x\| < 1$. This means that $x \in U = [-e,e]$ and so $-e \le \epsilon\hat{x} \le e$ for each $0 < \epsilon < 1$. Letting $\epsilon \uparrow 1$ and taking into account that $[\![-e,e]\!]$ is a norm closed subset of E^δ (see Problem 1.2.12), we get $-e \le \hat{x} = \lim_{\epsilon\uparrow 1} \epsilon\hat{x} \le e$. Hence, $U^\delta \subseteq [\![-e,e]\!]$. Consequently, $U^\delta = [\![-e,e]\!]$, and so E^δ is an AM-space with unit e too. ∎

Problem 3.1.13. *Assume that S and Ω are two compact metric spaces, μ is an arbitrary (finite) Borel measure on S, and that $K\colon S\times\Omega \to \mathbb{R}$ is a continuous function. Let also $T\colon C(S) \to C(\Omega)$ denote the* ***integral operator*** *with kernel K, i.e., T is defined by*

$$Tx(\omega) = \int_S K(s,\omega)x(s)\,d\mu(s)$$

for all $x \in C(S)$ and each $\omega \in \Omega$. Establish the following.

(a) *T is a compact operator.*

(b) *The modulus $|T|\colon C(S) \to C(\Omega)$ of the operator T (which exists according to Theorem 3.14) is also an integral operator satisfying $|T|x(\omega) = \int_S |K(s,\omega)|x(s)\,d\mu(s)$ for all $x \in C(S)$ and all $\omega \in \Omega$.*

(c) *The norm of T is given by*

$$\|T\|_\infty = \sup\{\|Tx\|_\infty\colon\ \|x\|_\infty \le 1\} = \sup_{\omega\in\Omega}\int_S |K(s,\omega)|\,d\mu(s)\,.$$

Solution: (a) Let d and ρ be the metrics on S and Ω, respectively, and let $\epsilon > 0$. Since K is a uniformly continuous function, there exists some $\delta > 0$ such that $d(s_1,s_2)+\rho(\omega_1,\omega_2) < \delta$ implies $|K(s_1,\omega_1) - K(s_2,\omega_2)| < \epsilon$. Now if $x \in C(S)$ satisfies $\|x\|_\infty \le 1$ and ω_1,ω_2 in Ω satisfy $\rho(\omega_1,\omega_2) < \delta$, then

$$\big|Tx(\omega_1) - Tx(\omega_2)\big| \le \int_S \big|K(s,\omega_1) - K(s,\omega_2)\big|\big|x(s)\big|\,d\mu(s) \le \epsilon\mu(S)\,.$$

This shows that the norm bounded set $\{Tx\colon\ \|x\|_\infty \le 1\}$ is an equicontinuous subset of $C(\Omega)$, and hence it is a norm totally bounded set. This proves that T is a compact operator.

(b) Fix $0 \le x \in C(S)$ and $\omega \in \Omega$. Clearly, $|T|x(\omega) \le \int_S |K(s,\omega)|x(s)\,d\mu(s)$. Denote by σ the sign function of $K(\cdot,\omega)$, i.e., $\sigma(s) = 1$ if $K(s,\omega) \ge 0$ and $\sigma(s) = -1$ if $K(s,\omega) < 0$, and then choose a sequence $\{x_n\} \subseteq C(S)$ such that $x_n(s) \to \sigma(s)$ for μ-almost all s. Replacing (if necessary) each function x_n by $[x_n \vee (-\mathbf{1})] \wedge \mathbf{1}$, we can assume that $\|x_n\|_\infty \le 1$ for each n. Since $|x_n(s)x(s)| \le x(s)$ for each $s \in S$ and all n, it follows from the Lebesgue Dominated Convergence Theorem that

$$\begin{aligned}
\int_S \big|K(s,\omega)\big|x(s)\,d\mu(s) &= \int_S K(s,\omega)\sigma(s)x(s)\,d\mu(s)\\
&= \lim_{n\to\infty}\int_S K(s,\omega)x_n(s)x(s)\,d\mu(s)\\
&\le |T|x(\omega)\,.
\end{aligned}$$

Hence, $|T|x(\omega) = \int_S \big|K(s,\omega)\big|x(s)\,d\mu(s)$ holds for each $\omega \in \Omega$.

(c) Clearly, $\|T\|_\infty \le \sup_{\omega\in\Omega}\int_S |K(s,\omega)|\,d\mu(s)$. For the converse inequality pick some $\omega_0 \in \Omega$ such that $\int_S |K(s,\omega_0)|\,d\mu(s) = \sup_{\omega\in\Omega}\int_S |K(s,\omega)|\,d\mu(s)$. Again, denote by σ the sign function of $K(\cdot,\omega_0)$, and then choose a sequence $\{x_n\} \subseteq C(S)$ satisfying $\|x_n\|_\infty \le 1$ for each n and $x_n(s) \to \sigma(s)$ for μ-almost all s. Now use the Lebesgue Dominated Convergence Theorem once more to see that

$$\begin{aligned}\int_S \big|K(s,\omega_0)\big|\,d\mu(s) &= \int_S K(s,\omega_0)\sigma(s)\,d\mu(s)\\ &= \lim_{n\to\infty}\int_S K(s,\omega_0)x_n(s)\,d\mu(s)\\ &\le \|T\|_\infty .\end{aligned}$$

Hence, $\|T\|_\infty = \max_{\omega\in\Omega}\int_S\big|K(s,\omega)\big|\,d\mu(s)$. ∎

Problem 3.1.14. *Show that every Markov operator carries order units to order units.*

Solution: Let $T\colon E$ be a Markov operator between two AM-spaces with units u and v, respectively. That is, T is positive and $Tu = v$. Let e by an arbitrary order unit in E. Then there exists some $\lambda > 0$ satisfying $u \le \lambda e$, and so (by the positivity of T) we have $v = Tu \le \lambda Te$. Since v is an order unit in F, this implies that Te is also an order unit in F. ∎

Problem 3.1.15. *Every Euclidean space $\mathbb{R}^n$ can be considered either as an AM-space with unit $e = (1,1,\ldots,1)$ or as an AL-space under the usual norms*

$$\|x\|_\infty = \max_{1\le i\le n}|x_i| \quad \textit{and} \quad \|x\|_1 = \sum_{i=1}^n |x_i|\,,$$

*respectively. An $m\times n$ matrix $A = [a_{ij}]$ with non-negative real entries is called **Markov** (resp. **stochastic**) if $\sum_{j=1}^n a_{ij} = 1$ for each $1\le i\le m$ (resp. $\sum_{i=1}^m a_{ij} = 1$ for each $1 \le j \le n$).*

Now let A be an $m\times n$ matrix with non-negative entries and consider A as an operator from $\mathbb{R}^n$ to $\mathbb{R}^m$. Show the following:

(a) *The positive operator $A\colon (\mathbb{R}^n, \|\cdot\|_\infty) \to (\mathbb{R}^m, \|\cdot\|_\infty)$ is a Markov operator if and only if the matrix A is a Markov matrix.*

(b) *The positive operator $A\colon (\mathbb{R}^n, \|\cdot\|_1) \to (\mathbb{R}^m, \|\cdot\|_1)$ is a stochastic operator if and only if the matrix A is a stochastic matrix.*

Solution: (a) Notice that $Ae = \big(\sum_{j=1}^n a_{1j}, \sum_{j=1}^n a_{2j}, \ldots, \sum_{j=1}^n a_{mj}\big)$. Thus, the positive operator $A\colon (\mathbb{R}^n, \|\cdot\|_\infty) \to (\mathbb{R}^m, \|\cdot\|_\infty)$ is a Markov operator if and only if $Ae = e$ or if and only if $\sum_{j=1}^n a_{ij} = 1$ for each $1 \le i \le m$. Clearly, this is equivalent to saying that A is a Markov matrix.

(b) Notice that for each $x \in \mathbb{R}^n_+$, we have

$$\|Ax\|_1 = \sum_{i=1}^m\sum_{j=1}^n a_{ij}x_j = \sum_{j=1}^n\Big[\sum_{i=1}^m a_{ij}\Big]x_j\,. \tag{$\star$}$$

Now if $A\colon (\mathbb{R}^n, \|\cdot\|_1) \to (\mathbb{R}^m, \|\cdot\|_1)$ defines a stochastic operator, then we have $\|Ax\|_1 = \|x\|_1$ for each $x \in \mathbb{R}^n_+$. Letting $x = e_j$, the jth unit vector in $\mathbb{R}^n$, we see that $\sum_{i=1}^m a_{ij} = 1$ holds for all $1 \le j \le n$, i.e., A is a stochastic matrix.

For the converse, assume that A is a stochastic matrix. Then it follows from $(\star)$ that $\|Ax\|_1 = \|x\|_1$ for all $x \in \mathbb{R}^n_+$. That is, A defines a stochastic operator. ∎

3.2. Complex Banach Lattices

Problem 3.2.1. *If E is a Banach lattice and $z = x + \imath y \in E_{\mathbf{c}}$, then show that*

$$|z| = \sup_{\theta\in[0,2\pi]} |x\cos\theta + y\sin\theta| = \sup_{\theta\in\mathbb{R}} |x\cos\theta + y\sin\theta| .$$

Solution: Notice that for each angle $\phi \in \mathbb{R}$ there exists an angle $\theta \in [0, 2\pi]$ such that $\cos\phi = \cos\theta$ and $\sin\phi = \sin\theta$. This implies

$$|z| = \sup_{\theta\in[0,2\pi]} [x\cos\theta + y\sin\theta] = \sup_{\theta\in\mathbb{R}} [x\cos\theta + y\sin\theta] .$$

If the suprema below exist, then clearly

$$\sup_{\theta\in[0,2\pi]} |x\cos\theta + y\sin\theta| = \sup_{\theta\in\mathbb{R}} |x\cos\theta + y\sin\theta| .$$

Now for each $\theta \in [0, 2\pi]$ we have

$$\begin{aligned} x\cos\theta + y\sin\theta &\le |x\cos\theta + y\sin\theta| = [x\cos\theta + y\sin\theta] \vee [-x\cos\theta - y\sin\theta] \\ &= [x\cos\theta + y\sin\theta] \vee \big[x\cos(\pi+\theta) + y\sin(\pi+\theta)\big] \le |z| , \end{aligned}$$

and consequently

$$|z| = \sup_{\theta\in[0,2\pi]} |x\cos\theta + y\sin\theta| = \sup_{\theta\in\mathbb{R}} |x\cos\theta + y\sin\theta| ,$$

as claimed. ∎

Problem 3.2.2. *If E is a Banach lattice, then for arbitrary $z, z_1, z_2 \in E_{\mathbf{c}}$ and $\lambda \in \mathbb{C}$ show that:*

(a) *$|z| \ge 0$, and $|z| = 0$ if and only if $z = 0$.*

(b) *$|\lambda z| = |\lambda| \cdot |z|$.*

(c) *$|z_1 + z_2| \le |z_1| + |z_2|$.*

Use the above properties to verify that the function $\|\cdot\|_{\mathbf{c}}\colon E_{\mathbf{c}} \to \mathbb{R}$, defined by $\|z\|_{\mathbf{c}} = \big\||z|\big\|$, is a norm.

Solution: (a) We proved in Problem 3.2.1 that for each $z = x + \imath y \in E_{\mathbf{c}}$ we have $|z| = \sup_{\theta\in\mathbb{R}} |x\cos\theta + y\sin\theta|$. This implies $|z| \ge 0$.

If $z = 0$, i.e., if $x = y = 0$, then $|z| = 0$ is trivially true. On the other hand, if $|z| = 0$, then $|x\cos\theta + y\sin\theta| = 0$ for each angle θ. Letting $\theta = 0$ and $\theta = \frac{\pi}{2}$, we get $|x| = |y| = 0$ or $x = y = 0$. Therefore, $|z| = 0$ implies $z = 0$.

Now notice that $\|z\|_{\mathbf{c}} = \big\||z|\big\| \geq 0$ for each $z \in E_{\mathbf{c}}$. Moreover,

$$\|z\|_{\mathbf{c}} = \big\||z|\big\| = 0 \iff |z| = 0 \iff z = 0\,.$$

(b) Let $\lambda = \alpha + \imath\beta \in \mathbb{C}$ and $z = x + \imath y \in E_{\mathbf{c}}$. Then $\lambda z = (\alpha x - \beta y) + \imath(\beta x + \alpha y)$. Also, we can assume that x, y belong to $C(\Omega)$, where Ω is some compact Hausdorff space. Therefore,

$$\begin{aligned} |\lambda z| &= = \big|(\alpha x - \beta y) + \imath(\beta x + \alpha y)\big| = \sqrt{(\alpha x - \beta y)^2 + (\beta x + \alpha y)^2} \\ &= \sqrt{(\alpha^2 + \beta^2)(x^2 + y^2)} = \sqrt{\alpha^2 + \beta^2}\sqrt{x^2 + y^2} \\ &= |\lambda| \cdot |z|\,. \end{aligned}$$

For the homogeneity of $\|\cdot\|_{\mathbf{c}}$ note that:

$$\|\lambda z\|_{\mathbf{c}} = \big\||\lambda z|\big\| = \big\||\lambda| \cdot |z|\big\| = |\lambda| \cdot \big\||z|\big\| = |\lambda| \cdot \|z\|_{\mathbf{c}}\,.$$

(c) Let $z_1 = x_1 + \imath y_1$ and $z_2 = x_2 + \imath y_2$. Then $z_1 + z_2 = x_1 + x_2 + \imath(y_1 + y_2)$, and so for each angle θ we have

$$(x_1 + x_2)\cos\theta + (y_1 + y_2)\sin\theta = (x_1\cos\theta + y_1\sin\theta) + (x_2\cos\theta + y_2\sin\theta) \leq |z_1| + |z_2|\,.$$

This implies $|z_1 + z_2| = \sup_{\theta\in\mathbb{R}}\big[(x_1 + x_2)\cos\theta + (y_1 + y_2)\sin\theta\big] \leq |z_1| + |z_2|$. Thus,

$$\|z_1 + z_2\|_{\mathbf{c}} = \big\||z_1 + z_2|\big\| \leq \big\||z_1| + |z_2|\big\| \leq \big\||z_1|\big\| + \big\||z_2|\big\| = \|z_1\|_{\mathbf{c}} + \|z_2\|_{\mathbf{c}}$$

holds for all $z_1, z_2 \in E_{\mathbf{c}}$. ∎

Problem 3.2.3. *Show that the complexification of $C_{\mathbb{R}}(\Omega)$ coincides with $C_{\mathbb{C}}(\Omega)$. Moreover, show that for each pair of functions $f, g \in C_{\mathbb{R}}(\Omega)$ we have $|f + \imath g| = \sqrt{f^2 + g^2}$ and hence $\|f + \imath g\|_{\mathbf{c}} = \|\sqrt{f^2 + g^2}\|_\infty$.*

Solution: Recall that $C_{\mathbb{C}}(\Omega)$ is the complex vector space consisting of all continuous complex-valued functions defined on Ω. Now if $f \in C_{\mathbb{C}}(\Omega)$, then there exist two (uniquely determined) functions $\mathrm{Re}f$ and $\mathrm{Im}f$ in $C_{\mathbb{R}}(\Omega)$ such that $f(\omega) = (\mathrm{Re}f)(\omega) + \imath(\mathrm{Im}f)(\omega)$ for each $\omega \in \Omega$. This means that we can define a mapping $T\colon C_{\mathbb{C}}(\Omega) \to (C_{\mathbb{R}}(\Omega))_{\mathbf{c}}$ via the formula

$$Tf = \mathrm{Re}f + \imath\mathrm{Im}f\,.$$

It is a routine matter to verify that T is a surjective isometry that leaves $C_{\mathbb{R}}(\Omega)$ invariant and such that $|Tf| = T|f|$ for all $f \in C_{\mathbb{C}}(\Omega)$.

By means of this isometry we can identify $C_{\mathbb{C}}(\Omega)$ with $(C_{\mathbb{R}}(\Omega))_{\mathbf{c}}$. The remaining properties follow from this identification. ∎

Problem 3.2.4. *Show that the complexification of $\mathbb{R}^n$ coincides with $\mathbb{C}^n$. Also, show that all norms on $\mathbb{C}^n$ (or $\mathbb{R}^n$) are equivalent.*

Solution: From Problem 3.2.3, it follows that $(\mathbb{R}^n)_{\mathbf{c}} = C_{\mathbb{C}}(\{1, 2, \ldots, n\}) = \mathbb{C}^n$.

We shall prove that all norms on $\mathbb{C}^n$ are equivalent. Let us denote by $\|\cdot\|_1$ the usual ℓ_1-norm, i.e., $\|\mathbf{z}\|_1 = \sum_{k=1}^n |z_k|$ for each $\mathbf{z} = (z_1, z_2, \ldots, z_n) \in \mathbb{C}^n$. An easy argument shows that a sequence $\{\mathbf{z}_k\}$, where $\mathbf{z}_k = \big(z_1^k, z_2^k, \ldots, z_n^k\big)$, satisfies $\|\mathbf{z}_k - \mathbf{z}\|_1 \to 0$ if and only if $z_i^k \to z_i$ holds in $\mathbb{C}$ for each $1 \leq i \leq n$.

Now let $\|\cdot\|$ be an arbitrary norm on $\mathbb{C}^n$. Assume that a sequence $\{\mathbf{z}_k\}$ in $\mathbb{C}^n$ satisfies $\|\mathbf{z}_k - \mathbf{z}\|_1 \to 0$. From

$$\mathbf{z}_k = \left(z_1^k, z_2^k, \ldots, z_n^k\right) = \sum_{i=1}^n z_i^k \mathbf{e}_i ,$$

where $\mathbf{e}_i$ is the standard i^{th} unit vector, the preceding remark and the inequality

$$\|\mathbf{z}_k - \mathbf{z}\| \le \sum_{i=1}^n |z_i^k - z_i| \cdot \|e_i\| ,$$

we see that $\|\mathbf{z}_k - \mathbf{z}\| \to 0$. This shows that the identity $I\colon (\mathbb{C}^n, \|\cdot\|_1) \to (\mathbb{C}^n, \|\cdot\|)$ is continuous. In particular, the $\|\cdot\|_1$-compact set $K = \{z \in \mathbb{C}^n\colon\ \|z\|_1 = 1\}$ is also a $\|\cdot\|$-compact subset of $\mathbb{C}^n$. The continuity of the norm function $z \mapsto \|z\|$ guarantees that $m = \min_{z\in K} \|z\| > 0$ and $M = \max_{z\in K} \|z\| < \infty$. Now using the identity $\|z\| = \|z\|_1 \cdot \|\frac{z}{\|z\|_1}\|$, we obtain that $m\|z\|_1 \le \|z\| \le M\|z\|_1$ for all $z \in \mathbb{C}^n$. This proves that $\|\cdot\|$ is equivalent to $\|\cdot\|_1$. Thus, all norms on $\mathbb{C}^n$ are equivalent. ∎

Problem 3.2.5. *Show that the complexification of the real $L_p(\mu)$ coincides with the complex $L_p(\mu)$. Also, show that for any pair of real functions f, g in $L_p(\mu)$ we have $|f + \imath g| = \sqrt{f^2 + g^2}$ and that $\|f + \imath g\|_{\mathbf{c}} = \|\sqrt{f^2 + g^2}\|_p$.*

Solution: The solution is similar to that of Problem 3.2.3. ∎

Problem 3.2.6. *Recall that $\|\cdot\|_1$- and $\|\cdot\|_\infty$-norms on $\mathbb{C}^n$ are defined via the formulas*

$$\|z\|_1 = \sum_{i=1}^n |z_i| \quad \textit{and} \quad \|z\|_\infty = \max_{1\le i\le n} |z_i| .$$

If $A = \left[a_{ij}\right]$ is an $n \times n$ matrix, then the $\|\cdot\|_1$- and the $\|\cdot\|_\infty$-norms of A are defined by

$$\|A\|_1 = \sup_{\|z\|_1\le 1} \|Az\|_1 \quad \textit{and} \quad \|A\|_\infty = \sup_{\|z\|_\infty\le 1} \|Az\|_\infty .$$

Show that:

$$\|A\|_1 = \max_{1\le j\le n} \sum_{i=1}^n |a_{ij}| \quad \textit{and} \quad \|A\|_\infty = \max_{1\le i\le n} \sum_{j=1}^n |a_{ij}| .$$

Solution: Let $A = [a_{ij}]$ be an $n \times n$ matrix with complex (or real) entries. Before proceeding with the solution, we need to recall a few definitions. If $a = \alpha + \imath\beta$ is a complex number, then its **conjugate** is the complex number $\overline{a} = \alpha - \imath\beta$. We have $a\overline{a} = |a|^2$. The **conjugate** of the matrix A is the matrix $A^* = \overline{A}^t$, where $\overline{A} = [\overline{a}_{ij}]$ and B^t denotes the transpose of any matrix $B = [b_{ij}]$ and is defined by $B^t = [b_{ji}]$.

In $\mathbb{C}^n$ the dot product of two vectors x and y is defined by

$$\langle x, y\rangle = \sum_{i=1}^n x_i \overline{y}_i .$$

If A is any $n \times n$ matrix, then for all $x, y \in \mathbb{C}^n$ we have the "duality" identity

$$\langle Ax, y\rangle = \langle x, A^* y\rangle = \langle x, \overline{A}^t y\rangle . \qquad (\star)$$

Moreover, if y is an arbitrary vector in $\mathbb{C}^n$, then

$$\|y\|_1 = \sup_{\|x\|_\infty \le 1} \big|\langle y, x\rangle\big| \quad \text{and} \quad \|y\|_\infty = \sup_{\|x\|_1 \le 1} \big|\langle y, x\rangle\big| . \qquad (\star\star)$$

Let us establish these formulas. Consider the ℓ_1-norm of y first. If $x \in \mathbb{C}^n$ satisfies $\|x\|_\infty \le 1$, then $|\langle y, x\rangle| = \big|\sum_{i=1}^n y_i \overline{x}_i\big| \le \sum_{i=1}^n |y_i||\overline{x}_i| \le \sum_{i=1}^n |y_i| = \|y\|_1$. On the other hand, if for each i we choose some angle ϕ_i such that $y_i e^{\imath\phi_i} = |y_i|$ (where $\imath$ is the imaginary unit!), then the vector $x = \big(e^{\imath\phi_1}, \ldots, e^{\imath\phi_n}\big)$ satisfies $\|x\|_\infty = 1$ and $|\langle y, x\rangle| = \sum_{i=1}^n |y_i| = \|y\|_1$. Now let us look at the ℓ_∞-norm of y. If $\|x\|_1 \le 1$, then we have $|\langle y, x\rangle| = \big|\sum_{i=1}^n y_i \overline{x}_i\big| \le \sum_{i=1}^n |y_i||\overline{x}_i| \le \|y\|_\infty \big[\sum_{i=1}^n |\overline{x}_i|\big] \le \|y\|_\infty$. Choose some k with $|y_k| = \|y\|_\infty$, and then pick an angle ϕ with $y_k e^{\imath\phi} = |y_k|$. If we let $x = e^{\imath\phi} e_k$, then $\|x\|_1 = 1$ and $|\langle y, x\rangle| = |y_k| = \|y\|_\infty$.

We now return to our problem. For any $n \times n$ matrix $B = [b_{ij}]$ let

$$\tau(B) = \max_{1 \le j \le n} \sum_{i=1}^n |b_{ij}| \quad \text{and} \quad \sigma(B) = \max_{1 \le i \le n} \sum_{j=1}^n |a_{ij}| .$$

Clearly,

$$\sigma(B) = \tau(B^t) = \tau(B^*) \quad \text{and} \quad \tau(B) = \sigma(B^t) = \sigma(B^*) .$$

First, we shall establish the formula for the norm $\|A\|_\infty$. If $\|z\|_\infty \le 1$ and $y = Az$, then

$$|y_i| = \Big|\sum_{j=1}^n a_{ij} z_j\Big| \le \sum_{j=1}^n |a_{ij}||z_j| \le \sum_{j=1}^n |a_{ij}| \le \sigma(A)$$

for each i, and so $\|A\|_\infty \le \sigma(A)$. Next, choose some natural number m such that $\sum_{j=1}^n |a_{mj}| = \sigma(A)$ and then for each j select an angle θ_j satisfying $a_{mj} e^{\imath\theta_j} = |a_{mj}|$. Clearly, the vector $z = (e^{\imath\theta_1}, \ldots, e^{\imath\theta_n})$ satisfies $\|z\|_\infty = 1$. Now, if we let $y = Az$, then $y_m = \sum_{j=1}^n a_{mj} e^{\imath\theta_j} = \sum_{j=1}^n |a_{mj}| = \sigma(A)$, proving that $\|A\|_\infty = \sigma(A)$.

We now compute the ℓ_1-norm of A. Using $(\star)$ and $(\star\star)$, we see that

$$\begin{aligned}
\|A\|_1 &= \sup_{\|z\|_1 \le 1} \|Az\|_1 = \sup_{\|z\|_1 \le 1} \Big[\sup_{\|x\|_\infty \le 1} |\langle Az, x\rangle| \Big] \\
&= \sup_{\|x\|_\infty \le 1} \Big[\sup_{\|z\|_1 \le 1} |\langle Az, x\rangle| \Big] \\
&= \sup_{\|x\|_\infty \le 1} \Big[\sup_{\|z\|_1 \le 1} |\langle z, A^* x\rangle| \Big] \\
&= \sup_{\|x\|_\infty \le 1} \Big[\sup_{\|z\|_1 \le 1} |\langle A^* x, z\rangle| \Big] = \sup_{\|x\|_\infty \le 1} \|A^* x\|_\infty \\
&= \sigma(A^*) = \tau(A) ,
\end{aligned}$$

and the proof is finished. ∎

Problem 3.2.7. *Let E be a real Banach lattice with complexification $E_{\mathbf{C}}$. A vector subspace J of $E_{\mathbf{C}}$ is said to be an **ideal** whenever $|z_1| \le |z_2|$ and $z_2 \in J$ imply $z_1 \in J$.*

(a) *Show that the ideals in* $E_{\mathbf{c}}$ *are precisely the vector subspaces of* $E_{\mathbf{c}}$ *of the form* $J = J_0 \oplus \imath J_0$*, where* J_0 *is an ideal in* E.

(b) *A bounded operator* $T: F \to F$ *on a (real or complex) Banach lattice is called **ideal irreducible** if there is no non-trivial closed ideal in* F *which is* T*-invariant. That is,* T *is ideal irreducible if and only if* $T(J) \subseteq J$ *and* J *a closed ideal in* F *imply either* $J = \{0\}$ *or* $J = F$.

If $T: E \to E$ *is an operator on a real Banach lattice, then show that* T *is ideal irreducible if and only if* $T_{\mathbf{c}}: E_{\mathbf{c}} \to E_{\mathbf{c}}$ *is likewise ideal irreducible.*

Solution: (a) Assume that J_0 is an ideal of E, and let $J = J_0 \oplus \imath J_0$. Clearly, J is a vector subspace of $E_{\mathbf{c}}$. Now assume that $|x_1 + \imath y_1| \leq |x + \imath y|$ and that $x + \imath y \in J$. From

$$|x_1| \leq |x_1 + \imath y_1| \leq |x + \imath y| \leq |x| + |y| \in J_0\,,$$

it follows that $x_1 \in J_0$, and similarly $y_1 \in J_0$. Hence, $x_1 + \imath y_1 \in J$, and so J is an ideal in $E_{\mathbf{c}}$.

For the converse, assume that J is an ideal in $E_{\mathbf{c}}$. Let

$$J_0 = \big\{x \in E\colon \text{ There exists some } y \in E \text{ with } x + \imath y \in J\big\}\,.$$

Clearly, J_0 is a vector subspace of E. Now if $x + \imath y \in J$, then from the equality $y + \imath(-x) = (-\imath)(x + \imath y) \in J$ it follows that $y \in J_0$. That is, $x + \imath y \in J$ implies $x, y \in J_0$ or $J \subseteq J_0 \oplus \imath J_0$. Moreover, if $x \in J_0$, then pick some $y \in E$ with $x + \imath y \in J$, and then note that $|x + \imath 0| = |x| \leq |x + \imath y|$, which implies $x = x + \imath 0 \in J$. In particular, $x \in J_0$ implies $\imath x \in J$. Thus, if $x, y \in J_0$, then $x + \imath y \in J$, and this shows that $J_0 \oplus \imath J_0 \subseteq J$. Therefore, $J = J_0 \oplus \imath J_0$.

(b) Start by observing that an ideal $J = J_0 \oplus \imath J_0$ in $E_{\mathbf{c}}$ is closed if and only if J_0 is a closed ideal in E. Next, assume that T is ideal irreducible, and let $J = J_0 \oplus \imath J_0$ be a closed $T_{\mathbf{c}}$-invariant ideal in $E_{\mathbf{c}}$. If $x \in J_0$, then $x + \imath 0 \in J$, and consequently $Tx + \imath 0 = T_{\mathbf{c}}(x + \imath 0) \in T(J) \subseteq J = J_0 \oplus \imath J_0$. This implies $Tx \in J_0$ and so J_0 is a closed T-invariant ideal in E. Since T is irreducible, either $J_0 = \{0\}$ or $J_0 = E$. This implies either $J = \{0\}$ or $J = E_{\mathbf{c}}$, i.e., $T_{\mathbf{c}}$ is an irreducible operator.

Finally, for the converse assume that $T_{\mathbf{c}}$ is an irreducible operator, and let J_0 be a closed T-invariant ideal in E. Then $J = J_0 \oplus \imath J_0$ is a closed ideal in $E_{\mathbf{c}}$, and from

$$T_{\mathbf{c}}(J) \subseteq T(J_0) \oplus \imath T(J_0) \subseteq J_0 \oplus \imath J_0 = J\,,$$

we infer that either $J = \{0\}$ or $J = E_{\mathbf{c}}$. This implies either $J_0 = \{0\}$ or $J_0 = E$, i.e., T is an irreducible operator. ∎

Problem 3.2.8. *Show that all conclusions regarding domination of operators stated in the results* 2.34–2.40 *are also true for complex Banach lattices.*

Solution: Recall (from Problem 1.1.9) that if $X_{\mathbf{c}}$ and $Y_{\mathbf{c}}$ are the complexifications of two real vector spaces, then every operator $T: X_{\mathbf{c}} \to Y_{\mathbf{c}}$ can be represented by a matrix of the form $\mathcal{T} = \begin{bmatrix} T & -S \\ S & T \end{bmatrix}$, where $T, S\colon X \to Y$ are uniquely determined linear operators.

Now assume that $\mathcal{T}\colon E_{\mathbf{c}} \to F_{\mathbf{c}}$ is an operator between two complex Banach lattices having the matrix representation $\mathcal{T} = \begin{bmatrix} T & -S \\ S & T \end{bmatrix}$. Then we have the following properties.

(1) *$\mathcal{T}$ is a compact operator (resp. weakly compact) if and only if S and T are both compact (resp. weakly compact) operators.*

(2) *If $\mathcal{T}$ is dominated by a positive operator R, then S and T are both dominated by R.*

For (1) we shall consider the case of compact operators only. Assume first that $\mathcal{T}$ is a compact operator and let $\{x_n\} \subseteq U_E$. If we put $z_n = x_n + \imath 0 \in E_{\mathbf{c}}$, then $\{z_n\} \subseteq U_{E_{\mathbf{c}}}$, and so (by passing to a subsequence and relabeling if necessary) we can assume without loss of generality that $\{\mathcal{T} z_n\}$ converges in $E_{\mathbf{c}}$. Now from

$$\mathcal{T} z_n = \begin{bmatrix} T & -S \\ S & T \end{bmatrix} \begin{bmatrix} x_n \\ 0 \end{bmatrix} = \begin{bmatrix} Tx_n \\ Sx_n \end{bmatrix} = Tx_n + \imath S x_n\,,$$

we see that both sequences $\{Tx_n\}$ and $\{Sx_n\}$ converge in F. This proves that S and T are both compact operators.

For the converse, assume that both S and T are compact operators, and let $\{z_n\} \subseteq U_{E_{\mathbf{c}}}$, where $z_n = x_n + \imath y_n$. Then

$$\mathcal{T} z_n = \begin{bmatrix} T & -S \\ S & T \end{bmatrix} \begin{bmatrix} x_n \\ y_n \end{bmatrix} = \begin{bmatrix} Tx_n - Sy_n \\ Sx_n + Ty_n \end{bmatrix} = Tx_n - Sy_n + \imath(Sx_n + Ty_n)\,. \qquad (\star)$$

Moreover, from $\|x_n\| \le \|z_n\|_{\mathbf{c}} \le 1$, $\|y_n\| \le \|z_n\|_{\mathbf{c}} \le 1$, and the compactness of the operators S and T, by passing to appropriate subsequences and relabeling, we can assume that the four sequences $\{Tx_n\}$, $\{Ty_n\}$, $\{Sx_n\}$, and $\{Sy_n\}$ are all norm convergent in F. From $(\star)$, we see that $\{\mathcal{T} z_n\}$ converges in $F_{\mathbf{c}}$, and so $\mathcal{T}$ is a compact operator.

For (2) assume that $\mathcal{T}$ is dominated by a positive operator R. If $x \in E$, then let $z = x + \imath 0$ and note that

$$|\mathcal{T} z| = \left| \begin{bmatrix} T & -S \\ S & T \end{bmatrix} \begin{bmatrix} x \\ 0 \end{bmatrix} \right| = |Tx + \imath Sx| \le R|z| = R|x|\,.$$

This implies $|Sx| \le R|x|$ and $|Tx| \le R|x|$ for each $x \in E$, so that R dominates both S and T.

To show how one can generalize the results 2.34–2.40 to complex Banach lattices, we shall prove only Corollary 2.35. To this end, let us assume that $\mathcal{T} = \begin{bmatrix} T & -S \\ S & T \end{bmatrix} : E_{\mathbf{c}} \to E_{\mathbf{c}}$ is an operator on a complex Banach lattice and T is dominated by a compact positive operator R. A direct computation shows that

$$\mathcal{T}^2 = \begin{bmatrix} T^2 - S^2 & -ST - TS \\ ST + TS & T^2 - S^2 \end{bmatrix} \quad \text{and}$$

$$\mathcal{T}^3 = \begin{bmatrix} T^3 - TS^2 - S^2T - STS & S^3 - ST^2 - T^2S - TST \\ -S^3 + ST^2 + T^2S + TST & T^3 - TS^2 - S^2T - STS \end{bmatrix}.$$

By part (2) above, both S and T are dominated by the compact positive operator R. So, by the (real) Theorem 2.34, all entries of the matrix $\mathcal{T}^3$ are compact operators from E to E. By part (1) above, $\mathcal{T}^3$ is a compact operator. ∎

Problem 3.2.9 (Wickstead). *Let $E = \mathbb{R}^2$ be equipped with the sup norm $\|\cdot\|_\infty$ (i.e., $\|(x,y)\|_\infty = \max\{|x|,|y|\}$) and let $F = \mathbb{R}^2$ be equipped with the ℓ_1-norm $\|\cdot\|_1$ (i.e., $\|(x,y)\|_1 = |x|+|y|$). Now consider the operator $T\colon E \to F$ defined by*

$$T(x,y) = (x+y, x-y)\,.$$

Show that T is a non-positive regular operator satisfying the inequalities $\|T\|_{\mathbf{c}} \geq 2\sqrt{2} > 2 = \|T\|$.

Solution: To see that T is a regular operator note that if $T_1, T_2\colon E \to F$ are defined via the formulas $T_1(x,y) = (x+y, x)$ and $T_2(x,y) = (0,y)$, then T_1 and T_2 are both positive operators and $T = T_1 - T_2$. (The regularity of T is, of course, automatic since here E is a finite dimensional space.) Now from

$$\begin{aligned}\big\|T(x,y)\big\|_1 &= \big\|(x+y, x-y)\big\|_1 = |x+y| + |x-y| \\ &= 2\max\{|x|,|y|\} = 2\big\|(x,y)\big\|_\infty,\end{aligned}$$

we see that $\|T\| = 2$.

On the other hand, we claim that $\|T\|_{\mathbf{c}} \geq 2\sqrt{2}$. To see this, consider the vector $z = (1,1) + \imath(1,-1)$. Then $|z| = (\sqrt{2}, \sqrt{2})$, and so $\|z\|_{\infty,\mathbf{c}} = \big\||z|\big\|_\infty = \sqrt{2}$. From

$$T_{\mathbf{c}}(z) = T(1,1) + \imath T(1,-1) = (2,0) + \imath(0,2)\,,$$

we get $\big|T_{\mathbf{c}}(z)\big| = (2,2)$. This implies $\big\|T_{\mathbf{c}}(z)\big\|_{1,\mathbf{c}} = \big\||T_{\mathbf{c}}(z)|\big\|_1 = 4$. Therefore,

$$\|T_{\mathbf{c}}z\|_{1,\mathbf{c}} = \big\||T_{\mathbf{c}}z|\big\|_1 = 4 = 2\sqrt{2}\,\|z\|_{\infty,\mathbf{c}}\,,$$

which proves that $\|T\|_{\mathbf{c}} \geq 2\sqrt{2} > 2 = \|T\|$. ∎

Problem 3.2.10. *Let H be a real Hilbert space with inner product $(\cdot,\cdot)$. Extending the inner product to $H_{\mathbf{C}}$ via the formula*

$$(z_1,z_2) = (x_1+\imath y_1, x_2+\imath y_2) = [(x_1,x_2)+(y_1,y_2)] + \imath[(y_1,x_2)-(x_1,y_2)]\,, \quad (\star)$$

we turn $H_{\mathbf{C}}$ into a (complex) Hilbert space. Show that the standard norm $\|\cdot\|_{\mathbf{C}}$ on $H_{\mathbf{C}}$ defined by

$$\|z\|_{\mathbf{C}} = \|x+\imath y\|_{\mathbf{C}} = \sup_{\theta\in\mathbb{R}} \|x\cos\theta + y\sin\theta\|$$

is not in general an inner product norm—and hence $\|\cdot\|_{\mathbf{C}}$ does not coincide with the norm induced by the above inner product.

Solution: We leave it for the reader to verify that the formula $(\star)$ for (z_1,z_2) indeed defines an inner product on $H_{\mathbf{C}}$.

Now we shall show that $\|\cdot\|_{\mathbf{c}}$ does not satisfy the Parallelogram Law. To this end, consider the real Hilbert space $H = L_2[0,1]$, and let $x = \chi_{[0,\frac{1}{2})}$ and $y = \chi_{[\frac{1}{2},1]}$. Put $u = x$ and $v = \imath y$. Then

$$\begin{aligned}\|u+v\|_{\mathbf{c}}^2 &= \sup_{\theta\in\mathbb{R}} \int_0^1 [\,\chi_{[0,\frac{1}{2})}\cos\theta + \chi_{[\frac{1}{2},1]}\sin\theta\,]^2\,dt \\ &= \sup_{\theta\in\mathbb{R}} [\,\tfrac{1}{2}\cos^2\theta + \tfrac{1}{2}\sin^2\theta\,] = \tfrac{1}{2}\,,\end{aligned}$$

$$\begin{aligned}\|u-v\|_{\mathbf{c}}^2 &= \sup_{\theta\in\mathbb{R}} \int_0^1 [\,\chi_{[0,\frac{1}{2})}\cos\theta - \chi_{[\frac{1}{2},1]}\sin\theta\,]^2\,dt \\ &= \sup_{\theta\in\mathbb{R}} [\,\tfrac{1}{2}\cos^2\theta + \tfrac{1}{2}\sin^2\theta\,] = \tfrac{1}{2}\,,\end{aligned}$$

$$\begin{aligned}\|u\|_{\mathbf{c}}^2 &= \int_0^1 [\chi_{[0,\frac{1}{2})}]^2\,dt = \tfrac{1}{2}\,, \\ \|v\|_{\mathbf{c}}^2 &= \int_0^1 [\chi_{[\frac{1}{2},1]}]^2\,dt = \tfrac{1}{2}\,.\end{aligned}$$

Hence, $\|u+v\|_{\mathbf{c}}^2 + \|u-v\|_{\mathbf{c}}^2 = 1 \neq 2 = 2(\|u\|_{\mathbf{c}}^2 + \|v\|_{\mathbf{c}}^2)$. This shows that the norm $\|\cdot\|_{\mathbf{c}}$ does not satisfy the Parallelogram Law and so it cannot be the norm obtained from an inner product. ∎

Problem 3.2.11. *Let X be a real Banach space and define $N\colon X\times X \to \mathbb{R}$ by:*

$$N(x,y) = \big(\|x\|^2 + \|y\|^2\big)^{\frac{1}{2}}\,.$$

Establish the following:

(1) *The real vector space $X\oplus X$ normed with N is a Banach space.*

(2) *The same function N (as defined above) now considered on the complexification $X_{\mathbf{c}} = X\oplus\imath X$, that is, $N(x+\imath y) = (\|x\|^2+\|y\|^2)^{\frac{1}{2}}$, is a norm on $X_{\mathbf{c}}$ if and only if X is a real Hilbert space.*

Solution: (1) Note that $\|x\|, \|y\| \le N(x,y)$ and $N(x,y) \le \|x\|+\|y\|$. Hence,

$$\tfrac{1}{2}(\|x\|+\|y\|) \le N(x,y) \le \|x\|+\|y\| \qquad (\star\star)$$

holds for all $(x,y)\in X\oplus X$.

Let us verify that N is a norm on $X\oplus X$. Clearly, $N(x,y)\ge 0$ for each $x\oplus y\in X\oplus X$, and from $(\star\star)$ it follows that $N(x,y)=0$ if and only if $x=y=0$.

For the homogeneity of N notice that if $\lambda\in\mathbb{R}$, then

$$N(\lambda(x,y)) = N(\lambda x,\lambda y) = [\,\|\lambda x\|^2 + \|\lambda y\|^2\,]^{\frac{1}{2}} = [\,\lambda^2(\|x\|^2+\|y\|^2)\,]^{\frac{1}{2}} = |\lambda|N(x,y)\,.$$

For the triangle inequality observe that

$$\begin{aligned}N((x_1,y_1)+(x_2,y_2)) &= N(x_1+x_2,y_1+y_2) = [\,\|x_1+y_1\|^2 + \|x_2+y_2\|^2\,]^{\frac{1}{2}} \\ &\le [\,(\|x_1\|+\|y_1\|)^2 + (\|x_2\|+\|y_2\|)^2\,]^{\frac{1}{2}} \\ &\le (\|x_1\|^2+\|x_2\|^2)^{\frac{1}{2}} + (\|y_1\|^2+\|y_2\|^2)^{\frac{1}{2}} \\ &= N(x_1,x_2) + N(y_1,y_2)\,.\end{aligned}$$

Finally, notice that it follows from $(\star\star)$ that a sequence $\{x_n \oplus y_n\}$ in $X \oplus X$ converges to $x \oplus y$ with respect to the norm N if and only if $\|x_n - x\| \to 0$ and $\|y_n - y\| \to 0$. This easily implies that $X \oplus X$ equipped with the norm N is a Banach space.

(2) Assume first that X is a real Hilbert space with inner product $(\cdot,\cdot)$. Then, by Problem 3.2.10, the complexification $X_{\mathbf{c}}$ of X is also a Hilbert space whose inner product is the extension of $(\cdot,\cdot)$ and is given by

$$(z_1, z_2) = (x_1 + \imath y_1, x_2 + \imath y_2) = [(x_1, x_2) + (y_1, y_2)] + \imath[(y_1, x_2) - (x_1, y_2)].$$

In particular, if $z = x + \imath y$, then

$$\|z\| = (z, z)^{\frac{1}{2}} = (\|x\|^2 + \|y\|^2)^{\frac{1}{2}} = N(x, y).$$

This shows that N is a norm on $X_{\mathbf{c}}$.

For the converse, assume that N is a norm on $X_{\mathbf{c}}$, and let $x, y \in X$ be arbitrary vectors. Then $(1+\imath)(x+\imath y) = (x-y)+\imath(x+y)$, and from the homogeneity of N we get $N(x-y, x+y) = N((1+\imath)(x+\imath y)) = |1+\imath|N(x,y)$ or $(\|x-y\|^2+\|x+y\|^2)^{\frac{1}{2}} = 2^{\frac{1}{2}}(\|x\|^2 + \|y\|^2)^{\frac{1}{2}}$. That is,

$$\|x - y\|^2 + \|x + y\|^2 = 2(\|x\|^2 + \|y\|^2),$$

for all $x, y \in X$. This shows that the norm of X satisfies the Parallelogram Law, and so X is a (real) Hilbert space. ∎

Problem 3.2.12. *Let $E_{\mathbf{c}}$ be the complexification of a real Banach lattice E and assume that $|z| \le u$ for some $z \in E_{\mathbf{c}}$ and $u \in E^+$. Show that for each $\epsilon > 0$ there exist positive vectors $u_1, \dots, u_n$ in E and complex scalars $\alpha_1, \dots, \alpha_n$ such that $|\alpha_i| \le 1$ holds for each $i = 1, \dots, n$, $\sum_{i=1}^n u_i = 1$, and $\left|z - \sum_{i=1}^n \alpha_i u_i\right| \le \epsilon u$.*

Solution: Assume $|z| \le u$ and let $\epsilon > 0$. By the Kakutani–Bohnenblust–Krein Representation Theorem 3.6, we identify the complex Banach lattice $E_u \oplus \imath E_u$ with some $C_{\mathbb{C}}(\Omega)$-space in such a way that u corresponds to the constant function one.

Since $z(\Omega)$ is a compact set, there exist points $\omega_1, \dots, \omega_n \in \Omega$ such that $z(\Omega) \subseteq \bigcup_{i=1}^n D\big(z(\omega_i), \epsilon\big)$. Consequently, if $V_i = z^{-1}\big(D\big(z(\omega_i), \epsilon\big)\big)$, then each V_i is a non-empty open subset of Ω and $\Omega = \bigcup_{i=1}^n V_i$. Next, pick a partition of unity $\{u_1, u_2, \dots, u_n\}$ subordinate to the open cover $\{V_1, \dots, V_n\}$. That is, each u_i is a continuous function from Ω to $[0,1]$, each u_i vanishes outside V_i and $\sum_{i=1}^n u_i = \mathbf{1}$; see [**7**, Theorem 10.9, p. 85]. Clearly, $\big|u_i(\omega)\big[z(\omega) - z(\omega_i)\big]\big| \le u_i(\omega)\epsilon$ holds for each i and all $\omega \in \Omega$. Therefore, if $\alpha_i = z(\omega_i)$, then $|\alpha_i| \le 1$ (and $\alpha_i \in \mathbb{R}$ if z is a real-valued function) and for each $\omega \in \Omega$, we have

$$\begin{aligned} \Big|z(\omega) - \sum_{i=1}^n \alpha_i u_i(\omega)\Big| &= \Big|\sum_{i=1}^n u_i(\omega)[z(\omega) - z(\omega_i)]\Big| \\ &\le \sum_{i=1}^n \big|u_i(\omega)\big[z(\omega) - z(\omega_i)\big]\big| \le \sum_{i=1}^n u_i(\omega)\epsilon = \epsilon. \end{aligned}$$

This shows that $\left|z - \sum_{i=1}^n \alpha_i u_i\right| \le \epsilon u$ holds in $E_{\mathbf{c}}$. ∎

3.3. The Center of a Banach Lattice

Problem 3.3.1. *Show that a positive operator $T\colon E \to E$ on a Riesz space is an orthomorphism if and only if $I+T$ is a lattice homomorphism.*

Solution: Let $T\colon E \to E$ be a positive operator on a Riesz space. Assume first that $I+T$ is a lattice homomorphism and let $x \wedge y = 0$ in E. Then we have

$$\begin{aligned} 0 \le x \wedge Ty &\le (x+Tx) \wedge (y+Ty) = \big[(I+T)x\big] \wedge \big[(I+T)y\big] \\ &= (I+T)(x \wedge y) = 0 \,. \end{aligned}$$

This implies $x \wedge Ty = 0$, and so T is an orthomorphism.

For the converse, we need the following lattice inequality:

- *If u, v and w are positive vectors in a Riesz space, then*

$$u \wedge (v+w) \le u \wedge v + u \wedge w \,.$$

To see this, let $x = u \wedge (v+w)$. Then $x \le v+w$ and $x \le u \le v+u$, and so $x \le (v+u) \wedge (v+w) = v + u \wedge w$. On the other hand, we have $x \le u \le u + u \wedge w$. Therefore, $u \wedge (v+w) = x \le (u + u \wedge w) \wedge (v + u \wedge w) = u \wedge v + u \wedge w$.

Now assume that T is an orthomorphism, and that $x \wedge y = 0$ holds in E. Then $x \wedge Ty = Tx \wedge y = Tx \wedge Ty = 0$, and hence

$$\begin{aligned} 0 &\le \big[(I+T)x\big] \wedge \big[(I+T)y\big] = (x+Tx) \wedge (y+Ty) \\ &\le (x+Tx) \wedge y + (x+Tx) \wedge Ty \\ &\le x \wedge y + Tx \wedge y + x \wedge Ty + Tx \wedge Ty = 0 \,. \end{aligned}$$

Therefore, $\big[(I+T)x\big] \wedge \big[(I+T)y\big] = 0$, and thus $I+T$ is a lattice homomorphism. ∎

Problem 3.3.2. *Use Theorem 1.17 to establish that if E is a Dedekind complete Banach lattice, then the lattice operations of $\mathcal{Z}(E)$ are given by*

$$(S \vee T)x = S(x) \vee T(x) \quad \text{and} \quad (S \wedge T)x = S(x) \wedge T(x)$$

for all $S, T \in \mathcal{Z}(E)$ and all $x \in E^+$.

Solution: Assume that E is a Dedekind complete Banach lattice. We need the following property: *If S and T are central operators, then so are $|T|$, $S \vee T$, and $S \wedge T$.*

To see this, pick some $\lambda > 0$ such that $|Tx| \le \lambda |x|$ for each $x \in E$, and note that for each $u \in E^+$ we have $|T|u = \sup_{|x| \le u} |Tx| \le \sup_{|x| \le u} \lambda |x| = \lambda u$. This implies $|T| \in \mathcal{Z}(E)$. To see that $S \vee T$ and $S \wedge T$ belong to $\mathcal{Z}(E)$, use the identities $S \vee T = \frac{1}{2}(S+T+|T-S|)$ and $S \wedge T = \frac{1}{2}(S+T-|T-S|)$.

Now let $S, T \in \mathcal{Z}(E)$. We shall establish the formula for the infimum only—the other formula follows from the identity $S \wedge T = S + T - S \vee T$. Assume first that S and T are also positive operators. Notice that if $y \wedge z = 0$, then $y \wedge Tz = 0$

and $Sy \wedge Tz = 0$ hold, in which case we have $Sy + Tz = Sy \vee Tz$. Thus, using Theorem 1.17, for each $x \in E^+$ we have

$$\begin{aligned}(S \vee T)x &= \sup\{Sy + Tz\colon\ y \wedge z = 0 \text{ and } y + z = x\}\\ &= \sup\{Sy \vee Tz\colon\ y \wedge z = 0 \text{ and } y + z = x\}\\ &\leq Sx \vee Tx\,.\end{aligned}$$

Since $(S \vee T)x \geq Sx \vee Tx$ is trivially true, we infer that $(S \vee T)x = Sx \vee Tx$. Consequently, $(S \wedge T)x = \frac{1}{2}(S + T - S \vee T)x = \frac{1}{2}(Sx + Tx - Sx \vee Tx) = Sx \wedge Tx$ also holds for each $x \in E^+$.

We now consider the general case. By the property stated at the beginning of the proof, we know that $0 \leq S - S \wedge T \in \mathcal{Z}(E)$ and $0 \leq T - S \wedge T \in \mathcal{Z}(E)$. So, from $(S - S \wedge T) \wedge (T - S \wedge T) = 0$ and the preceding case, it follows that for each $x \in E^+$ we have

$$\begin{aligned}0 &= \big[(S - S \wedge T) \wedge (T - S \wedge T)\big]x\\ &= (S - S \wedge T)x \wedge (T - S \wedge T)x\\ &= \big[Sx - (S \wedge T)x\big] \wedge \big[Tx - (S \wedge T)x\big]\\ &= Sx \wedge Tx - (S \wedge T)x\end{aligned}$$

or $(S \wedge T)x = Sx \wedge Tx$. ∎

Problem 3.3.3. *Show that every central operator $T\colon E \to E$ on a Banach lattice is principal ideal preserving, i.e., show that $T(E_x) \subseteq E_x$ holds for each $x \in E^+$.*

Solution: Let $T\colon E \to E$ be a central operator on a Banach lattice. Pick some $\lambda > 0$ such that $|Tx| \leq \lambda|x|$ holds for all $x \in E$. Next, fix some $u \in E^+$, and assume $y \in E_u$. Choose a scalar $\alpha > 0$ such that $|y| \leq \alpha u$, and note that $|Ty| \leq \lambda|y| \leq \alpha\lambda u$ holds. This implies $Ty \in E_u$, and so $T(E_u) \subseteq E_u$. Therefore, T leaves invariant every principal ideal. ∎

Problem 3.3.4. *Show that in a Banach lattice E:*

(1) *Its center $\mathcal{Z}(E)$ with the operator norm is an AM-space with unit the identity operator I.*

(2) *If E is also Dedekind complete, then the center $\mathcal{Z}(E)$ (as a Riesz space) coincides with the band generated by the identity operator I in $\mathcal{L}_r(E)$; and so*

$$\mathcal{L}_r(E) = \mathcal{Z}(E) \oplus I^{\mathrm{d}}\,.$$

Solution: (1) Let E be a Banach lattice. We consider $\mathcal{Z}(E)$ as a subset of $\mathcal{L}(E)$. We claim that $\mathcal{Z}(E)$ is a closed subset of $\mathcal{L}(E)$. To see this, assume that a sequence $\{T_n\}$ of $\mathcal{Z}(E)$ satisfies $T_n \to T$ in $\mathcal{L}(E)$. Let $|x| \wedge |y| = 0$ in E. Since each T_n is a central operator, we have $|x| \wedge |T_n x| = 0$ for each n. Now using the fact that the lattice operations are norm continuous, we see that $|x| \wedge |Ty| = 0$. This implies that T is an orthomorphism, and so (by Theorem 3.29) T is a central operator too. Thus, $\mathcal{Z}(E)$ equipped with the operator norm is a Banach space—and a unital (sub)algebra of $\mathcal{L}(E)$.

Next, notice that if $T \in \mathcal{Z}(E)$, then it follows from Theorem 3.30 that

$$\|T\| = \sup_{\|x\|\le 1} \|Tx\| = \sup_{\|x\|\le 1} \big\||T|(|x|)\big\| = \big\||T|\big\| = \|T\|_r\,.$$

This implies that the operator norm of T coincides with the r-norm of T, and so the operator norm is a lattice norm. Thus, $\mathcal{Z}(E)$ is a Banach lattice with the operator norm.

Clearly, the identity operator I is a unit for $\mathcal{Z}(E)$ (and I is both an algebraic unit and an order unit). So, in order to establish that $\mathcal{Z}(E)$ is a Banach lattice with unit I, we must show that $U_{\mathcal{Z}(E)} = [-I, I]$ holds. Clearly, $[-I, I] \subseteq U_{\mathcal{Z}(E)}$. So, what needs verification is the inclusion $U_{\mathcal{Z}(E)} \subseteq [-I, I]$. In order to establish this, we need the following lattice-algebraic property of the operators in $\mathcal{Z}(E)$.

(P) *If two operators $S, T \in \mathcal{Z}(E)$ satisfy $S \wedge T = 0$, then $ST = 0$.*

To see this, assume that for some $\lambda > 0$ we have $Sx \le \lambda x$ and $Tx \le \lambda x$ for all $x \in E^+$. Now note that if $x \in E^+$, then

$$0 \le (ST)x = S(Tx) \le S(\lambda x) = \lambda S(x) \quad \text{and} \quad 0 \le (ST)(x) = S(Tx) \le \lambda Tx\,,$$

and so $0 \le (ST)x \le \lambda(Sx \wedge Tx) = \lambda(S \wedge T)x = 0$. This implies $ST = 0$.

Now let $T \in U_{\mathcal{Z}(E)}$, i.e., $\|T\| \le 1$. To finish the proof, we must show that $|T| \le I$. For this, it suffices to establish that $|T| \le \alpha I$ holds for all $\alpha > 1$. If this is not the case, then $(|T| - \alpha I)^+ > 0$ holds in the Riesz space $\mathcal{Z}(E)$ for some $\alpha > 1$. So, there exists some $x \in E^+$ such that $y = (|T| - \alpha I)^+ x > 0$. But then, from the above property (P), we see that

$$\begin{aligned}(|T| - \alpha I)y &= [(|T| - \alpha I)^+ - (|T| - \alpha I)^-]y \\ &= (|T| - \alpha I)^+ y - [(|T| - \alpha I)^-(|T| - \alpha I)^+]x \\ &= (|T| - \alpha I)^+ y \ge 0\,,\end{aligned}$$

and so $|Ty| = |T|y \ge \alpha y > 0$. This implies $\|Ty\| \ge \alpha\|y\| > 0$ and so $\|T\| \ge \alpha > 1$, which is impossible. Hence, $|T| \le \alpha I$ holds for all $\alpha > 1$, and consequently $|T| \le I$.

(2) Now assume that E is a Dedekind complete Banach lattice. Clearly $\mathcal{Z}(E)$ is an ideal of E. As a matter of fact $\mathcal{Z}(E) = E_I$, the principal ideal generated by I in $\mathcal{L}_r(E)$. To see that $E_I = B_I$, let $0 \le T \in B_I$. Then, by Problem 1.2.6, we have $T \wedge nI \uparrow T$. Next, suppose $x \wedge y = 0$. Then $x \wedge \big[(T \wedge nI)y\big] = 0$ for each n, and so (by the order continuity of the lattice operations) we get $x \wedge Ty = 0$. Thus, T is an orthomorphism and so (by Theorem 3.29) $T \in \mathcal{Z}(E)$. Therefore, $\mathcal{Z}(E) = B_I$ holds true in this case. ∎

Problem 3.3.5. *If μ is a finite measure, then show—with appropriate identifications—that $\mathcal{Z}(L_p(\mu)) = L_\infty(\mu)$ for each $1 \le p \le \infty$.*

Solution: Note that if $u = \mathbf{1}$ (the constant function one), then the principal ideal E_u coincides with $L_\infty(\mu)$, i.e. $E_u = L_\infty(\mu)$. Moreover, $\overline{E}_u = L_p(\mu)$. Thus, by Theorem 3.34, $\mathcal{Z}\big(L_p(\mu)\big)$ can be identified with $L_\infty(\mu)$, where as in the proof of Theorem 3.34 every function $f \in L_\infty(\mu)$ is identified with the multiplication operator M_f defined by $M_f(g) = fg$ for each $g \in L_1(\mu)$. (For a more general result see Problem 5.2.7.) ∎

Problem 3.3.6. *Prove the second part of the Maeda–Ogasawara–Vulikh Representation Theorem* 3.35 *by establishing the following general result for Riesz spaces: If F is an order dense Riesz subspace of an Archimedean Riesz space E and F is Dedekind complete in its own right, then F is an ideal in E.*

Solution: Let F be an order dense Riesz subspace of an Archimedean Riesz space E such that F is Dedekind complete in its own right.

We claim that for each $x \in E^+$ the non-empty subset $D_x = \{y \in F\colon\ 0 \leq y \leq x\}$ of F satisfies $D_x \uparrow x$ in E. To see this, fix some $x \in E^+$ and assume that $y \leq z$ holds in E for all $y \in D_x$ and some $z \in E$. Clearly, $y \leq x \wedge z$ for all $y \in D_x$. Now suppose by way of contradiction that $x - x \wedge z > 0$. Pick some $u \in F$ satisfying $0 < u \leq x - x \wedge z$ and note that $u \in D_x$ and that $2u = u + u \leq (x - x \wedge z) + x \wedge z = x$ implies $2u \in D_x$. Now an easy inductive argument shows that $nu \leq x$ holds for all n, which contradicts the Archimedeaness of E. Thus, $x - x \wedge z = 0$ or $x = x \wedge z \leq z$. This shows that $D_x \uparrow x$ for each $x \in E^+$.

Next, we claim that if a non-empty subset A of F satisfies $s = \sup A$ in F, then $s = \sup A$ also holds in E. To see this, assume that $s = \sup A$ holds in F and let $a \leq t$ hold in E for all $a \in A$ and some $t \in E$. If $s \wedge t < s$ is true, then there exists some $v \in F$ satisfying $0 < v \leq s - s \wedge t$, and so for each $a \in A$ we have $a \leq s \wedge t \leq s - v < s$. Since $s - v \in F$, the latter contradicts the fact that $s = \sup A$ holds in F. Hence, $s = s \wedge t \leq t$ holds in E, proving that $s = \sup A$ is true in E.

Our next task is to show that F is an ideal of E. To see this, assume $0 \leq x \leq y$ holds with $x \in E$ and $y \in F$. By the first part of the proof, we know that $D_x \uparrow x$. Since $D_x \uparrow \leq y$ holds in F and F is a Dedekind complete Riesz space in its own right, there exists some $w \in F$ such that $D_x \uparrow w$ holds in F. Now by the second part of the above proof, we also have $D_x \uparrow w$ in E. This implies $x = w \in F$, so that F is an ideal of E.

Now consider the Maeda–Ogasawara–Vulikh Representation Theorem. We know that E is an order Riesz subspace of $C^\infty(Q)$, the order unit e of E corresponds to the constant function $\mathbf{1}$ on Q, and E is Dedekind complete. By the property of Riesz spaces proven above, E is an ideal of $C^\infty(Q)$ and, consequently, $C(Q) \subseteq E$. Next, notice that $C(Q)$ is an order dense Riesz subspace of $C^\infty(Q)$. Also, by virtue of Q being extremely disconnected, $C(Q)$ is Dedekind complete in its own right. So, by the above discussion, $C(Q)$ is an ideal of $C^\infty(Q)$, and clearly contains E. This shows that $E = C(Q)$. ∎

Problem 3.3.7. *Show that a central operator on a Banach lattice commutes with all band projections.*

Solution: Let $T\colon E \to E$ be a central operator on a Banach lattice and let $P\colon E \to E$ be a band projection. This means that $B = P(E)$ is a projection band, i.e., $B \oplus B^{\mathrm{d}} = E$. From Theorem 3.29 we know that $T(B) \subseteq B$ and $T(B^{\mathrm{d}}) \subseteq B^{\mathrm{d}}$. Now fix some $x \in E$ and write $x = y \oplus z$ with $y \in B$ and $z \in B^{\mathrm{d}}$. Then $Ty \in B$ and $Tz \in B^{\mathrm{d}}$, and so

$$\begin{aligned} (TP)x &= T(Px) = Ty = P(Ty) = P(Ty) + P(Tz) \\ &= P(Ty + Tz) = P(Tx) = (PT)x \end{aligned}$$

for each $x \in X$. Consequently, $TP = PT$, and therefore T commutes with every band projection. ∎

Problem 3.3.8. *This problem justifies the name "diagonal projection" for the band projection $P_{\mathcal{Z}}$ of $\mathcal{L}_r(E)$ onto the center of the Banach lattice E. Let $E = \mathbb{R}^n$, the Euclidean n-dimensional (Riesz) space. Then $\mathcal{L}(E) = \mathcal{L}_r(E)$, the vector space of all $n \times n$ matrices with real entries. If D is a* ***diagonal matrix*** *with diagonal entries, then we write $D = \operatorname{diag}(d_1, \ldots, d_n)$. Establish the following properties.*

(a) *If $A = [a_{ij}]$ and $B = [b_{ij}]$ are two $n \times n$ matrices, then $A \geq B$ holds in $\mathcal{L}_r(E)$ if and only if $a_{ij} \geq b_{ij}$ for all i and j.*

(b) *If $A = [a_{ij}]$ and $B = [b_{ij}]$ are two $n \times n$ matrices, then*
$$|A| = [\,|a_{ij}|\,], \quad A \vee B = [a_{ij} \vee b_{ij}] \quad \text{and} \quad A \wedge B = [a_{ij} \wedge b_{ij}]\,.$$

(c) *The center of E consists of all diagonal matrices.*

(d) *If $A = [a_{ij}]$ is an $n \times n$ matrix, then $P_{\mathcal{Z}}(A) = \operatorname{diag}(a_{11}, a_{22}, \ldots, a_{nn})$.*

Solution: (a) It suffices to show that an $n \times n$ matrix $A = [a_{ij}]$ defines a positive operator on $\mathbb{R}^n$ if and only if $a_{ij} \geq 0$ holds for each i and j. It should be clear that if A has non-negative entries, then it defines a positive operator on $\mathbb{R}^n$. For the converse, assume that A defines a positive operator on $\mathbb{R}^n$. Notice that this is equivalent to saying that $\sum_{j=1}^n a_{ij}x_j \geq 0$ for all $x_1, \ldots, x_n \geq 0$ and every $i = 1, \ldots, n$. Letting $x_j = 1$ and $x_k = 0$ for $k \neq j$, it yields $a_{ij} \geq 0$ for all i and j.

(b) We establish the formula for the absolute value only. So, let $A = [a_{ij}]$ be an $n \times n$ matrix and put $B = [|a_{ij}|]$. From part (a), it follows that $\pm A \leq B$. Now assume that another matrix $M = [m_{ij}]$ satisfies $A \leq M$ and $-A \leq M$. By part (a), we get $\pm a_{ij} \leq m_{ij}$ or $|a_{ij}| \leq m_{ij}$ for all i and j. This implies $B \leq M$, and thus $|A| = A \vee (-A) = B$.

(c) Let $D = \operatorname{diag}(d_1, \ldots, d_n)$ be a diagonal matrix and put $\lambda = \max_{1 \leq i \leq n} |d_i|$. Then
$$|Dx| = \big|(d_1x_1, \ldots, d_nx_n)\big| \leq \lambda\big(|x_1|, \ldots, |x_n|\big) = \lambda|x|$$
holds for each $x \in \mathbb{R}^n$. This shows that D is a central operator.

For the converse, assume that an $n \times n$ matrix $A = [a_{ij}]$ defines a central operator. This means that there exists some $\lambda > 0$ such that $|Ax| \leq \lambda|x|$ holds for all $x \in \mathbb{R}^n$. That is, we have $\big|\sum_{j=1}^n a_{ij}x_j\big| \leq \lambda|x_i|$ for all $x_1, \ldots, x_n \in \mathbb{R}^n$ and all $i = 1, \ldots, n$.

Now assume that $i \neq k$. Letting $x_k = 1$ and $x_m = 0$ for $m \neq k$, we see that
$$|a_{ik}| = \Big|\sum_{j=1}^n a_{ij}x_j\Big| \leq \lambda|x_i| = 0\,.$$

Thus, $a_{ik} = 0$ holds for all $i \neq k$, and this shows that A is a diagonal operator. Therefore, $\mathcal{Z}(\mathbb{R}^n)$ consists of all diagonal matrices.

(d) If $A = [a_{ij}]$, let $D = \operatorname{diag}(a_{11}, \ldots, a_{nn})$ and $B = A - D$. Then $D \in \mathcal{Z}(\mathbb{R}^n)$, $|B| \wedge I = 0$ (i.e., $B \in I^{\mathrm{d}}$), and $A = D + B$. This guarantees that $P_{\mathcal{Z}}(A) = D = \operatorname{diag}(a_{11}, \ldots, a_{nn})$. ∎

Problem 3.3.9. *Let E be a Dedekind complete Banach lattice. According to the Maeda–Ogasawara–Vulikh Theorem 3.35 we can consider E as an order dense Riesz subspace of $C^\infty(Q)$, where Q is the Stone space of E. Show that in this case $\mathcal{Z}(E) = C(Q)$.*

Solution: Observe first that since E is Dedekind complete, E is an ideal of $C^\infty(Q)$. To see this, assume that $0 \le x \le y$ holds in $C^\infty(Q)$ with $y \in E$. Since E is order dense in $C^\infty(Q)$, the set $D = \{z \in E\colon\ 0 \le z \le x\}$ satisfies $D \uparrow x$ in $C^\infty(Q)$. Since E is Dedekind complete and $z \le y$ holds in E for each $z \in D$, it follows that there exists some $u \in E$ such that $D \uparrow u$ holds in E. Now observe that the order denseness of E in $C^\infty(Q)$ implies that $D \uparrow u$ also holds in $C^\infty(Q)$. Hence, $x = u \in E$ must be the case, and so E is an ideal of $C^\infty(Q)$.

Fix $h \in C(Q)$ and consider the multiplication operator $T\colon C^\infty(Q) \to C^\infty(Q)$. If $x \in E$, then from the inequality $|Tx| = |h||x| \le \|h\|_\infty |x|$ and the fact that E is an ideal of $C^\infty(Q)$, it follows that $Tx \in E$. Thus, for every $h \in C(Q)$, the multiplication operator defined by h is a central operator on E.

For the converse, assume that $T\colon E \to E$ is a central operator. By the first part of Theorem 3.36 there exists some function $h \in C^\infty(Q)$ such that $Tx = hx$ holds for each $x \in E$. We shall finish the proof by showing that $h \in C(\Omega)$.

To see this, assume by way of contradiction that $h \notin C(\Omega)$. This means that for each n the open set $V = \big\{q \in Q\colon\ |h(q)| > n\big\}$ is non-empty. Thus, the clopen set $W = \overline{V}$ is non-empty, and so $0 < \chi_W \in C^\infty(Q)$. Since E is order dense in $C^\infty(Q)$, there exists some $y \in E$ such that $0 < x \le \chi_W$. This implies $|Tx| = |hx| \ge nx > 0$ and so $\|Tx\| \ge n\|x\|$ for each n. Hence $\|T\| \ge n$ for each $n \in \mathbb{N}$, which is impossible. This contradiction guarantees $h \in C(Q)$, as desired. ∎

Problem 3.3.10. *Recall that the **principal band** B_x generated by a vector x in a Riesz space E is the smallest band that contains x. From Problem 1.2.6 we know that $B_x = \{y \in E\colon\ |y| \wedge n|x| \uparrow |y|\}$. If E is Dedekind complete, then B_x is a projection band, i.e., $B_x \oplus B_x^{\mathrm{d}} = E$. We denote the band projection of E onto B_x by P_x.*

Let E be a Dedekind complete Banach lattice and let $T, Z\colon E \to E$ be two disjoint regular operators (i.e., $|T| \wedge |Z| = 0$) with $Z \ne 0$ and $Z \in \mathcal{Z}(E)$. Show that for each $0 < x \in E$ and each $0 < \epsilon < \|Z\|$ there exists a non-zero component a of x such that $P_a(|T|a) \le \epsilon a$.

Solution: A glance at Problem 3.3.4 guarantees that we can suppose that T and Z are positive operators with Z satisfying $0 < Z \le I$. Fix $0 < \epsilon < \|Z\|$ and let $x > 0$. Replacing E by B_x, we can assume that x is a weak unit. Let Q be the Stone space of E and identify E with the ideal $C(Q)$ of $C^\infty(Q)$, where x corresponds to the constant function $\mathbf{1}$ on Q. By Theorem 3.36, there exists some $h \in C(Q)$ satisfying $Z(z) = hz$ for all $z \in E$ and $\|Z\| = \|h\|_\infty$. Now notice that the set $V = \overline{\{q \in Q\colon\ h(q) > \epsilon\}}$ is a clopen subset of Q and so $y = \chi_V \in E$. From Theorem 1.17, we know that

$$0 = (T \wedge Z)y = \inf\{Tz + Z(y-z)\colon\ z \in E^+ \text{ and } z \wedge (y-z) = 0\}.$$

This implies that there exist a component χ_W of $y = \chi_V$ (where W is a clopen subset of Q) and some $q_0 \in V$ such that

$$T(\chi_W)(q_0) + h(q_0)\chi_{V\setminus W}(q_0) < \epsilon.$$

In view of $h(q) \geq \epsilon$ for each $q \in V$, we see that $q_0 \in W$. From the inequality $T(\chi_W)(q_0) < \epsilon$, we infer that there exists some clopen set $W_1 \subseteq W$ satisfying $0 \leq T(\chi_{W_1})(q) \leq T(\chi_W)(q) < \epsilon$ for each $q \in W_1$. If we let $a = \chi_{W_1} \neq 0$, then a is a component of $\mathbf{1}$ (which has been identified with x) and clearly $P_a(Ta) = T(\chi_{W_1})\chi_{W_1} \leq \epsilon\chi_{W_1} = \epsilon a$. ∎

3.4. The Predual of a Principal Ideal

Problem 3.4.1. *Establish the following useful result:* If Y and Z are closed subspaces of a normed space X, one of which is finite dimensional, then $Y + Z$ is a closed subspace of X.

Solution: Assume that Y is a closed subspace of X, Z is a finite dimensional subspace of X, and let $\pi\colon X \to X/Y$ denote the quotient map, i.e., $\pi(x) = [x]$ for each $x \in X$. Then $\pi(Z)$ is a finite dimensional subspace of the normed space X/Y, and so $\pi(Z)$ is a closed subspace of X/Y. Now note that $Y + Z = \pi^{-1}(\pi(Z))$. Since π is continuous, the latter guarantees that $Y + Z$ is a closed subspace of X. ∎

Problem 3.4.2. *Let E be a Riesz space and let J be an ideal in E. Establish the following:*

(a) *The set $C = \{[x]\colon\ x \in E^+\}$ is a (convex) cone in E/J.*

(b) *The vector space E/J ordered by the cone C is a Riesz space.*

(c) *The quotient map $x \mapsto [x]$ is a lattice homomorphism from E onto E/J.*

Solution: (a) It should be clear that $C + C \subseteq C$ and $\alpha C \subseteq C$ for all $\alpha \geq 0$. Now assume that $[x] \in C \cap (-C)$. This implies that there exist $x_1 \geq 0$ and $x_2 \geq 0$ such that $x_1 \in [x]$ and $-x_2 \in [x]$. Thus, $x_1 - x \in J$ and $-x_2 - x \in J$, and so $x_1 + x_2 = (x_1 - x) - (-x_2 - x) \in J$. The latter, in view of $0 \leq x_1 \leq x_1 + x_2$, implies $[x] = [x_1] = 0$. This shows that C is a cone in E/J.

The reader should notice that the inequality $[x] \leq [y]$ holds in E/J if and only if there exist $x_1 \in [x]$ and $y_1 \in [y]$ such that $x_1 \leq y_1$ in E. This means that whenever $[x] \leq [y]$ holds in E/J, we can assume without loss of generality that $x \leq y$.

(b) We claim that the lattice operations in E/J satisfy

$$[x] \vee [y] = [x \vee y], \quad [x] \wedge [y] = [x \wedge y], \quad \text{and} \quad |[x]| = [|x|].$$

In order to establish these formulas, it suffices to show that $[x]^+ = [x^+]$ for each $x \in E$. So, let $x \in E$. From $x^+ \geq x$ and $x^+ \geq 0$, it follows that $[x^+] \geq [x]$ and $[x^+] \geq 0$ are both true. To see that $[x^+]$ is the least upper bound of $[x]$ and 0, assume that some $[y] \in E/J$ satisfies $[y] \geq [x]$ and $[y] \geq 0$.

Pick $y_1 \in [y]$, $x_1 \in [x]$, and $z \in [y]$ such that $y_1 \geq x_1$ and $z \geq 0$. This yields

$$z + (y_1 - z) + (x - x_1) = x + (y_1 - x_1) \geq x\,,$$

and by letting $u = (y_1 - z) + (x - x_1) \in J$ we have $x \leq z + u$. Consequently, $x^+ \leq (z+u)^+ \leq z^+ + u^+ = z + u^+$. Taking into account that $u^+ \in J$ (since J is an ideal), we see that $[x^+] \leq [z + u^+] = [z] = [y]$. This shows that $[x]^+ = [x^+]$, and the validity of the formulas has been established.

(c) The formula $[x] \vee [y] = [x \vee y]$ established in (b) demonstrates that the quotient map is indeed a lattice homomorphism. ∎

Problem 3.4.3. *Let E be a normed Riesz space and let J be a closed ideal in E. Establish the following.*

(a) *The quotient norm is a lattice norm—and so E/J is a normed Riesz space.*

(b) *If E is a Banach lattice, then E/J is also a Banach lattice.*

Solution: (a) Assume $|[x]| \leq |[y]|$ or, equivalently, $[|x|] \leq [|y|]$. We must show $|||[x]||| \leq |||[y]|||$, where $|||\cdot|||$ denotes the quotient norm of E/J.

Replacing $|x|$ by $|x| \wedge |y|$ (and using the fact that the quotient map is a lattice homomorphism), we see that it suffices to show that $0 \leq x \leq y$ in E implies $|||[x]||| \leq |||[y]|||$. To this end, assume $0 \leq x \leq y$ in E.

Fix some $z \in [y]$ and let $u = |z| \wedge y$. From $|y - u| = |y - |z| \wedge y| \leq |y - z| \in J$, we get $u \in [y]$ and $0 \leq u \leq y$. Moreover, from $0 \leq x - u \wedge x = (x-u)^+ \leq (y-u)^+ \in J$, we obtain $u \wedge x \in [x]$. Therefore,

$$|||[x]||| \leq \|u \wedge x\| \leq \|u\| = |||z| \wedge y\| \leq \|\,|z|\,\| = \|z\|\,,$$

holds for all $z \in [y]$. So, $|||[x]||| \leq \inf_{z \in [y]} \|z\| = |||[y]|||$, and we have established that $|||\cdot|||$ is a lattice norm.

(b) This follows immediately from (a) and part (d) of Theorem 1.11. ∎

Problem 3.4.4. *Show that the norm completion of a normed Riesz space E is a Banach lattice containing E as a Riesz subspace.*

Solution: Let E be a normed Riesz space and let $\hat{E}$ denotes its norm completion. Also, let $C = \overline{E^+}$ be the norm closure of E^+ in $\hat{E}$. It should be clear that $C + C \subseteq C$ and $\alpha C \subseteq C$ for all $\alpha \geq 0$. Now assume that $x \in C \cap (-C)$. Pick two sequences $\{x_n\}$ and $\{y_n\}$ in E^+ satisfying $x_n \to x$ and $y_n \to -x$. Then from the relation $0 \leq x_n \leq x_n + y_n \to x - x = 0$ it follows that $x_n \to 0$. Thus, $x = 0$, and so C is a (convex) cone in $\hat{E}$. We shall denote the order induced by the cone C on $\hat{E}$ by $\leq_C$. That is, $x \leq_C y$ holds in $\hat{E}$ if and only if $y - x \in C$. Next, we shall show that $(\hat{E}, C)$ is a Riesz space.

Start by observing that from the inequality $|x^+ - y^+| \le |x - y|$ valid for all $x, y \in E$, it follows that the function $p\colon E \to E \subseteq \hat{E}$, defined by $p(x) = x^+$, is uniformly continuous. Therefore, p has a (uniformly) continuous extension to all of $\hat{E}$, which we shall denote by p again. We claim that $p(x) = x^+ = x \vee_C 0$ holds in E for each $x \in \hat{E}$. This will establish that $(\hat{E}, C)$ is a Riesz space.

To see this, fix $x \in \hat{E}$. Pick a sequence $\{x_n\} \subseteq E$ such that $x_n \to x$. Then we have $x_n^+ = p(x_n) \to p(x)$, and so $p(x) \ge_C 0$. On the other hand, from $x_n \le x_n^+$, it easily follows (by taking norm limits) that $x \le_C p(x)$. That is, $p(x)$ is a C-upper bound of x and 0. To establish that $p(x)$ is the least C-upper bound of x and 0, assume that some $z \in C$ satisfies $x \le_C z$. Choose sequences $\{u_n\}$ and $\{z_n\}$ in E^+ satisfying $u_n \to z - x$ and $z_n \to z$. Clearly, $z_n - u_n \to z - (z - x) = x$. Now, from $z_n - u_n \le z_n = z_n^+$, it follows that $(z_n - u_n)^+ \le z_n$ for each n, and so by taking norm limits, we see that $p(x) = \lim (z_n - u_n)^+ \le_C \lim z_n = z$. This shows that $x^+ = p(x)$ holds in $(\hat{E}, C)$, and so $(\hat{E}, C)$ is a Riesz space. Notice also the following: If $x, y \in \hat{E}$ and two sequences $\{x_n\}$ and $\{y_n\}$ in E satisfy $x_n \to x$ and $y_n \to y$, then

$$x \vee_C y = \lim_{n\to\infty} x_n \vee y_n \quad \text{and} \quad x \wedge_C y = \lim_{n\to\infty} x_n \wedge y_n\,. \tag{$\star$}$$

From the above formulas, it follows immediately that E is a Riesz subspace of $\hat{E}$.

To finish the proof, we must show that the norm of $\hat{E}$ is a lattice norm. To this end, let $|v|_C \le_C |w|_C$ in $\hat{E}$. Choose two sequences $\{v_n\}$ and $\{w_n\}$ in E such that $v_n \to v$ and $w_n \to w$. From $(\star)$, we get $|v_n| \to |v|_C$ and $|w_n| \to |w|_C$. Using $(\star)$ once more, we see that $|v_n| \wedge |w_n| \to |v|_C \wedge_C |w|_C = |v|_C$. Finally, using that $|v_n| \wedge |w_n| \le |w_n|$ implies $\|\,|v_n| \wedge |w_n|\,\| \le \|\,|w_n|\,\|$, by taking norm limits, we obtain $\|\,|v|_C\,\| \le \|\,|w|_C\,\|$. This completes the proof that $(\hat{E}, C)$ is a Banach lattice. ∎

Problem 3.4.5. *Let E be an arbitrary* ***M-space****, that is, E is a normed Riesz space such that $\|x \vee y\| = \max\{\|x\|, \|y\|\}$ holds for all $x, y \in E^+$. Show the following.*

(a) *The norm completion of E is an AM-space.*

(b) *If E has a unit e and the closed unit ball of E coincides with the order interval $[-e, e]$, then the norm completion of E is an AM-space with unit e.*

Solution: (a) Assume that E is an M-space, and let $\hat{E}$ denote its norm completion. By Problem 3.4.4, $\hat{E}$ is a Banach lattice containing E as a Riesz subspace. Now assume that $x \wedge y = 0$ holds in $\hat{E}$. Pick two sequences $\{x_n\}$ and $\{y_n\}$ in E satisfying $x_n \to x$ and $y_n \to y$. Since $(x_n - x_n \wedge y_n) \wedge (y_n - x_n \wedge y_n) = 0$ for each n and since E is an M-space, we conclude that

$$\|(x_n - x_n \wedge y_n) \vee (y_n - x_n \wedge y_n)\| = \max\bigl\{\|x_n - x_n \wedge y_n\|, \|y_n - x_n \wedge y_n\|\bigr\}\,. \tag{$\star\star$}$$

From $x_n \wedge y_n \to x \wedge y = 0$ and $(\star\star)$, we get $\|x \vee y\| = \max\{\|x\|, \|y\|\}$. This shows that $\hat{E}$ is an AM-space.

(b) Assume that E has a unit e, i.e., $U_E = [-e, e]$. Also, let

$$[\![-e, e]\!] = \bigl\{x \in \hat{E}\colon\ -e \le x \le e\bigr\}\,.$$

Clearly, $[\![-e,e]\!] \subseteq U_{\hat{E}}$. For the reverse inclusion, let $0 \neq x \in U_{\hat{E}}$, i.e., $\|x\| \leq 1$. Pick a sequence $\{x_n\}$ in E consisting of non-zero vectors such that $x_n \to x$. Let $y_n = \frac{x_n}{\|x_n\|} \in E$ and note that $\|y_n\| = 1$ for each n. This implies that $y_n \in [\![-e,e]\!]$ and $y_n \to \frac{x}{\|x\|}$. Since the order interval $[\![-e,e]\!]$ is norm closed, we infer that $-e \leq \frac{x}{\|x\|} \leq e$. That is, $-e \leq -\|x\|e \leq x \leq \|x\|e \leq e$. Hence, $U_{\hat{E}} \subseteq [\![-e,e]\!]$, and so $U_{\hat{E}} = [\![-e,e]\!]$. Therefore, $\hat{E}$ is an AM-space with unit e. ∎

Problem 3.4.6. *Let E be an arbitrary L-**space**, that is, E is a normed Riesz space such that $\|x+y\| = \|x\| + \|y\|$ holds for all $x, y \in E$ with $x \wedge y = 0$. Show that the norm completion of E is an AL-space.*

Solution: Assume that E is an L-space, and let $\hat{E}$ denote its norm completion. By Problem 3.4.4, $\hat{E}$ is a Banach lattice containing E as a Riesz subspace. Now assume that $x \wedge y = 0$ holds in $\hat{E}$. Pick two sequences $\{x_n\}$ and $\{y_n\}$ in E satisfying $x_n \to x$ and $y_n \to y$. Since $(x_n - x_n \wedge y_n) \wedge (y_n - x_n \wedge y_n) = 0$ holds for each n and since E is an L-space, we conclude that

$$\|(x_n - x_n \wedge y_n) + (y_n - x_n \wedge y_n)\| = \|x_n - x_n \wedge y_n\| + \|y_n - x_n \wedge y_n\| . \qquad (\star\star\star)$$

From $x_n \wedge y_n \to x \wedge y = 0$ and $(\star\star\star)$, we see that $\|x+y\| = \|x\| + \|y\|$. This shows that $\hat{E}$ is an AL-space. ∎

Problem 3.4.7. *Let X be a normed space and let $\hat{X}$ be its norm completion. For each $f \in X^*$, let $\hat{f}$ denote the unique continuous (linear) extension of f to $\hat{X}$. Show the following.*

(a) *The mapping $f \mapsto \hat{f}$ is a surjective linear isometry from X^* onto $\hat{X}^*$ (and so, subject to this linear isometry, $X^* = \hat{X}^*$).*

(b) *If X is a normed Riesz space, then $f \mapsto \hat{f}$ is a surjective lattice isometry from X^* onto $\hat{X}^*$.*

Solution: (a) Straightforward.

(b) To show that $f \mapsto \hat{f}$ is a lattice isometry, we need to verify only that this mapping is a lattice homomorphism. To see this, fix $f \in X^*$ and let $x \in X^+$. Then the modulus of f in X^* is given by

$$|f|(x) = \sup\{|f(y)|\colon\ y \in X \text{ and } |y| \leq x\}.$$

Similarly, the modulus of $\hat{f}$ in $\hat{X}^*$ satisfies

$$|\hat{f}|(x) = \sup\{|\hat{f}(y)|\colon\ y \in \hat{X} \text{ and } |y| \leq x\}.$$

Clearly, $|f|(x) \leq |\hat{f}|(x)$. For the reverse inequality, pick $y \in \hat{X}$ such that $|y| \leq x$. Choose a sequence $\{y_n\}$ in X satisfying $y_n \to y$, and then let $z_n = [y_n \vee (-x)] \wedge x \in X$ for each n. Clearly, $z_n \to y$ and $|z_n| \leq x$ holds for each n. This implies

$$|\hat{f}(y)| = \lim_{n\to\infty} |\hat{f}(z_n)| = \lim_{n\to\infty} |f(z_n)| \leq |f|(x)$$

for all $y \in X$ with $|y| \leq x$. It follows that

$$|\hat{f}|(x) = \sup\{|\hat{f}(y)|\colon\ y \in \hat{X} \text{ and } |y| \leq x\} \leq |f|(x),$$

and so $|f|(x) = |\hat{f}|(x)$ for all $x \in X^+$. The latter guarantees $\widehat{|f|} = |\hat{f}|$, which means that $f \mapsto \hat{f}$ is a lattice homomorphism. This completes the solution. ∎

Problem 3.4.8. *Let ρ be an arbitrary lattice seminorm on a Riesz space E, and define its* **null ideal** *by $N_\rho = \{x \in E\colon\ \rho(x) = 0\}$. Establish the following.*

(a) *N_ρ is indeed an ideal in E.*

(b) *The function $\hat{\rho}\colon E/N_\rho \to \mathbb{R}$ defined by $\hat{\rho}([x]) = \rho(x)$ is a lattice norm on E/N_ρ and $\hat{\rho}([x]) = \inf_{y\in[x]} \rho(y)$.*

(c) *If ρ is an M-seminorm (resp. an L-seminorm), then $\hat{\rho}$ is an M-norm (resp. an L-norm).*

Solution: (a) This is straightforward.

(b) First, let us establish that $\hat{\rho}$ is well defined. To this end, let $y \in [x]$, that is, $\rho(|x-y|) = 0$. Then from $|\rho(x) - \rho(y)| \le \rho(|x-y|) = 0$, we see that $\rho(x) = \rho(y)$, and so $\hat{\rho}$ is well defined.

Since $\rho(y) = \rho(x)$ holds for each $y \in [x]$, it follows that

$$\hat{\rho}([x]) = \rho(x) = \inf_{y\in[x]} \rho(y)\,.$$

To see that $\hat{\rho}$ is a lattice seminorm, assume $|[x]| \le |[y]|$, i.e., $[|x|] \le [|y|]$. We can suppose $|x| \le |y|$. This implies $\hat{\rho}([x]) = \rho(x) = \rho(|x|) \le \rho(|y|) = \rho(y) = \hat{\rho}([y])$.

Finally, to get that $\hat{\rho}$ is a lattice norm, notice that $\hat{\rho}([x]) = \rho(x) = 0$ implies $x \in N_\rho$ or $[x] = 0$.

(c) Assume first that ρ is a lattice M-seminorm, and let $[x]\wedge[y] = [x \wedge y] = 0$ in E/N_ρ. Replacing x by $x - x\wedge y$ and y by $y - x\wedge y$ and using that the quotient map is a lattice homomorphism, we can assume without loss of generality that $x \wedge y = 0$ holds in E. Now notice that

$$\hat{\rho}([x]\vee[y]) = \hat{\rho}([x\vee y]) = \rho(x\vee y) = \max\{\rho(x), \rho(y)\} = \max\{\hat{\rho}([x]), \hat{\rho}([y])\}\,.$$

This shows that $\hat{\rho}$ is an M-norm. If ρ is an L-seminorm, then a similar argument shows that $\hat{\rho}$ is an L-norm. ∎

Problem 3.4.9. *Complete the details of the proof of Lemma* 3.38.

Solution: Let us recall the background associated with Lemma 3.38. We consider a Banach lattice E and we fix a positive linear functional $0 < \phi \in E^*$.

As usual, E^*_ϕ denotes the ideal generated by $\phi \in E^*$, i.e.,

$$E^*_\phi = \big\{x^* \in E^*\colon\ \exists\,\lambda > 0 \text{ such that } |x^*| \le \lambda\phi\big\}\,.$$

The ideal E^*_ϕ with its $\|\cdot\|_\infty$-norm, defined by $\|x^*\|_\infty = \inf\{\lambda \ge 0\colon\ |x^*| \le \lambda\phi\}$, is an AM-space with unit ϕ.

The linear functional ϕ defines a lattice seminorm $\|\cdot\|_\phi$ on E via the formula

$$\|x\|_\phi = \phi(|x|)\,.$$

This is an L-seminorm in the sense that (besides being a lattice seminorm) it also satisfies $\|x+y\|_\phi = \|x\|_\phi + \|y\|_\phi$ for all $x, y \in E^+$. If we consider the **null ideal** of ϕ, i.e., the closed ideal $N_\phi = \{x \in E\colon\ \phi(|x|) = 0\}$, then the quotient seminorm induced on the quotient Riesz space E/N_ϕ by $\|\cdot\|_\phi$, i.e., the lattice seminorm

$$\|[x]\|_\phi = \inf_{y\in[x]} \|y\|_\phi = \phi\big(|x|\big)\,,$$

is in fact an L-norm. The norm completion of the normed Riesz space $(E/N_\phi, \|\cdot\|_\phi)$ is an AL-space, which we shall denote $E(\phi)$. Also, we shall denote the norm dual of $E(\phi)$ by $E(\phi)^*$. Keep in mind that $E(\phi)^*$ is an AM-space with unit e^* defined by the norm of $E(\phi)$, i.e., $e^*([x]) = \phi(x^+) - \phi(x^-)$ for each $x \in E$.

There is a natural mapping $\Phi\colon E^*_\phi \to E(\phi)^*$ or $x^* \mapsto \Phi_{x^*}$, defined by

$$\Phi_{x^*}([x]) = x^*(x)\,, \tag{$\star$}$$

for all $x^* \in E^*_\phi$ and all $x \in E$. To see that Φ is well defined, we argue as follows. Fix $x^* \in E^*_\phi$, and then select a scalar $\lambda > 0$ such that $|x^*| \le \lambda\phi$. Now assume that $y \in [x]$, i.e., $[y] = [x]$, or $\phi(|x-y|) = 0$. From

$$\big|x^*(x) - x^*(y)\big| \le |x^*|(|x-y|) \le \lambda\phi(|x-y|) = 0\,,$$

we see that $x^*(x) = x^*(y)$, and so ($\star$) is a well-defined linear functional on the quotient Riesz space E/N_ϕ. From

$$\big|\Phi_{x^*}([x])\big| = \big|x^*(x)\big| \le |x^*|(|x|) \le \lambda\phi(|x|) = \lambda\|[x]\|_\phi = \lambda e^*\big(|x|\big)\,, \tag{$\dagger$}$$

it follows that Φ_{x^*} is a continuous linear functional on E/N_ϕ. Therefore, Φ_{x^*} has a unique continuous linear extension to the norm completion $E(\phi)^*$ of E/N_ϕ, which we shall denote by Φ_{x^*} again. Lemma 3.38 now can be stated as follows.

- *The mapping $\Phi\colon E^*_\phi \to E(\phi)^*$ as defined by ($\star$) is a surjective lattice isometry.*

We shall verify the above claims by steps. It should be clear that Φ is a bounded (linear) operator.

STEP I: Φ *is a surjective operator.*

Let $\Psi \in E(\phi)^*$. So, there is some $\lambda > 0$ such that $|\Psi([x])| \le \lambda\|[x]\|_\phi = \lambda\phi(|x|)$ holds for each $x \in E$. Now if we consider the linear functional $x^*\colon E \to \mathbb{R}$, defined by $x^*(x) = \Psi([x])$, then from $|x^*(x)| \le \lambda\phi(|x|)$ for each $x \in E$, we see that $|x^*| \le \lambda\phi$ or $x^* \in E^*_\phi$. Moreover, note that $\Phi_{x^*} = \Psi$, and so Φ is surjective.

STEP II: Φ *is one-to-one.*

Assume that $\Phi_{x^*} = 0$. This means that $x^*(x) = 0$ for all $x \in E$, so that $x^* = 0$. Therefore, Φ is one-to-one.

STEP III: Φ *is an isometry.*

Fix $x^* \in E^*_\phi$. Then from

$$|x^*(x)| = |\Phi_{x^*}([x])| \le \|\Phi_{x^*}\| \cdot \|[x]\|_\phi = \|\Phi_{x^*}\|\phi(|x|)\,,$$

we get $|x^*| \le \|\Phi_{x^*}\|\phi$. This shows that $\|x^*\|_\infty = \inf\{\lambda \ge 0\colon\ |x^*| \le \lambda\phi\} \le \|\Phi_{x^*}\|$. On the other hand, from $|x^*| \le \|x^*\|_\infty\phi$, it follows that

$$|\Phi_{x^*}([x])| = |x^*(x)| \le |x^*|(|x|) \le \|x^*\|_\infty\phi(|x|) = \|x^*\|_\infty \cdot \|[x]\|_\phi$$

for each $x \in X$. This implies that $\|\Phi_{x^*}\| \le \|x^*\|_\infty$ is also true. Consequently, $\|\Phi_{x^*}\| = \|x^*\|_\infty$ holds for each $x^* \in E_\phi^*$, and so Φ is a linear isometry.

STEP IV: Φ *is a surjective lattice isomorphism.*

Let $x^* \ge 0$ and $[x] \ge 0$. If $0 \le y \in [x]$, then $\Phi_{x^*}([x]) = \Phi_{x^*}([y]) = x^*(y) \ge 0$. This implies $\Phi_{x^*} \ge 0$, and so Φ is a positive operator. On the other hand, if $\Phi_{x^*} \ge 0$, then for each $x \in E^+$ we have $x^*(x) = \Phi_{x^*}([x]) \ge 0$. That is, the inverse of Φ is also a positive operator. This guarantees that Φ is a (surjective) lattice isomorphism.[2] ∎

[2] Here we use the fact that a surjective one-to-one operator $T\colon F \to G$ between two Riesz spaces is a **lattice isomorphism** if and only if T and T^{-1} are both positive operators. To see this, let $x, y \in F$. From $x \le x \vee y$ and $y \le x \vee y$, we get $Tx \le T(x \vee y)$ and $Ty \le T(x \vee y)$ or $Tx \vee Ty \le T(x \vee y)$. The latter conclusion applied to T^{-1} yields $T^{-1}u \vee T^{-1}v \le T^{-1}(u \vee v)$ for all $u, v \in G$. Letting $u = Tx$ and $v = Ty$, we get $x \vee y \le T^{-1}(Tx \vee Ty)$, and by applying the positive operator T, we see that $T(x \vee y) \le Tx \vee Ty$. Hence, $T(x \vee y) = Tx \vee Ty$, and so T is a lattice isomorphism.

Chapter 4

Special Classes of Operators

4.1. Finite-rank Operators

Problem 4.1.1. *Let $f, f_1, \ldots, f_k$ be linear functionals on a vector space. Show that f lies in the linear span of $f_1, \ldots, f_k$ if and only if*

$$\bigcap_{i=1}^{k} \operatorname{Ker} f_i \subseteq \operatorname{Ker} f\,.$$

In particular, a collection $g_1, \ldots, g_n$ of linear functionals on a vector space is linearly dependent if and only if $\bigcap_{i\neq j} \operatorname{Ker} g_i \subseteq \operatorname{Ker} g_j$ for some j.

Solution: Let $f, f_1, \ldots, f_k$ be linear functionals defined on a real vector space V. (The arguments below show that the conclusion is true for vector spaces over any field.) Note first that if $f = \sum_{i=1}^{k} \lambda_i f_i$, then clearly $\bigcap_{i=1}^{k} \operatorname{Ker} f_i \subseteq \operatorname{Ker} f$.

For the converse, let $\bigcap_{i=1}^{k} \operatorname{Ker} f_i \subseteq \operatorname{Ker} f$. Consider the vector subspace of $\mathbb{R}^k$

$$W = \{(f_1(v), \ldots, f_k(v)) \in \mathbb{R}^k\colon\ v \in V\},$$

and then define the linear functional $\phi\colon W \to \mathbb{R}$ by $\phi(f_1(v), \ldots, f_k(v)) = f(v)$. Note that if $(f_1(v), \ldots, f_k(v)) = (f_1(u), \ldots, f_k(u))$, then $v - u \in \bigcap_{i=1}^{k} \operatorname{Ker} f_i$ and so by our hypothesis $f(v) = f(u)$, which shows that ϕ is well defined. The verification of the linearity of ϕ is straightforward. Now notice that ϕ extends linearly to all of $\mathbb{R}^k$. That is, there exist scalars $\lambda_1, \ldots, \lambda_k$ such that $\phi(t_1, \ldots, t_k) = \sum_{i=1}^{k} \lambda_i t_i$ holds for all $(t_1, \ldots, t_k) \in \mathbb{R}^k$. This easily implies $f = \sum_{i=1}^{k} \lambda_i f_i$. (Notice that this proof is applicable to vector spaces over any field.) ∎

Problem 4.1.2. *Show that an operator $T\colon V \to W$ between two vector spaces is a finite-rank operator if and only if there exist vectors $w_1, \ldots, w_n$ in W and linear functionals $g_1, \ldots, g_n$ on V such that $T = \sum_{i=1}^{n} g_i \otimes w_i$.*

Solution: Assume first that $T = \sum_{i=1}^{n} g_i \otimes w_i$. If U is the vector subspace of W generated by $w_1, \ldots, w_n$, then clearly $Tv \in U$ for all $v \in V$, i.e., $T(V) \subseteq U$. Since U is finite dimensional, $T(V)$ (the range of T) is likewise finite dimensional.

For the converse, assume that T is a finite-rank operator, i.e., the range $T(V)$ of T is finite dimensional. Fix a basis $w_1, \ldots, w_n$ of $T(V)$ and for each $v \in V$ write $Tv = \sum_{i=1}^{n} g_i(v) w_i$, where each g_i is a scalar-valued function defined on V. From

$$\begin{aligned}\sum_{i=1}^{n} g_i(\alpha v_1 + \beta v_2) w_i &= T(\alpha v_1 + \beta v_2) = \alpha T(v_1) + \beta T(v_2) \\ &= \sum_{i=1}^{n} [\alpha g_i(v_1) + \beta g_i(v_2)] w_i\,,\end{aligned}$$

it follows that $g_i(\alpha v_1 + \beta v_2) = \alpha g_i(v_1) + \beta g_i(v_2)$ for each i, all $v_1, v_2 \in V$, and all scalars α and β. Thus, each g_i is a linear functional and $T = \sum_{i=1}^{n} g_i \otimes w_i$. ∎

Problem 4.1.3. *Let $T\colon V \to W$ be an operator of rank k and let $w_1, \ldots, w_k$ be a basis of the range of T. Then there exist (uniquely determined) linear functionals $f_1, \ldots, f_k$ on V such that $T = \sum_{i=1}^{k} f_i \otimes w_i$. Show that $f_1, \ldots, f_k$ are linearly independent.*

Solution: The uniqueness of the linear functionals f_i should be immediate from the linear independence of the vectors $w_1, \ldots, w_k$. The existence and linearity of the f_i follow from Problem 4.1.2.

Now suppose by way of contradiction that $f_1, \ldots, f_n$ are not linearly independent. By rearranging, we can assume without loss of generality that there exist scalars $\lambda_2, \lambda_3, \ldots, \lambda_k$ such that $f_1 = \sum_{i=2}^{k} \lambda_i f_i$. This implies

$$\begin{aligned}T = \sum_{i=1}^{k} f_i \otimes w_i &= f_1 \otimes w_1 + \sum_{i=2}^{k} f_i \otimes w_i = \sum_{i=2}^{k} f_i \otimes (\lambda_i w_1) + \sum_{i=2}^{k} f_i \otimes w_i \\ &= \sum_{i=2}^{k} f_i \otimes (w_i + \lambda_i w_1)\,.\end{aligned}$$

This shows that rank of T is at most $k - 1$, a contradiction. Consequently, the linear functionals $f_1, \ldots, f_k$ are linearly independent. ∎

Problem 4.1.4. *Let $T\colon V \to V$ be a finite-rank operator on a vector space and assume that $T = \sum_{i=1}^{n} f_i \otimes v_i = \sum_{j=1}^{m} g_j \otimes u_j$ are two arbitrary representations of T, where the vectors v_i and u_j need not belong to the range of T. Show that*

$$\sum_{i=1}^{n} f_i(v_i) = \sum_{j=1}^{m} g_j(u_j)\,.$$

That is, the scalar $\operatorname{tr}(T) = \sum_{i=1}^{n} f_i(v_i)$ (called the ***trace*** *of T) is independent of the representation of T.*

Solution: Fix a basis $w_1, \ldots, w_k$ of the range of T and write $T = \sum_{r=1}^{k} h_r \otimes w_r$, where the linear functionals h_r are uniquely determined. It suffices to show that for an arbitrary representation $T = \sum_{i=1}^{n} f_i \otimes v_i$ we have $\sum_{i=1}^{n} f_i(v_i) = \sum_{r=1}^{k} g_r(w_r)$.

To this end, let $T = \sum_{i=1}^n f_i \otimes v_i$ be a representation of T. Assume first that the linear functionals $f_1, \ldots, f_n$ are linearly independent. Then for each $1 \le i \le n$ there exists some $x_i \in V$ such that $f_j(x_i) = \delta_{ij}$; see Problem 4.1.1. This implies $v_i \in T(V)$ for each i and so $v_i = \sum_{r=1}^k \lambda_r^i w_r$ for appropriate (uniquely determined) scalars λ_r^i. Hence, $T = \sum_{i=1}^n f_i \otimes v_i = \sum_{r=1}^k \big(\sum_{i=1}^n \lambda_r^i f_i\big) \otimes w_r$, and from this we infer that $h_r = \sum_{i=1}^n \lambda_r^i f_i$ for each $1 \le r \le k$. Consequently,

$$\sum_{i=1}^n f_i(v_i) = \sum_{i=1}^n f_i\Big(\sum_{r=1}^k \lambda_r^i w_r\Big) = \sum_{r=1}^k \Big(\sum_{i=1}^n \lambda_r^i f_i\Big)(w_r) = \sum_{r=1}^k h_r(w_r).$$

Next, suppose that $f_1, \ldots, f_n$ are linearly dependent. Then there exist linearly independent linear functionals $\theta_1, \ldots, \theta_\ell$ such that $f_i = \sum_{\nu=1}^\ell \mu_\nu^i \theta_\nu$ for appropriate scalars μ_ν^i. This implies $\sum_{i=1}^n f_i \otimes v_i = \sum_{\nu=1}^\ell \theta_\nu \otimes \big(\sum_{i=1}^n \mu_\nu^i v_i\big)$. By the preceding case, we have

$$\sum_{r=1}^k h_r(w_r) = \sum_{\nu=1}^\ell \theta_\nu\Big(\sum_{i=1}^n \mu_\nu^i v_i\Big) = \sum_{i=1}^n \Big(\sum_{\nu=1}^\ell \mu_\nu^i \theta_\nu\Big)(v_i) = \sum_{i=1}^n f_i(v_i),$$

and the solution of the problem is complete. ∎

Problem 4.1.5. *Show that the trace of a finite-rank operator on a Banach space is similarity invariant. That is, show that if $T\colon X \to X$ is a finite-rank operator on a Banach space, then $\operatorname{tr}(T) = \operatorname{tr}(S^{-1}TS)$ for each invertible bounded operator $S\colon X \to X$.*

Solution: Let $T\colon V \to V$ be a finite-rank operator on a vector space and let $T = \sum_{i=1}^n f_i \otimes u_i$ be a representation of T. If $S\colon V \to V$ is an invertible operator, then for each $v \in V$ we have

$$\begin{aligned} [S^{-1}TS]v &= S^{-1}\big(T(Sv)\big) = S^{-1}\Big(\sum_{i=1}^n f_i(Sv)u_i\Big) = \sum_{i=1}^n f_i(Sv)(S^{-1}u_i) \\ &= \Big[\sum_{i=1}^n (f_i \circ S) \otimes (S^{-1}u_i)\Big]v. \end{aligned}$$

This shows that $S^{-1}TS = \sum_{i=1}^n (f_i \circ S) \otimes (S^{-1}u_i)$. Now a glance at Problem 4.1.4 yields

$$\operatorname{tr}(S^{-1}TS) = \sum_{i=1}^n (f_i \circ S)(S^{-1}u_i) = \sum_{i=1}^n f_i\big(S(S^{-1}u_i)\big) = \sum_{i=1}^n f_i(u_i) = \operatorname{tr}(T),$$

as desired. ∎

Problem 4.1.6. *Let X be a Banach space and let $\{x_1^*, x_2^*, \ldots, x_n^*\} \subseteq X^*$ and $\{x_1, x_2, \ldots, x_n\} \subseteq X$ satisfy $x_i^*(x_j) = \delta_{ij}$ for all i and j. Show that the finite-rank operator $P = \sum_{i=1}^n x_i^* \otimes x_i$ has rank n, is a continuous projection, and has range the vector subspace generated by $\{x_1, x_2, \ldots, x_n\}$.*

Use this conclusion to show that every finite-dimensional subspace Y of X is ***complemented****, i.e., there exists a closed subspace Z of X satisfying $Y \cap Z = \{0\}$ and $Y \oplus Z = X$.*

Solution: For the first part notice that the condition $x_i^*(x_j) = \delta_{ij}$ for all i and j easily implies that both collections $\{x_1^*, x_2^*, \ldots, x_n^*\}$ and $\{x_1, x_2, \ldots, x_n\}$ are linearly independent. This implies that the range of the continuous operator $P = \sum_{i=1}^n x_i^* \otimes x_i$ is the n-dimensional vector subspace generated by the collection of vectors $\{x_1, x_2, \ldots, x_n\}$. Thus, P is a finite-rank operator of rank n. To see that P is a projection, note that for each $x \in X$ we have

$$\begin{aligned} P^2(x) &= P(Px) = P\Big(\sum_{i=1}^n x_i^*(x)x_i\Big) = \sum_{j=1}^n x_j^*\Big(\sum_{i=1}^n x_i^*(x)x_i\Big)x_j \\ &= \sum_{j=1}^n\sum_{i=1}^n x_i^*(x)x_j^*(x_i)x_j = \sum_{i=1}^n x_i^*(x)x_i = P(x)\,. \end{aligned}$$

Thus, $P^2 = P$, and so P is a projection.

For the second part, let $\{x_1, x_2, \ldots, x_n\}$ be a basis of the vector subspace Y of X. Pick linear functionals $x_1^*, x_2^*, \ldots, x_n^*$ on Y such that $x_i^*(x_j) = \delta_{ij}$ holds for all i and j. Since the norm topology of X induces the Euclidean topology on Y, each x_i^* is norm continuous. By the Hahn–Banach theorem, each x_i^* has a continuous extension to all of X, which we shall denote by x_i^* again. That is, we can assume that $x_i^* \in X^*$ for each i. Now consider the continuous projection $P = \sum_{i=1}^n x_i^* \otimes x_i$, and use the preceding conclusion to see that $P(X) = Y$. If we let $Z = (I - P)(X)$, then Z is a closed subspace of X satisfying $Y \cap Z = \{0\}$ and $Y \oplus Z = X$. ∎

Problem 4.1.7 (Grothendieck [**36**]). *For a Banach space X establish that the following statements are equivalent.*

(a) *X has the approximation property.*

(b) *For every Banach space Y the finite-rank operators $Y^* \otimes X$ are $\tau_c(Y, X)$-dense in $\mathcal{L}(Y, X)$.*

(c) *For every Banach space Y the finite-rank operators $X^* \otimes Y$ are $\tau_c(X, Y)$-dense in $\mathcal{L}(X, Y)$.*

Solution: (a) $\Longrightarrow$ (b) Fix a Banach space Y and let $T \in \mathcal{L}(Y, X)$. Also, let C be a compact subset of Y, and fix $\epsilon > 0$. Then $T(C)$ is a norm compact subset of X. Since X has the approximation property, there exists a finite-rank operator $S \in \mathcal{L}(X)$ such that $\|Sx - x\| < \epsilon$ holds for each $x \in T(C)$ or $\|STy - Ty\| < \epsilon$ for each $y \in C$. This (by taking into account that the operator $TS \in \mathcal{L}(Y, X)$ is a finite-rank operator) shows that T can be approximated uniformly on every compact subset of Y by finite-rank operators of $\mathcal{L}(Y, X)$. That is, $Y^* \otimes X$ is $\tau_c(Y, X)$-dense in $\mathcal{L}(Y, X)$.

(b) $\Longrightarrow$ (c) Let $T \in \mathcal{L}(X, Y)$. We can assume that $T \neq 0$. Fix a norm compact subset C of X and let $\epsilon > 0$. By our assumption, the finite-rank operators in $\mathcal{L}(X)$ are $\tau_c(X)$-dense in $\mathcal{L}(X) = \mathcal{L}(X, X)$. Consequently, there exists a finite-rank operator $S \in \mathcal{L}(X)$ satisfying $\|Sx - x\| < \frac{\epsilon}{\|T\|}$ for all $x \in C$. This implies $\|TSx - Tx\| \leq \|T\| \cdot \|Sx - x\| < \epsilon$ for all $x \in C$. Since the operator $TS \in \mathcal{L}(X, Y)$ has finite-rank, it follows that the vector space of all finite-rank operators $X^* \otimes Y$ is $\tau_c(X, Y)$-dense in $\mathcal{L}(X, Y)$.

(c) $\Longrightarrow$ (a) Letting $Y = X$, we see that the identity operator $I: X \to X$ belongs to the $\tau_c(X)$-closure of the finite-rank operators in $\mathcal{L}(X)$. Now a glance at Lemma 4.8 shows that X has the approximation property, and the solution is finished. ∎

Problem 4.1.8. *Let X be a Banach space with the approximation property. If Y is a Banach space and $T: Y \to X$ is a compact operator, then show that $T^*: X^* \to Y^*$ belongs to the norm closure of the finite-rank operators from X^* to Y^*, i.e., $T^* \in \overline{X^{**} \otimes Y^*}$.*

Solution: Assume that $T: Y \to X$ is a compact operator and fix $\epsilon > 0$. Since the Banach space X has the approximation property, there exists (by Theorem 4.12) a finite-rank operator $S = \sum_{i=1}^n y_n^* \otimes x_n$ in $\mathcal{L}(Y, X)$ such that $\|T - S\| < \epsilon$. Then $S^* = \sum_{i=1}^n x_n \otimes y_n^* \in \mathcal{L}(X^*, Y^*)$, where now x_n is considered as a vector in X^{**}, is a finite-rank operator, and $\big\|T^* - S^*\big\| = \|T - S\| < \epsilon$ holds. This shows that $T^* \in \overline{X^{**} \otimes Y^*}$. ∎

Problem 4.1.9. *Consider the rank-two operator $T = \mathbf{1} \otimes u - u \otimes \mathbf{1}$ on $L_1[0,1]$, where $\mathbf{1}(s) = 1$ and $u(s) = s$ for each $s \in [0,1]$. Also, recall that each function $v \in L_\infty[0,1]$ is viewed as a continuous linear functional on $L_1[0,1]$ via the formula $\langle x, v\rangle = \int_0^1 v(s)x(s)\,ds$. Prove the following.*

(a) *$Tx(t) = \int_0^1 (t-s)x(s)\,ds$ for all $x \in L_1[0,1]$ and each $t \in [0,1]$.*

(b) *The modulus of the operator T satisfies $|T|x(t) = \int_0^1 |s-t|x(s)\,ds$ for all $x \in L_1[0,1]$ and $t \in [0,1]$.*

(c) *$|T|$ is not a finite-rank operator.*

Solution: (a) Let $T = \mathbf{1} \otimes u - u \otimes \mathbf{1}$. If $x \in L_1[0,1]$, then for each $t \in [0,1]$ we have

$$\begin{aligned} Tx(t) &= [(\mathbf{1} \otimes u)(x)](t) - [(u \otimes \mathbf{1})(x)](t) \\ &= \Big[\int_0^1 x(s)\,ds\Big]u(t) - \Big[\int_0^1 u(s)x(s)\,ds\Big]\mathbf{1}(t) \\ &= \int_0^1 \big[u(t) - u(s)\big]x(s)\,ds \\ &= \int_0^1 (t-s)x(s)\,ds\,. \end{aligned}$$

(b) Repeat the solution of part (b) of Problem 3.1.13.

(c) For each k let $x_k(s) = (k+1)(k+2)s^k$, and note that

$$\begin{aligned}
|T|x_k(t) &= (k+1)(k+2)\int_0^1 |s-t|s^k\,ds \\
&= (k+1)(k+2)\int_0^t (t-s)s^k\,ds + (k+1)(k+2)\int_t^1 (s-t)s^k\,ds \\
&= \big[(k+2)ts^{k+1} - (k+1)s^{k+2}\big]\Big|_{s=0}^{s=t} + \big[(k+1)s^{k+2} - (k+2)ts^{k+1}\big]\Big|_{s=t}^{s=1} \\
&= \big[(k+2)t^{k+2} - (k+1)t^{k+2}\big] + \big[(k+1) - (k+2)t - (k+1)t^{k+2} + (k+2)t^{k+2}\big] \\
&= 2t^{k+2} - (k+2)t + k + 1\,.
\end{aligned}$$

Since the countable collection of polynomials $\{|T|x_1(t), |T|x_2(t), |T|x_3(t), \ldots\}$ is linearly independent, it follows that the range of $|T|$ cannot be finite dimensional. Hence, the modulus of the 2-rank operator T is not a finite-rank operator. ∎

Problem 4.1.10. *An operator $T\colon X \to Y$ between two Banach spaces is said to be a **nuclear operator** if there exist two sequences $\{x_n^*\} \subseteq X^*$ and $\{y_n\} \subseteq Y$ satisfying $\sum_{n=1}^\infty \|x_n^*\| \cdot \|y_n\| < \infty$ and*

$$T(x) = \sum_{n=1}^\infty x_n^*(x) y_n$$

for each $x \in X$, i.e., $T = \sum_{n=1}^\infty x_n^* \otimes y_n$ in $\mathcal{L}(X,Y)$. Show that:

(a) *Every nuclear operator is compact.*

(b) *If X and Y are Banach lattices, $\sum_{n=1}^\infty \|x_n^*\| < \infty$ and $\sum_{n=1}^\infty \|y_n\| < \infty$, then the nuclear operator $T = \sum_{n=1}^\infty x_n^* \otimes y_n$ has a compact modulus.*

Solution: (a) Let $M = \sum_{n=1}^\infty \|x_n^*\| \cdot \|y_n\| < \infty$. From

$$\sum_{n=1}^\infty \|x_n^*(x)y_n\| \le \sum_{n=1}^\infty \|x_n^*\| \cdot \|y_n\| \cdot \|x\| = M\|x\| < \infty\,,$$

it follows that the series defining $T(x)$ is norm convergent in Y for each $x \in X$ and that $\|T(x)\| \le M\|x\|$ holds. This implies that T is a bounded operator.

Now for each k consider the finite-rank operator $T_k\colon X \to Y$, defined by $T_k(x) = \sum_{n=1}^k x_n^*(x)y_n$. Then for each $x \in X$ with $\|x\| \le 1$ we have

$$\big\|T(x) - T_k(x)\big\| = \Big\|\sum_{n=k+1}^\infty x_n^*(x)y_n\Big\| \le \sum_{n=k+1}^\infty \|x_n^*\| \cdot \|y_n\| \xrightarrow[k\to\infty]{} 0\,.$$

This shows that the sequence $\{T_k - T\}$ converges uniformly to zero on U_X, which means that $T_k \to T$ in $\mathcal{L}(X,Y)$. Since each T_k is a compact operator, it follows that T is also a compact operator.

(b) Now assume that X and Y are Banach lattices, $C = \sum_{n=1}^\infty \|x_n^*\| < \infty$ and $K = \sum_{n=1}^\infty \|y_n\| < \infty$. This implies $\sum_{n=1}^\infty \|x_n^*\| \cdot \|y_n\| < \infty$, and so the formula $T = \sum_{n=1}^\infty x_n^* \otimes y_n$ defines a nuclear operator from X to Y.

Let $u = \sum_{n=1}^{\infty} |y_n| \in Y^+$. If $x \in X$, then

$$|Tx| = \Big|\sum_{n=1}^{\infty} x_n^*(x)y_n\Big| \le \sum_{n=1}^{\infty} |x_n^*(x)| \cdot |y_n| \le \sum_{n=1}^{\infty} |x_n^*(x)| \cdot |u| = C\|x\| \cdot |u|,$$

and so T carries X into the principal ideal Y_u. Moreover, if T_k is the finite-rank operator of part (a), then T_k is also a finite-rank operator from X to Y_u, and for each $x \in X$ with $\|x\| \le 1$ we have

$$\big|T(x) - T_k(x)\big| = \Big|\sum_{n=k+1}^{\infty} x_n^*(x)y_n\Big| \le \Big[\sum_{n=k+1}^{\infty} \|x_n^*\|\Big]u\,.$$

Thus, $\|T - T_k\|_\infty \le \sum_{n=k+1}^{\infty} \|x_n^*\|$ for each k, and this shows that $T\colon X \to Y_u$ (i.e., T as an operator from X to the AM-space Y_u) is also compact.

Now, by Theorem 3.14, the operator $T\colon X \to Y_u$ has a compact modulus that is given by the Riesz–Kantorovich formula: $|T|x = \sup\{Ty\colon\ y \in X \text{ and } |y| \le x\}$ for each $x \in X^+$. Since, Y_u is an ideal in Y, the latter implies that $|T|$ considered as an operator from X to Y is also the modulus of the operator $T\colon X \to Y$.

Since the $\|\cdot\|_\infty$-topology on Y_u is finer than the norm topology on Y_u, it follows that $|T|\colon X \to Y$ is likewise a compact operator. ∎

4.2. Multiplication Operators

Problem 4.2.1. *Prove the following implications:*

Order Unit $\Longrightarrow$ Quasi-interior Point $\Longrightarrow$ Weak Order Unit

Also show that in general no converse implication is true.

Solution: Let E be a Banach lattice and let $u > 0$. Assume first that u is an order unit. Then $E_u = E$, and so $\overline{E}_u = E_u = E$. That is, u is a quasi-interior point.

Now assume that u is a quasi-interior point of E. Then from Lemma 4.15(2), it follows that $\big\||x| \wedge nu - |x|\big\| \to 0$ for each $x \in E$. Since $|x| \wedge ne \uparrow$, Problem 1.2.12(c) guarantees that $|x| \wedge nu \uparrow |x|$ for each $x \in E$. The latter (in view of Theorem 1.27) shows that $B_u = E$, and so u is a weak unit of E.

To see that no converse implication holds true, note that the constant function $\mathbf{1}$ on $[0,1]$ is a quasi-interior point in $L_1[0,1]$ but fails to be an order unit. Finally, the function $u(t) = t$ is a weak unit in $C[0,1]$ but it is not a quasi-interior point. ∎

Problem 4.2.2. *Show that a Banach lattice E has a weak unit if and only if E has an at most countable maximal subset of pairwise disjoint vectors.*

Solution: Let E be a Banach lattice. Assume first that E has a weak unit, say e. Then the singleton $\{e\}$ is the required maximal subset of E. For the converse, let D be an at most countable maximal subset of pairwise disjoint vectors in E^+. We can assume that D is countable, say $D = \{x_1, x_2, \ldots\}$, and that $\|x_n\| = 1$ for each n. Since E is a Banach lattice, the series $e = \sum_{n=1}^{\infty} \frac{1}{2^n} x_n$ converges in E and $0 \leq \frac{1}{2^n} x_n \leq e$ holds for each n.

Now notice that if some vector $x \in E$ satisfies $x \perp e$, then $x \perp x_n$ for each n. Since D is maximal, we infer that $x = 0$. This shows that e is a weak unit of E. ∎

Problem 4.2.3. *Show that every separable Banach lattice has a quasi-interior point—and hence it also has a weak unit.*

Solution: Let E be a separable Banach space and fix a countable dense set $\{x_1, x_2, \ldots\}$. Choose a sequence $\{\alpha_n\}$ of positive weights such that the series $u = \sum_{n=1}^{\infty} \alpha_n |x_n|$ is norm convergent. For instance, for each n pick some $\alpha_n > 0$ such that $\alpha_n \|x_n\| < \frac{1}{2^n}$. Clearly, $0 \leq \alpha_n |x_n| \leq u$ for each n, and so $x_n \in E_u$ holds for each n. Therefore, $\{x_1, x_2, \ldots\} \subseteq E_u$ and from this we infer that $\overline{E_u} = E$. That is, u is a quasi-interior point. ∎

Problem 4.2.4. *In a Banach lattice with order continuous norm, show that a positive vector is a quasi-interior point if and only it is a weak unit.*

Solution: Assume that E is a Banach lattice with order continuous norm and let $e > 0$ be a weak unit. This means that $x \wedge ne \uparrow x$ holds in E for each $x \in E^+$, and so (by the order continuity of the norm) we have $\|x \wedge ne - x\| \to 0$ for each $x \in E^+$. Now a glance at Lemma 4.15 guarantees that e is a quasi-interior point. The converse was established in Problem 4.2.1. ∎

Problem 4.2.5. *Prove that the Banach lattice $\mathcal{M}[0,1]$, of all signed Borel measures on $[0,1]$ of bounded variation, does not have quasi-interior points.*

Solution: Notice first that $\mathcal{M}[0,1]$ as an AL-space has order continuous norm; see Corollary 3.7. So, by Problem 4.2.4, the notions of weak unit and quasi-interior point coincide.

Now assume by way of contradiction that the Banach lattice $\mathcal{M}[0,1]$ has a weak unit, say $0 < \mu \in \mathcal{M}[0,1]$. This guarantees that $\mu \wedge \nu = 0$ implies $\nu = 0$. This implies that $\mu(\{x\}) > 0$ for each $x \in [0,1]$. To see this, assume that $\mu(\{x\}) = 0$ for some $x \in [0,1]$. Then $\mu \wedge \delta_x = 0$, where δ_x is the Dirac measure supported at x, and so $\delta_x = 0$, which is impossible. If we let $E_n = \{x \in [0,1]\colon\ \mu(\{x\}) \geq \frac{1}{n}\}$, then it should be clear that each E_n is a finite set. But then $E = \bigcup_{n=1}^{\infty} E_n$ must be a countable set, a contradiction. This contradiction establishes that the AL-space $\mathcal{M}[0,1]$ has neither a quasi-interior point nor a weak unit. (The same proof shows that if Ω is an uncountable compact Hausdorff space, then $\mathrm{ca}(\Omega)$ is an AL-space without quasi-interior points.) ∎

Problem 4.2.6. *Show that if E is a Banach lattice, then its set of quasi-interior points is either empty or else it is dense in E^+.*

Solution: Let $A \subseteq E^+$ be the collection of all quasi-interior points of a Banach lattice E. Assume that $A \neq \emptyset$, and let $x \in E^+$. Fix some $e \in A$ and note that $\alpha e + x$ is a quasi-interior point for each $\alpha > 0$, i.e., $\alpha e + x \in A$. Now notice that $\lim_{\alpha \to 0^+} (\alpha e + x) = x$. This shows that $x \in \overline{A}$, and consequently $\overline{A} = E^+$. ∎

Problem 4.2.7. *If $T\colon E \to F$ is a strictly positive operator between two Banach lattices, then show that the adjoint operator $T^*\colon F^* \to E^*$ carries quasi-interior points to quasi-interior points.*

Solution: Assume that $T\colon E \to F$ is strictly positive, and let $e^* \in F^*$ be a quasi-interior point. Now let $0 < x \in E$. The strict positivity of T implies $Tx > 0$. Therefore, since e^* is a quasi-interior point of F^*, we have $T^*e^*(x) = e^*(Tx) > 0$. This implies that T^*e^* is a quasi-interior point of E^*; see Lemma 4.15. ∎

Problem 4.2.8. *Fix a function $\phi \in C(\Omega)$ (with Ω compact and Hausdorff) and let $\mathcal{A}_\phi$ denote the closed unital algebra generated by ϕ in $C(\Omega)$. That is,*

$$\mathcal{A}_\phi = \overline{\{p(\phi)\colon\ p \text{ is a polynomial}\}}\,.$$

Show that $\mathcal{A}_\phi$ is a Riesz subspace of $C(\Omega)$.

Solution: Consider the square root function $f(t) = \sqrt{t}$ on $[0,1]$. By the classical Stone–Weierstrass Approximation Theorem, the function f can be approximated uniformly by a sequence of polynomials, say $\{q_n\}$. That is, $\|q_n - f\|_\infty \to 0$.

Assume that x is a function of the form $x = p(\phi)$, where p is a polynomial. We can suppose without loss of generality that $\|x\|_\infty \le 1$ (otherwise, we replace x by the function $\frac{x}{\|x\|_\infty}$). Clearly, $r_n(x) = q_n(x^2) = q_n\big(p(\phi)^2\big) \in \mathcal{A}_\phi$ for each n and $\big\|r_n(x) - |x|\big\|_\infty = \|q_n(x^2) - f(x^2)\|_\infty \le \|q_n - f\|_\infty \to 0$. Thus, $\big\|r_n(x) - |x|\big\|_\infty \to 0$, and so $|x| \in \mathcal{A}_\phi$. From the inequality $\big\||y| - |z|\big\|_\infty \le \|y - z\|_\infty$ for all $y, z \in C(\Omega)$, we get $|x| \in \mathcal{A}_\phi$ for each $x \in \mathcal{A}_\phi$. This implies $\mathcal{A}_\phi$ is a Riesz subspace of $C(\Omega)$. ∎

Problem 4.2.9. *Let (Θ, Σ, μ) be a σ-finite measure space and let ϕ be in $L_\infty(\mu)$. Show that for the multiplication operator $M_\phi\colon L_p(\mu) \to L_p(\mu)$, where $1 \le p \le \infty$, the following statements are equivalent.*

(1) *The function ϕ is constant on some measurable set of positive measure.*

(2) *M_ϕ commutes with a rank-one positive operator.*

(3) *M_ϕ commutes with a rank-one operator.*

(4) *M_ϕ commutes with a non-zero finite-rank operator.*

Solution: (1) $\Longrightarrow$ (2) Fix a measurable subset A of Θ satisfying $0 < \mu(A) < \infty$ and $\phi(a) = c$ for each $a \in A$. Then it is easy to verify that M_ϕ commutes with the rank-one positive operator $K\colon L_p(\mu) \to L_p(\mu)$ defined by $Kx = \big(\int_A x\,d\mu\big)\chi_A$.

(2) $\Longrightarrow$ (3) $\Longrightarrow$ (4) Obvious.

(4) $\Longrightarrow$ (1) The proof of this implication is almost identical to that of Theorem 4.20. The arguments in the proof of Theorem 4.20 remain the same up to the verification of formula $(\star\star)$. After that we need to conclude that there exists some measurable set A with $0 < \mu(A) < \infty$ such that $p\big(\phi(a)\big) = 0$ for all $a \in A$. Recall that p has a finite number of zeros, say $\lambda_1, \ldots, \lambda_m$. So, if $A_i = \{a \in A\colon\ \phi(a) = \lambda_i\}$, then $A = \bigcup_{i=1}^m A_i$. Hence $\mu(A_i) > 0$ for some i, and thus $\phi = \lambda_i$ on this A_i. ∎

Problem 4.2.10. *Let $\tau\colon \Omega \to \Omega$ be a continuous mapping on a compact Hausdorff space Ω, and let $C_\tau\colon C(\Omega) \to C(\Omega)$ be the* **composition operator** *defined by $C_\tau(x) = x \circ \tau$. Establish the following.*

(a) *C_τ is a contractive positive operator satisfying $C_\tau(\mathbf{1}) = \mathbf{1}$.*

(b) *C_τ commutes with a rank-one positive operator.*

Solution: (a) Clearly, C_τ is a positive operator and

$$\big|[C_\tau(x)](\omega)\big| = \big|x(\tau(\omega))\big| \le \|x\|_\infty$$

for each $\omega \in \Omega$. Therefore, $\|C_\tau x\|_\infty = \sup_{\omega\in\Omega}\big|[C_\tau(x)](\omega)\big| \le \|x\|_\infty$. This shows that $\|C_\tau\| \le 1$, and so C_τ is a contraction. Moreover, note that

$$[C_\tau(\mathbf{1})](\omega) = \mathbf{1}(\tau(\omega)) = 1$$

for each $\omega \in \Omega$. This shows that $C_\tau(\mathbf{1}) = \mathbf{1}$ (and that $\|C_\tau\| = 1$).

(b) Let $\tau\colon \Omega \to \Omega$ be a continuous mapping. We claim that there exists some $0 < x^* \in C^*(\Omega)$ satisfying $C_\tau^* x^* = x^*$. To see this, consider the set $\mathcal{C} = \big\{x^* \in C_+^*(\Omega)\colon\ x^*(\mathbf{1}) = 1\big\}$ and note that $\mathcal{C}$ is non-empty, w^*-compact, and convex. Using (a), an easy argument shows that $C_\tau^*(\mathcal{C}) \subseteq \mathcal{C}$, and moreover $C_\tau^*\colon (\mathcal{C}, w^*) \to (\mathcal{C}, w^*)$ is continuous. Now invoke the Brouwer–Schauder–Tychonoff theorem (see, for instance [**4**, Corollary 14.51, p. 485]) to conclude that C_τ^* has a fixed point. Thus, there exists some $x^* \in \mathcal{C}$ with $C_\tau^* x^* = x^*$, i.e., $x^*(x) = x^*(x \circ \tau)$ holds for each $x \in C(\Omega)$. (Regarding this conclusion, see also Lemma 9.45.)

Next, fix some $x^* > 0$ such that $C_\tau^* x^* = x^*$ and consider the rank-one operator $K = x^* \otimes \mathbf{1}$. We claim that K and C_τ commute. Indeed, for each $x \in C(\Omega)$, on one hand we have

$$C_\tau K x = C_\tau(x^*(x)\mathbf{1}) = \big[x^*(x)\mathbf{1}\big] \circ \tau = x^*(x)\mathbf{1}\,,$$

and on the other hand

$$K C_\tau x = K(x \circ \tau) = x^*(x \circ \tau)\mathbf{1} = x^*(x)\mathbf{1}\,.$$

Thus, $KC_\tau = C_\tau K$. ∎

Problem 4.2.11. *Show that for a continuous map $\tau\colon \Omega \to \Omega$, where Ω is compact and Hausdorff, the following statements are equivalent.*

(1) *The map τ is surjective.*

(2) *The composition operator C_τ is an isometry.*

(3) *The composition operator C_τ is one-to-one.*

Solution: (1) $\Longrightarrow$ (2) Fix $x \in C(\Omega)$ and let $t \in \Omega$. Since τ is surjective, there exists some $\omega \in \Omega$ such that $\tau(\omega) = t$. So,

$$\big|x(t)\big| = \big|x\big(\tau(\omega)\big)\big| = \big|(C_\tau x)(\omega)\big| \le \|C_\tau x\|_\infty$$

holds for all $t \in \Omega$. This implies $\|x\|_\infty \le \sup_{t\in\Omega}\big|x(t)\big| \le \|C_\tau x\|_\infty$. Since, by Problem 4.2.10, C_τ is a contraction, we infer that $\|C_\tau x\|_\infty = \|x\|_\infty$. Thus, C_τ is an isometry.

(2) $\Longrightarrow$ (3) Obvious.

(3) $\Longrightarrow$ (1) Assume by way of contradiction that τ is not surjective. This means that there exists some $\omega_0 \in \Omega$ such that $\omega_0 \notin \tau(\Omega)$. Since Ω is compact and τ is continuous, we know that $\tau(\Omega)$ is compact, and hence closed. Now pick some function $x \in C(\Omega)$ such that $x(\omega_0) = 1$ and $x = 0$ on $\tau(\Omega)$. But then $C_\tau(x) = 0$, contradicting the fact that C_τ is one-to-one. Hence, τ is surjective. ∎

4.3. Lattice and Algebraic Homomorphisms

Problem 4.3.1. *According to Theorem 4.25, for any given lattice homomorphism $T\colon C(\Omega) \to C(Q)$ there exist a mapping $\tau\colon Q \to \Omega$ and some $w \in C(Q)$ such that for each $f \in C(\Omega)$ and each $q \in Q$ we have*

$$Tf(q) = w(q)f(\tau(q)).$$

Show that $w = T\mathbf{1}_\Omega$ and that the mapping τ is uniquely determined and continuous on the set $Q_0 = \{q \in Q\colon\ w(q) > 0\}$.

Solution: Clearly, $w = T\mathbf{1}_\Omega \in C(Q)$, and if $w(q) > 0$, then $\tau(q)$ is uniquely determined. Now assume that $q_\alpha \to q$ holds in Q_0. Then $w(q_\alpha) \to w(q) > 0$ and

$$w(q_\alpha)f(\tau(q_\alpha)) = Tf(q_\alpha) \to Tf(q) = w(q)f(\tau(q)).$$

So, $f(\tau(q_\alpha)) \to f(\tau(q))$ for each $f \in C(\Omega)$. This implies $\tau(q_\alpha) \to \tau(q)$, and hence τ is continuous on the set $Q_0 = \{q \in Q\colon\ w(q) > 0\}$. ∎

Problem 4.3.2. *Give an example of a lattice homomorphism which is not an algebraic homomorphism.*

Solution: Consider the operator $T\colon C[0,1] \to C[0,1]$ defined by $T(f) = 2f$. Notice that $f \wedge g = 0$ implies $T(f) \wedge T(g) = (2f) \wedge (2g) = 2(f \wedge g) = 0$, and so T is a lattice homomorphism. However, $T(\mathbf{1}^2) = T(\mathbf{1}) = 2 \cdot \mathbf{1} \neq (2 \cdot \mathbf{1})(2 \cdot \mathbf{1}) = T(\mathbf{1})T(\mathbf{1})$, and this proves that T is not an algebraic homomorphism. ∎

Problem 4.3.3. *Show that for a Markov operator $T\colon C(\Omega) \to C(Q)$ the following statements are equivalent.*

(1) *T is an algebraic homomorphism.*

(2) *T is a lattice homomorphism.*

(3) *T is a composition operator.*

Solution: (1) $\Longrightarrow$ (2) This is Theorem 4.26.

(2) $\Longrightarrow$ (3) This follows from Theorem 4.25 and the fact that $T\mathbf{1}_\Omega = \mathbf{1}_Q$.

(3) $\Longrightarrow$ (1) Assume that $Tf = f \circ \tau$ for some continuous function $\tau\colon Q \to \Omega$. Then for each $f, g \in C(\Omega)$ and each $q \in Q$ we have

$$\begin{aligned} [T(fg)](q) &= [(fg) \circ \tau](q) = (fg)(\tau(q)) = f(\tau(q)) \cdot g(\tau(q)) \\ &= [(Tf)(q)][(Tg)(q)] = [T(f)T(g)](q). \end{aligned}$$

That is, $T(fg) = T(f)T(g)$, and so T is an algebraic homomorphism. ∎

Problem 4.3.4. *Let Ω and Q be two compact Hausdorff spaces, and consider the convex set $\mathcal{C}$ of all Markov operators from $C(\Omega)$ to $C(Q)$, i.e.,*

$$\mathcal{C} = \{T \in \mathcal{L}(C(\Omega), C(Q))\colon\ T \geq 0 \text{ and } T\mathbf{1}_\Omega = \mathbf{1}_Q\}.$$

Show that a Markov operator T is a lattice homomorphism if and only if T is an extreme point of $\mathcal{C}$.

Solution: Assume first that T is an extreme point of $\mathcal{C}$. Fix a function $h \in C(\Omega)$ such that $\mathbf{1}_\Omega \leq h \leq 2 \cdot \mathbf{1}_\Omega$, and then define the operators $S, R\colon C(\Omega) \to C(Q)$ by

$$Sf = \tfrac{T(hf)}{T(h)} \quad \text{and} \quad Rf = 2Tf - Sf\,.$$

Notice that $S, R \in \mathcal{C}$ and $T = \frac{1}{2}S + \frac{1}{2}R$. Since T is an extreme point of $\mathcal{C}$, it follows that $T = S$. This implies $T(hf) = T(h)T(f)$ for each $f \in C(\Omega)$ and all $h \in C(\Omega)$ satisfying $\mathbf{1}_\Omega \leq h \leq 2 \cdot \mathbf{1}_\Omega$. Now consider any non-zero $g \in C(\Omega)$. Then the function $h = \mathbf{1}_\Omega + \frac{g}{\|g\|_\infty} \in C(\Omega)$ satisfies $\mathbf{1}_\Omega \leq h \leq 2 \cdot \mathbf{1}_\Omega$, and so $T\big((\mathbf{1}_\Omega + \frac{g}{\|g\|_\infty})f\big) = T\big(\mathbf{1}_\Omega + \frac{g}{\|g\|_\infty}\big)T(f)$ for all $f \in C(\Omega)$. This easily implies $T(gf) = T(g)T(f)$ for all $f \in C(\Omega)$ and all $g \in C(\Omega)$. Thus, T is an algebraic homomorphism, and hence a lattice homomorphism.

For the converse, assume that $T \in \mathcal{C}$ is a lattice homomorphism. By Theorem 4.25, there exists a continuous mapping $\tau\colon Q \to \Omega$ such that $Tf = f \circ \tau$ for each $f \in C(\Omega)$. Assume that $T = \alpha S + (1-\alpha)R$ holds with $S, R \in \mathcal{C}$ and $0 < \alpha < 1$. Clearly, $T^*, S^*, R^*\colon \mathrm{ca}(Q) \to \mathrm{ca}(\Omega)$, where $\mathrm{ca}(Q)$ and $\mathrm{ca}(\Omega)$ are the AL-spaces of all signed Borel measures on Q and Ω, respectively. Also, it should be noticed that the norm dual of every Markov operator carries probability measures to probability measures. Next, observe that for each $q \in Q$ and each $f \in C(\Omega)$, we have

$$\langle f, T^*\delta_q\rangle = \langle Tf, \delta_q\rangle = (Tf)(q) = f\big(\tau(q)\big) = \langle f, \delta_{\tau(q)}\rangle\,.$$

This shows that $T^*\delta_q = \delta_{\tau(q)}$, and consequently

$$\delta_{\tau(q)} = T^*\delta_q = \alpha S^*\delta_q + (1-\alpha)R^*\delta_q\,.$$

Since the extreme points of the convex set of all (Borel) probability measures on Ω are the Dirac measures (why?), it follows that $T^*\delta_q = S^*\delta_q = R^*\delta_q$ for each $q \in Q$. Therefore,

$$(Tf)(q) = (Sf)(q) = (Rf)(q)$$

for all $f \in C(\Omega)$ and each $q \in Q$. Therefore, $T = S = R$, and so T is an extreme point of the convex set $\mathcal{C}$. ∎

Problem 4.3.5. *Show that a positive operator $T\colon E \to F$ between two Banach lattices is a lattice homomorphism if and only if $T_{\mathbf{C}}\colon E_{\mathbf{C}} \to F_{\mathbf{C}}$ is a complex lattice homomorphism, i.e., $|T_{\mathbf{C}}z| = T|z|$ for all $z \in E_{\mathbf{C}}$.*

Solution: Let $z = x + \imath y \in E_{\mathbf{C}}$, and put $u = |x| + |y| \in E$ and $v = T|x| + T|y| = Tu \in F$. It is a routine matter to verify that $T(E_u) \subseteq F_v$, and so $T\colon E_u \to F_v$ is a Markov operator. Moreover, it is easy to see that in order to verify that $|T_{\mathbf{C}}z| = T|z|$, it suffices to consider T as an operator from E_u to F_v. From the Kakutani–Bohnenblust–Krein Representation Theorem, we can assume that

$T\colon C(\Omega) \to C(Q)$. Theorem 4.28 guarantees this T is a composition operator. That is, there exists a continuous function $\tau\colon Q \to \Omega$ such that

$$Tf(q) = f(\tau(q))$$

holds for all $f \in C(\Omega)$ and all $q \in Q$. In particular, note that for each $q \in Q$ we have

$$\begin{aligned} |T_{\mathbf{c}}z|(q) &= \big|Tx + \imath Ty\big|(q) = \sqrt{(Tx)^2 + (Ty)^2}\,(q) \\ &= \sqrt{[(Tx)(q)]^2 + [(Ty)(q)]^2} = \sqrt{[x(\tau(q))]^2 + [y(\tau(q))]^2} \\ &= \sqrt{x^2 + y^2}\,(\tau(q)) = |x + \imath y|(\tau(q)) \\ &= [T|z|](q)\,. \end{aligned}$$

That is, $|T_{\mathbf{c}}z| = T|z|$, as desired. ∎

Problem 4.3.6. *For a multiplicative linear functional ϕ on a Banach algebra $\mathcal{A}$ establish the following.*

(a) *If $a \in \mathcal{A}$ is an idempotent element (i.e., $a^2 = a$), then show that $\phi(a) = 0$ or $\phi(a) = 1$.*

(b) *Let p and q be two idempotent elements in $\mathcal{A}$. Then show that the linear functional ψ defined on $\mathcal{A}$ by $\psi(a) = \phi(qap)$ is also a multiplicative functional—even though the mapping $a \mapsto qap$ need not be an algebraic homomorphism.*

Solution: (a) If $a^2 = a$, then multiplicativity of ϕ yields $[\phi(a)]^2 = \phi(a^2) = \phi(a)$. This implies either $\phi(a) = 0$ or $\phi(a) = 1$.

(b) It should be clear that ψ is a linear functional. For the multiplicativity of ϕ notice that for any a and b we have

$$\psi(ab) = \phi(qabp) = \phi(q)\phi(a)\phi(b)\phi(p)\,.$$

Therefore, from $\phi(p) = [\phi(p)]^2$ and $\phi(q) = [\phi(q)]^2$, we get

$$\begin{aligned} \psi(ab) &= [\phi(q)]^2\phi(a)\phi(b)[\phi(p)]^2 = [\phi(q)\phi(a)\phi(p)][\phi(q)\phi(b)\phi(p)] \\ &= \phi(qap)\phi(qbp) = \psi(a)\psi(b)\,. \end{aligned}$$

This shows that ψ is also multiplicative. ∎

Problem 4.3.7. *Fix any Banach space X and let $Y = X \oplus X \oplus \cdots \oplus X$ be the direct sum of $n > 1$ copies of X; in particular, let $Y = \mathbb{C}^n$ or $Y = \mathbb{R}^n$ where $n > 1$. Show that the only multiplicative linear functional on the Banach algebra $\mathcal{L}(Y)$ is the zero functional.*

Solution: Let ϕ be a multiplicative linear functional on $\mathcal{L}(Y)$. Since $T = TI$ holds for each $T \in \mathcal{L}(X)$, it follows from the multiplicative property of ϕ that $\phi(T) = \phi(T)\phi(I)$. This shows that $\phi = 0$ if and only if $\phi(I) = 0$. So, we must show that $\phi(I) = 0$. Define the bounded operator $J\colon Y \to Y$ by

$$J(x_1, x_2, \ldots, x_n) = (x_n, x_1, x_2, \ldots, x_{n-1})$$

and note that $J^n = I$. Therefore, $\phi(I) = [\phi(J)]^n$, and so in order to establish that $\phi(I) = 0$, it suffices to prove that $\phi(J) = 0$.

To this end, for each $1 \le i < n$ define the bounded operator $T_i \in \mathcal{L}(Y)$ by $T_i(x_1, x_2, \ldots, x_n) = (0, \ldots, 0, x_i, 0, \ldots, 0)$, where the vector x_i occupies the $(i+1)$-position. Also, define $T_n \in \mathcal{L}(Y)$ by $T_n(x_1, x_2, \ldots, x_n) = (x_n, 0, 0, \ldots, 0)$. Clearly, $T_i^2 = 0$ for all $1 \le i \le n$, and therefore from $[\phi(T_i)]^2 = \phi(T_i^2) = 0$ we get $\phi(T_i) = 0$ for each i. Now note that $J = \sum_{i=1}^n T_i$ holds. This implies $\phi(J) = \sum_{i=1}^n \phi(T_i) = 0$, as desired. ∎

Problem 4.3.8. *For a Banach lattice E, a majorizing vector subspace X of E, and an arbitrary normed Riesz space F establish the following automatic continuity property:* Every positive operator $T\colon X \to F$ is continuous.

Solution: Let $\hat{F}$ be the norm completion of F. According to Problem 1.2.21 the Dedekind completion $\hat{F}^\delta$ of $\hat{F}$ is a Banach lattice that contains $\hat{F}$ as a Banach sublattice. Now consider $T\colon X \to F$ as an operator from X to $\hat{F}^\delta$, i.e., $T\colon X \to \hat{F}^\delta$. Since $\hat{F}^\delta$ is Dedekind complete and X majorizes E, it follows from the Kantorovich Extension Theorem 4.32 that the operator $T\colon X \to \hat{F}^\delta$ has a positive linear extension to all of E, say $\hat{T}\colon E \to \hat{F}^\delta$. By Theorem 1.31 the positive operator $\hat{T}$ is continuous. This implies that the operator $T\colon X \to F$ is likewise continuous. ∎

Problem 4.3.9 (Arendt [**13**]). *Let F be a Dedekind complete Riesz space and let $R\colon E \to H$ be a positive operator between two Riesz spaces. Then the positive operator R gives rise to a positive operator $T \mapsto TR$ from $\mathcal{L}_r(H, F)$ to $\mathcal{L}_r(E, F)$. Establish the following.*

(a) *If R is a lattice homomorphism, then $T \mapsto TR$ is interval preserving.*

(b) *If R is interval preserving, then $T \mapsto TR$ is a lattice homomorphism.*

Solution: (a) Assume that R is a lattice homomorphism. A straightforward verification shows that the mapping $T \mapsto TR$, from $\mathcal{L}_r(H, F)$ to $\mathcal{L}_r(E, F)$, is indeed a positive operator.[1] To verify that this positive operator is interval preserving, fix $0 \le T \in \mathcal{L}(H, F)$ and assume that some operator $S \in \mathcal{L}_r(E, F)$ satisfies $0 \le S \le TR$. We must show that there exists some operator $T_1 \in \mathcal{L}_r(H, F)$ satisfying $0 \le T_1 \le T$ and $T_1 R = S$.

We start by defining an operator $T_1\colon R(E) \to F$ by $T_1(Rx) = Sx$, where $R(E)$ is the range of the operator R in H. Observe that T_1 is a well-defined operator. Indeed, if $Rx = Ry$, then from

$$|Sx - Sy| \le S|x - y| \le TR|x - y| = T|Rx - Ry| = 0\,,$$

it follows that $Sx = Sy$. Next, notice that for each $x \in E$ we have

$$T_1(Rx) = Sx \le Sx^+ \le TR(x^+) = T(Rx)^+\,. \qquad (\star)$$

In addition, it is easy to see that the mapping $p\colon H \to F$, defined by $p(h) = T(h^+)$, is sublinear. So, by $(\star)$ and Theorem 4.31, there exists a linear extension of T_1 to

[1] Operators between vector spaces of operators are called **transformers**.

all of H (which we shall denote by T_1 again) satisfying $T_1 h \leq p(h) = T(h^+)$ for each $h \in H$. This implies $T_1 \leq T$, and moreover if $h \in H^+$, then

$$-T_1(h) = T_1(-h) \leq T((-h)^+) = T(0) = 0\,,$$

and so $T_1 h \geq 0$. Therefore, T_1 is a positive operator. This operator satisfies $0 \leq T_1 \leq T$ and $T_1 R = S$.

(b) Assume that R is interval preserving, and let $S, T \in \mathcal{L}_r(H, F)$. Now fix $x \in E^+$. From $[0, Rx] = R[0, x]$, it follows that two vectors $y, z \in [0, Rx]$ satisfy $y + z = Rx$ if and only if there exist vectors $u, v \in [0, x]$ such that $u + v = x$, $y = Ru$, and $z = Rv$. This implies

$$\begin{aligned}
\big[SR \vee TR\big](x) &= \sup\big\{SR(u) + TR(v)\colon\ u, v \in E^+ \text{ and } u + v = x\big\} \\
&= \sup\big\{Sy + Tz\colon\ y, z \in H^+ \text{ and } y + z = Rx\big\} \\
&= (S \vee T)(Rx) = \big[(S \vee T)R\big](x)\,.
\end{aligned}$$

Therefore, $SR \vee TR = (S \vee T)R$ in $\mathcal{L}_r(E, F)$, and this shows that $T \mapsto TR$ is a lattice homomorphism. ∎

Problem 4.3.10 (Aliprantis–Burkinshaw–Kranz [**9**]). *Let $R\colon H \to F$ be a positive operator between two Dedekind complete Riesz spaces and let E be an arbitrary Riesz space. For the positive operator $T \mapsto TR$, from $\mathcal{L}_r(H, F)$ to $\mathcal{L}_r(E, F)$, establish the following.*

(a) *If R is an order continuous lattice homomorphism, then $T \mapsto RT$ is a lattice homomorphism.*

(b) *If the operator $T \mapsto RT$ is a lattice homomorphism and the order dual $F^\sim$ is non-trivial, then R is a lattice homomorphism.*

Solution: (a) We shall use here the following characterization of the supremum of two operators:

- *If X and Y are two Riesz spaces with Y Dedekind complete, then for each pair of operators $A, B \in \mathcal{L}_r(X, Y)$ and each $x \in X^+$ we have:*

$$\Big\{\sum_{i=1}^{n} Ax_i \vee Bx_i\colon\ x_i \in E^+ \text{ and } \sum_{i=1}^{n} x_i = x\Big\} \uparrow (A \vee B)(x)\,.$$

(For a proof see [**6**, Theorem 1.16, p. 15].) Now assume that R is an order continuous lattice homomorphism, and let $S, T \in \mathcal{L}_r(E, H)$. Using the above formula, for each $x \in E^+$ we have

$$\begin{aligned}
R(S \vee T)(x) &= R\Big(\sup\Big\{\sum_{i=1}^{n} Sx_i \vee Tx_i\colon\ x_i \in E^+ \text{ and } \sum_{i=1}^{n} x_i = x\Big\}\Big) \\
&= \sup\Big\{R\Big(\sum_{i=1}^{n} Sx_i \vee Tx_i\Big)\colon\ x_i \in E^+ \text{ and } \sum_{i=1}^{n} x_i = x\Big\} \\
&= \sup\Big\{\sum_{i=1}^{n} RSx_i \vee RTx_i\colon\ x_i \in E^+ \text{ and } \sum_{i=1}^{n} x_i = x\Big\} \\
&= \big[(RS) \vee (RT)\big](x)\,.
\end{aligned}$$

Thus, $R(S \vee T) = (RS) \vee (RT)$, and so $T \mapsto RT$ is a lattice homomorphism.

(b) Assume that the transformer $T \mapsto RT$ is a lattice homomorphism. We claim that R is also a lattice homomorphism. To see this, fix some $0 < \phi \in F^\sim$ and then use Lemma 4.13 to obtain that for each $h \in H$ we have

$$\begin{aligned}\phi \otimes |Rh| &= |\phi \otimes Rh| = (\phi \otimes Rh) \vee (-(\phi \otimes Rh)) = R(\phi \otimes h) \vee R(\phi \otimes (-h)) \\ &= R[(\phi \otimes h) \vee (\phi \otimes (-h))] = R(\phi \otimes |h|) = \phi \otimes R|h| .\end{aligned}$$

This easily implies $R|h| = |Rh|$, and so R is a lattice homomorphism. ∎

Problem 4.3.11 (Kim [**48**]; Andô–Lotz [**55**]). *For an arbitrary positive operator $T\colon E \to F$ between two Banach lattices establish the following.*

(a) *If T is interval preserving, then its adjoint $T^*\colon F^* \to E^*$ is a lattice homomorphism.*

(b) *T is a lattice homomorphism if and only if T^* is interval preserving.*

Solution: (a) Let $T\colon E \to F$ be an interval preserving positive operator between two Banach lattices. Start by observing that $T^*y^* = y^* \circ T$ holds for each $y^* \in F^*$. By Problem 4.3.9(b), the operator $y^* \mapsto y^* \circ T = T^*y^*$ is a lattice homomorphism. That is, $T^*\colon F^* \to E^*$ is a lattice homomorphism.

(b) Assume first that $T\colon E \to F$ is interval preserving. By Problem 4.3.9(a), it follows that the operator $y^* \mapsto y^* \circ T = T^* \circ y^*$ (i.e., the operator $T^*\colon F^* \to E^*$) is interval preserving.

For the converse, assume that T^* is interval preserving. Then by part (a), the operator $T^{**}\colon E^{**} \to F^{**}$ is a lattice homomorphism. Since E is a Riesz subspace of E^{**} and $T^{**} = T$ on E, it easily follows that $T\colon E \to F$ itself is indeed a lattice homomorphism. ∎

Problem 4.3.12. *Let $E = \ell_\infty$ and let c denote the Riesz subspace of all convergent sequences. Show that the* ***limit functional*** *$\mathcal{L}im\colon c \to \mathbb{R}$, defined by $\mathcal{L}im(x) = \lim_{n\to\infty} x_n$ for each $x = (x_1, x_2, \ldots) \in c$, is a lattice homomorphism that has a lattice homomorphic extension to all of ℓ_∞.*

Solution: It should be clear that c is a majorizing Riesz subspace of ℓ_∞. In addition, for any $x = (x_1, x_2, \ldots) \in c$ we have

$$|\mathcal{L}im(x)| = \Big|\lim_{n\to\infty} x_n\Big| = \lim_{n\to\infty} |x_n| = \mathcal{L}im(|x|) .$$

This shows that the limit functional is a lattice homomorphism. By Corollary 4.36, $\mathcal{L}im$ has a lattice homomorphic extension to ℓ_∞. ∎

4.4. Fredholm Operators

Problem 4.4.1. *Show that a bounded operator $T\colon X \to Y$ between real Banach spaces is Fredholm if and only if the operator $T_{\mathbf{c}}\colon X_{\mathbf{c}} \to Y_{\mathbf{c}}$ is likewise Fredholm—in which case we have $i(T_{\mathbf{c}}) = i(T)$.*

Solution: The solution is based upon the following simple algebraic property:

- *A real vector space V is finite dimensional if and only if its complexification $V_{\mathbf{c}} = V \oplus \imath V$ is finite dimensional—in which case we have* $\dim V = \dim V_{\mathbf{c}}$.

To see this, assume first that V is finite dimensional. Let $e_1, e_2, \ldots, e_n$ be a basis of V and fix $v, u \in V$. Write $v = \sum_{j=1}^{n} \alpha_j e_j$ and $u = \sum_{j=1}^{n} \beta_j e_j$, and then note that

$$v + \imath u = \sum_{j=1}^{n} \alpha_j e_j + \imath \sum_{j=1}^{n} \beta_j e_j = \sum_{j=1}^{n} (\alpha_j + \imath \beta_j) e_j .$$

Since the collection of vectors $e_1, e_2, \ldots, e_n$ is also linearly independent in $V_{\mathbf{c}}$ (why?), the latter identity easily implies that $\{e_1, e_2, \ldots, e_n\}$ is also a basis in $V_{\mathbf{c}}$. So, $V_{\mathbf{c}}$ is finite dimensional and $\dim V_{\mathbf{c}} = \dim V$. For the converse, assume that $V_{\mathbf{c}}$ is finite dimensional and let $\{u_j + \imath w_j\}_{j=1}^{n}$ be a basis in $V_{\mathbf{c}}$. Now let $v \in V$. Then there exist complex numbers $\{\alpha_j + \imath \beta_j\}_{j=1}^{n}$ such that

$$v + \imath 0 = \sum_{j=1}^{n} (\alpha_j + \imath \beta_j)(u_j + \imath w_j) = \sum_{j=1}^{n} (\alpha_j u_j - \beta_j w_j) + \imath \Big[\sum_{j=1}^{n} (\alpha_j w_j + \beta_j u_j) \Big] .$$

This implies $v = \sum_{j=1}^{n} (\alpha_j u_j - \beta_j w_j)$, and so V is spanned by the finite collection of vectors $\{u_1, \ldots, u_n\} \cup \{w_1, \ldots, w_n\}$. Therefore, V is finite dimensional.

Now consider a bounded operator $T\colon X \to Y$ between real Banach spaces. Clearly, $T_{\mathbf{c}}(x + \imath y) = Tx + \imath Ty = 0$ if and only if $Tx = Ty = 0$ or equivalently, if and only if $x, y \in N(T)$. This implies $N(T_{\mathbf{c}}) = N(T) \oplus \imath N(T) = [N(T)]_{\mathbf{c}}$. The identity $R(T_{\mathbf{c}}) = R(T) \oplus \imath R(T) = [R(T)]_{\mathbf{c}}$ is also straightforward.

Next, assume that $T\colon X \to Y$ is a Fredholm operator. Pick a closed subspace V of X and a finite dimensional subspace W of Y with $X = N(T) \oplus V$ and $R(T) \oplus W = Y$. It is easy to see that the vector subspace $V_{\mathbf{c}}$ of $X_{\mathbf{c}}$ is closed, $X_{\mathbf{c}} = N(T_{\mathbf{c}}) \oplus V_{\mathbf{c}}$, and $Y_{\mathbf{c}} = R(T_{\mathbf{c}}) \oplus W_{\mathbf{c}}$. From the above property, we infer that the operator $T_{\mathbf{c}}$ is a Fredholm operator and that

$$i(T_{\mathbf{c}}) = n(T_{\mathbf{c}}) - d(T_{\mathbf{c}}) = \dim N(T_{\mathbf{c}}) - \dim W_{\mathbf{c}} = \dim N(T) - \dim W = i(T) .$$

For the converse, suppose that $T_{\mathbf{c}}$ is a Fredholm operator. From the property stated at the beginning of the solution, we have $\dim N(T) = \dim N(T_{\mathbf{c}}) < \infty$. Next, pick a finite dimensional subspace W_1 of $Y_{\mathbf{c}}$ such that

$$Y_{\mathbf{c}} = R(T_{\mathbf{c}}) \oplus W_1 = \big[R(T) \oplus \imath R(T)\big] \oplus W_1 , \tag{$\star$}$$

and let $U = \{x \in X\colon$ there exists some $z \in X$ such that $x + \imath z \in W_1\}$. Clearly, U is a vector subspace of X. Since W_1 is finite dimensional, an easy argument shows that U is also finite dimensional. Moreover, from ($\star$), it easily follows that $R(T) + U = Y$. Now a glance at Problem 4.4.3 guarantees the existence of a finite dimensional subspace U_1 of U such that $Y = R(T) \oplus U_1$. This implies $d(T) < \infty$. Therefore, T is a Fredholm operator and the solution is finished. ∎

Problem 4.4.2. *Prove Riesz's lemma without using the Hahn–Banach Extension Theorem: If Y is a proper closed subspace of a (real or complex)*

Banach space X, then for each $0 < \epsilon < 1$ there exists some unit vector $x \in X$ such that

$$\|y - x\| \geq 1 - \epsilon$$

holds for all $y \in Y$. Also, present an example of an infinite dimensional proper closed subspace Y of a Banach space X such that for each unit vector $x \in X \setminus Y$ we have $d(x, Y) < 1$.

Solution: Fix some $x_0 \in X \setminus Y$, and note that $d = d(x_0, Y) > 0$. Next, select some $y_0 \in Y$ with $\|x_0 - y_0\| < \frac{d}{1-\epsilon}$ and let $x = \frac{x_0 - y_0}{\|x_0 - y_0\|}$. Clearly, $\|x\| = 1$ and if $y \in Y$, then

$$\|x - y\| = \left\| \tfrac{x_0 - y_0}{\|x_0 - y_0\|} - y \right\| = \tfrac{\|x_0 - (y_0 + \|x_0 - y_0\| y)\|}{\|x_0 - y_0\|} \geq \tfrac{d}{\|x_0 - y_0\|} > 1 - \epsilon .$$

Now we present an example of an infinite dimensional closed vector subspace Y of a Banach space X such that $d(x, Y) < 1$ holds for each unit vector $x \in X$. We let $X = c$, the Banach space of all convergent sequences equipped with the norm defined by

$$\|x\| = \tfrac{1}{2}\|x\|_\infty + \sum_{n=1}^{\infty} \tfrac{1}{2^{n+1}} |x_n|$$

for each $x = (x_1, x_2, \ldots) \in c$. We also let $Y = c_0$, the closed vector subspace of c consisting of all sequences converging to zero.

Now take any unit vector $x \in X$. We claim that $d(x, c_0) < 1$. To see this, fix some k with $x_k \neq 0$, and note that the vector $y = (x_1, x_2, \ldots, x_k, 0, 0, \ldots) \in c_0$ satisfies

$$\|x - y\| = \tfrac{1}{2} \sup_{n>k} |x_n| + \sum_{n=k+1}^{\infty} \tfrac{1}{2^n} |x_n| < \tfrac{1}{2}\|x\|_\infty + \sum_{n=1}^{\infty} \tfrac{1}{2^{n+1}} |x_n| = 1 .$$

This shows that $d(x, c_0) \leq \|x - y\| < 1$. ∎

Problem 4.4.3. *This problem presents some elementary algebraic properties that have been used in the proofs of the results in Section 4.4. Establish the following.*

(a) *If W is a vector subspace of a vector space V, then W has an* ***algebraic complement****, i.e., there exists a vector subspace U of V satisfying $W \cap U = \{0\}$ and $W + U = V$.*

(b) *If U and W are two vector subspaces of a vector space V, then there exists a vector subspace W_1 of W satisfying $V \cap W_1 = \{0\}$ and $U + W_1 = U + W$.*

(c) *If $T\colon V \to W$ is a linear operator between two vector spaces, then the quotient vector space $V/N(T)$ is linearly isomorphic to $R(T)$.*

Solution: (a) Let W be a vector subspace of a vector space V. Fix a Hamel basis $\{e_i\}_{i \in I}$ of W and then choose a family $\{f_j\}_{j \in J}$ of linearly independent vectors in V such that the collection $\{e_i\colon\ i \in I\} \cup \{f_j\colon\ j \in J\}$ is a Hamel basis of V. (Zorn's lemma guarantees that such a family of linearly independent vectors exists.) Now let U be the vector subspace generated by the family $\{f_j\}_{j\in J}$. We claim that U is an algebraic complement of W.

To see this, assume first that $x \in W \cap U$. Then we there exist a family of scalars $\{\lambda_i\}_{i\in I}$ with $\lambda_i = 0$ for all but a finite number of i and a family of scalars $\{\mu_j\}_{j\in J}$ with $\mu_j = 0$ for all but a finite number of j satisfying $x = \sum_{i\in I} \lambda_i e_i = \sum_{j\in J} \mu_j f_j$. This implies $\sum_{i\in I} \lambda_i e_i + \sum_{j\in J}(-\mu_j)e_j = 0$. The linear independence of the set $\{e_i\colon\ i \in I\} \cup \{f_j\colon\ j \in J\}$ implies $\lambda_i = 0$ for all $i \in I$ and $\mu_j = 0$ for all $j \in J$. Therefore, $x = 0$, and so $W \cap U = \{0\}$.

To see that $W + U = V$, pick any $v \in V$. Since $\{e_i\colon\ i \in I\} \cup \{f_j\colon\ j \in J\}$ is a Hamel basis, there exist families of scalars $\{\lambda_i\}_{i\in I}$ with $\lambda_i = 0$ for all but a finite number of i and $\{\mu_j\}_{j\in J}$ with $\mu_j = 0$ for all but a finite number of j such that $v = \sum_{i\in I} \lambda_i e_i + \sum_{j\in J} \mu_j f_j$. If we let $w = \sum_{i\in I} \lambda_i e_i \in W$ and $u = \sum_{j\in J} \mu_j f_j \in U$, then $v = w + u$ holds. Therefore, $V = W \oplus U$.

(b) By part (a) there exist vector subspaces U_1 of U and W_1 of W such that $U = U_1 \oplus (W \cap U)$ and $W = W_1 \oplus (W \cap U)$. It is easy to check that $W_1 \cap U_1 = \{0\}$. So, $U + W = U_1 \oplus (W \cap U) \oplus W_1 = U \oplus W_1$.

(c) Consider the mapping $\hat{T}\colon V/N(T) \to R(T)$ defined by $\hat{T}([x]) = Tx$. Since $[x] = [y]$ is equivalent to $x - y \in N(T)$ or $Tx = Ty$, it follows that the mapping $\hat{T}$ is well defined. If $\hat{T}[x] = Tx = 0$, then $x \in N(T)$ or $[x] = 0$. That is, $\hat{T}$ is one-to-one. Moreover, $\hat{T}$ is linear and surjective. Therefore, $\hat{T}$ is an algebraic isomorphism, and so the vector spaces $V/N(T)$ and $R(T)$ are linearly isomorphic. ∎

Problem 4.4.4. *Let X be an n-dimensional vector space, let Y be an m-dimensional vector space, and let $T\colon X \to Y$ be an operator. If we let $r = \dim R(T)$, then show that $n(T) = n - r$ and $d(T) = m - r$; conclude from this that $i(T) = n - m$.*

Solution: Choose a vector subspace V of X and W of Y such that $X = N(T) \oplus V$ and $Y = R(T) \oplus W$. Since V and $R(T)$ are isomorphic, it follows that $n(T) + r = n$. Thus, $n(T) = n - r$. On the other hand, we know that

$$d(T) = \dim\,[Y/R(T)] = \dim Y - \dim\,[R(T)] = m - r\,.$$

Thus, $i(T) = n(T) - d(T) = n - r - (m - r) = n - m$. (See also Problem 2.1.8.) ∎

Problem 4.4.5. *For a bounded operator $T\colon X \to Y$ between Banach spaces define $\hat{T}\colon X/N(T) \to Y$ via the formula $\hat{T}[x] = Tx$. Let $\pi\colon X \to X/N(T)$ be the quotient map and $J\colon R(T) \to Y$ be the natural embedding. Now consider the scheme of bounded operators*

$$X \xrightarrow{\ \pi\ } X/N(T) \xrightarrow{\ \hat{T}\ } R(T) \xrightarrow{\ J\ } Y\,,$$

and note that $T = J\hat{T}\pi$. If T is a Fredholm operator, then show that all operators in the above scheme are Fredholm operators and compute their indices. Also, verify directly that $i(T) = i(J) + i(\hat{T}) + i(\pi)$.

Solution: Assume that $T\colon X \to Y$ is a Fredholm operator. We shall compute the indices of the introduced operators individually.

The operator $\pi\colon X \to X/N(T)$ is surjective and its null space is $N(T)$. So, $i(\pi) = n(\pi) - d(\pi) = n(T) - 0 = n(T)$. The operator $\hat{T}\colon X/N(T) \to R(T)$ is an isomorphism, and thus $i(\hat{T}) = 0$. Finally, the natural embedding $J\colon R(T) \to Y$ satisfies $N(J) = \{0\}$ and $R(J) = R(T)$. Therefore,

$$i(J) = \dim N(J) - \dim Y/R(J) = 0 - \dim Y/R(T) = -d(T)\,.$$

Consequently, we have

$$i(J) + i(\hat{T}) + i(\pi) = -d(T) + 0 + n(T) = n(T) - d(T) = i(T) = i(J\hat{T}\pi)\,,$$

which, of course, is the assertion of the Index Theorem. ∎

Problem 4.4.6. *If $K\colon X \to X$ is a compact operator on a Banach space, then show that the operator $I - K$ has a finite descent.*

Solution: Assume by way of contradiction that $R\big((I-K)^{n+1}\big)$ is a proper vector subspace of $R\big((I-K)^n\big)$ for each n. Since $(I-K)^n = I - K_1$ for some compact operator K_1, it follows from the proof of Lemma 4.45 that $R\big((I-K)^n\big)$ is a closed subspace for each n. So, by Riesz's Lemma 4.44, there exists a sequence $\{y_n\}$ in X satisfying $y_n \in R\big((I-K)^n\big)$ for each n and $\|y_n - y\| \geq \frac{1}{2}$ for each $y \in R\big((I-K)^{n+1}\big)$. Now if $1 \leq m < n$, then we have

$$Ky_m - Ky_n = y_m - \big[y_n + (I-K)y_m - (I-K)y_n\big]$$

and $y = y_n + (I-K)y_m - (I-K)y_n \in R\big((I-K)^{m+1}\big)$. This implies that for $n \neq m$ we have $\|Ky_m - Ky_n\| \geq \big\|y_m - y\big\| \geq \frac{1}{2}$. Consequently, the sequence $\{Ky_n\}$ cannot have any convergent subsequence, contrary to the compactness of K. This contradiction establishes that $I - K$ has a finite descent. ∎

Problem 4.4.7 (The Fredholm Alternative). *Let X be a Banach space and $K \in \mathcal{K}(X)$. Show that for each $\lambda \neq 0$ exactly one of the following two exclusive statements is true:*

(a) *For each $y \in X$ the equation $x - \lambda Kx = y$ has a unique solution.*

(b) *The equation $x - \lambda Kx = 0$ has a non-zero solution.*

When (b) *is true, show also that the equation $x - \lambda Kx = 0$ has a finite number of linearly independent solutions.*

Solution: Let us verify first that statements (a) and (b) are mutually exclusive. Assume that (a) and (b) are true simultaneously. If some $x \neq 0$ is a solution for the equation in (b), then letting $y = 0$ in (a), we see that 0 and x are two distinct solutions of (a), which is a contradiction. Next, we shall show that either (a) or (b) is true. We consider the following two cases.

CASE I: $\frac{1}{\lambda} \notin \sigma(K)$.

In this case, the operator $\frac{1}{\lambda} - K$ is invertible. So, the operator $I - \lambda K$ is likewise invertible. This implies that for each $y \in X$ the equation $y = x - \lambda Kx = (I - \lambda K)x$ has a unique solution, namely, $x = (I - \lambda K)^{-1}y$.

CASE II: $\frac{1}{\lambda} \in \sigma(K)$.

Since K is a compact operator, it follows that $\frac{1}{\lambda}$ is an eigenvalue of K, and so the equation $x - \lambda Kx = 0$ has a non-trivial solution. Now Lemma 4.45 guarantees that the non-trivial vector subspace

$$N_K\big(\tfrac{1}{\lambda}\big) = \big\{x \in X\colon\ Kx = \tfrac{1}{\lambda}x\big\} = \{x \in X\colon\ x - \lambda Kx = 0\}$$

is finite dimensional. ∎

Problem 4.4.8. *Let $T\colon X \to Y$ be a bounded operator between Banach spaces having a pseudoinverse $S\colon Y \to X$. Fix some $y \in Y$ and consider the equation*

$$Tx = y\,. \tag{$\star$}$$

Show that $(\star)$ has a solution if and only if Sy is a solution of $(\star)$, i.e, $T(Sy) = y$.

Solution: If $x = Sy$ is a solution of $(\star)$, then $(\star)$ has at least one solution. For the converse, assume that some $x \in X$ satisfies $Tx = y$. Then we have

$$y = Tx = (TST)x = TS(Tx) = TSy = T(Sy)\,.$$

That is, Sy is a solution of $(\star)$. ∎

Problem 4.4.9. *Show that a bounded operator between Hilbert spaces has a pseudoinverse if and only if it has a closed range.*

Solution: Let $T\colon H_1 \to H_2$ be a bounded operator between Hilbert spaces. If T has a pseudoinverse, then it follows from Lemma 4.52(2) that there exists a closed subspace W of H_1 such that $R(T) \oplus W = H_2$. But then, by Theorem 2.16, $R(T)$ is a closed subspace.

For the converse, assume that $R(T)$ is closed. Since, $N(T)$ and $R(T)$ are closed subspaces of the Hilbert spaces H_1 and H_2, respectively, it follows that they are both complemented—keep in mind that every closed subspace in a Hilbert space is complemented. This, coupled with Lemma 4.52, guarantees that T has a pseudoinverse. ∎

4.5. Strictly Singular Operators

Problem 4.5.1. *Show that a bounded operator $T\colon X \to Y$ between Banach spaces is strictly singular if and only if T is not bounded below on any infinite dimensional closed subspace of X.*

Solution: Since $\|Tx\| \le \|T\| \cdot \|x\|$ for each $x \in X$, it easily follows that if Z is a closed subspace of X, then $T\colon Z \to T(Z)$ is an isomorphism if and only if there exists some $c > 0$ such that $c\|z\| \le \|Tz\|$ for each $z \in Z$. That is, $T\colon Z \to T(Z)$ is an isomorphism if and only if T is bounded below on Z. Therefore, T is strictly singular if and only if T is not bounded below on any infinite dimensional closed subspace of X. ∎

Problem 4.5.2. *Let X_i $(i = 1, 2, \dots, n)$ and Y_j $(j = 1, 2, \dots, m)$ be all real (or all complex) Banach spaces and consider a bounded operator*

$$\mathcal{T}\colon X_1 \oplus X_2 \oplus \cdots \oplus X_n \to Y_1 \oplus Y_2 \oplus \cdots \oplus Y_m .$$

As in Problem 1.1.11, *it is easy to see that there exist (uniquely determined) bounded operators $T_{ji}\colon X_i \to Y_j$ $(i = 1, 2, \dots, n;\ j = 1, 2, \dots, m)$ such that $\mathcal{T}$ can be represented by the $m \times n$ matrix*

$$\mathcal{T} = \begin{bmatrix} T_{11} & T_{12} & \dots & T_{1n} \\ T_{21} & T_{22} & \dots & T_{2n} \\ \vdots & \vdots & \ddots & \vdots \\ T_{m1} & T_{m2} & \dots & T_{mn} \end{bmatrix}.$$

Show that $\mathcal{T}$ is strictly singular if and only if each T_{ji} is strictly singular.

Solution: For simplicity, let $X = X_1 \oplus X_2 \oplus \cdots \oplus X_n$ and $Y = Y_1 \oplus Y_2 \oplus \cdots \oplus Y_m$. Assume first that the bounded operator $\mathcal{T}\colon X \to Y$ is strictly singular, and assume by way of contradiction that some operator T_{ji} is not strictly singular. This means that there exists an infinite dimensional vector subspace V of X_i such that T_{ji} is bounded below on V, i.e., there exists some $c > 0$ such that $\|T_{ji}v\| \geq c\|v\|$ holds for all $v \in V$. If we let $\hat{V} = \{0\} \oplus \{0\} \oplus \cdots \oplus \{0\} \oplus V \oplus \{0\} \oplus \cdots \oplus \{0\}$, where V occupies the ith summand, then it is easy to see that $\hat{V}$ is an infinite dimensional vector subspace of X and $\|\mathcal{T}x\| \geq c\|x\|$ holds for each $x \in \hat{V}$. The latter contradicts the fact that $\mathcal{T}$ is strictly singular. Therefore, each T_{ji} is strictly singular.

For the converse, assume that each T_{ji} is a strictly singular operator. Let $P_i\colon X \to X$ be the natural projection of X with range X_i and define $R_j\colon Y_j \to Y$ by $Ry_j = 0 \oplus 0 \oplus \cdots \oplus 0 \oplus y_j \oplus 0 \oplus \cdots \oplus 0$, where y_j occupies the jth coordinate. Now consider the scheme of bounded operators $X \xrightarrow{P_i} X_i \xrightarrow{T_{ji}} Y_j \xrightarrow{R_j} Y$, and let $\mathcal{T}_{ij} = R_j T_{ji} P_i$. Clearly, $\mathcal{T}_{ji}(x_1 \oplus x_2 \oplus \cdots \oplus x_n) = 0 \oplus 0 \oplus \cdots \oplus 0 \oplus T_{ji}x_i \oplus 0 \oplus \cdots \oplus 0$, where the vector $T_{ji}x_i$ occupies the jth component. By Corollary 4.62, each $\mathcal{T}_{ji}$ is a strictly singular operator. Now from $\mathcal{T} = \sum_{i=1}^{n} \sum_{j=1}^{m} \mathcal{T}_{ji}$ and Corollary 4.62, we easily infer that $\mathcal{T}$ is a strictly singular operator. ∎

Problem 4.5.3. *Show that a bounded operator $T\colon X \to Y$ between two real Banach spaces is strictly singular if and only if $T_{\mathbf{c}}\colon X_{\mathbf{c}} \to Y_{\mathbf{c}}$ is strictly singular.*

Solution: Assume that $T_{\mathbf{c}}$ is strictly singular. If T is not strictly singular, then there exists an infinite dimensional closed subspace Z of X such that $T\colon Z \to T(Z)$ is an isomorphism. So, there exists some constant $\epsilon > 0$ satisfying $\epsilon\|z\| \leq \|Tz\|$ for each $z \in Z$. Clearly, $Z_{\mathbf{c}} = Z \oplus \imath Z$ is an infinite dimensional closed subspace of $X_{\mathbf{c}}$. Now if $a + \imath b \in Z_{\mathbf{c}}$, then for each angle θ we have

$$\begin{aligned} \big\|T_{\mathbf{c}}(a + \imath b)\big\| &= \big\|Ta + \imath Tb\big\| \geq \big\|(Ta)\cos\theta + (Tb)\sin\theta\big\| \\ &= \big\|T(a\sin\theta + b\cos\theta)\big\| \geq \epsilon\|a\sin\theta + b\cos\theta\| . \end{aligned}$$

This implies $\epsilon\|a+\imath b\| \leq \big\|T_{\mathbf{c}}(a+\imath b)\big\|$. That is, $T_{\mathbf{c}}$ restricted to $Z_{\mathbf{c}}$ is an isomorphism, which is impossible. Therefore, T is strictly singular.

For the converse, assume that T is strictly singular, and assume contrary to our claim that there exists an infinite dimensional vector subspace V of $X_{\mathbf{c}}$ and some $\delta > 0$ such that $\|T_{\mathbf{c}}z\| \geq \delta\|z\|$ holds for all $z \in V$. Now consider the vector subspace of $X \oplus X$ defined by

$$W = \big\{x_1 \oplus x_2 \colon\ x_1, x_2 \in X \text{ and } x_1 + \imath x_2 \in V\big\}.$$

Note that W is an infinite dimensional vector subspace of $X \oplus X$ and that

$$\begin{aligned}\big\|[T \oplus T](x_1 \oplus x_2)\big\| &= \|Tx_1\| + \|Tx_2\| \geq \|T_{\mathbf{c}}(x_1 + \imath x_2)\| \\ &\geq \delta\|x_1 + \imath x_2\| \geq \tfrac{\delta}{2}\|x_1 \oplus x_2\|\end{aligned}$$

holds for each $x_1 \oplus x_2 \in W$. This shows that the operator $T \oplus T\colon X \oplus X \to Y \oplus Y$ is strictly singular, contrary to the conclusion of Problem 4.5.2. Therefore, $T_{\mathbf{c}}$ must be strictly singular. (The proof of this part was communicated to the authors by Julio Flores.) ∎

Problem 4.5.4. *Prove the following property of ℓ_p-spaces that was used in Theorem 4.58: If X is an infinite dimensional vector subspace of any ℓ_p-space, then for each $m \in \mathbb{N}$ the vector subspace*

$$X_m = \big\{x = (x_1, x_2, \ldots) \in X \colon\ x_1 = x_2 = \cdots = x_m = 0\big\}$$

is also infinite dimensional.

Solution: The proof is based on the following property: *If Y is an infinite dimensional vector subspace of any ℓ_p-space, then for every $m \in \mathbb{N}$ the vector subspace $Y_m = \{y \in Y \colon\ y_m = 0\}$ is also infinite dimensional.*

The proof of this claim is as follows. Consider two arbitrary linearly independent vectors $u, v \in Y$. If $u_m = 0$ or $v_m = 0$, then either $u \in Y_m$ or $v \in Y_m$. If $u_m \neq 0$ and $v_m \neq 0$, then the vector $w = u_m v - v_m u \in Y_m$ is non-zero since u and v are linearly independent. Therefore, we have shown that if $u, v \in Y$ are linearly independent, then there exists a non-zero vector in the linear span of $\{u, v\}$ that belongs to Y_m. Now let $\{u_1, u_2, \ldots\} \subseteq Y$ be a countable linearly independent set. For each n, let w_n be a non-zero vector in the linear span of $\{u_{2n-1}, u_{2n}\}$ that lies in Y_m. It follows that $\{w_1, w_2, \ldots\}$ is a countable linearly independent subset of Y_m, and so Y_m is infinite dimensional.

We are now ready to prove the result using induction on m. For $m = 1$, the conclusion follows immediately from the above established property. Now assume that X_m is infinite dimensional for some $m \geq 1$. From

$$\begin{aligned}X_{m+1} &= \big\{x = (x_1, x_2, \ldots) \in X \colon\ x_1 = x_2 = \cdots = x_m = 0\big\} \\ &= \{x \in X_m \colon\ x_{m+1} = 0\}\end{aligned}$$

and the above established result, we infer that X_{m+1} is infinite dimensional. This completes the induction, and the desired conclusion has been established. ∎

Problem 4.5.5. *Assume that for a pair of Banach spaces (X, Y) we have the following:*

(a) *Each infinite dimensional closed subspace of X contains a closed vector subspace that is isomorphic to X.*

(b) *No closed vector subspace of Y is isomorphic to X.*

Show that each operator in $\mathcal{L}(X,Y)$ is strictly singular. (It is known that the pair (ℓ_1, ℓ_p), where $1 < p < \infty$, satisfies properties (a) *and* (b)*; see* [**52**, *Proposition* 2.a.2, p. 53].)

Solution: Let (X, Y) be a pair of Banach spaces satisfying properties (a) and (b), and assume that contrary to our claim some operator $T \in \mathcal{L}(X,Y)$ is not strictly singular. Then we can find an infinite dimensional closed vector subspace X_0 of X such that the restriction $T\colon X_0 \to Y$ is an isomorphism. By (a) there exists a closed vector subspace X_1 of X_0 that is isomorphic to X. Consequently, the closed vector subspace $T(X_1)$ of Y is isomorphic to X which contradicts (b). Hence, $\mathcal{L}(X,Y)$ consists of singular operators. ∎

Problem 4.5.6. *This problem presents an example of a strictly singular operator whose adjoint is not strictly singular. Establish the following.*

(i) *If Y is a separable Banach space, then there exists a bounded surjective operator $T\colon \ell_1 \to Y$ whose adjoint $T^*\colon Y^* \to \ell_\infty$ is a linear isometry.*

(ii) *If $T\colon \ell_1 \to \ell_p$, where $1 < p < \infty$, is any bounded operator satisfying* (i)*, then its adjoint operator $T^*\colon \ell_q \to \ell_\infty$, where $\frac{1}{p} + \frac{1}{q} = 1$, is not strictly singular.*

Solution: (i) Assume that Y is a separable Banach space. Fix a countable norm dense subset $\{y_1, y_2, \ldots\}$ in the closed unit ball U_Y of Y. Now define the linear operator $T\colon \ell_1 \to Y$ by $T(\lambda_1, \lambda_2, \ldots) = \sum_{n=1}^{\infty} \lambda_n y_n$. Observe that for each vector $x = (\lambda_1, \lambda_2, \ldots) \in \ell_1$ we have

$$\|Tx\| = \Big\|\sum_{n=1}^{\infty} \lambda_n y_n\Big\| \le \sum_{n=1}^{\infty} |\lambda_n| \cdot \|y_n\| \le \sum_{n=1}^{\infty} |\lambda_n| = \|x\|\,.$$

This implies $T(U_1) \subseteq U_Y$, where U_1 is the closed unit ball of ℓ_1. From $Te_n = y_n$, where $e_n \in U_1$ is the sequence having its nth coordinate equal to one and every other equal to zero, it follows that $\overline{T(U_1)} = U_Y$. Therefore, in view of the Open Mapping Theorem, the operator T is an open mapping, and hence T is surjective. In particular, we have $\|T\| = 1$. To see that $T^*\colon Y^* \to \ell_\infty$ is a linear isometry, note that for each $y^* \in Y^*$ we have

$$\|T^*y^*\| = \sup_{x \in U_1} \big|T^*y^*(x)\big| = \sup_{x \in U_1} \big|y^*(Tx)\big| = \sup_{y \in U_Y} \big|y^*(y)\big| = \|y^*\|\,.$$

(ii) Fix any $1 < p < \infty$. Then the pair of Banach spaces (ℓ_1, ℓ_p) satisfies properties (a) and (b) of Problem 4.5.5. So, if $T\colon \ell_1 \to \ell_p$ is any operator described in part (i) above, then T must be strictly singular, while its adjoint operator $T^*\colon Y^* \to \ell_\infty$ being a linear isometry is not strictly singular. ∎

Problem 4.5.7 (T. Oikhberg). *Show that there is an infinite dimensional Banach space X with a non-zero multiplicative linear functional on the Banach algebra $\mathcal{L}(X)$.*

Solution: Let X be an infinite dimensional **hereditarily indecomposable** Banach space, in brief X is an HI-space. This means that no infinite dimensional closed subspace of X is isomorphic to a direct sum of two infinite dimensional Banach spaces. The existence of the HI-spaces was established by W. T. Gowers and B. Maurey in [**32**]. Moreover, it was shown in [**32, 33**] that:

- *If X is an HI-space and $T \in \mathcal{L}(X)$, then there exist a scalar and a strictly singular operator $S \in \mathcal{L}(X)$ (both uniquely determined) such that T is of the form $T = \alpha I + S$, where I is the identity operator on X.*

Now define a function ϕ on $\mathcal{L}(X)$ by $\phi(T) = \phi(\alpha I + S) = \alpha$, where $T = \alpha I + S$ is the unique decomposition of T as a sum of a scalar multiple of I and a strictly singular operator. (The uniqueness of α and S follows from Corollary 4.62.) Clearly, ϕ is a non-zero function. Now let $T_1, T_2 \in \mathcal{L}(X)$ so that $T_1 = \alpha_1 I + S_1$ and $T_2 = \alpha_2 I + S_2$, where S_1 and S_2 are strictly singular operators. Then we have

$$\lambda T_1 = \lambda\alpha_1 + \lambda S_1\,, \quad T_1 + T_2 = (\alpha_1 + \alpha_2) + (S_1 + S_2)\,, \text{ and}$$

$$T_1 T_2 = \alpha_1\alpha_2 + (\alpha_1 S_1 + \alpha_2 S_2 + S_1 S_2)\,.$$

According to Corollary 4.62 these decompositions are the unique decompositions of the above operators as sums of multiples of the identity operator plus singular operators. This implies

$$\begin{aligned} \phi(\lambda T_1) &= \lambda\alpha_1 = \lambda\phi(T_1)\,, \\ \phi(T_1 + T_2) &= \alpha_1 + \alpha_1 = \phi(T_1) + \phi(T_2)\,, \text{ and} \\ \phi(T_1 T_2) &= \alpha_1\alpha_2 = \phi(T_1)\phi(T_2)\,, \end{aligned}$$

and so ϕ is a non-zero multiplicative linear functional on $\mathcal{L}(X)$. ∎

Problem 4.5.8. *Show that if $1 \le q < p < \infty$, then the natural embedding $J\colon L_p[0,1] \to L_q[0,1]$ is not a strictly singular operator.*

Solution: As indicated in the hint of the problem, the solution will be based upon the following famous classical inequality, known as **Khintchine's inequality**, regarding the standard sequence $\{r_n\}$ of Rademacher functions on $[0,1]$ that were introduced in Problem 3.1.9.

- *For each fixed $1 \le p < \infty$ there exist two constants $A_p > 0$ and $B_p > 0$ such that*

$$A_p\Big[\sum_{i=1}^{n} |a_n|^2\Big]^{\frac{1}{2}} \le \Big[\int_0^1 \Big|\sum_{i=1}^{n} a_n r_n(t)\Big|^p\,dt\Big]^{\frac{1}{p}} \le B_p\Big[\sum_{i=1}^{n} |a_n|^2\Big]^{\frac{1}{2}}$$

holds for all choices of the complex numbers $a_1, a_2, \ldots, a_n$.

For a proof of this inequality see [**23**, Theorem 1.10] or [**52**, Theorem 2.b.3].

This inequality implies the following. If X denotes the closed vector subspace generated by the sequence $\{r_n\}$ of Rademacher functions in $L_2[0,1]$, then for any $1 \le p < \infty$ the vector space X also coincides with the closed vector subspace generated by the sequence $\{r_n\}$ of Rademacher functions in $L_p[0,1]$ and that the two norms $\|\cdot\|_p$ and $\|\cdot\|_2$ are equivalent on X. In particular, if $r, s \in (0,\infty)$, then the two norms $\|\cdot\|_r$ and $\|\cdot\|_s$ are equivalent on X.

Now assume that p and q satisfy $1 \leq q < p < \infty$ and consider the natural embedding $J\colon L_p[0,1] \to L_q[0,1]$. By the above, J restricted to X is an isomorphism. This proves that the positive operator $J\colon L_p[0,1] \to L_q[0,1]$ is not strictly singular. ∎

Problem 4.5.9 (Hernández–Rodriguez-Salinas [**37**]). *A bounded operator $T\colon E \to X$, from a Banach lattice to a Banach space, is said to be* ***disjointly strictly singular*** *if for each disjoint sequence $\{x_n\}$ of non-zero vectors in E the restriction of T to the closed vector subspace generated by $\{x_n\}$ is not an isomorphism. Clearly, a strictly singular operator from a Banach lattice to a Banach space is disjointly strictly singular.*

Show that if p and q satisfy $1 \leq q < p < \infty$, then the natural embedding $J\colon L_p[0,1] \to L_q[0,1]$ is a disjointly strictly singular operator.

Solution: Let $\{x_n\}$ be an arbitrary disjoint sequence of non-zero vectors in L_p. Let Y be the closed linear span of this sequence in $L_p[0,1]$. A straightforward verification shows that Y is linearly isometric to ℓ_p. Similarly, the closed linear span Z of the sequence $\{x_n\}$ in $L_q[0,1]$ is linearly isometric to the Banach lattice ℓ_q. Since $1 \leq q < p < \infty$, we know from Theorem 4.58 that the embedding of ℓ_p to ℓ_q is strictly singular, and so $J|_Y$ cannot be an isomorphism. Therefore, the natural embedding $J\colon L_p[0,1] \to L_q[0,1]$ is disjointly strictly singular. ∎

Chapter 5

Integral Operators

5.1. The Basics of Integral Operators

Problem 5.1.1. *Assume that μ and ν are σ-finite measures. Show that if the kernels of two integral operators $S, T\colon E \to F$ satisfy $S(s,t) = T(s,t)$ for $\mu \times \nu$-almost all $(s,t) \in \mathcal{S} \times \mathcal{T}$, then $Sx = Tx$ for each $x \in E$.*

Solution: Assume that the kernels of two integral operators $S, T\colon E \to F$ satisfy $S(s,t) = T(s,t)$ for $\mu \times \nu$-almost all $(s,t) \in \mathcal{S} \times \mathcal{T}$, where μ and ν are σ-finite measures. That is, there exists a $\mu \times \nu$-null set D such that $S(s,t) = T(s,t)$ holds for all $(s,t) \notin D$. Fix some $x \in E$ and pick a ν-null subset B_1 of $\mathcal{T}$ such that for each $t \notin B_1$ both functions $S(\cdot,t)x(\cdot)$ and $T(\cdot,t)x(\cdot)$ belong to $L_1(\mu)$ and satisfy

$$Sx(t) = \int_{\mathcal{S}} S(s,t)x(s)\,d\mu(s) \quad \text{and} \quad Tx(t) = \int_{\mathcal{S}} T(s,t)x(s)\,d\mu(s)\,.$$

From $(\mu \times \nu)^*(D) = 0$ and the σ-finiteness of the measures, it follows from Fubini's Theorem 1.97 that there exists a ν-null subset B_2 such that for each $t \notin B_2$ the subset $A_t = \{s \in \mathcal{S}\colon\ (s,t) \in D\}$ of $\mathcal{S}$ is a μ-null set. Now note that if $B = B_1 \cup B_2$, then B is a ν-null set and, moreover, for each $t \notin B$ the functions $S(\cdot,t)x(\cdot)$ and $T(\cdot,t)x(\cdot)$ belong to $L_1(\mu)$ and satisfy $S(s,t)x(s) = T(s,t)x(s)$ for all s outside of the μ-null set A_t. This implies

$$Sx(t) = \int_{\mathcal{S}} S(s,t)x(s)\,d\mu(s) = \int_{\mathcal{S}} T(s,t)x(s)\,d\mu(s) = Tx(t)$$

for all $t \notin B$. Therefore, $Sx = Tx$ for each $x \in E$. This means that the operators S and T are identical, i.e. $S = T$. ∎

Problem 5.1.2. *Let $(\mathcal{S}, \Sigma_1, \mu)$ and $(\mathcal{T}, \Sigma_2, \nu)$ be two σ-finite measure spaces, and consider a $\mu \times \nu$-measurable function $K\colon \mathcal{S} \times \mathcal{T} \to \mathbb{R}$. Assume that a function $f \in L_0(\mu)$ satisfies $K(\cdot,t)f(\cdot) \in L_1(\mu)$ for ν-almost all $t \in \mathcal{T}$. Show* *that the function $g\colon \mathcal{T} \to \mathbb{R}$ defined ν-almost everywhere by the formula*

$$g(t) = \int_{\mathcal{S}} K(s,t) f(s)\, d\mu(s)$$

is ν-measurable, i.e., $g \in L_0(\nu)$.

Solution: Assume that for some $f \in L_0(\mu)$ we have $K(\cdot,t)f(\cdot) \in L_1(\mu)$ for ν-almost all $t \in \mathcal{T}$. So, there exists a ν-null subset B_0 of $\mathcal{T}$ such that $K(\cdot,t)f(\cdot) \in L_1(\mu)$ for all $t \notin B_0$. Without loss of generality, we can assume that $f(s) \in \mathbb{R}$ for each $s \in \mathcal{S}$.

By Lemma 1.90 there exists a sequence $\{K_n\}$ of $\mu \times \nu$-integrable functions satisfying $|K_n(s,t)| \le |K(s,t)|$ and $K_n(s,t) \to K(s,t)$ for all (s,t) in $\mathcal{S} \times \mathcal{T}$. Since $|K_n(s,t)f| \le |K(s,t)f|$ and since each K_n is $\mu \times \nu$-integrable for each $n \in \mathbb{N}$, Fubini's Theorem 1.97 guarantees the existence of some ν-null set B_n such that for each $t \notin B_n$ the function $s \mapsto K_n(s,t)f(s)$ is μ-integrable and the formula $g_n(t) = \int_{\mathcal{S}} K_n(s,t)f(s)\, d\mu(s)$ defines a ν-integrable function. If $B = \bigcup_{n=0}^{\infty} B_n$, then B is a ν-null set and from the Lebesgue Dominated Convergence Theorem it follows that $g_n(t) \to g(t)$ holds for each $t \notin B$. This shows that g is a measurable function, i.e., $g \in L_0(\nu)$. ∎

Problem 5.1.3. *Let $T(\cdot,\cdot)$ be a $\mu \times \nu$-measurable function such that:*

(a) $\int_{\mathcal{S}} |T(s,t)|\, d\mu(s) \le 1$ *for ν-almost all $t \in \mathcal{T}$.*

(b) $\int_{\mathcal{T}} |T(s,t)|\, d\nu(t) \le 1$ *for μ-almost all $s \in \mathcal{S}$.*

Prove that the integral operator T with kernel $T(\cdot,\cdot)$ is a bounded operator from $L_2(\mu)$ to $L_2(\nu)$.

Solution: Fix $x \in L_2(\mu)$, and formally let $y(t) = \int_{\mathcal{S}} T(s,t)x(s)\, d\mu(s)$. We shall verify that the formula defining y indeed gives a function that belongs to $L_2(\nu)$. To this end, apply Hölder's inequality in connection with the first condition on the kernel to see that for ν-almost all $t \in \mathcal{T}$ we have:

$$\begin{aligned}
\int_{\mathcal{S}} |T(s,t)x(s)|\, d\mu(s) &= \int_{\mathcal{S}} \big[|T(s,t)||x(s)|^2\big]^{\frac{1}{2}} \cdot |T(s,t)|^{\frac{1}{2}}\, d\mu(s) \\
&\le \Big[\int_{\mathcal{S}} |T(s,t)||x(s)|^2\, d\mu(s)\Big]^{\frac{1}{2}} \cdot \Big[\int_{\mathcal{S}} |T(s,t)|\, d\mu(s)\Big]^{\frac{1}{2}} \\
&\le \Big[\int_{\mathcal{S}} |T(s,t)||x(s)|^2\, d\mu(s)\Big]^{\frac{1}{2}}.
\end{aligned}$$

Therefore, we get

$$\begin{aligned}
\int_{\mathcal{T}} |y(t)|^2\, d\nu(t) &\le \int_{\mathcal{T}} \Big[\int_{\mathcal{S}} |T(s,t)x(s)|\, d\mu(s)\Big]^2 d\nu(t) \\
&\le \int_{\mathcal{T}} \Big[\int_{\mathcal{S}} |T(s,t)||x(s)|^2\, d\mu(s)\Big] d\nu(t).
\end{aligned}$$

Changing the order of integration (by virtue of Tonelli's theorem) and using the second condition on the kernel, we finally obtain

$$\int_{\mathcal{T}} |y(t)|^2\, d\nu(t) \le \int_{\mathcal{S}} \Big[\int_{\mathcal{T}} |T(s,t)|\, d\nu(t)\Big] \cdot |x(s)|^2\, d\mu(s) \le \int_{\mathcal{S}} |x(s)|^2\, d\mu(s) = \|x\|_2^2.$$

This proves that the expression defining y is indeed a function that belongs to $L_2(\nu)$. Moreover, it follows that the integral operator with kernel $T(\cdot,\cdot)$ defines a bounded operator from $L_2(\mu)$ to $L_2(\nu)$ satisfying $\|T\| \le 1$.

Remark: General conditions on the kernel $T(\cdot,\cdot)$ guaranteeing that the integral operator generated by this kernel belongs to $\mathcal{L}(L_p(\mu), L_q(\nu))$ for some $p, q \in [1,\infty]$ can be found in [**45**, Chapter XI]. See also Problem 5.1.10. ∎

Problem 5.1.4. ([**20**, Section 4.1]) *Let $T\colon L_2(\mu) \to L_2(\nu)$ be an integral operator. Assume that there exists a subset $B \in \Sigma_2$ such that $\nu(B) = 0$ and for each $x \in L_2(\mu)$ we have $\int_{\mathcal{S}} |T(s,t)x(s)|\,d\mu(s) < \infty$ for each $t \notin B$. That is, the exceptional set in the definition of the integral operator T is independent of the "input" function x.*

*Prove that $\int_{\mathcal{S}} |T(s,t)|^2\,d\mu(s) < \infty$ for ν-almost all $t \in \mathcal{T}$. (Operators that satisfy this condition are known as **Carleman operators**.)*

Solution: Fix any point $t \in \mathcal{T} \setminus B$ and define the positive linear functional $\phi_t\colon L_2(\mu) \to \mathbb{R}$ via the formula $\phi_t(x) = \int_{\mathcal{S}} |T(s,t)|x(s)\,d\mu(s)$. The hypothesis $\int_{\mathcal{S}} |T(s,t)x(s)|\,d\mu(s) < \infty$ for each $t \notin B$ and each $x \in L_2(\mu)$ implies that ϕ_t is a well-defined positive linear functional. It follows that ϕ_t is a continuous linear functional on $L_2(\mu)$. Therefore, by the Riesz Representation Theorem, this linear functional is represented by a unique function in $L_2(\mu)$. This implies that $T(\cdot,t) \in L_2(\mu)$ for each $t \notin B$. ∎

Problem 5.1.5. *Assume that $(\mathcal{S}, \Sigma_1, \mu)$ and $(\mathcal{T}, \Sigma_2, \nu)$ are two σ-finite measure spaces, and let $1 < p, q < \infty$ satisfy $\frac{1}{p} + \frac{1}{q} = 1$. Show that an integral operator $T\colon L_p(\mu) \to L_0(\nu)$ maps the closed unit ball of $L_p(\mu)$ to an order bounded subset of $L_0(\nu)$ (i.e., there exists a function $g \in L_0(\nu)$ such that $|Tx| \le g$ for each $x \in L_p(\mu)$ with $\|x\|_p \le 1$) if and only if there exists a ν-null set $B \in \Sigma_2$ such that*

$$\int_{\mathcal{S}} |T(s,t)|^q\,d\mu(s) < \infty$$

for each $t \in \mathcal{T} \setminus B$.

In particular, an integral operator $T\colon L_2(\mu) \to L_2(\nu)$ is a Carleman operator if and only if T maps the closed unit ball of $L_2(\mu)$ into an order bounded subset of $L_0(\nu)$.

Solution: We shall denote by U the closed unit ball of the Banach space $L_p(\mu)$, i.e., $U = \{x \in L_p(\mu)\colon \|x\|_p \le 1\}$. Suppose first that there exists a ν-null set $B \in \Sigma_2$ such that

$$\int_{\mathcal{S}} |T(s,t)|^q\,d\mu(s) < \infty$$

for each $t \in \mathcal{T} \setminus B$. In view of this condition, the function

$$g(t) = \Big[\int_{\mathcal{S}} [T(s,t)]^q\,d\mu(s)\Big]^{\frac{1}{q}}$$

is finite for each $t \notin B$, and in view of Problem 5.1.2 the function g is ν-measurable. Thus, $g \in L_0(\nu)$. We claim that g satisfies the required property. To see this, fix $x \in U$. Then there exists a ν-null set B_1 such that

$$Tx(t) = \int_{\mathcal{S}} T(s,t)x(s)\,d\mu(s)$$

for each $t \notin B_1$. Therefore, by Hölder's inequality, for each $t \notin B \cup B_1$ we have

$$|Tx(t)| \le \Big[\int_{\mathcal{S}} [T(s,t)]^q\,d\mu(s)\Big]^{\frac{1}{q}} \cdot \Big[\int_{\mathcal{S}} |x(s)|^p\,d\mu(s)\Big]^{\frac{1}{p}} = g(t)\|x\|_p \le g(t)\,.$$

That is, $|Tx| \le g$ holds for each $x \in U$.

For the converse, assume that there exists a function $g \in L_0(\nu)$ such that $|Tx| \le g$ for each $x \in U$. Without loss of generality, we can assume that $g(t) < \infty$ for each $t \in \mathcal{T}$. If we assume that the measure μ is separable,[1] then the proof is considerably easier. We shall present this simpler proof first and then we shall provide the general proof.

So, assume that the measure μ is separable. In this case, $L_p(\mu)$ is separable. So, there exists a countable dense subset $\{x_1, x_2, \ldots\}$ of U. Now for each n pick a ν-null set B_n such that

$$\big|Tx_n(t)\big| = \Big|\int_{\mathcal{S}} T(s,t)x_n(s)\,d\mu(s)\Big| \le g(t)$$

holds for each $t \notin B_n$. Therefore, for each t that does not belong to the ν-null set $\bigcup_{n=1}^{\infty} B_n$ we have

$$\Big[\int_{\mathcal{S}} [T(s,t)]^q d\mu(s)\Big]^{\frac{1}{q}} = \sup_{n\in\mathbb{N}} \Big|\int_{\mathcal{S}} T(s,t)x_n(s)\,d\mu(s)\Big| = \sup_{n\in\mathbb{N}} \big|Tx_n(t)\big| \le g(t) < \infty\,.$$

Thus, T satisfies the desired condition.

For the general case, assume that μ is an arbitrary σ-finite measure. As before, the function $T(s,t)$ is the kernel of the operator T. We shall say that a non-empty subset D of $L_0(\mu)$ and the kernel $T(s,t)$ are **compatible** if there exists a null set $B_0 \in \Sigma_2$ such that for each $t \notin B_0$ and for each $x \in D$:

(1) The function $s \mapsto T(s,t)x(s)$ is μ-measurable and its extended Lebesgue integral exists.

(2) The function x_t, defined by $x_t(s) = |x(s)|\big[\operatorname{Sgn} T(s,t)\big]$ if $T(s,t) \neq 0$ and $x_t(s) = 0$ if $T(s,t) = 0$, belongs to D, i.e., $x_t \in D$.

Note that conditions (1) and (2) are both independent of the representative chosen from the equivalence class $[x]$ which, in accordance with our agreement, is denoted by x rather than $[x]$. In addition, for each $t \notin B_0$ the (extended) value of the integral $\int_{\mathcal{S}} T(s,t)x(s)\,d\mu(s)$ is also independent of the representative from the equivalence class $[x]$. Using Fubini's theorem, it is easy to see that if $T(\cdot,\cdot)$ is non-negative, Δ is a solid subset of $L_0(\mu)$ and $D = \Delta \cap L_0(\mu)^+$, then (D, T) is an example of a compatible pair.

[1] Recall that a measure μ is **separable** if and only if $L_p(\mu)$ is a separable Banach space; for details see [**4**, Section 12.5].

Now let (D,T) be an arbitrary compatible pair. Then an extended function $y\colon \mathcal{T}\setminus B_0 \to [-\infty,\infty]$ can be defined via the formula

$$y(t) = \sup\Big\{\int_{\mathcal{S}} T(s,t)x(s)\,d\mu(s)\colon x \in D\Big\}. \qquad (\star)$$

It should be noticed that y is a function and not an equivalence class. If the set D is not countable, then the ν-measurability of the function y, which is the pointwise supremum of ν-measurable functions, is far from being obvious. Nevertheless, according to Y. I. Gribanov (see [**45**, Theorem XI.1.2]) the following result is true.

- (Gribanov's theorem) *If (D,T) is a compatible pair, then the function y is ν-measurable. Moreover, there exists a sequence $\{f_n\}$ in D such that*

$$y(t) = \sup_{n\in\mathbb{N}} \int_{\mathcal{S}} T(s,t)f_n(s)\,d\mu(s) \qquad (\star\star)$$

for almost all $\nu \in \mathcal{T}$.

Now we are ready to prove the converse of our statement—without assuming that μ is separable. Consider the $\mu\times\nu$-measurable function $|T(\cdot,\cdot)|$. As noted above, the pair $(D,|T(\cdot,\cdot)|)$, where $D = U^+$ is the positive part of the closed unit ball U in $L_p(\mu)$, is compatible. Let $B_0 \in \Sigma_2$ be the corresponding ν-null set. By Gribanov's theorem, the function

$$y(t) = \sup_{x\in D} \int_{\mathcal{S}} |T(s,t)|x(s)\,d\mu(s)$$

is ν-measurable and there is a sequence $\{x_n\}$ in D such that

$$y(t) = \sup_{n\in\mathbb{N}} \int_{\mathcal{S}} |T(s,t)|x_n(s)\,d\mu(s)$$

for all $t \in \mathcal{T}\setminus B_1$, where B_1 is some ν-null set in $\in \Sigma_2$. This obviously guarantees that $y \le g$, and so $y(t) \le g(t)$ for all $t \in \mathcal{T}\setminus B_2$, where B_2 is some ν-null set in Σ_2. (As a matter of fact, it is easy to see that y is the supremum of the set $\{|T|x\colon\ x \in D\}$ in the Dedekind complete Riesz space $L_0(\nu)$.)

Since $T(\cdot,\cdot)$ is a $\mu\times\nu$-measurable function, Fubini's theorem guarantees that there exists a ν-null set $B_3 \in \Sigma_2$ such that for each $t \notin B_3$ the function $T(\cdot,t)$ is μ-measurable. Finally, let us verify that the μ-measurable function $|T(\cdot,t)|$ belongs to $L_q(\mu)$ for each $t \in \mathcal{T}\setminus(B_0\cup B_1\cup B_2\cup B_3)$. To this end, note that a μ-measurable function $f\colon \mathcal{S}\to\mathbb{R}$ belongs to $L_q(\mu)$ if and only if

$$\|f\|_q = \sup_{x\in U} \int_{\mathcal{T}} f(s)x(s)\,d\mu(s) = \sup_{x\in D} \int_{\mathcal{T}} |f(s)|x(s)\,d\mu(s) < \infty .$$

Applying this to the function $|T(\cdot,t)|$, we get

$$\begin{aligned}\Big[\int_{\mathcal{T}} [T(s,t)]^q\,d\mu(s)\Big]^{\frac{1}{q}} &= \sup_{x\in U} \int_{\mathcal{T}} |T(s,t)|x(s)\,d\mu(s)\\ &= \sup_{x\in D} \int_{\mathcal{T}} |T(s,t)|x(s)\,d\mu(s)\\ &= y(t) \le g(t) < \infty .\end{aligned}$$

This establishes that T satisfies the desired condition. ∎

Problem 5.1.6. *Give an example of an integral operator on $L_2[0,1]$ which is not a Carleman operator—where $[0,1]$ is equipped with its Lebesgue measure.*

Solution: This example can be found in [**20**, Section 4.1.3]. Let g be an arbitrary non-negative function in $L_1[0,1]$ such that $\|g\|_1 = \frac{1}{2}$ and $g \notin L_2[0, \frac{1}{2}]$ (for instance, let $g(s) = \frac{1}{4\sqrt{s}}$), and consider on $[0,1] \times [0,1]$ the function $T(s,t) = g(|s-t|)$. Clearly, $T(\cdot,\cdot)$ is a measurable function. To show that the integral operator T with kernel $T(s,t)$ belongs to $\mathcal{L}(L_2[0,1])$, we will verify that the kernel satisfies the conditions given in Problem 5.1.3. Indeed, for each $t \in [0,1]$ we have

$$\int_0^1 |T(s,t)|\, ds = \int_0^t g(t-s)\, ds + \int_t^1 g(s-t)\, ds \le 2\int_0^1 g(s)\, ds = 1\,.$$

By symmetry, $\int_0^1 |T(s,t)|\, dt \le 1$ also holds for each $s \in [0,1]$.

Finally, for each $t \in [0,1]$ we have

$$\begin{aligned}
\int_0^1 |T(s,t)|^2\, ds &= \int_0^t g^2(t-s)\, ds + \int_t^1 g^2(s-t)\, ds \\
&= \int_0^t g^2(u)\, du + \int_0^{1-t} g^2(u)\, du \\
&\ge \int_0^{\frac{1}{2}} g^2(u)\, du = \infty\,.
\end{aligned}$$

This shows that the kernel T does not satisfy the Carleman condition at each point $t \in [0,1]$.

Remark: In view of Problem 5.1.4, the above example demonstrates that in general the exceptional set in the definition of an integral operator depends on the function the operator is acting upon. ∎

Problem 5.1.7. *Let $\{T_n\}$ be an increasing sequence of regular integral operators and let T be another regular integral operator. Show that $T_n \uparrow T$ holds in $\mathcal{L}_r(E,F)$ if and only if $T_n(s,t) \uparrow T(s,t)$ holds for $\mu \times \nu$-almost all (s,t) in $C_E \times C_F$.*

Solution: Without loss of generality we can assume that E and F are order dense ideals in $L_0(\mu)$ and $L_0(\nu)$, respectively—and so we have $C_E = \mathcal{S}$ and $C_F = \mathcal{T}$. Suppose first that $T_n \uparrow T$ in $\mathcal{L}_r(E,F)$. From Theorem 5.5, it follows that there exists a $\mu \times \nu$-null set A such that $T_n(s,t) \uparrow \le T(s,t)$ holds for each $(s,t) \notin A$. Replacing $T_n(\cdot,\cdot)$ by $T_n(\cdot,\cdot)\chi_A$ and $T(\cdot,\cdot)$ by $T(\cdot,\cdot)\chi_A$, respectively, we can assume that $T_n(s,t) \uparrow \le T(s,t)$ holds for each (s,t). Let $K(s,t) = \lim_{n\to\infty} T_n(s,t)$ for each $(s,t) \in \mathcal{S} \times \mathcal{T}$. So, K is a $\mu \times \nu$-measurable function, and since for each $x \in E^+$ we have $T_n x(t) = \int_{\mathcal{S}} T_n(s,t)x(s)\, d\mu(s) \uparrow \int_{\mathcal{S}} K(s,t)x(s)\, d\mu(s)$ and $T_n x(t) \uparrow Tx(t)$ for ν-almost all $t \in \mathcal{T}$, we infer that $Tx(t) = \int_{\mathcal{S}} K(s,t)x(s)\, d\mu(s)$ for ν-almost all t. From the uniqueness of the kernel of the operator T (see Corollary 5.7), we get $T(s,t) = K(s,t)$ for $\mu \times \nu$-almost all (s,t). Hence, $T_n(s,t) \uparrow T(s,t)$ for $\mu \times \nu$-almost all $(s,t) \in \mathcal{S} \times \mathcal{T}$.

For the converse, assume that $T_n(s,t) \uparrow T(s,t)$ for $\mu\times\nu$-almost all (s,t); without loss of generality we can assume that $T_n(s,t) \uparrow T(s,t)$ for all (s,t). Now if $x \in E^+$, then for each $t \in \mathcal{T}$ we have $T_n(s,t)x(s) \uparrow T(s,t)x(s)$ for μ-almost all $s \in \mathcal{S}$. So, by the Lebesgue Dominated Convergence Theorem, we get

$$T_n x(t) = \int_{\mathcal{S}} T_n(s,t)x(s)\,d\mu(s) \uparrow \int_{\mathcal{S}} T(s,t)x(s)\,d\mu(s) = Tx(t)$$

for all $t \in \mathcal{T}$. This shows that $T_n \uparrow T$ in $\mathcal{L}_r(E,F)$. ∎

Problem 5.1.8. *Let E be an ideal in $L_0(\mu)$, where $(\mathcal{S},\Sigma,\mu)$ is a non-atomic σ-finite measure space. If $T\colon E \to E$ is a regular integral operator, then show that T is disjoint from the identity operator, i.e., $|T| \wedge I = 0$.*

Solution: Assume that $(\mathcal{S},\Sigma,\mu)$ is a non-atomic σ-finite measure space and that $T\colon E \to E$ is a regular integral operator, where E is an ideal in $L_0(\mu)$. Without loss of generality we can suppose that E is order dense and that the kernel of the operator T satisfies $T(s,t) \geq 0$ for all $(s,t) \in \mathcal{S} \times \mathcal{T}$. Since $\mu \times \mu$ is also a σ-finite measure, there exists a sequence $\{\phi_n\}$ of $\mu \times \mu$-step functions such that $0 \leq \phi_n(s,t) \uparrow T(s,t)$ for all (s,t) in $\mathcal{S} \times \mathcal{T}$.

In order to establish that $I \perp T$, it suffices to show (according to Problem 5.1.7) that $I \perp T$ holds true when the kernel of T is of the form $T(s,t) = \chi_A(s,t)$ for some $\mu \times \mu$-measurable subset A of $\mathcal{S} \times \mathcal{S}$ with $(\mu \times \mu)^*(A) < \infty$. So, suppose that $T(s,t) = \chi_A(s,t)$ for some $\mu\times\mu$-measurable subset A of $\mathcal{S}\times\mathcal{S}$ with $(\mu \times \mu)^*(A) < \infty$ and all $(s,t) \in \mathcal{S} \times \mathcal{S}$.

Since $\mathbf{1} = \chi_{\mathcal{S}}$ belongs to $L_0(\mu)$, there exists (according to Corollary 1.91) a disjoint sequence $\{V_n\}$ in Σ satisfying $\chi_{V_n} \in E$ and $\mu(V_n) < \infty$ for each n, and $\mathcal{S} = \bigcup_{n=1}^{\infty} V_n$. If we let $\mathcal{S}_n = \bigcup_{i=1}^{n} V_i$, then $\chi_{\mathcal{S}_n} \in E$ and $\mu(\mathcal{S}_n) < \infty$ hold for each n. Moreover, $\chi_{\mathcal{S}_n} \uparrow \mathbf{1}$ in $L_0(\mu)$. In particular, note that if $x \in L_0^+(\mu)$, then the sequence $\{x \wedge n\chi_{\mathcal{S}_n}\}$ in E satisfies $0 \leq x \wedge n\chi_{\mathcal{S}_n} \uparrow x$ in $L_0(\mu)$.

Now fix some n. Since the measure μ is non-atomic, for each natural number k there exist pairwise disjoint μ-measurable subsets $B_1^n, \ldots, B_k^n$ of $\mathcal{S}$ satisfying $\mathcal{S}_n = \bigcup_{i=1}^{k} B_i^n$ and $\mu(B_i^n) = \frac{\mu(\mathcal{S}_n)}{k}$ for each $i = 1, \ldots, k$. For each $1 \leq i \leq k$ let $C_i^n = \mathcal{S}_n \setminus \bigcup_{j \neq i} B_j^n$. Clearly, $\chi_{B_i^n}, \chi_{C_i^n} \in E$ for each $i = 1, \ldots, k$.

If for an arbitrary subset D of $\mathcal{S} \times \mathcal{S}$ we let $D_t = \{s \in \mathcal{S}\colon\ (s,t) \in D\}$, then

$$\begin{aligned} T\chi_{B_i^n}(t) &= \int_{\mathcal{S}} \chi_A(s,t)\chi_{B_i^n}\,d\mu(s) = \int_{\mathcal{S}} \chi_{A_t \cap B_i^n}(s)\,d\mu(s) \\ &= \mu\big(A_t \cap B_i^n\big) \leq \mu(B_i^n) = \tfrac{\mu(\mathcal{S}_n)}{k} \end{aligned}$$

for μ-almost all $t \in \mathcal{S}$. This implies that

$$0 \leq \big[(T \wedge I)\chi_{\mathcal{S}_n}\big](t) \leq \big[T\chi_{B_i^n} + I\chi_{C_i^n}\big](t) = \tfrac{\mu(\mathcal{S}_n)}{k} + \chi_{C_i^n}(t)$$

for μ-almost all t and all $i = 1, \ldots, k$. This implies

$$0 \leq \big[(T \wedge I)\chi_{\mathcal{S}_n}\big](t) \leq \tfrac{\mu(\mathcal{S}_n)}{k}$$

for all k and μ-almost all $t \in \mathcal{S}$. Consequently, $\big[(T \wedge I)\chi_{\mathcal{S}_n}\big](t) = 0$ holds for μ-almost all t, or $(T \wedge I)\chi_{\mathcal{S}_n} = 0$ for each n.

Finally, let $x \in E^+$. From $0 \leq (T\wedge I)(x \wedge n\chi_{\mathcal{S}_n}) \leq (T\wedge I)(n\chi_{\mathcal{S}_n}) = 0$, we easily get $(T\wedge I)(x \wedge n\chi_{\mathcal{S}_n}) = 0$ for each n. Since $0 \leq x \wedge n\chi_{\mathcal{S}_n} \uparrow x$ in E and $T\wedge I$ is order continuous, we infer that $(T \wedge I)(x) = 0$ for each $x \in E^+$. That is, $T \wedge I = 0$. ∎

Problem 5.1.9 (Abramovich–Aliprantis–Zame [**3**]). *Show that for an arbitrary Dedekind complete AM-space E with unit e the following two statements are equivalent.*

(a) *There exists a probability measure space (Ω, Σ, π) such that E is lattice isometric to $L_\infty(\pi)$ with e corresponding to the constant function $\mathbf{1}$ on Ω.*

(b) *E admits a strictly positive order continuous linear functional.*

Proof. (a) $\Longrightarrow$ (b) If (Ω, Σ, π) is a probability measure space, then $L_\infty(\pi)$ is Dedekind complete with order unit $\mathbf{1}$, and integration against π yields a strictly positive order continuous linear functional on $L_\infty(\pi)$. Moreover, if $T\colon E \to L_\infty(\pi)$ is a surjective lattice isometry with $T(e) = \mathbf{1}$, then it should be clear that the mapping $\phi\colon E \to \mathbb{R}$, defined by $\phi(x) = \int T(x)\, d\pi$, is a strictly positive order continuous linear functional on E satisfying $\phi(e) = 1$.

(b) $\Longrightarrow$ (a) Let $\phi\colon E \to \mathbb{R}$ be an order continuous strictly positive linear functional. Replacing ϕ by $\frac{\phi}{\phi(e)}$, we can suppose that $\phi(e) = 1$. Next, define $|||\cdot|||\colon E \to \mathbb{R}$ by the formula $|||y||| = \phi\big(|y|\big)$. It is easy to check that $|||\cdot|||$ is an L-norm on E, which (in view of the order continuity of ϕ) is also order continuous. Let $\hat{E}$ be the completion of $(E, |||\cdot|||)$. From Problem 3.4.6 we know that $\hat{E}$ is an AL-space.

Next, we claim that E is an ideal in $\hat{E}$. Clearly, E is a Riesz subspace of $\hat{E}$. We must verify that whenever $0 \le z \le y$ holds with $y \in E$ and $z \in \hat{E}$, then $z \in E$. So, assume $0 \le z \le y \in E$ with $y \in E$. Since E is norm dense in its completion $\hat{E}$, there exists a sequence $\{z_n\} \subseteq E$ converging to z; that is, $|||z_n - z||| \to 0$. Since $z \le y$ and the mapping $u \mapsto u \wedge y$ is $|||\cdot|||$-continuous, we can assume (by replacing $\{z_n\}$ with $\{z_n \wedge y\}$) that $0 \le z_n \le y$ holds for each n. Also, by passing to a subsequence (if necessary), we can suppose that $|||z_n - z||| < \frac{1}{n2^n}$ for each n. Now let

$$w_{k,m} = \bigvee_{i=k}^{k+m} z_i\,,$$

and note that $0 \le w_{k,m} \le y$ for all k and m. Since E is Dedekind complete, there exists some $w_k \in E$ satisfying $w_{k,m} \uparrow_m w_k$ in E, and since $|||\cdot|||$ is order continuous on E, it follows that $|||w_{k,m} - w_k||| \xrightarrow[m\to\infty]{} 0$. From

$$|w_{k,m} - z| = \Big|\bigvee_{i=k}^{k+m} z_i - z\Big| = \Big|\bigvee_{i=k}^{k+m} (z_i - z)\Big| \le \bigvee_{i=k}^{k+m} |z_i - z| \le \sum_{i=k}^{k+m} |z_i - z|\,,$$

we see that $|||w_{k,m} - z||| \le \sum_{i=k}^{k+m} |||z_i - z||| \le \frac{1}{k}$, and so by the order continuity of the norm $|||\cdot|||$, we get $|||w_k - z||| = \lim_{m\to\infty} |||w_{k,m} - z||| \le \frac{1}{k}$ for each k. In particular, $\lim_{k\to\infty} |||w_k - z||| = 0$. Clearly, the sequence $\{w_k\}$ is decreasing and bounded from below by zero. Using once more the Dedekind completeness of E, it follows that $w_k \downarrow w$ holds in E for some $w \in E$. Now from the order continuity of the norm $|||\cdot|||$ on E, we infer that $|||w_k - w||| \to 0$. But then $|||z - w||| = \lim_{k\to\infty} |||z - w_k||| = 0$, proving that $z = w \in E$.

The preceding conclusion also implies that e is a weak unit in $\hat{E}$. To see this, assume that $u \wedge e = 0$ holds in $\hat{E}$ for some $u \in \hat{E}$. Pick a sequence $\{u_n\}$ of E^+ such that $|||u_n - u||| \to 0$. Now let $v_n = u_n \wedge u$ and note that (since E is an ideal of $\hat{E}$) we have $v_n \in E$ and $0 \le v_n \le u$ for each n and that $|||v_n - u||| \to 0$. Clearly, $v_n \wedge e = 0$ for each n, and since e is an order unit of E, we infer that $v_n = 0$ for each n. This implies $u = 0$ and it confirms that e is a weak unit of $\hat{E}$. Next, notice that by the Kakutani–Bohnenblust–Nakano Representation Theorem 3.5, we can find a probability measure space (Ω, Σ, π) and a surjective lattice isometry $T\colon \hat{E} \to L_1(\pi)$ such that $T(e) = \mathbf{1}$.

Since T is a surjective lattice isomorphism of $\hat{E}$ onto $L_1(\pi)$, it sends norm dense ideals of $\hat{E}$ to norm dense ideals in $L_1(\pi)$, and hence $T(E)$ must be a norm dense ideal in $L_1(\pi)$. Furthermore, $T(E) = L_\infty(\pi)$ since $T(e) = \mathbf{1}$. Finally, from $\|x\| = \inf\{\lambda \ge 0\colon\ |x| \le \lambda e\}$ and the definition of the $\|\cdot\|_\infty$ norm in $L_\infty(\pi)$, it follows that $T\colon \big(E, \|\cdot\|\big) \to \big(L_\infty(\pi), \|\cdot\|_\infty\big)$ is also a surjective lattice isometry. ■

Problem 5.1.10. *Let $(\mathcal{S}, \Sigma_1, \mu)$ and $(\mathcal{T}, \Sigma_2, \nu)$ be two arbitrary measure spaces, and assume that $1 < p, q < \infty$ satisfy $\frac{1}{p} + \frac{1}{q} = 1$. If a function $T(\cdot,\cdot) \in L_p(\mu \times \nu)$, then show that the formula*

$$Tx(t) = \int_{\mathcal{S}} T(s,t)x(s)\,d\mu(s)$$

*defines a regular integral operator from $L_q(\mu)$ to $L_p(\nu)$ such that its norm satisfies $\|T\| \le \|T(\cdot,\cdot)\|_p$. (When $p = q = 2$, such an operator T is called a **Hilbert–Schmidt operator**.)*

Solution: From Fubini's Theorem 1.97 we know that $T(\cdot,t) \in L_p(\mu)$ for ν-almost all $t \in \mathcal{T}$. So, if $x \in L_q(\mu)$, then $T(\cdot,t)x(\cdot) \in L_1(\mu)$ for ν-almost all $t \in \mathcal{T}$. This implies that the formula

$$Tx(t) = \int_{\mathcal{S}} T(s,t)x(s)\,d\mu(s)$$

defines a measurable function, i.e., $Tx \in L_0(\nu)$. We claim that $Tx \in L_p(\nu)$. To see this, notice that Hölder's inequality implies

$$\begin{aligned}
\int_{\mathcal{T}} |Tx(t)|^p\,d\nu(t) &= \int_{\mathcal{T}} \Big| \int_{\mathcal{S}} T(s,t)x(s)\,d\mu(s)\Big|^p\,d\nu(t)\\
&\le \int_{\mathcal{T}} \Big(\Big[\int_{\mathcal{S}} |T(s,t)|^p\,d\mu(s)\Big]^{\frac{1}{p}} \|x\|_q\Big)^p\,d\nu(t)\\
&= \Big(\int_{\mathcal{T}}\Big[\int_{\mathcal{S}} |T(s,t)|^p\,d\mu(s)\Big]\,d\nu(t)\Big)\cdot \|x\|_q^p\,.
\end{aligned}$$

So, $\|Tx\|_p \le \|T(\cdot,\cdot)\|_p \cdot \|x\|_q$. This (according to Theorem 5.11) shows that T is a regular integral operator from $L_q(\mu)$ to $L_p(\nu)$ and that $\|T\| \le \|T(\cdot,\cdot)\|_p$. ■

Problem 5.1.11. *Show that Hilbert–Schmidt operators are Carleman operators.*

Solution: Let $T\colon L_2(\mu) \to L_2(\nu)$ be a Hilbert–Schmidt operator. Then, according to the definition of a Hilbert–Schmidt operator (see Problem 5.1.10), its kernel $T(\cdot,\cdot)$ belongs to $L_2(\mu \times \nu)$. That is,

$$\|T(\cdot,\cdot)\|_2^2 = \iint_{\mathcal{S}\times\mathcal{T}} |T(s,t)|^2 \, d(\mu \times \nu)(s,t) < \infty .$$

Applying Fubini's theorem to this integral, we obtain that $\int_{\mathcal{S}} |T(s,t)|^2 \, d\mu(s) < \infty$ for ν-almost all $t \in \mathcal{T}$. This shows that T is a Carleman operator; see the definition of a Carleman operator in Problem 5.1.4. ∎

Problem 5.1.12. *Assume that $(\mathcal{S}, \Sigma_1, \mu)$ are $(\mathcal{T}, \Sigma_2, \nu)$ are two σ-finite measure spaces, and assume that $(\mathcal{S}, \Sigma_1, \mu)$ is* ***atomic****—which in this case means that $\mathcal{S}$ is a union of at most countably many atoms.*

Show that a linear operator $T\colon E \to L_0(\nu)$, where E is an ideal in $L_0(\mu)$, is an integral operator if and only if T is order continuous.

Proof. The order continuity of each integral operator $T\colon E \to L_0(\nu)$ has been established in Corollary 5.12 and it is true for any measure space $(\mathcal{S}, \Sigma_1, \mu)$. It is the converse statement that requires the extra assumption that $(\mathcal{S}, \Sigma_1, \mu)$ is atomic. Since μ is σ-finite, we can represent $\mathcal{S}$ as a union $\bigcup_{n\in\mathcal{N}} A_n$ of atoms of positive measure, where $\mathcal{N}$ is at most countable. Clearly, each $x \in E$ can be written as $x = \sum_{n\in\mathcal{N}} \lambda_n \chi_{A_n}$, where λ_n are scalars and the series converges μ-almost everywhere. Since T is order continuous, it follows that $Tx = \sum_{n\in\mathcal{N}} \lambda_n T\chi_{A_n}$, where again the series converges ν-almost everywhere.

For each $s \in A_n$ and $t \in \mathcal{T}$ let $T(s,t) = \frac{1}{\mu(A_n)} T\chi_{A_n}(t)$. It is obvious that the function $T(\cdot,\cdot)$ is $\mu \times \nu$-measurable. Let us verify that $T(\cdot,\cdot)$ is the kernel of the operator T. For each $x \in E$ we have

$$\begin{aligned}\int_{\mathcal{S}} T(s,t)x(s)\, d\mu(s) &= \sum_{n\in\mathcal{N}} \tfrac{1}{\mu(A_n)} \big[T\chi_{A_n}(t)\big]\lambda_n \mu(A_n) \\ &= \sum_{n\in\mathcal{N}} \lambda_n T\chi_{A_n}(t) = Tx(t) .\end{aligned}$$

This shows that T is an integral operator. ∎

5.2. Abstract Integral Operators

Problem 5.2.1. *Let E and G be two Banach function spaces associated with $L_0(\pi)$. Show that $E \cap G$ is also a Banach function space—which is order dense if both E and F are order dense ideals.*

Solution: Observe first that if E and F are order dense ideals, then their intersection $E \cap G$ is also order dense. Indeed, if $0 < x \in L_0^+(\pi)$, pick some $y \in E$ with $0 < y \le x$, and then select some $z \in G$ such that $0 < z \le y$. Hence, $z \in E \cap G$ and $0 < z \le x$.

Let $\|\cdot\|_E$ and $\|\cdot\|_G$ denote the lattice norms of E and G, respectively, and define the function $\|\cdot\|\colon E\cap G\to[0,\infty)$ via the formula

$$\|x\| = \|x\|_E + \|x\|_G\,.$$

A straightforward verification shows that $\|\cdot\|$ is a lattice norm on the ideal $E\cap G$. We claim that $(E\cap G, \|\cdot\|)$ is a Banach lattice. To see this, assume that $\{x_n\}\subseteq E\cap G$ is a $\|\cdot\|$-Cauchy sequence. This implies that $\{x_n\}$ is both a $\|\cdot\|_E$-Cauchy sequence and a $\|\cdot\|_G$-Cauchy sequence. Since E and G are Banach function spaces, there exist $x\in E$ and $y\in G$ such that $\|x_n-x\|_E\to 0$ and $\|x_n-y\|_G\to 0$.

Now, according to Problem 1.2.13, there exist a subsequence $\{x_{k_n}\}$ of $\{x_n\}$, a vector $u\in E^+$, and a vector $v\in G^+$ such that

$$|x_{k_n}-x|\le \tfrac{1}{n}u \quad\text{and}\quad |x_{k_n}-y|\le \tfrac{1}{n}v$$

for each n. This implies $x=y$. Letting $w=x=y$, we obtain $w\in E\cap G$ and

$$\|x_n-w\| = \|x_n-w\|_E+\|x_n-w\|_G = \|x_n-x\|_E+\|x_n-y\|_G \longrightarrow 0\,.$$

This shows that $(E\cap G, \|\cdot\|)$ is a Banach function space. ∎

Problem 5.2.2. *Let G be an ideal in a Riesz space E, let F be a Dedekind complete Riesz space, and let $T\colon G\to F$ be a positive operator. Assuming that T has a regular extension to all of E, establish the following.*

(a) *The operator T has a minimal positive extension to all of E. That is, there exists a positive extension $T_G\colon E\to F$ of T such that if $S\colon E\to F$ is any positive extension of T, then $T_G\le S$.*

(b) *The minimal extension T_G of T is given by the formula*

$$T_G(x)=\sup\{Ty\colon\ y\in G \ \textit{and}\ 0\le y\le x\}$$

for each $x\in E^+$. In particular, $T_G(x)=0$ for each $x\in G^{\rm d}$.

(c) *If T is order continuous, then T_G is also order continuous.*

Proof. Let $S\colon E\to F$ be a regular extension of the positive operator $T\colon G\to F$. The fact that G is an ideal in E implies that for each $x\in G^+$ we have

$$\begin{aligned} S^+(x) &= \sup\bigl\{Sy\colon\ y\in E \text{ and } 0\le y\le x\bigr\} = \sup\bigl\{Sy\colon\ y\in G \text{ and } 0\le y\le x\bigr\}\\ &= \sup\bigl\{Ty\colon\ y\in G \text{ and } 0\le y\le x\bigr\} = Tx\,. \end{aligned}$$

This shows that S^+ is a positive extension of T to all of E.

(a) Let $\mathcal{C}=\bigl\{R\in\mathcal{L}_r^+(E,F)\colon\ R=T \text{ on } G\bigr\}$, i.e., $\mathcal{C}$ is the set of all positive linear extensions of T from G to all of E. By the above, $\mathcal{C}\ne\emptyset$. Now if $R_1,R_2\in\mathcal{C}$, then for each $x\in G^+$ we have

$$\begin{aligned} (R_1\wedge R_2)x &= \inf\bigl\{R_1y+R_2(x-y)\colon\ y\in E \text{ and } 0\le y\le x\bigr\}\\ &= \inf\bigl\{R_1y+R_2(x-y)\colon\ y\in G \text{ and } 0\le y\le x\bigr\}\\ &= \inf\bigl\{Ty+T(x-y)\colon\ y\in G \text{ and } 0\le y\le x\bigr\}\\ &= Tx\,. \end{aligned}$$

This shows that $R_1 \wedge R_2 \in \mathcal{C}$, and so $\mathcal{C}$ is directed downward. Since $\mathcal{C}$ consists of positive operators and $\mathcal{L}_r(E,F)$ is Dedekind complete, there exists a positive operator $T_G \in \mathcal{L}_r(E,F)$ such that $\mathcal{C} \downarrow T_G$. From $\{Rx\colon\ R \in \mathcal{C}\} \downarrow T_G x$ for each $x \in E^+$ and $Rx = Tx$ for each $x \in G$, it follows that $T_G x = Tx$ for all $x \in G$, i.e., T_G is a positive linear extension of T, and obviously $T_G = \min \mathcal{C}$.

(b) Define the mapping $T_m\colon E^+ \to F^+$ via the formula

$$T_m x = \sup\{Ty\colon\ y \in G \text{ and } 0 \le y \le x\}.$$

We claim that T_m is an additive function. To see this, let $u, v \in E^+$. If $y_1, y_2 \in G$ satisfy $0 \le y_1 \le u$ and $0 \le y_2 \le v$, then $0 \le y_1 + y_2 \le u + v$ and so

$$Ty_1 + Ty_2 = T(y_1 + y_2) \le T_m(u + v).$$

This implies $T_m u + T_m v \le T_m(u+v)$. Now assume that $y \in G$ satisfies $0 \le y \le u+v$. By the Riesz Decomposition Property there exist vectors $y_1, y_2 \in E^+$ satisfying $0 \le y_1 \le u$, $0 \le y_2 \le v$, and $y = y_1 + y_2$. Since G is an ideal, the latter implies $y_1, y_2 \in G$, and so $Ty = Ty_1 + Ty_2 \le T_m u + T_m v$ for all $0 \le y \le u+v$. Consequently, $T_m(u+v) \le T_m u + T_m v$. Thus, $T_m(u+v) = T_m u + T_m v$, and so T_m is additive.

By Theorem 1.15, the formula $T_m(x) = T_m(x^+) - T_m(x^-)$ extends T_m (uniquely) to a positive operator on all of E. We shall show that $T_G = T_m$. To see this, notice first that since T_m is a positive extension of T, we have $T_G \le T_m$. On the other hand, if $x \in E^+$, then for each $y \in G$ with $0 \le y \le x$ we have $Ty = T_G y \le T_G x$, and so

$$T_m x = \sup\{Ty\colon\ y \in G \text{ and } 0 \le y \le x\} \le T_G x.$$

Therefore, $T_m \le T_G$ is also true, and thus $T_G = T_m$.

If $0 \le x \in G^{\mathrm{d}}$, then $y \in G$ and $0 \le y \le x$ imply $y = 0$. This shows that $T_G x = 0$ for each $x \in G^{\mathrm{d}}$.

(c) Assume that $T\colon G \to F$ is also order continuous and let $0 \le x_\alpha \uparrow x$ in E. Clearly, $T_G x_\alpha \uparrow \le T_G x$. To see that $T_G x$ is the least upper bound of the net $\{T_G x_\alpha\}$ assume that $T_G x_\alpha \le u$ for all α. Fix $y \in G$ with $0 \le y \le x$ and note that $0 \le y \wedge x_\alpha \uparrow y$ in E (and, of course, in G). The order continuity of T implies $T(y \wedge x_\alpha) \uparrow Ty$. So, from

$$T(y \wedge x_\alpha) = T_G(y \wedge x_\alpha) \le T_G x_\alpha \le u,$$

we get $Ty \le u$. Consequently, $T_G x = \sup\{Ty\colon\ y \in G \text{ and } 0 \le y \le x\} \le u$. This shows $T_G x_\alpha \uparrow T_G x$, and the order continuity of T_G has been established. ■

Problem 5.2.3. *By means of Nakano's theorem (see Problem 1.2.22), show directly that if E is a Banach function space associated with a σ-finite measure π, then its order continuous dual E_{n}^* has weak units.*

Solution: Let E be a Banach function space associated with a σ-finite measure π. For each $f \in E_{\mathrm{n}}^*$ let $C_f = N_f^{\mathrm{d}} = \{x \in E\colon\ |f|(|x|) = 0\}^{\mathrm{d}}$ be the **carrier** of f. We claim that each pairwise disjoint subset of E_{n}^* is countable. To see this, let $\mathcal{C}$ be a pairwise disjoint subset of E_{n}^* consisting of non-zero linear functionals. According to Nakano's theorem (see Problem 1.2.22), for every pair $f, g \in \mathcal{C}$ with $f \neq g$ we have $C_f \cap C_g = \{0\}$. Since C_f is a band of E, there exists a unique set $D_f \in \Sigma$ with $\pi(D_f) > 0$ such that $C_f = \big\{x \in E\colon\ x = 0 \ \pi\text{-a.e. on } \Omega \setminus D_f\big\}$;

see Theorem 1.92. Notice that the mapping $f \mapsto D_f$ is one-to-one and that the condition $C_f \cap C_g = \{0\}$ is equivalent to $D_f \cap D_g = \emptyset$. Since π is σ-finite, the family $\{D_f\colon\ f \in \mathcal{C}\}$ is at most countable. This implies that $\mathcal{C}$ is at most countable.

Now assume that $\mathcal{C} = \{f_1, f_2, \ldots\}$ is a countable maximal pairwise disjoint family in $E^*_{\rm n}$ satisfying $\|f_n\| = 1$ for each n; by Zorn's lemma such a family exists. Since $E^*_{\rm n}$ is a Banach lattice (in fact, it is a band in E^*), it follows that the functional $f = \sum_{n=1}^{\infty} \frac{1}{2^n}|f_n| \in E^*_{\rm n}$ is a weak unit in $E^*_{\rm n}$. (If $\mathcal{C}$ is finite, we let $f = \sum_{g\in\mathcal{C}} |g|$.) ∎

Problem 5.2.4. *Let E be a normed Riesz space. Show that $E^*_{\rm n}$ separates the points of E if and only if for each $0 < x \in E$ there exists some $0 < \phi \in E^*_{\rm n}$ satisfying $\phi(x) > 0$.*

Solution: If $E^*_{\rm n}$ separates the points of E and $0 < x \in E$, then there exists some $\phi \in E^*_{\rm n}$ such that $\phi(x) \neq 0$, and so either $\phi^+(x) > 0$ or $\phi^-(x) > 0$.

Conversely, assume that for each $0 < x \in E$ there exists some $0 < \phi \in E^*_{\rm n}$ satisfying $\phi(x) > 0$ and let $y \in E$ be a non-zero vector. Pick $0 < \phi \in E^*_{\rm n}$ such that $\phi(|y|) > 0$, and then use the formula

$$0 < \phi(|y|) = |y|(\phi) = \sup\{\psi(y)\colon\ \psi \in E^* \text{ and } |\psi| \le \phi\}$$

to conclude that for some $\psi \in E^*_{\rm n}$ satisfying $|\psi| \le \phi$ we have $\psi(y) \neq 0$. ∎

Problem 5.2.5. *Assume that (Ω, Σ, π) is a σ-finite measure space and let E be an order dense function space associated with $L_0(\pi)$. Show that a function $g \in E^\sim_{\rm n}$ satisfies $g = 0$ if and only if $\int_\Omega g(\omega)x(\omega)\,d\pi(\omega) = 0$ for each $x \in E$. (This guarantees the uniqueness of the function g in the Representation Theorem 5.26.)*

Solution: Assume that a function $g \in E^\sim_{\rm n}$ satisfies $\int_\Omega g(\omega)x(\omega)\,d\pi(\omega) = 0$ for each $x \in E$. If $g \neq 0$, then there exists some $\epsilon > 0$ such that the π-measurable set $A = \{\omega \in \Omega\colon\ |g(\omega)| \ge \epsilon\}$ has positive measure. Since E is order dense in $L_0(\pi)$, there exists some $y \in E$ satisfying $0 < y \le \epsilon\chi_A \le |g|$. Now if $x = (\operatorname{Sign} g)y$, then $x \in E$ and

$$0 = \int_\Omega g(\omega)x(\omega)\,d\pi(\omega) = \int_\Omega |g(\omega)|y(\omega)\,d\pi(\omega) \ge \epsilon \int_A y(\omega)\,d\pi(\omega) > 0\,,$$

which is absurd. This contradiction establishes that $g = 0$ π-a.e. ∎

Problem 5.2.6. *Let (Ω, Σ, π) be a σ-finite measure space. Show that the order continuous dual of $L_\infty(\pi)$ coincides with $L_1(\pi)$, i.e., $L_\infty(\pi)^*_{\rm n} = L_1(\pi)$.*

Solution: Let $E = L_\infty(\pi)^*_{\rm n}$. By Theorem 5.26, E is an order dense Banach function space associated with $L_0(\pi)$. Clearly, $L_1(\pi) \subseteq E$. Now assume that $f \in E$. This means that $\int_\Omega f(\omega)x(\omega)\,d\pi(\omega)$ is a real number for each x in $L_\infty(\pi)$. In particular, if $x = \operatorname{Sign} f$, then $x \in L_\infty(\pi)$, and so $\int_\Omega |f|\,d\pi = \int_\Omega f(\omega)x(\omega)\,d\pi(\omega) \in \mathbb{R}$. This shows that $f \in L_1(\pi)$, and thus $E \subseteq L_1(\pi)$. Hence, $L_\infty(\pi)^*_{\rm n} = L_1(\pi)$. ∎

Problem 5.2.7. *Let (Ω, Σ, π) be a σ-finite measure space. If E is an order dense Banach function space in $L_0(\pi)$, then show that the center of E coincides with $L_\infty(\pi)$, i.e., $\mathcal{Z}(E) = L_\infty(\pi)$.*

Solution: Start by observing that the Maeda–Ogasawara–Vulikh completion of E can be identified with $L_0(\pi)$ equipped with its (π-a.e.) pointwise multiplication; see Theorem 3.35. Now fix some operator $0 \le T \in \mathcal{Z}(E)$. Then, according to Theorem 3.36, there exists a function $0 \le f \in L_0(\pi)$ such that

$$T(g) = fg$$

for each $g \in E$. To finish the solution, we must show that $f \in L_\infty(\pi)$.

To this end, assume by way of contradiction that $\|f\|_\infty = \infty$. This implies that for each n the measurable set $A_n = \{\omega \in \Omega\colon\ f(\omega) \ge n\}$ has positive measure. Replacing (if necessary) the set A_n by some of its subsets in Σ, we can assume that $0 < \pi(A_n) < \infty$ and $\chi_{A_n} \in E$ hold for each n. Then $T\chi_{A_n} = f\chi_{A_n} \ge n\chi_{A_n}$, and so $\|T\| \ge n$ for each n, a contradiction. Therefore, f belongs to $L_\infty(\pi)$, and from this we easily conclude that $\mathcal{Z}(E) = L_\infty(\pi)$. ∎

Problem 5.2.8. *For an order bounded sequence $\{x_n\}$ in a Banach function space with order continuous norm consider the following convergence properties:*

$$\text{(a) } x_n \xrightarrow{o} x \qquad \text{(b) } x_n \xrightarrow{p} x \qquad \text{(c) } x_n \xrightarrow{\pi} x \qquad \text{(d) } \|x_n - x\| \longrightarrow 0\,.$$

Show that (a) $\iff$ (b), (c) $\iff$ (d), *and* (b) $\Longrightarrow$ (c). *Does* (d) $\Longrightarrow$ (b)*?*

Solution: (a) $\iff$ (b) The equivalence of (a) and (b) is established in Problem 1.6.4. (This proof is also valid in any function space.)

(b) $\Longrightarrow$ (c) This is Corollary 1.83.

(c) $\iff$ (d) Let $\{x_n\}$ be a sequence in a Banach function space E such that $|x_n| \le u$ for each n and some $u \in E$. Assume first that $x_n \xrightarrow{\pi} x$. If $\|x_n - x\| \not\to 0$, then there exist a subsequence $\{y_n\}$ of $\{x_n\}$ and some $\epsilon > 0$ satisfying $\|y_n - x\| \ge \epsilon$ for each n. From $x_n \xrightarrow{\pi} x$, it follows that there exists a subsequence $\{z_n\}$ of $\{y_n\}$ satisfying $z_n \xrightarrow{p} x$; see Theorem 1.82. By the equivalence of (a) and (b), we have $z_n \xrightarrow{o} x$. But then the order continuity of the norm implies $\|z_n - x\| \to 0$, which contradicts $\|z_n - x\| \ge \epsilon$ for each n. Hence, $x_n \xrightarrow{\pi} x$ implies $\|x_n - x\| \to 0$.

For the converse, assume that $\|x_n - x\| \to 0$, and let $\{y_n\}$ be a subsequence of $\{x_n\}$. Since $\|y_n - x\| \to 0$, it follows from Lemma 5.21 that $\{y_n\}$ has a subsequence that converges π-a.e. to x. This implies that $x_n \xrightarrow{\pi} x$ is true; see Theorem 1.82.

(d) $\not\Longrightarrow$ (b) Let $E = L_1[0,1]$. Now consider the sequence of closed intervals

$$[0,1],\ [0,\tfrac{1}{2}],\ [\tfrac{1}{2},1],\ [0,\tfrac{1}{3}],\ [\tfrac{1}{3},\tfrac{2}{3}],\ [\tfrac{2}{3},1],\ [0,\tfrac{1}{4}],\ [\tfrac{1}{4},\tfrac{2}{4}],\ [\tfrac{2}{4},\tfrac{3}{4}],\ [\tfrac{3}{4},1],\ [0,\tfrac{1}{5}],\ldots\ .$$

Also, let $\{x_n\}$ be the sequence of the characteristic functions of these intervals. Then it is easy to see that $0 \le x_n \le \mathbf{1}$ for each n, $\|x_n\|_1 \to 0$, but $x_n(\omega) \not\to 0$ for each $\omega \in [0,1]$. ∎

Problem 5.2.9. *If $\{x_n\}$ is an order bounded positive sequence in some Banach function space E in $L_0(\pi)$, then prove that $x_n \xrightarrow{\pi} 0$ is equivalent to $x_n \xrightarrow{\sigma(E,E^*_n)} 0$, i.e., to the property: $\int_\Omega x_n(\omega)y(\omega)\,d\pi(\omega) \to 0$ for each $y \in E^*_n$.*

Is this equivalence true if $\{x_n\}$ is not assumed to be a positive sequence?

Solution: Assume that E is a Banach function space, and let $\{x_n\}$ be a positive sequence satisfying $0 \le x_n \le x$ for each n and some $x \in E^+$. It is not difficult to see that $x_n \xrightarrow{\sigma(E,E_n^*)} 0$ is equivalent to $\int_A x_n(\omega)\,d\pi(\omega) \to 0$ for each $A \in \Sigma$ with $\chi_A \in E$ and $\pi(A) < \infty$.

Suppose first that $x_n \xrightarrow{\pi} 0$ is true, and assume by way of contradiction that there is some $A \in \Sigma$ satisfying $\chi_A \in E$, $\pi(A) < \infty$, and $\int_A x_n(\omega)\,d\pi(\omega) \not\to 0$. Therefore, there exists a subsequence $\{y_n\}$ of $\{x_n\}$ and some $\delta > 0$ such that $\int_A x_n(\omega)\,d\pi(\omega) \ge \delta$ for each n. Since $x_n \xrightarrow{\pi} 0$, there exists a subsequence $\{z_n\}$ of $\{y_n\}$ satisfying $z_n \xrightarrow{p} 0$. But then the Lebesgue Dominated Convergence Theorem guarantees that $\int_A z_n(\omega)\,d\pi(\omega) \longrightarrow 0$, a contradiction. So, $x_n \xrightarrow{\sigma(E,E_n^*)} 0$.

For the converse, assume that $x_n \xrightarrow{\sigma(E,E_n^*)} 0$. Fix a set $A \in \Sigma$ of finite measure with $\chi_A \in E$, and for each $\epsilon > 0$ and n let $A_n(\epsilon) = \{\omega \in A\colon\ x_n(\omega) \ge \epsilon\}$. From $\epsilon\chi_{A_n(\epsilon)} \le x_n$, we get $0 \le \epsilon\pi(A_n(\epsilon)) \le \int_A x_n(\omega)\,d\pi(\omega)$. Since $\int_A x_n(\omega)\,d\pi(\omega) \to 0$, the latter implies $\pi(A_n(\epsilon)) \to 0$. Thus, $x_n \xrightarrow{\pi} 0$.

The equivalence is false if the sequence $\{x_n\}$ is not positive. For an example, let $E = L_\infty[0,1]$ and consider the sequence of Rademacher functions $\{r_n\}$ as introduced in Problem 3.1.9. We have $E_n^* = L_1[0,1]$ and $r_n \xrightarrow{w} 0$; see the solution of Problem 3.1.9. However, the sequence $\{r_n\}$ does not converge in measure to zero since no subsequence of $\{r_n\}$ converges pointwise almost everywhere to zero. ∎

Problem 5.2.10. *If E and F are Banach function spaces, then show that $(E_n^* \otimes F)^{dd}$ is a principal band in $\mathcal{L}_r(E,F)$.*

Solution: We can assume that E and F are order dense ideals. Let e be a weak unit in F and ϕ be a weak unit in E_n^*. (The existence of these weak units is guaranteed by Corollary 5.22 and Theorem 5.26.) Clearly, the operator $\phi \otimes e$ belongs to $(E_n^* \otimes F)^{dd}$. Assume now that some elements $0 \le x^* \in E_n^*$ and $0 \le x \in E$ satisfy $(x^* \otimes x) \wedge (\phi \otimes e) = 0$. From $0 \le (x^* \wedge \phi) \otimes (x \wedge e) \le (x^* \otimes x) \wedge (\phi \otimes e)$, we see that $(x^* \wedge \phi) \otimes (x \wedge e) = 0$. This implies that either $x^* \wedge \phi = 0$ or $x \wedge e = 0$. So, either $x^* = 0$ or $x = 0$, i.e., $x^* \otimes x = 0$. This shows that $\phi \otimes e$ is a weak unit in $(E_n^* \otimes F)^{dd}$, and the conclusion follows. ∎

Problem 5.2.11. *Show using Theorem 5.28 instead of Theorem 5.30 that if E is a Banach function space with order continuous norm associated with a σ-finite measure μ, then every continuous operator $T\colon E \to L_\infty(\nu)$ is a regular integral operator. That is, $\mathcal{L}(E, L_\infty(\nu)) = \mathcal{L}_\kappa(E, L_\infty(\nu))$.*

Solution: Since $L_\infty(\nu)$ is a Dedekind complete AM-space with unit, Theorem 3.9 guarantees that $\mathcal{L}(E, L_\infty(\nu)) = \mathcal{L}_r(E, L_\infty(\nu))$. We can assume that E is order dense in $L_\infty(\mu)$. Fix any weak unit $\phi \in E_n^*$; hence $\phi(s) > 0$ for μ-almost all $s \in \mathcal{S}$. As in the solution of Problem 5.2.10, we see that the operator $\phi \otimes \mathbf{1}_\mathcal{T}$ is a weak unit in $\mathcal{L}_\kappa(E, L_\infty(\nu))$. Therefore, to establish the identity $\mathcal{L}(E, L_\infty(\nu)) = \mathcal{L}_\kappa(E, L_\infty(\nu))$, it suffices (in view of Theorem 5.28) to show that the zero operator is the only operator $T\colon E \to L_\infty(\nu)$ that satisfies $T \wedge (\phi \otimes \mathbf{1}_\mathcal{T}) = 0$. So, let $T\colon E \to L_\infty(\nu)$ be a positive operator satisfying $T \wedge (\phi \otimes \mathbf{1}_\mathcal{T}) = 0$.

Since E has order continuous norm, it follows that $T = 0$ if and only if $T\chi_A = 0$ for each measurable subset $A \subseteq \Sigma_1$ of finite measure satisfying $\chi_A \in E$. Assume, by way of contradiction, that there exists some A as above and such that $T\chi_A > 0$. Then there exists some $\epsilon > 0$ such that $v = (T\chi_A - \epsilon)^+ > 0$. By the continuity of T there exists some $\delta > 0$ such that:

$$f \in E \text{ and } \|f\| < \delta \text{ imply } |Tf(s)| < \epsilon \text{ for } \mu\text{-almost all } s \in \mathcal{S}. \qquad (\star)$$

The order continuity of the norm in E implies that there is some $0 < \eta < \mu(A)$ such that if $B \subseteq A$ and $\mu(B) < \eta$, then $\|\chi_B\| < \delta$. To see this, assume by way of contradiction that this is not the case. This means that there exists some sequence $\{B_n\}$ such that $B_n \subseteq A$ for each n, $\mu(B_n) \to 0$, and $\|\chi_{B_n}\| \geq \delta$ for each n. By passing to a subsequence, we can assume that $\mu(B_n) < \frac{1}{2^n}$ holds for each n. Now if $C_n = \bigcup_{i=n}^{\infty} B_i$, then from $\mu(C_n) \leq \sum_{i=n}^{\infty} \mu(B_i) \leq 2^{1-n}$ it follows that $\chi_{C_n} \downarrow 0$ holds in E. Since E has order continuous norm, the latter implies $\|\chi_{C_n}\| \downarrow 0$. But then from $0 \leq \chi_{B_n} \leq \chi_{C_n}$ we obtain $\|\chi_{B_n}\| \to 0$, contrary to $\|\chi_{B_n}\| \geq \delta$ for each n. This establishes the existence of η.

Since ϕ is a weak unit in E_{n}^*, Problem 1.6.9 guarantees the existence of some real number $\alpha > 0$ such that $C \in \Sigma_1$, $C \subseteq A$ and $\mu(C) \geq \eta$ imply $\int_C \phi\, d\mu \geq \alpha$. Pick any $B \in \Sigma_1$ with $B \subseteq A$. If $\mu(A \setminus B) < \eta$, then $\|\chi_{A\setminus B}\| < \delta$, and $(\star)$ yields

$$0 \leq T\chi_{A\setminus B} = T\chi_A - T\chi_B < \epsilon \mathbf{1}_{\mathcal{T}},$$

whence $T\chi_B \geq (T\chi_A - \epsilon)^+ = v$. On the other hand, if $\mu(A \setminus B) \geq \eta$, then we have $\int_{A\setminus B} \phi\, d\mu \geq \alpha$. Thus, if $w = v \wedge (\alpha \mathbf{1}_{\mathcal{T}})$, then $0 < w \in F$, and if $B \in \Sigma_1$ satisfies $B \subseteq A$, then $T\chi_B + (\phi \otimes \mathbf{1}_{\mathcal{T}})\chi_{A\setminus B} = T\chi_B + \big(\int_{A\setminus B} \phi\, d\mu\big)\mathbf{1}_{\mathcal{T}} \geq w$. But then Theorem 1.17 implies

$$\begin{aligned} 0 < w &\leq \inf\{T\chi_B + (\phi \otimes \mathbf{1}_{\mathcal{T}})\chi_{A\setminus B} \colon\ B \in \Sigma_1 \text{ and } B \subseteq A\} \\ &= [T \wedge (\phi \otimes \mathbf{1}_{\mathcal{T}})](\chi_A) = 0, \end{aligned}$$

which is impossible. This contradiction shows that $T\chi_A = 0$, and hence $T = 0$, as desired. ∎

Problem 5.2.12. *Show that Dunford's Theorem (Corollary 5.31) is not valid for continuous operators between L_1-spaces.*

Solution: According to Problem 5.1.8 the identity operator $I\colon L_1[0,1] \to L_1[0,1]$ is disjoint from every regular integral operator. So, if I were a regular integral operator, then I would coincide with the zero operator, which is absurd. Hence, not every positive operator between L_1-spaces is an integral operator. (The preceding argument establishes really the following result: *If $T\colon L_1[0,1] \to L_1[0,1]$ is an integral operator, then $I + T$ is not an integral operator.*) ∎

Problem 5.2.13. *Assume that $(\mathcal{S}, \Sigma_1, \mu)$, $(\mathcal{T}, \Sigma_2, \nu)$, and $(\mathcal{Q}, \Sigma_3, \theta)$ are σ-finite measure spaces and that E, F, and G are ideals in $L_0(\mu)$, $L_0(\nu)$, and $L_0(\pi)$, respectively. Assume also that in the scheme $E \xrightarrow{S} F \xrightarrow{T} G$ the operators S and T are both regular and order continuous.*

If either S or T is an integral operator, then show that TS is likewise an integral operator.

Solution: Let $\{x_n\} \subseteq E$ be an order bounded sequence satisfying $x_n \xrightarrow{\pi} 0$. According to Theorem 5.30 it suffices to show that $TSx_n \xrightarrow{p} 0$.

Consider first the case when S is an integral operator. Since S is regular, the sequence $\{Sx_n\}$ is order bounded in F and, since S is an integral operator, $Sx_n \xrightarrow{p} 0$. Hence, by Corollary 1.86, $Sx_n \xrightarrow{o} 0$ in F, and so (by the order continuity of T) we have $TSx_n \xrightarrow{o} 0$ in G. In particular, $TSx_n \xrightarrow{p} 0$. Thus, in this case, TS is indeed an integral operator.

Assume now that T is an integral operator. Since S is regular, the sequence $\{Sx_n\}$ is order bounded in F, and we claim that $Sx_n \xrightarrow{\nu} 0$ holds in F. If not, then there exist some $B \in \Sigma_2$ with $\nu(B) < \infty$, some $\epsilon > 0$, and some subsequence $\{y_n\}$ of $\{x_n\}$ such that for each n we have

$$\nu\big(\{t \in B\colon \ |Sy_n(t)| \geq \epsilon\}\big) \geq \epsilon\,. \tag{$\star\star$}$$

Now, according to Theorem 1.82, there exists a subsequence $\{z_n\}$ of $\{y_n\}$ such that $z_n \xrightarrow{p} 0$. But then, by Corollary 1.86, we have $z_n \xrightarrow{o} 0$, and so by the order continuity of S, we get $Sz_n \xrightarrow{o} 0$. That is, $Sz_n \xrightarrow{p} 0$. This (according to Corollary 1.83) implies $Sz_n \xrightarrow{\nu} 0$, contrary to ($\star\star$). Consequently, $Sx_n \xrightarrow{\nu} 0$. Since T is an integral operator, it follows from Theorem 5.30 that $TSx_n \xrightarrow{p} 0$. Using Theorem 5.30 once more, we get that TS is an integral operator.

Regarding this problem, two remarks are in order.

(1) When S is an integral operator, the assumption $T \in \mathcal{L}_{\rm n}(F,G)$ can be replaced by the weaker assumption $T \in \mathcal{L}_{\rm n}(F, L_0(\pi))$, and when T is an integral operator we do not need the assumption that $T\colon F \to G$ is regular.

(2) In general, the composition of integral operators may fail to be an integral operator. We refer to [**49**] for an example of two compact integral operators on $L_2[0,1]$ whose composition is not an integral operator. ∎

Problem 5.2.14. *Assume that $(\mathcal{S}, \Sigma_1, \mu)$ and $(\mathcal{T}, \Sigma_2, \nu)$ are two σ-finite measure spaces with $(\mathcal{S}, \Sigma_1, \mu)$ non-atomic. Also, suppose that E and F are ideals in $L_0(\mathcal{S}, \Sigma_1, \mu)$ and $L_0(\mathcal{T}, \Sigma_2, \nu)$, respectively, and let $S\colon E \to F$ be a regular integral operator.*

Let F_0 denote the order ideal in F generated by the range space of the operator S and assume that there is an operator $T\colon F_0 \to E$ such that $TS = I_E$, the identity operator on E (in a sense T is a right inverse of the operator S). Show that T cannot be an integral operator.

Solution: Assume, contrary to our claim, that T is an integral operator. Therefore, T is order continuous as an operator from F_0 to $L_0(\mu)$. By the hypotheses $S\colon E \to F$ is a regular integral operator and, therefore, S considered as an operator from E to F_0 is also a regular integral operator. Hence, by Problem 5.2.13 and the remark at the end of it, the composition TS is again an integral operator. However, TS is the identity operator and according to Problem 5.1.8 the identity operator cannot be an integral operator, since $(\mathcal{S}, \Sigma_1, \mu)$ is non-atomic. ∎

Problem 5.2.15. *Assume that $(\mathcal{S}, \Sigma_1, \mu)$, $(\mathcal{T}, \Sigma_2, \nu)$, and $(\mathcal{Q}, \Sigma_3, \theta)$ are three σ-finite measure spaces and that E, F, and G are function spaces in $L_0(\mu)$, $L_0(\nu)$, and $L_0(\theta)$, respectively. Suppose that in the scheme of operators $E \xrightarrow{S} F \xrightarrow{T} G$ both S and T are regular integral operators with kernels $S(\cdot,\cdot)$ and $T(\cdot,\cdot)$, respectively.*

Show that the composition operator $R = TS$ is an integral operator whose kernel is given by the **convolution** *of the two kernels $S(\cdot,\cdot)$ and $T(\cdot,\cdot)$, i.e.,*

$$R(s,q) = \int_{\mathcal{T}} S(s,t)T(t,q)\,d\nu(t)\,.$$

Solution: Since T and S are regular integral operators, they are both order continuous. So, by Problem 5.2.13, the operator $R = TS$ is an integral operator. Let us calculate the kernel of R. Without loss of generality, we can assume that both operators S and T are positive and, hence, their kernels $S(\cdot,\cdot)$ and $T(\cdot,\cdot)$ are non-negative functions. For each $x \in E^+$ we have

$$Sx(t) = \int_{\mathcal{S}} S(s,t)x(s)\,d\mu(s)$$

for ν-almost all $t \in \mathcal{T}$, and therefore

$$[TSx](q) = \int_{\mathcal{T}} T(t,q)\Big[\int_{\mathcal{S}} S(s,t)x(s)\,d\mu(s)\Big]\,d\nu(t)$$

for θ-almost all $q \in \mathcal{Q}$. Since all functions are non-negative, an application of Tonelli's theorem yields that

$$[TSx](q) = \int_{\mathcal{S}}\Big[\int_{\mathcal{T}} S(s,t)T(t,q)\,d\nu(t)\Big]x(s)\,d\mu(s)$$

and that the function $s \mapsto \big[\int_{\mathcal{T}} S(s,t)T(t,q)\,d\nu(t)\big]x(s)$ is integrable for θ-almost all $q \in \mathcal{Q}$. Consequently, $R(s,q) = \int_{\mathcal{T}} S(s,t)T(t,q)\,d\nu(t)$ is the required kernel for the operator R. ∎

Problem 5.2.16. *Let $(\mathcal{S}, \Sigma_1, \mu)$ and $(\mathcal{T}, \Sigma_2, \nu)$ be two σ-finite measure spaces with $(\mathcal{S}, \Sigma_1, \mu)$ non-atomic, and let $0 < K(\cdot,\cdot) \in L_0(\mu \times \nu)$. Do there exist a function $0 < f \in L_0(\mu)$ and a subset $B \in \Sigma_2$ such that:*

(a) *$\nu(B) > 0$.*

(b) *For each $t \in B$ we have $K(s,t) \geq f(s)$ for μ-almost all $s \in \mathcal{S}$?*

Solution: We claim that the existence of a function f and a set B with the required properties is equivalent to the statement that there exists an $\epsilon > 0$ such that the set $U_\epsilon = \{(s,t)\colon\ K(s,t) \geq \epsilon\}$ includes a non-trivial rectangle $A \times B$, that is, $(\mu \times \nu)(A \times B \setminus U_\epsilon) = 0$ for some $A \in \Sigma_1$ with $\mu(A) > 0$ and some $B \in \Sigma_2$ with $\nu(B) > 0$.

Indeed, assume first that there is an $\epsilon > 0$ such that $(\mu \times \nu)(A \times B \setminus U_\epsilon) = 0$ for some $A \in \Sigma_1$ with $\mu(A) > 0$ and some $B \in \Sigma_2$ with $\nu(B) > 0$. Assume also that $\mu(A) < \infty$. Let $C = A \times B \setminus U_\epsilon$ and let C_t denote the t-section of C for $t \in \mathcal{T}$. Since $(\mu \times \nu)(C) = 0$, Fubini's theorem implies that $\mu(C_t) = 0$ for ν-almost all $t \in \mathcal{T}$. We will assume that $\mu(C_t) = 0$ for each $t \in B$; otherwise we will remove from B

the ν-null set of the exceptional points t. We claim that the function $f = \epsilon\chi_A$ is as required. Indeed, for each $t \in B$, if $s \notin C_t$, then clearly $K(s,t) \geq \epsilon\chi_A(s)$. Since $\mu(C_t) = 0$, this means that $K(s,t) \geq f(s)$ for μ-almost all $s \in \mathcal{S}$, as claimed.

For the converse, assume that there exists a function $0 < f \in L_0(\mu)$ and a subset $B \in \Sigma_2$ such that $\nu(B) > 0$ and for each $t \in B$ the inequality $K(s,t) \geq f(s)$ holds for μ-almost all $s \in \mathcal{S}$. Let us denote this set by $\mathcal{S}_t$.

Since $f > 0$, we can find an $\epsilon > 0$ such that the set $A = \{s \in \mathcal{S}\colon\ f(s) \geq \epsilon\}$ satisfies $\mu(A) > 0$. We will verify that $(\mu \times \nu)(C) = 0$, where $C = A \times B \setminus U_\epsilon$. For each $t \in \mathcal{T}$ let C_t denote the t-section of C. We claim that $\mu(C_t) = 0$ for each $t \in \mathcal{T}$ (and hence $(\mu \times \nu)(C) = 0$ by Fubini's theorem). Indeed, if $t \notin B$, then $C_t = \emptyset$. On the other hand, if $t \in B$, then $C_t \subseteq \mathcal{S} \setminus \mathcal{S}_t$ and so again $\mu(C_t) = 0$. Thus, indeed, the rectangle $A \times B$ is contained in U_ϵ.

To answer our question in the negative, it remains to recall that in general there exist measurable subsets of $\mathcal{S} \times \mathcal{T}$ of positive measure that include no non-trivial rectangles. (See, for instance, Exercises 136.12 and 136.13 in [**83**], where $\mathcal{S} = \mathcal{T} = [a,b]$.) ∎

Problem 5.2.17 (Benedek–Panzone [**16**]). *Let $(\mathcal{S}, \Sigma_1, \mu)$ and $(\mathcal{T}, \Sigma_2, \nu)$ be two σ-finite measure spaces. For each $p \in [1,\infty)$ we denote by $L_{p,\infty}(\mu \times \nu)$ (or simply by $L_{p,\infty}$) the collection of all functions $f \in L_0(\mu \times \nu)$ such that*

$$\|f\|_{p,\infty} = \operatorname*{ess\,sup}_{s \in \mathcal{S}} \left[\int_{\mathcal{T}} |f(s,t)|^p \, d\nu(t) \right]^{\frac{1}{p}} < \infty \,.$$

As usual, the equivalent functions are identified. Similarly, we denote by $L_{p,1}(\mu \times \nu)$ (or simply by $L_{p,1}$) the collection of all functions $f \in L_0(\mu \times \nu)$ such that

$$\|f\|_{p,1} = \int_{\mathcal{S}} \left[\int_{\mathcal{T}} |f(s,t)|^p \, d\nu(t) \right]^{\frac{1}{p}} d\mu(s) < \infty \,.$$

*The non-negative real numbers $\|f\|_{p,\infty}$ and $\|f\|_{p,1}$ are called the **mixed norms** of f. Show that:*

(a) *The function $\|\cdot\|_{p,\infty}$ is well defined, i.e., if $f, g \in L_{p,\infty}$ satisfy $f = g$ $\mu \times \nu$-a.e., then $\|f\|_{p,\infty} = \|g\|_{p,\infty}$.*

(b) *The collections $L_{p,\infty}$ and $L_{p,1}$ (with their mixed norms) are both Banach function spaces associated with $L_0(\mu \times \nu)$.*

(c) *If $1 < p, q < \infty$ satisfy $\frac{1}{p} + \frac{1}{q} = 1$, then $L_{p,1}^* = L_{q,\infty}$.*

Solution: (a) Let $f, g \in L_{p,\infty}$ satisfy $f = g$ $\mu \times \nu$-a.e. This means that there exists a $\mu \times \nu$-null set A such that $f(s,t) = g(s,t)$ holds for all $(s,t) \notin A$. Since μ and ν are σ-finite measures, for μ-almost all $s \in \mathcal{S}$ the function $s \mapsto \nu^*(A_s)$ is ν-integrable, where $A_s = \{t \in \mathcal{T}\colon\ (s,t) \in A\}$, and $\int_{\mathcal{S}} \nu^*(A_s)\, d\mu(s) = (\mu \times \nu)^*(A) = 0$. This implies that $\nu^*(A_s) = 0$ for μ-almost all $s \in \mathcal{S}$. So, for μ-almost all $s \in \mathcal{S}$ we have

$$\int_{\mathcal{T}} [f(s,t)]^p \, d\nu(t) = \int_{\mathcal{T} \setminus A_s} [f(s,t)]^p \, d\nu(t) = \int_{\mathcal{T} \setminus A_s} [g(s,t)]^p \, d\nu(t) = \int_{\mathcal{T}} [g(s,t)]^p \, d\nu(t) \,,$$

and from this we easily get $\|f\|_{p,\infty} = \|g\|_{p,\infty}$.

(b) We shall verify that $L_{p,\infty}$ is a Banach function space; the case of $L_{p,1}$ can be done similarly. Obviously, if $f \in L_{p,\infty}$ and λ is a scalar, then $\lambda f \in L_{p,\infty}$. To verify that $L_{p,\infty}$ is closed under addition, let $f, g \in L_{p,\infty}$. Using the triangle inequality of the norm in $L_p(\nu)$, for μ-almost all $s \in \mathcal{S}$ we have

$$\begin{aligned}\Big[\int_{\mathcal{T}} |f(s,t)+g(s,t)|^p \, d\nu(t)\Big]^{\frac{1}{p}} &\leq \Big[\int_{\mathcal{T}} |f(s,t)|^p \, d\nu(t)\Big]^{\frac{1}{p}} + \Big[\int_{\mathcal{T}} |g(s,t)|^p \, d\nu(t)\Big]^{\frac{1}{p}} \\ &\leq \|f\|_{p,\infty} + \|g\|_{p,\infty}\,.\end{aligned}$$

This implies

$$\|f+g\|_{p,\infty} = \operatorname*{ess\,sup}_{s\in\mathcal{S}} \Big[\int_{\mathcal{T}} |f(s,t)+g(s,t)|^p \, d\nu(t)\Big]^{\frac{1}{p}} \leq \|f\|_{p,\infty} + \|g\|_{p,\infty}\,.$$

Therefore, $f+g \in L_{p,\infty}$ and $\|\cdot\|_{p,\infty}$ satisfies the triangle inequality. The verification of the other two axioms of the norm is straightforward and is omitted.

If $|f| \leq |g|$, then clearly $\|f\|_{p,\infty} \leq \|g\|_{p,\infty}$. This implies that $\|\cdot\|_{p,\infty}$ is a lattice norm on $L_{p,\infty}$ and that $L_{p,\infty}$ is an ideal in $L_0(\mu\times\nu)$. In particular, $L_{p,\infty}$ is a Dedekind complete Riesz space.

Next, we shall verify that $L_{p,\infty}$ is a Banach lattice. According to Problem 1.2.16, it suffices to show that every increasing norm Cauchy sequence in $L^+_{p,\infty}$ is norm convergent. To this end, assume that $\{f_n\}$ is a $\|\cdot\|_{p,\infty}$-Cauchy sequence in $L_{p,\infty}$ satisfying $0 \leq f_n \uparrow$. Without loss of generality, we can suppose that $0 \leq f_n(s,t) \leq f_{n+1}(s,t)$ holds in $\mathbb{R}$ for all n and each $(s,t) \in \mathcal{S}\times\mathcal{T}$. Let $f(s,t) = \lim_{n\to\infty} f_n(s,t)$, and so $f\colon \mathcal{S}\times\mathcal{T} \to [0,\infty]$.

Now fix some $M > 0$ such that

$$\|f_n\|_{p,\infty} = \operatorname*{ess\,sup}_{s\in\mathcal{S}} \Big[\int_{\mathcal{T}} [f_n(s,t)]^p \, d\nu(t)\Big]^{\frac{1}{p}} \leq M$$

for each n. This implies that we have $\int_{\mathcal{T}} [f_n(s,t)]^p \, d\nu(t) \uparrow\leq M$ for μ-almost all $s \in \mathcal{S}$. By Beppo Levi's classical theorem, we infer that for μ-almost all $s \in \mathcal{S}$ we have $f(s,\cdot) \in L_p(\nu)$ and that $\int_{\mathcal{T}} [f(s,t)]^p \, d\nu(t) \leq M$. This implies $\operatorname*{ess\,sup}_{s\in\mathcal{S}} \big[\int_{\mathcal{T}} [f(s,t)]^p \, d\nu(t)\big]^{\frac{1}{p}} \leq M$. Therefore, if we can prove that $f(s,t) < \infty$ for $\mu\times\nu$-almost all $(s,t) \in \mathcal{S}\times\mathcal{T}$, then it will follow that $f \in L_{p,\infty}$.

To see that $f(s,t) < \infty$ for $\mu\times\nu$-almost all $(s,t) \in \mathcal{S}\times\mathcal{T}$, assume by way of contradiction that this is not the case. Then there exists a $\mu\times\nu$-measurable subset A of $\mathcal{S}\times\mathcal{T}$ with $0 < (\mu\times\nu)^*(A) < \infty$ satisfying $f(s,t) = \infty$ for each $(s,t) \in A$. By Fubini's Theorem 1.97, the function $s \mapsto \nu^*(A_s)$, where $A_s = \{t \in \mathcal{T}\colon\ (s,t) \in A\}$, is ν-integrable and $(\mu\times\nu)^*(A) = \int_{\mathcal{S}} \nu^*(A_s) \, d\mu(s)$. In particular, the μ-measurable set $B = \{s \in \mathcal{S}\colon\ \nu^*(A_s) > 0\}$ satisfies $\mu^*(B) > 0$. But then for each $s \in B$ we have

$$\int_{\mathcal{T}} [f(s,t)]^p \, d\nu(t) \geq \int_{A_s} [f(s,t)]^p \, d\nu(t) = \infty\,,$$

which is impossible. Therefore, $f \in L_{p,\infty}$.

We shall finish the proof of the completeness of the normed Riesz space $L_{p,\infty}$ by proving that $\|f_n - f\|_{p,\infty} \to 0$. To this end, let $\epsilon > 0$. Fix some k such that $\|f_m - f_n\|_{p,\infty} < \epsilon$ for all $m,n \geq k$. In particular, for μ-almost all $s \in \mathcal{S}$ we have $\int_{\mathcal{T}} |f_m(s,t) - f_n(s,t)|^p \, d\nu(t) < \epsilon$ for all $n,m \geq k$. From $f_m(s,t) \uparrow_m f(s,t)$, it follows that for μ-almost all $s \in \mathcal{S}$ we have $\int_{\mathcal{T}}[f(s,t) - f_n(s,t)]^p \, d\nu(t) < \epsilon$ for all $n \geq k$. This implies $\|f - f_n\|_{p,\infty} \leq \epsilon$ for all $n \geq k$. That is, $f_n \to f$ in $L_{p,\infty}$.

(c) Let $1 < p < \infty$ and let q be its conjugate index, i.e., $\frac{1}{p} + \frac{1}{q} = 1$. We shall prove that $L^*_{p,1} = L_{q,\infty}$.

First, we shall establish that the norm on $L_{p,1}$ is order continuous. To this end, assume that $x_n \downarrow 0$ holds in $L_{p,1}$. This implies $x_n(s,t) \downarrow 0$ for $\mu \times \nu$-almost all $(s,t) \in \mathcal{S} \times \mathcal{T}$. Therefore, for μ-almost each $s \in \mathcal{S}$ we have $x_n(s,\cdot) \downarrow 0$ in $L_0(\nu)$, and so the Lebesgue Dominated Convergence Theorem yields

$$\|x_n(s,\cdot)\|_p = \Big[\int_{\mathcal{T}} [x_n(s,t)]^p \, d\nu(t)\Big]^{\frac{1}{p}} \downarrow 0$$

for μ-almost all $s \in \mathcal{S}$. This and the hypothesis that the functions $\|x_n(s,\cdot)\|_p$ belong to $L_1(\mu)$ imply (again by the Lebesgue Dominated Convergence Theorem) that

$$\|x_n\|_{p,1} = \int_{\mathcal{S}} \Big[\int_{\mathcal{T}} [x_n(s,t)]^p \, d\nu(t)\Big]^{\frac{1}{p}} d\mu(s) = \int_{\mathcal{S}} \|x_n(s,\cdot)\|_p \, d\mu(s) \downarrow 0 \,.$$

Therefore, $\|\cdot\|_{p,1}$ is order continuous.

The order continuity of the norm $\|\cdot\|_{p,1}$ guarantees that $L^*_{p,1} = (L_{p,1})^*_{\mathrm{n}}$. That is, each bounded linear functional ϕ on $L_{p,1}$ is order continuous and hence allows an integral representation by Theorem 5.26.

Our next step is to establish the inclusion $L_{q,\infty} \subseteq L^*_{p,1}$. Take any function $g \in L_{q,\infty}$, and let us verify that $xg \in L_1(\mu \times \nu)$ for each $x \in L_{p,1}$. So, fix $x \in L_{p,1}$. Applying Hölder's inequality, we see that

$$\begin{aligned}
\int_{\mathcal{S}} \Big[\int_{\mathcal{T}} |x(s,t)g(s,t)| \, d\nu(t)\Big] d\mu(s) &\leq \int_{\mathcal{S}} \|x(s,\cdot)\|_p \cdot \|g(s,\cdot)\|_q \, d\mu(s) \\
&\leq \operatorname*{ess\,sup}_{s \in \mathcal{S}} \|g(s,\cdot)\|_q \cdot \int_{\mathcal{S}} \|x(s,\cdot)\|_p \, d\mu(s) \\
&= \|g\|_{q,\infty} \cdot \|x\|_{p,1} \,.
\end{aligned}$$

Now an easy application of Tonelli's Theorem 1.98 shows that $xg \in L_1(\mu \times \nu)$. So, g defines a bounded linear functional on $L_{p,1}$, i.e., $g \in L^*_{p,1}$, and $\|g\| \leq \|g\|_{q,\infty}$, where $\|g\|$ denotes the norm of g as a bounded linear functional on $L_{p,1}$. We shall show now that in actuality we have equality. That is, we claim that

$$\|g\| = \|g\|_{q,\infty} \,. \tag{$\star$}$$

To this end, fix $\epsilon > 0$ and suppose that $\|g\|_{q,\infty} = 1$. Thus, $\int_{\mathcal{T}} |g(s,t)|^q \, d\nu(t) \leq 1$ for μ-almost all $s \in \mathcal{S}$. Without loss of generality we can assume that $g(s,t) \geq 0$ for all $(s,t) \in \mathcal{S} \times \mathcal{T}$. According to the definition of the space $L_{q,\infty}$, we can find some $A \in \Sigma_1$ such that $0 < \mu(A) < \infty$ and $\|g(s,\cdot)\|_q = \big[\int_{\mathcal{T}} g^q(s,t) \, d\nu(t)\big]^{\frac{1}{q}} > 1 - \epsilon$ for each $s \in A$. Consider the function $x(s,t) = \frac{1}{\mu(A)} \chi_A(s)[g(s,t)]^{q-1}$. From

$$\begin{aligned}
\int_{\mathcal{S}} \Big[\int_{\mathcal{T}} [x(s,t)]^p \, d\nu(t)\Big]^{\frac{1}{p}} d\mu &= \tfrac{1}{\mu(A)} \int_A \Big[\int_{\mathcal{T}} [g(s,t)]^{(q-1)p} \, d\nu(t)\Big]^{\frac{1}{p}} d\mu(s) \\
&= \tfrac{1}{\mu(A)} \int_A \Big[\int_{\mathcal{T}} [g(s,t)]^q \, d\nu(t)\Big]^{\frac{1}{p}} d\mu(s) \\
&\leq 1 \,,
\end{aligned}$$

it follows that $x \in L_{p,1}$ and that $\|x\|_{p,1} \le 1$. Consequently,

$$\begin{aligned}\|g\| \ge g(x) &= \int_{\mathcal{S}\times\mathcal{T}} xg\,d(\mu\times\nu) = \tfrac{1}{\mu(A)}\int_{\mathcal{S}\times\mathcal{T}} \chi_A(s)[g(s,t)]^q\,d(\mu\times\nu)(s,t)\\ &= \tfrac{1}{\mu(A)}\int_A\int_{\mathcal{T}}[g(s,t)]^q\,d\nu(t)\,d\mu(s) > (1-\epsilon)^q\,.\end{aligned}$$

Since $\epsilon > 0$ is arbitrary, we get $\|g\| \ge 1$ and ($\star$) has been established.

The inclusion $L_{q,\infty} \subseteq L^*_{p,1} = (L_{p,1})^*_{\rm n}$ guarantees that $L_{q,\infty}$ is an order dense ideal in $(L_{p,1})^*_{\rm n}$. To establish the equality $L_{q,\infty} = L^*_{p,1}$, take any $0 < g \in L^*_{p,1}$. Since the measure $\mu\times\nu$ is σ-finite, there exists a sequence $\{g_n\} \in L_{q,\infty}$ satisfying $0 \le g_n(s,t) \uparrow g(s,t)$ for $\mu\times\nu$-almost all $(s,t) \in \mathcal{S}\times\mathcal{T}$. So, for μ-almost all $s \in \mathcal{S}$ we have $0 \le g_n(s,t) \uparrow g(s,t)$ for ν-almost all $t \in \mathcal{T}$. This implies

$$\Big[\int_{\mathcal{T}}[g_n(s,t)]^q\,d\nu(t)\Big]^{\frac{1}{q}} \uparrow \Big[\int_{\mathcal{T}}[g(s,t)]^q\,d\nu(t)\Big]^{\frac{1}{q}}$$

for μ-almost all $s \in \mathcal{S}$, and consequently (from Lemma 1.101) we get

$$\|g_n\|_{q,\infty} = \operatorname*{ess\,sup}_{s\in\mathcal{S}}\Big[\int_{\mathcal{T}}[g_n(s,t)]^q\,d\nu(t)\Big]^{\frac{1}{q}} \uparrow \operatorname*{ess\,sup}_{s\in\mathcal{S}}\Big[\int_{\mathcal{T}}[g(s,t)]^q\,d\nu(t)\Big]^{\frac{1}{q}}.$$

The function g belongs to $L_{q,\infty}$ if (and only if) the expression on the right-hand side is finite, or, equivalently, if the sequence $\{\|g_n\|_{q,\infty}\}$ is bounded. But $g_n \uparrow g$ in $L^*_{p,1}$ implies that $\|g_n\| \uparrow \|g\|$ and, as we proved above, $\|g_n\|_{q,\infty} = \|g_n\|$. This proves that the sequence $\{\|g_n\|_{q,\infty}\}$ is bounded from above by $\|g\|$, and consequently $g \in L_{q,\infty}$. The solution is finished. ∎

Problem 5.2.18. *The result of this problem can be found in many books; see for instance* [**24, 28, 45, 46**]. *For the notation used here see Problem* 5.2.17.

Assume that $(\mathcal{S},\Sigma_1,\mu)$ and $(\mathcal{T},\Sigma_2,\nu)$ are two σ-finite measure spaces. Show that a $\mu\times\nu$-measurable function $T(\cdot,\cdot)\colon \mathcal{S}\times\mathcal{T}\to\mathbb{R}$ defines, via the usual formula

$$Tx(t) = \int_{\mathcal{S}} T(s,t)x(s)\,d\mu(s)\,,$$

an integral operator from $L_1(\mu)$ to $L_p(\nu)$ for some $1 < p < \infty$ if and only if $T(\cdot,\cdot)$ belongs to $L_{p,\infty}$. Moreover, prove that in this case the integral operator T determined by the kernel $T(\cdot,\cdot)$ is a regular operator and its norm is given by $\|T\| = \|T(\cdot,\cdot)\|_{p,\infty}$.

Solution: Assume first that $T(\cdot,\cdot) \in L_{p,\infty}$ for some $1 < p < \infty$. We must show that the integral operator T generated by the function $T(\cdot,\cdot)$ maps $L_1(\mu)$ to $L_p(\nu)$. That is, we must show that for each $x \in L_1(\mu)$ the formula $\int_{\mathcal{S}} T(s,t)x(s)\,d\mu(s)$ defines a function that belongs to $L_p(\nu)$. Without loss of generality we can suppose that $T(s,t) \ge 0$ holds for all $(s,t) \in \mathcal{S}\times\mathcal{T}$. Pick a non-negative function $x \in L_1(\mu)$ and let

$$y(t) = \int_{\mathcal{S}} T(s,t)x(s)\,d\mu(s)\,.$$

Now let $1 < q < \infty$ satisfy $\frac{1}{p} + \frac{1}{q} = 1$. From Hölder's inequality, we get

$$\begin{aligned} y(t) &= \int_S T(s,t)x(s)\,d\mu(s) = \int_S \big[[T(s,t)]^p x(s)\big]^{\frac{1}{p}} [x(s)]^{1-\frac{1}{p}}\,d\mu(s) \\ &\le \Big[\int_S [T(s,t)]^p x(s)\,d\mu(s)\Big]^{\frac{1}{p}} \cdot \Big[\int_S [x(s)]^{q(1-\frac{1}{p})}\,d\mu(s)\Big]^{\frac{1}{q}} \\ &= \Big[\int_S [T(s,t)]^p x(s)\,d\mu(s)\Big]^{\frac{1}{p}} \cdot \|x\|_1^{\frac{1}{q}}. \end{aligned}$$

Therefore, using Tonelli's theorem, we obtain that

$$\begin{aligned} \|y\|_p^p &= \int_T |y(t)|^p\,d\nu(t) \le \int_T \Big(\Big[\int_S [T(s,t)]^p x(s)\,d\mu(s)\Big] \cdot \|x\|_1^{\frac{p}{q}}\Big)\,d\nu(t) \\ &= \Big[\int_S \Big(\int_T [T(s,t)]^p\,d\nu(t)\Big) x(s)\,d\mu(s)\Big] \|x\|_1^{\frac{p}{q}} \\ &\le \|T(\cdot,\cdot)\|_{p,\infty}^p \cdot \|x\|_1^p < \infty. \end{aligned}$$

This shows that $y \in L_p(\nu)$, and so the $\mu \times \nu$-measurable function $T(\cdot,\cdot)$ defines an integral operator $T\colon L_1(\mu) \to L_p(\nu)$ that satisfies $\|T\| \le \|T(\cdot,\cdot)\|_{p,\infty}$.

To establish the reverse inequality $\|T\| \ge \|T(\cdot,\cdot)\|_{p,\infty}$, consider the adjoint operator $T^*\colon L_q(\nu) \to L_\infty(\mu)$. Since both Banach lattices $L_1(\mu)$ and $L_p(\nu)$ have order continuous norms, Theorem 5.29 guarantees that T^* is also an integral operator whose kernel $T^*(t,s)$ satisfies $T^*(t,s) = T(s,t)$.

Fix some $\epsilon > 0$ and then find some set $A \in \Sigma_1$ such that $0 < \mu(A) < \infty$ and $\big[\int_T [T(s,t)]^p\,d\nu(t)\big]^{\frac{1}{p}} \ge \|T(\cdot,\cdot)\|_{p,\infty} - \epsilon$ for each $s \in A$. In particular, we have $\operatorname{esssup}_{s\in S}\big[\int_T [T(s,t)]^p\,d\nu(t)\big]^{\frac{1}{p}} \ge \|T(\cdot,\cdot)\|_{p,\infty} - \epsilon$. Now let U denote the closed unit ball of $L_q(\nu)$ and let U^+ be its positive part, i.e., $U^+ = U \cap L_q^+(\nu)$. Clearly, the set $T^*(U^+) = \big\{\int_T T(\cdot,t)\psi(t)\,d\nu(t)\colon \psi \in U^+\big\}$ is a norm bounded subset of $L_\infty^+(\mu)$, and hence it is also order bounded. In particular, since $L_\infty(\mu)$ is Dedekind complete, $\sup T^*(U^+)$ exists in $L_\infty(\mu)$. Now using part (2) of Lemma 1.101, we see that

$$\begin{aligned} \|T^*\| &= \sup_{\psi\in U^+} \|T^*\psi\|_\infty = \sup_{\psi\in U^+} \Big\|\int_T T(\cdot,t)\psi(t)\,d\nu(t)\Big\|_\infty \\ &= \Big\|\sup_{\psi\in U^+} \int_T T(\cdot,t)\psi(t)\,d\nu(t)\Big\|_\infty = \Big\|\int_T [T(\cdot,t)]^p\,d\nu(t)\Big\|_\infty \\ &= \operatorname*{ess\,sup}_{s\in S} \Big[\int_T [T(s,t)]^p\,d\nu(t)\Big]^{\frac{1}{p}} \\ &\ge \|T(\cdot,\cdot)\|_{p,\infty} - \epsilon. \end{aligned}$$

Since $\|T\| = \|T^*\|$ and $\epsilon > 0$ is arbitrary, we have $\|T\| \ge \|T(\cdot,\cdot)\|_{p,\infty}$, and so indeed $\|T\| = \|T(\cdot,\cdot)\|_{p,\infty}$.

For the converse, assume that a $\mu \times \nu$-measurable function $T(\cdot,\cdot)$ defines an integral operator from $L_1(\mu)$ to $L_p(\nu)$. We want to show that the kernel $T(\cdot,\cdot)$ belongs to $L_{p,\infty}$. This follows immediately from the previous estimate by considering the dual operator for which we have

$$\|T\| = \|T^*\| \ge \|T(\cdot,\cdot)\|_{p,\infty}.$$

Hence, $T(\cdot,\cdot) \in L_{p,\infty}$. ∎

Problem 5.2.19 (Lozanovsky [**54**]). *Dunford's theorem stated in part* (2) *of Corollary* 5.33 *asserts that: If* $(\mathcal{S}, \Sigma_1, \mu)$ *and* $(\mathcal{T}, \Sigma_2, \nu)$ *are* σ*-finite measure spaces and* $1 < p < \infty$*, then every bounded operator* $T\colon L_1(\mu) \to L_p(\nu)$ *is a regular integral operator. Present an alternate proof of Dunford's theorem using Problem* 5.2.18.

Solution: Let $T\colon L_1(\mu) \to L_p(\nu)$ be a bounded operator, where $1 < p < \infty$. As usual, we let $1 < q < \infty$ satisfy $\frac{1}{p} + \frac{1}{q} = 1$. By virtue of Theorem 3.9, we can assume that T is a positive operator and assume also that $\|T\| \le 1$.

Let Σ_1^f be the collection of all subsets A of Σ_1 with $0 < \mu(A) < \infty$. For each set $A \in \Sigma_1^f$ let $\mathcal{U}_A$ denote the collection of all non-negative functions $K \in L_{p,\infty} = L_{q,1}^*$ satisfying $\|K\|_{p,\infty} \le 1$ and

$$T\chi_A(t) = \int_{\mathcal{S}} K(s,t)\chi_A(s)\, d\mu(s) \tag{$\star$}$$

for ν-almost all $t \in \mathcal{T}$.

We shall prove that there exists some $\mu \times \nu$-measurable function $K_0(\cdot,\cdot)$ that belongs to each $\mathcal{U}_A$. If this is established, then it will follow from $(\star)$ that for each μ-step function ϕ we have $T\phi(t) = \int_{\mathcal{S}} K_0(s,t)\phi(s)\, d\mu(s)$ for ν-almost all $t \in \mathcal{T}$. Then, using the fact that the step functions are norm (and also order) dense in $L_1(\mu)$, a continuity argument will guarantee that for each $x \in L_1(\mu)$ we must have

$$Tx(t) = \int_{\mathcal{S}} K_0(s,t)x(s)\, d\mu(s)$$

for ν-almost all $t \in \mathcal{T}$. This proves that T is an integral operator with kernel K_0.

We claim that the family of sets $\{\mathcal{U}_A\}_{A\in\Sigma_1^f}$ has the finite intersection property. To see this, let $A_1, \ldots, A_n \in \Sigma_1^f$, and then pick pairwise disjoint μ-measurable sets $C_1, \ldots, C_k \in \Sigma_1^f$ such that each A_i is a finite union of some subcollection of the sets $C_1, \ldots, C_k$. If we define the function $K\colon \mathcal{S} \times \mathcal{T} \to \mathbb{R}$ by

$$K(s,t) = \sum_{i=1}^{k} \tfrac{1}{\mu(C_i)} \chi_{C_i}(s)(T\chi_{C_i})(t)\,,$$

then it is not difficult to check that $K \in \mathcal{U}_{A_j}$ for each $j = 1, \ldots, n$. Therefore, the family $\{\mathcal{U}_A\}_{A\in\Sigma_1^f}$ has the finite intersection property.

To finish the proof, it suffices to show that each $\mathcal{U}_A$ is compact in the weak* topology $\sigma(L_{p,\infty}, L_{q,1})$. To see this, assume that a net $\{K_\alpha\}$ in some $\mathcal{U}_A$ satisfies $K_\alpha \xrightarrow{w^*} K$ in $L_{p,\infty}$. Obviously $K \ge 0$ and $\|K\|_{p,\infty} \le 1$. It is easy to see that for each set $B \in \Sigma_2$ with $\nu(B) < \infty$ the function $\phi(s,t) = \chi_A(s)\chi_B(t)$ belongs to $L_{q,1}$. This implies

$$\begin{aligned}
\int_B \Big[\int_A K(s,t)\, d\mu(s)\Big]\, d\nu(t) &= \int_{\mathcal{S}\times\mathcal{T}} K\phi\, d(\mu\times\nu) = \lim_\alpha \int_{\mathcal{S}\times\mathcal{T}} K_\alpha \phi\, d(\mu\times\nu) \\
&= \lim_\alpha \int_B \Big[\int_A K_\alpha(s,t)\, d\mu(s)\Big]\, d\nu(t) \\
&= \int_B T\chi_A(t)\, d\nu(t)
\end{aligned}$$

for all $B \in \Sigma_2$ with $\nu(B) < \infty$. But then Lemma 1.96 yields

$$T\chi_A(t) = \int_A K(s,t)\,d\mu(s) = \int_S K(s,t)\chi_A(s)\,d\mu(s)$$

for ν-almost all $t \in \mathcal{T}$. That is, $K \in \mathcal{U}_A$, and the solution is finished. ∎

Problem 5.2.20. *Let E and F be two Banach function spaces with F having order continuous norm. Show that every regular integral operator from E to F carries order bounded subsets of E to norm totally bounded subsets of F. That is, show that if $x \in E^+$, then for each regular integral operator $T\colon E \to F$ the set $T[0,x]$ is norm totally bounded.*

Solution: Let E and F be as above and fix $x \in E^+$. According to Problem 2.3.9 the collection of operators

$$\mathcal{B} = \big\{T \in \mathcal{L}_r(E,F)\colon\ T[0,x] \text{ is norm totally bounded}\big\}$$

is a band in $\mathcal{L}_r(E,F)$.

From Theorem 5.25 we know that E is a normal function space. But then, from Theorem 5.28, we have $\mathcal{L}_\kappa(E,F) = (E_n^\sim \otimes F)^{dd}$. Since every finite-rank operator is a compact operator, it is easy to see that $E_n^\sim \otimes F \subseteq \mathcal{B}$. Consequently, $\mathcal{L}_\kappa(E,F) = (E_n^\sim \otimes F)^{dd} \subseteq \mathcal{B}$, and the solution is finished. ∎

5.3. Conditional Expectations and Positive Projections

Problem 5.3.1. *Let (Ω,Σ,μ) be a probability space, and assume that $\mathcal{A}$ is a σ-subalgebra of Σ. Show that for each function $f \in L_\infty(\mathcal{A})$ we have $\|f\|_{\mathcal{A},\infty} = \|f\|_{\Sigma,\infty}$.*

Solution: Assume that $\mathcal{A}$ is a σ-subalgebra of Σ, and let $f \in L_\infty(\mathcal{A})$. Clearly, $f \in L_\infty(\Sigma)$ and:

$$\begin{aligned}\|f\|_{\mathcal{A},\infty} &= \inf\big\{M>0\colon\ A=\{\omega\in\Omega\colon\ |f(\omega)|>M\}\in\mathcal{A} \text{ and } \mu(A)=0\big\},\\ \|f\|_{\Sigma,\infty} &= \inf\big\{M>0\colon\ A=\{\omega\in\Omega\colon\ |f(\omega)|>M\}\in\Sigma \text{ and } \mu(A)=0\big\}.\end{aligned}$$

Since for each scalar c we have $\{\omega \in \Omega\colon\ |f(\omega)| > c\} \in \mathcal{A}$, it follows that the collections where the above infima are taken coincide. Consequently, the two norms are the same, i.e., $\|f\|_{\mathcal{A},\infty} = \|f\|_{\Sigma,\infty}$. ∎

Problem 5.3.2. *This problem shows that every Banach lattice with order continuous norm and a weak unit can be represented by a Banach function space that "lives" between L_∞ and L_1.*

Let E be a Banach lattice with order continuous norm and a weak unit e. Show that there exists a probability measure space (Ω,Σ,μ) and an ideal F in $L_1(\Sigma)$ containing the constants, i.e., $L_\infty(\Sigma) \subseteq F \subseteq L_1(\mu)$, and a surjective lattice isomorphism $T\colon E \to F$ such that $T(e) = \mathbf{1}$.

Solution: For the solution we shall need the following property of Banach lattices with order continuous norms.

- *If a Banach lattice E has order continuous norm, then for each $x > 0$ there exists a positive linear functional on E that is strictly positive on E_x, the ideal generated by x.*

We begin by presenting a proof of this result. Fix $x > 0$. For each $0 < f \in E^*$, put $N_f = \{y \in E\colon\ f(|y|) = 0\}$ and $C_f = N_f^{\mathrm{d}}$. Clearly, f is strictly positive on C_f and, since f is order continuous, N_f is a band in E.

Let $J = \bigcup_{f>0} C_f$. Since $0 \le f \le g$ implies $C_f \subseteq C_g$, it easily follows that J is an ideal in E. We claim that J is order dense. To see this, let $0 \le z \in J^{\mathrm{d}}$, i.e., $z \perp C_f$ for all $0 < f \in E^*$. That is, $z \in C_f^{\mathrm{d}} = (N_f)^{\mathrm{dd}} = N_f$ for all $0 < f \in E^*$. Thus, $f(z) = 0$ holds for all $0 < f \in E^*$, and so $f(z) = 0$ for all $f \in E^*$. This implies $z = 0$ or $J^{\mathrm{d}} = \{0\}$, and so J is order dense in E. Since E has order continuous norm, J is also norm dense in E.

Next, we claim that J is norm closed. To see this, assume that a sequence $\{y_n\} \subseteq J$ satisfies $y_n \to y$ in E. For each n pick some $0 < f_n \in E^*$ with $\|f_n\| = 1$ such that $y_n \in C_{f_n}$, and let $f = \sum_{n=1}^{\infty} \frac{1}{2^n} f_n$. Clearly, $0 < f \in E^*$. From $N_f \subseteq N_{f_n}$ for each n, it follows that $C_{f_n} \subseteq C_f$ for each n. Since C_f (being a band) is norm closed, we see that $y \in C_f \subseteq J$. Thus, J is norm closed and consequently (since it is norm dense) $J = E$. In particular, there exists some $0 < g \in E^*$ such that $x \in C_g$, and from this we see that g is strictly positive on E_x.

Now let E be a Banach lattice with order continuous norm and a weak unit e. By the preceding part, there exists a positive linear functional $\phi\colon E \to \mathbb{R}$ that is strictly positive on the ideal E_e. Since e is a weak unit, it follows that ϕ is a strictly positive linear functional on E. We can assume that $\phi(e) = 1$.

Next, we consider on E the L-norm defined by $|||x||| = \phi(|x|)$ and, as in the solution of (b) $\Longrightarrow$ (a) of Problem 5.1.9, we can show that E is an ideal in the norm completion $\hat{E}$ of $(E, |||\cdot|||)$ and that e is a weak unit in $\hat{E}$. Since $(E, |||\cdot|||)$ is an AL-space, it follows from the Kakutani–Bohnenblust–Nakano Representation Theorem 3.5 that we can identify $\hat{E}$ with an $L_1(\mu)$-space over some probability measure space (Ω, Σ, μ) with e corresponding to the function $\mathbf{1}$. Since E is an ideal in $\hat{E}$ and $e = \mathbf{1} \in E$, it easily follows that $L_\infty(\mu) \subseteq E \subseteq L_1(\mu)$. ∎

Problem 5.3.3. *Show that for every σ-subalgebra $\mathcal{A}$ of Σ the conditional expectation operator $\mathcal{E}(\cdot|\mathcal{A})\colon L_1(\Sigma) \to L_1(\Sigma)$ is a strictly positive order continuous contractive projection with range $L_1(\mathcal{A})$. That is, show that:*

(1) *If $f \in L_1(\Sigma)$ and $f > 0$, then $\mathcal{E}(f|\mathcal{A}) > 0$ in $L_1(\mathcal{A})$.*

(2) *If $f_n \downarrow 0$ in $L_1(\Sigma)$, then $\mathcal{E}(f_n|\mathcal{A}) \downarrow 0$ in $L_1(\mathcal{A})$.*[2]

(3) *The operator $\mathcal{E}(\cdot|\mathcal{A})$ has norm one, i.e., $\|\mathcal{E}(\cdot|\mathcal{A})\|_1 = 1$.*

[2] Recall (see Theorem 1.80 and the paragraph thereafter) that since we deal here with a finite measure space, each ideal E in $L_0(\Sigma)$ satisfies the countable sup property. Accordingly, a positive operator $T\colon E \to E$ is order continuous if and only if $f_n \downarrow 0$ implies $Tf_n \downarrow 0$.

(4) *If* $g \in L_1(\mathcal{A})$, *then* $\mathcal{E}(g|\mathcal{A}) = g$, *and so*

$$\mathcal{E}(\cdot|\mathcal{A})^2 = \mathcal{E}(\cdot|\mathcal{A}) \quad \textit{and} \quad \mathcal{E}(\mathbf{1}|\mathcal{A}) = \mathbf{1}\,.$$

(5) *The operator* $\mathcal{E}(\cdot|\mathcal{A})$ *leaves* $L_\infty(\Sigma)$ *invariant and* $\|\mathcal{E}(\cdot|\mathcal{A})\|_\infty = 1$.

Solution: (1) Assume that a function $f \in L_1(\Sigma)$ satisfies $f > 0$. Then for each $A \in \mathcal{A}$ we have $\int_A \mathcal{E}(f|\mathcal{A})\,d\mu = \int_A f\,d\mu \geq 0$ in $L_1(\mathcal{A})$. This implies (see Problem 1.6.14) that $\mathcal{E}(f|\mathcal{A}) \geq 0$. Moreover, from $\int_\Omega \mathcal{E}(f|\mathcal{A})\,d\mu = \int_\Omega f\,d\mu > 0$, we infer that $\mathcal{E}(f|\mathcal{A}) > 0$ in $L_1(\mathcal{A})$. Therefore, the operator $\mathcal{E}(\cdot|\mathcal{A})$ is strictly positive.

(2) Let $f_n \downarrow 0$ in $L_1(\Sigma)$ and assume that some function $g \in L_1(\mathcal{A})$ satisfies $0 < g \leq \mathcal{E}(f_n|\mathcal{A})$ in $L_1(\mathcal{A})$ for each n. Using the Lebesgue Dominated Convergence Theorem, we see that

$$0 < \int_\Omega g\,d\mu \leq \int_\Omega \mathcal{E}(f_n|\mathcal{A})\,d\mu = \int_\Omega f_n\,d\mu \downarrow 0\,,$$

which is impossible. Hence, $g = 0$ and thus $\mathcal{E}(f_n|\mathcal{A}) \downarrow 0$.

(3) Assume that $f \in L_1(\Sigma)$ satisfies $\|f\|_1 \leq 1$. The positivity of $\mathcal{E}(\cdot|\mathcal{A})$ yields

$$\big\|\mathcal{E}(f|\mathcal{A})\big\|_1 = \int_\Omega \big|\mathcal{E}(f|\mathcal{A})\big|\,d\mu \leq \int_\Omega \mathcal{E}(|f|\,|\mathcal{A})\,d\mu = \int_\Omega |f|\,d\mu = \|f\|_1 \leq 1\,.$$

The latter implies that the operator $\mathcal{E}(\cdot|\mathcal{A})\colon L_1(\Sigma) \to L_1(\Sigma)$ is bounded and that $\big\|\mathcal{E}(\cdot|\mathcal{A})\big\|_1 \leq 1$. Taking into account that $\mathcal{E}(\mathbf{1}|\mathcal{A}) = \mathbf{1}$, we get $\big\|\mathcal{E}(\cdot|\mathcal{A})\big\|_1 = 1$.

(4) and (5) These are obvious. ∎

Problem 5.3.4. *Assume that* $\{\Omega_i\}_{i\in I}$ *is an at most countable* ***partition*** *of a set* Ω. *That is,* I *is at most countable,* $\Omega_i \neq \emptyset$ *for each* $i \in I$, $\Omega_i \cap \Omega_j = \emptyset$ *for* $i \neq j$, *and* $\bigcup_{i\in I} \Omega_i = \Omega$. *Establish the following.*

(a) *The* σ*-algebra generated by the family* $\{\Omega_i\}_{i\in I}$ *is given by*

$$\mathcal{A} = \big\{\textstyle\bigcup_{i\in J} \Omega_i\colon\ J \subseteq I\big\}\,.$$

(b) *A function* $f\colon \Omega \to \mathbb{R}$ *is* $\mathcal{A}$*-measurable if and only if it is of the form* $f = \sum_{i\in I} \alpha_i \chi_{\Omega_i}$.

(c) *If* Ω *is at most countable, then every* σ*-algebra of subsets of* Ω *is generated by a unique at most countable partition of* Ω.

(d) *Now suppose that* (Ω, Σ, μ) *is a probability measure space and that* $\{\Omega_i\}_{i\in I}$ *is an at most countable measurable partition of* Ω, *i.e, besides* $\{\Omega_i\}_{i\in I}$ *being a partition of* Ω, *we also have* $\Omega_i \in \Sigma$ *for each* i.

If $\mu(\Omega_i) > 0$ *for each* $i \in I$, *then for any* $f \in L_1(\Omega, \Sigma, \mu)$ *we have*

$$\mathcal{E}(f|\mathcal{A}) = \sum_{i\in I} \big[\tfrac{1}{\mu(\Omega_i)} \textstyle\int_{\Omega_i} f\,d\mu\big]\chi_{\Omega_i}\,.$$

Solution: (a) Since every element of $\mathcal{A}$ belongs to the σ-algebra generated by the family $\{\Omega_i\}_{i\in I}$, it suffices to verify that $\mathcal{A}$ itself is a σ-algebra. This claim follows easily from the identities

$$\Omega = \bigcup_{i\in I}\Omega_i \quad \text{and} \quad \Big[\bigcup_{i\in J}\Omega_i\Big]^{\mathrm{c}} = \bigcup_{i\in I\setminus J}\Omega_i$$

and the fact that $\mathcal{A}$ is closed under arbitrary unions.

(b) If a function $f\colon \Omega \to \mathbb{R}$ is of the form $f = \sum_{i\in I}\alpha_i\chi_{\Omega_i}$, then it should be clear that f is $\mathcal{A}$-measurable.

For the converse, suppose that f is $\mathcal{A}$-measurable. To establish that f is of the form $f = \sum_{i\in I}\alpha_i\chi_{\Omega_i}$, it suffices to show that f is constant on each Ω_i. To see this, assume by way of contradiction that there exist some $i \in I$ and two distinct points $\omega_1, \omega_2 \in \Omega_i$ such that $f(\omega_1) \neq f(\omega_2)$. Now notice that the sets

$$\Theta_1 = \big\{\omega \in \Omega_i\colon f(\omega) = f(\omega_1)\big\} \quad \text{and} \quad \Theta_2 = \big\{\omega \in \Omega_i\colon f(\omega) = f(\omega_2)\big\}$$

belong to $\mathcal{A}$ and they are both proper non-empty subsets of Ω_i, which is impossible. This proves that f must be of the desired form.

(c) Assume that Ω is at most countable and that $\mathcal{B}$ is a σ-algebra of subsets of Ω. Let us show first that there is at most one partition of Ω that generates $\mathcal{B}$.

To see this, assume that the two partitions $\{\Omega_i\}_{i\in I}$ and $\{\Theta_j\}_{j\in J}$ of Ω generate $\mathcal{B}$. Fix some $i_0 \in I$. Then there exists a non-empty subset J_1 of J such that $\Omega_{i_0} = \bigcup_{j\in J_1}\Theta_j$. Now fix any index $j \in J_1$. There exists a non-empty subset I_1 of I such that $\Theta_j = \bigcup_{i\in I_1}\Omega_i \subseteq \Omega_{i_0}$. Since the family $\{\Omega_i\}_{i\in I}$ is pairwise disjoint, it follows that $I_1 = \{i_0\}$. This implies that J_1 must be a singleton too, say $J_1 = \{j\}$. Consequently, $\Omega_{i_0} = \Theta_j$. Thus, each Ω_i must be some Θ_j and similarly every Θ_j must equal some Ω_i. In other words, the two partitions $\{\Omega_i\}_{i\in I}$ and $\{\Theta_j\}_{j\in J}$ coincide, and so $\mathcal{B}$ is generated by at most one partition, which is necessarily at most countable.

Next, we shall verify that $\mathcal{B}$ is generated by a partition. For each $\omega \in \Omega$ let

$$B_\omega = \bigcap\{U \in \mathcal{B}\colon\ \omega \in U\}\,.$$

We claim that if $B_{\omega_1} \cap B_{\omega_2} \neq \emptyset$, then necessarily $B_{\omega_1} = B_{\omega_2}$. Indeed, take any $U \in \mathcal{B}$ such that $\omega_1 \in U$. If $\omega_2 \notin U$, then $\omega_2 \in U^{\mathrm{c}}$, and therefore (keeping in mind that the set U^{c} belongs to $\mathcal{B}$) we get that $B_{\omega_1} \cap B_{\omega_2} \subseteq U \cap U^{\mathrm{c}} = \emptyset$, contrary to our assumption. This proves that $\omega_2 \in U$ and consequently $B_{\omega_2} \subseteq B_{\omega_1}$. By symmetry we have $B_{\omega_1} \subseteq B_{\omega_2}$, whence $B_{\omega_1} = B_{\omega_2}$. This establishes that the collection $\{B_\omega\colon\ \omega \in \Omega\}$ is a partition of Ω.

To finish the proof, it remains to be shown that each B_ω belongs to $\mathcal{B}$. To see this, let $A_\omega = \Omega \setminus B_\omega$. Since Ω is at most countable, A_ω is likewise at most countable. Now notice that for each $s \in A_\omega$ there exists some $U_s \in \mathcal{B}$ with $\omega \in U_s$ and $s \notin U_s$. In particular, we have $B_\omega \subseteq \bigcap_{s\in A_\omega}U_s$. Now let $t \in \bigcap_{s\in A_\omega}U_s$, i.e., $t \in U_s$ for each $s \in A_\omega$. If $t \notin B_\omega$, then $t \in A_\omega$ and so $t \notin U_t$, a contradiction. Hence, $t \in B_\omega$, and thus $\bigcap_{s\in A_\omega}U_s \subseteq B_\omega$. Consequently, $B_\omega = \bigcap_{s\in A_\omega}U_s \in \mathcal{B}$, as desired.

(d) Assume that the function f belongs to $L_1(\Omega, \Sigma, \mu)$. By part (b) the function $\sum_{i\in I}\big[\frac{1}{\mu(\Omega_i)}\int_{\Omega_i} f\,d\mu\big]\chi_{\Omega_i}$ is $\mathcal{A}$-measurable. Now for each $A = \bigcup_{i\in J}\Omega_i \in \mathcal{A}$ we have that

$$\begin{aligned}\int_A \sum_{i\in I}\big[\tfrac{1}{\mu(\Omega_i)}\textstyle\int_{\Omega_i} f\,d\mu\big]\chi_{\Omega_i}\,d\mu &= \int_\Omega \Big[\sum_{i\in I}\big[\tfrac{1}{\mu(\Omega_i)}\textstyle\int_{\Omega_i} f\,d\mu\big]\chi_{\Omega_i}\Big]\chi_{\bigcup_{j\in J}\Omega_j}\,d\mu\\ &= \int_\Omega \sum_{i\in J}\big[\tfrac{1}{\mu(\Omega_i)}\textstyle\int_{\Omega_i} d\mu\big]\chi_{\Omega_i}\,d\mu\\ &= \sum_{i\in J}\int_{\Omega_i} f\,d\mu = \int_{\bigcup_{i\in J}\Omega_i} f\,d\mu\\ &= \int_A f\,d\mu\,.\end{aligned}$$

This shows that $\mathcal{E}(f|\mathcal{A}) = \sum_{i\in I}\big[\frac{1}{\mu(\Omega_i)}\int_{\Omega_i} f\,d\mu\big]\chi_{\Omega_i}$, as claimed. ∎

Problem 5.3.5. *For an arbitrary convex function $\psi\colon \mathbb{R}\to\mathbb{R}$, any function $f \in L_1(\Sigma)$ for which $\psi\circ f \in L_1(\Sigma)$, and any σ-subalgebra $\mathcal{A}$ of Σ establish the following inequalities—known as* ***Jensen's inequalities:***

(a) $\psi\big(\int f\,d\mu\big) \le \int \psi\circ f\,d\mu$.

(b) $\psi\big(\mathcal{E}(f|\mathcal{A})\big) \le \mathcal{E}(\psi\circ f|\mathcal{A})$ *in* $L_1(\mathcal{A})$.

Solution: Assume that the function $f \in L_1(\Omega, \Sigma, \mu)$ satisfies $\psi\circ f \in L_1(\mu)$.

(a) It is well known that at each point $a \in \mathbb{R}$ the convex function ψ has left and right derivatives and the inequality $\psi'_r(a) \le \psi'_\ell(a)$ holds. Moreover, if m satisfies $\psi'_r(a) \le m \le \psi'_\ell(a)$, then the straight line with slope m passing through $\big(a, \psi(a)\big)$ supports that graph of ψ at a, i.e.,

$$m(x-a) + \psi(a) \le \psi(x) \qquad (\star)$$

holds for each $x \in \mathbb{R}$; see Figure 1.

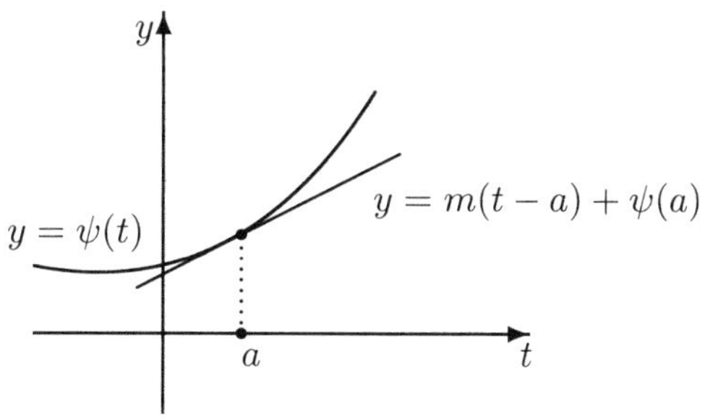

Figure 1. The support property of a convex function

In particular, letting $a = \int f\,d\mu$ and choosing m as above, we have

$$m\Big[f(\omega) - \int f\,d\mu\Big] + \psi\Big(\int f\,d\mu\Big) \le \psi(f(\omega))$$

for μ-almost all $\omega \in \Omega$. Integrating yields

$$m\Big[\int f\,d\mu - \int f\,d\mu\Big] + \psi\Big(\int f\,d\mu\Big) \le \int \psi(f)\,d\mu\,,$$

or $\psi(\int f\,d\mu) \le \int \psi \circ f\,d\mu$.

(b) Assume that $\mathcal{A}$ is a σ-subalgebra of Σ. Without loss of generality we can assume that both $f(\omega)$ and $\mathcal{E}(f|\mathcal{A})(\omega)$ are finite for each $\omega \in \Omega$. Let

$$D = \big\{(\alpha,\beta) \in \mathbb{R}^2\colon\ \alpha t + \beta \le \psi(t) \text{ for all } t\big\}\,.$$

It follows from $(\star)$ that $D \neq \emptyset$ and that for each $t \in \mathbb{R}$ we have

$$\sup\big\{\alpha t + \beta\colon\ (\alpha,\beta) \in D\big\} = \psi(t)\,.$$

Since $\mathbb{R}^2$ is separable, there exists a countable subset D_0 of D that is dense in D. In particular, notice that

$$\sup\big\{\alpha t + \beta\colon\ (\alpha,\beta) \in D_0\big\} = \psi(t)\,. \tag{$\star\star$}$$

Now fix $(\alpha,\beta) \in D_0$. Substituting $f(\omega)$ for t in $\alpha t + \beta \le \psi(t)$, we get

$$\alpha f(\omega) + \beta \le \psi\big(f(\omega)\big)$$

for each $\omega \in \Omega$. This inequality, in conjunction with $\mathcal{E}(\mathbf{1}|\mathcal{A}) = \mathbf{1}$ and the monotonicity and linearity of the conditional expectation operator $\mathcal{E}(\cdot|\mathcal{A})$, yields

$$\alpha\mathcal{E}(f|\mathcal{A}) + \beta = \mathcal{E}(\alpha f + \beta|\mathcal{A}) \le \mathcal{E}(\psi \circ f|\mathcal{A})$$

in $L_1(\mathcal{A})$. So, for each $(\alpha,\beta) \in D_0$ there exists some μ-null set $A_{\alpha,\beta} \in \mathcal{A}$ such that

$$\mathcal{E}(\alpha f + \beta|\mathcal{A})(\omega) \le \mathcal{E}(\psi \circ f|\mathcal{A})(\omega)$$

for each $\omega \notin A_{\alpha,\beta}$. If we let $A = \bigcup_{(\alpha,\beta)\in D} A_{\alpha,\beta}$, then $A \in \mathcal{A}$, $\mu(A) = 0$ and

$$\alpha\mathcal{E}(f|\mathcal{A})(\omega) + \beta \le \mathcal{E}(\psi \circ f|\mathcal{A})(\omega)$$

for all $\omega \notin A$ and all $(\alpha,\beta) \in D_0$. Taking the supremum over D_0 and using $(\star\star)$, we obtain

$$\psi(\mathcal{E}(f|\mathcal{A})(\omega)) \le \mathcal{E}(\psi \circ f|\mathcal{A})(\omega)$$

for each $\omega \notin A$. Therefore, $\psi\big(\mathcal{E}(f|\mathcal{A})\big) \le \mathcal{E}(\psi \circ f|\mathcal{A})$ in $L_1(\mathcal{A})$.

It is worth pointing out that we can get the conclusion of part (a) from part (b). Indeed, if we take $\mathcal{A} = \big\{\emptyset, \Omega\big\}$, then $L_1(\mathcal{A})$ consists of the constant functions on Ω. Moreover, it is easy to see that $\mathcal{E}(f|\mathcal{A}) = \big(\int f\,d\mu\big)\mathbf{1}$ and $\mathcal{E}(\psi(f)|\mathcal{A}) = \big(\int \psi \circ f\,d\mu\big)\mathbf{1}$. Consequently, part (b) implies $\psi(\int f\,d\mu) \le \int \psi \circ f\,d\mu$. ∎

Problem 5.3.6. *Give an example of an ideal E in $L_1(\Sigma)$ such that E is a Banach lattice in its own right and E contains a non-trivial norm closed order dense ideal in E.*

Solution: Let $\Omega = [0,1]$ and let Σ be the σ-algebra of all Borel subsets of $[0,1]$. Choose some $u \in L_1(\Sigma) \setminus L_\infty(\Sigma)$ such that $u \ge \mathbf{1}$; for instance let $u(t) = \frac{1}{\sqrt{t}}$. Now consider the principal ideal E in $L_1(\Sigma)$ generated by the function u and equip E with the uniform norm $\|\cdot\|_u$. We know that $(E, \|\cdot\|_u)$ is a Dedekind complete Banach lattice. Moreover, it should be clear that $L_\infty(\Sigma)$ is a non-trivial order dense ideal in E.

We claim that the $\|\cdot\|_u$-distance from u to $L_\infty(\Sigma)$ is strictly positive. To see this, let $y \in L_\infty(\Sigma)$ and assume that some $\lambda > 0$ satisfies $|u-y| \le \lambda u$. This implies $(1-\lambda)u \le y$ and so if $0 \le \lambda < 1$ were true, then y could not belong to $L_\infty(\Sigma)$, a contradiction. Hence, $\lambda \ge 1$. Consequently, $\|u-y\|_u = \inf\{\lambda\colon\ |u-y| \le \lambda u\} \ge 1$. In particular, we have

$$d\big(u, L_\infty(\Sigma)\big) = \inf\big\{\|u-y\|_u\colon\ y \in L_\infty(\Sigma)\big\} \ge 1\,.$$

Finally, let F be the norm closure of $L_\infty(\Sigma)$ in $(E, \|\cdot\|_u)$. Then F is a norm closed proper ideal in E. Obviously, this ideal is order dense in E. ∎

Problem 5.3.7. *Show that the conclusions of Lemma* 5.42 *do not hold in general for the remaining values of the indices.*

Solution: (a) We shall present an example of a contractive projection T on $L_\infty[0,1]$ such that $T\mathbf{1} = \mathbf{1}$ but the adjoint operator T^* does not fix $\mathbf{1}$.

Pick a multiplicative non-zero positive linear functional ψ on $L_\infty[0,1]$. (Since $L_\infty[0,1]$, as an AM-space with unit, can be represented as a $C(K)$-space in view of the Kakutani–Bohnenblust–Krein Theorem 3.6, non-zero multiplicative positive linear functionals exist.) In particular, $\psi(\mathbf{1}) = 1$ and (by Theorem 4.30) we have $\|\psi\| = 1$. Consider next the rank-one operator $T = \psi \otimes \mathbf{1}$ on $L_\infty[0,1]$, that is, $Tx = \psi(x)\mathbf{1}$ for each $x \in L_\infty[0,1]$. Clearly, the operator T is a positive contractive projection and $T\mathbf{1} = \mathbf{1}$. Now notice that the adjoint operator $T^* = \mathbf{1} \otimes \psi$ satisfies $T^*(\mathbf{1}) = [\mathbf{1} \otimes \psi](\mathbf{1}) = \langle \mathbf{1}, \mathbf{1}\rangle\psi = \psi \ne \mathbf{1}$, that is, T^* does not fix the function $\mathbf{1}$.

(b) We will present here a contractive operator T on $L_2[0,1]$ such that $T\mathbf{1} = \mathbf{1}$ but T is not positive. First consider the three-dimensional Euclidean space $\mathbb{R}^3$ and the straight line L passing through the origin in the direction of the vector $(1,1,1)$. Let R be the rotation of $\mathbb{R}^3$ about L through a "small" positive angle θ, say $\theta = 20^{\mathrm{o}}$. Clearly, R is a linear isometry, R fixes the vector $(1,1,1)$ and R is not positive.

Next, consider the sets $A = \big[0, \frac{1}{3}\big)$, $B = \big[\frac{1}{3}, \frac{2}{3}\big)$, and $C = \big[\frac{2}{3}, 1\big]$, and let

$$Y = \big\{c_1\chi_A + c_2\chi_B + c_3\chi_C\colon\ c_1, c_2, c_3 \in \mathbb{R}\big\}\,.$$

It should be clear that Y is a closed vector sublattice of $L_2[0,1]$ such that $\mathbf{1} \in Y$. Now equip $\mathbb{R}^3$ with the ℓ_2-norm and consider the lattice isometry $J\colon \mathbb{R}^3 \to Y$ defined by $J(x_1,x_2,x_3) = 3x_1\chi_A + 3x_2\chi_B + 3x_3\chi_C$. Also, let Q denote the orthogonal projection from $L_2[0,1]$ onto Y. If we consider the scheme of bounded operators

$$L_2[0,1] \xrightarrow{\ Q\ } Y \xrightarrow{\ J^{-1}\ } \mathbb{R}^3 \xrightarrow{\ R\ } \mathbb{R}^3 \xrightarrow{\ J\ } Y$$

and let $T = JRJ^{-1}Q\colon L_2[0,1] \to L_2[0,1]$, then it should be clear that T is a contractive operator such that $T\mathbf{1} = \mathbf{1}$ but T is not positive. For instance, notice that $T\chi_C \notin L_2^+[0,1]$. ∎

Problem 5.3.8. *Show that the conditional expectation operator $\mathcal{E}(\cdot|\mathcal{A})$ satisfies the following property:* If a function $f \in L_1(\Sigma)$ has its support in some set $A \in \mathcal{A}$, then the function $\mathcal{E}(f|\mathcal{A})$ also has its support in A.

Solution: Assume first that $0 \le f \le \chi_A$. Then $0 \le \mathcal{E}(f|\mathcal{A}) \le \mathcal{E}(\chi_A|\mathcal{A}) = \chi_A$ holds in $L_1(\mathcal{A})$, and so $\operatorname{Supp}\mathcal{E}(f|\mathcal{A}) \subseteq A$. For the general case, let a function $f \in L_1(\Sigma)$ satisfy $\operatorname{Supp} f \subseteq A$. From $|f| \wedge n\chi_A \uparrow |f|$ and the order continuity of the operator

$\mathcal{E}(\cdot|\mathcal{A})$, it follows that $\mathcal{E}(|f| \wedge n\chi_A|\mathcal{A}) \uparrow \mathcal{E}(|f|\,|\mathcal{A})$ in $L_1(\mathcal{A})$. So, by the preceding case, we see that $\operatorname{Supp}\mathcal{E}(f|\mathcal{A}) \subseteq \operatorname{Supp}\mathcal{E}(|f|\,|\mathcal{A}) = \bigcup_{n=1}^{\infty} \operatorname{Supp}(|f| \wedge n\chi_A) \subseteq A$. ∎

Problem 5.3.9 (Dodds–Huijsmans–de Pagter [**26**]). *Suppose that E is an ideal in $L_1(\Sigma)$ containing the constants, and let $0 \le \phi \in E_n^\sim$. Assume also that for some σ-subalgebra $\mathcal{A}$ of Σ the operator $P\colon L_1(\Sigma) \to L_1(\Sigma)$, defined by $Px = \mathcal{E}(\phi x|\mathcal{A})$, leaves E invariant.*

Show that P is a projection if and only if there exists a subset $A \in \mathcal{A}$ such that $\operatorname{Supp}\phi \subseteq A$ *and $\mathcal{E}(\phi|\mathcal{A}) = \chi_A$ in $L_1(\mathcal{A})$.*

Solution: By the definition of the operator P we have $P\mathbf{1} = \mathcal{E}(\phi|\mathcal{A})$. Assume that $P^2 = P$. Using the averaging property of the conditional expectation operator, we obtain

$$\begin{aligned}\mathcal{E}(\phi|\mathcal{A}) &= P\mathbf{1} = P^2\mathbf{1} = P(\mathcal{E}(\phi|\mathcal{A})) = \mathcal{E}(\phi\mathcal{E}(\phi|\mathcal{A})|\mathcal{A}) \\ &= \mathcal{E}(\phi|\mathcal{A})\mathcal{E}(\phi|\mathcal{A}) = \mathcal{E}(\phi|\mathcal{A})^2 .\end{aligned}$$

This guarantees that $\mathcal{E}(\phi|\mathcal{A})$ is a characteristic function of some set $A \in \mathcal{A}$. To verify that $\operatorname{Supp}\phi \subseteq A$, suppose by way of contradiction that $\operatorname{Supp}\phi$ is not a subset of A. This means that $\phi\chi_{A^c}$ is a non-zero positive function. By Problem 5.3.8 we have $\operatorname{Supp}\mathcal{E}(\phi\chi_{A^c}|\mathcal{A}) \subseteq A^c$. Now from $\chi_{A^c} \le \mathbf{1}$ and the positivity of P we get $\mathcal{E}(\phi\chi_{A^c}|\mathcal{A}) = P(\chi_{A^c}) \le P(\mathbf{1}) = \mathcal{E}(\phi|\mathcal{A}) = \chi_A$. That is, $\operatorname{Supp}\mathcal{E}(\phi\chi_{A^c}|\mathcal{A}) \subseteq A$. This implies that the set $\operatorname{Supp}\mathcal{E}(\phi\chi_{A^c}|\mathcal{A}) \in \mathcal{A}$ has μ-measure zero, and so $\mathcal{E}(\phi\chi_{A^c}|\mathcal{A}) = 0$. In turn, the latter implies $\phi\chi_{A^c} = 0$, which is a contradiction.

For the converse, assume that $\operatorname{Supp}\phi \subseteq A$ and $\mathcal{E}(\phi|\mathcal{A}) = \chi_A$. We must show that the operator P is a projection. Start by observing that for each $x \in E$ we have

$$\begin{aligned}P^2x &= P(\mathcal{E}(\phi x|\mathcal{A})) = \mathcal{E}(\phi\mathcal{E}(\phi x|\mathcal{A})|\mathcal{A}) \\ &= \mathcal{E}(\phi|\mathcal{A})\mathcal{E}(\phi x|\mathcal{A}) = \chi_A\mathcal{E}(\phi x|\mathcal{A}) .\end{aligned}$$

Notice that for each $x \in E$ we have $\operatorname{Supp}\phi x \subseteq \operatorname{Supp}\phi$. Hence, from Problem 5.3.8, it follows that $\operatorname{Supp}\mathcal{E}(\phi x|\mathcal{A}) \subseteq A$, and so $\chi_A\mathcal{E}(\phi x|\mathcal{A}) = \mathcal{E}(\phi x|\mathcal{A}) = Px$. This proves that $P^2 = P$. ∎

Problem 5.3.10 (Moy [**58**]; Dodds–Huijsmans–de Pagter [**26**]). *This problem is a modification of Theorem* 5.45 *for linear operators that are not necessarily projections. Let E be an ideal in $L_1(\Sigma)$ containing the constants, and let $T\colon E \to E$ be a linear operator such that:*

(a) *If $x_n \xrightarrow{o} 0$ in E, then $\|Tx_n\|_1 \to 0$.*
(b) *T leaves $L_\infty(\Sigma)$ invariant.*
(c) *For each $h \in L_\infty(\Sigma)$ and each $g \in E$ we have $T(hTg) = ThTg$.*

Then there exists a σ-subalgebra $\mathcal{A}$ of Σ and a function $\phi \in E_n^\sim$ so that $Tx = \mathcal{E}(\phi x|\mathcal{A})$ for each $x \in E$.

Solution: Let

$$Y = \bigl\{h \in L_\infty(\Sigma)\colon\ T(hf) = hTf \ \text{ for all } \ f \in L_\infty(\Sigma)\bigr\},$$

and note that Y is a vector subspace of $L_\infty(\Sigma)$ such that $\mathbf{1} \in Y$. We claim that Y is an order closed subalgebra of $L_\infty(\Sigma)$. If this claim is true, then by Corollary 5.41 there exists a σ-subalgebra $\mathcal{A}$ of Σ such that $Y = L_\infty(\mathcal{A})$.

The fact that Y is a subalgebra follows immediately from condition (c). To show that Y is order closed in $L_\infty(\Sigma)$, assume $\{h_n\} \subseteq Y$, $h \in L_\infty(\Sigma)$ and $h_n \xrightarrow{o} h$ in $L_\infty(\Sigma)$. We must verify that $h \in Y$. To this end, fix any $f \in L_\infty(\Sigma)$. Since $h_n \xrightarrow{o} h$ in $L_\infty(\Sigma)$, we certainly have $h_n f \xrightarrow{o} hf$ and $h_n Tf \xrightarrow{o} hTf$ in E. In view of (a) we have $\|T(h_n f) - T(hf)\|_1 \to 0$, and consequently $\|h_n T(f) - T(hf)\|_1 \to 0$, as $h_n \in Y$ implies $T(h_n f) = h_n Tf$ for each n. The latter shows that $T(hf) = hTf$ for each $f \in L_\infty(\Sigma)$, and thus $h \in Y$.

Next, we claim that $T(E) \subseteq Y$. To see this, we shall establish first that $T\big(L_\infty(\Sigma)\big) \subseteq Y$. So, let $x \in L_\infty(\Sigma)$. By (b), the function Tx belongs to $L_\infty(\Sigma)$. Thus, in view of (c), for each $f \in L_\infty(\Sigma)$ we have $T\big((Tx)f\big) = TxTf$. This shows that $Tx \in Y$, and thus $T\big(L_\infty(\Sigma)\big) \subseteq Y$. Now, since $L_\infty(\Sigma)$ is order dense in E and Y order closed in $L_\infty(\Sigma)$, condition (a) easily implies that $Tx \in Y$ for each $x \in E$.

To complete the solution, we shall consider two cases. Assume first that $E = L_\infty(\Sigma)$. Define the linear functional $\Phi\colon L_\infty(\Sigma) \to \mathbb{R}$ by $\Phi(x) = \int_\Omega Tx\,d\mu$. Condition (a) guarantees that $\Phi \in L_\infty(\Sigma)^*_{\rm n} = L_1(\Sigma)$. Therefore, there exists a function $\phi \in L_1(\Sigma)$ such that $\Phi(x) = \int_\Omega \phi x\,d\mu$ for each $x \in L_\infty(\Sigma)$. Using this and the fact that for each set $A \in \mathcal{A}$ the function χ_A belongs to Y, for each $A \in \mathcal{A}$ and $x \in E$ we obtain

$$\int_A \phi x\,d\mu = \int_\Omega \phi x \chi_A\,d\mu = \Phi(\phi x) = \int_\Omega T(x\chi_A)\,d\mu = \int_\Omega \chi_A Tx\,d\mu = \int_A Tx\,d\mu\,.$$

Since Tx is $\mathcal{A}$-measurable, we can conclude that $Tx = \mathcal{E}(\phi x|\mathcal{A})$.

The general case of an arbitrary E can be easily reduced to the previous one. Indeed, consider the restriction of T to $L_\infty(\Sigma)$. We denote this restriction by T_∞. It should be clear that T_∞ also satisfies conditions (a), (b) and (c), and consequently, by the previous part, there are a σ-algebra $\mathcal{A}$ of Σ and a function $\phi \in L_1(\Sigma)$ such that $T_\infty(x) = \mathcal{E}(\phi x|\mathcal{A})$ for each $x \in L_\infty(\Sigma)$. This and condition (a) imply immediately (since $L_\infty(\Sigma)$ is an order dense ideal in E) that the representation

$$Tx = \mathcal{E}(\phi x|\mathcal{A})$$

is valid on all of E. In particular, the obtained representation of T implies that $\phi x \in L_1(\Sigma)$ for each $x \in E$. This shows that the function ϕ belongs to $E^\sim_{\rm n}$. ∎

Problem 5.3.11. *Give an example of a Markov projection on $L_1[0,1]$ that is not bistochastic. Show that such a projection cannot be contractive.*

Solution: A Markov projection which is not bistochastic can be found in Example 5.56. We present another example below.

Let $\Omega = [0,1]$, let Σ be the σ-algebra of all Lebesgue measurable subsets of $[0,1]$, and let μ be the Lebesgue measure on $[0,1]$. Also, we let $\mathcal{A}$ denote the σ-subalgebra generated by the two sets $A_1 = \big[0, \frac{1}{2}\big]$ and $A_2 = \big[\frac{1}{2}, 1\big]$. We now consider the operator $P\colon L_1(\Sigma) \to L_1(\Sigma)$ defined by the formula $Px = \mathcal{E}(\phi x|\mathcal{A})$, where $\phi = 2\chi_{[0,\frac{1}{4}]} + 2\chi_{[\frac{3}{4},1]}$. A glance at Problem 5.3.4 shows that $\mathcal{E}(\phi|\mathcal{A}) = \mathbf{1}$, and so in view of Problem 5.3.9 we know that P is a positive projection on $L_1[0,1]$ such that $P\mathbf{1} = \mathbf{1}$.

It remains to show that the adjoint operator $P^\sim$ does not fix the function $\mathbf{1}$. To this end, consider the function $x = \chi_{[\frac{1}{4},\frac{1}{2}]}$, and note that $\phi x = 0$. This implies $Px = \mathcal{E}(\phi x|\mathcal{A}) = 0$, and so we have $\int Px\,d\mu = 0 \neq \frac{1}{4} = \int x\,d\mu$. By Lemma 5.48, the adjoint operator $P^\sim$ does not fix the constant function $\mathbf{1}$.

This projection P cannot be contractive since, otherwise, Lemma 5.42(a) would guarantee that the adjoint operator $P^\sim$ must fix the function $\mathbf{1}$, a contradiction. Of course, it can also be shown directly that $\|P\| > 1$. ∎

Corollary 5.52 was generalized to $L_p(\Sigma)$-spaces for $1 < p < \infty$ with $p \neq 2$ by T. Andô [**10**]. Under the assumption that the contractive projection is positive, the restriction $p \neq 2$ is redundant. We present below a very simple proof of this result.

Problem 5.3.12. *For a vector subspace Y of some $L_p(\Sigma)$ with $1 \le p < \infty$ establish the following.*

(a) *If Y is a closed vector sublattice, then Y has a weak unit.*

(b) *Y is a closed vector sublattice if and only if it is the range of a contractive positive projection on $L_p(\Sigma)$.*

Solution: (a) Let Y be a closed vector sublattice of an $L_p(\Sigma)$ with $1 \le p < \infty$. By Zorn's lemma there exists a maximal pairwise disjoint family $\{x_\lambda\}_{\lambda\in\Lambda}$ of functions in Y. If $D_\lambda \in \Sigma$ denotes the support of x_λ, i.e., $D_\lambda = \{\omega \in \Omega\colon\ x_\lambda(\omega) \neq 0\}$, then $\{D_\lambda\}_{\lambda\in\Lambda}$ is a μ-a.e. pairwise disjoint family of subsets of Σ. Since μ is a probability measure, it follows that the set I is at most countable. We can assume that $\Lambda = \mathbb{N}$ and that $\|x_n\|_p = 1$ for each n. Let $x = \sum_{n=1}^\infty \frac{1}{2^n}|x_n| \in Y$, and we claim that x is a weak unit in Y. To see this, let $y \in Y$ satisfy $x \perp y = 0$. If $y \neq 0$, then $y \neq x_n$ and $y \perp x_n$ holds for each n, and so the family $\{y, x_1, x_2, x_3, \ldots\}$ is pairwise disjoint and includes $\{x_1, x_2, \ldots\}$ properly, contrary to the maximality property of $\{x_1, x_2, \ldots\}$. Hence, $y = 0$ and so x is a weak unit in Y.

(b) Assume first that Y is the range of a contractive positive projection on $L_p(\Sigma)$. Since the norm of $L_p(\Omega, \Sigma, \mu)$ is strictly monotone, it follows from Lemma 5.43(b) that Y is a closed vector sublattice of $L_p(\Sigma)$.

For the converse, assume that Y is a closed vector sublattice of $L_p(\Sigma)$. By part (a), we know that Y has weak units. Let $u \in Y_+$ be a weak unit in Y with $\|u\|_p = 1$. Replacing Ω with $\operatorname{Supp} u$, we can assume without loss of generality that $\operatorname{Supp} u = \Omega$.

Now define a new probability measure ν on Σ via the formula

$$\nu(A) = \int_A u^p\,d\mu\,.$$

Clearly, μ and ν are equivalent measures and the Randon–Nikodym derivative of ν with respect to μ equals u^p, i.e., $\frac{d\nu}{d\mu} = u^p$. This implies that the mapping $J\colon L_p(\Omega, \Sigma, \nu) \to L_p(\Omega, \Sigma, \mu)$, defined by

$$J(f) = f\left(\tfrac{d\nu}{d\mu}\right)^{\frac{1}{p}} = fu\,,$$

is a surjective lattice isometry; see [**8**, Problem 37.11, p. 358].

It follows that, $Y_1 = J^{-1}(Y)$ is a closed vector sublattice of $L_p(\Omega, \Sigma, \nu)$ and since $J^{-1}g = g/u$ for each $g \in L_p(\Omega, \Sigma, \mu)$, we see that $\mathbf{1} \in Y_1$. Therefore, by Lemma 5.40, there exists a σ-subalgebra $\mathcal{A}$ of Σ such that $Y_1 = L_p(\Omega, \mathcal{A}, \nu)$ and consequently, by Corollary 5.52, the conditional expectation operator $\mathcal{E}(\cdot|\mathcal{A})$ on $L_1(\Omega, \Sigma, \nu)$ acts as a positive contractive projection from $L_p(\Omega, \Sigma, \nu)$ onto Y_1. To finish the proof, notice that the operator $P = J\mathcal{E}(\cdot|\mathcal{A})J^{-1}\colon L_p(\Sigma, \mu) \to L_p(\Sigma, \mu)$ is a positive contractive projection with range Y.

It is interesting to point out that the above result almost characterizes the L_p-spaces. Namely, T. Andô [**11**] proved that:

- *If every closed vector sublattice of a Banach lattice E is the range of a positive contractive projection, then E is lattice isometric to an L_p- or a $c_0(\Gamma)$-space.*

Later, L. Tzafriri [**81**] and J. Lindenstrauss and L. Tzafriri [**51**] established an isomorphic version of this result. ∎

Problem 5.3.13. *According to Problem* 5.3.12(b), *if a Markov projection P on some $L_p(\Sigma)$ with $1 \le p < \infty$ is contractive, then its range is a vector sublattice of $L_p(\Sigma)$. Show that the converse is not true. That is, give an example of a Markov projection P on $L_p(\Sigma)$ such that its range is a vector sublattice of $L_p(\Sigma)$ but $\|P\| > 1$.*

Solution: We shall exhibit an example of a Markov projection P on $L_p(\Sigma)$ such that its range is a vector sublattice of $L_p(\Sigma)$ but $\|P\| > 1$. Let $\Omega = [0,1]$, let Σ be the Borel σ-algebra, let μ be the Lebesgue measure, and let $\mathcal{A} = \{\emptyset, \Omega\}$ be the trivial σ-subalgebra of Σ. So,

$$\mathcal{E}(x|\mathcal{A}) = \Big[\int_0^1 x(\omega)\,d\mu(\omega)\Big]\mathbf{1}$$

holds for each $x \in L_1(\Sigma)$. Now let $\phi = 2\chi_{[0,1/2]}$ and note that $\mathcal{E}(\phi|\mathcal{A}) = \mathbf{1}$. This implies that the formula $Px = \mathcal{E}(\phi x|\mathcal{A})$ defines a Markov projection whose range Y is the one-dimensional subspace spanned by the constant function one. In particular, Y is a vector sublattice of $L_p(\Sigma)$ for each $1 \le p \le \infty$.

However, for each $1 \le p < \infty$ the projection P is not a contraction, i.e., $\|P\| > 1$. To see this, consider the function $x = 2^{\frac{1}{p}}\chi_{[0,1/2]}$ and note that $\|x\|_p = 1$. An easy calculation shows that $Px = 2^{\frac{1}{p}} \cdot \mathbf{1}$, whence $\|P\| \ge \|Px\| = 2^{\frac{1}{p}} > 1$. ∎

Problem 5.3.14. *Show that a projection on a Hilbert space is an orthogonal projection if and only if it is a contraction.*

Solution: Let $P\colon H \to H$ be any projection on a (real or complex) Hilbert space. Assume first that P is also an orthogonal projection, that is, the decomposition $H = \operatorname{Ker}(P) \oplus R(P)$ satisfies $\operatorname{Ker}(P) \perp R(P)$. Now for each $x \in H$ write $x = x_1 + x_2$ with $x_1 \in \operatorname{Ker}(P)$ and $x_2 \in R(P)$ and note that

$$\|Px\|^2 = \|x_2\|^2 \le \|x_1\|^2 + \|x_2\|^2 = \|x_1 + x_2\|^2 = \|x\|^2.$$

Thus, $\|Px\| \le \|x\|$ for each $x \in H$ and this proves that P is a contraction.

For the converse, assume that P is a contraction. Let $a \in \operatorname{Ker}(P)$ and $b \in R(P)$ and let $(a,b) = \alpha + \beta\imath \in \mathbb{C}$ be the inner product of the vectors a and b. Then for each scalar λ we have

$$\|\lambda b\|^2 = \|P(a+\lambda b)\|^2 \le \|P\|^2 \cdot \|a+\lambda b\|^2 \le \|a\|^2 + 2\lambda\alpha + \|\lambda b\|^2 .$$

Therefore, $\|a\|^2 + 2\lambda\alpha \ge 0$ for each $\lambda \in \mathbb{R}$, and from this it easily follows that $\alpha = 0$.

Observing that $(\imath a, b) = -\beta + \alpha\imath$ and applying the preceding conclusion to the vectors $\imath a \in \operatorname{Ker}(P)$ and $b \in R(P)$, we get $\beta = 0$. Thus, $(a,b) = 0$ holds true for all $a \in \operatorname{Ker}(P)$ and $b \in R(P)$, i.e., $\operatorname{Ker}(P) \perp R(P)$. This shows that P is an orthogonal projection. ∎

Problem 5.3.15. *Give an alternative proof of Lemma* 5.38 *by employing Theorem* 5.57*. That is, show that for each σ-subalgebra $\mathcal{A}$ of Σ and for each $1 \le p \le \infty$ the operator $\mathcal{E}(\cdot|\mathcal{A})$ leaves the space $L_p(\Sigma)$ invariant and is contractive on this space.*

Solution: We need to consider only $1 < p < \infty$. Let q be the index conjugate to p. Take any function $x \in L_p(\Sigma)$ and consider the function $\mathcal{E}(x|\mathcal{A})$. Recall that this function is $\mathcal{A}$-measurable, and therefore its $\|\cdot\|$-norm can be calculated as follows:

$$\begin{aligned}
\|\mathcal{E}(x|\mathcal{A})\|_p &= \sup\{\langle \mathcal{E}(x|\mathcal{A}), y\rangle :\ y \in L_q(\mathcal{A}), \|y\|_q \le 1\} \\
&= \sup\{\langle \mathcal{E}(x|\mathcal{A}), y\rangle :\ y \in L_\infty(\mathcal{A}), \|y\|_q \le 1\} \\
&= \sup\{\langle x, \mathcal{E}(y|\mathcal{A})^*\rangle :\ y \in L_\infty(\mathcal{A}), \|y\|_q \le 1\} \\
&= \sup\{\langle x, \mathcal{E}(y|\mathcal{A})\rangle :\ y \in L_\infty(\mathcal{A}), \|y\|_q \le 1\} \\
&\le \sup\{\langle x, y\rangle :\ y \in L_\infty(\Sigma), \|y\|_q \le 1\} = \|x\|_p .
\end{aligned}$$

This establishes both that $\mathcal{E}(x|\mathcal{A}) \in L_p(\mathcal{A})$ and that $\|\mathcal{E}(\cdot|\mathcal{A})\|_p \le 1$.

Remark. Formally speaking, there is a circular argument in proving Lemma 5.38 by using Theorem 5.57 since our proof of Theorem 5.57 depends upon this lemma, where we used that each conditional expectation operator acts as a contractive operator on $L_2(\Sigma)$. However, the reference to Lemma 5.38 can be avoided by invoking Theorem 5.54. ∎

5.4. Positive Projections and Lattice-subspaces

Problem 5.4.1. *If P is a positive projection on an $L_p(\mu)$-space for some $1 \le p < \infty$, then show that its range is again order isomorphic (up to an equivalent norm) to an $L_p(\nu)$-space for some measure space $(\Omega_1, \Sigma_1, \nu)$.*

Solution: In view of Theorem 5.60 the range Y of the operator P is a closed lattice-subspace of $L_p(\mu)$ which is lattice isomorphic to a closed vector sublattice F of $L_p(\Omega, \Sigma, \mu)$. But each closed vector sublattice of an L_p-space is an abstract L_p-space and it is lattice isometric to an $L_p(\nu)$-space for an appropriate measure space $(\Omega_1, \Sigma_1, \nu)$; see [**5**, Theorem 10.12, p. 71] .

Some remarks are in order regarding this problem.

(1) If we replace the assumption "a positive projection" by "a contractive projection," then a corresponding version of the above result is also true. For $p = 1$ this was established by A. Grothendieck [**35**] and for the remaining values of p by L. Tzafriri [**80**].

(2) We do not claim in this problem that Y is order isomorphic to the original $L_p(\mu)$-space. This is not true in general. However, we have the following result: *If a positive projection $P\colon \ell_p \to \ell_p$, $1 \le p < \infty$, has an infinite dimensional range, then the range of P is order isomorphic to ℓ_p.* ∎

Problem 5.4.2. *Let $\{x_{ij}\colon\ i \in I = \{1, 2, \ldots, n\};\ j \in J = \{1, 2, \ldots, m\}\}$ be a finite subset of vectors in a vector lattice. Then we have*

$$\bigwedge_{\tau\in J^I}\bigvee_{i=1}^{n} x_{i,\tau(i)} = \bigvee_{i=1}^{n}\bigwedge_{j=1}^{m} x_{ij} \quad \textit{and} \quad \bigvee_{\tau\in I^J}\bigwedge_{j=1}^{m} x_{\tau(j),j} = \bigwedge_{i=1}^{n}\bigvee_{j=1}^{m} x_{ij}\,.$$

Solution: The solution will be based on the two (infinite) distributive laws that are valid in vector lattices. We state and prove them for later use in the solution.

(D1) *If a non-empty subset D in a vector lattice E has a supremum, then for each vector $a \in E$ the set $D \wedge a = \{x \wedge a\colon\ x \in D\}$ also has a supremum and* $\sup(D \wedge a) = (\sup D) \wedge a$.

(D2) *If a non-empty subset D in a vector lattice E has an infimum, then for each vector $a \in E$ the set $D \vee a = \{x \vee a\colon\ x \in D\}$ also has an infimum and* $\inf(D \vee a) = (\inf D) \vee a$.

To prove (D1), let $s = \sup D$ and fix an arbitrary vector $a \in E$. Then for each $x \in D$ we have $x \wedge a \le s \wedge a$. This shows that $s \wedge a$ is an upper bound for the set $D \wedge a$. To see that $s \wedge a$ is the least upper bound of $D \wedge a$, assume that a vector y satisfies $x \wedge a \le y$ for each $x \in D$. We have

$$x + a - s \vee a \le x + a - x \vee a = x \wedge a \le y\,,$$

and so $x \le y - a + s \vee a$ for each $x \in D$. Consequently, $s \le y - a + s \vee a$, and thus $s \wedge a = s + a - s \vee a \le y$. The proof of (D2) is similar.

Let us return to our problem. We shall verify the first formula only. The second formula can be proven in the same manner. Start by observing that

$$\bigwedge_{\tau\in J^I}\bigvee_{i=1}^{n} x_{i,\tau(i)} = \bigwedge_{j_1=1}^{m}\bigwedge_{j_2=1}^{m}\cdots\bigwedge_{j_n=1}^{m}\big[x_{1,j_1} \vee x_{2,j_2} \vee \cdots \vee x_{n,j_n}\big]\,.$$

We claim that the right-hand side of this identity is equal to $\bigvee_{i=1}^{n}\bigwedge_{j=1}^{m} x_{ij}$. This will be proved by induction on n. For $n = 1$ and any vectors $\{x_{11}, x_{12}, \ldots, x_{1m}\}$ we have $\bigwedge_{j_1=1}^{m} x_{1,j_1} = \bigvee_{i=1}^{1}\bigwedge_{j=1}^{m} x_{ij}$, and so the desired identity is true for $n = 1$. Now assume that for some $n \ge 1$ and for all finite subsets of vectors of the form $\{x_{ij}\colon\ i = 1, 2, \ldots, n;\ j = 1, 2, \ldots, m\}$ we have

$$\bigwedge_{j_1=1}^{m}\bigwedge_{j_2=1}^{m}\cdots\bigwedge_{j_n=1}^{m}\big[x_{1,j_1} \vee x_{2,j_2} \vee \cdots \vee x_{n,j_n}\big] = \bigvee_{i=1}^{n}\bigwedge_{j=1}^{m} x_{ij}\,. \tag{$\star$}$$

To prove our claim for $n+1$ consider an arbitrary finite subset of vectors $\{x_{ij}\colon\ i=1,\ldots,n+1;\ j=1,\ldots,m\}$. To simplify the computations below, let $x=\bigwedge_{j_{n+1}=1}^{m} x_{n+1,j_{n+1}}$. Using the induction hypothesis $(\star)$ and the distributive laws, we see that

$$\begin{aligned}
\bigvee_{i=1}^{n+1}\bigwedge_{j=1}^{m} x_{ij} &= \Big[\bigvee_{i=1}^{n}\bigwedge_{j=1}^{m} x_{ij}\Big]\bigvee x\\
&= \Big[\bigwedge_{j_1=1}^{m}\bigwedge_{j_2=1}^{m}\cdots\bigwedge_{j_n=1}^{m}\big[x_{1,j_1}\vee x_{2,j_2}\vee\cdots\vee x_{n,j_n}\big]\Big]\bigvee x\\
&= \bigwedge_{j_1=1}^{m}\bigwedge_{j_2=1}^{m}\cdots\bigwedge_{j_n=1}^{m}\big[x_{1,j_1}\vee x_{2,j_2}\vee\cdots\vee x_{n,j_n}\vee x\big]\\
&= \bigwedge_{j_1=1}^{m}\bigwedge_{j_2=1}^{m}\cdots\bigwedge_{j_n=1}^{m}\Big[\bigwedge_{j_{n+1}=1}^{m} x_{1,j_1}\vee x_{2,j_2}\vee\cdots\vee x_{n,j_n}\vee x_{n+1,j_{n+1}}\Big]\\
&= \bigwedge_{j_1=1}^{m}\bigwedge_{j_2=1}^{m}\cdots\bigwedge_{j_n=1}^{m}\bigwedge_{j_{n+1}=1}^{m}\Big[x_{1,j_1}\vee x_{2,j_2}\vee\cdots\vee x_{n,j_n}\vee x_{n+1,j_{n+1}}\Big].
\end{aligned}$$

Thus, $(\star)$ is true for $n+1$, and the solution is finished. ∎

Problem 5.4.3. *If Y is a vector subspace of a vector lattice E, then the vector sublattice of E generated by Y coincides with $Y^{\wedge\vee}$ and also with $Y^{\vee\wedge}$. That is,*

$$\mathcal{R}(Y)=Y^{\wedge\vee}=Y^{\vee\wedge}.$$

Solution: A glance at Problem 5.4.2 shows that $Y^{\wedge\vee}=Y^{\vee\wedge}$. Also, it should be clear that $Y^{\wedge\vee}\subseteq\mathcal{R}(Y)$. Therefore, in order to establish the identity $Y^{\wedge\vee}=\mathcal{R}(Y)$, it suffices to show that $Y^{\wedge\vee}$ is a vector sublattice of E.

To see this, let $x=\bigvee_{i=1}^{n}\bigwedge_{j=1}^{m} x_{ij}$ and $y=\bigvee_{p=1}^{k}\bigwedge_{q=1}^{\ell} y_{pq}$ be two vectors in $Y^{\wedge\vee}$. Then

$$x+y=\bigvee_{i=1}^{n}\bigwedge_{j=1}^{m} x_{ij}+\bigvee_{p=1}^{k}\bigwedge_{q=1}^{\ell} y_{pq}=\bigvee_{i=1}^{n}\bigvee_{p=1}^{k}\bigwedge_{j=1}^{m}\bigwedge_{q=1}^{\ell}(x_{ij}+y_{pq})\in Y^{\wedge\vee}.$$

Now let α be a scalar. If $\alpha\geq 0$, then

$$\alpha x=\alpha\Big[\bigvee_{i=1}^{n}\bigwedge_{j=1}^{m} x_{ij}\Big]=\bigvee_{i=1}^{n}\bigwedge_{j=1}^{m}\alpha x_{ij}\in Y^{\wedge\vee},$$

and for $\alpha<0$ we have

$$\alpha x=\alpha\Big[\bigvee_{i=1}^{n}\bigwedge_{j=1}^{m} x_{ij}\Big]=\bigwedge_{i=1}^{n}\bigvee_{j=1}^{m}\alpha x_{ij}\in Y^{\vee\wedge}=Y^{\wedge\vee}.$$

The above show that $Y^{\wedge\vee}$ is a vector space. Finally, from

$$x\vee y=\Big[\bigvee_{i=1}^{n}\bigwedge_{j=1}^{m} x_{ij}\Big]\bigvee\Big[\bigvee_{p=1}^{k}\bigwedge_{q=1}^{\ell} y_{pq}\Big]\in Y^{\wedge\vee},$$

we conclude that $Y^{\wedge\vee}$ is indeed a vector sublattice of E, as desired. ∎

Problem 5.4.4. *Show that a vector subspace of a finite dimensional vector lattice is a lattice-subspace if and only if it is the range of a positive projection.*

Solution: Let Y be a vector subspace of some $\mathbb{R}^n$. If Y is the range of a positive projection, then Y is a lattice-subspace by Theorem 5.59.

For the converse, assume that Y is a lattice-subspace. Then the identity operator $I\colon Y \to Y$ is a positive operator and Y (as being finite dimensional) is Dedekind complete. By the Kantorovich Extension Theorem 4.32, the operator I has a positive linear extension $P\colon B \to Y$, where B is the ideal generated by Y in $\mathbb{R}^n$. Since B is closed, B is a band in the Dedekind complete Riesz space $\mathbb{R}^n$. Write $\mathbb{R}^n = B \oplus B^{\mathrm{d}}$, and let P_B denote the band projection of $\mathbb{R}^n$ onto B. Now notice that the operator $PP_B\colon \mathbb{R}^n \to \mathbb{R}^n$ is a positive projection whose range is Y. ∎

Problem 5.4.5. *For a vector subspace Y of some $C(K)$-space establish the following.*

(a) *Y separates the points of K if and only if $\mathcal{R}(Y)$ separates the points of K.*

(b) *Assume that the induced cone $Y_+ = Y \cap [C(K)]^+$ is generating in Y, i.e., $Y = Y_+ - Y_+$. Then Y majorizes $C(K)$ if and only if $\mathcal{R}(Y)$ majorizes $C(K)$.*

(c) *Assume that:*

(i) *Y is a Dedekind complete lattice-subspace. (In particular, this is so if Y is a finite dimensional lattice-subspace.)*

(ii) *Y separates the points of K and $\mathbf{1} \in Y$.*

Then the Miyajima projection $P\colon \mathcal{R}(Y) \to \mathcal{R}(Y)$ extends uniquely to a positive projection on $C(K)$ with range Y.

Solution: (a) If Y separates the points of K, then $\mathcal{R}(Y)$ automatically separates the points of K since $Y \subseteq R(Y)$. For the converse, assume that $\mathcal{R}(Y)$ separates the points of K. If Y does not separate the points of K, then there exist two distinct points $t_1, t_2 \in K$ such that $x(t_1) = x(t_2)$ for each $x \in C(K)$.

Now let $y \in \mathcal{R}(Y)$ and choose $x_{ij} \in Y$ $(i = 1, \ldots, n;\ j = 1, \ldots, m)$ such that $y = \bigvee_{i=1}^{n} \bigwedge_{j=1}^{m} x_{ij}$. Since the lattice operations in $C(K)$ are pointwise, we have

$$y(t_1) = \bigvee_{i=1}^{n} \bigwedge_{j=1}^{m} x_{ij}(t_1) = \bigvee_{i=1}^{n} \bigwedge_{j=1}^{m} x_{ij}(t_2) = y(t_2)\,.$$

That is, $y(t_1) = y(t_2)$ for each $y \in \mathcal{R}(Y)$. In other words, $\mathcal{R}(Y)$ does not separate the points of K, a contradiction. This shows that Y separates the points of K.

(b) If Y majorizes $C(K)$, then $\mathcal{R}(X)$ also majorizes $C(K)$. For the converse, assume that $\mathcal{R}(Y)$ majorizes $C(K)$. Let $x \in C(K)$ and pick some $y \in \mathcal{R}(Y)$ such that $x \leq y$. Now choose functions $x_{ij} \in Y$ $(i = 1, 2, \ldots, n;\ j = 1, 2, \ldots, m)$ such that $y = \bigvee_{i=1}^{n} \bigwedge_{j=1}^{m} x_{ij}$. Since Y_+ is generating, for each pair (i, j) there exists some $z_{ij} \in Y_+$ such that $y_{ij} \leq z_{ij}$. If we let $z = \sum_{i=1}^{n} \sum_{j=1}^{m} z_{ij}$, then $z \in Y_+$ and it is easy to see that $x \leq y \leq z$. This shows that Y is majorizing $C(K)$.

(c) Since $\mathcal{R}(Y)$ includes the constant function $\mathbf{1}$ and separates the points of K, it follows from the classical Stone–Weierstrass theorem that $\mathcal{R}(Y)$ is norm dense in $C(K)$. In particular, if the operator $P\colon \mathcal{R}(Y) \to \mathcal{R}(Y)$ extents to a positive operator $\hat{P}\colon C(K) \to C(K)$, then $\hat{P}$ (as a positive operator defined on a Banach lattice) is continuous. Since $\hat{P}$ agrees with P on $\mathcal{R}(Y)$, it is uniquely determined. To see that $P\colon \mathcal{R}(Y) \to \mathcal{R}(Y)$ does have a positive linear extension to all of $C(K)$, note that the operator $P\colon \mathcal{R}(Y) \to Y$ satisfies all the assumptions of the Kantorovich Extension Theorem 4.32. So, according to this theorem, a positive linear extension of $P\colon \mathcal{R}(Y) \to Y$ to a positive linear operator $\hat{P}\colon C(K) \to Y$ exists. This extension is a positive projection on $C(K)$ with range Y that extends the Miyajima projection $P\colon \mathcal{R}(Y) \to \mathcal{R}(Y)$. ∎

Problem 5.4.6. *Let $E = C[0,1]$ and let Y be the vector subspace generated by the functions $\mathbf{1}$ and s, where $s(t) = t$ for each $t \in [0,1]$. That is, Y consists of all linear functions. Establish the following.*

(a) *Y is a lattice-subspace of E.*

(b) *$\mathcal{R}(Y)$ consists of all piecewise linear continuous functions.*

(c) *The Miyajima projection $P\colon \mathcal{R}(Y) \to \mathcal{R}(Y)$ is given by*

$$Px(t) = tx(1) + (1-t)x(0)$$

for each $x \in \mathcal{R}(Y)$ and each $t \in [0,1]$.

(d) *If we define $Q\colon C[0,1] \to C[0,1]$ by $Qx(t) = tx(1) + (1-t)x(0)$ for each $x \in C[0,1]$ and each $t \in [0,1]$, then Q is the only extension of the Miyajima projection to a positive projection on all of $C[0,1]$.*

(e) *There are infinitely many bounded projections on E with range Y.*

Solution: Recall that a function $x \in C[0,1]$ is said to be **piecewise linear** if there exist a partition $0 = a_0 < a_1 < \cdots < a_n = 1$ of the interval $[0,1]$ and constants m_i, b_i $(i = 1, 2, \ldots, n)$ such that $x(t) = m_i t + b_i$ for each $a_{i-1} \le t \le a_i$. If $x(t) = mt + b$ for each $t \in [0,1]$, then x is called a **linear function**.

(a) It should be clear that the vector space Y generated by the functions s and $\mathbf{1}$ is precisely the vector space of all linear functions. If $x_1(t) = m_1 t + b_1$ and $x_2(t) = m_2 t + b_2$ are two linear functions, then the linear function whose graph is the line joining the points $\big(0, \max\{x_1(0), x_2(0)\}\big) = \big(0, \max\{b_1, b_2\}\big)$ and $\big(1, \max\{x_1(1), x_2(1)\}\big) = \big(1, \max\{m_1+b_1, m_2+b_2\}\big)$ is the least upper bound of x_1 and x_2 in Y. That is, the linear function

$$y(t) = \big[\max\{m_1 + b_1, m_2 + b_2\} - \max\{b_1, b_2\}\big]t - \max\{b_1, b_2\}$$

satisfies $y = x_1 \curlyvee x_2$ in Y. This shows that Y is a lattice-subspace of $C[0,1]$.

(b) We shall denote by L the collection of all piecewise linear functions. Clearly, L is a vector sublattice of $C[0,1]$ and $Y \subseteq L$. Therefore, $\mathcal{R}(Y) \subseteq L$. So, what needs verification is the inclusion $L \subseteq \mathcal{R}(Y)$.

For each $0 \le a < b \le 1$ and each $m \in \mathbb{R}$, we consider the continuous function $u_{a,b,m}\colon [0,1] \to \mathbb{R}$ defined by

$$u_{a,b,m}(t) = \begin{cases} 0 & \text{if } 0 \le t < a \\ m(t-a) & \text{if } a \le t < b \\ m(b-a) & \text{if } b \le t \le 1\,. \end{cases}$$

Now notice that if x is a typical piecewise linear function as defined at the beginning of the solution, then

$$x = b_1 + u_{0,a_1,m_1} + u_{a_1,a_2,m_2} + u_{a_2,a_3,m_3} + \cdots + u_{a_{n-1},1,m_n}\,.$$

Therefore, in order to establish that $L \subseteq \mathcal{R}(Y)$, it suffices to show that each function $u_{a,b,m}$ belongs to $\mathcal{R}(Y)$. To this end, consider the functions $y_1, y_2 \in C[0,1]$ defined by

$$\begin{aligned} y_1(t) &= \big[m(s-\alpha\mathbf{1})^+\big](t) = \begin{cases} 0 & \text{if } 0 \le t < a \\ m(t-a) & \text{if } a \le t \le 1\,, \end{cases} \\ y_2(t) &= \big[-m(s-\beta\mathbf{1})^+\big](t) = \begin{cases} 0 & \text{if } 0 \le t < b \\ -m(t-a) & \text{if } b \le t \le 1\,. \end{cases} \end{aligned}$$

Clearly, $y_1, y_2 \in \mathcal{R}(X)$ and they satisfy $y_1 + y_2 = u_{a,b,m}$. Hence, each function of the form $u_{a,b,m}$ belongs to $\mathcal{R}(X)$, and thus $\mathcal{R}(Y) = L$.

(c) The mapping $P\colon \mathcal{R}(Y) \to \mathcal{R}(Y)$ defined by

$$Px(t) = tx(1) + (1-t)x(0)$$

for each $x \in \mathcal{R}(Y)$ and each $t \in [0,1]$ is clearly a positive projection on $\mathcal{R}(Y)$ with range Y. By the uniqueness of the Miyajima projection, P must coincide with the Miyajima projection.

(d) The operator $Q\colon C[0,1] \to C[0,1]$, defined by $Qx(t) = tx(1) + (1-t)x(0)$ for each $x \in C[0,1]$ and each $t \in [0,1]$, is a positive linear extension of the Miyajima projection to all of $C[0,1]$. Since any such extension must be continuous and $\mathcal{R}(Y)$ is norm dense in $C[0,1]$, it follows that Q is the only extension of the Miyajima projection to a positive projection on all of $C[0,1]$.

(e) Fix any two points α and β in $[0,1]$ such that $0 < \alpha < \beta < 1$. For each $x \in C[0,1]$ let Tx be the linear function whose graph is the line joining the points $(\alpha, x(\alpha))$ and $(\beta, x(\beta))$, i.e.,

$$Tx(t) = \tfrac{x(\beta)-x(\alpha)}{\beta-\alpha}\,(t-\alpha) + x(\alpha)\,.$$

Then T is a continuous linear projection on $C[0,1]$ with range Y. However, it is easy to see that the projection T is not positive. ∎

Problem 5.4.7. *If Y is a majorizing lattice-subspace of a Banach lattice, then show that the Miyajima projection $P\colon \mathcal{R}(Y) \to \mathcal{R}(Y)$ is continuous.*

Solution: We consider $\mathcal{R}(Y)$ equipped with the lattice norm of E. Then, according to Theorem 5.59(5), there exists a lattice norm $|||\cdot|||$ on Y satisfying $\|x\| \le |||x|||$ for each $x \in Y$.

According to Problem 4.3.8 the operator $P\colon (\mathcal{R}(Y),\|\cdot\|) \to (Y, |||\cdot|||)$ is continuous. Therefore, there exists some $M > 0$ such that $|||Px||| \leq M\|x\|$ for each $x \in \mathcal{R}(Y)$. This implies $\|Px\| \leq M\|x\|$ for each $x \in \mathcal{R}(Y)$. The obtained inequality shows that the Miyajima projection $P\colon (\mathcal{R}(Y),\|\cdot\|) \to (\mathcal{R}(Y),\|\cdot\|)$ is continuous. ∎

Problem 5.4.8. *This problem presents an application of lattice-subspaces to harmonic analysis. Recall that a C^2-function $u\colon \mathcal{O} \to \mathbb{R}$, where $\mathcal{O}$ is an open subset of $\mathbb{R}^2$, is said to be* ***harmonic*** *if it satisfies* ***Laplace's equation*** $\nabla^2 u = \frac{\partial^2 u}{\partial x^2} + \frac{\partial^2 u}{\partial y^2} = 0$ *at every point of $\mathcal{O}$.*

Now let $D = \{(x,y)\colon\ x^2 + y^2 < 1\}$ be the open unit disk. We shall denote by $\mathcal{H}$ the vector space of all continuous functions defined on the closed unit disk $\overline{D} = \{(x,y)\colon\ x^2 + y^2 \leq 1\}$ and harmonic on D. The classical Dirichlet theorem states that if f is a continuous function on the unit circle $\partial D = \{(x,y)\colon\ x^2+y^2 = 1\}$, then there exists a unique continuous extension $\widehat{f}$ of f to all of $\overline{D}$ which is harmonic on D. The harmonic extension $\widehat{f}$ in polar coordinates is given by ***Poisson's classical formula***

$$\widehat{f}(r,\theta) = \frac{1}{2\pi}\int_0^{2\pi} \frac{f(\cos\phi,\sin\phi)(1-r^2)}{1-2r\sin(\theta-\phi)+r^2}\, d\phi,\ \ (r,\theta) \in [0,1) \times [0,2\pi].$$

Establish the following.

(a) *A function $f \in C(\overline{D})$ satisfies $\widehat{f_{\partial D}} \geq 0$ if and only if $f_{\partial D} \geq 0$, where $f_{\partial D}$ denotes the restriction of the function f to ∂D.*

(b) *The mapping $Q\colon C(\overline{D}) \to C(\overline{D})$, defined by $Qf = \widehat{f_{\partial D}}$, is a positive projection whose range is $\mathcal{H}$.*

(c) *$\mathcal{H}$ is a lattice-subspace of $C(\overline{D})$ and Q is the unique extension of the Miyajima projection to $C(\overline{D})$.*

(d) *$\mathcal{H}$ is an AM-space with unit $\mathbf{1}$, the constant function one, and the mapping $f \mapsto \widehat{f}$, from $C(\partial D)$ to $\mathcal{H}$, is a surjective lattice isometry.*

Solution: (a) Assume first that $\widehat{f}(x,y) \geq 0$ for all $(x,y) \in D$. Since $\widehat{f}$ extends continuously to the boundary ∂D, we have $\widehat{f}(x,y) \geq 0$ for each $(x,y) \in \partial D$. Hence, from $\widehat{f} = f$ on ∂D, we see that $f_{\partial D} \geq 0$.

For the converse, assume that $f_{\partial D} \geq 0$, i.e., $f(x,y) \geq 0$ for each $(x,y) \in \partial D$. Fix some $(r,\theta) \in D$. From $2r\cos(\theta-\phi) \leq 2r$, we obtain

$$1 - 2r\cos(\theta-\phi) + r^2 \geq 1 - 2r + r^2 = (1-r)^2 > 0\,.$$

For each $\phi \in [0,2\pi]$ we have $f(\cos\phi,\sin\phi) \geq 0$, and so $\frac{f(\cos\phi,\sin\phi)(1-r^2)}{1-2r\sin(\theta-\phi)+r^2} \geq 0$ for each $\theta \in [0,2\pi]$. Now a glance at the Poisson formula guarantees that $\widehat{f}(r,\theta) \geq 0$ for each $(r,\theta) \in [0,1) \times [0,2\pi]$.

(b) The linearity and positivity of Q are easy consequences of the Poisson formula and part (a). The identity $Q^2 f = Qf$ for each $f \in C(\overline{D})$ follows immediately from the uniqueness property of the Dirichlet theorem. The fact that Q is the identity on $\mathcal{H}$ (and hence its range is $\mathcal{H}$) also follows easily from Dirichlet's theorem.

(c) Since $\mathcal{H}$ is the range of the positive projection Q, Theorem 5.59 guarantees that $\mathcal{H}$ is a lattice-subspace of $C(\overline{D})$.

To see that the positive projection Q is uniquely determined, notice first that $\mathbf{1} \in \mathcal{H}$ and that $\mathcal{H}$ separates the points of $\overline{D}$. For instance, let $(x_1, y_1), (x_2, y_2) \in D$ satisfy $(x_1, y_1) \neq (x_2, y_2)$. For instance, if $x_1 \neq x_2$, then the function $g(x, y) = x$ satisfies $g \in \mathcal{H}$ and $g(x_1, y_1) \neq g(x_2, y_2)$. From the Stone–Weierstrass theorem, we infer that the vector sublattice $\mathcal{R}(\mathcal{H})$ generated by $\mathcal{H}$ is norm dense in $C(\overline{D})$. Recall also that the Miyajima projection $P\colon \mathcal{R}(\mathcal{H}) \to \mathcal{R}(\mathcal{H})$ is a uniquely determined positive projection on $\mathcal{R}(\mathcal{H})$ with range $\mathcal{H}$. So, since every positive projection on $C(\overline{D})$ is continuous, we infer that Q is the unique extension of P to a positive projection on $C(\overline{D})$ with range $\mathcal{H}$.

(d) The proof will be based upon the following well-known **maximum modulus principle** for harmonic functions.

- *If* $u \in \mathcal{H}$, *then* $\max\limits_{(x,y)\in\overline{D}} |u(x, y)| = \max\limits_{(x,y)\in\partial D} |u(x, y)|$.

Hence if $f \in C(\overline{D})$ satisfies $\|f\|_\infty \leq 1$, then $\|Qf\|_\infty \leq 1$. Since $Q\mathbf{1} = \mathbf{1}$, it easily follows that the projection Q has norm one.

From Theorem 5.59(6), we know that under the norm

$$|||h||| = \big\|Q|h|\big\|_\infty$$

the lattice-subspace $\mathcal{H}$ is an AM-space with unit, where the unit is the function $Q\mathbf{1} = \mathbf{1}$. Now notice that for each $h \in \mathcal{H}$ we have

$$\|h\|_\infty = \big\||h|\big\|_\infty = \big\||Qh|\big\|_\infty \leq \big\|Q|h|\big\|_\infty = |||h||| \leq \|Q\| \cdot \|h\|_\infty = \|h\|_\infty .$$

This shows that $|||h||| = \|h\|_\infty$ for each $h \in \mathcal{H}$. In particular, the lattice-subspace $\mathcal{H}$ with the induced norm from $C(\overline{D})$ is an AM-space. Moreover, it should be clear that the closed unit ball of $\mathcal{H}$ coincides with the closed interval $[-\mathbf{1}, \mathbf{1}]$. That is, $(\mathcal{H}, \|\cdot\|_\infty)$ is an AM-space with unit $\mathbf{1}$.

From the above conclusions it is easy to see that the mapping $f \mapsto \widehat{f}$, from $C(\partial D)$ to $\mathcal{H}$, is a surjective lattice isometry. ∎

Chapter 6

Spectral Properties

6.1. The Spectrum of an Operator

Problem 6.1.1. *Establish the following identity—known as the* **second resolvent identity***. If $S, T \in \mathcal{L}(X)$ and $\lambda \in \rho(T) \cap \rho(S)$, then*

$$R(\lambda, S) - R(\lambda, T) = R(\lambda, S)(S - T)R(\lambda, T)\,.$$

Solution: For each $\lambda \in \rho(T) \cap \rho(S)$ we have

$$\begin{aligned} R(\lambda, S)&(S-T)R(\lambda,T) \\ &= R(\lambda,S)\big[(S-\lambda)-(T-\lambda)\big]R(\lambda,T) \\ &= R(\lambda,S)(S-\lambda)R(\lambda,T) - R(\lambda,S)(T-\lambda)R(\lambda,T) \\ &= -(\lambda-S)^{-1}(\lambda-S)R(\lambda,T) + R(\lambda,S)(\lambda-T)(\lambda-T)^{-1} \\ &= R(\lambda,S) - R(\lambda,T)\,, \end{aligned}$$

as desired. ∎

Problem 6.1.2. *Generalize Corollary 6.8 by proving that the derivatives of the resolvent function of an operator $T \in \mathcal{L}(X)$ are given by the formulas*

$$\frac{d^n R(\lambda, T)}{d\lambda^n} = (-1)^n n! \big[R(\lambda, T)\big]^{n+1}$$

for $n = 1, 2, \dots$.

Solution: By Theorem 6.6, we know that if $\lambda_0 \in \rho(T)$ satisfies $|\lambda - \lambda_0| < \frac{1}{\|R(\lambda_0)\|}$, then

$$R(\lambda, T) = \sum_{n=0}^{\infty}(-1)^n(\lambda-\lambda_0)^n R(\lambda_0)^{n+1} = \sum_{n=0}^{\infty}\big[(-1)^n R(\lambda_0)^{n+1}\big](\lambda-\lambda_0)^n\,.$$

Now a glance at Theorem 1.73(3) shows that

$$\frac{d^n R(\lambda_0,T)}{d\lambda^n} = n!\big[(-1)^n R(\lambda_0)^{n+1}\big] = (-1)^n n!\big[R(\lambda_0)\big]^{n+1}$$

holds for each $n = 1, 2, \ldots$. ∎

Problem 6.1.3. *If X is a Banach space and $T\colon X \to X$ is a bounded operator, then show that $\lim\limits_{\lambda\to\infty} \lambda R(\lambda,T) = I$.*

Solution: Note that for each $|\lambda| > \|T\|$, we have

$$\lambda R(\lambda,T) = \lambda\big(\tfrac{I}{\lambda}+\tfrac{T}{\lambda^2}+\tfrac{T^2}{\lambda^3}+\cdots\big) = I+\tfrac{T}{\lambda}\big(I+\tfrac{T}{\lambda}+\tfrac{T^2}{\lambda^2}+\cdots\big) = I+\tfrac{T}{\lambda}\big(I-\tfrac{T}{\lambda}\big)^{-1}. \quad (\star)$$

From $\lim_{\lambda\to\infty}(I-\frac{T}{\lambda}) = I$ and Problem 2.1.11, we get $\lim_{\lambda\to\infty}\big(I-\frac{T}{\lambda}\big)^{-1} = I$. Hence, $(\star)$ implies $\lim_{\lambda\to\infty}\lambda R(\lambda,T) = I$.

An alternative solution can be obtained as follows. As above, for each $\lambda \in \mathbb{C}$ with $|\lambda| > \|T\|$ we have $\lambda R(\lambda,T) = I + \sum_{n=1}^{\infty}\frac{T^n}{\lambda^n}$. Now, a glance at Problem 1.5.6 shows that $\lim_{\lambda\to\infty}\sum_{n=1}^{\infty}\frac{T^n}{\lambda^n} = 0$. This implies $\lim_{\lambda\to\infty}\lambda R(\lambda,T) = I$. ∎

Problem 6.1.4. *For any operator $T \in \mathcal{L}(X)$ show that $[R(\lambda,T)]^* = R(\lambda,T^*)$ holds for each $\lambda \in \rho(T)$.*

Solution: Let $\lambda \in \rho(T)$. By Theorem 6.14 we have

$$\big[R(\lambda,T)\big]^* = \big[(\lambda-T)^{-1}\big]^* = \big[(\lambda-T)^*\big]^{-1} = (\lambda-T^*)^{-1} = R(\lambda,T^*),$$

and we are done. ∎

Problem 6.1.5. *Let $T \in \mathcal{L}(X)$. If $\lambda_0 \in \rho(T)$, then show that $R(\lambda_0,T)$ cannot be quasinilpotent.*

Solution: Start by observing that an operator $S \in \mathcal{L}(X)$ is invertible if and only if $0 \notin \sigma(S)$. In particular, an invertible operator cannot be quasinilpotent.

Now assume that $T \in \mathcal{L}(X)$ and let $\lambda_0 \in \rho(T)$. Therefore, $\lambda_0 - T$ is invertible. This implies that $R(\lambda_0,T) = (\lambda_0-T)^{-1}$ is also invertible, and thus, by the above, $R(\lambda_0,T)$ cannot be a quasinilpotent operator. ∎

Problem 6.1.6. *Show directly that if $T\colon X \to X$ is a bounded operator on a (real or complex) Banach space, then the limit $\lim_{n\to\infty}\|T^n\|^{\frac{1}{n}}$ always exists in $\mathbb{R}$ and*

$$\lim_{n\to\infty}\|T^n\|^{\frac{1}{n}} = \inf_n \|T^n\|^{\frac{1}{n}}.$$

Solution: Let $\alpha = \inf_n \|T^n\|^{\frac{1}{n}} \geq 0$. From the inequality $\alpha \leq \|T^n\|^{\frac{1}{n}}$ we see that $\liminf_{n\to\infty}\|T^n\|^{\frac{1}{n}} \geq \alpha$.

Now let $\epsilon > 0$ and then fix some k such that $\|T^k\|^{\frac{1}{k}} < \alpha+\epsilon$. For any positive integer n write $n = km + r$, where $m \geq 0$ and $0 \leq r < k$. Note that

$$\begin{aligned}\|T^n\| &= \|T^r(T^k)^m\| \leq \|T\|^r\cdot\|T^k\|^m = \|T\|^r\cdot(\|T^k\|^{\frac{1}{k}})^{km}\\ &= \|T\|^r\cdot(\alpha+\epsilon)^{n-r} \leq M(\alpha+\epsilon)^n,\end{aligned}$$

where $M = \max_{0\le r<k}\big(\frac{\|T\|}{\alpha+\epsilon}\big)^r$. So, $\|T^n\|^{\frac{1}{n}} \le M^{\frac{1}{n}}(\alpha+\epsilon)$, from which it follows $\limsup_{n\to\infty}\|T^n\|^{\frac{1}{n}} \le \alpha+\epsilon$ for all $\epsilon>0$, or $\limsup_{n\to\infty}\|T^n\|^{\frac{1}{n}} \le \alpha$. This implies $\alpha = \limsup_{n\to\infty}\|T^n\|^{\frac{1}{n}} = \liminf_{n\to\infty}\|T^n\|^{\frac{1}{n}}$. So, $\alpha = \lim_{n\to\infty}\|T^n\|^{\frac{1}{n}}$. ∎

Problem 6.1.7. *For any operator $T \in \mathcal{L}(X)$ show that*

$$r(T) = \inf\big\{\alpha > 0\colon\ \tfrac{\|T^n\|}{\alpha^n} \to 0\big\}.$$

Solution: Consider the set $A = \big\{\alpha > 0\colon\ \frac{\|T^n\|}{\alpha^n} \to 0\big\}$ and let $\alpha_0 = \inf A$. We must show that $r(T) = \alpha_0$.

Take any λ and β such that $r(T) < \beta < \lambda$. Since $r(T) = \lim_{n\to\infty}\|T^n\|^{\frac{1}{n}}$, it follows that there exists some $k \in \mathbb{N}$ such that $\|T^n\|^{\frac{1}{n}} < \beta$ for all $n > k$. This implies $\frac{\|T^n\|}{\beta^n} < 1$ for all $n > k$, and so $\frac{\|T^n\|}{\lambda^n} = \frac{\|T^n\|}{\beta^n}\cdot\big(\frac{\beta}{\lambda}\big)^n \to 0$. Thus, $\lambda \in A$ for each $\lambda > r(T)$, and consequently $\alpha_0 = \inf A \le r(T)$.

To see that $\alpha_0 = r(T)$, assume by way of contradiction that $\alpha_0 < r(T)$ and fix some β satisfying $\alpha_0 < \beta < r(T)$. Clearly, $\beta \in A$. However, since $\beta < r(T)$, it follows that $\beta < \|T^n\|^{\frac{1}{n}}$ for all n, and so $1 < \frac{\|T^n\|}{\beta^n} \to 0$, which is impossible. This contradiction establishes that $r(T) = \alpha_0$. ∎

Problem 6.1.8. *If $S, T\colon X \to X$ are two commuting bounded operators on a Banach space, then show that $r(ST) \le r(S)r(T)$.*

Solution: Assume that two operators $S, T \in \mathcal{L}(X)$ commute, i.e., $ST = TS$. Then we have $(ST)^n = S^nT^n$ for each n, and so

$$\|(ST)^n\|^{\frac{1}{n}} = \|S^nT^n\|^{\frac{1}{n}} \le \|S^n\|^{\frac{1}{n}}\cdot\|T^n\|^{\frac{1}{n}}$$

holds for each n. Therefore,

$$\begin{aligned} r(ST) &= \lim_{n\to\infty}\|(ST)^n\|^{\frac{1}{n}} \le \lim_{n\to\infty}\big[\,\|S^n\|^{\frac{1}{n}}\cdot\|T^n\|^{\frac{1}{n}}\,\big] \\ &= \big[\lim_{n\to\infty}\|S^n\|^{\frac{1}{n}}\,\big]\cdot\big[\lim_{n\to\infty}\|T^n\|^{\frac{1}{n}}\,\big] = r(S)r(T), \end{aligned}$$

as claimed. ∎

Problem 6.1.9. *Let $T\colon X \to X$ be a bounded operator and assume that for some non-zero complex number α the series $\sum_{n=0}^{\infty}\alpha^{-(n+1)}T^nx$ converges in norm for each $x \in X$. Show that $\alpha \in \rho(T)$.*

Solution: Assume that the operator T and the scalar α satisfy the stated properties. Clearly, $\lim\limits_{n\to\infty}\alpha^{-(n+1)}T^{n+1}x = 0$ for each $x \in X$.

The series $Sx = \sum_{n=0}^{\infty}\alpha^{-(n+1)}T^nx$ defines an operator from X to X. If for each n we consider the bounded operator $S_n = \sum_{i=0}^{n}\alpha^{-(i+1)}T^i$ on X, then $S_nx \to Sx$ for all $x \in X$. By the Uniform Boundedness Principle, $S \in \mathcal{L}(X)$. Now for each

$x \in X$ we have

$$\begin{aligned}[(\alpha - T)S]x &= \lim_{n\to\infty}\Big[(\alpha - T)\sum_{i=0}^{n}\alpha^{-(i+1)}T^i x\Big]\\ &= \lim_{n\to\infty}\Big[\sum_{i=0}^{n}\alpha^{-i}T^i x - \sum_{i=0}^{n}\alpha^{-(i+1)}T^{i+1}x\Big]\\ &= \lim_{n\to\infty}\big[x - \alpha^{-(n+1)}T^{n+1}x\big] = x\,.\end{aligned}$$

That is, $(\alpha - T)S = I$. Similarly, $S(\alpha - T) = I$. Therefore, $\alpha - T$ is an invertible operator or $\alpha \in \rho(T)$. ■

Problem 6.1.10. *Let $T \in \mathcal{L}(X)$. Show that $r(T) = \lim_{n\to\infty}\|T^n\|^{\frac{1}{n}} = 0$ if and only if $\lim_{n\to\infty}\|T^n x\|^{\frac{1}{n}} = 0$ for each $x \in X$.*

Solution: If $\lim_{n\to\infty}\|T^n\|^{\frac{1}{n}} = 0$ and $x \in X$, then from

$$\|T^n x\|^{\frac{1}{n}} \le \big(\|T^n\|\cdot\|x\|\big)^{\frac{1}{n}} = \|T^n\|^{\frac{1}{n}}\cdot\|x\|^{\frac{1}{n}} \longrightarrow 0\,,$$

it follows that $\|T^n x\|^{\frac{1}{n}} \to 0$.

For the converse, assume that $\|T^n x\|^{\frac{1}{n}} \to 0$ holds for each $x \in X$. Fix a complex number $\alpha \neq 0$ and let $x \in X$. Since $\|T^n x\|^{\frac{1}{n}} \to 0$, there exists some $n_0 \in \mathbb{N}$ such that $\|T^n x\|^{\frac{1}{n}} < \frac{|\alpha|}{2}$ for all $n \ge n_0$. This implies $\big\|\alpha^{-(n+1)}T^n x\big\| \le \frac{1}{|\alpha|}\cdot\frac{1}{2^n}$ for all $n \ge n_0$. In particular, the series $\sum_{n=0}^{\infty}\alpha^{-(n+1)}T^n x$ converges in X for each $x \in X$.

By Problem 6.1.9, $\alpha \in \rho(T)$ for each $\alpha \neq 0$. This and the fact that $\sigma(T) \neq \emptyset$ imply that $\rho(T) = \mathbb{C}\setminus\{0\}$ or $\sigma(T) = \{0\}$. So, $\lim_{n\to\infty}\|T^n\|^{\frac{1}{n}} = r(T) = 0$. ■

Problem 6.1.11. *If X is a Banach space and $T \in \mathcal{L}(X)$, then show that*

$$\sigma(T^k) = \{\lambda^k\colon\ \lambda \in \sigma(T)\}$$

for each k (and conclude from this that $r(T^k) = [r(T)]^k$).

Also establish that an operator $T \in \mathcal{L}(X)$ satisfies $\|T^n\| \to 0$ if and only if $r(T) < 1$.

Solution: In this proof, we shall employ the following simple algebraic property (P):

(P) *If $A, B, C \in \mathcal{L}(X)$ satisfy $CA = BC = I$, then $A = B = C^{-1}$.*

To see this, note that $A = IA = (BC)A = B(CA) = BI = B$.

Now let $T \in \mathcal{L}(X)$ and fix some natural number k. Assume first that $\lambda \in \sigma(T)$. In this case, we claim that $\lambda^k \in \sigma(T^k)$. Indeed, if $\lambda^k \notin \sigma(T^k)$, then there exists some $S \in \mathcal{L}(X)$ such that $(\lambda^k - T^k)S = S(\lambda^k - T^k) = I$. Since

$$\lambda^k - T^k = (\lambda - T)\big(\lambda^{k-1} + \lambda^{k-2}T + \cdots + \lambda T^{k-2} + T^{k-1}\big)\,,$$

it follows that the pair of operators $A = \big(\lambda^{k-1} + \lambda^{k-2}T + \cdots + \lambda T^{k-2} + T^{k-1}\big)S$ and $B = S\big(\lambda^{k-1} + \lambda^{k-2}T + \cdots + \lambda T^{k-2} + T^{k-1}\big)$ satisfies $(\lambda - T)A = B(\lambda - T) = I$. Now property (P) implies $\lambda \in \rho(T)$, which is a contradiction. Hence, $\lambda^k \in \sigma(T^k)$, and so $\{\lambda^k\colon\ \lambda \in \sigma(T)\} \subseteq \sigma(T^k)$.

For the converse inclusion, fix some $\mu \in \sigma(T^k)$ and consider the polynomial $p(z) = \mu - z^k$ with factorization

$$\mu - z^k = (-1)^k (r_1 - z)(r_2 - z) \cdots (r_k - z),$$

where $r_1, r_2, \ldots, r_k$ are the roots of the polynomial p counted with their multiplicities. In particular, $\mu = r_i^k$ holds for each i. The above factorization yields

$$\mu - T^k = (-1)^k (r_1 - T)(r_2 - T) \cdots (r_k - T).$$

Now notice that if each operator $r_i - T$ is invertible, then it follows from the above factorization that $\mu - T^k$ is also an invertible operator, which is impossible. Hence, for some $1 \le i \le k$ the operator $r_i - T$ is not invertible, i.e., $r_i \in \sigma(T)$. This implies $\mu = r_i^k$, and so $\mu \in \{\lambda^k \colon \lambda \in \sigma(T)\}$. Thus, $\sigma(T^k) \subseteq \{\lambda^k \colon \lambda \in \sigma(T)\}$ is also true, and so

$$\sigma(T^k) = \{\lambda^k \colon \lambda \in \sigma(T)\}.$$

For the last part assume first that $r(T) < 1$. Fix some $\delta > 0$ such that $r(T) < \delta < 1$ and note that $\lim_{n\to\infty} \|T^n\|^{\frac{1}{n}} = r(T)$ implies that there exists some n_0 such that $\|T^n\|^{\frac{1}{n}} < \delta$ or $\|T^n\| < \delta^n$ for each $n \ge n_0$. Since $\delta^n \to 0$, we see that $\|T^n\| \to 0$. For the converse assume $\|T^n\| \to 0$. Then $\|T^k\| < 1$ for some k, and so $r(T) = \inf_n \|T^n\|^{\frac{1}{n}} \le \|T^k\|^{\frac{1}{k}} < 1$. ∎

Problem 6.1.12. *If $T \colon X \to X$ is an invertible bounded operator on a Banach space, then show that $\sigma(T^{-1}) = \{\lambda^{-1} \colon \lambda \in \sigma(T)\}$.*

Solution: Start by observing that since both T and T^{-1} are invertible, the number zero is neither in $\sigma(T)$ nor in $\sigma(T^{-1})$.

Now let $\mu \in \sigma(T^{-1})$. Put $\lambda = \frac{1}{\mu}$ and assume by way of contradiction that $\lambda \in \rho(T)$. This means that there exists a bounded operator $S \in \mathcal{L}(X)$ satisfying $S(\lambda - T) = (\lambda - T)S = I$. Then

$$I = S(\lambda - T) = STT^{-1}(\lambda - T) = ST(\lambda T^{-1} - I) = \big(-\tfrac{1}{\mu}ST\big)\big(\mu - T^{-1}\big),$$

and similarly $\big(\mu - T^{-1}\big)\big(-\frac{1}{\mu}TS\big) = (\lambda - T)S = I$. Now invoking property (P) stated at the beginning of the solution to Problem 6.1.11, we see that $\mu \notin \sigma\big(T^{-1}\big)$, which is a contradiction. Hence, $\lambda = \frac{1}{\mu} \in \sigma(T)$ and $\mu = \lambda^{-1}$. Therefore, we have $\sigma\big(T^{-1}\big) \subseteq \{\lambda^{-1} \colon \lambda \in \sigma(T)\}$.

Now, by the symmetry of the situation, if $\lambda \in \sigma(T) = \sigma\big((T^{-1})^{-1}\big)$, then there exists some $\mu \in \sigma\big(T^{-1}\big)$ such that $\mu^{-1} = \lambda$. Thus, $\lambda^{-1} = \mu \in \sigma\big(T^{-1}\big)$. That is, $\{\lambda^{-1} \colon \lambda \in \sigma(T)\} \subseteq \sigma\big(T^{-1}\big)$. This proves that $\sigma\big(T^{-1}\big) = \{\lambda^{-1} \colon \lambda \in \sigma(T)\}$. ∎

Problem 6.1.13. *If an operator $T \in \mathcal{L}(X)$ satisfies $\|T^k\| < 1$ for some k, then show that $\lim_{n\to\infty} \|T^n\| = 0$.*

Solution: If an operator $T \in \mathcal{L}(X)$ satisfies $\|T^k\| < 1$ for some k, then

$$r(T) = \inf_n \|T^n\|^{\frac{1}{n}} \le \|T^k\|^{\frac{1}{k}} < 1.$$

So, by Problem 6.1.11, $\|T^n\| \to 0$. ∎

Problem 6.1.14. *Let Ω be a compact Hausdorff space and let μ be a Borel measure on Ω. Assume that either $E = C(\Omega)$ or E is a Banach function space that is order dense in $L_0(\mu)$. If $\phi\colon \Omega \to \mathbb{C}$ is a continuous function, then show that the spectrum of the multiplication operator $M_\phi\colon E \to E$ coincides with the range of ϕ, i.e., $\sigma(M_\phi) = R_\phi = \{\phi(\omega)\colon\ \omega \in \Omega\}$.*

Solution: We start by observing the following properties:

- *The inverse operator M_ϕ^{-1} exists if and only if $\phi(\omega) \neq 0$ for each $\omega \in \Omega$, in which case we have $M_\phi^{-1} = M_{\phi^{-1}}$.*
- *The identity $\lambda - M_\phi = M_{\lambda-\phi}$ holds for each $\lambda \in \mathbb{C}$.*

By the above properties, $\lambda - M_\phi$ is invertible if and only if $\phi(\omega) \neq \lambda$ for each $\omega \in \Omega$. Since ϕ is continuous, the latter happens if and only if $\lambda \notin R_\phi$. This establishes the desired equality $\sigma(M_\phi) = R_\phi$. ∎

Problem 6.1.15. *For a non-empty compact subset K of $\mathbb{C}$ establish the following properties.*

(a) *There exists a bounded operator $T\colon \ell_2 \to \ell_2$ such that $\sigma(T) = K$.*

(b) *If $S\colon C_{\mathbb{C}}(K) \to C_{\mathbb{C}}(K)$ is the multiplication operator defined by $(Sx)(z) = zx(z)$, then $\sigma(S) = K$.*

Solution: (a) Let K be a non-empty compact subset of complex numbers. Assume first that K is a finite set, say $K = \{\lambda_1, \lambda_2, \ldots, \lambda_n\}$. Then the operator defined on $\mathbb{C}^n$ via the diagonal matrix $A = \operatorname{diag}(\lambda_1, \lambda_2, \ldots, \lambda_n)$ defines a bounded operator whose spectrum is precisely K.

Now assume that K is an infinite set. Let $\{\lambda_1, \lambda_2, \ldots\}$ be a countable dense subset of K, and define the operator $T\colon \ell_2 \to \ell_2$ via the formula

$$T(x_1, x_2, \ldots) = (\lambda_1 x_1, \lambda_2 x_2, \ldots)\,.$$

If e_n denotes the sequence whose n^{th} coordinate is equal to one and every other is zero, then $Te_n = \lambda_n e_n$ is obviously true for each n. This shows that $\lambda_n - T$ is not one-to-one, and so $\lambda_n - T$ is not invertible. Therefore, $\{\lambda_1, \lambda_2, \ldots\} \subseteq \sigma(T)$. Since $\sigma(T)$ is a compact set, it follows that $K \subseteq \sigma(T)$.

To see that $K \subseteq \sigma(T)$, let $\lambda \notin K$ and then choose some $\epsilon > 0$ such that $|\lambda - z| \geq \epsilon$ holds for all $z \in K$. This implies

$$\|(\lambda - T)x\| = \big\|\big((\lambda - \lambda_1)x_1, (\lambda - \lambda_2)x_2, \ldots\big)\big\| = \Big[\sum_{n=1}^{\infty} |\lambda - \lambda_n|^2 |x_n|^2\Big]^{\frac{1}{2}} \geq \epsilon\|x\|$$

for all $x \in \ell_2$. That is, $\lambda - T$ is a bounded below operator. On the other hand, if $y = (y_1, y_2, \ldots) \in \ell_2$, then the vector $z = \big(\frac{y_1}{\lambda - \lambda_1}, \frac{y_2}{\lambda - \lambda_2}, \ldots\big)$ also belongs to ℓ_2 and satisfies $(\lambda - T)z = y$. This shows that $\lambda - T$ is also surjective, and hence an invertible operator. In other words, we have shown that $\lambda \notin K$ implies $\lambda \notin \sigma(T)$. The above properties establish that $\sigma(T) = K$.

(b) If $\phi(z) = z$, then according to Problem 6.1.14 the spectrum of the multiplication operator $S = M_\phi$ satisfies $\sigma(S) = \{\phi(z)\colon\ z \in K\} = K$. (See also Problem 7.1.11.) ∎

Problem 6.1.16. *If an invertible operator $T\colon X \to X$ on a Banach space satisfies $\|T^n\| \le M < \infty$ for all integers n (such an operator is called a* ***Cesaro operator*** *or a* ***double power bounded operator****), then show that $\sigma(T) \subseteq \{\lambda \in \mathbb{C}\colon\ |\lambda| = 1\}$.*

Solution: Here is a solution different than the one suggested by the hint. The given conditions imply

$$r(T) = \lim_{n\to\infty} \|T^n\|^{\frac{1}{n}} \le \lim_{n\to\infty} M^{\frac{1}{n}} = 1\,,$$

and similarly $r\big(T^{-1}\big) \le 1$.

Now let $\lambda \in \sigma(T)$. Then $|\lambda| \le r(T) \le 1$. On the other hand, by Problem 6.1.12, we know that $\frac{1}{\lambda} \in \sigma\big(T^{-1}\big)$. This implies $\frac{1}{|\lambda|} \le r\big(T^{-1}\big) \le 1$ or $|\lambda| \ge 1$. Consequently, $|\lambda| = 1$, and thus $\sigma(T) \subseteq \big\{\lambda \in \mathbb{C}\colon\ |\lambda| = 1\big\}$. ∎

Problem 6.1.17. *Let $T_i\colon X_i \to X_i$ $(i = 1, \ldots, n)$ be bounded operators on Banach spaces. If $X = X_1 \oplus \cdots \oplus X_n$ is the direct sum Banach space, then show that:*

(a) *The* ***direct sum operator*** *$T = T_1 \oplus \cdots \oplus T_n\colon X \to X$, defined by*

$$T(x_1 \oplus \cdots \oplus x_n) = T_1x_1 \oplus \cdots \oplus T_nx_n\,,$$

is invertible if and only if each T_i is invertible—in which case the inverse operator T^{-1} satisfies $T^{-1} = T_1^{-1} \oplus T_2^{-1} \oplus \cdots \oplus T_n^{-1}$.

(b) *The spectrum of the direct sum operator T is the union of the spectra of the operators T_i, i.e., $\sigma\big(T_1 \oplus \cdots \oplus T_n\big) = \bigcup_{i=1}^n \sigma(T_i)$.*

Solution: (a) If each T_i is invertible, then the direct computation shows that the operator $S = T_1^{-1} \oplus \cdots \oplus T_n^{-1}$ satisfies $ST = TS = I$ and so $S = T^{-1}$.

Conversely, assume that T^{-1} exists. Let $y_i \in X_i$ and consider the vector $y = 0 \oplus \cdots \oplus 0 \oplus y_i \oplus 0 \oplus \cdots \oplus 0$, where the y_i occupies the ith position. To see that T_i is one-to-one, assume $T_iy_i = 0$. Then $Ty = 0$, and so $y = 0$ or $y_i = 0$, so that T_i is one-to-one. To show that T_i is surjective, let $T^{-1}y = x = x_1 \oplus \cdots \oplus x_n$. Then from $y = TT^{-1}y = Tx = T_1x_1 \oplus \cdots \oplus Tx_n$, we get $T_ix_i = y_i$. This shows that T_i is indeed surjective—and so each $T_i\colon X_i \to X_i$ is invertible.

(b) A scalar $\lambda \in \sigma(T)$ if and only if $\lambda - T = (\lambda - T_1) \oplus \cdots \oplus (\lambda - T_n)$ is not invertible on X. In view of (a) this is equivalent to saying that $\lambda \in \bigcup_{i=1}^n \sigma(T_i)$. ∎

Problem 6.1.18. *Show that an operator $T\colon X \to X$ on a complex Banach space is nilpotent if and only if it is quasinilpotent and has finite descent.*

Solution: Clearly, if T is nilpotent, then T is also quasinilpotent and has finite descent. For the converse assume that T is quasinilpotent and has finite descent. Let $p = \delta(T) < \infty$. To finish the proof, it suffices to show that $R(T^p) = \{0\}$. To this end, assume by way of contradiction that $R(T^p) \ne \{0\}$.

Suppose first that T has finite ascent. This implies that $\alpha(T) = p$ and that $X = N(T^p) \oplus R(T^p)$. In particular, $R(T^p)$ is a closed T-invariant subspace of X on which T is invertible. The latter conclusion guarantees that the spectrum of the restriction of T to $R(T^p)$ has a non-zero point. Consequently, from the decomposition $T = T|_{N(T^p)} \oplus T|_{R(T^p)}$, it follows that the spectrum of T has a non-zero point (see Problem 6.1.17), which contradicts the fact that T is quasinilpotent.

Now assume that T has infinite ascent. According to Theorem 2.14, we can renorm $R(T^p)$ so that it becomes a Banach space and, moreover, the restriction S of T to $R(T^p)$ is continuous. Since p is the descent of T, it follows that S is also surjective. The fact that T has infinite ascent implies $N(T) \cap R(T^p) \neq \{0\}$; see part (d) of Problem 2.2.8. So, the surjective operator $S\colon R(T^p) \to R(T^p)$ is not invertible. But then, by Theorem 6.20, S (and hence T) has a non-zero eigenvalue. Hence, the spectrum of T has a non-zero point, which is a contradiction. ∎

Problem 6.1.19. *If $S, T \in \mathcal{L}(X)$, then show that*

$$\sigma(ST) \cup \{0\} = \sigma(TS) \cup \{0\}.$$

Conclude from this that:

(a) $\sigma(ST) = \{0\}$ *if and only if* $\sigma(TS) = \{0\}$.

(b) $r(ST) = r(TS)$.

Give an example of two operators $S, T \in \mathcal{L}(X)$ such that $\sigma(ST) \neq \sigma(TS)$.

Solution: The solution will be based upon the following algebraic property.

($\star$) *If $A, B, C \in \mathcal{L}(X)$ satisfy $(I - AB)C = C(I - AB) = I$, then*

$$(I + BCA)(I - BA) = (I - BA)(I + BCA) = I.$$

We verify the first identity. From $C(I - AB) = I$, we get $CAB = C - I$, and so

$$\begin{aligned}(I + BCA)(I - BA) &= I + BCA - BA - BCABA\\ &= I + BCA - BA - B(C - I)A\\ &= I + BCA - BA - BCA + BA = I.\end{aligned}$$

(a) It follows from ($\star$) that if $1 \notin \sigma(AB)$, then $1 \notin \sigma(BA)$. Now let $\lambda \in \sigma(ST)$ be non-zero. Assume by way of contradiction that $\lambda \notin \sigma(TS)$. This means that there exists a bounded operator $D\colon X \to X$ such that $(\lambda - TS)D = D(\lambda - TS) = I$. Rewriting this identity, we get

$$\left(I - \tfrac{1}{\lambda}TS\right)(\lambda D) = (\lambda D)\left(I - \tfrac{1}{\lambda}TS\right) = I.$$

Letting $A = \frac{1}{\lambda}T$, $B = S$, and $C = \lambda D$ in ($\star$), we see that

$$(I + SCT)\left(I - \tfrac{1}{\lambda}ST\right) = \left(I - \tfrac{1}{\lambda}ST\right)(I + SCT) = I.$$

Rewriting the latter identity, we get

$$\left[\tfrac{1}{\lambda}(I + SCT)\right](\lambda - ST) = (\lambda - ST)\left[\tfrac{1}{\lambda}(I + SCT)\right] = I.$$

Therefore, $\lambda \notin \sigma(ST)$, which is a contradiction. Thus, $\sigma(ST) \cup \{0\} \subseteq \sigma(TS) \cup \{0\}$. By the symmetry of the situation, we have $\sigma(TS) \cup \{0\} \subseteq \sigma(ST) \cup \{0\}$. Hence, $\sigma(ST) \cup \{0\} = \sigma(TS) \cup \{0\}$.

(b) Note that

$$r(ST) = \max_{\lambda \in \sigma(ST)} |\lambda| = \max_{\lambda \in \sigma(ST) \cup \{0\}} |\lambda| = \max_{\lambda \in \sigma(TS) \cup \{0\}} |\lambda| = \max_{\lambda \in \sigma(TS)} |\lambda| = r(TS).$$

For a counterexample consider the operators $S, T\colon \ell_1 \to \ell_1$ defined by

$$S(x_1, x_2, x_3, \ldots) = (x_2, x_3, \ldots) \quad \text{and} \quad T(x_1, x_2, x_3, \ldots) = (0, x_1, x_2, x_3, \ldots).$$

Notice that $ST = I$, and so $\sigma(ST) = \{1\}$. On the other hand, we have

$$TS(x, 0, 0, 0, \ldots) = (0, 0, 0, \ldots),$$

and so TS is non-invertible, i.e., $0 \in \sigma(TS)$. So, $\sigma(ST) = \{1\} \neq \{0, 1\} = \sigma(TS)$. (See also Problems 6.2.9 and 8.2.7.) ∎

Problem 6.1.20. *Let $T \in \mathcal{L}(X)$ be an operator and let $z_0 \in \mathbb{C}$ be a point such that every $\lambda \in \sigma(T) \setminus \{z_0\}$ is an isolated point of $\sigma(T)$. Show that $\sigma(T)$ is at most countable and that if $\sigma(T)$ is countable, say $\sigma(T) = \{\lambda_1, \lambda_2, \ldots\}$, then $\lambda_n \to z_0$.*

Solution: Assume that a bounded operator $T \in \mathcal{L}(X)$ satisfies the stated property. Clearly, $\sigma(T)$ must be at most countable.

Next, suppose that $\sigma(T)$ is countable, say $\sigma(T) = \{\lambda_1, \lambda_2, \ldots\}$. Now for each $\epsilon > 0$ let $C_\epsilon = \big\{\lambda \in \sigma(T)\colon\ |\lambda - z_0| \geq \epsilon\big\}$. If some C_ϵ is an infinite set, then it must have an accumulation point, say μ. It follows that μ is an accumulation point of $\sigma(T)$ satisfying $\mu \in \sigma(T) \setminus \{z_0\}$, contrary to our hypothesis. Therefore, each C_ϵ is a finite set and from this we easily infer that $\lambda_n \to z_0$. ∎

6.2. Special Points of the Spectrum

Problem 6.2.1. *Show that an operator $T \in \mathcal{L}(X)$ is not one-to-one if and only if zero is an eigenvalue of T.*

Solution: If 0 is an eigenvalue of T, then there exists some non-zero vector x such that $Tx = 0x = 0$. This shows that T is not one-to-one.

Conversely, if T is not one-to-one, then there exists some non-zero vector x such that $Tx = 0 = 0x$. Therefore, 0 is an eigenvalue of T. ∎

Problem 6.2.2. *Let $T\colon V \to V$ be a linear operator on a vector space. Show that a collection of eigenvectors of T any two of which correspond to distinct eigenvalues is a linearly independent set.*

Solution: Let $T\colon V \to V$ be a linear operator on a vector space. The proof is by induction. For $n = 1$, the conclusion should be obvious. For the induction step, assume that the result is true for any collection of n eigenvectors of T corresponding to distinct eigenvalues. Now let $v_1, \ldots, v_n, v_{n+1}$ be $n+1$ eigenvectors of T corresponding to the distinct eigenvalues $\lambda_1, \ldots, \lambda_n, \lambda_{n+1}$, respectively. To establish that the vectors $v_1, \ldots, v_n, v_{n+1}$ are linearly independent, we must show that if

$$\alpha_1 v_1 + \cdots + \alpha_n v_n + \alpha_{n+1} v_{n+1} = 0 \qquad (\star)$$

holds, then necessarily $\alpha_1 = \alpha_2 = \cdots = \alpha_n = \alpha_{n+1} = 0$.

So, assume that $(\star)$ is true. This implies

$$\begin{aligned}\alpha_1\lambda_1 v_1 + \cdots + \alpha_n\lambda_n v_n + \alpha_{n+1}\lambda_{n+1}v_{n+1} \\ &= \alpha_1 Tv_1 + \cdots + \alpha_n Tv_n + \alpha_{n+1}Tv_{n+1} \\ &= T(\alpha_1 v_1 + \cdots + \alpha_n x_n + \alpha_{n+1}v_{n+1}) \\ &= T(0) = 0\,.\end{aligned}$$

Multiplying $(\star)$ by λ_{n+1} and subtracting it from the preceding equation, we get

$$\alpha_1(\lambda_1 - \lambda_{n+1})v_1 + \alpha_2(\lambda_2 - \lambda_{n+1})v_2 + \cdots + \alpha_n(\lambda_n - \lambda_{n+1})v_n = 0\,.$$

Hence, our induction hypothesis yields $\alpha_i(\lambda_i - \lambda_{n+1}) = 0$ for $i = 1, \ldots, n$. Since $\lambda_i - \lambda_{n+1} \neq 0$ for each i, we infer that $\alpha_1 = \alpha_2 = \cdots = \alpha_n = 0$. From $(\star)$ we get $\alpha_{n+1}v_{n+1} = 0$ or $\alpha_{n+1} = 0$. Therefore, $\alpha_i = 0$ for all i, and so the vectors $v_1, \ldots, v_n, v_{n+1}$ are linearly independent. ∎

Problem 6.2.3. *Let $T\colon X \to X$ be a bounded operator on a Banach space. Show that if for some $\lambda \in \mathbb{C}$ there exists a sequence $\{x_n\}$ of unit vectors such that $\|\lambda x_n - Tx_n\| \to 0$, then $\lambda \in \sigma(T)$, i.e., show that $\sigma_a(T) \subseteq \sigma(T)$.*

Solution: If $\lambda \in \sigma_a(T)$, then the operator $\lambda - T$ is not bounded below, and so it cannot be invertible. Hence, $\lambda \in \sigma(T)$. ∎

Problem 6.2.4. *Show that a complex number λ is an approximate eigenvalue of an operator $T \in \mathcal{L}(X)$ if and only if there exists a sequence $\{x_n\}$ in X satisfying $x_n \not\to 0$ and $\lambda x_n - Tx_n \to 0$.*

Solution: Assume that a sequence $\{x_n\}$ in X satisfies $x_n \not\to 0$ and $\lambda x_n - Tx_n \to 0$. Since $x_n \not\to 0$, there exist some $\epsilon > 0$ and a subsequence $\{y_n\}$ of $\{x_n\}$ such that $\|y_n\| \geq \epsilon$ for all n. Now let $z_n = \frac{y_n}{\|y_n\|}$ and note that $\|z_n\| = 1$ for all n. Moreover, from

$$\|\lambda z_n - Tz_n\| = \frac{\|\lambda y_n - Ty_n\|}{\|y_n\|} \leq \frac{\|\lambda y_n - Ty_n\|}{\epsilon} \longrightarrow 0\,,$$

it follows that $\lambda z_n - Tz_n \to 0$. This shows that $\lambda \in \sigma_a(T)$. ∎

Problem 6.2.5. *Let $T\colon X \to X$ be a bounded below operator on a Banach space. If $\gamma > 0$ satisfies $\|Tx\| \geq \gamma\|x\|$ for each $x \in X$, then show that*

$$\sigma_a(T) \cap \{\lambda \in \mathbb{C}\colon\ |\lambda| < \gamma\} = \emptyset\,.$$

In particular, if $T\colon X \to X$ is an isometry, then $\sigma_a(T) \subseteq \{\lambda \in \mathbb{C}\colon\ |\lambda| = 1\}$.

Solution: Assume that $\|Tx\| \geq \gamma\|x\|$ holds for each $x \in X$ and some $\gamma > 0$. If $|\lambda| < \gamma$ and $x \in X$ satisfies $\|x\| = 1$, then

$$\|Tx - \lambda x\| \geq \|Tx\| - |\lambda| \cdot \|x\| \geq \gamma - |\lambda| > 0\,.$$

This shows that λ cannot belong to the approximate point spectrum of the operator T. That is, $\sigma_a(T) \cap \{\lambda \in \mathbb{C}\colon\ |\lambda| < \gamma\} = \emptyset$.

If T is an isometry, then $\|Tx\| = \|x\|$ for all $x \in X$. By the above, the approximate point spectrum of T is disjoint from the open unit disk. Since $\|T^n\| = 1$ for each n, we have $r(T) = 1$, and thus $\sigma_a(T)$ must lie on the unit circle. (Since T need not be invertible, this conclusion does not follow from Problem 6.1.16.) ∎

Problem 6.2.6. *Let $T\colon X \to X$ be a non-invertible bounded operator on a complex Banach space. If some $\gamma > 0$ satisfies $\|Tx\| \geq \gamma\|x\|$ for each $x \in X$, then show that*

$$D(0,\gamma) = \{\lambda \in \mathbb{C}\colon\ |\lambda| < \gamma\} \subseteq \sigma_r(T)\,.$$

Solution: First of all note that since T is not invertible, $0 \in \sigma(T)$. Moreover, since T is bounded below, $0 \in \sigma_r(T)$. Let $\beta = \sup\{\alpha \geq 0\colon\ D(0,\alpha) \subseteq \sigma_r(T)\}$. Clearly, $D(0,\beta) \subseteq \sigma_r(T) \subseteq \sigma(T)$, and so $\overline{D(0,\beta)} \subseteq \sigma(T)$. To finish the proof, it suffices to show that $\beta \geq \gamma$.

To establish this claim, assume by way of contradiction that $\beta < \gamma$. From the definition of β it follows that there exists a sequence of complex numbers $\{\lambda_n\}$ satisfying $\beta < |\lambda_n| < \beta + \frac{\gamma-\beta}{2n}$ and $\lambda_n \notin \sigma_r(T)$ for each n. By passing to a subsequence, we can assume that $\lambda_n \to \lambda$ for some $\lambda \in \mathbb{C}$. Clearly, $|\lambda| = \beta$, and from $\overline{D(0,\beta)} \subseteq \sigma(T)$, we get $\lambda \in \sigma(T)$. Now notice that

$$\|(\lambda_n - T)x\| \geq \|Tx\| - \|\lambda_n x\| \geq \gamma\|x\| - |\lambda_n|\cdot\|x\| \geq \tfrac{\gamma-\beta}{2}\|x\| \qquad (\star)$$

holds for each $x \in X$. Since each operator $\lambda_n - T$ has dense range, it follows from Lemma 2.8 that each $\lambda_n - T$ is an invertible operator. But then, by Theorem 2.5, the operator $\lambda - T$ is itself invertible, a contradiction. This contradiction completes the solution. ∎

Problem 6.2.7. *If $P \in \mathcal{L}(X)$ is a non-trivial projection (that is, if $P^2 = P$ and $P \neq 0, I$), then show that $\sigma(P) = \sigma_p(P) = \{0,1\}$.*

Solution: Assume that $P \in \mathcal{L}(X)$ is a non-trivial bounded projection. Let $Y = P(X)$ and $Z = (I-P)(X)$. Then Y and Z are non-trivial closed vector subspaces of X and $X = Y \oplus Z$ holds. If $y \in X$ and $z \in Z$ are both non-zero, then

$$Py = y \quad \text{and} \quad Pz = 0 = 0z\,.$$

Therefore, 0 and 1 are eigenvalues of P.

Now fix $\lambda \in \mathbb{C}$ such that $\lambda \neq 0$ and $\lambda \neq 1$. Notice that if $x = y+z \in X = Y \oplus Z$, then

$$(\lambda - P)x = \lambda x - Px = (\lambda - 1)y + \lambda z\,.$$

So, if $(\lambda - P)x = 0$, then $(1-\lambda)y = 0$ and $\lambda z = 0$. Since $\lambda \neq 0, 1$, we get $y = z = 0$ or $x = 0$. Thus, $\lambda - P$ is one-to-one.

Now if $x = y + z \in X$, then the vector $u = \frac{y}{\lambda-1} + \frac{z}{\lambda}$ satisfies $(\lambda - P)u = x$. Hence, $\lambda - P$ is a surjective one-to-one operator, and is thus invertible. Therefore, $\rho(P) = \{\lambda \in \mathbb{C}\colon\ \lambda \neq 0, 1\}$ or $\sigma_p(T) = \sigma(T) = \{0,1\}$. ∎

Problem 6.2.8. *Let $S \in \mathcal{L}(X)$ and $T \in \mathcal{L}(Y)$ be two bounded operators, where X and Y are two complex Banach spaces. The operator S and T are said to be* ***similar*** *whenever there exists an invertible operator R in $\mathcal{L}(X,Y)$ satisfying $T = RSR^{-1}$. In other words, S and T are similar if there exists an isomorphism (i.e., a surjective, one-to-one, and continuous operator) $R\colon X \to Y$ such that the diagram shown in Figure 1 commutes.*

If S and T are similar operators, then show that

$$\sigma(S) = \sigma(T), \quad \sigma_p(S) = \sigma_p(T), \quad \sigma_a(S) = \sigma_a(T), \quad \text{and} \quad \sigma_r(S) = \sigma_r(T).$$

$$\begin{array}{ccc} X & \xrightarrow{R} & Y \\ {\scriptstyle S}\downarrow & & \downarrow{\scriptstyle T} \\ X & \xrightarrow{R} & Y \end{array}$$

Figure 1

Solution: By the symmetry of the situation, it suffices to prove that

$$\sigma(S) \subseteq \sigma(T), \quad \sigma_p(S) \subseteq \sigma_p(T), \quad \text{and} \quad \sigma_a(S) \subseteq \sigma_a(T)$$

hold. We shall do this in each case separately.

(a) $\sigma_p(S) \subseteq \sigma_p(T)$.

Let $\lambda \in \sigma_p(S)$. Pick some non-zero vector $x \in X$ such that $Sx = \lambda x$. Since R is an isomorphism, we have $Rx \neq 0$, and from the commuting diagram we get $T(Rx) = RSx = \lambda(Rx)$. This shows that $\lambda \in \sigma_p(T)$, and so $\sigma_p(S) \subseteq \sigma_p(T)$.

(b) $\sigma_a(S) \subseteq \sigma_a(T)$.

Let $\lambda \in \sigma_a(S)$. Pick a sequence of unit vectors $\{x_n\}$ so that $\lambda x_n - Sx_n \to 0$. Since the operator R (as being an isomorphism) is bounded below, there exists some $\delta > 0$ such that $\|Rx_n\| \geq \delta$ for each n. Now note that

$$(\lambda - T)Rx_n = \lambda Rx_n - T(Rx_n) = \lambda Rx_n - RSx_n = R(\lambda x_n - Sx_n) \to 0.$$

This implies (see Problem 6.2.4) that $\lambda \in \sigma_a(T)$, and so $\sigma_a(S) \subseteq \sigma_a(T)$.

(c) $\sigma(S) \subseteq \sigma(T)$.

Let $\lambda \in \sigma(S)$. Assume $\lambda - S$ is a non-surjective one-to-one bounded operator. By part (a), we know that $\lambda - T$ is one-to-one. We claim that $\lambda - T$ is not surjective. To see this, assume by way of contradiction that $\lambda - T$ is surjective, and let $x \in X$ be an arbitrary vector. Pick some $y_0 \in Y$ such that $(\lambda - T)y_0 = Rx$. Since R is surjective, there exists some $x_0 \in X$ such that $y_0 = Rx_0$. Therefore,

$$Rx = (\lambda - T)y_0 = (\lambda - T)Rx_0 = \lambda Rx_0 - TRx_0 = \lambda Rx_0 - RSx_0 = R(\lambda - S)x_0.$$

Since R is an isomorphism, it follows that $(\lambda - S)x_0 = x$. This shows that $\lambda - S$ is surjective, which is a contradiction. Hence, $\lambda - T$ is not surjective, and so $\lambda \in \sigma(T)$, i.e., $\sigma(S) \subseteq \sigma(T)$.

(d) $\sigma_r(S) \subseteq \sigma_r(T)$.

Let $\lambda \in \sigma_r(S)$. This means that $\lambda - S$ is one-to-one and $\overline{(\lambda - S)X} \neq X$. To see that $\lambda - T$ is one-to-one, assume that $(\lambda - T)y = 0$ for some $y \in Y$. Pick some $x \in X$ such that $y = Rx$, and note that $(\lambda - S)x = [R^{-1}(\lambda - T)R]x = R^{-1}(\lambda - T)y = 0$. Since $\lambda - S$ is one-to-one, it follows that $x = 0$, and so $y = Rx = 0$.

On the other hand, if $\overline{(\lambda - T)Y} = Y$, then

$$\begin{aligned}\overline{(\lambda - S)X} &= \overline{[R^{-1}(\lambda - T)]R(X)} = \overline{R^{-1}(\lambda - T)Y}\\ &= R^{-1}\big(\overline{(\lambda - T)Y}\big) = R^{-1}(Y) = X\,,\end{aligned}$$

which is a contradiction. Hence, $\overline{(\lambda - T)Y} \neq Y$, and so $\lambda \in \sigma_r(T)$. ∎

Problem 6.2.9. *For two operators $S, T \in \mathcal{L}(X)$ establish the following.*

(a) *A non-zero complex number is an eigenvalue of ST if and only if it is an eigenvalue of TS.*

(b) *If X is finite dimensional, then show that $\sigma(ST) = \sigma(TS)$; compare this with Problem 6.1.19.*

Solution: (a) Assume that for some $\lambda \neq 0$ and some $x \neq 0$ we have $STx = \lambda x$. This guarantees $Tx \neq 0$. Now note that

$$TS(Tx) = T(STx) = T(\lambda x) = \lambda Tx\,.$$

Therefore, λ is an eigenvalue of TS. By symmetry, the converse is also true.

(b) Since X is a finite dimensional complex Banach space, we can identify X with some $\mathbb{C}^n$. Also, we can view S and T as $n \times n$ (complex) matrices. In this case, it should be noted that the spectrum of any operator consists only of eigenvalues.

From part (a), we know that the matrices ST and TS have the same non-zero eigenvalues. Now notice that 0 is an eigenvalue of the matrix ST if and only if $\det(ST) = \det(S)\det(T) = \det(TS) = 0$. Since $\det(ST) = 0$ if and only if $\det(TS) = 0$, we see that $0 \in \sigma(ST)$ if and only if $0 \in \sigma(TS)$. Therefore, $\sigma(ST) = \sigma(TS)$ must be true in this case. (However, as we saw in Problem 6.1.19, the two sets $\sigma(ST)$ and $\sigma(TS)$ need not coincide in general.) ∎

6.3. The Resolvent of a Positive Operator

Problem 6.3.1. *Assume that $\sum_{n=1}^{\infty} |z_n|$ exists in a Banach lattice E, where $z_n = x_n + \imath y_n \in E_{\mathbf{c}}$ for each n. Show that $\sum_{n=1}^{\infty} z_n$ exists in $E_{\mathbf{c}}$ and $\big|\sum_{n=1}^{\infty} z_n\big| \leq \sum_{n=1}^{\infty} |z_n|$, where the series are assumed to converge in norm.*

Solution: If k and θ are fixed, then

$$\begin{aligned}\Big(\sum_{n=1}^{k} x_n\Big)\cos\theta + \Big(\sum_{n=1}^{k} y_n\Big)\cos\theta &= \sum_{n=1}^{k}(x_n\cos\theta + y_n\sin\theta)\\ &\leq \sum_{n=1}^{k}|z_n| \leq \sum_{n=1}^{\infty}|z_n|.\end{aligned}$$

Hence, $(\sum_{n=1}^{\infty} x_n)\cos\theta + (\sum_{n=1}^{\infty} y_n)\sin\theta \leq \sum_{n=1}^{\infty}|z_n|$ holds for all θ. This implies

$$\Big|\sum_{n=1}^{\infty} z_n\Big| = \sup_{\theta\in\mathbb{R}}\Big[\Big(\sum_{n=1}^{\infty} x_n\Big)\cos\theta + \Big(\sum_{n=1}^{\infty} y_n\Big)\sin\theta\Big] \leq \sum_{n=1}^{\infty}|z_n|\,,$$

as desired. ∎

Problem 6.3.2. *If $S, T\colon E \to E$ are positive operators, then show that*

$$r(S+T) \geq \max\{r(S), r(T)\}\,.$$

Solution: Let $S, T\colon E \to E$ be two positive operators. Then $0 \leq S \leq S+T$ holds, and so $0 \leq S^n \leq (S+T)^n$ for each n. This implies $\|S^n\| \leq \|(S+T)^n\|$ for each n, and so

$$r(S) = \lim_{n\to\infty} \|S^n\|^{\frac{1}{n}} \leq \lim_{n\to\infty} \|(S+T)^n\|^{\frac{1}{n}} = r(S+T)\,.$$

Similarly, $r(T) \leq r(S+T)$, and so $\max\{r(S), r(T)\} \leq r(S+T)$. ∎

Problem 6.3.3. *Let $T\colon E \to E$ be a positive operator on a Banach lattice. If T is σ-order continuous (resp. order continuous), then show that $R(\lambda, T)$ is also σ-order continuous (resp. order continuous) for each $\lambda > r(T)$.*

Solution: We prove the σ-order continuous case only. So, assume that $T\colon E \to E$ is a σ-order continuous positive operator on a Banach lattice. If $x_n \downarrow 0$ in E, then $Tx_n \downarrow 0$, and so $T^2x_n = T(Tx_n) \downarrow 0$. This shows that T^2 is a σ-order continuous operator. By induction, T^k is a σ-order continuous operator for each $k = 0, 1, \ldots$. So, if $\lambda > r(T)$ is fixed, then each positive operator $S_k = \sum_{n=1}^{k} \lambda^{-(n+1)} T^k$ is σ-order continuous. Moreover, $S_k \uparrow R(\lambda, T)$ and $\|S_k - R(\lambda, T)\| \to 0$ in $\mathcal{L}(E)$.

Now let $y_n \downarrow 0$ in E, and assume that $0 \leq y \leq R(\lambda, T)y_n \downarrow$. For each k and n we have

$$0 \leq y \leq R(\lambda, T)y_n = \big[R(\lambda, T) - S_k\big]y_n + S_k y_k \leq \big[R(\lambda, T) - S_k\big]y_1 + S_k x_n\,.$$

Since $S_k x_n \downarrow_n 0$ for each k, the above inequality implies $0 \leq y \leq \big[R(\lambda, T) - S_k\big]x_1$ for each k. Hence, $0 \leq \|y\| \leq \big\|\big[R(\lambda, T) - S_k\big]x_1\big\| \to 0$, and so $\|y\| = 0$ or $y = 0$. This shows that $R(\lambda, T)y_n \downarrow 0$, proving that $R(\lambda, T)$ is a σ-order continuous positive operator. ∎

Problem 6.3.4. *Let $T\colon E \to E$ be a positive operator on a Banach lattice. If a scalar $\alpha > 0$ satisfies $\alpha x \leq Tx$ for some $x > 0$ or $\alpha x^* \leq T^*x^*$ for some $0 < x^* \in E^*$, then show that $\alpha \leq r(T)$.*

Solution: Assume that $\alpha x \leq Tx$ holds for some $\alpha > 0$ and $x > 0$. Since T is a positive operator, an easy inductive argument shows that $\alpha^n x \leq T^n x$ for each n. This implies $\alpha^n \leq \|T^n\|$ or $\alpha \leq \|T^n\|^{\frac{1}{n}}$ for each n. Consequently, we have $\alpha \leq \lim_{n\to\infty} \|T^n\|^{\frac{1}{n}} = r(T)$. A similar argument applies when $\alpha x^* \leq T^*x^*$. ∎

Problem 6.3.5. *Let $T\colon X \to X$ be a bounded operator on a complex Banach space and let $\rho_\infty(T)$ denote the unbounded (open) component in $\mathbb{C}$ of the resolvent $\rho(T)$. Establish the following.*

(a) *If $\sigma(T)$ is at most countable (in particular, if every non-zero point in $\sigma(T)$ is an isolated point), then $\rho_\infty(T) = \rho(T)$.*

(b) *A closed subspace Y of the Banach space X is T-invariant if and only if Y is $R(\lambda, T)$-invariant for each $\lambda \in \rho_\infty(T)$.*

(c) *If a closed subspace Y of X is T-invariant and $T|_Y$ is the restriction of T to Y, then $\rho_\infty(T) \subseteq \rho(T|_Y)$ and $R(\lambda, T|_Y) = R(\lambda, T)|_Y$ for each $\lambda \in \rho_\infty(T)$.*

(d) *If $\rho_\infty(T) = \rho(T)$, then for every T-invariant closed subspace Y of X we have $\sigma(T|_Y) \subseteq \sigma(T)$.*

(e) *Assume that an open annulus $A_{R_1,R_2}(\lambda_0)$ lies entirely in $\rho_\infty(T)$ and that $R(\lambda, T) = \sum_{n=-\infty}^{\infty}(\lambda - \lambda_0)^n A_n$ is the Laurent series expansion of the resolvent in $A_{R_1,R_2}(\lambda_0)$. If Y is a T-invariant closed subspace of X, then Y is also A_n-invariant for each $n \in \mathbb{Z}$.*

Solution: (a) Assume that $\sigma(T)$ is at most countable. We shall show that $\rho(T)$ is arcwise connected. To see this, let $\lambda, \mu \in \rho(T)$ with $\lambda \neq \mu$. We claim that there exists a straight line passing through λ that lies entirely in $\rho(T)$. Otherwise, if every straight line L_1 passing through λ intersects $\sigma(T)$, then $\sigma(T)$ would be uncountable, which is impossible. Similarly, there exists a straight line L_2 passing through μ that is not parallel to L_1 and lies entirely in $\rho(T)$. If z is the intersection of L_1 and L_2, then the continuous path consisting of the line segments from λ to z and from z to μ lies entirely in $\rho(T)$ and joins λ with μ . This shows that $\rho(T)$ is arcwise connected—and hence $\rho(T)$ is connected. In particular, $\rho_\infty(T) = \rho(T)$.

(b) Let Y be a closed subspace of X. Assume first that Y is T-invariant. Fix any $y \in Y$ and $y^* \in Y^\perp$, and consider the analytic function $f\colon \rho_\infty(T) \to \mathbb{C}$ defined by

$$f(\lambda) = y^*\big(R(\lambda, T)y\big).$$

From the Neumann series, it is easy to see that for each $\lambda \in \mathbb{C}$ with $|\lambda| > r(T)$ the vector subspace Y is also $R(\lambda, T)$-invariant. So, $R(\lambda, T)y \in Y$ for all $|\lambda| > r(T)$. Since $y^* \in Y^\perp$, we obtain $f(\lambda) = y^*\big(R(\lambda, T)y\big) = 0$ for each $|\lambda| > r(T)$. The analyticity of f and the connectedness of $\rho_\infty(T)$ imply $y^*\big(R(\lambda, T)y\big) = 0$ for each $\lambda \in \rho_\infty(T)$, all $y^* \in Y^\perp$ and all $y \in Y$. Now an easy application of the Hahn–Banach theorem shows that $R(\lambda, T)y \in Y$ for all $\lambda \in \rho_\infty(T)$ and all $y \in Y$. That is, Y is $R(\lambda, T)$-invariant for each $\lambda \in \rho_\infty(T)$.

For the converse, assume that Y is $R(\lambda, T)$-invariant for every $\lambda \in \rho_\infty(T)$. According to Problem 6.1.3 we have $\lim_{\lambda\to\infty} \lambda R(\lambda, T) = I$. Therefore,

$$\lim_{\lambda\to\infty} T\big(\lambda R(\lambda, T)\big) = T. \qquad (\star)$$

Now note that if $\lambda \in \rho(T)$, then

$$T\big(\lambda R(\lambda, T)\big) = \lambda T R(\lambda, T) = -\lambda(\lambda - T - \lambda)R(\lambda, T) = \lambda^2 R(\lambda, T) - \lambda.$$

So, for each $y \in Y$ we have $\big[T\big(\lambda R(\lambda, T)\big)\big]y = \lambda^2 R(\lambda, T)y - \lambda y \in Y$. Now a glance at $(\star)$ shows that $Ty \in Y$ for each $y \in Y$. So, Y is T-invariant.

(c) Assume that Y is a closed T-invariant subspace of X, and let $\lambda \in \rho_\infty(T)$. By part (b), Y is also $R(\lambda, T)$-invariant. From $(\lambda - T)R(\lambda, T) = R(\lambda, T)(\lambda - T) = I_X$, it follows that $(\lambda - T)R(\lambda, T)y = R(\lambda, T)(\lambda - T)y = y$ for each $y \in Y$. This shows that $R(\lambda, T)|_Y$ is the inverse of the operator $\lambda \; T|_Y \in \mathcal{L}(Y)$. Therefore, $\lambda \in \rho(T|_Y)$ and $R(\lambda, T|_Y) = R(\lambda, T)|_Y$.

(d) From part (c) we have $\rho(T) = \rho_\infty(T) \subseteq \rho(T|_Y)$. Taking complements yields $\sigma(T|_Y) \subseteq \sigma(T)$.

(e) Assume that $A_{R_1,R_2}(\lambda_0) = \{\lambda \in \mathbb{C}\colon\ R_1 < |\lambda - \lambda_0| < R_2\}$ lies entirely in $\rho_\infty(T)$. Fix some R such that $R_1 < R < R_2$ and let C_R denote the positively oriented circle centered at λ_0 with radius R. According to Laurent's Theorem 1.78 for each $n \in \mathbb{Z}$ we have

$$A_n = \tfrac{1}{2\pi\imath} \int_{C_R} \tfrac{R(\lambda,T)}{(\lambda-\lambda_0)^{n+1}}\, d\lambda\,.$$

Now fix $y \in Y$. Part (b) guarantees that $R(\lambda, T)y \in Y$ for each $\lambda \in \rho_\infty(T)$, and so from Problem 1.5.3 we get

$$A_n y = \tfrac{1}{2\pi\imath} \int_{C_R} \tfrac{R(\lambda,T)y}{(\lambda-\lambda_0)^{n+1}}\, d\lambda \in Y$$

for each $n \in \mathbb{Z}$. So, Y is A_n-invariant for each $n \in \mathbb{Z}$, and the solution is finished. ∎

Problem 6.3.6. *Give an example of a positive operator $T\colon E \to E$ on a Banach lattice and some T-invariant closed ideal J in E such that:*

(a) *There is some $\lambda \in \rho(T)$ for which J is not $R(\lambda, T)$-invariant.*

(b) *The spectrum $\sigma(T|_J)$ is not included in $\sigma(T)$, where $T|_J$ denotes the restriction of T to J.*

Solution: Let $E = \ell_2(\mathbb{Z})$, the Banach lattice of all square summable sequences defined on the integers $\mathbb{Z}$. We let $T\colon E \to E$ be the right-shift operator. That is,

$$T(\ldots, x_{-2}, x_{-1}, \underbrace{x_0}_{0}, x_1, x_2, \ldots) = (\ldots, x_{-3}, x_{-2}, \underbrace{x_{-1}}_{0}, x_0, x_1, x_2, \ldots)$$

for each $(\ldots, x_{-2}, x_{-1}, x_0, x_1, x_2, \ldots) \in \ell_2(\mathbb{Z})$. It is easy to see that T is a surjective lattice isometry; and so $r(T) = \|T\| = 1$. The inverse of T is the left-shift $S\colon E \to E$ defined by

$$S(\ldots, x_{-2}, x_{-1}, \underbrace{x_0}_{0}, x_1, x_2, \ldots) = (\ldots, x_{-1}, x_0, \underbrace{x_1}_{0}, x_2, x_3, \ldots)\,.$$

We claim that $\sigma(T) = \Gamma = \{\lambda \in \mathbb{C}\colon\ |\lambda| = 1\}$. To see this, consider first some $\lambda \in \sigma(T)$. Then on one hand $|\lambda| \le \|T\| = 1$, and on the other hand (from Problem 6.1.12) we have $\left|\frac{1}{\lambda}\right| \le \|S\| = 1$ or $|\lambda| \ge 1$. Thus, $|\lambda| = 1$, and so $\sigma(T) \subseteq \Gamma$.

For the reverse inclusion, let $\lambda \in \Gamma$. We claim that the vector $y \in \ell_2(\mathbb{Z})$ defined by $y_n = 0$ if $n \neq 0$ and $y_0 = 1$ does not belong to the range of the operator $\lambda - T$. To establish this, assume by way of contradiction that there exists some vector $x = (\ldots, x_{-2}, x_{-1}, x_0, x_1, x_2, \ldots) \in \ell_2(\mathbb{Z})$ such that $(\lambda - T)x = y$. That is, $\lambda x_n - x_{n-1} = y_n$ for each $n \in \mathbb{Z}$. It follows that $x_n = \frac{1}{\lambda^n} x_0$ for each $n \ge 1$, and from $|x_0| = |x_n| \to 0$, we see that $x_0 = 0$. This implies $x_{-1} = -1$ and $x_{-n} = -\lambda^n$ for each $n \ge 1$. But then we have $1 = |x_{-n}| \to 0$, which is impossible. This establishes that $y \notin R(\lambda - T)$, and so $\lambda \in \sigma(T)$. Thus, $\Gamma \subseteq \sigma(T)$ is also true, and so $\sigma(T) = \Gamma$.

Since T is invertible, $0 \notin \sigma(T)$. Now let

$$J = \ell_2 = \{x \in \ell_2(\mathbb{Z})\colon\ x_{-i} = 0 \ \text{for all}\ i \ge 1\}\,.$$

Clearly, J is T-invariant and the restriction of T to J is the familiar forward shift. From Example 6.21 we know that $\sigma(T|_J) = \{z \in \mathbb{C}\colon\ |z| \leq 1\}$. This implies $\sigma(T|_J) \not\subseteq \sigma(T)$. Finally, notice that $0 \in \rho(T)$ and $R(0,T) = -S$. Consequently, the closed ideal J of $\ell_2(\mathbb{Z})$ is T-invariant but not $R(0,T)$-invariant. (Compare this problem with Problem 6.3.5.) ∎

6.4. Functional Calculus

Problem 6.4.1. *Show directly (i.e., by avoiding the Spectral Mapping Theorem) that if $p(\lambda) = \sum_{k=0}^n a_k\lambda^k$ is a polynomial, then for every bounded operator $T\colon X \to X$ on a Banach space we have*

$$\sigma(p(T)) = p(\sigma(T)) = \{p(\lambda)\colon\ \lambda \in \sigma(T)\}.$$

Solution: Let $p(\lambda) = \sum_{k=0}^n a_k\lambda^k$ be a polynomial—with real or complex coefficients. For each $\mu \in \mathbb{C}$ we have the factorization

$$\mu - p(T) = (-1)^{n+1}a_n(\lambda_1 - T)(\lambda_2 - T)\cdots(\lambda_n - T), \tag{$\star$}$$

where $\lambda_1, \lambda_2, \ldots, \lambda_n$ are the roots (counted with their multiplicities) of the polynomial $q(\lambda) = \mu - p(\lambda)$.

Assume that $\mu \in \sigma(p(T))$, i.e., the operator $\mu - p(T)$ is non-invertible. In view of $(\star)$, there exists some root λ_i of the polynomial $q(\lambda)$, i.e., $\mu = p(\lambda_i)$, such that the operator $\lambda_i - T$ is non-invertible, that is, $\lambda_i \in \sigma(T)$. This implies $\mu \in \{p(\lambda)\colon\ \lambda \in \sigma(T)\}$, and so $\sigma(p(T)) \subseteq \{p(\lambda)\colon\ \lambda \in \sigma(T)\}$.

Now let $\alpha \in \sigma(T)$, and put $\mu = p(\alpha)$. In this case, we claim that the operator $\mu - p(T)$ cannot be invertible. Indeed, if $\mu - p(T)$ is invertible, then it follows from $(\star)$ that each $\lambda_i - T$ is invertible. Since $\alpha = \lambda_i$ for some i, we get $\alpha \notin \sigma(T)$, which is a contradiction. Therefore, we have $\mu = p(\alpha) \in \sigma(p(T))$, and consequently $\{p(\lambda)\colon\ \lambda \in \sigma(T)\} \subseteq \sigma(p(T))$. Thus, $\sigma(p(T)) = \{p(\lambda)\colon\ \lambda \in \sigma(T)\} = p(\sigma(T))$. ∎

Problem 6.4.2. *Assume that f is an analytic function defined on a neighborhood V of ∞, that is, the function f is defined and is analytic on an open set of the form $V = \{\lambda \in \mathbb{C}\colon\ |\lambda| > R\}$ for some $R > 0$. Suppose also that f vanishes at infinity. If $\mathcal{C}$ is a Jordan contour in V surrounding the spectrum of an operator $T \in \mathcal{L}(X)$, then show that $\int_{\mathcal{C}} f(\lambda)R(\lambda,T)\,d\lambda = 0$.*

Solution: Assume that f satisfies the stated properties. Then, according to Cauchy's Integral Theorem 1.70, there exists some $r_0 > \|T\|$ such that

$$A = \int_{\mathcal{C}} f(\lambda)R(\lambda,T)\,d\lambda = \int_{C_r} f(\lambda)R(\lambda,T)\,d\lambda$$

holds for all $r > r_0$, where C_r is the positively oriented circle centered at zero with radius r.

Now let $\epsilon > 0$. Pick some $r_1 > r_0$ such that $|f(\lambda)| < \epsilon$ for all $\lambda \in \mathbb{C}$ with $|\lambda| > r_1$. So, according to Theorem 6.3, for each $r > r_1$ we have

$$\|A\| = \left\| \int_{C_r} f(\lambda)R(\lambda,T)\,d\lambda \right\| < \epsilon \cdot 2\pi r \cdot \tfrac{1}{r-\|T\|} = 2\pi\epsilon \tfrac{r}{r-\|T\|}\,.$$

Letting $r \to \infty$, we get $\|A\| \leq 2\pi\epsilon$ for each $\epsilon > 0$. That is, $\|A\| = 0$ or $A = 0$, as claimed. ∎

Problem 6.4.3. *Let $T\colon X \to X$ be a bounded operator on a Banach space. Fix any $f \in \mathcal{F}(T)$ and let $\mathcal{C}$ be a Jordan contour surrounding $\sigma(T)$ in the domain of analyticity of f. Show that*

$$f(T)x = \tfrac{1}{2\pi \imath} \int_{\mathcal{C}} f(\lambda)R(\lambda,T)x\,d\lambda$$

holds for each $x \in X$. In particular, if $Tx = \alpha x$ for some $x \in X$ and a scalar $\alpha \in \mathbb{C}$, then show that $f(T)x = f(\alpha)x$.

Solution: Fix some $f \in \mathcal{F}(T)$ and let $\mathcal{C}$ be a Jordan contour in the domain of analyticity of f surrounding $\sigma(T)$. The formula

$$f(T)x = \tfrac{1}{2\pi \imath} \int_{\mathcal{C}} f(\lambda)R(\lambda,T)x\,d\lambda$$

follows immediately from Problem 1.5.3.

Now assume that $Tx = \alpha x$. From the identity $(\lambda - T)x = (\lambda - \alpha)x$, it follows that $R(\lambda,T)x = (\lambda - T)^{-1}x = \frac{1}{\lambda-\alpha}x$ for each $\lambda \in \mathcal{C}$. So, from the preceding case and Theorem 1.73 we get

$$\begin{aligned} f(T)x &= \tfrac{1}{2\pi \imath} \int_{\mathcal{C}} f(\lambda)R(\lambda,T)x\,d\lambda = \tfrac{1}{2\pi \imath} \int_{\mathcal{C}} \tfrac{f(\lambda)}{\lambda-\alpha}x\,d\lambda \\ &= \left[\tfrac{1}{2\pi \imath} \int_{\mathcal{C}} \tfrac{f(\lambda)}{\lambda-\alpha}\,d\lambda\right]x = f(\alpha)x\,, \end{aligned}$$

as desired. ∎

Problem 6.4.4. *Let $T \in \mathcal{L}(X)$, let $f \in \mathcal{F}(T)$, and let $g \in \mathcal{F}(f(T))$. Establish that the composition $g \circ f$ belongs to $\mathcal{F}(T)$ and that $(g \circ f)(T) = g(f(T))$.*

Solution: If f is a constant function, then the conclusion is trivial. So, we can assume that f is non-constant. In this case, f is an open mapping.

Now let V and U be open neighborhoods of $\sigma(T)$ and $\sigma\big(f(T)\big)$ on which f and g are analytic, respectively. According to the Spectral Mapping Theorem 6.31, we have $\sigma\big(f(T)\big) = f\big(\sigma(T)\big)$, and thus $f(V)$ is an open neighborhood of $\sigma\big(f(T)\big)$. Replacing U by $U_1 = U \cap f(V)$ and V by $f^{-1}(U_1)$, we can suppose without loss of generality that $f(V) \subseteq U$. Therefore, $g \circ f \in \mathcal{F}(T)$.

Pick a Jordan contour $\mathcal{C}_2$ in U surrounding $\sigma\big(f(T)\big)$, and let W be the open set satisfying $\sigma(f(T)) \subseteq W \subseteq \overline{W} \subseteq U$ whose boundary is $\mathcal{C}_2$. Clearly, $\sigma(T) \subseteq f^{-1}(W)$. Replacing V by $f^{-1}(W)$, we can also assume that $f(V) \cap \mathcal{C}_2 = \emptyset$. Finally, fix any Jordan contour $\mathcal{C}_1$ in V surrounding $\sigma(T)$.

Fix $\mu \in \mathcal{C}_2$, and note that the function $h\colon V \to \mathbb{C}$ defined by $h(\lambda) = \frac{1}{\mu - f(\lambda)}$ is analytic, i.e., $h \in \mathcal{F}(T)$. From $h(\lambda)[\mu - f(\lambda)] = 1$ for each $\lambda \in V$, it follows that $h(T)[\mu - f(T)] = [\mu - f(T)]h(T) = I$. Therefore, $h(T) = [\mu - f(T)]^{-1}$, and so

$$h(T) = \tfrac{1}{2\pi\imath} \int_{\mathcal{C}_1} \tfrac{R(\lambda,T)}{\mu - f(\lambda)}\, d\lambda = [\mu - f(T)]^{-1} = R(\mu, f(T))$$

for each $\mu \in \mathcal{C}_2$. This implies

$$\begin{aligned} g\big(f(T)\big) &= \tfrac{1}{2\pi\imath} \int_{\mathcal{C}_2} g(\mu) R\big(\mu, f(T)\big)\, d\mu \\ &= -\tfrac{1}{4\pi^2} \int_{\mathcal{C}_2} \int_{\mathcal{C}_1} \tfrac{g(\mu)R(\lambda,T)}{\mu - f(\lambda)}\, d\lambda d\mu \\ &= -\tfrac{1}{4\pi^2} \int_{\mathcal{C}_1} \int_{\mathcal{C}_2} \tfrac{g(\mu)R(\lambda,T)}{\mu - f(\lambda)}\, d\mu d\lambda \\ &= \tfrac{1}{2\pi\imath} \int_{\mathcal{C}_1} \Big[\tfrac{1}{2\pi\imath} \int_{\mathcal{C}_2} \tfrac{g(\mu)}{\mu - f(\lambda)}\, d\mu \Big] R(\lambda, T)\, d\lambda \\ &= \tfrac{1}{2\pi\imath} \int_{\mathcal{C}_1} g\big(f(\lambda)\big) R(\lambda, T)\, d\lambda \\ &= (g \circ f)(T)\,, \end{aligned}$$

as claimed. ∎

Problem 6.4.5. *Let $T\colon X \to X$ be a bounded operator on a Banach space, and let V be an open neighborhood of $\sigma(T)$. Assume that a sequence $\{f_n\}$ of analytic functions defined on V converges uniformly to some function $f\colon V \to \mathbb{C}$. Show that $f \in \mathcal{F}(T)$ and that $\|f_n(T) - f(T)\| \to 0$.*

Solution: Since the sequence $\{f_n\}$ of analytic functions converges uniformly to f on V, it follows that f is analytic on V and so $f \in \mathcal{F}(T)$. This can be easily seen from Morera's theorem; see Problem 1.5.2. Indeed, if $\mathcal{C}$ is any Jordan contour that lies in V, then from

$$0 = \int_{\mathcal{C}} f_n(\lambda)\, d\lambda \longrightarrow \int_{\mathcal{C}} f(\lambda)\, d\lambda\,,$$

we see that $\int_{\mathcal{C}} f(\lambda)\, d\lambda = 0$, and so by Morera's theorem f is analytic on V.

Now fix $\epsilon > 0$ and consider any Jordan contour $\mathcal{C}$ in V surrounding $\sigma(T)$. Put $M = \max_{\lambda \in \mathcal{C}} \|R(\lambda, T)\|$, and pick some n_0 such that $|f_n(\lambda) - f(\lambda)| < \frac{2\pi\epsilon}{M\ell(\mathcal{C})}$ for all $n \geq n_0$, where $\ell(\mathcal{C})$ denotes the length of the contour $\mathcal{C}$. Now note that for each $n \geq n_0$ we have

$$\big\|f_n(T) - f(T)\big\| = \tfrac{1}{2\pi} \Big\| \int_{\mathcal{C}} [f_n(\lambda) - f(\lambda)] R(\lambda, T)\, d\lambda \Big\| \leq \tfrac{1}{2\pi} \cdot \tfrac{2\pi\epsilon}{M\ell(\mathcal{C})} \cdot M\ell(\mathcal{C}) = \epsilon\,.$$

This shows that $\lim_{n\to\infty} \|f_n(T) - f(T)\| = 0$. ∎

Problem 6.4.6. *Let $T\colon X \to X$ be a bounded operator on a Banach space, and let σ be a spectral set of T. Also, let (Y, Z) be the pair of reducing subspaces as determined in Theorem 6.34. If $h \in \mathcal{F}(T)$, then show that*

$$h(T)|_Y = h(T|_Y)\,.$$

Solution: If $\sigma = \emptyset$ or $\sigma = \sigma(T)$, the conclusion is trivial. So, assume that σ is a non-trivial spectral set. Let us recall the "background" for this problem. Fix a non-empty spectral set σ such that $\sigma' = \sigma(T) \setminus \sigma$ is also non-empty. If we let $2\delta = d(\sigma, \sigma') > 0$, then the two non-empty open sets $V_\sigma = \bigcup_{\lambda\in\sigma} B(\lambda, \delta)$ and $V_{\sigma'} = \bigcup_{\lambda\in\sigma'} B(\lambda, \delta)$ satisfy

$$\sigma \subseteq V_\sigma, \quad \sigma' \subseteq V_{\sigma'}, \quad \text{and} \quad V_\sigma \cap V_{\sigma'} = \emptyset\,.$$

That is, if σ is a spectral set, then there exist disjoint open neighborhoods V_σ and $V_{\sigma'}$ of σ and σ', respectively.

Next, consider the functions $f_\sigma, f_{\sigma'} \colon V_\sigma \cup V_{\sigma'} \to \mathbb{C}$ defined by

$$f_\sigma(\lambda) = \begin{cases} 1 & \text{if } \lambda \in V_\sigma \\ 0 & \text{if } \lambda \in V_{\sigma'} \end{cases} \quad \text{and} \quad f_{\sigma'}(\lambda) = \begin{cases} 0 & \text{if } \lambda \in V_\sigma \\ 1 & \text{if } \lambda \in V_{\sigma'}\,. \end{cases}$$

Clearly, $f_\sigma, f_{\sigma'} \in \mathcal{F}(T)$, and

$$(f_\sigma)^2 = f_\sigma\,, \quad (f_{\sigma'})^2 = f_{\sigma'}\,, \quad f_\sigma f_{\sigma'} = 0\,, \quad \text{and} \quad f_\sigma + f_{\sigma'} = \mathbf{1}\,.$$

So, by Theorem 6.29, we obtain

$$\begin{aligned} \big[f_\sigma(T)\big]^2 = f_\sigma(T)\,, \ \big[f_{\sigma'}(T)\big]^2 &= f_{\sigma'}(T)\,, \ f_\sigma(T) f_{\sigma'}(T) = 0\,, \ \text{and} \\ f_\sigma(T) + f_{\sigma'}(T) &= I\,. \end{aligned}$$

It follows that $f_\sigma(T)$ and $f_{\sigma'}(T)$ are both projections on X satisfying

$$X = R\big(f_\sigma(T)\big) \oplus R\big(f_{\sigma'}(T)\big)\,.$$

Moreover, the Spectral Mapping Theorem implies

$$\sigma\big(f_\sigma(T)\big) = f_\sigma\big(\sigma(T)\big) = \{0,1\} \quad \text{and} \quad \sigma\big(f_{\sigma'}(T)\big) = f_{\sigma'}\big(\sigma(T)\big) = \{0,1\}\,.$$

This shows that $f_\sigma(T)$ and $f_{\sigma'}(T)$ are both proper bounded projections. If we let $Y = R(f_\sigma(T))$ and $Z = R(f_{\sigma'}(T))$, then (Y, Z) is a reducing pair for T.

Now fix some $h \in \mathcal{F}(T)$. We must show that $h(T)|_Y = h(T|_Y)$. To this end, pick a Jordan contour $\mathcal{C}_1$ in V_σ surrounding σ and a Jordan contour $\mathcal{C}_2$ in $V_{\sigma'}$ surrounding σ'. Let $\mathcal{C}$ be the union of $\mathcal{C}_1$ and $\mathcal{C}_2$. So, $\mathcal{C}$ is a Jordan contour surrounding $\sigma(T)$. Now for each $y \in Y$ we have

$$\begin{aligned} h(T)(y) &= \big[h(T) f_\sigma(T)\big] y \\ &= \Big[\tfrac{1}{2\pi\imath} \int_{\mathcal{C}} h(\lambda) f_\sigma(\lambda) R(\lambda, T)\, d\lambda\Big] y \\ &= \Big[\tfrac{1}{2\pi\imath} \int_{\mathcal{C}_1} h(\lambda) R(\lambda, T)\, d\lambda\Big] y \\ &= \Big[\tfrac{1}{2\pi\imath} \int_{\mathcal{C}_1} h(\lambda) R(\lambda, T|_Y)\, d\lambda\Big] y \\ &= h(T|_Y) y\,. \end{aligned}$$

This shows that $h(T)|_Y = h(T|_Y)$ holds. ∎

Problem 6.4.7. *Let X be a Banach space. Show that there do not exist operators $S, T \in \mathcal{L}(X)$ such that $ST - TS = I$, the identity operator on X.*[1]

Solution: Fix $S, T \in \mathcal{L}(X)$ and assume by way of contradiction that $ST - TS = I$. To see this, start by observing that from Problem 6.1.19 we know that

$$\sigma(ST) \setminus \{0\} = \sigma(TS) \setminus \{0\}. \tag{$\star$}$$

The identity $ST - TS = I$ implies $ST = I + TS$, and so using Problem 6.4.1, we see that

$$\sigma(ST) = 1 + \sigma(TS). \tag{$\star\star$}$$

From ($\star$) and ($\star\star$), it follows that $\sigma(ST) \neq \{0\}$ and $\sigma(TS) \neq \{0\}$.

Now let λ be a non-zero point in the spectrum of ST. We claim that λ is a negative integer. To see this, notice that ($\star$) implies $\lambda \in \sigma(TS)$ and so ($\star\star$) yields $1 + \lambda \in \sigma(ST)$. Similarly, if $1 + \lambda \neq 0$, then $2 + \lambda \in \sigma(ST)$. Since, $\sigma(ST)$ is a compact set, there must exist some $k \in \mathbb{N}$ such that $k + \lambda = 0 \in \sigma(ST)$. This implies that λ is a negative integer. Let $\lambda_0 = \min[\mathbb{Z} \cap \sigma(ST)]$. Consequently, by ($\star\star$), the negative integer $\lambda_0 - 1$ satisfies $\lambda_0 - 1 \in \sigma(TS)$, and so $\lambda_0 - 1 \in \sigma(ST)$, which is a contradiction. This establishes that the identity operator cannot be of the form $ST - TS$ for any $S, T \in \mathcal{L}(X)$. ∎

Problem 6.4.8. *Let $T\colon X \to X$ be a bounded operator and let σ be a spectral set of T. Also, let (Y_σ, Z_σ) be the unique T-reducing pair of subspaces such that $X = Y_\sigma \oplus Z_\sigma$, $\sigma(T|_{Y_\sigma}) = \sigma$, and $\sigma(T|_{Z_\sigma}) = \sigma(T) \setminus \sigma$; see Theorem 6.34. If we let $T_\sigma = T|_{Y_\sigma}$, then show that*

$$\sigma_p(T_\sigma) = \sigma \cap \sigma_p(T), \quad \sigma_r(T_\sigma) = \sigma \cap \sigma_r(T), \quad \textit{and} \quad \sigma_c(T_\sigma) = \sigma \cap \sigma_c(T).$$

Solution: We shall verify each identity separately.

I: $\sigma_p(T_\sigma) = \sigma \cap \sigma_p(T)$.

Let $\lambda \in \sigma_p(T_\sigma)$. Then there exists some non-zero vector $y \in Y_\sigma$ satisfying $T_\sigma y = Ty = \lambda y$. This implies $\lambda \in \sigma_p(T)$, and so $\sigma_p(T_\sigma) \subseteq \sigma \cap \sigma_p(T)$.

For the reverse inclusion, let $\lambda \in \sigma \cap \sigma_p(T)$. Pick a non-zero vector $x \in X$ satisfying $Tx = \lambda x$, and write $x = y + z$ with $y \in Y_\sigma$ and $z \in Z_\sigma$. Clearly, $Ty + Tz = \lambda y + \lambda z$ or $Ty - \lambda y = \lambda z - Tz \in Y_\sigma \cap Z_\sigma = \{0\}$. This implies $Ty = \lambda y$ and $Tz = \lambda z$. If $y = 0$, then $z \neq 0$, and so $\lambda \in \sigma(T|_{Z_\sigma}) = \sigma(T) \setminus \sigma$, which contradicts $\lambda \in \sigma$. Hence, $y \neq 0$ and $T_\sigma y = Ty = \lambda y$ implies $\lambda \in \sigma_p(T_\sigma)$. Therefore, $\sigma \cap \sigma_p(T) \subseteq \sigma_p(T_\sigma)$, and consequently $\sigma_p(T_\sigma) = \sigma \cap \sigma_p(T)$.

II: $\sigma_r(T_\sigma) = \sigma \cap \sigma_r(T)$.

Let $\lambda \in \sigma_r(T_\sigma)$, and so $\overline{(\lambda - T_\sigma)(Y_\sigma)} \neq Y_\sigma$. We claim that $\overline{(\lambda - T)(X)} \neq X$. To see this, assume by way of contradiction that $\overline{(\lambda - T)(X)} = X$ holds true and fix $y \in Y_\sigma$. Pick a sequence $\{x_n\}$ of X such that $(\lambda - T)x_n \to y$. Then we have $\{(\lambda - T)P_\sigma(T)x_n\} \subseteq (\lambda - T_\sigma)(Y_\sigma)$ and $(\lambda - T)P_\sigma(T)x_n \to P_\sigma(T)y = y$. This implies

[1] If $S, T \in \mathcal{L}(X)$, then the operator $[S, T] = ST - TS$ is called the **commutator** of S and T. In this terminology, the conclusion of this problem can be stated as follows: *The identity operator on any Banach space cannot be a commutator.*

$y \in \overline{(\lambda - T_\sigma)(Y_\sigma)}$, and consequently $\overline{(\lambda - T_\sigma)(Y_\sigma)} = Y_\sigma$, which is impossible. This establishes that $\overline{(\lambda - T)(X)} \neq X$, and so $\lambda \in \sigma_r(T)$. Therefore, $\sigma_r(T_\sigma) \subseteq \sigma \cap \sigma_r(T)$.

Now assume that $\lambda \in \sigma \cap \sigma_r(T)$. This implies $\overline{(\lambda - T)(X)} \neq X$ and that the operator $\lambda - T_{Z_\sigma}: Z_\sigma \to Z_\sigma$ is invertible. Now if $\overline{(\lambda - T_\sigma)(Y_\sigma)} = Y_\sigma$, then the latter conclusion implies that $\overline{(\lambda - T)(X)} = X$, which is impossible. Hence, $\overline{(\lambda - T_\sigma)(Y_\sigma)} \neq Y_\sigma$ must be true, in which case we infer that $\lambda \in \sigma_r(T_\sigma)$. Therefore, $\sigma \cap \sigma_r(T) \subseteq \sigma_r(T_\sigma)$ is also valid, and thus $\sigma_r(T_\sigma) = \sigma \cap \sigma_r(T)$.

III: $\sigma_c(T_\sigma) = \sigma \cap \sigma_c(T)$.

This can be proven in a similar manner as II above. ∎

Problem 6.4.9. *Let λ_0 be an isolated point in the spectrum of an operator $T \in \mathcal{L}(X)$. Show that the resolvent of T has an essential singularity at λ_0 if and only if either the ascent or the descent of $\lambda_0 - T$ is infinite.*

Solution: By Theorem 6.39 we know that λ_0 is a pole if and only if both $\alpha(\lambda_0 - T)$ and $\delta(\lambda_0 - T)$ are finite. This implies the conclusion. ∎

Problem 6.4.10. *If $T \in \mathcal{L}(X)$ and $\lambda_0 \in \sigma(T)$ is a pole of order k of the resolvent function $R(\cdot, T)$ around λ_0, then show that:*

(1) *λ_0 is an eigenvalue of T.*

(2) *$Y_{\lambda_0} = N\big((\lambda_0 - T)^k\big)$ and $Z_{\lambda_0} = R\big((\lambda_0 - T)^k\big)$.*

Solution: (1) Since λ_0 is a pole of order k of the resolvent of T around λ_0, it follows from Theorem 6.39 that $(T - \lambda_0)^{k-1}A_{-1} \neq 0$ and $(T - \lambda_0)^k A_{-1} = 0$. So, if x is any vector in X with $(T - \lambda_0)^{k-1}A_{-1}x \neq 0$, then the non-zero vector $y = (T - \lambda_0)^{k-1}A_{-1}x$ satisfies $(T - \lambda_0)y = (T - \lambda_0)^k A_{-1}x = 0$. This implies $Ty = \lambda_0 y$, and so λ_0 is an eigenvalue of the operator T.

(2) According to Theorem 6.39 we have $\alpha(\lambda_0 - T) = \delta(\lambda_0 - T) = k$. If we let $Y = N\big((\lambda_0 - T)^k\big)$ and $Z = R\big((\lambda_0 - T)^k\big)$, then by Theorem 2.23 both Y and Z are T-invariant closed subspaces satisfying $X = Y \oplus Z$, and the operator $\lambda_0 - T$ is quasinilpotent on Y and invertible on Z.

The quasinilpotence of $\lambda_0 - T$ on Y implies that $\sigma(\lambda_0 - T|_Y) = \{0\}$. Invoking the Spectral Mapping Theorem, we get $\sigma(T|_Y) = \{\lambda_0\}$. Now from the identity $T = T|_Y \oplus T|_Z$, we obtain $\sigma(T) = \sigma(T|_Y) \cup \sigma(T|_Z)$; see Problem 6.1.17. Since $\lambda_0 \notin \sigma(T|_Z)$ and $\sigma(T|_Y) = \{\lambda_0\}$, we see that $\sigma(T|_Z) = \sigma(T) \setminus \{\lambda_0\}$. Now a glance at Theorem 6.34 guarantees that $Y = Y_{\lambda_0}$ and $Z = Z_{\lambda_0}$. ∎

Problem 6.4.11. *Let $T \in \mathcal{L}(X)$ and let λ_0 be an isolated point in the spectrum $\sigma(T)$ such that every neighborhood of λ_0 contains points that are not eigenvalues of T. Show that λ_0 is a pole of the resolvent $R(\cdot, T)$ if and only if $\lambda_0 - T$ has finite descent.*

Solution: If λ_0 is a pole of the resolvent $R(\cdot, T)$, then it follows from Theorem 6.39 that $\delta(\lambda_0 - T) < \infty$.

For the converse, let $p = \delta(\lambda_0 - T) < \infty$ and assume by way of contradiction that $\alpha(\lambda_0 - T) = \infty$. In particular, we have $N\big((\lambda_0 - T)^{p+1}\big) \neq \{0\}$. This implies that the surjective operator $\lambda_0 - T\colon R\big((\lambda_0 - T)^p\big) \to R\big((\lambda_0 - T)^p\big)$ is not invertible. Now a glance at Theorem 6.20 implies that there exists some $\epsilon > 0$ such that every $\lambda \in D(\lambda_0, \epsilon)$ is an eigenvalue of T, which is a contradiction. Hence, $\alpha(\lambda_0 - T) < \infty$. Now invoke Theorem 6.39 to conclude that λ_0 is a pole of $R(\cdot, T)$. ∎

Problem 6.4.12. *If a quasinilpotent operator $T\colon X \to X$ on a Banach space satisfies $p = \delta(T) < \infty$, then show that $T^p = 0$.*

Solution: Assume that $T \in \mathcal{L}(X)$ is quasinilpotent, i.e., $\sigma(T) = \{0\}$. From Problem 6.4.11, we know that 0 is a pole of the resolvent $R(\cdot, T)$. But then, from Theorem 6.39, we obtain that 0 is a pole of order p. If we consider the spectral set $\sigma = \{0\}$, then it follows from Problem 6.4.10 that $Y_0 = N\big((-T)^p\big) = N(T^p) = X$ and $Z_0 = \{0\}$. Consequently, $T^p = 0$. ∎

Problem 6.4.13. *For a spectral set σ of an operator $T \in \mathcal{L}(X)$ establish the following.*

(a) *If $0 \notin \sigma$, then there exists an operator $S \in \mathcal{L}(X)$ such that*

$$P_\sigma(T) = TS = ST\,.$$

(b) *If $0 \in \sigma$, then there exists an operator $S \in \mathcal{L}(X)$ such that*

$$P_\sigma(T) = I + TS = I + ST\,.$$

Solution: According to Theorem 6.34 there is a unique T-reducing pair (Y_σ, Z_σ) such that $\sigma\big(T|_{Y_\sigma}\big) = \sigma$, $\sigma\big(T|_{Z_\sigma}\big) = \sigma(T) \setminus \sigma$, and $P_\sigma(T)$ is the projection of X onto Y_σ along Z_σ.

(a) Assume $0 \notin \sigma$. This means that the operator $T|_{Y_\sigma}\colon Y_\sigma \to Y_\sigma$ is invertible. Let $R\colon Y_\sigma \to Y_\sigma$ be the inverse of the operator $T|_{Y_\sigma}$, and consider the operator $S = RP_\sigma(T) \in \mathcal{L}(X)$. Then for each $x = y_\sigma + z_\sigma \in Y_\sigma \oplus Z_\sigma = X$ we have

$$\begin{aligned} TSx &= TRP_\sigma(T)(y_\sigma + z_\sigma) = TRy_\sigma = y_\sigma = P_\sigma(T)x\,, \text{ and} \\ STx &= RP_\sigma(T)T(y_\sigma + z_\sigma) = RP_\sigma(T)(Ty_\sigma + Tz_\sigma) = RTy_\sigma = y_\sigma = P_\sigma(T)x\,. \end{aligned}$$

This shows that $P_\sigma(T) = ST = TS$.

(b) If $0 \in \sigma(T)$, then $0 \notin \sigma' = \sigma(T) \setminus \sigma$. So, according to part (a), there exists an operator $R \in \mathcal{L}(X)$ such that $P_{\sigma'}(T) = I - P_\sigma(T) = RT = TR$. If we let $S = -R$, then we have $P_\sigma(T) = I + TS = I + ST$. ∎

Problem 6.4.14. *Let σ be a spectral set of an operator $T \in \mathcal{L}(X)$. If there exists some $r > 0$ such that*

$$\sigma \subseteq \big\{\lambda \in \mathbb{C}\colon\ |\lambda| < r\big\} \quad \textit{and} \quad \sigma(T) \setminus \sigma \subseteq \big\{\lambda \in \mathbb{C}\colon\ |\lambda| > r\big\}\,,$$

then show that the range of the spectral projection $P_\sigma(T)$ is given by

$$R\big(P_\sigma(T)\big) = \big\{x \in X\colon\ \tfrac{T^n x}{r^n} \to 0\big\}\,.$$

Solution: Let σ be a spectral set of an operator $T \in \mathcal{L}(X)$ and let $r > 0$ be such that $\sigma \subseteq \big\{\lambda \in \mathbb{C}\colon\ |\lambda| < r\big\}$ and $\sigma(T) \setminus \sigma \subseteq \big\{\lambda \in \mathbb{C}\colon\ |\lambda| > r\big\}$. Also let (Y_σ, Z_σ) be the T-reducing pair, where Y_σ is the range of the projection $P_\sigma(T)$.

From $\sigma(T|_{Y_\sigma}) = \sigma$, it follows that $r(T|_{Y_\sigma}) < r$. Consequently, $\frac{(T|_{Y_\sigma})^n}{r^n} \to 0$ in $\mathcal{L}(Y_\sigma)$; see Problem 6.1.11. In particular, $\frac{(T|_{Y_\sigma})^n x}{r^n} \to 0$ for each $x \in Y_\sigma$, and so

$$Y_\sigma \subseteq \{x \in X\colon\ \tfrac{T^n x}{r^n} \to 0\}.$$

For the reverse inclusion, assume that some $x \in X$ satisfies $\frac{T^n x}{r^n} \to 0$ in X. If we choose an n_0 such that $\left\|\frac{T^n x}{r^n}\right\| < 1$ for all $n \geq n_0$, then $\|T^n x\|^{\frac{1}{n}} < r$ for each $n \geq n_0$. This implies $\limsup_{n\to\infty} \sqrt[n]{\|T^n x\|} \leq r$, and from Theorem 1.77 we see that the series $\sum_{n=0}^{\infty} \frac{T^n x}{\lambda^{n+1}}$ converges for each $\lambda \in \mathbb{C}$ with $|\lambda| > r$.

Next, choose some $R > r$ such that $\sigma(T) \setminus \sigma \subseteq \{\lambda \in \mathbb{C}\colon\ |\lambda| > R\}$ and let $A = \{\lambda \in \mathbb{C}\colon\ r < |\lambda| < R\}$. Clearly, $A \cap \sigma(T) = \emptyset$ and for each $\lambda \in A$ we have

$$\begin{aligned}(\lambda - T)\sum_{n=0}^{\infty} \tfrac{T^n x}{\lambda^{n+1}} &= (\lambda - T)\lim_{m\to\infty}\sum_{n=0}^{m} \tfrac{T^n x}{\lambda^{n+1}} = \lim_{m\to\infty}(\lambda - T)\sum_{n=0}^{m} \tfrac{T^n x}{\lambda^{n+1}} \\ &= \lim_{m\to\infty}\left(x - \tfrac{T^{m+1}x}{\lambda^{m+1}}\right) = x\,.\end{aligned}$$

Since $A \subseteq \rho(T)$, it follows that $\lambda - T$ is one-to-one for each $\lambda \in A$, and so for each $\lambda \in A$ we have $R(\lambda, T)x = \sum_{n=0}^{\infty} \frac{T^n x}{\lambda^{n+1}}$. Using the formula for the residue of this series, we get

$$x = T^0 x = \tfrac{1}{2\pi i}\int_{C_R} R(\lambda, T)x\, d\lambda = \left[\tfrac{1}{2\pi i}\int_{C_R} R(\lambda, T)\, d\lambda\right]x = P_\sigma(T)x \in Y_\sigma\,.$$

Consequently, $\{x \in X\colon\ \frac{T^n x}{r^n} \to 0\} \subseteq Y_\sigma$, and so $\{x \in X\colon\ \frac{T^n x}{r^n} \to 0\} = Y_\sigma$, as claimed. ■

Problem 6.4.15 (R. Melton). *For an entire function $f\colon \mathbb{C} \to \mathbb{C}$ having the series expansion $f(\lambda) = \sum_{n=0}^{\infty} a_n \lambda^n$ the following statements are equivalent.*

(a) *Each coefficient a_n of the expansion is non-negative.*

(b) *For each positive operator T on any Banach lattice the operator $f(T) = \sum_{n=1}^{\infty} a_n T^n$ is positive.*

(c) *There exists an infinite dimensional Banach lattice E such that $f(K)$ is positive for each positive compact operator K on E.*

Solution: It should be clear that (a) $\Longrightarrow$ (b) $\Longrightarrow$ (c). What needs verification is the implication (c) $\Longrightarrow$ (a). So, assume that (c) is true. The solution will be based upon the following three properties.

(1) *If every pairwise disjoint subset of an Archimedean Riesz space E is finite, then E is finite dimensional.*

(2) *Let $\{x_n\}$ be a disjoint sequence in a Banach lattice such that the series $\sum_{n=1}^{\infty} x_n$ is norm convergent. Then $x_n \geq 0$ holds for each n if and only if $\sum_{n=1}^{\infty} x_n \geq 0$.*

(3) *If x is a non-zero positive vector in a Banach lattice E, then there exists a positive linear functional $x^* \in E^*$ satisfying $x^*(x) = 1$ and $x^*(y) = 0$ for all $y \perp x$.*

We shall provide proofs to these claims below.

(1) We start by observing that for any $x > 0$ there exists an atom u such that $0 < u \leq x$. Indeed, if x is not an atom, then there exist two vectors $0 < u_1^1 \leq x$ and $0 < u_2^1 \leq x$ such that $u_1^1 \wedge u_2^1 = 0$. If neither of them is an atom, then there exist two vectors $0 < u_1^2 \leq u_1^1$ and $0 < u_2^2 \leq u_1^1$ such that $u_1^2 \wedge u_2^2 = 0$. Clearly, the vectors u_2^1, u_1^2, u_2^2 are pairwise disjoint. If neither u_1^2 nor u_2^2 is an atom, then we can continue this process. The process must terminate in a finite number of steps and produce a desired atom. Otherwise, if the process continues, then we can construct a countable disjoint sequence of non-zero vectors, contradicting our assumption.

Since any two atoms are linearly dependent or disjoint, it follows from our assumption that there exists a maximal finite number of pairwise disjoint atoms, say $u_1, u_2, \ldots, u_n$. We claim that the linear span V of $\{u_1, u_2, \ldots, u_n\}$ coincides with E.

To see this, let $x \in E^+$ be in the ideal generated by $\{u_1, u_2, \ldots, u_n\}$; we can assume that $0 \leq x \leq \sum_{i=1}^n u_i$. For each $1 \leq i \leq n$ pick some $\lambda_i \geq 0$ satisfying $x \wedge u_i = \lambda_i u_i$, and note that

$$x = x \wedge \Big(\sum_{i=1}^{n} u_i\Big) = x \wedge \Big(\bigvee_{i=1}^{n} u_i\Big) = \bigvee_{i=1}^{n} x \wedge u_i = \bigvee_{i=1}^{n} \lambda_i u_i = \sum_{i=1}^{n} \lambda_i u_i\,.$$

This shows that V coincides with the ideal generated by $\{u_1, u_2, \ldots, u_n\}$. Now an easy argument shows that the band generated by $\{u_1, u_2, \ldots, u_n\}$ also coincides with V; see also the solution to Problem 2.3.2. Finally, to see that $V = E$, it suffices to show that $V^{\mathrm{d}} = \{0\}$. If there exists some $0 < z \in V^{\mathrm{d}}$, then by the first part there exists an atom u such that $0 < u \leq z$. However, the latter leads to a contradiction since the collection of pairwise disjoint atoms $\{u_1, u_2, \ldots, u_n, u\}$ contradicts the maximality property of $\{u_1, u_2, \ldots, u_n\}$.

(2) If $x_n \geq 0$ for each $n \in \mathbb{N}$, then clearly $\sum_{n=1}^{\infty} x_n \geq 0$. Now assume that $\sum_{n=1}^{\infty} x_n \geq 0$. From $\sum_{n=1}^{\infty} x_n = \big|\sum_{n=1}^{\infty} x_n\big| = \sum_{n=1}^{\infty} |x_n|$, it follows that $\sum_{n=1}^{\infty} \big(|x_n| - x_n\big) = 0$. Taking into account that $|x_n| - x_n \geq 0$ for each n, the latter shows that $|x_n| - x_n = 0$ or $x_n = |x_n| \geq 0$ for each n.

(3) Pick any $y^* \in E^*$ with $y^*(x) = 1$. According to Problem 5.2.2 there exists a positive linear functional $x^* \in E^*$ such that $x^*(z) = y^*(z)$ for all $z \in E_x$ and $x^*(y) = 0$ for all $y \perp x$.

We now return to the solution of the problem. Since E is infinite dimensional, property (1) guarantees the existence of a pairwise disjoint sequence of positive unit vectors. By property (3) above, for each n there exists some positive linear functional $x_n^* \in E^*$ satisfying $x_n^*(x_n) = 1$ and $x_n^*(x_m) = 0$ for all $n \neq m$. Choose a sequence $\{\lambda_n\}$ of positive scalars satisfying $0 < \lambda_n < 1$ for each n and such that the formula

$$K = \sum_{n=1}^{\infty} \lambda_n \big(x_n^* \otimes x_{n+1}\big)$$

defines an operator on E. Clearly, K is a positive compact operator. Moreover, note that $Kx_1 = \lambda_1 x_2 \leq x_2$. An easy inductive argument shows that $K^n x_1 \leq x_{n+1}$ for each n.

By our hypothesis, the operator

$$f(K) = \sum_{n=0}^{\infty} a_n K^n$$

is positive. In particular, $0 \le f(K)x_1 = \sum_{n=0}^{\infty} a_n K^n x_1 \le a_0 x_1 + \sum_{n=1}^{\infty} \lambda_n a_n x_{n+1}$. Now according to property (2) above, we must have $a_n x_{n+1} \ge 0$ for each $n \ge 0$. This implies $a_n \ge 0$ for each $n \ge 0$, and the solution is finished. ∎

Chapter 7

Some Special Spectra

7.1. The Spectrum of a Compact Operator

Problem 7.1.1. *Show that the only compact operators with closed range are the finite-rank operators.*

Solution: Let $T\colon X \to Y$ be a compact operator between two Banach spaces, and assume that its range $R(T)$ is closed. The surjective operator $\hat{T}\colon X/N(T) \to R(T)$, defined by $\hat{T}[x] = Tx$, is an isomorphism. In particular, from $\hat{T}(U_{X/N(T)}) = T(U_X)$ and the compactness of T, it follows that the closed unit ball $U_{X/N(T)}$ of $X/N(T)$ is compact. Therefore, the vector space $X/N(T)$ is finite dimensional. This implies that $R(T)$ is finite dimensional, and so T is a finite-rank operator. ∎

Problem 7.1.2. *For each* $1 \le p \le \infty$ *define the integral positive operator* $T\colon L_p[0,1] \to L_p[0,1]$ *by*

$$Tf(t) = \int_0^1 (s^2 + t^2) f(s)\, ds\,,$$

where the interval $[0,1]$ *is equipped with its Lebesgue measure. Show that* T *is a finite-rank integral operator and find its spectrum.*

Solution: To see that T is a finite-rank operator, note that

$$Tf(t) = \int_0^1 (s^2 + t^2) f(s)\, ds = \Big[\int_0^1 f(s)\, ds\Big] t^2 + \int_0^1 s^2 f(s)\, ds\,.$$

This shows that T is, in fact, an operator of rank 2.

If we let $u_1(t) = t^2$ and $u_2(t) = 1$ for each $t \in [0,1]$ and $x_1^*(f) = \int_0^1 f(s)\, ds$ and $x_2^*(f) = \int_0^1 s^2 f(s)\, ds$ for each $f \in L_p[0,1]$, then $T = x_1^* \otimes u_1 + x_2^* \otimes u_2$. So, according to Theorem 7.6, the non-zero eigenvalues of T are the non-zero eigenvalues of the matrix

$$A = \begin{bmatrix} x_1^*(u_1) & x_1^*(u_2) \\ \\ x_2^*(u_1) & x_2^*(u_2) \end{bmatrix} = \begin{bmatrix} \int_0^1 t^2\,dt & \int_0^1 dt \\ \\ \int_0^1 t^4\,dt & \int_0^1 t^2\,dt \end{bmatrix} = \begin{bmatrix} \frac{1}{3} & 1 \\ \\ \frac{1}{5} & \frac{1}{3} \end{bmatrix}.$$

The characteristic polynomial of A is: $p_A(\lambda) = \det(A - \lambda I) = \lambda^2 - \frac{2}{3}\lambda - \frac{4}{45}$. Solving the equation $p_A(\lambda) = 0$ yields $\lambda = \frac{5\pm 3\sqrt{5}}{15}$. So, according to Theorem 7.6, the spectrum of the operator T is $\sigma(T) = \bigl\{0, \frac{5+3\sqrt{5}}{15}, \frac{5-3\sqrt{5}}{15}\bigr\}$.

The reader can verify that the eigenvalue $\lambda_1 = \frac{5+3\sqrt{5}}{15}$ corresponds to the eigenfunction (unique up to a constant multiple) $f_1(s) = 5s^2 + \sqrt{5}$ and the eigenvalue $\lambda_2 = \frac{5-3\sqrt{5}}{15}$ corresponds to the eigenfunction $f_2(s) = 5s^2 - \sqrt{5}$. ∎

Problem 7.1.3. *Consider the* **Volterra operator** $V\colon L_p[0,1] \to L_p[0,1]$ $(1 \le p < \infty)$ *defined by*

$$Vf(t) = \int_0^t f(s)\,ds\,.$$

Show that V is quasinilpotent, i.e., $\sigma(V) = \{0\}$, by proving that for each $\lambda \neq 0$ and each $g \in L_p[0,1]$ the equation $(\lambda - V)f = g$ has a unique solution given by

$$f(s) = \tfrac{1}{\lambda^2} e^{\frac{s}{\lambda}} \int_0^s g(\xi) e^{-\frac{\xi}{\lambda}}\,d\xi + \tfrac{1}{\lambda} g(s)\,.$$

Solution: Let $1 \le p < \infty$. Fix a non-zero complex number λ and define the operator $S\colon L_p[0,1] \to L_p[0,1]$ via the formula

$$Sf(t) = \tfrac{1}{\lambda^2} e^{\frac{t}{\lambda}} \int_0^t f(\xi) e^{-\frac{\xi}{\lambda}}\,d\xi + \tfrac{1}{\lambda} f(t)\,.$$

It should be clear that S is indeed a linear operator. What needs verification is that S maps $L_p[0,1]$ into $L_p[0,1]$ and that it is a bounded operator.

To see this, fix $f \in L_p[0,1]$, and define $g\colon [0,1] \to \mathbb{R}$ by

$$g(t) = \tfrac{1}{\lambda^2} e^{\frac{t}{\lambda}} \int_0^t f(\xi) e^{-\frac{\xi}{\lambda}}\,d\xi\,.$$

Clearly, g is a continuous function (see, for instance, [**8**, Problem 22.6, p. 176]), and so g belongs to $L_p[0,1]$. Moreover, if $M = \frac{1}{|\lambda|^2} e^{\frac{2}{|\lambda|}}$, then for each $t \in [0,1]$ it follows from Hölder's inequality that $|g(t)| \le M \int_0^1 |f(\xi)|\,d\xi \le M\|f\|_p$. Hence, $\|g\|_p \le M\|f\|_p$. Now from $Sf = g + \frac{1}{\lambda} f$, we obtain

$$\|Sf\|_p \le \|g\|_p + \tfrac{1}{|\lambda|}\|f\|_p \le \bigl(M + \tfrac{1}{|\lambda|}\bigr)\|f\|_p\,.$$

This establishes that S is a bounded operator.

Next, we claim that $(\lambda - V)Sf = S(\lambda - V)f = f$ for each function $f \in L_p[0,1]$. Since the continuous functions are norm dense in $L_p[0,1]$ (here we use the fact that

$1 \le p < \infty$), it suffices to prove that $(\lambda - V)Sf = S(\lambda - V)f = f$ holds true for each function $f \in C[0,1]$. So, let $f \in C[0,1]$. Then for each $t \in [0,1]$ we have

$$\begin{aligned}
\big[(\lambda - V)Sf\big](t) &= \tfrac{1}{\lambda}e^{\frac{t}{\lambda}} \int_0^t f(\xi)e^{-\frac{\xi}{r\lambda}}\,d\xi + f(t) \\
&\quad - \int_0^t \Big[\tfrac{1}{\lambda^2}e^{\frac{s}{\lambda}} \int_0^s f(\xi)e^{-\frac{\xi}{\lambda}}\,d\xi + \tfrac{1}{\lambda}f(s)\Big]\,ds \\
&= \tfrac{1}{\lambda}e^{\frac{t}{\lambda}} \int_0^t f(\xi)e^{-\frac{\xi}{\lambda}}\,d\xi + f(t) \\
&\quad - \int_0^t \Big[\int_0^s f(\xi)e^{-\frac{\xi}{\lambda}}\,d\xi\Big]\,d\big(\tfrac{1}{\lambda}e^{\frac{s}{\lambda}}\big) - \tfrac{1}{\lambda}\int_0^t f(s)\,ds \\
&= \tfrac{1}{\lambda}e^{\frac{t}{\lambda}} \int_0^t f(\xi)e^{-\frac{\xi}{\lambda}}\,d\xi + f(t) - \Big[\tfrac{1}{\lambda}e^{\frac{s}{\lambda}} \int_0^s f(\xi)e^{-\frac{\xi}{\lambda}}\,d\xi\Big]\Big|_{s=0}^{s=t} \\
&\quad + \int_0^t \tfrac{1}{\lambda}e^{\frac{s}{\lambda}} f(s)e^{-\frac{s}{\lambda}}\,ds - \tfrac{1}{\lambda}\int_0^t f(s)\,ds \\
&= \tfrac{1}{\lambda}e^{\frac{t}{\lambda}} \int_0^t f(\xi)e^{-\frac{\xi}{\lambda}}\,d\xi + f(t) - \tfrac{1}{\lambda}e^{\frac{t}{\lambda}} \int_0^t f(\xi)e^{-\frac{\xi}{\lambda}}\,d\xi \\
&\quad + \tfrac{1}{\lambda}\int_0^t f(s)\,ds - \tfrac{1}{\lambda}\int_0^t f(s)\,ds \\
&= f(t)\,.
\end{aligned}$$

Similarly, for each $f \in C[0,1]$ and each $t \in [0,1]$ we have

$$\begin{aligned}
\big[S(\lambda - V)f\big](t) &= \lambda Sf(t) - S(Vf)(t) \\
&= \tfrac{1}{\lambda}e^{\frac{t}{\lambda}} \int_0^t f(\xi)e^{-\frac{\xi}{\lambda}}\,d\xi + f(t) \\
&\quad - \tfrac{1}{\lambda^2}e^{\frac{t}{\lambda}} \int_0^t \Big[\int_0^\xi f(s)\,ds\Big]e^{-\frac{\xi}{\lambda}}\,d\xi - \tfrac{1}{\lambda}\int_0^t f(s)\,ds \\
&= \tfrac{1}{\lambda}e^{\frac{t}{\lambda}} \int_0^t f(\xi)e^{-\frac{\xi}{\lambda}}\,d\xi + f(t) \\
&\quad + \tfrac{1}{\lambda}e^{\frac{t}{\lambda}} \int_0^t \Big[\int_0^\xi f(s)\,ds\Big]\,d\big(e^{-\frac{\xi}{\lambda}}\big) - \tfrac{1}{\lambda}\int_0^t f(s)\,ds \\
&= \tfrac{1}{\lambda}e^{\frac{t}{\lambda}} \int_0^t f(\xi)e^{-\frac{\xi}{\lambda}}\,d\xi + f(t) + \Big[\tfrac{1}{\lambda}e^{\frac{t}{\lambda}}e^{-\frac{\xi}{\lambda}} \int_0^\xi f(s)\,ds\Big]\Big|_{\xi=0}^{\xi=t} \\
&\quad - \tfrac{1}{\lambda}e^{\frac{t}{\lambda}} \int_0^t f(\xi)e^{-\frac{\xi}{\lambda}}\,d\xi - \tfrac{1}{\lambda}\int_0^t f(s)\,ds \\
&= \tfrac{1}{\lambda}e^{\frac{t}{\lambda}} \int_0^t f(\xi)e^{-\frac{\xi}{\lambda}}\,d\xi + f(t) + \tfrac{1}{\lambda}\int_0^t f(s)\,ds \\
&\quad - \tfrac{1}{\lambda}e^{\frac{t}{\lambda}} \int_0^t f(\xi)e^{-\frac{\xi}{\lambda}}\,d\xi - \tfrac{1}{\lambda}\int_0^t f(s)\,ds \\
&= f(t)\,.
\end{aligned}$$

The above equations show that $S(\lambda - V)f = (\lambda - V)Sf = f$ holds for each $f \in L_p[0,1]$. Therefore, $S = (\lambda - V)^{-1}$, and the desired conclusion follows. ∎

Problem 7.1.4. *Show directly (i.e., without using Problem 7.1.3) that the Volterra operator has no eigenvalues—and conclude from this that the Volterra operator is quasinilpotent.*

Solution: Consider the Volterra operator $V\colon L_p[0,1] \to L_p[0,1]$ $(1 \le p < \infty)$ defined by

$$Vf(t) = \int_0^t f(s)\,ds\,.$$

First, let us show that T does not have 0 as an eigenvalue, i.e., that T is one-to-one. To see this, assume $Tf = 0$ or

$$\int_0^t f(s)\,ds = 0 \quad \text{for all} \quad t \in [0,1]\,. \tag{$\star$}$$

Then for each subinterval (closed, open, half-closed, or half-open) $[a,b]$ of $[0,1]$ we have $\int_a^b f(t)\,dt = \int_0^b f(t)\,dt - \int_0^a f(t)\,dt = 0$. Since every open subset of $[0,1]$ is an at most countable disjoint union of intervals in $[0,1]$, it follows that $\int_O f(t)\,dt = 0$ for each open subset O of $[0,1]$. Since the Lebesgue measure is regular, the latter implies $\int_A f(t)\,dt = 0$ for every Lebesgue measurable subset A of $[0,1]$.

Now consider the two Lebesgue measurable sets $B = \{x \in [0,1]\colon\ f(x) > 0\}$ and $C = \{x \in [0,1]\colon\ f(x) < 0\}$. From $\int_B f(t)\,dt = \int_C f(t)\,dt = 0$, it follows that B and C are both Lebesgue null sets. This implies $f = 0$ a.e., and so T is one-to-one.

Next, assume that for some $\lambda \neq 0$ and some $f \in L_p[0,1]$ we have

$$\int_0^t f(s)\,ds = \lambda f(t) \tag{$\star\star$}$$

for almost all $t \in [0,1]$. Since the function $F(t) = \int_0^t f(s)\,ds$ is (absolutely) continuous (see, for instance [**8**, Problem 22.6, p. 176]) and $\lambda \neq 0$, it follows from $(\star\star)$ that f can be assumed to be a continuous function. But then from $(\star\star)$ and the Fundamental Theorem of Calculus, it also follows that the function f is differentiable and $\lambda f'(t) = f(t)$ for each $t \in [0,1]$. Solving this differential equation yields $f(t) = Ce^{\frac{t}{\lambda}}$. Now one more glance at $(\star\star)$ shows that $f(0) = 0$. This implies $C = 0$ or $f = 0$. Therefore, V does not have any non-zero eigenvalues.

Since the Volterra operator is a compact operator (see Example 7.8), it follows from Theorem 7.3 that $\sigma(V) = \{0\}$. That is, the Volterra operator is quasinilpotent. ∎

Problem 7.1.5. *Let $T\colon E \to E$ be a positive contraction on a Banach lattice. Show that 1 belongs to the approximate point spectrum of T if and only if $r(T) = 1$.*

Solution: Let $T\colon E \to E$ be a positive operator which is also a contraction, i.e., $\|T\| \le 1$. Then $\|T^n\| \le \|T\|^n \le 1$, and so $r(T) \le \|T^n\|^{\frac{1}{n}} \le 1$.

Assume that $1 \in \sigma_a(T)$. This implies $1 \le r(T)$, and hence $r(T) = 1$. For the converse, suppose that $r(T) = 1$. By Theorem 7.9, we know that 1 belongs to the spectrum of T (and hence, as a boundary point, it also belongs to the approximate point spectrum of T). ∎

Problem 7.1.6. *Show that the hypothesis $r(T) > 0$ cannot be dropped from the Krein–Rutman Theorem* 7.10.

Solution: Let $V: L_1[0,1] \to L_1[0,1]$ be the standard **Volterra operator** defined by $Vf(t) = \int_0^t f(s)\,ds$ for each $f \in L_1[0,1]$ and all $t \in [0,1]$. According to Problem 7.1.4, V is a compact quasinilpotent positive operator, i.e., $r(T) = 0$. We claim that 0 cannot be an eigenvalue of the operator V.

To see this, assume that a function $f \in L_1[0,1]$ satisfies $Vf(t) = \int_0^t f(s)\,ds = 0$ for all $t \in [0,1]$. This implies $\int_a^b f(s)\,ds = 0$ for all $a, b \in [0,1]$, and from this we conclude that $f(s) = 0$ for almost all $s \in [0,1]$. Consequently, 0 cannot be an eigenvalue of V. (See also Problem 7.1.4.) ∎

Problem 7.1.7. *Show that the compactness hypothesis of the operator in the Krein–Rutman Theorem* 7.10 *cannot be dropped.*

Solution: Consider the shift operator $T: \ell_2 \to \ell_2$ defined by

$$T(x_1, x_2, \ldots) = (0, x_1, x_2, \ldots)\,.$$

Then T is a non-compact positive operator—in fact, it is a lattice isometry. So, $\|T^n\| = 1$ holds for each n, and so $r(T) = \lim_{n\to\infty} \|T^n\|^{\frac{1}{n}} = 1$.

Now note that if $T(x_1, x_2, \ldots) = (0, x_1, x_2, \ldots) = (x_1, x_2, \ldots)$, then it is easy to see that $x_i = 0$ for each i. Hence, $x = (x_1, x_2, \ldots) = 0$, and this shows that $r(T) = 1$ cannot be an eigenvalue of T. ∎

Problem 7.1.8. *Let $T: E \to E$ be a positive operator on a Banach lattice and let p be a polynomial with non-negative coefficients. Show that $r(p(T)) = p(r(T))$.*

Solution: Assume that a polynomial p has non-negative coefficients, and let $T: E \to E$ be a positive operator. Then the operator $p(T)$ is also positive, and so, by Theorem 7.9, we know that $r\big(p(T)\big) \in \sigma\big(p(T)\big)$. At the same time, from the Spectral Mapping Theorem or Problem 6.4.1, we know that $p\big(\sigma(T)\big) = \sigma\big(p(T)\big)$. Since the polynomial p has non-negative coefficients, we have $\big|p(\lambda)\big| \le p\big(|\lambda|\big) \le p\big(r(T)\big)$ for each $\lambda \in \sigma(T)$. Therefore,

$$r\big(p(T)\big) = \max\big\{|\mu| \colon\ \mu \in \sigma\big(p(T)\big)\big\} = \max\big\{\big|p(\lambda)\big| \colon\ \lambda \in \sigma(T)\big\} = p\big(r(T)\big)\,,$$

as claimed. ∎

Problem 7.1.9. *Prove the following generalization of the Krein–Rutman Theorem* 7.10. *If a power compact positive operator $T: E \to E$ satisfies $r(T) > 0$, then there exists some vector $x > 0$ such that $Tx = r(T)x$.*

Solution: Assume that $r = r(T) > 0$ and that T^k is compact for some $k > 1$. By Theorem 7.10, there exists some $x > 0$ such that $T^k x = r^k x$. Now consider the operator $S = \sum_{i=0}^{k-1} r^i T^{k-1-i}$, and note that the vector $y = Sx$ satisfies

$$y = Sx = \sum_{i=0}^{k-1} r^i T^{k-1-i} x = \sum_{i=0}^{k-2} r^i T^{k-1-i} x + r^{k-1} x \ge r^{k-1} x > 0\,.$$

Now we have

$$\begin{aligned} Ty - ry &= TSx - rSx = (T-r)Sx = (T-r)\Big[\sum_{i=0}^{k-1} r^i T^{k-1-i}\Big]x \\ &= (T^k - r^k)x = 0\,. \end{aligned}$$

That is, $Ty = ry$, and so r is an eigenvalue having a positive eigenvector. ∎

Problem 7.1.10. *Prove Lemma 4.18 using the Krein–Rutman theorem. That is, show by using the Krein–Rutman theorem that if Ω is a compact Hausdorff space without isolated points, then the only weakly compact multiplication operator on $C(\Omega)$ is the zero operator.*

Solution: Assume that Ω does not have isolated points and suppose by way of contradiction that there exists a non-zero weakly compact multiplication operator M_ϕ on $C(\Omega)$. Then $M_\phi M_\phi = M_{\phi^2}$ is a non-zero compact positive multiplication operator; see [**6**, p. 337]. So, replacing ϕ by ϕ^2, we can suppose that $\phi > 0$ and that M_ϕ is compact. Clearly,

$$r = r(M_\phi) = \lim_{n\to\infty} \|M_\phi^n\|^{\frac{1}{n}} = \lim_{n\to\infty} \|M_{\phi^n}\|^{\frac{1}{n}} = \lim_{n\to\infty} (\|\phi^n\|_\infty)^{\frac{1}{n}} = \|\phi\|_\infty > 0\,.$$

Consequently, by the Krein–Rutman Theorem 7.10 there exists some $0 < x \in C(\Omega)$ such that $M_\phi(x) = \phi x = rx$, or $(\phi - r)x = 0$. So, if $V = \{\omega \in \Omega\colon\ x(\omega) > 0\}$, then V is a non-empty open subset of Ω and $\phi(\omega) = r$ for all for all $\omega \in \overline{V}$.

Now let $\{f_n\} \subseteq C(\overline{V})$ be any sequence in $C(\overline{V})$ satisfying $\|f_n\|_\infty \le 1$ for each n. By Tietze's extension theorem, we can assume that each f_n is defined on all of Ω. Now the compactness of M_ϕ guarantees the existence of a subsequence $\{g_n\}$ of $\{f_n\}$ and a function $f \in C(\Omega)$ such that $\|M_\phi(g_n) - f\|_\infty \to 0$. This implies that $\{g_n\}$ converges uniformly to $\frac{1}{r}f$ on $\overline{V}$ or that $g_n \to \frac{1}{r}f$ holds in $C(\overline{V})$. This shows that the closed unit ball of $C(\overline{V})$ is compact, and hence $C(\overline{V})$ is a finite dimensional space. In turn, this implies that $\overline{V}$ (and hence the open set V) must be a finite set. But then Ω has isolated points, which is a contradiction. Consequently, there are no non-zero weakly compact multiplication operators on $C(\Omega)$. ∎

Problem 7.1.11. *Show that there exist a Banach space X and a non-empty compact subset K of $\mathbb{C}$ such that $\sigma(T) \neq K$ for each $T \in \mathcal{L}(X)$.*

Solution: Let X be the *HI*-space introduced in Problem 4.5.7. As explained there, each operator $T \in \mathcal{L}(X)$ is of the form $T = \lambda I + S$, where λ is a constant and S is a strictly singular operator. Therefore, $\sigma(T) = \lambda + \sigma(S)$, and the spectrum of T is at most countable since by Theorem 7.11 the spectrum of S is at most countable. Consequently, if K is an arbitrary uncountable compact subset of $\mathbb{C}$, then no operator in $\mathcal{L}(X)$ can have K as its spectrum. (In connection with this problem see also Problem 6.1.15.) ∎

Problem 7.1.12. *Assume that $S \in \mathcal{L}(X)$ is strictly singular and that an operator $T \in \mathcal{L}(X)$ has an at most countable spectrum. Show that the spectrum of $S+T$ is at most countable and that zero and the points of $\sigma(T)$ are the only possible accumulation points of $\sigma(S+T)$.*

Solution: The solution is a modification of the proof of Theorem 7.11. For any $\lambda \neq 0$ that does not belong to $\sigma(T)$ the operator $\lambda - T$ is invertible and hence Fredholm of index $i(\lambda - T) = 0$. Therefore, by Theorem 4.63, the operator $\lambda - T - S$ is also Fredholm of index zero. Now Lemma 4.55(3) guarantees that there exists some open disk $D(\lambda)$ centered at λ such that for each $\mu \in D(\lambda)$ the operator $\mu - T - S$ is Fredholm and the functions $n(\mu - T - S)$ and $d(\mu - T - S)$ are constant on the set $D(\lambda) \setminus \{\lambda\}$.

Fix a non-zero $\lambda \notin \sigma(T)$. We claim that there exists a half-ray that starts at λ and does not meet the set $\sigma(T) \cup \{0\}$. Indeed, otherwise, every half-ray starting at λ and not containing zero would intersect the spectrum of T, and so $\sigma(T)$ would be uncountable, which is a contradiction. Consequently, we can find some λ_1 such that $|\lambda_1| > \|T + S\|$ and the line segment $L = [\lambda, \lambda_1]$ joining λ and λ_1 contains neither zero nor any point from $\sigma(T)$.

Since the operator $\mu - T - S$ is invertible for $|\mu| > \|T + S\|$, it follows that $n(\mu - T - S) = 0$ and $d(\mu - T - S) = 0$ for each μ that is close enough to λ_1, and hence, by our choice of the disk $D(\lambda_1)$, we have $n(\mu - T - S) = d(\mu - T - S) = 0$ for every $\mu \in D(\lambda_1)$.

The compactness and connectedness of L guarantee that there exists a finite number of disks $D(\xi_1), \dots, D(\xi_k)$ that cover L and satisfy $D(\xi_j) \cap D(\xi_{j+1}) \neq \emptyset$ for $1 \leq j \leq k-1$; see also Problem 1.5.7. Using this chain of disks connecting the point λ_1 with λ and keeping in mind that the functions $n(\mu - T - S)$ and $d(\mu - T - S)$ are constant on each set $D(\xi_i) \setminus \{\xi_i\}$, we can conclude that $n(\mu - T - S)$ and $d(\mu - T - S)$ are constant on the union $D = \bigcup_{i=1}^{k} [D(\xi_i) \setminus \{\xi_i\}]$. Moreover, since $n(\mu - T - S) = d(\mu - T - S) = 0$ for each $\mu \in D(\lambda_1) \setminus \{\lambda_1\}$, it follows that $n(\mu - T - S) = d(\mu - T - S) = 0$ for each $\mu \in D$. This implies, in particular, that the operator $\mu - T - S$ is invertible for each $\mu \in D(\lambda) \setminus \{\lambda\}$ and so if λ belongs to $\sigma(T + S)$, then λ is an isolated point of $\sigma(T + S)$.

To finish the proof, it suffices to show that for every $\epsilon > 0$ the compact set

$$A_\epsilon = \{\lambda \in \sigma(S + T) \colon \ |\lambda - \mu| \geq \epsilon \text{ for all } \mu \in \sigma(T) \cup \{0\}\}$$

is finite. Indeed, since A_ϵ is compact and disjoint from $\sigma(T) \cup \{0\}$, the previous arguments guarantee the existence of a finite number of points $\mu_1, \dots, \mu_n$ such that each of them is outside of $\sigma(T)$ and the disks $D(\mu_i)$ cover A_ϵ. This implies that $A_\epsilon \subseteq \{\mu_1, \dots, \mu_n\}$. That is, A_ϵ is finite, and the solution is finished. ∎

Problem 7.1.13 (T. Oikhberg and V. Troitsky). *Let ϕ be a non-zero multiplicative functional on the Banach algebra $\mathcal{L}(X)$, where X is an arbitrary Banach space with* $\dim X > 1$. *Show that:*

(a) *For each $T \in \mathcal{L}(X)$ we have $\phi(T) \in \sigma(T)$.*

(b) *If for $T \in \mathcal{L}(X)$ there is a reducing pair of closed subspaces (V, W) such that* $\dim V < \infty$ *and T vanishes on W, then $\phi(T) = 0$.*

(c) *If $T \in \mathcal{L}(X)$ is a strictly singular operator (in particular, if T is compact), then $\phi(T) = 0$.*

Solution: (a) Since ϕ is non-zero, we know that $\phi(I) = 1$. Fix $T \in \mathcal{L}(X)$ and pick any $\lambda \in \rho(T)$. So, there exists an operator $S \in \mathcal{L}(X)$ such that $(\lambda - T)S = I$.

This implies that $[\lambda - \phi(T)]\phi(S) = \phi(I) = 1$, and hence $\lambda - \phi(T) \neq 0$. That is, $\phi(T) \neq \lambda$ for each $\lambda \in \rho(T)$. Consequently, $\phi(T) \in \mathbb{C} \setminus \rho(T) = \sigma(T)$.

(b) If $\dim X < \infty$, then the result is a consequence of Problem 4.3.7. So, let $\dim X = \infty$. Replacing, if necessary, V by a larger subspace, we can assume without loss of generality that $\dim V > 1$.

Consider the collection $\mathcal{B}$ of all operators $S \in \mathcal{L}(X)$ that leave V invariant and vanish on W. Clearly, the functional $\psi = \phi|_{\mathcal{B}}$ is multiplicative on $\mathcal{B}$. Furthermore, the algebra $\mathcal{B}$ is obviously isomorphic to $\mathcal{L}(V)$ and $T \in \mathcal{B}$. Since V is finite dimensional and $\dim V > 1$, it follows from Problem 4.3.7 that $\psi = 0$. Thus, $\phi(T) = \psi(T) = 0$.

(c) Let $T \in \mathcal{L}(X)$ be a strictly singular operator. From Theorem 7.11 we know that the spectrum $\sigma(T)$ is at most countable and that every non-zero point in $\sigma(T)$ is isolated and is an eigenvalue of finite (geometric) multiplicity.

Therefore for each $\epsilon > 0$ the set $A_\epsilon = \{\lambda \in \sigma(T)\colon\ |\lambda| > \epsilon\}$ is finite and consists of isolated points of the spectrum $\sigma(T)$. Thus, A_ϵ is a spectral set of T. Consider the spectral subspaces V and W of X corresponding to A_ϵ and $B_\epsilon = \sigma(T) \setminus A_\epsilon$, respectively. Theorem 6.34 guarantees that $V \oplus W = X$ and also that $\sigma(T|_V) = A_\epsilon$ and $\sigma(T|_W) = B_\epsilon$.

Consider the bounded operator $T_1\colon X \to X$ defined by $T_1 v = Tv$ for $v \in V$ and $T_1 w = 0$ for $w \in W$. Clearly, the pair (V, W) reduces the operator T_1. Since $\dim V < \infty$, part (b) applies and it follows that $\phi(T_1) = 0$.

Finally, consider the operator $T_2 = T - T_1$. In view of the properties of the pair of the spectral subspaces (V, W), the spectrum of T_2 is contained in the set B_ϵ and therefore, by (a), we have $\phi(T_2) \in B_\epsilon$, whence $|\phi(T_2)| \leq \epsilon$. At the same time we have $\phi(T_2) = \phi(T) - \phi(T_1) = \phi(T)$. That is, we have established that $|\phi(T)| \leq \epsilon$, and hence $\phi(T) = 0$ since $\epsilon > 0$ is arbitrary. ∎

7.2. Turning Approximate Eigenvalues into Eigenvalues

Problem 7.2.1. *Show that if X is an arbitrary Banach space, then for each $x = (x_1, x_2, \dots) \in \ell_\infty(X)$ the norm of the vector $[x] \in \widehat{X}$ is given by*

$$\big\|[x]\big\| = \inf_{y \in [x]} \|y\|_\infty = \limsup_{n \to \infty} \|x_n\|\,.$$

Solution: Let X be an arbitrary Banach space and let $x = (x_1, x_2, \dots) \in \ell_\infty(X)$. Put $\alpha = \limsup_{n \to \infty} \|x_n\|$. If $y = (y_1, y_2, \dots) \in [x]$, then $z_n = x_n - y_n \to 0$, and so from $\|x_n\| = \|y_n + z_n\| \leq \|y_n\| + \|z_n\| \leq \|y\|_\infty + \|z_n\|$, we get

$$\alpha = \limsup_{n \to \infty} \|x_n\| \leq \limsup_{n \to \infty} \big[\|y\|_\infty + \|z_n\|\big] = \|y\|_\infty\,.$$

Hence, $\alpha \leq \inf_{y \in [x]} \|y\|_\infty = \big\|[x]\big\|$.

Now let $\epsilon > 0$ be fixed. Then there exists some k such that $\|x_n\| < \alpha + \epsilon$ holds for all $n \geq k$. If we consider the vector $y = (y_1, y_2, \dots) \in \ell_\infty(X)$ defined by

$y_i = 0$ if $1 \le i < k$ and $y_i = x_i$ if $i \ge k$, then $y \in [x]$ and $\|y\|_\infty \le \alpha + \epsilon$. This implies $\|[x]\| \le \|y\|_\infty \le \alpha + \epsilon$. Since $\epsilon > 0$ is arbitrary, $\|[x]\| \le \|y\|_\infty \le \alpha$. Hence, $\|[x]\| \le \|y\|_\infty = \alpha$, as desired. ∎

Problem 7.2.2. *Assume that $\{X_n\}$ is a sequence of real Banach spaces. Show that (with the obvious identifications) for each $1 \le p \le \infty$ we have*

$$\ell_p((X_1)_{\mathbf{c}} \oplus (X_2)_{\mathbf{c}} \oplus \cdots) = \ell_p(X_1 \oplus X_2 \oplus \cdots) \oplus \imath\ell_p(X_1 \oplus X_2 \oplus \cdots).$$

Also show that

$$c_0\big((X_1)_{\mathbf{c}} \oplus (X_2)_{\mathbf{c}} \oplus \cdots\big) = c_0(X_1 \oplus X_2 \oplus \cdots) \oplus \imath c_0(X_1 \oplus X_2 \oplus \cdots).$$

Solution: We establish the result for the ℓ_p-sums when $1 \le p < \infty$. To this end, let $z = (x_1 + \imath y_1, x_2 + \imath y_2, \ldots) \in \ell_p((X_1)_{\mathbf{c}} \oplus (X_2)_{\mathbf{c}} \oplus \cdots)$. Put $x = (x_1, x_2, \ldots)$ and $y = (y_1, y_2, \ldots)$, and note that both sequences x and y belong to $\ell_p(X_1 \oplus X_2 \oplus \cdots)$. This allows us to define a mapping

$$\mathcal{T}\colon \ell_p((X_1)_{\mathbf{c}} \oplus (X_2)_{\mathbf{c}} \oplus \cdots) \to \ell_p(X_1 \oplus X_2 \oplus \cdots) \oplus \imath\ell_p(X_1 \oplus X_2 \oplus \cdots)$$

via the formula $\mathcal{T}(z) = x + \imath y$. We claim that $\mathcal{T}$ is a surjective (topological) isomorphism. For the additivity of $\mathcal{T}$ note that if $z = (x_1 + \imath y_1, x_2 + \imath y_2, \ldots)$ and $u = (a_1 + \imath b_1, a_2 + \imath b_2, \ldots)$, then we have

$$\begin{aligned} \mathcal{T}(z+u) &= (x_1 + a_1, x_2 + a_2, \ldots) + \imath(y_1 + b_1, y_2 + b_2, \ldots) \\ &= (x + \imath y) + (a + \imath b) = \mathcal{T}(z) + \mathcal{T}(u). \end{aligned}$$

For the homogeneity of $\mathcal{T}$ let $\lambda = \alpha + \imath\beta$, and note that

$$\begin{aligned} \mathcal{T}(\lambda z) &= \mathcal{T}\big((\alpha + \imath\beta)(x_1 + \imath y_1), (\alpha + \imath\beta)(x_2 + \imath y_2), \ldots\big) \\ &= \mathcal{T}\big(\alpha x_1 - \beta y_1 + \imath(\beta x_1 + \alpha y_1), \alpha x_2 - \beta y_2 + \imath(\beta x_2 + \alpha y_2), \ldots\big) \\ &= (\alpha x_1 - \beta y_1, \alpha x_2 - \beta y_2, \ldots) + \imath(\beta x_1 + \alpha y_1, \beta x_2 + \alpha y_2, \ldots) \\ &= (\alpha x - \beta y) + \imath(\beta x + \alpha y) = (\alpha + \imath\beta)(x + \imath y) \\ &= \lambda \mathcal{T} z. \end{aligned}$$

We leave it to the reader to verify that $\mathcal{T}$ is surjective and one-to-one.

Finally, we shall show that $\mathcal{T}$ is continuous. To see this, let

$$z = (x_1 + \imath y_1, x_2 + \imath y_2, \ldots) \in \ell_p((X_1)_{\mathbf{c}} \oplus (X_2)_{\mathbf{c}} \oplus \cdots),$$

and consider the vectors $x = (x_1, x_2, \ldots)$ and $y = (y_1, y_2, \ldots)$. Then for each $\theta \in \mathbb{R}$ we have

$$\begin{aligned} \|x\cos\theta + y\sin\theta\|_p &= \Big(\sum_{n=1}^{\infty} \|x_n\cos\theta + y_n\sin\theta\|^p\Big)^{\frac{1}{p}} \\ &\le \Big(\sum_{n=1}^{\infty} \|x_n + \imath y_n\|^p\Big)^{\frac{1}{p}} = \|z\|_p. \end{aligned}$$

Hence, the norm of $\mathcal{T}(z) = x + \imath y$ in $\ell_p(X_1 \oplus X_2 \oplus \cdots) \oplus \imath\ell_p(X_1 \oplus X_2 \oplus \cdots)$ satisfies

$$\|\mathcal{T}(z)\| = \sup_{\theta\in\mathbb{R}} \|x\cos\theta + y\sin\theta\|_p \le \|z\|_p.$$

On the other hand, we know that $\|\mathcal{T}(z)\| > \frac{1}{2}(\|x\|_p + \|y\|_p) \ge \frac{1}{2}\|z\|_p$. That is,

$$\tfrac{1}{2}\|z\|_p \le \|\mathcal{T}(z)\| \le \|z\|_p$$

holds for each $z \in \ell_p((X_1)_{\mathbf{c}} \oplus (X_2)_{\mathbf{c}} \oplus \cdots)$, proving that T is a surjective topological isomorphism. ∎

Problem 7.2.3. *Let $\{X_n\}$ be a sequence of real (or complex) Banach spaces, and assume that $1 < p, q < \infty$ satisfy $\frac{1}{p} + \frac{1}{q} = 1$. Show that*

$$(X_1 \oplus X_2 \oplus \cdots)_p^* = (X_1^* \oplus X_2^* \oplus \cdots)_q\,,$$

where the equality is understood subject to the natural duality

$$\langle x, x^* \rangle = \sum_{n=1}^{\infty} \langle x_n, x_n^* \rangle$$

for $x = (x_1, x_2, \ldots) \in (X_1 \oplus X_2 \oplus \cdots)_p$, $x^* = (x_1^*, x_2^*, \ldots) \in (X_1^* \oplus X_2^* \oplus \cdots)_q$.

Similarly, show that

$$\begin{aligned} (X_1 \oplus X_2 \oplus \cdots)_1^* &= (X_1^* \oplus X_2^* \oplus \cdots)_\infty \quad \text{and} \\ (X_1 \oplus X_2 \oplus \cdots)_0^* &= (X_1^* \oplus X_2^* \oplus \cdots)_1\,. \end{aligned}$$

Solution: We shall establish the result for $1 < p, q < \infty$ only. This will be done by showing that the mapping

$$F\colon (X_1^* \oplus X_2^* \oplus \cdots)_q \to (X_1 \oplus X_2 \oplus \cdots)_p^*\,,$$

defined by $Fx^*(x_1, x_2, \ldots) = \sum_{n=1}^{\infty} \langle x_n, x_n^* \rangle = \sum_{n=1}^{\infty} x_n^*(x_n)$, is a surjective linear isometry, where $x^* = (x_1^*, x_2^*, \ldots) \in (X_1^* \oplus X_2^* \oplus \cdots)_q$.

To this end, fix $x^* = (x_1^*, x_2^*, \ldots) \in (X_1^* \oplus X_2^* \oplus \cdots)_q$. Then for each vector $x = (x_1, x_2, \ldots)$ in $(X_1 \oplus X_2 \oplus \cdots)_p$ we have $|x_n^*(x_n)| \le \|x_n^*\| \cdot \|x_n\|$, and so for each k we get the estimate

$$\Big|\sum_{n=1}^{k} \langle x_n, x_n^* \rangle\Big| \le \sum_{n=1}^{k} \|x_n^*\| \cdot \|x_n\| \le \Big(\sum_{n=1}^{\infty} \|x_n^*\|^q\Big)^{\frac{1}{q}} \cdot \Big(\sum_{n=1}^{\infty} \|x_n\|^p\Big)^{\frac{1}{p}} = \|x^*\|_q \cdot \|x\|_p\,.$$

This implies that the formula $Fx^*(x) = \sum_{n=1}^{\infty} \langle x_n, x_n^* \rangle$ defines a continuous linear functional on $(X_1 \oplus X_2 \oplus \cdots)_p$ and

$$\|Fx^*\| \le \|x^*\|_q\,. \qquad (\star)$$

Moreover, it should be clear that $Fx^* + Fy^* = F(x^* + y^*)$ and $F(\lambda x^*) = \lambda Fx^*$, that is, F is a linear operator.

Now let $f \in (X_1 \oplus X_2 \oplus \cdots)_p^*$. For each n define $x_n^*\colon X_n \to \mathbb{R}$ by

$$x_n^*(x_n) = f(0, 0, \ldots, 0, x_n, 0, 0, \ldots)\,,$$

where the vector $x_n \in X_n$ occupies the nth coordinate. Then $x_n^* \in X_n^*$, and moreover $f(x_1, x_2, \ldots) = \sum_{n=1}^{\infty} x_n^*(x_n)$ holds for each vector $x = (x_1, x_2, \ldots)$ in $(X_1 \oplus X_2 \oplus \cdots)_p$. Next, fix $0 < \epsilon < 1$. For each n choose some $y_n \in X_n$ satisfying $\|y_n\| = 1$ and $x_n^*(y_n) \ge \epsilon \|x_n^*\|$. Put $z_n = \|x_n^*\|^{q-1} y_n$, and note that for each k we have

$$\begin{aligned}\epsilon\Big(\sum_{n=1}^{k}\|x_n^*\|^q\Big) &= \sum_{n=1}^{k}\|x_n^*\|^{q-1}\epsilon\|x_n^*\| \le \sum_{n=1}^{k}\|x_n^*\|^{q-1}x_n^*(y_n)\\ &= f(z_1,z_2,\dots,z_k,0,0,0,\dots)\\ &\le \|f\|\Big(\sum_{n=1}^{k}\|z_n\|^p\Big)^{\frac{1}{p}} = \|f\|\Big(\sum_{n=1}^{k}\|x_n^*\|^q\Big)^{\frac{1}{p}}.\end{aligned}$$

Consequently, $\epsilon\big(\sum_{n=1}^{k}\|x_n^*\|^q\big)^{\frac{1}{q}} \le \|f\|$ holds for each k and each $0<\epsilon<1$. It follows that $x^* = (x_1^*, x_2^*, \dots) \in (X_1^* \oplus X_2^* \oplus \cdots)_q$, $f = Fx^*$, and $\|x^*\|_q \le \|Fx^*\|$. Now use $(\star)$ to conclude that $\|Fx^*\| = \|x^*\|_q$. Therefore, F is a surjective linear isometry—which is also a lattice isometry if each X_n is a Banach lattice. ∎

Problem 7.2.4. *Let $\{X_n\}$ be a sequence of all real or all complex Banach spaces, and consider the Banach space $X = (X_1 \oplus X_2 \oplus \cdots)_p$ for some $1 \le p < \infty$. For each $\mathcal{T} = (T_1, T_2, \dots) \in (\mathcal{L}(X_1) \oplus \mathcal{L}(X_2) \oplus \cdots)_\infty$ define the mapping $\mathcal{F}_\mathcal{T}\colon X \to X$ by*

$$\mathcal{F}_\mathcal{T}(x_1, x_2, \dots) = \big(T_1x_1, T_2x_2, \dots\big).$$

Establish the following.

(a) *$\mathcal{F}_\mathcal{T}$ is a bounded operator on X, i.e., $\mathcal{F}_\mathcal{T} \in \mathcal{L}(X)$.*

(b) *The mapping $\mathcal{T} \mapsto \mathcal{F}_\mathcal{T}$, from $(\mathcal{L}(X_1) \oplus \mathcal{L}(X_2) \oplus \cdots)_\infty$ to $\mathcal{L}(X)$, is a linear isometry—and therefore for each $1 \le p < \infty$ the Banach space $(\mathcal{L}(X_1) \oplus \mathcal{L}(X_2) \oplus \cdots)_\infty$ is a closed subspace of $\mathcal{L}\big((X_1 \oplus X_2 \oplus \cdots)_p\big)$.*

Solution: (a) Fix $1 \le p < \infty$, pick $\mathcal{T} = (T_1, T_2, \dots) \in (\mathcal{L}(X_1) \oplus \mathcal{L}(X_2) \oplus \cdots)_\infty$, and put $M = \sup_n \|T_n\| = \|\mathcal{T}\|_\infty < \infty$. If $x = (x_1, x_2, \dots) \in X$, then note that

$$\big(\|\mathcal{F}_\mathcal{T}x\|_p\big)^p = \sum_{n=1}^{\infty}\|T_nx_n\|^p \le \sum_{n=1}^{\infty}\|T_n\|^p\|x_n\|^p \le M^p\sum_{n=1}^{\infty}\|x_n\|^p = M^p\|x\|^p < \infty.$$

This implies, $\|\mathcal{F}_\mathcal{T}x\|_p \le M\|x\|_p < \infty$, and so $\mathcal{F}_\mathcal{T}$ is indeed a mapping from X to X. It should be obvious that $\mathcal{F}_\mathcal{T}\colon X \to X$ is a bounded operator satisfying $\|\mathcal{F}_\mathcal{T}\| \le M = \|\mathcal{T}\|_\infty$.

(b) To see that $\|\mathcal{F}_\mathcal{T}\| = \|\mathcal{T}\|_\infty$, let $x_n \in X_n$ be an arbitrary vector. Denote by $\mathbf{x}_n$ the vector in X whose n^{th} component is x_n and all others are zero. Then we have

$$\|T_nx_n\| = \big(\|T_nx_n\|^p\big)^{\frac{1}{p}} = \|\mathcal{F}_\mathcal{T}\mathbf{x}_n\|_p \le \|\mathcal{F}_\mathcal{T}\| \cdot \|\mathbf{x}_n\|_p = \|\mathcal{F}_\mathcal{T}\| \cdot \|x_n\|.$$

Therefore, $\|T_n\| \le \|\mathcal{F}_\mathcal{T}\|$ for each n, and from this we infer that $\|\mathcal{T}\|_\infty \le \|\mathcal{F}_\mathcal{T}\|$ is also true. Consequently, $\|\mathcal{F}_\mathcal{T}\| = \|\mathcal{T}\|_\infty$. This shows that the mapping $\mathcal{T} \mapsto \mathcal{F}_\mathcal{T}$, from the Banach space $(\mathcal{L}(X_1) \oplus \mathcal{L}(X_2) \oplus \cdots)_\infty$ to $\mathcal{L}(X)$, is a linear isometry. ∎

Problem 7.2.5. *Verify that the mapping $x \mapsto \dot{x}$, from X to $\widehat{X}$ defined by the formula $\dot{x} = [(x, x, x, \dots)]$, is a linear isometry—which is also a lattice isometry if X is in addition a Banach lattice.*

Solution: Let $J\colon X \to \widehat{X}$ be defined by $J(x) = \dot{x} = [(x,x,x,\ldots)]$. For the additivity of J note that if $x, y \in X$, then

$$\begin{aligned} J(x+y) &= [(x+y, x+y, \ldots)] = [(x,x,\ldots) + (y,y,\ldots)] \\ &= [(x,x,\ldots)] + [(y,y,\ldots)] = J(x) + J(y)\,. \end{aligned}$$

To see that J is homogeneous, let $x \in X$ and λ be an arbitrary scalar. Then we have

$$J(\lambda x) = [(\lambda x, \lambda x, \ldots)] = [\lambda(x,x,\ldots)] = \lambda[(x,x,\ldots)] = \lambda J(x)\,.$$

To see that J is an isometry, notice that for each $x \in X$ we have

$$\|J(x)\| = \big\|[(x,x,\ldots)]\big\| = \limsup_{n\to\infty} \|x\| = \|x\|\,.$$

Finally, assume that X is a Banach lattice. Then the lattice operations of $\widehat{X}$ are given by

$$[x] \vee [y] = [(x_1 \vee y_1, x_2 \vee y_2, \ldots)] \quad \text{and} \quad [x] \wedge [y] = [(x_1 \wedge y_1, x_2 \wedge y_2, \ldots)]\,.$$

Therefore, if $x, y \in X$, then

$$J(x \vee y) = [(x \vee y, x \vee y, \ldots)] = \dot{x} \vee \dot{y} = J(x) \vee J(y)\,.$$

This proves that when X is a Banach lattice, J is a lattice isometry. ∎

Problem 7.2.6. *If X is a real Banach lattice, then show that a bounded operator $T\colon X \to X$ is positive if and only if the operator $\widehat{T}\colon \widehat{X} \to \widehat{X}$ is positive.*

Solution: Let $T\colon X \to X$ be a bounded operator on a (real) Banach lattice. Assume first that T is positive and let $[x] \geq 0$ in $\widehat{E}$. Then there exists some vector $y = (y_1, y_2, \ldots) \in [x]$ such that $y_n \geq 0$ for each n. Since T is positive, $Ty_n \geq 0$ likewise holds for each n. This implies

$$\widehat{T}[x] = [(Tx_1, Tx_2, \ldots)] = [(Ty_1, Ty_2, \ldots)] \geq 0\,,$$

and so $\widehat{T}$ is a positive operator.

For the converse, assume that $\widehat{T}$ is a positive operator, and take any $x \in X^+$. Then we have $\widehat{T}\dot{x} = [(Tx, Tx, \ldots)] \geq 0$. Consequently, there exists some sequence $y = (y_1, y_2, \ldots) \in [(Tx, Tx, \ldots)]$ satisfying $y_n \geq 0$ for each n. It follows that $y_n \to Tx$. Now a glance at Problem 1.2.12 guarantees that $Tx \geq 0$. This shows that T is a positive operator. ∎

Problem 7.2.7. *If $T\colon X \to X$ is a surjective operator on a Banach space, then show that $\widehat{T}\colon \widehat{X} \to \widehat{X}$ is also surjective. Conversely, if T has a closed range and $\widehat{T}$ is surjective, then show that T is also surjective.*

Solution: Assume that $T\colon X \to X$ is surjective. By Corollary 2.15 there is a constant $c > 0$ such that for every $y \in Y$ there exists some $x \in X$ satisfying $\|x\| \leq c\|y\|$ and $Tx = y$. Now let $[y] = [(y_1, y_2, \ldots)] \in \widehat{X}$. For each n pick some $x_n \in X$ satisfying $\|x_n\| \leq c\|y_n\| \leq c\|y\|_\infty$ and $Tx_n = y_n$. Therefore, if we let $x = (x_1, x_2, \ldots)$, then $x \in \ell_\infty(X)$ and

$$\widehat{T}[x] = [(Tx_1, Tx_2, \ldots)] = [y]\,.$$

This shows that $\widehat{T}$ is surjective.

For the converse, assume that the operator $T\colon X \to X$ has a closed range and that $\widehat{T}$ is surjective. Fix some $y \in X$. Since $\widehat{T}$ is surjective, there exists some $x = (x_1, x_2, \ldots) \in \ell_\infty(X)$ such that $\widehat{T}[x] = [(Tx_1, Tx_2, \ldots)] = \dot{y}$. This implies $Tx_n \to y$. Since the range of T is closed, we infer that $y \in R(T)$. Therefore, $R(T) = X$, and so T is surjective. ∎

Problem 7.2.8. *Show that a positive operator $T\colon E \to E$ on a real Banach lattice is a lattice homomorphism if and only if $\widehat{T}\colon \widehat{E} \to \widehat{E}$ is likewise a lattice homomorphism.*

Solution: Assume first that T is a lattice homomorphism and let $x = (x_1, x_2, \ldots)$ and $y = (y_1, y_2, \ldots)$ belong to $\ell_\infty(E)$. Then we have

$$\begin{aligned}\widehat{T}\big([x] \vee [y]\big) &= \widehat{T}\big([(x_1 \vee y_1, x_2 \vee y_2, \ldots)]\big)\\ &= [(T(x_1 \vee y_1), T(x_2 \vee y_2), \ldots)] = [(Tx_1 \vee Ty_1, Tx_2 \vee Ty_2, \ldots)]\\ &= [(Tx_1, Tx_2, \ldots)] \vee [(Ty_1, Ty_2, \ldots)] = \widehat{T}\big([x]\big) \vee \widehat{T}\big([y]\big)\,.\end{aligned}$$

This shows that $\widehat{T}$ is a lattice homomorphism.

For the converse, suppose that $\widehat{T}$ is a lattice homomorphism. If $x, y \in E$, then

$$\begin{aligned}[(Tx \vee Ty, Tx \vee Ty, \ldots)] &= [(Tx, Tx, \ldots)] \vee [(Ty, Ty, \ldots)]\\ &= \widehat{T}\dot{x} \vee \widehat{T}\dot{y} = \widehat{T}(\dot{x} \vee \dot{y})\\ &= \widehat{T}\big([(x \vee y, x \vee y, \ldots)]\\ &= [(T(x \vee y), T(x \vee y), \ldots)]\,.\end{aligned}$$

This implies $T(x \vee y) = Tx \vee Ty$, and so T is a lattice homomorphism. ∎

Problem 7.2.9. *Let X be an arbitrary real Banach space with complexification $X_{\mathbf{C}} = X \oplus \imath X$. Establish the following:*

(a) *A sequence $z = (x_1 + \imath y_1, x_2 + \imath y_2, \ldots)$ in $X_{\mathbf{C}}$ is bounded, i.e., $z \in \ell_\infty(X_{\mathbf{C}})$, if and only if the sequences $x = (x_1, x_2, \ldots)$ and $y = (y_1, y_2, \ldots)$ in X are both bounded, i.e., $x, y \in \ell_\infty(X)$. In this case, write $z = x + \imath y$.*

(b) *Show that the mapping $\mathcal{T}\colon \widehat{X_{\mathbf{C}}} \to \widehat{X} \oplus \imath\widehat{X}$, defined by*

$$\mathcal{T}\big([x + \imath y]\big) = [x] + \imath[y]$$

for all $x = (x_1, x_2, \ldots)$ and $y = (y_1, y_2, \ldots)$ in $\ell_\infty(X)$, is a surjective (topological) isomorphism—which is a lattice isomorphism if X is also a Banach lattice.

Solution: (a) This is a consequence of the following fact: If $a + \imath b \in X \oplus \imath X$, then

$$\tfrac{1}{2}\big(\|a\| + \|b\|\big) \le \|a + \imath b\| \le \|a\| + \|b\|\,.$$

(b) First, let us establish that $\mathcal{T}$ is well defined. To this end, let the sequence $z = a + \imath b = (a_1 + \imath b_1, a_2 + \imath b_2, \ldots)$ belong to $[x + \imath y]$. This means that the sequence $a + \imath b$ is norm bounded, and $\|a_n + \imath b_n - (x_n + \imath y_n)\| = \|(a_n - x_n) + \imath(b_n - y_n)\| \to 0$.

From the inequalities of part (a), it follows that the sequences $a = (a_1, a_2, \ldots)$ and $b = (b_1, b_2, \ldots)$ are both bounded, i.e., $a, b \in \ell_\infty(X)$, $\|a_n - x_n\| \to 0$, and $\|b_n - y_n\| \to 0$. Thus, $a \in [x]$ and $b \in [y]$, and so

$$[a] + \imath[b] = [x] + \imath[y]\,,$$

holds true, proving that $\mathcal{T}$ is well defined.

We now verify the conditions needed to guarantee that $\mathcal{T}$ is a linear isomorphism. For the additivity of $\mathcal{T}$, note that

$$\begin{aligned} \mathcal{T}\big([x + \imath y] + [a + \imath b]\big) &= \mathcal{T}\big([(x+a) + \imath(y+b)]\big) = [x+a] + \imath[y+b] \\ &= \big([x] + \imath[y]\big) + \big([a] + \imath[b]\big) = \mathcal{T}\big([x+\imath y]\big) + \mathcal{T}\big([a + \imath b]\big)\,. \end{aligned}$$

The homogeneity of $\mathcal{T}$ can be proven as follows:

$$\begin{aligned} \mathcal{T}\big(\lambda[x + \imath y]\big) &= \mathcal{T}([\lambda x + \imath\lambda y]) = [\lambda x] + \imath[\lambda y] = \lambda[x] + \imath\lambda[y] \\ &= \lambda([x] + \imath[y]) = \lambda\mathcal{T}([x + \imath y])\,. \end{aligned}$$

To see that $\mathcal{T}$ is surjective, note that if $[x] + \imath[y] \in \widehat{X} \oplus \imath\widehat{X}$, then $x = (x_1, x_2, \ldots)$ and $y = (y_1, y_2, \ldots)$ are both bounded sequences in X, and moreover the bounded sequence $x + \imath y$ (see part (a)) in $X_{\mathbf{C}}$ satisfies

$$\mathcal{T}([x + \imath y]) = [x] + \imath[y]\,.$$

To prove that $\mathcal{T}$ is one-to-one, let $T\big([x + \imath y]\big) = [x] + \imath[y] = 0$. This means that $[x] = [y] = 0$ or that x and y are null sequences in X. This implies that the sequence $[x + \imath y]$ in $\ell_\infty(X_{\mathbf{C}})$ is also a null sequence, and so $[x + \imath y] = 0$ in $\widehat{X}_{\mathbf{C}}$. So, $\mathcal{T}$ is one-to-one too.

Finally, for the continuity of $\mathcal{T}$ note that:

$$\begin{aligned} \|\mathcal{T}([x + \imath y])\| &= \|[x] + \imath[y]\| = \sup_{\theta\in\mathbb{R}} \|[x]\cos\theta + [y]\sin\theta\| \\ &= \sup_{\theta\in\mathbb{R}} \|[(x_1\cos\theta + y_1\sin\theta, x_2\cos\theta + y_2\sin\theta, \ldots)]\| \\ &= \sup_{\theta\in\mathbb{R}} \big[\limsup_{n\to\infty} \|x_n\cos\theta + y_n\sin\theta\|\,\big] \\ &\le \limsup_{n\to\infty} \|x_n + \imath y_n\| = \|[x + \imath y]\|\,. \end{aligned}$$

Therefore, $\mathcal{T}$ is continuous, and consequently by the Open Mapping Theorem, the operator $\mathcal{T}$ is a topological isomorphism. ∎

Problem 7.2.10. *Let $T\colon X \to X$ be a bounded operator on a real Banach space, and let $\widehat{T}\colon \widehat{X} \to \widehat{X}$ be its extension to $\widehat{X}$. If $X_{\mathbf{C}} = X \oplus \imath X$ is the complexification of X, then in Problem 7.2.9 we defined the mapping $\mathcal{T}\colon \widehat{X}_{\mathbf{C}} \to \widehat{X} \oplus \imath\widehat{X}$ by*

$$\mathcal{T}([x + \imath y]) = [x] + \imath[y]$$

for all $x = (x_1, x_2, \ldots)$ and $y = (y_1, y_2, \ldots)$ in $\ell_\infty(X)$ and proved that it was a surjective (topological) isomorphism. Establish the following properties.

(a) *The diagram shown in Figure 1 is commutative.*

(b) *The spectra of the operators T, $\widehat{T}_{\mathbf{C}}$, and $\widehat{T}+\imath\widehat{T}$ coincide. That is,*

$$\sigma(T)=\sigma(\widehat{T}_{\mathbf{C}})=\sigma(\widehat{T}+\imath\widehat{T}),\quad \sigma_p(T)=\sigma_p(\widehat{T}_{\mathbf{C}})=\sigma_p(\widehat{T}+\imath\widehat{T}),\quad \textit{and}$$

$$\sigma_a(T)=\sigma_a(\widehat{T}_{\mathbf{C}})=\sigma_a(\widehat{T}+\imath\widehat{T})\,.$$

(c) *If X is a Banach lattice and T is a lattice homomorphism, then*

$$\big|(\widehat{T})_{\mathbf{C}}(z)\big|=\widehat{T}|z|$$

holds for all $z\in(\widehat{X})_{\mathbf{C}}$.

Solution: (a) Note that if $[z]=[x+\imath y]\in\widehat{X}_{\mathbf{C}}$, then we have

$$\begin{aligned}(\widehat{T}+\imath\widehat{T})\mathcal{T}([z]) &= (\widehat{T}+\imath\widehat{T})([x]+\imath[y])=\widehat{T}([x])+\imath\widehat{T}([y])\\ &= [(Tx_1,Tx_2,\ldots)]+\imath[(Ty_1,Ty_2,\ldots)]\,,\end{aligned}$$

and

$$\begin{aligned}\mathcal{T}\widehat{T}_{\mathbf{C}}([z]) &= \mathcal{T}[(T_{\mathbf{C}}(x_1+\imath y_1),T_{\mathbf{C}}(x_2+\imath y_2),\ldots)]\\ &= \mathcal{T}[(Tx_1+\imath Ty_1,Tx_2+\imath Ty_2,\ldots)]\\ &= [(Tx_1,Tx_2,\ldots)]+\imath[(Ty_1,Ty_2,\ldots)]\,.\end{aligned}$$

So, $(\widehat{T}+\imath\widehat{T})\mathcal{T}=\mathcal{T}\widehat{T}_{\mathbf{C}}$. That is, the diagram in Figure 1 is indeed commutative.

(b) From Problem 6.2.8, we know that

$$\sigma(\widehat{T}_{\mathbf{C}})=\sigma(\widehat{T}+\imath\widehat{T}),\quad \sigma_p(\widehat{T}_{\mathbf{C}})=\sigma_p(\widehat{T}+\imath\widehat{T}),\quad\text{and}\quad \sigma_a(\widehat{T}_{\mathbf{C}})=\sigma_a(\widehat{T}+\imath\widehat{T})\,.$$

The other equalities follow from Corollary 7.17.

(c) Assume that T is a lattice homomorphism. Then, by Problem 7.2.8, the operator $\widehat{T}\colon\widehat{X}\to\widehat{X}$ is also a lattice homomorphism. The desired conclusion now follows from Theorem 4.29. ∎

$$\begin{array}{ccc}\widehat{X_{\mathrm{C}}} & \xrightarrow{\ \mathcal{T}\ } & (\widehat{X})_{\mathrm{C}}\\ {\scriptstyle\widehat{T_{\mathrm{C}}}}\big\downarrow & & \big\downarrow{\scriptstyle(\widehat{T})_{\mathrm{C}}}\\ \widehat{X_{\mathrm{C}}} & \xrightarrow{\ \mathcal{T}\ } & (\widehat{X})_{\mathrm{C}}\end{array}$$

Figure 1

Problem 7.2.11. *If $T\colon X\to X$ is a bounded operator on a Banach space, then for every free ultrafilter $\mathcal{U}$ on $\mathbb{N}$ we have:*

(1) $\sigma(T)=\sigma(T_{\mathcal{U}})$.

(2) $\sigma_a(T)=\sigma_a(T_{\mathcal{U}})=\sigma_p(T_{\mathcal{U}})$.

Solution: (1) Fix some $\lambda\in\mathbb{C}$. According to part (c) of Problem 1.4.9 we know that the operator $\lambda-T$ is invertible if and only if the operator $(\lambda-T)_{\mathcal{U}}=\lambda-T_{\mathcal{U}}$ is invertible—in which case we have $(\lambda-T_{\mathcal{U}})^{-1}=\big[(\lambda-T)^{-1}\big]_{\mathcal{U}}$.

From the above conclusion, we infer immediately that $\rho(T)=\rho(T_{\mathcal{U}})$. This is, of course, equivalent to $\sigma(T)=\sigma(T_{\mathcal{U}})$.

(2) We shall show first that $\sigma_a(T) = \sigma_p(T_{\mathcal{U}})$. To this end, let $\lambda \in \sigma_a(T)$. Then there exists some sequence $\{x_n\}$ in X satisfying $\|x_n\| = 1$ for each n and $Tx_n - \lambda x_n \to 0$. So, if we let $x = (x_1, x_2, \ldots)$, then $[x] \neq 0$ and $T_{\mathcal{U}}[x] = \lambda[x]$. Thus, $\lambda \in \sigma_p(T_{\mathcal{U}})$, and so $\sigma_a(T) \subseteq \sigma_p(T_{\mathcal{U}})$.

For the reverse inequality, let $\lambda \in \sigma_p(T_{\mathcal{U}})$ and pick some $x = (x_1, x_2, \ldots)$ in $\ell_\infty(X)$ with $\|[x]\| = \lim_{\mathcal{U}} \|x_n\| = 1$ and $T_{\mathcal{U}}[x] = \lambda[x]$ or $\lim_{\mathcal{U}}[Tx_n - \lambda x_n] = 0$. Therefore, for each k the set

$$A_k = \left\{n \in \mathbb{N}\colon \ \|Tx_n - \lambda x_n\| < \tfrac{1}{k} \ \text{ and } \ \|x_n\| > \tfrac{1}{2}\right\}$$

belongs to $\mathcal{U}$, and so it is an infinite subset of $\mathbb{N}$. Moreover, we have $A_{k+1} \subseteq A_k$ for each k. Now if we choose a sequence of natural numbers $\{m_k\}$ satisfying $m_k \in A_k$ and $m_k < m_{k+1}$ for each k, then $x_{m_k} \not\to 0$ and $Tx_{m_k} - \lambda x_{m_k} \to 0$. This shows that $\lambda \in \sigma_a(T)$, and so $\sigma_p(T_{\mathcal{U}}) \subseteq \sigma_a(T)$. Consequently, $\sigma_a(T) = \sigma_p(T_{\mathcal{U}})$.

Since, $\sigma_p(T_{\mathcal{U}}) \subseteq \sigma_a(T_{\mathcal{U}})$ is always true, what remains to be proven is the inclusion $\sigma_a(T_{\mathcal{U}}) \subseteq \sigma_p(T_{\mathcal{U}}) = \sigma_a(T)$. To this end, let $\lambda \in \sigma_a(T_{\mathcal{U}})$. This means that $\lambda - T_{\mathcal{U}} = (\lambda - T)_{\mathcal{U}}$ is not bounded below on $X_{\mathcal{U}}$. Now if λ does not belong to $\sigma_a(T)$, then $\lambda - T$ is bounded below on X. But then it follows from part (c) of Problem 1.4.9 that $(\lambda - T)_{\mathcal{U}}$ must be bounded below on $X_{\mathcal{U}}$, a contradiction. Hence, $\lambda \in \sigma_a(T)$ and so $\sigma_a(T_{\mathcal{U}}) \subseteq \sigma_a(T)$ is true, as desired. ∎

7.3. The Spectrum of a Lattice Homomorphism

Problem 7.3.1. *Let $G = \{z_1, \ldots, z_p\}$ be a subset of the unit circle Γ consisting of p distinct numbers. If G is closed under multiplication, then show that*

$$G = \{z \in \mathbb{C}\colon \ z^p = 1\} = \{e^{\frac{2\pi k}{p}\imath}\colon \ k = 0, 1, \ldots, p-1\}.$$

Solution: Fix any $z \in G$ and consider $p+1$ elements $z, z^2, \ldots, z^{p+1}$. Each of them lies in G and so $z^m = z^\ell$ for some $1 \leq m < \ell \leq p+1$. This implies $z^n = 1$ for some $1 \leq n \leq p$. In particular, $1 \in G$ and from $zz^{n-1} = z^n = 1$, we see that G is a subgroup of Γ. Next we will show that $z^p = 1$. To this end let $\nu = \min\{n \in \mathbb{N}\colon \ 1 \leq n \leq p \ \text{ and } \ z^n = 1\}$. Clearly the set $H = \{1, z, \ldots, z^{\nu-1}\}$ is a subgroup of G and so p is divisible by ν, and thus $z^p = 1$.

The above proves that the solutions of the equation $\xi^p = 1$ are precisely the p elements of the group G. Therefore $G = \{e^{\frac{2\pi k}{p}\imath}\colon \ k = 0, 1, \ldots, p-1\}$. ∎

Problem 7.3.2. *Fix an angle $\theta \in \mathbb{R}$ and let $G = \{e^{\imath n\theta}\colon \ n \in \mathbb{N}\}$. Establish the following.*

(a) *If θ is a rational multiple of 2π, then G is a finite subgroup of the unit circle Γ (and so G consists of roots of unity).*

(b) *If θ is not a rational multiple of 2π, then G is dense in Γ.*

Solution: Clearly G is closed under multiplication.

(a) Assume that the angle θ is a rational multiple of 2π, that is, $\theta = 2\pi\frac{k}{m}$ for some $k \in \mathbb{N}$ and $m \in \mathbb{Z}$. Consequently $e^{m\theta\imath} = e^{2k\pi\imath} = 1$. Now if $n \in \mathbb{N}$, then we can write $n = \ell k + r$, where $0 \le r < k$, and so $e^{n\theta\imath} = (e^{k\theta\imath})^{\ell} e^{r\theta\imath} = e^{r\theta\imath}$. This shows that G is a finite subset of Γ. Since G is closed under multiplication, it follows from Problem 7.3.1 that G is a subgroup of Γ consisting of roots of unity.

(b) Suppose that the angle θ is not a rational multiple of 2π. In this case, for any distinct $n, m \in \mathbb{N}$ we have $e^{n\theta\imath} \ne e^{m\theta\imath}$. Indeed, otherwise, $e^{n\theta\imath} = e^{m\theta\imath}$ would imply that $e^{(n-m)\theta\imath} = 1$, and so $(n-m)\theta = 2\pi q$ for some $q \in \mathbb{Z}$ or, equivalently, $\theta = \frac{q}{n-m}\cdot 2\pi$, which is impossible. In particular, G is an infinite set.

Next fix any $k \in \mathbb{N}$ and for each $1 \le j \le k$ let

$$\Gamma_j = \left\{ e^{x\imath}\colon\ \tfrac{2\pi}{k}(j-1) \le x < \tfrac{2\pi}{k} j \right\}.$$

Clearly $\Gamma = \bigcup_{j=1}^{k} \Gamma_j$ and so, since G is infinite, there exist some $1 \le j \le k$ and some $n, m \in \mathbb{N}$ with $n > m$ such that $e^{n\theta\imath}, e^{m\theta\imath} \in \Gamma_j$. Pick $x_1, x_2 \in [\frac{2\pi}{k}(j-1), \frac{2\pi}{k}j)$ such that $e^{n\theta\imath} = e^{x_1\imath}$ and $e^{m\theta\imath} = e^{x_2\imath}$. Finally, let $x = x_1 - x_2$ and note that $|x| < \frac{2\pi}{k}$ and $e^{x\imath} = e^{(n-m)\theta\imath} \in G$. Consequently, $e^{\ell x\imath} \in G$ for each $\ell \in \mathbb{N}$. This implies that for each $1 \le j \le k$ there exists some $\ell \in \mathbb{N}$ such that $e^{\ell x\imath} \in \Gamma_j$. Since $k \in \mathbb{N}$ is arbitrary, we infer that G is dense in Γ. ∎

Problem 7.3.3 (Kronecker's theorem)**.** *Let G be an arbitrary subgroup of the unit circle Γ. Show that either G is dense in Γ or else it consists of roots of unity, i.e., $G = \{e^{\frac{2\pi k}{p}\imath}\colon\ k = 0, 1, \ldots, p-1\}$ for some $p \in \mathbb{N}$.*

Solution: The conclusion follows immediately from Problems 7.3.1 and 7.3.2. ∎

Problem 7.3.4. *Establish the following result that was used in the proof of Theorem 7.20.* If μ is a regular Borel probability measure on a compact Hausdorff space Ω and some function $z \in C_{\mathbb{C}}(\Omega)$ satisfies $\int_\Omega z(t)\,d\mu(t) = 1$ and $|z| = \mathbf{1}$, then $z(t) = 1$ for each t in the support of μ.

Solution: Replacing Ω by $\operatorname{Supp}\mu$, we can assume without loss of generality that $\operatorname{Supp}\mu = \Omega$. Now suppose that a function $z = x + \imath y \in C_{\mathbb{C}}(\Omega)$ satisfies $|z| = \mathbf{1}$ and $\int_\Omega z(t)\,d\mu(t) = \int_\Omega x(t)\,d\mu(t) + \imath \int_\Omega y(t)\,d\mu(t) = 1$. Clearly,

$$\int_\Omega x(t)\,d\mu(t) = 1 \quad \text{and} \quad \int_\Omega y(t)\,d\mu(t) = 0\,.$$

From $|z| = \mathbf{1}$, it follows that $x \le \mathbf{1}$. If $x(t_0) < 1$ holds for some $t_0 \in \Omega$, then there exists an open neighborhood V of t_0 such that $x(t) < 1$ for each $t \in V$. Taking into account that $\mu(V) > 0$, the latter implies

$$1 = \int_\Omega x(t)\,d\mu(t) = \int_V x(t)\,d\mu(t) + \int_{\Omega\setminus V} x(t)\,d\mu(t) < \mu(V) + \mu(\Omega\setminus V) = 1\,,$$

which is impossible. Hence, $x = \mathbf{1}$, and from $\mathbf{1} = |z|^2 = x^2 + y^2 = \mathbf{1} + y^2$, we get $y = 0$. Therefore, $z = x + \imath y = x = \mathbf{1}$. ∎

Problem 7.3.5. *Let $T\colon E \to E$ be an interval preserving positive operator. If a complex number $\lambda = |\lambda| e^{\imath\theta}$ belongs to the residual spectrum of T, then* *show that for each integer n the complex number $|\lambda|e^{\imath n\theta}$ is an eigenvalue of the adjoint operator T^*.*

Solution: By Theorem 6.19 we know that λ is an eigenvalue of T^*. Also, by Theorem 1.35, $T^*: E^* \to E^*$ is a lattice homomorphism. According to Corollary 7.22, the point spectrum of T^* is cyclic, and the desired conclusion follows. ∎

Problem 7.3.6. *Assume that for some continuous function $\phi\colon \Omega \to \mathbb{R}$ on a compact Hausdorff space the multiplication operator $M_\phi\colon C(\Omega) \to C(\Omega)$ is a lattice homomorphism.*

Show directly (i.e., without using Theorem 7.23) that the spectrum of the multiplication operator M_ϕ is cyclic.

Solution: Note that the multiplication operator M_ϕ is a lattice homomorphism if and only if ϕ is a non-negative function. Now a glance at Problem 6.1.14 shows that the spectrum of M_ϕ coincides with the range of ϕ, a subset of $[0, \infty)$. This implies that the spectrum of M_ϕ is cyclic. ∎

Problem 7.3.7. *Let $T\colon X \to X$ be a bounded operator on a Banach space. If the approximate point spectrum of T is rotation invariant, then show that the spectrum of T is also rotation invariant.*

Solution: Assume $\lambda = |\lambda|e^{\imath\theta_0} \in \sigma(T)$, and let $C = \{z \in \mathbb{C}\colon\ |z| = |\lambda|\}$. We must show that $C \subseteq \sigma(T)$. If $C \cap \sigma_a(T) \neq \emptyset$, then $C \subseteq \sigma_a(T)$ by our hypothesis. So, assume that $C \cap \sigma_a(T) = \emptyset$.

In this case, $\lambda \in \sigma(T) \setminus \sigma_a(T)$. By Theorem 6.18 we know that $\sigma(T) \setminus \sigma_a(T)$ is an open subset of $\mathbb{C}$. Therefore, there exists some $\epsilon > 0$ such that $\theta_0 - \epsilon < \theta < \theta_0 + \epsilon$ implies $|\lambda|e^{\imath\theta} \in \sigma(T) \setminus \sigma_a(T)$. Now let

$$\Delta = \big\{\delta \in [0, 2\pi]\colon\ |\lambda|e^{\imath\theta} \in \sigma(T) \setminus \sigma_a(T) \text{ for each } \theta \in (\theta_0 - \delta, \theta_0 + \delta)\big\}.$$

Since $\epsilon \in \Delta$, we see that $\Delta \neq \emptyset$. Let $\delta_0 = \sup \Delta$. Since $\sigma(T)$ is closed, we have $|\lambda|e^{\imath(\theta_0-\delta_0)}, |\lambda|e^{\imath(\theta_0+\delta_0)} \in \sigma(T)$. From $C \cap \sigma_a(T) = \emptyset$, it follows that $|\lambda|e^{\imath(\theta_0-\delta_0)}, |\lambda|e^{\imath(\theta_0+\delta_0)} \in \sigma(T) \setminus \sigma_a(T)$.

We claim that $\delta_0 = 2\pi$. To see this, assume that $\delta_0 < 2\pi$. Since $\sigma(T) \setminus \sigma_a(T)$ is open, then—as above—we can find some $\eta \in \Delta$ such that $\delta_0 < \eta < 2\pi$, which is a contradiction. So, $\delta_0 = 2\pi$, which implies $C \subseteq \sigma(T)$, and the proof is finished. ∎

7.4. The Order Spectrum of an Order Bounded Operator

Problem 7.4.1. *Let $T\colon V \to V$ be an invertible operator on a vector space, and let $T^{-1}\colon V \to V$ be its inverse. Assume that W is a T-invariant subspace of V such that $T\colon W \to W$ is also invertible. Then show that T^{-1} leaves W invariant and that the inverse of $T|_W$ (the restriction of T to W) is the restriction of T^{-1} to W, i.e., $(T|_W)^{-1} = T^{-1}|_W$ holds.*

Solution: Let $S: W \to W$ be the inverse of $T: W \to W$, i.e., $S = (T|_W)^{-1}$. If $w \in W$, then we have

$$T(T^{-1}w) = T(Sw) = w\,.$$

Since T is one-to-one, it follows that $T^{-1}w = Sw$ for each $w \in W$. That is, T^{-1} also leaves the vector subspace W invariant and $T^{-1}|_W = (T|_W)^{-1}$ holds. ∎

Problem 7.4.2. *Describe the Banach lattice $\check{E}$ if $E = \mathbb{R}$.*

Solution: Assume that $E = \mathbb{R}$, so that $E^A = \mathbb{R}^A$. Now notice that a function $f \in \mathbb{R}^A$ belongs to $\check{E}$ if and only if there exists some $x \in \mathbb{R}$ such that $|f(\alpha)| \leq x$ holds for each $\alpha \in A$. This means that $\check{E}$ consists of all bounded real-valued functions on α, that is, $\check{E} = \ell_\infty(\alpha)$. ∎

Problem 7.4.3. *Let α be an arbitrary index set. Show that a Riesz space E is Dedekind complete if and only if the Riesz space E^A is Dedekind complete.*

Solution: Let E be a Riesz space. Assume first that E is Dedekind complete, and let $D \subseteq E^A$ be non-empty and bounded from above. That is, there exists some $g \in E^A$ such that $f \leq g$ holds for each $f \in D$. This is equivalent to $f(\alpha) \leq g(\alpha)$ for all $f \in D$ and each $\alpha \in A$. So, if we fix $\alpha \in A$, then the non-empty subset $\{f(\alpha)\colon\ f \in D\}$ of E is bounded from above in E. Since E is Dedekind complete, for each $\alpha \in A$ there exists some $h(\alpha) \in E$ such that $\sup_{f\in D} f(\alpha) = h(\alpha)$ holds in E. Thus, we have defined a function $h\colon A \to E$ (i.e., $h \in E^A$), and we claim that $h = \sup D$ holds in E^A.

Clearly, $f \leq h$ holds for each $f \in D$, i.e., h is an upper bound of the set D in E^A. To see that h is the least upper bound of the set D, assume that some $\theta \in E^A$ satisfies $f \leq \theta$ for each $f \in D$. So, for each $\alpha \in A$ we have $f(\alpha) \leq \theta(\alpha)$ in E for all $f \in D$. This implies $h(\alpha) = \sup_{f\in D} f(\alpha) \leq \theta(\alpha)$ for each $\alpha \in A$, that is, $h \leq \theta$. Therefore, $h = \sup D$ holds in E^A, and consequently E^A is a Dedekind complete Riesz space.

For the converse, suppose that E^A is a Dedekind complete Riesz space, and let D be a non-empty bounded from above subset of E. Pick some $x \in E$ such that $d \leq x$ holds for each $d \in D$. It follows that the non-empty subset $\check{D} = \{\check{d}\colon\ d \in D\}$ of E^A is bounded from above by $\check{x}$. So, according to our hypothesis, $f = \sup \check{D}$ exists in E^A. In particular, if $\alpha_0 \in A$ is any fixed index, then $d = \check{d}(\alpha_0) \leq f(\alpha_0)$ holds in E for all $d \in D$, i.e., $f(\alpha_0) \in E$ is an upper bound of D. To finish the solution, we shall show that $f(\alpha_0) = \sup D$ holds in E.

To this end, assume that some vector $y \in E$ satisfies $d \leq y$ for each $d \in D$. Then $\check{y}$ is an upper bound for the set $\check{D}$. This implies $f \leq \check{y}$ in E^A, and in particular $f(\alpha_0) \leq \check{y}(\alpha_0) = y$. Thus, $f(\alpha_0) = \sup D$ holds in E, proving that E is Dedekind complete. ∎

Problem 7.4.4. *Let E be a normed Riesz space, and define the function $|||\cdot|||\colon \check{E} \to \mathbb{R}$ by*

$$|||f||| = \inf\bigl\{\|x\|\colon\ x \in E \quad \textit{and} \quad |f| \leq \check{x}\bigr\} \qquad (\star)$$

for each $f \in \check{E}$. Establish the following.

(a) $|||\cdot|||$ *is a lattice norm that extends* $\|\cdot\|$ *to* $\check{E}$.

(b) *If* E *is a Banach lattice, then* $\check{E}$ *equipped with the lattice norm given by* $(\star)$ *is a Banach lattice containing* E *as a Banach sublattice.*

Solution: (a) We shall verify first that the function $|||\cdot|||$ is a norm.

Clearly, $|||f||| \geq 0$ holds for each $f \in \check{E}$ and $|||0||| = 0$. Now assume that $|||f||| = 0$. This means that there exists some sequence $\{x_n\} \subseteq E^+$ satisfying $|f| \leq \check{x}_n$ for each n and $\|x_n\| \to 0$. From $|f(\alpha)| \leq \check{x}_n(\alpha) = x_n$, we see that $\|f(\alpha)\| \leq \|x_n\|$ for all n and each $\alpha \in A$. This implies $\|f(\alpha)\| = 0$ for each $\alpha \in A$, and so $f = 0$.

For the homogeneity of $|||\cdot|||$, let $f \in \check{E}$ and let λ be a non-zero scalar. Clearly, $|\lambda f| \leq \check{x}$ if and only if $|f| \leq \frac{\check{x}}{|\lambda|}$, and so

$$\begin{aligned} |||\lambda f||| &= \inf\{\|x\|\colon\ x \in E \text{ and } |\lambda f| \leq \check{x}\} \\ &= |\lambda| \inf\{\|\tfrac{x}{\lambda}\|\colon\ x \in E \text{ and } |f| \leq \tfrac{\check{x}}{|\lambda|}\} \\ &= |\lambda| \inf\{\|y\|\colon\ y \in E \text{ and } |f| \leq \check{y}\} \\ &= |\lambda| \cdot |||f|||\,. \end{aligned}$$

For the triangle inequality, assume $|f| \leq \check{x}$ and $|g| \leq \check{y}$ with $x, y \in E^+$. Then we have $|f+g| \leq |f|+|g| \leq \check{x}+\check{y} = (x+y)\check{}$, and so $|||f+g||| \leq \|x+y\| \leq \|x\|+\|y\|$. Taking infima yields $|||f+g||| \leq |||f||| + |||g|||$.

Next, we shall establish that $|||\cdot|||$ is a lattice norm. To this end, assume that $|f| \leq |g|$ holds in $\check{E}$. Now if $z \in E^+$ satisfies $|g| \leq \check{z}$, then $|f| \leq \check{z}$ is also true, and so $|||f||| \leq \|z\|$ holds. This implies $|||f||| \leq \inf\{\|z\|\colon\ z \in E \text{ and } |g| \leq \check{z}\} = |||g|||$. That is, the norm $|||\cdot|||$ is a lattice norm.

Finally, let us show that $|||\cdot|||$ is an extension of $\|\cdot\|$. To this end, let $x \in E$. From $|\check{x}| = |x|\check{}$ and $|x| \in E$, we see that $|||\check{x}||| \leq \|\,|x|\,\| = \|x\|$. On the other hand, if $y \in E$ satisfies $|\check{x}| \leq \check{y}$, then $|x| \leq y$, and so $\|x\| = \|\,|x|\,\| \leq \|y\|$. This implies $\|x\| \leq \inf\{\|y\|\colon\ y \in E \text{ and } |\check{x}| \leq \check{y}\} = |||\check{x}|||$. Hence, $|||\check{x}||| = \|x\|$ for each $x \in E$, and thus $|||\cdot|||$ extends the norm $\|\cdot\|$ of E to $\check{E}$.

(b) We only need to show that $|||\cdot|||$ is a complete norm. To this end, let $\{f_n\}$ be a $|||\cdot|||$-Cauchy sequence of $\check{E}$. By passing to a subsequence if necessary, we can assume without loss of generality that $|||f_{n+1} - f_n||| < \frac{1}{2^{n+1}}$ holds for each n. Now for each n pick some vector $x_n \in E^+$ satisfying $|f_{n+1} - f_n| \leq \check{x}_n$ and $\|x_n\| < \frac{1}{2^{n+1}}$. Since E is a Banach lattice, the series $y_n = \sum_{k=n}^{\infty} x_k$ is norm convergent in E and $\|y_n\| \leq \frac{1}{2^n}$ holds for each n.

From $|f_{n+1}(\alpha) - f_n(\alpha)| \leq x_n$ for each $\alpha \in A$ and all n, it easily follows that $\{f_n(\alpha)\}$ is a Cauchy sequence in E for each $\alpha \in A$. Since E is a Banach space, $f_n(\alpha) \to f(\alpha)$ holds in E for each $\alpha \in A$. So, we have defined a function $f\colon A \to E$. We shall finish the solution by proving that $f \in \check{E}$ and that $|||f_n - f||| \to 0$.

We first show that $f \in \check{E}$. Start by observing that for each $\alpha \in A$ we have

$$\begin{aligned} |f_n(\alpha)| &= |f_n(\alpha) - f_1(\alpha)| + |f_1(\alpha)| = \Big|\sum_{k=1}^{n-1}(f_{k+1} - f_k)\Big| + |f_1(\alpha)| \\ &\le \sum_{k=1}^{n-1} |f_{k+1} - f_k| + |f_1(\alpha)| \le \sum_{k=1}^{\infty} x_k + |f_1(\alpha)| \\ &= y_1 + |f_1(\alpha)|\,. \end{aligned}$$

Letting $n \to \infty$, we get $|f(\alpha)| \le y_1 + |f_1(\alpha)|$ for each $\alpha \in A$, and then using that $f_1 \in \check{E}$, we easily infer that $f \in \check{E}$. Finally, for $m > n$ we have

$$|f_m(\alpha) - f_n(\alpha)| = \Big|\sum_{k=n}^{m-1}[f_{k+1}(\alpha) - f_k(\alpha)]\Big| \le \sum_{k=n}^{m-1} |f_{k+1}(\alpha) - f_k(\alpha)| \le \sum_{k=n}^{\infty} \check{x}_k = \check{y}_n\,.$$

Letting $m \to \infty$, we obtain $|f(\alpha) - f_n(\alpha)| \le \check{y}_n$ for each n and all indices $\alpha \in A$. This implies $|||f_n - f||| \le \|y_n\| < \frac{1}{2^n}$, and so $|||f_n - f||| \to 0$. Therefore, $(\check{E}, |||\cdot|||)$ is a Banach lattice. ∎

Problem 7.4.5. *Let $S, T\colon E \to E$ be two order bounded operators on a Riesz space. If $(S+\imath T)\check{}\colon \check{E}_{\mathbf{c}} \to \check{E}_{\mathbf{c}}$ is invertible, then show that the operator $S+\imath T\colon E_{\mathbf{c}} \to E_{\mathbf{c}}$ is also invertible and that its inverse is order bounded and satisfies $[(S+\imath T)^{-1}]\check{} = [(S+\imath T)\check{}]^{-1} = (\check{S}+\imath\check{T})^{-1}$.*

Solution: Let $S, T\colon E \to E$ be two order bounded operators on a Riesz space such that the operator $(S+\imath T)\check{} = \check{S}+\imath\check{T} : \check{E}_{\mathbf{c}} \to \check{E}_{\mathbf{c}}$ is invertible. We shall show first that the operator $S+\imath T\colon E_{\mathbf{c}} \to E_{\mathbf{c}}$ is also invertible.

To show $S+\imath T$ is one-to-one, let $(S+\imath T)(x+\imath y) = Sx - Ty + \imath(Tx+Sy) = 0$. This easily implies

$$(\check{S}+\imath\check{T})(\check{x}+\imath\check{y}) = [(S+\imath T)(x+\imath y)]\check{} = 0\,,$$

and so (given that $\check{S}+\imath\check{T}\colon \check{E}_{\mathbf{c}} \to \check{E}_{\mathbf{c}}$ is invertible) we get $\check{x}+\imath\check{y} = 0$. This implies $x+\imath y = 0$, proving that the operator $S+\imath T\colon E_{\mathbf{c}} \to E_{\mathbf{c}}$ is one-to-one.

To see that the operator $S+\imath T$ is surjective, let $x+\imath y \in E_{\mathbf{c}}$. Then $\check{x}+\imath\check{y} \in \check{E}_{\mathbf{c}}$, and since $\check{S}+\imath\check{T}$ is surjective, there exists some $f+\imath g \in \check{E}_{\mathbf{c}}$ such that

$$(\check{S}+\imath\check{T})(f+\imath g) = \check{S}f - \check{T}g + \imath(\check{T}f+\check{S}g) = \check{x}+\imath\check{y}\,.$$

Now if $\alpha_0 \in A$ is any fixed index, then

$$S(f(\alpha_0)) - T(g(\alpha_0)) = x \quad \text{and} \quad T(f(\alpha_0)) + S(g(\alpha_0)) = y\,.$$

So, if $a = f(\alpha_0)$ and $b = g(\alpha_0)$, then $a+\imath b \in E_{\mathbf{c}}$ and

$$(S+\imath T)(a+\imath b) = Sa - Tb + \imath(Ta+Sb) = x+\imath y\,.$$

This shows that $S+\imath T\colon E_{\mathbf{c}} \to E_{\mathbf{c}}$ is also a surjective operator.

The above show that $S+\imath T\colon E_{\mathbf{c}} \to E_{\mathbf{c}}$ is invertible. Let $S_1, T_1\colon E \to E$ be two linear operators such that $S_1+\imath T_1$ is the inverse of $S+\imath T\colon E_{\mathbf{c}} \to E_{\mathbf{c}}$. That is, we have $(S+\imath T)(S_1+\imath T_1) = (S_1+\imath T_1)(S+\imath T) = \mathcal{I}\,(= I+\imath I)$. This implies

$$(\check{S}+\imath\check{T})(\check{S}_1+\imath\check{T}_1) = (\check{S}_1+\imath\check{T}_1)(\check{S}+\imath\check{T}) = \check{\mathcal{I}}\,(= \check{I}+\imath\check{I})\,.$$

So, the operator $\check{S}+\imath\check{T}\colon (E^A)_{\mathbf{c}} \to (E^A)_{\mathbf{c}}$ is invertible and $(\check{S}+\imath\check{T})^{-1} = \check{S}_1+\imath\check{T}_1$.

Since $\check{S} + \imath\check{T}$ is invertible on the subspace $E_{\mathbf{c}}$ of $(E^A)_{\mathbf{c}}$, it follows that $\check{A} + \imath\check{B}$ leaves invariant $E_{\mathbf{c}}$; see Problem 7.4.1. This means that $\check{S}_1$ and $\check{T}_1$ leave $\check{E}$ invariant, and so (by Theorem 7.31) S_1 and T_1 are both order bounded operators. Therefore, the inverse $S_1 + \imath T_1$ of the operator $S + \imath T\colon E_{\mathbf{c}} \to E_{\mathbf{c}}$ is an order bounded operator.

Finally, one more glance at Problem 7.4.1 guarantees that the inverse of the operator $(S + \imath T)\check{}\colon \check{E}_{\mathbf{c}} \to \check{E}_{\mathbf{c}}$ is the restriction of $\check{S}_1 + \imath\check{T}_1$ to $\check{E}_{\mathbf{c}}$. This implies

$$[(S + \imath T)\check{}\,]^{-1} = (\check{S} + \imath\check{T})^{-1} = \check{S}_1 + \imath\check{T}_1 = (S_1 + \imath T_1)\check{} = [(S + \imath T)^{-1}]\check{}\,,$$

and the solution is finished. ∎

Problem 7.4.6. *If T is an order bounded operator on a Dedekind complete Banach lattice, then show that $\sigma_o(T) = \sigma_o(\check{T}) = \sigma(\check{T})$.*

Solution: Let $T\colon E \to E$ be an order bounded operator on a Dedekind complete Banach lattice, and let $T_{\mathbf{c}}\colon E_{\mathbf{c}} \to E_{\mathbf{c}}$ be its extension to the complexification of E. Then for any complex number $\lambda = a + \imath b$ we have

$$(\lambda - T_{\mathbf{c}})\check{} = [(a - T) + \imath(b - T)]\check{} = (a - T)\check{} + \imath(b - T)\check{} = a - \check{T} + \imath(b - \check{T})\,.$$

Now, from Problem 7.4.5, it follows that the operator $\lambda - T_{\mathbf{c}}\colon E_{\mathbf{c}} \to E_{\mathbf{c}}$ has an order bounded inverse if and only if the operator $\lambda - \check{T}_{\mathbf{c}}\colon \check{E}_{\mathbf{c}} \to \check{E}_{\mathbf{c}}$ has likewise an order bounded inverse—and also if and only if $\lambda - \check{T}_{\mathbf{c}}\colon \check{E}_{\mathbf{c}} \to \check{E}_{\mathbf{c}}$ is invertible. These equivalent statements simply show that $\sigma_o(T) = \sigma_o(\check{T}) = \sigma(\check{T})$. ∎

Problem 7.4.7. *Establish the following elementary properties that were used in Example 7.36. Consider the matrix $M = \begin{bmatrix} \cos\phi & -\sin\phi \\ \sin\phi & \cos\phi \end{bmatrix}$ acting on $\mathbb{R}^2$, where ϕ is an arbitrary angle. Then:*

(a) *For each $n \in \mathbb{N}$ we have $M^n = \begin{bmatrix} \cos n\phi & -\sin n\phi \\ \sin n\phi & \cos n\phi \end{bmatrix}$.*

(b) *The modulus of M^n is the matrix $|M^n| = \begin{bmatrix} |\cos n\phi| & |\sin n\phi| \\ |\sin n\phi| & |\cos n\phi| \end{bmatrix}$.*

(c) *The symmetric matrix $|M^n|$ has eigenvalues $|\cos n\phi| \pm |\sin n\phi|$.*

(d) *The Euclidean norm of $|A^n|$ is $\||M^n|\| = |\cos n\phi| + |\sin n\phi|$.*

Solution: (a) This proof is by induction. Assume that the formula is valid for some $n \geq 1$. Then we have

$$\begin{aligned} M^{n+1} &= M^n M = \begin{bmatrix} \cos n\phi & -\sin n\phi \\ \sin n\phi & \cos n\phi \end{bmatrix} \begin{bmatrix} \cos\phi & -\sin\phi \\ \sin\phi & \cos\phi \end{bmatrix} \\ &= \begin{bmatrix} \cos\phi\cos n\phi - \sin\phi\sin n\phi & -(\sin\phi\cos n\phi + \cos\phi\sin n\phi) \\ \sin\phi\cos n\phi + \cos\phi\sin n\phi & \cos\phi\sin n\phi - \sin\phi\sin n\phi \end{bmatrix} \\ &= \begin{bmatrix} \cos[(n+1)\phi] & -\sin[(n+1)\phi] \\ \sin[(n+1)\phi] & \cos[(n+1)\phi] \end{bmatrix}. \end{aligned}$$

(b) This is a special case of Problem 8.3.1. We have

$$|M^n| = \begin{bmatrix} |\cos n\phi| & |-\sin n\phi| \\ |\sin n\phi| & |\cos n\phi| \end{bmatrix} = \begin{bmatrix} |\cos n\phi| & |\sin n\phi| \\ |\sin n\phi| & |\cos n\phi| \end{bmatrix}.$$

(c) If we let $a = |\cos n\phi|$ and $b = |\sin n\phi|$, then the characteristic polynomial of $|M^n|$ is

$$\det\big[\lambda I - |M^n|\big] = \det\begin{bmatrix}\lambda - a & -b\\ -b & \lambda - a\end{bmatrix} = (\lambda - a)^2 - b^2 .$$

Therefore, the eigenvalues of $|M^n|$ are the solutions of the quadratic equation

$$(\lambda - a)^2 - b^2 = (\lambda - a + b)(\lambda - a - b) = 0 .$$

Solving this equation yields: $\lambda = a \pm b = |\cos n\phi| \pm |\sin n\phi|$.

(d) Since $|M^n|$ is a real symmetric matrix, its Euclidean norm coincides with its spectral radius; see Problem 8.1.9. So, we have $\||M^n|\| = |\cos n\phi| + |\sin n\phi|$. ∎

7.5. The Essential Spectrum of a Bounded Operator

Problem 7.5.1. *For the identity operator I on a Banach space X establish the following:*

(a) *X is infinite dimensional if and only if $\big\|[I]\big\| = 1$.*

(b) *X is finite dimensional if and only if $\big\|[I]\big\| = 0$.*

Solution: (a) If X is finite dimensional, then $\mathcal{L}(X) = \mathcal{K}(X)$. Consequently, $\mathfrak{C}(X) = \{0\}$, the trivial Banach algebra. So, $\big\|[I]\big\| = 1$ implies that X must be infinite dimensional.

For the converse, assume that X is infinite dimensional. Then $I \notin \mathcal{K}(X)$, and so $\big\|[I]\big\| > 0$. From $0 < \big\|[I]\big\| = \big\|[I]^2\big\| \le \big\|[I]\big\|^2$, we see that $\big\|[I]\big\| \ge 1$. Since $\big\|[I]\big\| \le \|I\| \le 1$ is obviously true, we infer that $\big\|[I]\big\| = 1$.

(b) This equivalence follows immediately from the preceding part. ∎

Problem 7.5.2. *Let $\mathcal{A}$ be a unital Banach algebra, and let $G(\mathcal{A})$ denote the non-empty collection of all invertible elements of $\mathcal{A}$. Show that $G(\mathcal{A})$ is an open subset of $\mathcal{A}$ and that the mapping $x \mapsto x^{-1}$, from $G(\mathcal{A})$ to $G(\mathcal{A})$, is continuous—in particular, $G(\mathcal{A})$ is a topological group.*

Solution: Assume first that some element $a \in \mathcal{A}$ satisfies $\|a\| < 1$. From the inequality $\|a^n\| \le \|a\|^n$ it follows that the series $b = \sum_{n=0}^{\infty} a^n$ is norm convergent in $\mathcal{A}$. Now the continuity of the product map $(x, y) \mapsto xy$ implies that

$$(e - a)b = (e - a)\sum_{n=0}^{\infty} a^n = \sum_{n=0}^{\infty} a^n - \sum_{n=0}^{\infty} a^{n+1} = e ,$$

where e denotes the unit of $\mathcal{A}$. Similarly, $b(e - a) = e$. This shows that the element $e - a$ is invertible and $(e - a)^{-1} = \sum_{n=0}^{\infty} a^n$.

Now let any $a \in G(\mathcal{A})$, i.e., assume that a is an invertible element of $\mathcal{A}$. Pick some element $x \in \mathcal{A}$ such that $\|x - a\| < \frac{1}{\|a^{-1}\|}$. Then from

$$\|e - xa^{-1}\| = \|(a - x)a^{-1}\| \le \|x - a\| \cdot \|a^{-1}\| < 1$$

and the preceding conclusion, it follows that $xa^{-1} = e - (e - xa^{-1})$ is an invertible element. That is, there is some $y \in \mathcal{A}$ such that $(xa^{-1})y = y(xa^{-1}) = e$. This implies $x(a^{-1}y) = e$, and thus x has a right inverse. Similarly, x has a left inverse, and this guarantees that x is invertible,[1] that is, $x \in G(\mathcal{A})$. Consequently, $B(a, \frac{1}{\|a^{-1}\|}) \subseteq G(\mathcal{A})$, and so $G(\mathcal{A})$ is an open subset of $\mathcal{A}$. Since $e \in G(\mathcal{A})$, the open set $G(\mathcal{A})$ is automatically non-empty.

Finally, let us show that the mapping $x \mapsto x^{-1}$, from $G(\mathcal{A})$ to $G(\mathcal{A})$, is continuous. To this end, fix $a \in G(\mathcal{A})$. Then by the above, $B(a, \frac{1}{\|a^{-1}\|}) \subseteq G(\mathcal{A})$. Now let $y \in B(a, \frac{1}{\|a^{-1}\|})$. Clearly, $\|y-a\| \cdot \|a^{-1}\| < 1$ or $1 - \|y-a\| \cdot \|a^{-1}\| > 0$. From the identity

$$\begin{aligned} y^{-1} - a^{-1} &= a^{-1}(a-y)y^{-1} = a^{-1}(a-y)[(y^{-1}-a^{-1}) + a^{-1}] \\ &= a^{-1}(a-y)(y^{-1}-a^{-1}) + a^{-1}(a-y)a^{-1}, \end{aligned}$$

it follows that $\|y^{-1}-a^{-1}\| \le \|a^{-1}\| \cdot \|a-y\| \cdot \|y^{-1}-a^{-1}\| + \|a^{-1}\| \cdot \|a-y\| \cdot \|a^{-1}\|$. Therefore, $(1 - \|a^{-1}\| \cdot \|a-y\|)\|y^{-1}-a^{-1}\| \le \|a^{-1}\| \cdot \|a-y\| \cdot \|a^{-1}\|$, and so

$$\|y^{-1} - a^{-1}\| \le \tfrac{\|a^{-1}\|^2}{1-\|a^{-1}\|\cdot\|a-y\|}\, \|y-a\|$$

holds for all $y \in B(a, \frac{1}{\|a^{-1}\|})$. This implies that the map $x \mapsto x^{-1}$ is norm continuous at each element $a \in G(\mathcal{A})$. ∎

Problem 7.5.3. *Let $S\colon \ell_p \to \ell_p$, where $1 \le p < \infty$, be the forward shift operator defined by*

$$S(x_1, x_2, x_3, \ldots) = (0, x_1, x_2, x_3, \ldots).$$

Show that the essential spectrum of S coincides with the unit circle, that is,

$$\sigma_{\text{ess}}(S) = \Gamma = \{\lambda \in \mathbb{C}\colon\ |\lambda| = 1\}.$$

Solution: Clearly, S is an isometry, and so $\|S\| = 1$. From Example 6.21 we know that $\sigma(S) = \sigma(S^*) = \{\lambda \in \mathbb{C}\colon\ |\lambda| \le 1\}$. Hence, $\sigma_{\text{ess}}(S) \subseteq \{\lambda \in \mathbb{C}\colon\ |\lambda| \le 1\}$. We claim that:

(a) For any $\lambda \in \mathbb{C}$ the operator $\lambda - S$ is one-to-one, i.e., $N(\lambda - S) = \{0\}$.

(b) For any $\lambda \in \mathbb{C}$ with $|\lambda| < 1$ the operator $\lambda - S$ is bounded below, and hence $\lambda - S$ has closed range.

(c) For each complex number λ with $|\lambda| < 1$ the null space of the adjoint operator $(\lambda - S)^* = \lambda - S^*$ is $\{\alpha u\colon\ \alpha \in \mathbb{C}\}$, where $u = (1, \lambda, \lambda^2, \ldots) \in \ell_p$.

We verify each claim separately below.

(a) If $x = (x_1, x_2, \ldots) \in N(\lambda - S)$, then we have $\lambda x_1 = 0$ and $\lambda x_{i+1} - x_i = 0$ for each $i \ge 1$. If $\lambda = 0$, then $x_i = 0$ for each $i \ge 1$, i.e., $x = 0$. If $\lambda \ne 0$, then $x_1 = 0$ and $x_{i+1} = \frac{1}{\lambda} x_i$ for each $i \ge 1$, and so $x = 0$ in this case too. Thus, $N(\lambda - S) = \{0\}$.

(b) If $\lambda \in \mathbb{C}$ satisfies $|\lambda| < 1$, then for each $x \in \ell_p$ we have

$$\|(\lambda - S)x\| \ge \|Sx\| - \|\lambda x\| = \|x\| - |\lambda| \cdot \|x\| = (1 - |\lambda|)\|x\|.$$

[1] If $uv = wu = e$ in a unital algebra with unit e, then $w = we = w(uv) = (wu)v = ev = v$.

(c) From Example 6.21 we know that $S^*\colon \ell_q \to \ell_q$ is given by

$$S^*(x_1, x_2, x_3, \ldots) = (x_2, x_3, x_4, \ldots).$$

So, if $x \in N(\lambda - S^*)$, then $\lambda x_i - x_{i+1} = 0$ for each $i \geq 1$. This implies $x_{i+1} = \lambda^i x_1$ for each $i \geq 1$, and so $x = x_1(1, \lambda, \lambda^2, \ldots)$. On the other hand, if $x = \alpha(1, \lambda, \lambda^2, \ldots)$, then it is easy to see that $(\lambda - S^*)x = 0$. Therefore, $N(\lambda - S^*) = \{\alpha u\colon\ \alpha \in \mathbb{C}\}$.

In particular, (b) and (c) show that for each $\lambda \in \mathbb{C}$ with $|\lambda| < 1$ the operator $\lambda - S$ is a Fredholm operator (of index -1), and so the operator $\lambda - S$ is essentially invertible. This implies that $\sigma_{\text{ess}}(S) \subseteq \Gamma$.

Now let $\lambda \in \Gamma = \partial\sigma(S) \subseteq \sigma_a(S)$. Since the operator $\lambda - S$ is not bounded below and $N(\lambda - S) = \{0\}$, it follows that $\lambda - S$ does not have closed range. This implies that $\lambda - S$ is not a Fredholm operator. Consequently, the operator $\lambda - S$ is not essentially invertible, and so $\lambda \in \sigma_{\text{ess}}(S)$. Thus, $\Gamma \subseteq \sigma_{\text{ess}}(S)$ is also true, and hence $\sigma_{\text{ess}}(S) = \Gamma$. ∎

Problem 7.5.4. *Show that if an operator $T\colon X \to X$ on an infinite dimensional Banach space satisfies $\sigma(T) = \{\lambda_0\}$, then $\sigma_{\text{ess}}(T) = \{\lambda_0\}$.*

Solution: We know that $\sigma_{\text{ess}}(T) \subseteq \sigma(T)$. If $\lambda_0 \notin \sigma_{\text{ess}}(T)$, then $\sigma_{\text{ess}}(T) = \emptyset$, a contradiction. ∎

Problem 7.5.5. *Show that the power compact operators on a Banach space are precisely the essentially nilpotent operators.*

Deduce from this that every bounded operator on a Banach lattice that is dominated by a compact positive operator is essentially nilpotent.

Solution: Let $T\colon X \to X$ be a bounded operator on a Banach space. Then T^n is a compact operator if and only if the equality $[T]^n = [T^n] = 0$ holds in the Calkin algebra $\mathfrak{C}(X)$. That is, T is power compact if and only if $[T]$ is essentially nilpotent.

Now let $T\colon E \to E$ be a bounded operator on a Banach lattice that is dominated by a compact positive operator. By Corollary 2.35 we know that T^3 is a compact operator, and so T is essentially nilpotent. ∎

Problem 7.5.6. *Let λ_0 be an isolated point in the spectrum of an operator $T \in \mathcal{L}(X)$. Show without using Theorem 7.44 that if the spectral projection $P_{\lambda_0}(T)$ is of finite-rank, then λ_0 is a pole of the resolvent of T.*

Solution: Let $X_1 = R(P_{\lambda_0}(T))$ and $X_2 = N(P_{\lambda_0}(T))$. According to our hypothesis, X_1 is finite dimensional, $X = X_1 \oplus X_2$, and the closed subspaces X_1 and X_2 are both invariant under T. Moreover, if $T_i = T|_{X_i}$ (the restriction of T to X_i), then $\sigma(T_1) = \{\lambda_0\}$ and $\sigma(T_2) = \sigma(T) \setminus \{\lambda_0\}$. Also, for brevity write $P = P_{\lambda_0}$ and $Q = I - P$.

Notice that the resolvent operator satisfies

$$R(\lambda, T) = R(\lambda, T_1) \oplus R(\lambda, T_2) = R(\lambda, T)|_{X_1} \oplus R(\lambda, T)|_{X_2}.$$

From $\lambda_0 \notin \sigma(T_2)$, it follows that $R(\lambda, T_2) = R(\lambda, T)|_{X_2} = R(\lambda, T)Q$ is analytic around λ_0. Furthermore, since X_1 is finite dimensional, all singularities of $R(\lambda, T_1)$ are poles; see Corollary 6.41. So, λ_0 is a pole of $R(\lambda, T_1) = R(\lambda, T)|_{X_1} = R(\lambda, T)P$. This implies that λ_0 is also a pole of $R(\lambda, T) = R(\lambda, T)P + R(\lambda, T)Q$. ∎

Problem 7.5.7. *Give an example of an operator $T \in \mathcal{L}(X)$ and some point $\lambda \in \sigma(T) \setminus \sigma_{\rm ess}(T)$ that is not an eigenvalue of T.*

Solution: Let $S\colon \ell_2 \to \ell_2$ be the forward shift, i.e., $S(x_1, x_2, \dots) = (0, x_1, x_2, \dots)$. The spectrum of S coincides with the closed unit disk; see Example 6.21. According to Problem 7.5.3, we have $\sigma_{\rm ess}(S) = \Gamma$, the unit circle. Clearly, $0 \notin \sigma_{\rm ess}(S)$.

If $T\colon \ell_2 \to \ell_2$ is the forward shift, then $\pi(T)$ is the inverse of $\pi(S)$ in $\mathfrak{C}(X)$. This implies that $0 \in \sigma(S) \setminus \sigma_{\rm ess}(S) = \{\lambda \in \mathbb{C}\colon\ |\lambda| < 1\}$. Now notice that 0 is not an eigenvalue of S. ∎

Problem 7.5.8. *Show that every strictly singular operator on an infinite dimensional Banach space is essentially quasinilpotent.*

Solution: Assume that $T \in \mathcal{L}(X)$ is a strictly singular operator on an infinite dimensional Banach space. According to Theorem 7.48, in order to establish that T is essentially quasinilpotent, we must show that each non-zero point $\lambda \in \sigma(T)$ is an isolated point of $\sigma(T)$ and that the spectral projection $P_\lambda(T)$ has finite-rank.

To this end, let $\lambda \in \sigma(T) \setminus \{0\}$. From Theorem 7.11 we know that λ is an isolated point of $\sigma(T)$. Next, according to Problem 6.4.13, there exists an operator $S \in \mathcal{L}(X)$ such that $P_{\lambda_0}(T) = TS = ST$. This implies that the operator $P_{\lambda_0}(T)$ in $\mathcal{L}(X)$ is strictly singular. Since $P_{\lambda_0}(T)$ is the identity operator on its range, it follows that $P_{\lambda_0}(T)$ has finite-rank, and the solution is finished. ∎

Problem 7.5.9. *Establish the following properties regarding the measure of non-compactness.*

(a) *For each subset A of a metric space we have $\chi(A) = \chi(\overline{A})$.*

(b) *A subset A of a metric space is relatively compact if and only if $\chi(A) = 0$.*

(c) *A bounded operator $T \in \mathcal{L}(X)$ is compact if and only if $\chi(T) = 0$.*

(d) *If $T \in \mathcal{L}(X,Y)$ and $S \in \mathcal{L}(Y,Z)$, then $\chi(ST) \le \chi(S)\chi(T)$.*

Solution: (a) Let A be a subset of a metric space (M,d). If $\chi(A) = \infty$, then $\chi(\overline{A}) = \infty$ is also true. So, we can assume that $\chi(A) < \infty$. Pick any $r > \chi(A)$ and then choose points $x_1, \dots, x_n$ in M such that $A \subseteq \bigcup_{i=1}^n B(x_i, r)$. Since the set $C(x_i, r) = \{y \in M\colon\ d(x_i, y) \le r\}$ is a closed set and satisfies $B(x_i, r) \subseteq C(x_i, r)$, it follows that for each $\epsilon > 0$ we have

$$\overline{A} \subseteq \overline{\textstyle\bigcup_{i=1}^n B(x_i, r)} = \bigcup_{i=1}^n \overline{B(x_i, r)} \subseteq \bigcup_{i=1}^n C(x_i, r) \subseteq \bigcup_{i=1}^n B(x_i, r+\epsilon)\,.$$

This implies $\chi(\overline{A}) \le r + \epsilon$ for each $r > \chi(A)$ and all $\epsilon > 0$. Thus, $\chi(\overline{A}) \le \chi(A)$. Since $\chi(A) \le \chi(\overline{A})$ is obviously true, we easily conclude that $\chi(A) = \chi(\overline{A})$.

(b) It is easy to see that for a subset A of a metric space the condition $\chi(A) = 0$ is equivalent to saying that A is a totally bounded subset. That is, a subset A of a metric space A is relatively compact if and only if $\chi(A) = 0$.

(c) This follows immediately from (b).

(d) Assume $T \in \mathcal{L}(X,Y)$ and $S \in \mathcal{L}(Y,Z)$. Fix $r > \chi(T)$ and $\rho > \chi(S)$. Pick $y_1, \ldots, y_n \in Y$ such that $T(U_X) \subseteq \bigcup_{i=1}^{n}(y_i + rU_Y)$. Similarly, choose $z_1, \ldots, z_m \in Z$ such that $S(U_Y) \subseteq \bigcup_{j=1}^{m}(z_j + \rho U_Z)$, and note that

$$\begin{aligned} ST(U_X) &= S\Big(\bigcup_{i=1}^{n}[y_i + rU_Y]\Big) = \bigcup_{i=1}^{n} S(y_i + rU_Y) \\ &= \bigcup_{i=1}^{n}[Sy_i + rS(U_Y)] \subseteq \bigcup_{i=1}^{n}\bigcup_{j=1}^{m}\big[Sy_i + rz_j + r\rho U_Z\big]. \end{aligned}$$

This implies $\chi(ST) \le r\rho$ for all $r > \chi(T)$ and $\rho > \chi(S)$. So, $\chi(ST) \le \chi(S)\chi(T)$. ∎

Problem 7.5.10. *If X is a Banach space, then show that the measure of non-compactness $\chi\colon \mathcal{L}(X) \to \mathbb{R}$ is a seminorm. Also, show that this seminorm induces a norm on the Calkin algebra $\mathfrak{C}(X)$.*

Solution: Recall that if $T \in \mathcal{L}(X)$, then the measure $\chi(T)$ of non-compactness of T is defined by

$$\chi(T) = \inf\big\{r > 0\colon\ \exists\ x_1, \ldots, x_n \text{ such that } T(U_X) \subseteq \textstyle\bigcup_{i=1}^{n} B(x_i, r)\big\}.$$

In particular, we have $\chi(T) \ge 0$ for each $T \in \mathcal{L}(X)$. We now verify the remaining desired properties of the function $\chi\colon \mathcal{L}(X) \to \mathbb{R}$.

1. *The homogeneity, i.e., $\chi(\lambda T) = |\lambda|\chi(T)$ for all $T \in \mathcal{L}(X)$ and all scalars λ.*

The verification of the homogeneity property will be based upon the identity $\mu B(0,r) = B(0,|\mu|r)$ for each $r > 0$ and μ. Fix $T \in \mathcal{L}(X)$ and a scalar $\lambda \neq 0$.

If $r > \chi(T)$, pick vectors $x_1, \ldots, x_n$ such that $T(U_X) \subseteq \bigcup_{i=1}^{n}\big[x_i + B(0,r)\big]$ and note that $\lambda T(U_X) \subseteq \bigcup_{i=1}^{n}\big[\lambda x_i + \lambda B(0,r)\big] = \bigcup_{i=1}^{n}\big[\lambda x_i + B(0,|\lambda|r)\big]$. This implies $\chi(\lambda T) \le |\lambda| r$ for each $r > \chi(T)$, and from this it follows that $\chi(\lambda T) \le |\lambda|\chi(T)$.

Now let $\rho > \chi(\lambda T)$. Choose $y_1, \ldots, y_k$ such that $\lambda T(U_X) \subseteq \bigcup_{i=j}^{k}\big[y_j + B(0,\rho)\big]$. This implies $T(U_X) \subseteq \bigcup_{i=j}^{k}\big[\frac{y_j}{\lambda} + \frac{1}{\lambda}B(0,\rho)\big] = \bigcup_{i=j}^{k}\big[\frac{y_j}{\lambda} + B\big(0, \frac{\rho}{|\lambda|}\big)\big]$. Consequently, $\chi(T) \le \frac{\rho}{|\lambda|}$ holds for each $\rho > \chi(\lambda T)$, and so $\chi(T) \le \frac{\chi(\lambda T)}{|\lambda|}$ or $|\lambda|\chi(T) \le \chi(\lambda T)$. Therefore, $\chi(\lambda T) = |\lambda|\chi(T)$. For $\lambda = 0$, the preceding identity should be obvious.

2. *The triangle inequality, i.e., $\chi(S+T) \le \chi(S) + \chi(T)$ for all $S, T \in \mathcal{L}(X)$.*

Let $S, T \in \mathcal{L}(X)$. Fix $r > \chi(S)$ and $\rho > \chi(T)$. Pick vectors $x_1, \ldots, x_n$ and $y_1, \ldots, y_k$ such that $S(U_X) \subseteq \bigcup_{i=1}^{n}\big[x_i + B(0,r)\big]$ and $T(U_X) \subseteq \bigcup_{j=1}^{k}\big[y_j + B(0,\rho)\big]$. It follows that

$$\begin{aligned} (S+T)(U_X) &\subseteq S(U_X) + T(U_X) \subseteq \bigcup_{i=1}^{n}\bigcup_{j=1}^{k}\big[x_i + y_j + B(0,r) + B(0,\rho)\big] \\ &= \bigcup_{i=1}^{n}\bigcup_{j=1}^{k}\big[x_i + y_j + B(0, r+\rho)\big]. \end{aligned}$$

This implies $\chi(S+T) \le r + \rho$ for all $r > \chi(S)$ and all $\rho > \chi(T)$. Consequently, $\chi(S+T) \le \chi(S) + \chi(T)$.

3. *If* $S - T$ *is a compact operator, then* $\chi(S) = \chi(T)$.

Let $S - T = K$, a compact operator. From $S = T + K$, the triangle inequality, and Problem 7.5.9(c), it follows that $\chi(S) = \chi(T + K) \leq \chi(T) + \chi(K) = \chi(T)$. Similarly, from $T = S + (-K)$ we get $\chi(T) \leq \chi(S)$. Therefore, $\chi(S) = \chi(T)$.

A consequence of (3) and the other properties above is that the function $\chi\colon \mathcal{L}(X) \to \mathbb{R}$ when restricted to the Calkin algebra $\mathfrak{C}(X)$ is a norm. ∎

Chapter 8

Positive Matrices

8.1. The Banach Lattices $M_n(\mathbb{R})$ and $M_n(\mathbb{C})$

Problem 8.1.1. *If A is an $n \times n$ matrix with real or complex entries, then show that*

$$(Ax, y) = (x, \overline{A}^t y)$$

for all $x, y \in \mathbb{C}^n$. In particular, show that A is Hermitian if and only if $(Ax, y) = (x, Ay)$ for all $x, y \in \mathbb{C}^n$.

Solution: Let $A = [a_{ij}]$ be an $n \times n$ matrix with complex entries. Fix $x, y \in \mathbb{C}^n$, and note that

$$\begin{aligned} Ax &= \Big(\sum_{j=1}^n a_{1j}x_j, \sum_{j=1}^n a_{2j}x_j, \ldots, \sum_{j=1}^n a_{nj}x_j\Big), \quad \text{and} \\ \overline{A}^t y &= \Big(\sum_{i=1}^n \overline{a}_{i1}y_i, \sum_{i=1}^n \overline{a}_{i2}y_i, \ldots, \sum_{i=1}^n \overline{a}_{in}y_i\Big). \end{aligned}$$

Consequently, we have

$$(Ax, y) = \sum_{i=1}^n \Big[\sum_{j=1}^n a_{ij}x_j\Big]\overline{y}_i = \sum_{j=1}^n \Big[\sum_{i=1}^n a_{ij}\overline{y}_i\Big]x_j = \sum_{j=1}^n x_j\Big[\overline{\textstyle\sum_{i=1}^n \overline{a}_{ij}y_i}\Big] = (x, \overline{A}^t y).$$

Finally, note that $(Ax, y) - (x, Ay) = (x, (\overline{A}^t - A)y)$ holds for all $x, y \in \mathbb{C}^n$. So, if $(Ax, y) = (x, Ay)$ or $(Ax, y) - (x, Ay) = 0$ for all $x, y \in \mathbb{C}^n$, then $(x, (\overline{A}^t - A)y) = 0$ for all $x, y \in \mathbb{C}^n$. Letting, $x = (\overline{A}^t - A)y$, we get $((\overline{A}^t - A)y, (\overline{A}^t - A)y) = 0$ for all $y \in \mathbb{C}^n$. This implies $(\overline{A}^t - A)y = 0$ for all $y \in \mathbb{C}^n$, and so $\overline{A}^t = A$. ∎

Problem 8.1.2. *Show that for an $n \times n$ matrix A the following statements are equivalent.*

(a) *A is unitary.*

(b) *The rows of A form an orthonormal basis in* $\mathbb{C}^n$.

(c) *The columns of A form an orthonormal basis in* $\mathbb{C}^n$.

(d) *A is* ***inner product preserving****, i.e.,* $(Ax, Ay){=}(x, y)$ *for* $x, y \in \mathbb{C}^n$.

(e) *A is Euclidean* ***norm preserving****, i.e.,* $\|Ax\| = \|x\|$ *for* $x, y \in \mathbb{C}^n$.

(f) *A carries an orthonormal basis to an orthonormal basis.*

Solution: Clearly, (a), (b), and (c) are equivalent.

(a) $\Longrightarrow$ (d) Note that $(Ax, Ay) = (\overline{A}^t Ax, y) = (x, y)$.

(d) $\Longrightarrow$ (e) We have $\|Ax\|^2 = (Ax, Ax) = (x, x) = \|x\|^2$, and so $\|Ax\| = \|x\|$ for all $x \in \mathbb{C}^n$.

(e) $\Longrightarrow$ (a) The proof of this implication will be based upon the following property of Hermitian operators.

(P) *A Hermitian $n \times n$ matrix H is equal to zero if and only if $(Hx, x) = 0$ for all $x \in \mathbb{C}^n$.*

To establish (P), assume that a Hermitian $n \times n$ matrix H satisfies $(Hx, x) = 0$ for all $x \in \mathbb{C}^n$. Then for all $x, y \in \mathbb{C}^n$ we have

$$\begin{aligned} 0 &= (H(x + \imath y), x + \imath y) = (Hx, x) - \imath(Hx, y) + \imath(Hy, x) + (H(\imath y), \imath y) \\ &= \imath\big[(Hy, x) - (Hx, y)\big] = \imath\big[(Hy, x) - (x, Hy)\big] = \imath\big[(Hy, x) - \overline{(Hy, x)}\,\big] \\ &= 2\imath \operatorname{Im}(Hy, x)\,. \end{aligned}$$

Consequently, $\operatorname{Im}(Hy, x) = 0$ for all $x, y \in \mathbb{C}^n$. Replacing y by $\imath y$, we also have $\operatorname{Im}(H(\imath y), x) = \operatorname{Re}(Hy, x) = 0$. So, $(Hy, x) = 0$ for all $x, y \in \mathbb{C}^n$ and this implies that $H = 0$.

To see that (e) implies (a), note that $\|Ax\| = \|x\|$ is equivalent to the identity $(Ax, Ax) = (x, x)$ or $((\overline{A}^t A - I)x, x) = 0$ for each $x \in \mathbb{C}^n$. Since $\overline{A}^t A - I$ is a Hermitian matrix, it follows from (P) that $\overline{A}^t A - I = 0$ (and also that $A\overline{A}^t - I = 0$). That is, $\overline{A}^t A = A\overline{A}^t = I$, and this shows that A is a unitary matrix.

(d) $\Longrightarrow$ (f) Let $\{e_1, \ldots, e_n\}$ be an orthonormal basis. Then $\|Ae_i\| = \|e_i\| = 1$ for each i, and $(Ae_i, Ae_j) = (e_i, e_j) = 0$ for $i \neq j$. This implies that $\{Ae_1, \ldots, Ae_n\}$ is an orthonormal basis.

(f) $\Longrightarrow$ (e) Assume that A carries an orthonormal basis $\{e_1, \ldots, e_n\}$ to an orthonormal basis $\{Ae_1, \ldots, Ae_n\}$. Then for each $x = \sum_{i=1}^n \lambda_i e_i$ we have

$$\|x\|^2 = \sum_{i=1}^n |\lambda_i|^2 = \|Ax\|^2\,,$$

and so $\|Ax\| = \|x\|$ holds for each $x \in \mathbb{C}^n$. ∎

Problem 8.1.3. *As in the case of operators, two $n \times n$ matrices A and B are said to be* ***similar*** *if there exists an invertible matrix C such that $A = C^{-1}BC$. Recall also that the* ***trace*** $\operatorname{tr}(A)$ *of an $n \times n$ matrix $A = [a_{ij}]$*

is the sum of its diagonal entries, i.e., $\operatorname{tr}(A) = \sum_{i=1}^{n} a_{ii}$. *Establish the following.*

(i) *Two similar matrices have the same characteristic polynomial.*

(ii) *This part deals with the trace of matrices.*

(a) *Show that if A and B are two arbitrary $n \times n$ matrices, then we have* $\operatorname{tr}(AB) = \operatorname{tr}(BA)$.

(b) *Show that the trace of an $n \times n$ matrix A coincides with the sum of its eigenvalues. That is, if $\lambda_1, \ldots, \lambda_n$ are the eigenvalues of A (counted with their multiplicities), then* $\operatorname{tr}(A) = \sum_{i=1}^{n} \lambda_i$.

(c) *Show that two similar $n \times n$ matrices have the same trace, i.e., the trace of an $n \times n$ matrix is* similarity invariant.

(d) *Consider an $n \times n$ matrix A with column vectors $c_1, \ldots, c_n$. Also, let e_i denote the ith unit vector in $\mathbb{C}^n$ and consider e_i as a linear functional acting on $\mathbb{C}^n$ (or $\mathbb{R}^n$) via the formula* $e_i(x) = (x, e_i) = x_i$.

Show that A, as an operator on $\mathbb{C}^n$ (or $\mathbb{R}^n$), can be identified with the finite-rank operator $T = \sum_{i=1}^{n} e_i \otimes c_i$.

(e) *Using the notation of part* (b), *show that* $\operatorname{tr}(A) = \sum_{i=1}^{n} e_i(c_i)$.

Solution: (i) Assume $A = C^{-1}BC$. Then we have

$$\begin{aligned} p_A(\lambda) &= \det(\lambda I - A) = \det\bigl(C^{-1}(\lambda I - B)C\bigr) = \det\bigl(C^{-1}\bigr)\det(\lambda I - B)\det(C) \\ &= [\det(C)]^{-1}\det(\lambda I - B)\det(C) = \det(\lambda I - B) = p_B(\lambda)\,. \end{aligned}$$

(ii) (a) Observe that the $(i,i)^{\text{th}}$ entry of the matrix AB is $\sum_{k=1}^{n} a_{ik}b_{ki}$ and the $(k,k)^{\text{th}}$ entry of BA is $\sum_{i=1}^{n} b_{ki}a_{ik}$. So,

$$\operatorname{tr}(AB) = \sum_{i=1}^{n}\sum_{k=1}^{n} a_{ik}b_{ki} = \sum_{k=1}^{n}\sum_{i=1}^{n} a_{ik}b_{ki} = \sum_{k=1}^{n}\sum_{i=1}^{n} b_{ki}a_{ik} = \operatorname{tr}(BA)\,.$$

(b) From the expansion formula of the determinant, we see that

$$\begin{aligned} p_A(\lambda) &= \det(\lambda I - A) \\ &= \lambda^n + a_{n-1}\lambda^{n-1} + \cdots + a_1\lambda + a_0 \\ &= \lambda^n - \operatorname{tr}(A)\lambda^{n-1} + \cdots + a_1\lambda + (-1)^n\det(A)\,. \end{aligned}$$

On the other hand, we also have

$$\begin{aligned} p_A(\lambda) &= (\lambda - \lambda_1)(\lambda - \lambda_1)\cdots(\lambda - \lambda_n) \\ &= \lambda^n - \Bigl(\sum_{i=1}^{n}\lambda_i\Bigr)\lambda^{n-1} + \cdots + a_1\lambda + (-1)^n\prod_{i=1}^{n}\lambda_i\,. \end{aligned}$$

The above equations show that

$$\operatorname{tr}(A) = \sum_{i=1}^{n}\lambda_i \quad \text{and} \quad \det(A) = \prod_{i=1}^{n}\lambda_i\,.$$

(c) Let A and B be two similar $n \times n$ matrices. By part (i), we know that $p_A(\lambda) = p_B(\lambda)$. So, A and B have the same eigenvalues (together with their multiplicities), say $\lambda_1, \ldots, \lambda_n$. Now (b) above implies $\operatorname{tr}(A) = \sum_{i=1}^n \lambda_i = \operatorname{tr}(B)$.

(d) Note that if $x = (x_1, x_2, \ldots, x_n) \in \mathbb{C}^n$, then

$$T(x) = \sum_{i=1}^{n} e_i \otimes c_i(x) = \sum_{i=1}^{n} e_i(x) c_i = \sum_{i=1}^{n} x_i c_i = Ax\,,$$

where in the last expression x is considered as a column vector.

(e) Clearly, $\sum_{i=1}^n a_{ii} = \sum_{i=1}^n (c_i)_i = \sum_{i=1}^n e_i(c_i)$. (See also Problem 4.1.4.) ∎

Problem 8.1.4. *Show that each eigenvalue of a unitary matrix has modulus one.*

Solution: Let U be a unitary $n \times n$ matrix, i.e., $U^{-1} = \overline{U}^t$. Then with respect to the Euclidean norm we have

$$\|Ux\|^2 = (Ux, Ux) = (\overline{U}^t Ux, x) = (x, x) = \|x\|^2\,,$$

or $\|Ux\| = \|x\|$ for each $x \in \mathbb{C}^n$. This shows that U defines a surjective isometry. By Problem 6.2.5, it follows that every eigenvalue of U lies on the unit circle.

A direct proof goes as follows. Assume that a unit vector $z \in \mathbb{C}^n$ satisfies $Uz = \lambda z$, then

$$|\lambda|^2 = \lambda\overline{\lambda}(z, z) = (\lambda z, \lambda z) = (Uz, Uz) = (\overline{U}^t Uz, z) = (z, z) = 1\,,$$

and so $|\lambda| = 1$. ∎

Problem 8.1.5. *Show that every Hermitian matrix has real eigenvalues and that eigenvectors corresponding to distinct eigenvalues are orthogonal.*

Solution: If A is a Hermitian matrix and a unit vector $z \in \mathbb{C}^n$ satisfies $Az = \lambda z$, then

$$\overline{\lambda} = \overline{\lambda}(z, z) = (z, \lambda z) = (z, Az) = (\overline{A}^t z, z) = (Az, z) = (\lambda z, z) = \lambda(z, z) = \lambda\,.$$

That is, every eigenvalue of A is a real number.

Assume that two non-zero vectors $z, w \in \mathbb{C}^n$ satisfy $Az = \lambda_1 z$ and $Aw = \lambda_2 w$ with $\lambda_1 \neq \lambda_2$. From the preceding part, we know that λ_1 and λ_2 are real numbers. Now note that

$$\begin{aligned}(\lambda_1 - \lambda_2)(z, w) &= \lambda_1(z, w) - \lambda_2(z, w) = (\lambda_1 z, w) - (z, \overline{\lambda}_2 w)\\ &= (\lambda_1 z, w) - (z, \lambda_2 w) = (Az, w) - (z, Aw)\\ &= (Az, w) - (Az, w) = 0\,.\end{aligned}$$

This implies $(z, w) = 0$, and so z and w are orthogonal vectors. ∎

Problem 8.1.6. *Recall that a square matrix A is called* ***diagonalizable*** *if there exists an invertible $n \times n$ matrix R (called a* ***diagonalizing matrix*** *for A) such that the matrix $R^{-1}AR$ is diagonal. Establish the following.*

(a) *A matrix A is diagonalizable if and only if A has n linearly independent eigenvectors.*

(b) *Every Hermitian matrix is diagonalizable by a unitary matrix.*

Solution: (a) Assume that an invertible matrix R diagonalizes the matrix A, that is, $R^{-1}AR = D$ or, equivalently, $AR = RD$. From Problem 8.1.3(i), we know that $D = \text{diag}(\lambda_1, \lambda_2, \ldots, \lambda_n)$, where $\lambda_1, \lambda_2, \ldots, \lambda_n$ are the eigenvalues of A (counted with their multiplicities). If $c_1, c_2, \ldots, c_n$ denote the column vectors of the matrix R, then note that the equation $AR = RD$ can be written in the form

$$\begin{bmatrix} | & | & \cdots & | \\ Ac_1 & Ac_2 & \cdots & Ac_n \\ | & | & \cdots & | \end{bmatrix} = \begin{bmatrix} | & | & \cdots & | \\ \lambda_1 c_1 & \lambda_2 c_2 & \cdots & \lambda_n c_n \\ | & | & \cdots & | \end{bmatrix}. \qquad (\star)$$

That is, $Ac_i = \lambda_i c_i$ for $i = 1, 2, \ldots, n$. Since the (column) vectors $c_1, c_2, \ldots, c_n$ are linearly independent, it follows that A has n linearly independent eigenvectors.

Conversely, assume that $Ac_i = \lambda_i c_i$ for each i and that the vectors $c_1, c_2, \ldots, c_n$ are linearly independent. Then the matrix R with columns $c_1, c_2, \ldots, c_n$ is invertible and, in view of $(\star)$, it satisfies $AR = RD$ or $R^{-1}AR = \text{diag}(\lambda_1, \lambda_2, \ldots, \lambda_n)$.

(b) In order to solve this part, we need the following property that is also useful in many other contexts.

- **(Closed Subspaces Have Orthogonal Complements)** *If V is a closed subspace of a Hilbert space $\mathcal{H}$, then $V \oplus V^\perp = \mathcal{H}$.*

To see this, note first that the Pythagorean theorem guarantees that $V \oplus V^\perp$ is a closed subspace. Take any $x \in \mathcal{H}$ and consider the unique nearest to x vector $v \in V$. By Problem 1.1.18(f) this vector exists and $x - v \perp V$. Therefore, we have $x = v + (x - v)$ with $v \in V$ and $x - v \in V^\perp$. Hence, $V \oplus V^\perp = \mathcal{H}$.

We now return to our problem. So, let A be a Hermitian matrix. From Problem 8.1.5, we know that eigenvectors corresponding to distinct eigenvalues of A are orthogonal. We shall prove first that A is diagonalizable. To see this, let V be the vector subspace generated in $\mathbb{C}^n$ by the collection of all eigenvectors of A. It should be clear that $Av \in V$ for all $v \in V$, i.e., V is A-invariant. We claim that $V = \mathbb{C}^n$.

To see this, assume by way of contradiction that $V \neq \mathbb{C}^n$. Then, by ($\bullet$) above, we have $V \oplus V^\perp = \mathbb{C}^n$, and so the vector subspace

$$V^\perp = \{z \in \mathbb{C}^n \colon \ z \perp v = 0 \ \text{ for all } \ v \in V\}$$

of $\mathbb{C}^n$ is different from $\{0\}$. Observe that $V^\perp$ is A-invariant. Indeed, if $z \in V^\perp$, then note that for each $v \in V$ we have $(Az, v) = (z, Av) = 0$, that is, $Az \in V^\perp$. But then, since $V^\perp$ is finite dimensional, A must have a non-zero eigenvector lying in $V^\perp$, say $z \in V^\perp$. This implies, $z \in V$ and so $z = 0$, which is a contradiction. Hence, $V = \mathbb{C}^n$. So, A has n linearly independent eigenvectors, and thus it is diagonalizable.

To complete the solution, we need to show that A has n orthonormal eigenvectors. To this end, let $\lambda_1, \ldots, \lambda_k$ be the distinct eigenvalues of A, and let $V_i = \{z \in \mathbb{C}^n \colon \ Az - \lambda_i z\}$. By Problem 8.1.5, we know that $V_i \perp V_j$ if $i \neq j$ and $V_1 \oplus V_2 \oplus \cdots \oplus V_k = \mathbb{C}^n$. If $E_i = \{e_1^i, \ldots, e_{m_i}^i\}$ is an orthonormal basis of V_i, then $E = \bigcup_{i=1}^{k} E_i$ is an orthonormal basis of $\mathbb{C}^n$ consisting of eigenvectors of A. So, if U is the $n \times n$ matrix whose columns are the vectors of the orthonormal basis E, then U is a unitary matrix that diagonalizes A. ∎

Problem 8.1.7. *An $n \times n$ matrix A is said to be* ***positive semidefinite*** *whenever $(Az, z) \geq 0$ for all $z \in \mathbb{C}^n$. Establish the following:*

(a) *If A is a positive semidefinite matrix, then A has non-negative eigenvalues.*

(b) *A Hermitian matrix A is positive semidefinite if and only if it has non-negative eigenvalues.*

Solution: (a) Let A be a positive semidefinite matrix. If λ is an eigenvalue of A and $z \in \mathbb{C}^n$ is a unit vector satisfying $Az = \lambda z$, then note that

$$\lambda = \lambda(z, z) = (\lambda z, z) = (Az, z) \geq 0 .$$

(b) Assume that A is a Hermitian matrix. Then, according to Problem 8.1.6(b), the matrix A can be diagonalized by a unitary matrix. That is, there exists a unitary matrix U satisfying $\overline{U}^t AU = \operatorname{diag}(\lambda_1, \ldots, \lambda_n)$, where $\lambda_1, \lambda_2, \ldots, \lambda_n$ are the eigenvalues of A counted with their multiplicities. Now letting $z = Uy$, we see that

$$(Az, z) = (AUy, Uy) = (\overline{U}^t AUy, y) = \sum_{i=1}^{n} \lambda_i |y_i|^2 \geq 0$$

for all $y \in \mathbb{C}^n$. From this, it should be easy to establish that A is positive semidefinite if and only if $\lambda_i \geq 0$ for each i. ∎

Problem 8.1.8. *For each k let $A_k = [a_{ij}^{(k)}]$ be an $n \times n$ matrix. It is easy to see that the sequence of matrices $\{A_k\}$ converges to a matrix $A = [a_{ij}]$ in the Banach lattice $M_n(\mathbb{C})$ if and only if $\lim_{k\to\infty} a_{ij}^{(k)} = a_{ij}$ for each i and j.*

Show that an $n \times n$ matrix A satisfies $A^k \to 0$ in $M_n(\mathbb{C})$ if and only if every eigenvalue of A has modulus less than one.

Solution: Assume first that every eigenvalue of A has modulus less than one. Then we have

$$\lim_{k\to\infty} \|A^k\|^{\frac{1}{k}} = r(A) = \max\{|\lambda| \colon \ \lambda \text{ is an eigenvalue of } A\} < 1 .$$

So, if we fix some real number δ satisfying $r(A) < \delta < 1$, then there exists some k_0 such that $\|A^k\|^{\frac{1}{k}} < \delta$ or $\|A^k\| < \delta^k$ for all $k \geq k_0$. This implies $A^k \to 0$.

For the converse, suppose that $A^k \to 0$ and let $\lambda \in \mathbb{C}$ be an eigenvalue of A. Pick some unit vector $z \in \mathbb{C}^n$ such that $Az = \lambda z$, and note that $A^k z = \lambda^k z$ for each k. In particular, we have

$$|\lambda|^k = \|\lambda^k z\| = \|A^k z\| \leq \|A^k\| .$$

This implies $\lim_{k\to\infty} |\lambda|^k = 0$, and from this we infer that $|\lambda| < 1$ for each eigenvalue λ of the matrix A. (Compare this solution with that of Problem 6.1.11.) ∎

Problem 8.1.9. *Consider a matrix $A \in M_n(\mathbb{C})$. If λ_1 is the (non-negative) largest eigenvalue of the Hermitian matrix $\overline{A}^t A$, then show that $\|A\| = \sqrt{\lambda_1}$. In particular, if A is Hermitian, then show that its Euclidean norm coincides with its spectral radius.*

Solution: Since $(\overline{A}^t Az, z) = (Az, Az) = \|Az\|^2 \geq 0$ holds for all $z \in \mathbb{C}^n$, we infer that the Hermitian matrix $\overline{A}^t A$ is positive semidefinite. Thus, by Problem 8.1.7, the matrix $\overline{A}^t A$ has non-negative eigenvalues, say $\lambda_1 \geq \lambda_2 \geq \cdots \geq \lambda_n \geq 0$. Moreover, since $\overline{A}^t A$ is Hermitian, it is (according to Problem 8.1.6) diagonalizable by some unitary matrix, say U. That is, there exists a unitary matrix U such that $\overline{U}^t\overline{A}^t AU = \operatorname{diag}(\lambda_1, \lambda_2, \ldots, \lambda_n)$.

Fix a unit vector $x \in \mathbb{C}^n$ and let $y = U^{-1}x$. Clearly, $\|y\| = 1$. Then we have

$$\begin{aligned}\|Ax\|^2 &= (Ax, Ax) = (x, \overline{A}^t Ax) = (Uy, \overline{A}^t AUy)\\ &= (y, \overline{U}^t\overline{A}^t AUy) = \sum_{i=1}^n \lambda_i |y_i|^2 \leq \sum_{i=1}^n \lambda_1 |y_i|^2 = \lambda_1 .\end{aligned}$$

This implies $\|A\| = \sup_{\|x\|=1} \|Ax\| \leq \sqrt{\lambda_1}$.

On the other hand, there exists some unit vector $z \in \mathbb{C}^n$ satisfying $\overline{A}^t Az = \lambda_1 z$, and so

$$\|A\|^2 \geq \|Az\|^2 = (Az, Az) = (\overline{A}^t Az, z) = (\lambda_1 z, z) = \lambda_1(z, z) = \lambda_1 .$$

Therefore, $\|A\| \geq \sqrt{\lambda_1}$.

For the last part, assume that A is Hermitian, i.e., $\overline{A}^t = A$. So, in this case, we have $\overline{A}^t A = A^2$. Since $\sigma(A^2) = \{\lambda^2\colon\ \lambda \in \sigma(A)\}$ by the Spectral Mapping Theorem, we see that $\|A\| = \max\{|\lambda|\colon\ \lambda \in \sigma(A)\} = r(A)$. ∎

Problem 8.1.10. *Show that an $n \times n$ matrix is quasinilpotent if and only if it is nilpotent.*

Solution: Let A be an $n \times n$ matrix. If A is nilpotent, then $A^k = 0$ for some k, and so $A^n = 0$ for all $n \geq k$. Therefore, $r(A) = \lim_{n\to\infty} \|A^n\|^{\frac{1}{n}} = 0$.

For the converse, assume that A is quasinilpotent. That is, $r(A) = 0$, or equivalently $\sigma(A) = \{0\}$. This implies $\lambda_1 = \lambda_2 = \cdots = \lambda_n = 0$, and hence the characteristic polynomial of A is $p_A(\lambda) = \lambda^n$. By the Cayley–Hamilton Theorem 8.3, we have $A^n = 0$, so that A is a nilpotent matrix. (See also Problem 6.1.18.) ∎

Problem 8.1.11. *Let $A = [a_{ij}]$ be an $n \times n$ matrix with complex entries. If we let $a_{ij} = b_{ij} + \imath c_{ij}$ and consider the real matrices $B = [b_{ij}]$ and $C = [c_{ij}]$, then we can write $A = B + \imath C$ and view A as a vector in the complexification $M_n(\mathbb{R}) \oplus \imath M_n(\mathbb{R}) = M_n(\mathbb{C})$ of the real Banach lattice $M_n(\mathbb{R})$.*

Show that the modulus of $A = B + \imath C$ in $M_n(\mathbb{R}) \oplus \imath M_n(\mathbb{R})$ is the positive matrix $|A| = \big[\,|a_{ij}|\,\big]$.

Solution: We know that if $a, b \in \mathbb{R}$, then $\sup_{\theta\in\mathbb{R}}(a\cos\theta + b\sin\theta) = \sqrt{a^2 + b^2}$; see the solution to Problem 1.1.7. Also, we know that a non-empty family of $n \times n$ real matrices $\{D_\lambda\}_{\lambda\in\Lambda}$, where $D_\lambda = \left[d_{ij}^\lambda\right]$, satisfies $\sup_{\lambda\in\Lambda} D_\lambda = D = [d_{ij}]$ in $M_n(\mathbb{R})$ if and only if $\sup_{\lambda\in\Lambda} d_{ij}^\lambda = d_{ij}$ for all i and j, i.e., $\sup_{\lambda\in\Lambda} D_\lambda = \left[\sup_{\lambda\in\Lambda} d_{ij}^\lambda\right]$.

Now the modulus of A in $M_n(\mathbb{R}) \oplus \imath M_n(\mathbb{R})$ is given by

$$\begin{aligned} |A| &= \sup_{\theta\in\mathbb{R}} \big(B\cos\theta + C\sin\theta\big) = \sup_{\theta\in\mathbb{R}} \big[\, b_{ij}\cos\theta + c_{ij}\sin\theta \,\big] \\ &= \Big[\sup_{\theta\in\mathbb{R}} (b_{ij}\cos\theta + c_{ij}\sin\theta)\Big] = \Big[\sqrt{(b_{ij})^2 + (c_{ij})^2}\,\Big] \\ &= \big[\,|a_{ij}|\,\big], \end{aligned}$$

as desired. ∎

Problem 8.1.12. *Prove the Cayley–Hamilton Theorem* 8.3. *That is, show that every square matrix satisfies its characteristic polynomial.*

Solution: Let $p_A(\lambda) = \lambda^n + a_{n-1}\lambda^{n-1} + \cdots + a_1\lambda + a_0$ be the characteristic polynomial of an $n \times n$ matrix A. The proof employs the well-known formula

$$C(\mathrm{cof}C)^t = (\mathrm{cof}C)^t C = \det(C)I$$

for an $n \times n$ matrix $C = [c_{ij}]$. The cofactor matrix $\mathrm{cof}C = \left[\mathrm{cof}c_{ij}\right]$ is the $n \times n$ matrix whose $(i,j)^{\text{th}}$ entry is given by $\mathrm{cof}c_{ij} = (-1)^{i+j}\det(C_{ij})$, where C_{ij} is the $(n-1)\times(n-1)$ matrix that is obtained from C by deleting its ith row and j^{th} column. So, for each $\lambda \in \mathbb{C}$ we have

$$(\lambda I - A)\big[\mathrm{cof}(\lambda I - A)\big]^t = \det(\lambda - I)I = p_A(\lambda)I\,. \qquad (\star)$$

Now notice that each entry of $\mathrm{cof}(\lambda I - A)$ is a polynomial of degree at most $n-1$. This implies that the matrix $\big[\mathrm{cof}(\lambda I - A)\big]^t$ is of the form $\sum_{k=0}^{n-1} \lambda^k B_k$, where the B_k are fixed $n \times n$ matrices. Rewriting $(\star)$, we get

$$\begin{aligned} &\lambda^n B_{n-1} + \lambda^{n-1}(B_{n-2} - AB_{n-1}) + \cdots + \lambda(B_0 - AB_1) - AB_0 \\ &\qquad\qquad = \lambda^n I + a_{n-1}\lambda^{n-1} I + a_{n-2}\lambda^{n-2} I + \alpha_1\lambda I + a_0 I\,. \end{aligned}$$

Since this holds for each $\lambda \in \mathbb{C}$, it follows that:

$$\begin{aligned} I &= B_{n-1}\,, \\ a_{n-1}I &= B_{n-2} - AB_{n-1}\,, \\ a_{n-2}I &= B_{n-3} - AB_{n-2}\,, \\ &\;\;\vdots \\ a_1 I &= B_0 - AB_1\,, \\ a_0 I &= -AB_0\,. \end{aligned}$$

Multiplying the above equations by $A^n, A^{n-1}, A^{n-2}, \ldots, A$, and I, respectively, we get

$$\begin{aligned} A^n &= A^n B_{n-1}\,, \\ a_{n-1}A^{n-1} &= A^{n-1}B_{n-2} - A^n B_{n-1}\,, \\ a_{n-2}A^{n-2} &= A^{n-2}B_{n-3} - A^{n-1}B_{n-2}\,, \\ &\vdots \\ a_1 A &= AB_0 - A^2 B_1\,, \\ a_0 I &= -AB_0\,, \end{aligned}$$

and then adding yields $A^n + a_{n-1}A^{n-1} + a_{n-2}A^{n-2} + \cdots + a_1 A + a_0 I = 0$. For a different proof of the Cayley–Hamilton theorem see Problem 8.2.10. ∎

8.2. Operators on Finite Dimensional Spaces

Problem 8.2.1. *Show that for an operator $T\colon X \to X$ on a finite dimensional vector space the following statements are equivalent.*

(a) *T is invertible.*

(b) *T is one-to-one.*

(c) *T is surjective.*

Solution: (a) $\Longrightarrow$ (b) Obvious.

(b) $\Longrightarrow$ (c) If T is one-to-one and $\{e_1, \ldots, e_n\}$ is a basis of X, then $Te_1, \ldots, Te_n$ are n linearly independent vectors, and so they must be a basis of X. This implies that T is a surjective operator.

(c) $\Longrightarrow$ (a) Assume that X is n-dimensional and that the operator $T\colon X \to X$ is surjective. If $n = 1$, then T is clearly invertible. So, assume that $n > 1$. To finish the solution, we must show that T is one-to-one.

To this end, assume by way of contradiction that $Tx = 0$ for some $x \neq 0$. Choose vectors $x_2, \ldots, x_n$ such that $\{x, x_2, x_3, \ldots, x_n\}$ is a basis of X. If V is the $(n-1)$-dimensional vector space generated by the linearly independent vectors $x_2, \ldots, x_n$, then the linear operator $T\colon V \to X$ is also surjective. This implies $n = \dim(X) \leq \dim(V) = n - 1$, which is a contradiction. Hence, T is one-to-one, and consequently it is invertible. ∎

Problem 8.2.2. *A square matrix A is said to be* ***upper triangularizable*** *if there exists an invertible matrix C (called an* ***upper triangularizing matrix*** *for A) such that $C^{-1}AC$ is an upper triangular matrix.*

Show that every matrix A is upper triangularizable and that if C is an upper triangularizing matrix for A, then the diagonal entries of $C^{-1}AC$ consist of the eigenvalues of A (counted with their multiplicities).

Solution: Let $A = [a_{ij}]$ be an arbitrary $n \times n$ matrix with complex entries. Also, let $\{e_1, e_2, \dots, e_n\}$ be the standard orthonormal basis of $\mathbb{C}^n$. Then A defines an operator $T\colon \mathbb{C}^n \to \mathbb{C}^n$ via the formula $T(x) = Ax$ for each $x \in \mathbb{C}^n$. So, for each k we have

$$T(e_k) = Ae_k = \sum_{i=1}^{n} a_{ik} e_i \,. \tag{$\dagger$}$$

By Theorem 8.4 there exist T-invariant subspaces $V_1 \subseteq V_2 \subseteq \cdots \subseteq V_n = \mathbb{C}^n$ such that $\dim(V_i) = i$ for each i. In particular, there exist linearly independent vectors $f_1, f_2, \dots, f_n$ such that $\{f_1, f_2, \dots, f_i\}$ is a basis for V_i for each i. If we consider the basis $\{f_1, f_2, \dots, f_n\}$ of $\mathbb{C}^n$, then we can write

$$T(f_j) = \sum_{k=1}^{n} b_{kj} f_k \tag{$\dagger\dagger$}$$

for all $j = 1, 2, \dots, n$. Since $T(f_j) \in V_j$ for each j and $\{f_1, \dots, f_j\}$ is a basis of V_j, it follows that $b_{ij} = 0$ for $i > j$. That is, the matrix $B = [\,b_{ij}\,]$ is upper triangular. Let

$$B = \begin{bmatrix} b_{11} & \star & \star & \star & \star \\ 0 & b_{22} & \star & \star & \star \\ \vdots & \vdots & \ddots & \vdots & \vdots \\ 0 & 0 & 0 & b_{n-1n-1} & \star \\ 0 & 0 & 0 & 0 & b_{nn} \end{bmatrix}.$$

Next, for each j choose scalars c_{ij} $(i = 1, \dots, n)$ such that

$$f_j = \sum_{i=1}^{n} c_{ij} e_i \,,$$

and note that the matrix $C = [c_{ij}]$ is invertible. It follows from ($\dagger$) that

$$T(f_j) = \sum_{k=1}^{n} c_{kj} T(e_k) = \sum_{k=1}^{n} c_{kj} \Big(\sum_{i=1}^{n} a_{ik} e_i \Big) = \sum_{i=1}^{n} \Big(\sum_{k=1}^{n} a_{ik} c_{kj} \Big) e_i \,,$$

and from ($\dagger\dagger$) we get

$$T(f_j) = \sum_{k=1}^{n} b_{kj} f_k = \sum_{k=1}^{n} b_{kj} \Big(\sum_{i=1}^{n} c_{ik} e_i \Big) = \sum_{i=1}^{n} \Big(\sum_{k=1}^{n} c_{ik} b_{kj} \Big) e_i \,.$$

Therefore, $\sum_{k=1}^{n} a_{ik} c_{kj} = \sum_{k=1}^{n} c_{ik} b_{kj}$ for all i and j. In matrix notation this is equivalent to $AC = CB$ or $C^{-1}AC = B$, which shows that the matrix A is indeed upper triangularizable! To finish the proof now look at part (i) of Problem 8.1.3. ∎

Problem 8.2.3. *Assume that a matrix $A \in M_n(\mathbb{C})$ has the distinct eigenvalues $\lambda_1, \dots, \lambda_k$ with multiplicities $m_1, \dots, m_k$, respectively. If $q(\cdot)$ is a polynomial, then show that the characteristic polynomial of the $n \times n$ matrix $q(A)$ is*

$$p_{q(A)}(\lambda) = \big[\lambda - q(\lambda_1)\big]^{m_1} \big[\lambda - q(\lambda_2)\big]^{m_2} \cdots \big[\lambda - q(\lambda_k)\big]^{m_k} \,.$$

Solution: Let A be an $n \times n$ matrix having distinct eigenvalues $\lambda_1, \ldots, \lambda_k$ with multiplicities $m_1, \ldots, m_k$, respectively. By Problem 8.2.2, we know that the matrix A is similar to an upper triangular matrix, and so we can suppose that A is itself an upper triangular matrix. In particular, the diagonal entries of A are the eigenvalues λ_i each appearing m_i times. It is easy to see that $q(A)$ is also an upper triangular matrix having diagonal entries $q(\lambda_1), \ldots, q(\lambda_k)$ with each element $q(\lambda_i)$ appearing m_i times. This implies the desired formula. ∎

Problem 8.2.4. *This problem describes the minimal polynomial of an operator $T\colon X \to X$ on an n-dimensional complex Banach space.*

(a) *Show that the spectrum of T is finite and consists of eigenvalues. Moreover, if we let $\sigma(T) = \{\lambda_1, \ldots, \lambda_k\}$, then show that $k \le n$.*

(b) *Show that there is a non-zero polynomial $p(\lambda)$ such that $p(T) = 0$.*

(c) *Prove that some root of the polynomial $p(\lambda)$ in* (b) *must belong to the spectrum of T; and so the spectrum of T is non-empty.*

(d) *Assume that $\pi(\lambda)$ is a polynomial of minimal degree and leading coefficient one satisfying $\pi(T) = 0$. Show that*

$$\pi(\lambda) = (\lambda - \lambda_1)^{\alpha_1} \cdots (\lambda - \lambda_k)^{\alpha_k},$$

where, as in (a), *$\sigma(T) = \{\lambda_1, \ldots, \lambda_k\}$ and each α_i is the ascent of the operator $\lambda_i - T$.* (*This establishes that the polynomial $\pi(\lambda)$ is uniquely determined; it is referred to as the* ***minimal polynomial*** *of the operator T.*)

(e) *Show that a polynomial $q(\lambda)$ satisfies $q(T) = 0$ if and only if $q(\lambda)$ is divisible by the minimal polynomial $\pi(\lambda)$ of T. In particular, prove that two polynomials $\psi(\lambda)$ and $\phi(\lambda)$ satisfy $\psi(T) = \phi(T)$ if and only if there exists a polynomial $\theta(\lambda)$ such that $\phi(\lambda) = \psi(\lambda) + \theta(\lambda)\pi(\lambda)$.*

Solution: (a) Here we shall use the following simple fact: *An operator on a finite dimensional vector space is invertible if and only if it is one-to-one.* Thus, a complex number λ belongs to the spectrum of T if and only if $\lambda - T$ is not one-to-one or, equivalently, if and only if λ is an eigenvalue of T. Therefore, the spectrum of T consists of eigenvalues.

Now notice that eigenvectors corresponding to distinct eigenvalues are linearly independent; see Problem 6.2.2. Since the maximum number of linearly independent vectors in X is n, it follows that $\sigma(T)$ consists of at most n complex numbers.

(b) Let $\{x_1, \ldots, x_n\}$ be a basis of X. Since for each i the $n+1$ vectors $x_i, Tx_i, \ldots, T^n x_i$ must be linearly dependent, there exists a non-zero polynomial $p_i(\lambda)$ such that $p_i(T)x_i = 0$. If $p(\lambda) = p_1(\lambda)p_2(\lambda)\cdots p_n(\lambda)$, then $p(\lambda)$ is a non-zero polynomial satisfying $p(T)x_i = 0$ for each i. Since $\{x_1, \ldots, x_n\}$ is a basis, the latter implies $p(T)x = 0$ for each $x \in X$, i.e., $p(T) = 0$.

(c) Assume that $p(\lambda)$ is a non-zero polynomial satisfying $p(T) = 0$, and write

$$p(\lambda) = a(\lambda - r_1)^{k_1} \cdots (\lambda - r_m)^{k_m},$$

where $r_1, \ldots, r_m$ are the distinct roots of $p(\lambda)$. If $r_i \notin \sigma(T)$ for each $1 \leq i \leq m$, then the operator $T - r_i = -(r_i - T)$ is invertible for each $1 \leq i \leq m$. This implies that the operator $0 = p(T) = a(T - r_1)^{k_1} \cdots (T - r_m)^{k_m}$ is also invertible, which is impossible. Hence, $r_i \in \sigma(T)$ must hold for some i, and consequently $\sigma(T)$ is non-empty. (Of course, we already know that the spectrum of T is non-empty, but the above proof is direct and elementary.)

(d) Let $\pi(\lambda)$ be a non-zero polynomial of minimal degree with leading coefficient one satisfying $\pi(T) = 0$. Write

$$\pi(\lambda) = (\lambda - r_1)^{k_1} \cdots (\lambda - r_m)^{k_m},$$

where $r_1, \ldots, r_m$ are the distinct roots of $\pi(\lambda)$. Also, let α_i denote the ascent of the operator $r_i - T$. Consequently, $N((r_i - T)^{\alpha_i}) = N((r_i - T)^k)$ for all $k \geq \alpha_i$.

We claim that $k_i \leq \alpha_i$ for each i. Indeed, if $\alpha_i < k_i$ holds for some i, then from

$$(T - r_i)^{k_i} \Big[\prod_{j \neq i} (T - r_j)^{k_j} x \Big] = \pi(T)x = 0,$$

it follows that the non-zero polynomial $q(\lambda) = (\lambda - r_i)^{\alpha_i} \big[\prod_{j \neq i} (\lambda - r_j)^{k_j}\big]$ has degree less than the degree of $\pi(\lambda)$ and satisfies $q(T)x = (T - r_i)^{\alpha_i} \big[\prod_{j \neq i} (T - r_j)^{k_j} x\big] = 0$ for each $x \in X$. That is, $q(T) = 0$. This contradicts the minimality property of $\pi(\lambda)$, and so $k_i \leq \alpha_i$ holds for each i.

Next, we claim that $k_i = \alpha_i$ for each i. To see this, assume by way of contradiction that $k_i < \alpha_i$ for some i; without loss of generality we can suppose that $i = m$. Fix some $x \in N((r_m - T)^{k_m + 1})$ such that $y = (r_m - T)^{k_m} x \neq 0$. This implies $(r_m - T)y = 0$ or $Ty = r_m y$, and consequently

$$\begin{aligned} 0 = \pi(T)x &= \Big[\prod_{i=1}^{m-1} (r_i - T)^{k_i} \Big] (r_m - T)^{k_m} x = \prod_{i=1}^{m-1} (r_i - T)^{k_i} y \\ &= \Big[\prod_{i=1}^{m-1} (r_i - r_m)^{k_i} \Big] y \neq 0, \end{aligned}$$

which is a contradiction. This contradiction establishes that $k_i = \alpha_i$ for each i.

It remains to be shown that $\sigma(T) = \{r_1, \ldots, r_m\}$. Indeed, if $r_i \notin \sigma(T)$ for some i, then $r_i - T$ is invertible, and so the polynomial $q_1(\lambda) = \prod_{j \neq i} (\lambda - r_j)^{k_j}$ has degree less than the degree of $\pi(\lambda)$ and satisfies $q_1(T) = 0$, which is impossible. Thus, every r_i belongs to the spectrum of T. Finally, let $r \in \sigma(T)$ and fix some non-zero vector $z \in X$ with $Tz = rz$. If $r \neq r_i$ for each i, then we have $\pi(T)z = \pi(r)z \neq 0$, which is also a contradiction. This establishes that $\sigma(T) = \{r_1, \ldots, r_m\}$.

(e) Let q be a non-zero polynomial. If $q(\lambda) = p(\lambda)\pi(\lambda)$, then $q(T) = 0$. For the converse assume that $q(T) = 0$. Let $\{r_1, \ldots, r_k\} \cup \{r_{k+1}, \ldots, r_n\}$ be the distinct roots of the polynomial $q(\lambda)$, where $\{r_1, \ldots, r_k\} \subseteq \sigma(T)$ and $\{r_{k+1}, \ldots, r_n\} \subseteq \rho(T)$. Assume that each root r_i has multiplicity m_i. Thus,

$$q(\lambda) = a \Big[\prod_{i=1}^{k} (\lambda - r_i)^{m_i} \Big] \cdot \Big[\prod_{i=k+1}^{n} (\lambda - r_i)^{m_i} \Big] = q_1(\lambda) q_2(\lambda),$$

where a is the leading coefficient of the polynomial q, $q_1(\lambda) = \prod_{i=1}^{k}(\lambda - r_i)^{m_i}$ and $q_2(\lambda) = a\prod_{i=k+1}^{n}(\lambda - r_i)^{m_i}$. Since the operator $q_2(T) = a\prod_{i=k+1}^{n}(T - r_i)^{m_i}$ is invertible, it follows that $q_1(\lambda)$ is a non-zero polynomial satisfying $q_1(T) = 0$. Now for each $1 \le i \le k$ let $\beta_i = \min\{m_i, \alpha_i\}$, where α_i is the ascent of the operator $r_i - T$, and consider the non-zero polynomial $q_3(\lambda) = \prod_{i=1}^{k}(\lambda - r_i)^{\beta_i}$. Clearly, $q_1(\lambda) = q_3(\lambda)p(\lambda)$ holds for some non-zero polynomial $p(\lambda)$. Since the degree of the polynomial $q_3(\lambda)$ is less than or equal to the degree of the minimal polynomial $\pi(\lambda)$ and $q_3(T) = 0$, it follows that $q_3(\lambda) = \pi(\lambda)$. Thus, $q(\lambda) = q_2(\lambda)p(\lambda)\pi(\lambda)$. This shows that $q(\lambda)$ is divisible by $\pi(\lambda)$.

Finally, notice that two polynomials $\phi(\lambda)$ and $\psi(\lambda)$ satisfy $\phi(T) = \psi(T)$ if and only if $\psi(T) - \phi(T) = 0$. By the preceding part, this is equivalent to having $\psi(\lambda) - \phi(\lambda) = \theta(\lambda)\pi(\lambda)$ for some polynomial $\theta(\lambda)$. Therefore, $\phi(T) = \psi(T)$ if and only if there exists a polynomial $\theta(\lambda)$ satisfying $\psi(T) = \phi(T) + \theta(T)\pi(T)$. ∎

Problem 8.2.5. *Let $T\colon X \to X$ be an operator on a finite dimensional vector space. If a pair of vector subspaces (Y, Z) reduces the operator T, then show that $p_T(\lambda) = p_{T|_Y}(\lambda)p_{T|_Z}(\lambda)$.*

Solution: Let $T\colon X \to X$ be an operator on a finite dimensional complex vector space and let (Y, Z) be a pair of vector subspaces that reduces T. Fix a basis $\{y_1, \dots, y_k\}$ of Y and a basis $\{z_1, \dots, z_m\}$ of Z. Clearly, $\{y_1, \dots, y_k, z_1, \dots, z_m\}$ is a basis of X. Moreover, we have

$$Ty_j = \sum_{i=1}^{k} a_{ij}y_i \ \ (j = 1, \dots, k) \quad \text{and} \quad Tz_p = \sum_{r=1}^{m} b_{rp}z_r \ \ (p = 1, \dots, m).$$

Therefore, the matrices representing the operators $T\colon Y \to Y$ and $T\colon Z \to Z$ with respect to the bases $\{y_1, \dots, y_k\}$ of Y and $\{z_1, \dots, z_m\}$ of Z are given by

$$A = \begin{bmatrix} a_{11} & a_{12} & \cdots & a_{1k} \\ a_{21} & a_{22} & \cdots & a_{2k} \\ \vdots & \vdots & \ddots & \vdots \\ a_{k1} & a_{k2} & \cdots & a_{kk} \end{bmatrix} \quad \text{and} \quad B = \begin{bmatrix} b_{11} & b_{12} & \cdots & b_{1m} \\ b_{21} & b_{22} & \cdots & b_{2m} \\ \vdots & \vdots & \ddots & \vdots \\ b_{m1} & b_{m2} & \cdots & b_{mm} \end{bmatrix}.$$

It follows that the matrix representing the operator $T\colon X \to X$ with respect to the basis $\{y_1, \dots, y_k, z_1, \dots, z_m\}$ is the block diagonal $(k+m) \times (k+m)$ matrix $C = \begin{bmatrix} A & 0 \\ 0 & B \end{bmatrix}$. This implies that $\lambda I - C = \begin{bmatrix} \lambda I_k - A & 0 \\ 0 & \lambda I_m - B \end{bmatrix}$, and so

$$p_T(\lambda) = \det(\lambda I - C) = \det(\lambda I_k - A) \cdot \det(\lambda I_m - B) = p_{T|_Y}(\lambda)p_{T|_Z}(\lambda),$$

as claimed. ∎

Problem 8.2.6. *Let $T\colon X \to X$ be an operator on an n-dimensional vector space. If two $n \times n$ matrices A and B represent T with respect to two different bases of X, then show that A and B are similar.*

Solution: Assume that the two $n \times n$ matrices $A = [a_{ij}]$ and $B = [b_{ij}]$ represent the operator T with respect to the bases $\{e_1, e_2, \dots, e_n\}$ and $\{f_1, f_2, \dots, f_n\}$, respectively. That is,

$$T(e_k) = Ae_k = \sum_{i=1}^{n} a_{ik}e_i \quad \text{for } k = 1, 2, \ldots, n\,, \tag{$\star$}$$

and

$$T(f_j) = Bf_j = \sum_{k=1}^{n} b_{kj}f_k \quad \text{for } j = 1, 2, \ldots, n\,. \tag{$\star\star$}$$

Next, for each j choose scalars c_{ij} $(i = 1, 2, \ldots, n)$ such that

$$f_j = \sum_{i=1}^{n} c_{ij}e_i\,,$$

and note that the matrix $C = [c_{ij}]$ is invertible. Using ($\star$), we get

$$T(f_j) = \sum_{k=1}^{n} c_{kj}T(e_k) = \sum_{k=1}^{n} c_{kj}\Big(\sum_{i=1}^{n} a_{ik}e_i\Big) = \sum_{i=1}^{n}\Big(\sum_{k=1}^{n} a_{ik}c_{kj}\Big)e_i\,,$$

and from ($\star\star$) we obtain

$$T(f_j) = \sum_{k=1}^{n} b_{kj}f_k = \sum_{k=1}^{n} b_{kj}\Big(\sum_{i=1}^{n} c_{ik}e_i\Big) = \sum_{i=1}^{n}\Big(\sum_{k=1}^{n} c_{ik}b_{kj}\Big)e_i\,.$$

Therefore, $\sum_{k=1}^{n} a_{ik}c_{kj} = \sum_{k=1}^{n} c_{ik}b_{kj}$ for all i and j. In matrix notation this is equivalent to $AC = CB$ or $A = CBC^{-1}$, which establishes that the matrices A and B are similar. ∎

Problem 8.2.7. *If $S, T\colon X \to X$ are two operators on a finite dimensional vector space, then show that ST and TS have the same characteristic polynomial—and conclude from this that if A and B are two $n \times n$ square matrices, then AB and BA have the same characteristic polynomial.*

Solution: Assume first that S is also a projection, i.e., $S^2 = S$. If $Y = S(X)$ and $Z = (I - S)(X)$, then $X = Y \oplus Z$, S is the identity operator on Y, and S vanishes on Z. Assume that Y is a k-dimensional vector subspace. Then there exists a basis $\mathcal{B}$ of X such that S has the matrix representation

$$S = \begin{bmatrix} I_k & 0 \\ 0 & 0 \end{bmatrix},$$

where I_k is the $k \times k$ identity matrix and the "diagonal" 0 is the $(n-k) \times (n-k)$ zero square matrix. Now assume that $T = [t_{ij}]$ is the matrix representation of T with respect to the basis $\mathcal{B}$, and write the matrix T in the block form

$$T = \begin{bmatrix} C & D \\ U & V \end{bmatrix},$$

where the size of C is $k \times k$ and that of V is $(n-k) \times (n-k)$. Now note that

$$ST = \begin{bmatrix} C & D \\ 0 & 0 \end{bmatrix} \quad \text{and} \quad TS = \begin{bmatrix} C & 0 \\ U & 0 \end{bmatrix}.$$

This implies that for each $\lambda \neq 0$ we have

$$p_{ST}(\lambda) = \det \begin{bmatrix} \lambda I_k - C & -D \\ 0 & \lambda I_{n-k} \end{bmatrix} = \det \begin{bmatrix} \lambda I_k - C & 0 \\ 0 & \lambda I_{n-k} \end{bmatrix} = \lambda^{n-k} p_C(\lambda).$$

Similarly, for each $\lambda \neq 0$, we have $p_{TS}(\lambda) = \lambda^{n-k} p_C(\lambda)$. So, $p_{ST}(\lambda) = p_{TS}(\lambda)$ for each λ. That is, ST and TS have the same characteristic polynomial.

We now consider the general case. Let $W = S(X)$, the range of S, and let $\{Sx_1, \dots, Sx_k\}$ be a basis of W. Pick vectors $y_{k+1}, \dots, y_n$ such that the set of vectors $\{Sx_1, \dots, Sx_k, y_{k+1}, \dots, y_n\}$ is a basis of X. Since the vectors $Sx_1, \dots, Sx_k$ are linearly independent, the vectors $x_1, \dots, x_k$ are also linearly independent. So, we can find vectors $x_{k+1}, \dots, x_n$ such that $\{x_1, \dots, x_k, x_{k+1}, \dots, x_n\}$ is a basis of X. Next, consider the linear operator $R\colon X \to X$ defined by $RSx_i = x_i$ for $1 \leq i \leq k$ and $Ry_i = x_i$ for $k+1 \leq i \leq n$. Clearly, R is an invertible operator, and an easy argument shows that $(RS)^2 = RS$, i.e., RS is a projection.

By the first part $(RS)(TR^{-1}) = RSTR^{-1}$ and $(TR^{-1})(RS) = TS$ have the same characteristic polynomial. Since $RSTR^{-1}$ and ST have the same characteristic polynomial, we conclude that ST and TS have the same characteristic polynomial. (See also Problems 6.1.19 and 6.2.9.) ∎

Problem 8.2.8. *Let V be a finite dimensional vector space. If A is a non-empty subset of V, then its* ***annihilator*** *$A^\perp$ is the vector subspace of V^* defined by*

$$A^\perp = \{v^* \in V^*\colon\ v^*(a) = 0 \ \textit{for all}\ a \in A\}.$$

Similarly, the ***annihilator*** *of any non-empty subset B of V^* is the vector subspace of V defined by $B^\perp = \{v \in V\colon\ b^*(v) = 0 \ \textit{for all}\ b^* \in B\}$.*

Assume that $T\colon X \to Y$ is an operator between two finite dimensional vector spaces. As usual, we define its adjoint operator $T^*\colon Y^* \to X^*$ via the duality identity

$$\langle Tx, y^* \rangle = \langle x, T^*y^* \rangle = y^*(Tx)$$

for all $x \in X$ and all $y^* \in Y^*$. Establish the following:

(a) $[R(T)]^\perp = \operatorname{Ker}(T^*)$.

(b) $[\operatorname{Ker}(T)]^\perp = R(T^*)$.

(c) $\dim R(T^*) = \dim R(T)$.

Solution: (a) For this identity note that:

$$\begin{aligned} y^* \in [R(T)]^\perp &\iff \langle Tx, y^* \rangle = \langle x, T^*y^* \rangle = 0 \text{ for each } x \in X \\ &\iff T^*y^* = 0 \\ &\iff y^* \in \operatorname{Ker}(T^*). \end{aligned}$$

(See also Theorem 2.13.)

(b) Fix $x^* \in R(T^*)$, and then pick some $y^* \in Y^*$ such that $x^* = T^*y^*$. Now for each $x \in \operatorname{Ker}(T)$ we have $x^*(x) = T^*y^*(x) = y^*(Tx) = 0$. This shows that $x^* \in [\operatorname{Ker}(T)]^\perp$, and so $R(T^*) \subseteq [\operatorname{Ker}(T)]^\perp$.

Now let $x^* \in [\mathrm{Ker}\,(T)]^\perp$, and assume by way of contradiction that $x^* \notin R(T^*)$. Then, by the Separation Theorem, there exists some $x \in X$ satisfying $x^*(x) \neq 0$ and $\langle x, T^*y^* \rangle = \langle Tx, y^* \rangle = 0$ for all $y^* \in Y^*$. It follows that $Tx = 0$ or that $x \in \mathrm{Ker}\,(T)$. But then $x^* \in [\mathrm{Ker}\,(T)]^\perp$ implies $x^*(x) = 0$, which is a contradiction. Therefore, $x^* \in [\mathrm{Ker}\,(T)]^\perp$ implies $x^* \in R(T^*)$ or $[\mathrm{Ker}\,(T)]^\perp \subseteq R(T^*)$. Consequently, $[\mathrm{Ker}\,(T)]^\perp = R(T^*)$.

(c) Let $n = \dim X = \dim X^*$. Fix a basis $\{x_1, \ldots, x_p\}$ in $\mathrm{Ker}\,(T)$, and then choose vectors $x_{p+1}, \ldots, x_n$ such that $\{x_1, \ldots, x_p, x_{p+1}, \ldots, x_n\}$ is a basis of X. Next, select vectors $x_j^* \in X^*$ $(j = 1, \ldots, n)$ satisfying $x_j^*(x_i) = \delta_{ij}$ for all i and j. Clearly, $\{x_1^*, x_2^*, \ldots, x_n^*\}$ is a basis of X^*.

If $x = \sum_{i=1}^n \lambda_i x_i$, then $Tx = \sum_{i=p+1}^n \lambda_i T x_i$. This shows that

$$\dim R(T) \leq n - p\,. \tag{$\star$}$$

Furthermore, we have $\{x_{p+1}^*, \ldots, x_n^*\} \subseteq [\mathrm{Ker}\,(T)]^\perp = R(T^*)$, where the last identity holds true by virtue of part (b). Consequently,

$$n - p \leq \dim R(T^*)\,. \tag{$\star\star$}$$

From $(\star)$ and $(\star\star)$, we see that $\dim R(T) \leq \dim R(T^*)$. Applying this inequality to the operator $T^*\colon Y^* \to X^*$ and taking into account that $T^{**} = T$, we get

$$\dim R(T) \leq \dim R(T^*) \leq \dim R(T^{**}) = \dim R(T)\,.$$

Therefore, $\dim R(T) = \dim R(T^*)$ holds true. (Regarding this problem, see also Theorem 2.13.) ∎

Problem 8.2.9. *If $T\colon X \to Y$ is an operator between finite dimensional vector spaces, then show that*

$$\dim \mathrm{Ker}(T) + \dim \mathrm{R}(T) = \dim X\,.$$

Solution: A glance at the inequalities $(\star)$ and $(\star\star)$ in the solution of the preceding problem yields

$$\dim R(T) = \dim R(T^*) = n - p\,,$$

where $n = \dim X$ and $p = \dim \mathrm{Ker}(T)$. Therefore,

$$\dim \mathrm{Ker}(T) + \dim \mathrm{R}(T) = p + (n-p) = n = \dim X\,.$$

The result can be obtained directly from $(\star)$ (and thus avoiding the use of the Separation Theorem). The only thing that is needed here is to observe that the vectors $Tx_{p+1}, \ldots, Tx_n$ (introduced at the beginning of the solution of part (c) in the preceding problem) are linearly independent. To see this, notice that if $\sum_{i=p+1}^n \lambda_i T x_i = T\big(\sum_{i=p+1}^n \lambda_i x_i\big) = 0$ holds true, then $\sum_{i=p+1}^n \lambda_i x_i \in \mathrm{Ker}(T)$, and so we must have $\sum_{i=p+1}^n \lambda_i x_i = \sum_{i=1}^p \mu_i x_i$ for appropriate scalars μ_i. Since the set of vectors $\{x_1, \ldots, x_p, x_{p+1}, \ldots, x_n\}$ is a basis of X, the latter implies $\lambda_i = 0$ for each i with $p+1 \leq i \leq n$. This guarantees that $\{Tx_{p+1}, \ldots, Tx_n\}$ is a basis of $R(T)$ and $n - p = \dim R(T)$. ∎

Problem 8.2.10. *Use Theorem 8.9 to present an alternate proof of the Cayley–Hamilton Theorem 8.3.*

Solution: Let $T\colon X \to X$ be an operator on a finite dimensional complex vector space. Also, let $\lambda_1, \ldots, \lambda_k$ be the distinct eigenvalues of T with multiplicities $m_1, \ldots, m_k$, respectively. According to Theorem 8.9, we can decompose X in the form $X = X_1 \oplus X_2 \oplus \cdots \oplus X_k$, where each vector subspace X_i has dimension m_i, is T-invariant, and the operator $T - \lambda_i\colon X_i \to X_i$ is nilpotent. In particular, we have $(T - \lambda_i)^{m_i} = 0$ on X_i.

Now if $S_i = \prod_{j \neq i}(T - \lambda_j)^{m_j}$, then $p_T(T) = \prod_{r=1}^{k}(T - \lambda_r)^{m_i} = S_i(T - \lambda_i)^{m_i}$ for each i. So, if $x = x_1 + x_2 + \cdots + x_k \in X_1 \oplus X_2 \oplus \cdots \oplus X_k = X$, then

$$p_T(T)x = \sum_{i=1}^{k} p_T(T)x_i = \sum_{i=1}^{k} S_i(T - \lambda)^{m_i} x_i = \sum_{i=1}^{k} S_i(0) = 0\,.$$

Therefore, $p_T(T) = 0$. In particular, if A is any matrix representing the operator T, then $p_T(A) = 0$ is true, proving the Cayley–Hamilton theorem. ∎

Problem 8.2.11. *Show that every square matrix is similar to its transpose.*

Solution: Let A be an $n \times n$ square matrix, and let $\lambda_1, \ldots, \lambda_k$ be the distinct eigenvalues of A with multiplicities $m_1, \ldots, m_k$, respectively. As usual, in terms of the basis $\{e_1, \ldots, e_n\}$ of the standard unit vectors, the matrix A defines an operator $T\colon \mathbb{C}^n \to \mathbb{C}^n$ via the formula $T(x) = Ax$ for all $x \in \mathbb{C}^n$. According to Theorem 8.9, there exist T-invariant subspaces $X_1, \ldots, X_k$ of X such that:

(1) $\dim(X_i) = m_i$ for each $i = 1, 2, \ldots, k$.

(2) $\mathbb{C}^n = X_1 \oplus X_2 \oplus \cdots \oplus X_k$.

(3) The operator $T - \lambda_i\colon X_i \to X_i$ is nilpotent for each $i = 1, 2, \ldots, k$.

For each $1 \le i \le k$ let $\{f_1^i, f_2^i, \ldots, f_{m_i}^i\}$ be a basis for X_i. Clearly, the set of vectors

$$\mathcal{B} = \{f_1^1, f_2^1, \ldots, f_{m_1}^1, f_1^2, f_2^2, \ldots, f_{m_2}^2, \ldots, f_k^1, f_k^1, \ldots, f_{m_k}^k\}$$

is a basis for $\mathbb{C}^n$. Since the operator T leaves each X_i invariant, it is easy to see that the matrix representing the operator T with respect to the basis $\mathcal{B}$ is a bock-diagonal matrix of the form $\operatorname{diag}(A_1, A_2, \ldots, A_k)$, where each A_i is an $m_i \times m_i$ square matrix. In particular, it follows that the $n \times n$ matrices A and $B = \operatorname{diag}(A_1, A_2, \ldots, A_k)$ are similar. This implies that A^t and

$$B^t = [\operatorname{diag}(A_1, A_2, \ldots, A_k)]^t = \operatorname{diag}(A_1^t, A_2^t, \ldots, A_k^t)$$

are similar matrices.

Next, let $S\colon \mathbb{C}^n \to \mathbb{C}^n$ be the operator defined via the matrix

$$B^t = \operatorname{diag}(A_1^t, A_2^t, \ldots, A_k^t)$$

with respect to the basis $\mathcal{B}$. Then S leaves each X_i invariant and $S - \lambda_i\colon X_i \to X_i$ is quasinilpotent for each $i = 1, \ldots, k$. This follows from the fact each $T - \lambda_i$ has the same property which guarantees that the matrix $A_i - \lambda_i$ (and hence $A_i^t - \lambda_i$) is quasinilpotent. This implies that the matrices B and B^t have a common Jordan form, which guarantees that B and B^t are similar matrices. Thus, we have shown that $A \sim B \sim B^t \sim A^t$, and so the matrix A is similar to its transpose A^t. ∎

Problem 8.2.12. *Let $T\colon Y \to Y$ be a bounded operator on a Banach space. A vector $y \in Y$ is said to be* **cyclic** *(or more precisely T-***cyclic***) if the vector subspace generated by $\{y, Ty, T^2y, T^3y, \ldots\}$ is dense in Y. For an operator $T\colon X \to X$ on an n-dimensional vector space establish the following.*

(a) *A vector $x \in X$ is T-cyclic if and only if $\{x, Tx, \ldots, T^{n-1}x\}$ is a basis of X.*

(b) *The operator T has a cyclic vector if and only if T has a matrix representation of the form:*

$$\begin{bmatrix} 0 & 0 & 0 & \cdots & 0 & 0 & -a_0 \\ 1 & 0 & 0 & \cdots & 0 & 0 & -a_1 \\ 0 & 1 & 0 & \cdots & 0 & 0 & -a_2 \\ \vdots & \vdots & \vdots & \ddots & \vdots & \vdots & \vdots \\ 0 & 0 & 0 & \cdots & 0 & 0 & -a_{n-3} \\ 0 & 0 & 0 & \cdots & 1 & 0 & -a_{n-2} \\ 0 & 0 & 0 & \cdots & 0 & 1 & -a_{n-1} \end{bmatrix} \qquad (\star)$$

(c) *If x is a T-cyclic vector and $T^n x = -a_0x - a_1Tx - \cdots - a_{n-1}T^{n-1}x$, then the characteristic polynomial of T is given by*

$$p_T(\lambda) = \lambda^n + a_{n-1}\lambda^{n-1} + a_{n-2}\lambda^{n-2} + \cdots + a_1\lambda + a_0\,.$$

Solution: (a) Since every subspace of a finite dimensional space is itself finite dimensional (and hence closed for the Euclidean topology), it follows that a vector $x \in X$ is T-cyclic if and only if the vector subspace generated by its **orbit** $\{x, Tx, T^2x, \ldots\}$ is all of X.

Now fix a vector $x \in X$. If $\{x, Tx, \ldots, T^{n-1}x\}$ is a basis of X, then it should be clear that the vector subspace generated by $\{x, Tx, T^2x, \ldots\}$ coincides with X. For the converse, assume that the vector subspace generated by $\{x, Tx, T^2x, \ldots\}$ coincides with X. If $\{x, Tx, \ldots, T^{n-1}x\}$ is not a basis of X, then it is easy to see that there exists some $0 \le k \le n-1$ such that T^kx lies in the span of the k vectors $x, Tx, \ldots, T^{k-1}x$. This implies that the vector subspace generated by $\{x, Tx, T^2x, \ldots\}$ has dimension at most $k < n$, and so it cannot coincide with X, which is a contradiction. Hence, if x is T-cyclic, then $\{x, Tx, \ldots, T^{n-1}x\}$ must be a basis of X.

(b) Assume that the operator T has a cyclic vector, say x. By (a) this implies that $\{x, Tx, \ldots, T^{n-1}x\}$ is a basis of X. If we let $b_1 = x, b_2 = Tx, \ldots, b_n = T^{n-1}x$, then with respect to this basis we have:

$$\begin{aligned} Tb_1 &= b_2 = (0,1,0,0,\ldots,0)\,, \\ Tb_2 &= b_3 = (0,0,1,0,\ldots,0)\,, \\ &\ \vdots \\ Tb_{n-1} &= b_n = (0,0,0,0,\ldots,1)\,, \\ Tb_n &= (-a_0, -a_1, -a_2, -a_3, \ldots, -a_{n-1})\,. \end{aligned}$$

These vectors written as columns yield the matrix representing the operator T with respect to the basis $\{b_1, b_2, \dots, b_n\}$. This matrix is of the desired form $(\star)$.

For the converse, assume that the operator T has a matrix representation of the form $(\star)$. This means that there exists a basis $\{b_1, b_2, \dots, b_n\}$ such that:

$$\begin{aligned} Tb_1 &= (0,1,0,0,\dots,0) = b_2\,, \\ T^2b_1 = Tb_2 &= (0,0,1,0,\dots,0) = b_3\,, \\ T^3b_1 = Tb_3 &= (0,0,0,1,\dots,0) = b_4\,, \\ &\ \vdots \\ T^{n-1}b_1 = Tb_{n-1} &= (0,0,0,0,\dots,1) = b_nv\,, \\ T^nb_1 = Tb_n &= (-a_0,-a_1,-a_2,-a_3,\dots,-a_{n-1})\,. \end{aligned}$$

So, if $x = b_1$, then $\{x, Tx, \dots, T^{n-1}x\} = \{b_1, b_2, \dots, b_n\}$ is a basis of X. By part (a), this guarantees that x is a T-cyclic vector.

(c) Assume that x is a cyclic vector for the operator T. Then, according to part (a), the set $\{x, Tx, \dots, T^{n-1}x\}$ is a basis of X, and so (by part (b)) the operator T has the matrix representation with respect to this basis given by $(\star)$, where $T^nx = -a_0x - a_1Tx - \cdots - a_{n-1}T^{n-1}x$. So, what needs verification is the fact that the characteristic polynomial of an $n \times n$ square matrix A of the form $(\star)$ is given by $p_A(\lambda) = \lambda^n + a_{n-1}\lambda^{n-1} + a_{n-2}\lambda^{n-2} + \cdots + a_1\lambda + a_0$.

The proof is by induction. It is straightforward to prove the validity of the claim for $n = 1$ and $n = 2$. So, for the induction step, we assume that the claim is true for all $n \times n$ square matrices of the form $(\star)$. Now assume that $A \in M_{n+1}(\mathbb{C})$ is of the form $(\star)$, and let $-a_0, -a_1, \dots, -a_n$ be the $n+1$ scalars of the $(n+1)^{\text{th}}$ column of A. Therefore, the characteristic polynomial of A is:

$$p_A(\lambda) = \det(\lambda I - A) = \det \begin{bmatrix} \lambda & 0 & 0 & \cdots & 0 & 0 & a_0 \\ -1 & \lambda & 0 & \cdots & 0 & 0 & a_1 \\ 0 & -1 & \lambda & \cdots & 0 & 0 & a_2 \\ \vdots & \vdots & \vdots & \ddots & \vdots & \vdots & \vdots \\ 0 & 0 & 0 & \cdots & \lambda & 0 & a_{n-2} \\ 0 & 0 & 0 & \cdots & -1 & \lambda & a_{n-1} \\ 0 & 0 & 0 & \cdots & 0 & -1 & \lambda + a_n \end{bmatrix}$$

If we expand this determinant with respect to the first row, then we have

$$p_A(\lambda) = \lambda \det B - a_0(-1)^{n+1} \det U\,,$$

where $\det B$ is the characteristic polynomial of a matrix of the form $(\star)$ having last column with scalars $-a_1, -a_2, \dots, -a_n$, and U is an $n \times n$ upper triangular matrix having each diagonal entry equal to -1. Clearly, $\det U = (-1)^n$ and our induction hypothesis implies that $\det B = \lambda^n + a_n\lambda^{n-1} + a_n\lambda^{n-1} + \cdots + a_2\lambda + a_1$. Therefore,

$$\begin{aligned} p_A(\lambda) &= \lambda\bigl(\lambda^n + a_n\lambda^{n-1} + a_{n-1}\lambda^{n-2} + \cdots + a_2\lambda + a_1\bigr) - a_0(-1)^{n+1}(-1)^n \\ &= \lambda^{n+1} + a_n\lambda^n + a_{n-1}\lambda^{n-1} + \cdots + a_1\lambda + a_0\,. \end{aligned}$$

This completes the induction, and the solution of the problem is finished. ∎

8.3. Matrices with Non-negative Entries

Problem 8.3.1. *Let $A = [a_{ij}]$ be an $n \times n$ matrix with real entries. Show that A (as a regular operator on $\mathbb{R}^n$) has a modulus and that $|A| = \big[|a_{ij}|\big]$.*

Solution: We have observed before that an $n \times n$ matrix $C = [c_{ij}]$ with real entries defines a positive operator on $\mathbb{R}^n$ if and only if $c_{ij} \geq 0$ for all i and j. This implies that if $B = [b_{ij}]$ and $C = [c_{ij}]$ are two $n \times n$ matrices with real entries, then $B \geq C$ if and only if $b_{ij} \geq c_{ij}$ for all i and j.

Now let $A = [a_{ij}]$ be an $n \times n$ matrix with real entries, and let $B = [|a_{ij}|]$. By the above $B \geq A$ and $B \geq -A$ both hold. Now if another $n \times n$ matrix $C = [c_{ij}]$ with real entries satisfies $C \geq A$ and $C \geq -A$, then $c_{ij} \geq a_{ij}$ and $c_{ij} \geq -a_{ij}$ for all i and j. Therefore, $c_{ij} \geq |a_{ij}|$ for all i and j, and so $C \geq B$. Consequently, $|A| = A \vee (-A) = \big[|a_{ij}|\big]$ in the Dedekind complete Riesz space $M_n(\mathbb{R})$. (See also Problem 8.1.11.) ∎

Problem 8.3.2. *Show (by means of an example) that although the spectral radius of a (non-zero) positive matrix is an eigenvalue, it need not have a strictly positive eigenvector.*

Solution: Consider the 2×2 positive matrix $A = \begin{bmatrix} 1 & 0 \\ 0 & 0 \end{bmatrix}$. Clearly, $\sigma(A) = \{1, 0\}$, and so $r(A) = 1$. Now an easy computation shows that (up to a scalar multiple) the vector $(1, 0)$ is the only positive eigenvector of A corresponding to the eigenvalue 1. This vector is not strictly positive. (Notice that this conclusion does not contradict the Perron–Frobenius Theorem 8.26 since the matrix A is not irreducible.) ∎

Problem 8.3.3. *Assume that a non-zero positive matrix A commutes with a strongly positive matrix. Show that if $\lambda_A > 0$ and $Ax \geq \lambda_A x$ for some $x > 0$, then $Ax = \lambda_A x$.*

Solution: Assume that A is an $n \times n$ non-zero positive matrix that commutes with an $n \times n$ strongly positive matrix B. Also, suppose that $\lambda_A > 0$ and that $Ax \geq \lambda_A x$ holds for some $x > 0$. To see that $Ax = \lambda_A x$ is true, assume by way of contradiction that $Ax > \lambda_A x$.

Let $y = \frac{1}{\lambda_A}(Ax - \lambda_A x) > 0$. Then $Ax = \lambda_A(x + y) > 0$, and since B is strongly positive we have $B(x+y) \gg 0$ and $By \gg 0$. Since A is a non-zero positive matrix, we have $BAy = ABy > 0$, and so $Ay > 0$ is true. This in turn implies that $ABy = BAy \gg 0$. Now note that

$$AB(x + y) = ABx + ABy \gg ABx = BAx = \lambda_A B(x + y)\,.$$

In particular, there exists some $\epsilon > 0$ satisfying $AB(x + y) \geq (\lambda_A + \epsilon)B(x + y)$. This implies $\lambda_A + \epsilon \leq \lambda_A$, which is impossible. Hence, $Ax = \lambda_A x$. ∎

Problem 8.3.4 (Frobenius). *For each i let $r_i = \sum_{j=1}^n a_{ij}$ (the sum of all elements in row i) and put $\rho = \min_{1 \leq i \leq n} r_i$ and $R = \max_{1 \leq i \leq n} r_i$. Show that $\rho \leq r(A) \leq R$.*

Solution: Pick some $x > 0$ such that $Ax = r(A)x$, and then choose some k such that $x_k = \max_{1\leq i\leq n} x_i > 0$. Then we have

$$r(A)x_k = (Ax)_k = \sum_{j=1}^{n} a_{kj}x_j \leq \Big(\sum_{j=1}^{n} a_{kj}\Big)x_k \leq Rx_k\,.$$

This implies $r(A) \leq R$.

Now if $e = (1, 1, \ldots, 1)$, then for each $1 \leq i \leq n$ we have $\rho \leq \sum_{j=1}^{n} a_{ij} = (Ae)_i$. So, $\rho e \leq Ae$, and consequently $\rho \leq \lambda_A = r(A)$ is also true. ∎

Problem 8.3.5 (Gershgorin [**31**]). *This exercise presents a simple but very useful "approximation" of the spectrum of an arbitrary matrix. Let $A = [a_{ij}]$ be an $n \times n$ matrix and let $r_i = \sum_{j\neq i}^{n} |a_{ij}|$ and $C_i = \{z \in \mathbb{C}\colon\ |z - a_{ii}| \leq r_i\}$ for each i. Show that $\sigma(A) \subseteq \bigcup_{i=1}^{n} C_i$.*

Solution: Fix an eigenvalue λ of A, and then pick some $x = (x_1, \ldots, x_n) \neq 0$ satisfying $Ax = \lambda x$. Choose some k such that $|x_k| = \max_{1\leq i\leq n} |x_i| > 0$. From $\lambda x_k = \sum_{j=1}^{n} a_{kj}x_j$, we get $(\lambda - a_{kk})x_k = \sum_{j\neq k} a_{kj}x_j$. This implies

$$|\lambda - a_{kk}| = \Big|\sum_{j\neq k} a_{kj}\tfrac{x_j}{x_k}\Big| \leq \sum_{j\neq k} |a_{kj}|\big|\tfrac{x_j}{x_k}\big| \leq \sum_{j\neq k} |a_{kj}| = r_k\,,$$

and so $\lambda \in C_k$. Consequently, $\sigma(A) \subseteq \bigcup_{i=1}^{n} C_i$. ∎

Problem 8.3.6. *If an arbitrary $n \times n$ matrix $A = [a_{ij}] \in M_n(\mathbb{C})$ satisfies $|a_{ii}| > \sum_{j\neq i} |a_{ij}|$ for each i, then show that A is invertible.*

Solution: Notice that Problem 8.3.5 implies that $\lambda = 0$ does not belong to the spectrum of A. Thus, A is invertible. ∎

Problem 8.3.7. *A non-negative $n \times n$ matrix $A = [a_{ij}]$ is said to be a **Markov** matrix if the sum of the elements in each row equals one, i.e., if $\sum_{j=1}^{n} a_{ij} = 1$ for each i. (This is equivalent, of course, to saying that A defines a Markov operator on $\mathbb{R}^n$.) Show in four different ways that every Markov matrix A satisfies $r(A) = 1$ by using:*

(a) *Problem* 8.3.4,

(b) *Lemma* 8.14(4),

(c) *Gershgorin's theorem (see Problem* 8.3.5*), and*

(d) *the spectral radius formula.*

Solution: (a) Following the notation of Problem 8.3.4, we have $r_i = \sum_{j=1}^{n} a_{ij} = 1$ for each i. Therefore, $r = \min_{1\leq i\leq n} r_i = 1$ and $R = \max_{1\leq i\leq n} r_i = 1$ for each $i = 1, 2, \ldots, n$. This implies $1 = r \leq r(A) \leq R = 1$, and so $r(A) = 1$.

(b) If $e = (1, 1, \ldots, 1)$, then $Ae = e$. Now a glance at Lemma 8.14(4) yields $r(A) = 1$.

(c) We know that $r(A)$ is an eigenvalue of A. Therefore, by Problem 8.3.5, there exists some $1 \le i \le n$ such that

$$r(A) - a_{ii} \le \big|r(A) - a_{ii}\big| \le \sum_{j \ne i} a_{ij} = 1 - a_{ii}\,.$$

This implies $r(A) \le 1$. Since $Ae = e$, we conclude that $r(A) = 1$.

(d) Equip $\mathbb{R}^n$ with the sup norm $\|\cdot\|$, i.e., $\|x\| = \max_{1 \le i \le n} |x_i|$, and consider A as a positive operator on $\mathbb{R}^n$. From $Ae = e$, we see that $A^k e = e$ for each k, and so $\|A^k\| = \|A^k e\| = \|e\| = 1$ for each k. This implies $r(A) = \lim_{k\to\infty} \big\|A^k\big\| = 1$. ∎

Problem 8.3.8. *Let A be an $n \times n$ non-negative matrix. Then:*

(a) *A defines a lattice homomorphism on $\mathbb{R}^n$ if and only if each row has at most one non-zero entry.*

(b) *A is interval preserving if and only if every column of A has at most one non-zero entry.*

Solution: Let $A = [a_{ij}]$ be a non-negative matrix.

(a) Assume first that A defines a lattice homomorphism on $\mathbb{R}^n$ equipped with its standard ordering. Since the (column) unit vectors $e_1, \ldots, e_n$ are pairwise disjoint, it follows that the non-negative (column) vectors $Ae_1, \ldots, Ae_n$ are also pairwise disjoint. But Ae_j is simply the jth column of A, and so the matrix A has pairwise disjoint columns. This implies that A has at most one non-zero entry in each row.

For the converse, assume that A has at most one non-zero entry in each row. Now consider a row i of A. If A has a positive entry in row i, then let j_i be the column where this happens, i.e., $a_{ij_i} > 0$. If row i does not have a positive entry, let $j_i = 1$. Note that $Ax = (a_{1j_1}x_{j_1}, a_{2j_2}x_{j_2}, \ldots, a_{nj_n}x_{j_n})$. So, if $x \wedge y = 0$, i.e., $x_i \wedge y_i = \min\{x_i, y_i\} = 0$ for each i, then

$$\begin{aligned}
&Ax \wedge Ay \\
&\quad = \big(a_{1j_1}x_{j_1}, a_{2j_2}x_{j_2}, \ldots, a_{nj_n}x_{j_n}\big) \wedge \big(a_{1j_1}y_{j_1}, a_{2j_2}y_{j_2}, \ldots, a_{nj_n}y_{j_n}\big) \\
&\quad = \big((a_{1j_1}x_{j_1}) \wedge (a_{1j_1}y_{j_1}), (a_{2j_2}x_{j_2}) \wedge (a_{2j_2}y_{j_2}), \ldots, (a_{nj_n}x_{j_n}) \wedge (a_{nj_n}y_{j_n})\big) \\
&\quad = \big(a_{1j_1}(x_1 \wedge y_1), a_{2j_2}(x_2 \wedge y_2), \ldots, a_{nj_n}(x_n \wedge y_n)\big) = 0\,.
\end{aligned}$$

This shows that A defines a lattice homomorphism on $\mathbb{R}^n$.

(b) The direct proof of this claim is laborious. The easiest way to see its validity is by using part (a) in conjunction with Theorem 1.35 (see also Problem 4.3.11) asserting that: *The matrix A defines a lattice homomorphism if and only if its transpose A^t defines an interval preserving operator.* ∎

Problem 8.3.9. *If an $n \times n$ positive matrix $A = [a_{ij}]$ satisfies $\sigma(A) = \{1\}$, then show that $a_{ii} = 1$ for all i. Use this and the preceding exercise to show that if a positive matrix A with $\sigma(A) = \{1\}$ defines either a lattice homomorphism or an interval preserving operator, then $A = I$, the identity matrix.*

Solution: Assume that a non-negative $n\times n$ matrix $A = [a_{ij}]$ satisfies $\sigma(A) = \{1\}$. Then the matrix $B = (A - I)^2$ satisfies $\sigma(B) = \{0\}$ or $\lambda_1 = \lambda_2 = \cdots = \lambda_n = 0$, where $\lambda_1, \lambda_2, \ldots, \lambda_n$ are the eigenvalues of the matrix B. In particular, we have $\operatorname{tr}(B) = \sum_{i=1}^n \lambda_i = 0$. Clearly, $b_{ii} = (a_{ii} - 1)^2 + \sum_{k\neq i} a_{ik}a_{ki}$ for each i. Since A has non-negative entries and $\sum_{i=1}^n b_{ii} = \operatorname{tr}(B) = 0$, it follows that $(a_{ii} - 1)^2 = 0$ for each i. That is, $a_{ii} = 1$ for $i = 1, \ldots, n$. (The same formula for the trace of B also shows that for each i we have $a_{ik}a_{ki} = 0$ for all $k \neq i$.)

Now assume that, in addition to $\sigma(A) = \{1\}$, the matrix A defines either a lattice homomorphism or an interval preserving operator. Then, according to Problem 8.3.8, the matrix A either cannot have more than one non-zero entry in each row or else it has no more than one non-zero entry in each column. This shows that in both cases A must coincide with the identity matrix. ∎

Problem 8.3.10. *Let A be an $n \times n$ positive matrix. If A is nilpotent, i.e., $r(A) = 0$, then show that there exists some vector $x > 0$ such that $Ax = 0$.*

Solution: Assume that $A > 0$ and $A^p = 0$ holds for some $p \geq 2$. Now assume by way of contradiction that for each $x > 0$ we have $Ax > 0$. That is, A is a strictly positive matrix. This easily implies that $A^p = 0$ is likewise a strictly positive matrix, which is impossible. Thus, there must exist some $x > 0$ such that $Ax = 0$. ∎

8.4. Irreducible Matrices

Problem 8.4.1. *For a vector subspace J of $\mathbb{C}^n$ show that the following statements are equivalent.*

(a) *J is an ideal, i.e., $J = J_0 \oplus \imath J_0$, where J_0 is an ideal in $\mathbb{R}^n$.*

(b) *J is a band, i.e., $J = B \oplus \imath B$, where B is a band in $\mathbb{R}^n$.*

(c) *There exists some non-empty subset $\mathcal{I}$ of $\{1, 2, \ldots, n\}$ such that*
$$J = \big\{(z_1, \ldots, z_n) \in \mathbb{C}^n \colon \ z_i = 0 \ \text{ for all } \ i \in \mathcal{I}\big\}.$$

Solution: The solution is based upon the structure of ideal and bands in $\mathbb{R}^n$. We shall establish the following basic property.

- *A vector subspace V of $\mathbb{R}^n$ is an ideal if and only if it is a band—and also, if and only if there exists a subset $\mathcal{I}$ of $\{1, 2, \ldots, n\}$ such that*
$$V = \big\{x \in \mathbb{R}^n \colon \ x_i = 0 \ \text{for all } \ i \in \mathcal{I}\big\}.$$

To establish this basic property, let V be an ideal in $\mathbb{R}^n$. If $V = \{0\}$ or $V = \mathbb{R}^n$, then take $\mathcal{I} = \{1, 2, \ldots, n\}$ or $\mathcal{I} = \emptyset$, respectively. So, assume that V is a proper subspace of $\mathbb{R}^n$. We define the set $\mathcal{I}$ as follows:
$$\mathcal{I} = \big\{i \in \{1, 2, \ldots, n\} \colon \ x_i = 0 \ \text{ for all } \ x \in V\big\}.$$
We also let $W = \big\{x \in \mathbb{R}^n \colon \ x_i = 0 \ \text{ for all } \ i \in \mathcal{I}\big\}$. It should be clear that W is an ideal in $\mathbb{R}^n$ and that $V \subseteq W$. We claim that $V = W$.

To see this, let $x \in W$. Notice that if $i \notin \mathcal{I}$, then there exists some $y \in V$ such that $y_i \neq 0$. From the inequalities $0 \le e_i \le \frac{1}{|y_i|}|y|$ and the fact that V is an ideal in $\mathbb{R}^n$, we infer that each unit vector e_i belongs to V for each $i \notin \mathcal{I}$. But then we have $x = \sum_{i=1}^n x_i e_i = \sum_{i \notin \mathcal{I}} x_i e_i \in V$. This shows that $W \subseteq V$, and so $V = W$.

The claim that W is also a band follows from the simple fact $x^\alpha \xrightarrow{o} x$ in $\mathbb{R}^n$ implies $x_i^\alpha \to x_i$ in $\mathbb{R}$ for each coordinate i.

Now the equivalence of (a) and (b) should be obvious. For the equivalence of (a) with (c) notice that if $J_0 = \{x \in \mathbb{R}^n : \ x_i = 0 \text{ for all } i \in \mathcal{I}\}$, then

$$J = \{(z_1, \ldots, z_n) \in \mathbb{C}^n : \ z_i = 0 \ \text{ for all } \ i \in \mathcal{I}\} = J_0 \oplus \imath J_0 \,.$$

This completes the solution of the problem. ∎

Problem 8.4.2. *Show that an $n \times n$ positive matrix A is irreducible if and only if A^t is irreducible.*

Solution: Notice that for any $n \times n$ matrix A we have $(I+A^t)^{n-1} = [(I+A)^{n-1}]^t$. Also, observe that a matrix $B \in M_n(\mathbb{R})$ is strongly positive if and only B^t is strongly positive. Now use statement (3) of Theorem 8.18 to see that

$$\begin{aligned} A \text{ is irreducible} \quad &\iff \quad (I+A)^{n-1} \text{ is strongly positive} \\ &\iff \quad (I+A^t)^{n-1} \text{ is strongly positive} \\ &\iff \quad A^t \text{ is irreducible}\,, \end{aligned}$$

as claimed. ∎

Problem 8.4.3. *For positive $n \times n$ matrices A and B establish the following.*

(a) *If A^k is irreducible for some $k \ge 1$, then A is irreducible.*

(b) *If A is irreducible, then $A+B$ is irreducible.*

(c) *If A and B are irreducible, then AB and BA need not be irreducible.*

Solution: (a) Assume that A^k is irreducible for some power $k \ge 1$. Now let J be an ideal in $\mathbb{C}^n$ such that $A(J) \subseteq J$. Then $A^k(J) \subseteq J$ is also true. So, by the irreducibility of A^k, we have either $J = \{0\}$ or $J = \mathbb{C}^n$. Therefore, A is an irreducible matrix.

(b) Assume that A is irreducible and that B is a non-negative $n \times n$ matrix. Take any ideal J such that $(A+B)(J) \subseteq J$. Since $0 \le A \le A+B$, it follows that $A(J) \subseteq J$. The irreducibility of A implies that either $J = \{0\}$ or $J = \mathbb{C}^n$. So, the matrix $A+B$ is irreducible.

(c) Let $A = B = \begin{bmatrix} 0 & 1 \\ 1 & 0 \end{bmatrix}$. Then $AB = BA = \begin{bmatrix} 1 & 0 \\ 0 & 1 \end{bmatrix} = I$. This shows that A and B are both irreducible and that $AB = BA$ (the identity matrix) is not an irreducible matrix. ∎

Problem 8.4.4. *Show that a non-negative matrix A is irreducible if and only if the matrix $I + A$ defines a Krein operator.*

Solution: Recall that a positive operator $T\colon E \to E$ on an AM-space with unit is said to be a **Krein operator** if for each $x > 0$ there exists some $k \geq 1$ such that $T^k x$ is an order unit. Also, keep in mind that the order units of $\mathbb{R}^n$ are simply the strictly positive vectors in $\mathbb{R}^n$, and that $\mathbb{R}^n$ with the sup norm is an AM-space with unit $e = (1, 1, \ldots, 1)$.

Now assume that A is an irreducible matrix and let $x > 0$. According to statement (3) of Theorem 8.18 the matrix $(I + A)^{n-1}$ is strongly positive. This implies $(I + A)^{n-1}x \gg 0$, and so $I + A$ is a Krein operator.

For the converse, suppose that $I + A$ is a Krein operator, and let an ideal J in $\mathbb{C}^n$ satisfy $A(J) \subseteq J$. We claim that J must be a trivial ideal, that is, either $J = \{0\}$ or $J = \mathbb{C}^n$. To see this, assume that $J \neq \{0\}$, and fix some $0 < x \in J$. It is easy to see that $(I + A)^k x \in J$ for each $k \geq 1$. Since $I + A$ is a Krein operator, there exists some $k \geq 1$ such that $(I + A)^k x$ is an order unit. Therefore, J contains an order unit, and since J is an ideal, it follows that $J = \mathbb{C}^n$. This shows that A is an irreducible matrix. ∎

Problem 8.4.5. *Use Problem 8.4.4 and Theorem 9.42 to present an alternate proof of the fact that the spectral radius of a positive irreducible matrix has a unique (up to scalar multiples) strictly positive eigenvector.*

Solution: Let A be an irreducible positive matrix and assume that $Ax = \lambda_0 x$ for some $\lambda_0 \in \mathbb{C}$ and some $x > 0$. Since (according to Lemma 8.17) A does not have zero columns, it follows that the vector Ax has at least one positive coordinate, say the ith one. This implies $\lambda_0 x_i \in \mathbb{R}$ and $\lambda_0 x_i > 0$. Hence, λ_0 is a positive eigenvalue of A.

Next, notice that $(I + A)x = (1 + \lambda_0)x$. Since $I + A$ defines a Krein operator on $\mathbb{R}^n$ (see Problem 8.4.4), it follows from Theorem 9.42 that:

(1) $1 + \lambda_0 = r(I + A) = 1 + r(A)$.

(2) The vector x is strictly positive and (apart from scalar multiples) is the only positive eigenvector of $I + A$.

So, (1) implies $\lambda_0 = r(A)$, and (2) shows that the vector x is strictly positive and (apart from scalar multiples) is the only positive eigenvector of A. ∎

Problem 8.4.6. *Consider a positive irreducible matrix $A = [a_{ij}]$ and assume that $A_t = [a_{ij}(t)]$ is a positive matrix-valued function defined for each t in some subinterval (a, b) of $\mathbb{R}$ such that for some $t_0 \in (a, b)$ we have $\lim_{t \to t_0} a_{ij}(t) = a_{ij}$ for each (i, j). Establish that $\lim_{t \to t_0} r(A_t) = r(A)$.*

Solution: To show that $\lim_{t \to t_0} r(A_t) = r(A)$, we must show that if $t_n \to t_0$, then there is a subsequence $\{s_n\}$ of $\{t_n\}$ such that $\lim_{n \to \infty} r(A_{s_n}) = r(A)$.

So, fix a sequence $\{t_n\} \subseteq (a, b)$ such that $t_n \to t_0$. By Theorem 8.11, for each n there exists a positive unit vector x_n such that $A_{t_n} x_n = r(A_{t_n}) x_n$. Since the closed unit ball of $\mathbb{R}^n$ is compact, there exists a subsequence $\{x_{k_n}\}$ of $\{x_n\}$ such that $y_n = x_{k_n} \to y > 0$. Let $s_n = t_{k_n}$, and note that

$$\lim_{n \to \infty} r(A_{s_n}) y_n = \lim_{n \to \infty} A_{s_n} y_n = Ay > 0\,,$$

where $Ay > 0$ is true by Corollary 8.21. This implies that there is some $\lambda_0 \geq 0$ such that $\lim_{n\to\infty} r(A_{s_n}) = \lambda_0$, and so $Ay = \lambda_0 y$. By Corollary 8.23, $\lambda_0 = \lambda_A = r(A)$. That is, we have $\lim_{n\to\infty} r(A_{s_n}) = r(A)$, and thus $\lim_{t\to t_0} r(A_t) = r(A)$. ∎

8.5. The Perron–Frobenius Theorem

Problem 8.5.1. *Show that n arbitrary complex numbers $z_1, \ldots, z_n$ satisfy $\left|\sum_{i=1}^n z_i\right| = \sum_{i=1}^n |z_i|$ if and only if there exists some angle θ such that $z_i = |z_i|e^{\imath\theta}$ for each $i = 1, \ldots, n$.*

Solution: If $z_i = |z_i|e^{\imath\theta}$ for each $i = 1, \ldots, n$, then it should be clear that $|\sum_{i=1}^n z_i| = \sum_{i=1}^n |z_i|$. We shall establish the converse by induction. First we shall establish the conclusion for $n = 2$.

To this end, assume that two complex numbers $z_1 = |z_1|e^{\imath\theta_1}$ and $z_2 = |z_2|e^{\imath\theta_2}$, where $0 \leq \theta_1, \theta_2 < 2\pi$, satisfy $|z_1 + z_2| = |z_1| + |z_2|$. Squaring both sides, we get

$$|z_1 + z_2|^2 = (z_1 + z_2)(\overline{z}_1 + \overline{z}_2) = (|z_1| + |z_2|)^2$$

or $|z_1|^2 + |z_2|^2 + z_1\overline{z}_2 + \overline{z}_1 z_2 = |z_1|^2 + |z_2|^2 + 2|z_1| \cdot |z_2|$. Simplifying the latter expression yields $e^{\imath(\theta_1-\theta_2)} + e^{\imath(\theta_2-\theta_2)} = 2$. Letting $z = e^{\imath(\theta_1-\theta_2)}$, we get $z + \frac{1}{z} = 2$ or $z^2 - 2z + 1 = (z-1)^2 = 0$. This implies $z = e^{\imath(\theta_1-\theta_2)} = 1$, and so $\theta_1 = \theta_2$. Therefore, our conclusion is true for $n = 2$.

For the induction step, suppose the claim is true for some n, and assume that the non-zero complex numbers $z_1, \ldots, z_n, z_{n+1}$ satisfy $\left|\sum_{i=1}^{n+1} z_i\right| = \sum_{i=1}^{n+1} |z_i|$. From the inequalities

$$\sum_{i=1}^{n+1} |z_i| = \left|\sum_{i=1}^{n+1} z_i\right| = \left|\Big(\sum_{i=1}^{n} z_i\Big) + z_{n+1}\right| \leq \left|\sum_{i=1}^{n} z_i\right| + |z_{n+1}| \leq \sum_{i=1}^{n+1} |z_i|, \qquad (\star)$$

we see that $\left|\big(\sum_{i=1}^n z_i\big) + z_{n+1}\right| = \left|\sum_{i=1}^n z_i\right| + |z_{n+1}|$. So, according to the case $n = 2$, there exists an angle $0 \leq \theta < 2\pi$ such that $\sum_{i=1}^n z_i = \left|\sum_{i=1}^n z_i\right|e^{\imath\theta}$ and $z_{n+1} = |z_{n+1}|e^{\imath\theta}$.

A glance at $(\star)$ also shows that $\left|\sum_{i=1}^n z_i\right| = \sum_{i=1}^n |z_i|$. So, according to our induction hypothesis, there exists an angle $0 \leq \phi < 2\pi$ such that $z_i = |z_i|e^{\imath\phi}$ for each $i = 1, \ldots, n$. Now note that

$$\Big(\sum_{i=1}^{n} |z_i|\Big)e^{\imath\theta} = \left|\sum_{i=1}^{n} z_i\right|e^{\imath\theta} = \sum_{i=1}^{n} z_i = \sum_{i=1}^{n} |z_i|e^{\imath\phi} = \Big(\sum_{i=1}^{n} |z_i|\Big)e^{\imath\phi}.$$

This implies $e^{\imath\theta} = e^{\imath\phi}$. Hence, $\phi = \theta$, and the induction is complete. ∎

Problem 8.5.2. *Consider the polynomials:*

$$\text{(i) } \lambda^3 - 1, \quad \text{(ii) } \lambda^3 - \lambda, \quad \textit{and} \quad \text{(iii) } \lambda^4 + \lambda^2 - 1.$$

Which of the above polynomials are the characteristic polynomials of positive irreducible matrices?

Solution: (i) Consider the matrix $A = \begin{bmatrix} 0 & 1 & 0 \\ 0 & 0 & 1 \\ 1 & 0 & 0 \end{bmatrix}$, and notice that

$$A + A^2 + A^3 = \begin{bmatrix} 0 & 1 & 0 \\ 0 & 0 & 1 \\ 1 & 0 & 0 \end{bmatrix} + \begin{bmatrix} 0 & 0 & 1 \\ 1 & 0 & 0 \\ 0 & 1 & 0 \end{bmatrix} + \begin{bmatrix} 1 & 0 & 0 \\ 0 & 1 & 0 \\ 0 & 0 & 1 \end{bmatrix} = \begin{bmatrix} 1 & 1 & 1 \\ 1 & 1 & 1 \\ 1 & 1 & 1 \end{bmatrix}.$$

By part (4) of Theorem 8.18 the positive matrix A is irreducible. Now a simple calculation shows that

$$p_A(\lambda) = \det \begin{bmatrix} \lambda & -1 & 0 \\ 0 & \lambda & -1 \\ -1 & 0 & \lambda \end{bmatrix} = \lambda^3 - 1.$$

(ii) Let $B = \begin{bmatrix} 0 & \frac{\sqrt{2}}{2} & 0 \\ \frac{\sqrt{2}}{2} & 0 & \frac{\sqrt{2}}{2} \\ 0 & \frac{\sqrt{2}}{2} & 0 \end{bmatrix} = \frac{\sqrt{2}}{2} \begin{bmatrix} 0 & 1 & 0 \\ 1 & 0 & 1 \\ 0 & 1 & 0 \end{bmatrix}$. Then we have

$$B^2 = \tfrac{1}{2} \begin{bmatrix} 1 & 0 & 1 \\ 0 & 2 & 0 \\ 1 & 0 & 1 \end{bmatrix},$$

and so $B + B^2$ is a strongly positive matrix. This implies that B is an irreducible matrix. Now notice that

$$p_B(\lambda) = \det \begin{bmatrix} \lambda & -\frac{\sqrt{2}}{2} & 0 \\ -\frac{\sqrt{2}}{2} & \lambda & -\frac{\sqrt{2}}{2} \\ 0 & -\frac{\sqrt{2}}{2} & \lambda \end{bmatrix} = \lambda^3 - \lambda.$$

(iii) Let $p(\lambda) = \lambda^4 + \lambda^2 - 1$. Solving the quadratic $(\lambda^2)^2 + \lambda^2 - 1 = 0$ yields $\lambda^2 = \frac{-1 \pm \sqrt{5}}{2}$. Consequently, the roots of the polynomial $p(\lambda)$ are

$$\lambda_1 = \sqrt{\tfrac{-1+\sqrt{5}}{2}}, \quad \lambda_2 = -\sqrt{\tfrac{-1+\sqrt{5}}{2}}, \quad \lambda_3 = \sqrt{\tfrac{1+\sqrt{5}}{2}}\imath, \quad \text{and} \quad \lambda_4 = -\sqrt{\tfrac{1+\sqrt{5}}{2}}\imath.$$

Therefore, $r(A) = |\lambda_3| = |\lambda_4| = \sqrt{\frac{1+\sqrt{5}}{2}}$. Since $r(A)$ is not a root of $p(\lambda)$, it follows that the polynomial $p(\lambda) = \lambda^4 + \lambda^2 - 1$ cannot be the characteristic polynomial of any irreducible positive matrix. (In fact, it cannot be the characteristic polynomial of any positive matrix!) ∎

Problem 8.5.3. *Consider the following Markov matrix*

$$A = \begin{bmatrix} 0 & 0 & \frac{1}{2} & \frac{1}{2} \\ 1 & 0 & 0 & 0 \\ 0 & 1 & 0 & 0 \\ 0 & 0 & 1 & 0 \end{bmatrix}.$$

Show that A is irreducible and satisfies $\sigma_{\text{per}}(A) = \{1\}$.

Solution: Since A is a Markov matrix, we know that $r(A) = 1$; see Problem 8.3.7. Now the patient reader should verify the following computation:

$$A^{10} = \frac{1}{16}\begin{bmatrix} 6 & 6 & 3 & 1 \\ 2 & 6 & 6 & 2 \\ 4 & 2 & 6 & 4 \\ 8 & 4 & 2 & 2 \end{bmatrix}.$$

This, combined with condition (4) of Theorem 8.18, guarantees that A is a positive irreducible matrix. A glance at statement (2) of Theorem 8.29 shows that A is a primitive matrix. So, $\sigma_{\text{per}}(A) = \{r(A)\} = \{1\}$.

Next, let us discuss the eigenvalues of A. Again, a direct computation yields

$$p_A(\lambda) = \det\begin{bmatrix} \lambda & 0 & -\frac{1}{2} & -\frac{1}{2} \\ -1 & \lambda & 0 & 0 \\ 0 & -1 & \lambda & \lambda \\ 0 & 0 & -1 & \lambda \end{bmatrix} = \lambda^4 - \tfrac{\lambda}{2} - \tfrac{1}{2} = (\lambda - 1)\big(\lambda^3 + \lambda^2 + \lambda + \tfrac{1}{2}\big).$$

This polynomial has the real roots $\lambda_1 = 1$ and $-1 < \lambda_2 < -\frac{1}{2}$. The other two roots λ_3 and λ_4 are complex conjugates with modulus less than one. Indeed, if we look at the function $f(\lambda) = \lambda^3 + \lambda^2 + \lambda + \frac{1}{2}$, then $f'(\lambda) = 3\lambda^2 + 2\lambda + 1 > 0$ for each λ, and so (since its discriminant $b^2 - 4ac = 4 - 12 = -8$ is negative) $f(\lambda)$ is strictly increasing. This implies that $p(\lambda)$ has only two real roots. ∎

Problem 8.5.4. *This exercise illustrates the Perron–Frobenius theorem. Consider the positive matrix*

$$A = \begin{bmatrix} 0 & 0 & 0 & 0 & 1 & 0 & 0 & 0 \\ 0 & 0 & 0 & 0 & 0 & 1 & 0 & 0 \\ 0 & 0 & 0 & 0 & 0 & 0 & 1 & 0 \\ 0 & 0 & 0 & 0 & 0 & 0 & 0 & 1 \\ 1 & 1 & 1 & 0 & 0 & 0 & 0 & 0 \\ 1 & 1 & 0 & 1 & 0 & 0 & 0 & 0 \\ 1 & 0 & 1 & 1 & 0 & 0 & 0 & 0 \\ 0 & 1 & 1 & 1 & 0 & 0 & 0 & 0 \end{bmatrix}.$$

(a) *Show that A is irreducible.*

(b) *Show that the characteristic polynomial of A is*

$$p_A(\lambda) = (\lambda^2 - 3)(\lambda - 1)^2(\lambda + 1)^2(\lambda^2 + 1).$$

Obtain from this that on the circle of radius one there are six eigenvalues: 1 with multiplicity two, -1 with multiplicity two, $\imath$ and $-\imath$.

(c) *Show that the spectrum of A is rotation invariant under* 180°.

Solution: A direct computation shows that if $x = (x_1, x_2, x_3, x_4, x_5, x_6, x_7, x_8)$ is an arbitrary vector in $\mathbb{R}^8$, then

$$Ax = \big(x_5, x_6, x_7, x_8, x_1 + x_2 + x_3, x_1 + x_2 + x_4, x_1 + x_3 + x_4, x_2 + x_3 + x_4\big). \quad (\dagger)$$

(a) Let $J \neq \mathbb{C}^n$ be an ideal that is invariant under A, i.e., $A(J) \subseteq J$. There exists a non-empty subset $\mathcal{I}$ of $\{1, 2, \ldots, 8\}$ such that J consists of all vectors $x \in \mathbb{C}^n$ satisfying $x_i = 0$ for each $i \in \mathcal{I}$. Now assume that $1 \in \mathcal{I}$. From (†), we see that $5 \in \mathcal{I}$. Otherwise, if $5 \notin \mathcal{I}$, then the vector e_5 belongs to J while $Ae_5 = e_1 \notin J$. Once we are convinced that 5 belongs to $\mathcal{I}$, it follows from (†) that $2, 3 \in \mathcal{I}$, and then $6, 7 \in \mathcal{I}$, and eventually $\mathcal{I} = \{1, 2, \ldots, 8\}$. This shows that in this case $J = \{0\}$.

Now notice if $2 \in \mathcal{I}$, then $6 \in \mathcal{I}$, and so $1 \in \mathcal{I}$, which (by the preceding argument) implies $J = \{0\}$. Similarly, it is easy to see that once $\mathcal{I}$ is assumed to be non-empty, then $1 \in \mathcal{I}$, and so $J = \{0\}$. This shows that A is an irreducible positive matrix.

(b) The computation of the characteristic polynomial is left as an exercise for the reader.

(c) From the characteristic polynomial, we infer that $r(A) = \sqrt{3}$. Therefore, $\sigma_{\text{per}}(A) = \{\sqrt{3}, -\sqrt{3}\}$. In other words, the eigenvalues on the peripheral spectrum of A are spaced apart by 180°. By statement (5) of the Perron–Frobenius Theorem 8.26, the spectrum of A is rotation invariant under the angle 180°. ∎

Problem 8.5.5. *Let A be a primitive matrix and let $\left(\frac{A}{r}\right)^n \to S$, where $r = r(A)$. Show that each column of S is an eigenvector corresponding to the eigenvalue r.*

Solution: Let c_j denote the jth column of the matrix S. Clearly, $c_j = Se_j$. From Corollary 8.30, we know that the vector

$$c_j = Se_j = \lim_{n\to\infty} \left(\tfrac{A}{r}\right)^n e_j$$

is an eigenvector of A corresponding to the eigenvalue r, i.e., $Ac_j = rc_j$. ∎

Problem 8.5.6. *Show that an $n \times n$ positive matrix is primitive if and only if it defines a Krein operator.*

Solution: Let A be an $n \times n$ positive matrix. If A is a primitive matrix, then (according to condition (2) of Theorem 8.29) there exists some $m \in \mathbb{N}$ such that A^m is strongly positive. So, $A^m x \gg 0$ holds for each $x > 0$. This proves that A defines a Krein operator. (Keep in mind that in $\mathbb{R}^n$ the order units are precisely the strictly positive vectors.)

For the converse, assume that A defines a Krein operator. Note first that A is irreducible. To see this, assume that an ideal $J \neq \{0\}$ is A-invariant, i.e., $A(J) \subseteq J$. Fix some $0 < x \in J$, and then pick some k such that $A^k x \gg 0$. It follows that $A^k x \in J$, and thus the ideal J contains order units. This implies $J = \mathbb{C}^n$, and hence A is an irreducible matrix. Next, for each j pick up some $m_j \in \mathbb{N}$ such that $A^{m_j} e_j \gg 0$.

To finish the solution, we shall prove that if $m = m_1 + m_2 + \cdots + m_n$, then A^m is strongly positive. Indeed, since A as an irreducible matrix has no zero rows, it follows that $A^m e_j = A^{m-m_j}(A^{m_j} e_j) \gg 0$. Since $A^m e_j$ is the jth column of A^m, we see that A^m has positive entries. Now condition (2) of Theorem 8.29 guarantees that A is a primitive matrix. ∎

Problem 8.5.7. *We have two glasses; call them* 1 *and* 2*. Glass* 1 *contains* 100 *grams of water and glass* 2 *contains* 100 *grams of wine. We transfer* 10 *grams from glass* 1 *into glass* 2 *and stir the mixture very well. Subsequently, we transfer* 10 *grams from glass* 2 *to glass* 1 *and again stir the mixtures in both glasses very well. We repeat this process ad infinitum.*

If w_n^1 and w_n^2 denote the amounts (in grams) of wine in glasses 1 *and* 2*, respectively, after the nth stage of this process, show that*

$$\lim_{n\to\infty} w_n^1 = \lim_{n\to\infty} w_n^2 = 50\,.$$

Solution: We follow the notation given in the hint. So, at stage $n-1$ we have w_{n-1}^1 grams of wine in glass 1 and w_{n-1}^2 grams of wine in glass 2. (The water-wine mixtures are stirred very well so that the mixtures in both glasses have uniform densities.) Now we perform the two parts required at the nth step. First, we transfer ten grams of the mixture from glass 1 to glass 2, and then stir the mixtures in both glasses very well—so that the mixtures again have uniform densities. After this transfer, the amounts of wine in both glasses are:

$$W_1 = w_{n-1}^1 - \tfrac{10}{100} w_{n-1}^1 = \tfrac{9}{10} w_{n-1}^1 \quad \text{and} \quad W_2 = w_{n-1}^2 + \tfrac{10}{100} w_{n-1}^1 = \tfrac{1}{10} w_{n-1}^1 + w_{n-1}^2\,.$$

Next, we transfer ten grams from the (well-stirred) 110-gram mixture of glass 2 to glass 1 (and afterwards stir both mixtures very well). Then the amounts of wine in the two glasses are:

$$\begin{aligned} w_n^1 &= W_1 + \tfrac{10}{110} W_2 = \tfrac{9}{10} w_{n-1}^1 + \tfrac{1}{11} w_{n-1}^2 + \tfrac{1}{110} w_{n-1}^1 = \tfrac{10}{11} w_{n-1}^1 + \tfrac{1}{11} w_{n-1}^2 \\ w_n^2 &= W_2 - \tfrac{10}{110} W_2 = \tfrac{10}{11} W_2 = \tfrac{10}{11}\big(\tfrac{1}{10} w_{n-1}^1 + w_{n-1}^2\big) = \tfrac{1}{11} w_{n-1}^1 + \tfrac{10}{11} w_{n-1}^2\,. \end{aligned}$$

These equations can be written in the following matrix form:

$$\begin{bmatrix} w_n^1 \\ w_n^2 \end{bmatrix} = \begin{bmatrix} \frac{10}{11} & \frac{1}{11} \\ \frac{1}{11} & \frac{10}{11} \end{bmatrix} \begin{bmatrix} w_{n-1}^1 \\ w_{n-1}^2 \end{bmatrix}$$

Given that $w_0^1 = 0$ and $w_0^2 = 100$, the preceding recursive formula yields

$$\begin{bmatrix} w_n^1 \\ w_n^2 \end{bmatrix} = \begin{bmatrix} \frac{10}{11} & \frac{1}{11} \\ \frac{1}{11} & \frac{10}{11} \end{bmatrix}^n \begin{bmatrix} 0 \\ 100 \end{bmatrix}.$$

Now notice that the positive matrix $A = \begin{bmatrix} \frac{10}{11} & \frac{1}{11} \\ \frac{1}{11} & \frac{10}{11} \end{bmatrix}$ is a Markov primitive matrix. This implies that $r(A) = \frac{10}{11} + \frac{1}{11} = 1$ and $Ae = e$, where $e = (1,1)$; see Problem 8.3.7. Therefore, according to Corollary 8.30, we have

$$\begin{bmatrix} w_n^1 \\ w_n^2 \end{bmatrix} = \begin{bmatrix} \frac{10}{11} & \frac{1}{11} \\ \frac{1}{11} & \frac{10}{11} \end{bmatrix}^n \begin{bmatrix} 0 \\ 100 \end{bmatrix} \longrightarrow c \begin{bmatrix} 1 \\ 1 \end{bmatrix},$$

where $c > 0$ is a constant. This implies $w_n^1 \to c$ and $w_n^2 \to c$. Since $w_n^1 + w_n^2 = 100$ holds at each stage n, it follows that $2c = 100$ or $c = 50$. Hence, $w_n^1 \to 50$ and $w_n^2 \to 50$. ∎

Chapter 9

Irreducible Operators

9.1. Irreducible and Expanding Operators

Problem 9.1.1. *Verify the "Implication Scheme" of Theorem* 9.6 :

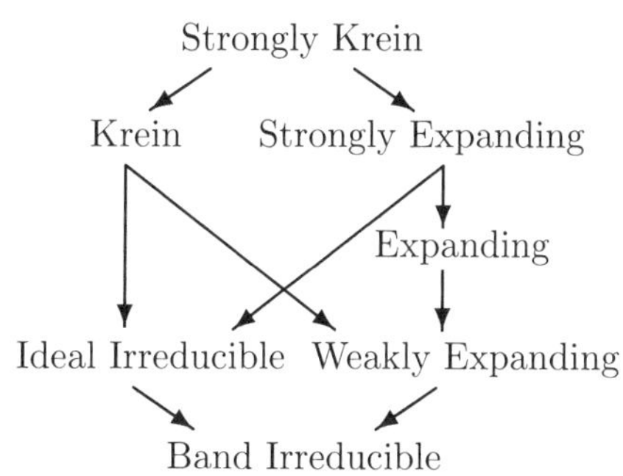

Solution: In the implications

$$\textit{Strongly Krein} \Longrightarrow \textit{Krein Operator} \Longrightarrow \textit{Ideal Irreducible} \Longrightarrow \textit{Band Irreducible}$$

only the implication "*Krein Operator* $\Rightarrow$ *Ideal Irreducible*" needs verification. To this end, let $T\colon E \to E$ be a Krein operator on an AM-space with unit. Also, let $J \neq \{0\}$ be a closed T-invariant ideal in E. Fix some $0 < x \in J$, and then pick some $n \in \mathbb{N}$ such that $T^n x$ is a strong unit. From the T-invariance of J, it follows that $T^n x \in J$, and from this we infer that $J = E$. Therefore, T is ideal irreducible. To see that an ideal irreducible operator need not be a Krein operator, consider the operator $T\colon \mathbb{R}^2 \to \mathbb{R}^2$ defined by $T(x, y) = (y, x)$.

In the implications

$$\begin{aligned} \textit{Strongly Expanding} &\Longrightarrow \textit{Expanding Operator} \\ &\Longrightarrow \textit{Weakly Expanding} \\ &\Longrightarrow \textit{Band Irreducible}\,, \end{aligned}$$

only the implication "*Weakly Expanding* $\Rightarrow$ *Band Irreducible*" needs verification. This can be established with an argument similar to the one in the preceding case.

The implication "*Strongly Krein* $\Rightarrow$ *Strongly Expanding*" should be obvious. The implication "*Krein Operator* $\Rightarrow$ *Weakly Expanding*" follows from the fact that a strong unit is always a weak unit. Finally, we shall verify the implication:

$$\textit{Strongly Expanding} \implies \textit{Ideal Irreducible}\,.$$

To see this, assume that $T\colon E \to E$ is a strongly expanding operator on a Banach lattice, and let a closed ideal $J \neq \{0\}$ be T-invariant. Fix some $0 < x \in J$, and note that the vector $u = Tx \in J$ is a quasi-interior point. So, from $E_u \subseteq J$, it follows that $E = \overline{E}_u \subseteq J$ or $J = E$. This shows that T is ideal irreducible. ∎

Problem 9.1.2. *Show that every weakly expanding operator is strictly positive.*

Solution: Let $T\colon E \to E$ be a weakly expanding operator, and let $x > 0$. Since T is weakly expanding, there exists some $n \in \mathbb{N}$ such that $T^n x$ is a weak unit, and so $T^n x > 0$. Now from $T^{n-1}(Tx) = T^n x > 0$, it follows that $Tx > 0$. Thus, T is strictly positive. ∎

Problem 9.1.3. *Let $S, T\colon E \to E$ be two arbitrary positive operators on a Banach lattice. If $0 \le S \le T$ and S is band (resp. ideal) irreducible, then show that T is likewise band (resp. ideal) irreducible.*

Solution: Assume that S is ideal irreducible, and let $T(J) \subseteq J$ for some closed ideal J. If $0 < x \in J$, then from $0 \le Sx \le Tx \in J$ it follows that $Sx \in J$. That is, $S(J) \subseteq J$. Since S is ideal irreducible, either $J = \{0\}$ or $J = E$. This shows that T is ideal irreducible.

Exactly the same arguments show that if S is band irreducible, then T is also a band irreducible operator. ∎

Problem 9.1.4. *If S and T are two positive operators on a Banach lattice and one of them is band (resp. ideal) irreducible, then show that $S + T$ is band (resp. ideal) irreducible.*

Solution: It follows from Problem 9.1.3 by observing that $0 \le S \le S + T$. ∎

Problem 9.1.5. *Give an example of a band irreducible positive operator T that is not a supremum of a family of ideal irreducible positive operators dominated by T.*

Solution: Let T be any band irreducible positive operator that is not ideal irreducible (for instance, take T in Example 9.7). Then each operator S satisfying $0 < S \le T$ is not ideal irreducible—otherwise, by Problem 9.1.4, T itself would be ideal irreducible. This, of course, implies the desired conclusion. ∎

Problem 9.1.6. *We know that each central operator leaves every ideal invariant while an ideal irreducible operator does not leave invariant any nontrivial closed ideal. Therefore, the following conjecture seems plausible:* If S is a central positive operator and T is an ideal irreducible positive operator, then $S \wedge T = 0$. *Disprove this conjecture.*

Solution: Let R be any ideal irreducible positive operator and consider $T = R+I$. Then T is also ideal irreducible. If $S = I$, then S is a central operator, and obviously $S \wedge T = S$. ∎

Problem 9.1.7. *If a positive operator T on a Banach lattice is either strongly expanding, or expanding, or strongly Krein and S is a strictly positive operator, then show that TS also possesses the corresponding property of T.*

Solution: Let $S, T \colon E \to E$ be two positive operators on a Banach lattice with S strictly positive. Fix some $x > 0$, and note that $Sx > 0$.

(a) Assume that T is strongly expanding. Then $TSx = T(Sx)$ is a quasi-interior point. This shows that TS is strongly expanding.

(b) Suppose that T is expanding. Then $TSx = T(Sx)$ is a weak unit. This shows that TS is expanding.

(c) Assume that T is a strong Krein operator. Then $TSx = T(Sx)$ is a strong unit. This shows that TS is a strong Krein operator. ∎

Problem 9.1.8. *If a Banach lattice E has order continuous norm, then show that a positive operator on E is expanding if and only if it is strongly expanding.*

Solution: The claim follows from the following fact: *In a Banach lattice with order continuous norm a positive vector is a quasi-interior point if and only if it is a weak unit.*

Indeed, if $e > 0$ is a weak unit, then $x \wedge ne \uparrow x$ holds for each $x \in E^+$, and so (by the order continuity of the norm) $\|x - x \wedge ne\| \to 0$ for each $x \in E^+$. Therefore, e is a quasi-interior point. The converse is obvious. ∎

Problem 9.1.9. *Show that a positive operator on a Banach lattice with order continuous norm is band irreducible if and only if it is ideal irreducible.*

Solution: This claim follows immediately from the following result: *If a Banach lattice has order continuous norm, then every closed ideal is a band.* ∎

Problem 9.1.10. *If $T \colon E \to E$ is a σ-order continuous band irreducible positive operator, then show that:*

(a) *T is strictly positive.*

(b) *T carries weak units to weak units.*

(c) *E has weak units.*

Solution: (a) Let $x > 0$, and let B_x denote the band generated by x in E. Now assume by way of contradiction that $Tx = 0$. Since every $y \in B^+$ satisfies $y \wedge nx \uparrow y$ and $T(y \wedge nx) = 0$ holds for each n, it follows from the σ-order continuity of T that $Ty = 0$ for each $y \in B_x^+$. Consequently, $Ty = 0$ for each $y \in B_x$. This implies $T(B_x) = \{0\} \subseteq B_x$, i.e., that B_x is T-invariant, and so the band irreducibility of

T guarantees that $B_x = E$. But then $T = 0$ which is impossible. Hence, $x > 0$ implies $Tx > 0$, and so T is strictly positive.

(b) Let $u > 0$ be a weak unit in E, i.e., $B_u = E$. Since, by part (a), T is strictly positive, it follows that $v = Tu > 0$, and so $B_v \neq \{0\}$. Now fix some $x > 0$, and note that $x \wedge nu \uparrow x$ implies $T(x \wedge nu) \uparrow Tx$. From $T(x \wedge nu) \leq nv$, it follows that $Tx \in B_v$. So, B_v is T-invariant, and the band irreducibility of T guarantees that $B_v = E$. That is, $v = Tu$ is a weak unit in E.

(c) Fix some $y > 0$ and consider the vector $u = \sum_{n=0}^{\infty} \frac{1}{2^n} \cdot \frac{T^n y}{\|T\|^n} > 0$. We claim that the principal ideal E_u is T-invariant. Indeed, if $|x| \leq \lambda u$, then

$$|Tx| \leq T|x| \leq \lambda Tu = \lambda \sum_{n=0}^{\infty} \frac{1}{2^n} \cdot \frac{T^{n+1}y}{\|T\|^n} = 2\lambda\|T\| \sum_{n=0}^{\infty} \frac{1}{2^{n+1}} \cdot \frac{T^{n+1}y}{\|T\|^{n+1}} \leq 2\lambda\|T\|u\,,$$

and so $Tx \in E_u$. Therefore, $T(E_u) \subseteq E_u \subseteq B_u$. Since T is σ-order continuous, it easily follows that $T(B_u) \subseteq B_u$. Now taking into account that T is band irreducible and $B_u \neq \{0\}$, we see that $B_u = E$. This shows that u is a weak unit in E, and consequently E has weak units. ∎

Problem 9.1.11. *For a positive operator $T\colon E \to E$ on a Banach lattice establish the following properties.*

(a) *If T^k is ideal (resp. band) irreducible for some $k > 1$, then T is also ideal (resp. band) irreducible.*

(b) *If T is ideal irreducible, then T^2 need not be ideal irreducible.*

Solution: (a) Assume that T^k is ideal irreducible for some $k > 1$, and let J be a closed ideal in E such that $T(J) \subseteq J$. Then $T^k(J) \subseteq J$ also holds, and so, by the ideal irreducibility of T^k, it follows that $J = \{0\}$ or $J = E$. This shows that T is ideal irreducible.

A similar argument shows that if T^k is band irreducible for some $k > 1$, then T itself must be band irreducible.

(b) Consider the positive operator $T\colon \mathbb{R}^2 \to \mathbb{R}^2$ defined by $T(x, y) = (y, x)$. Clearly, $T^2 = I$, and so T^2 is neither ideal nor band irreducible.

We claim that T is ideal irreducible. To see this, let $J \neq \{0\}$ be an ideal in $\mathbb{R}^2$ which is T-invariant. Let $0 < (x, y) \in J$. Then either $x > 0$ or $y > 0$. From $T(x, y) = (y, x) \in J$, we see that $0 \ll (x + y, x + y) = (x, y) + (y, x) \in J$. This implies $J = E$, and so T is ideal irreducible. ∎

Problem 9.1.12. *For a Markov operator $T\colon E \to E$ on an AM-space with unit e establish the following.*

(a) *There exists some $0 < \phi \in E^*$ satisfying $T^*\phi = \phi$. (Note that if $E = C(\Omega)$ and $e = \mathbf{1}$, then ϕ is a regular Borel measure on Ω, known as an* ***invariant measure*** *for T.)*

(b) *If $0 < \phi \in E^*$ satisfies $T^*\phi = \phi$, then the rank-one positive operator $S = \phi \otimes e$ satisfies $ST = TS = S$.*

Solution: (a) Consider the w^*-compact convex set

$$\mathcal{C} = \{\psi \in E^* \colon \ \psi \geq 0 \text{ and } \psi(e) = 1\}.$$

Then the adjoint operator $T^* \colon E^* \to E^*$ is w^*-continuous and satisfies $T^*(\mathcal{C}) \subseteq \mathcal{C}$. (Indeed, if $\psi \in \mathcal{C}$, then $(T^*\psi)(e) = \psi(Te) = \psi(e) = 1$, and so $T^*\psi \in \mathcal{C}$.) Thus, by the classical Tychonoff Fixed Point Theorem, there exists some $\phi \in \mathcal{C}$ such that $T^*\phi = \phi$.

(b) Notice that for each $x \in E$ we have:

$$\begin{aligned} STx &= S(Tx) = (\phi \otimes e)(Tx) = \phi(Tx)e = [(T^*\phi)x]e = \phi(x)e = Sx\,, \text{ and} \\ TSx &= T(Sx) = T(\phi(x)e) = \phi(x)Te = \phi(x)e = Sx. \end{aligned}$$

Therefore, $ST = TS = S$. (For related results see Lemma 9.45 and Corollary 9.46.) ∎

Problem 9.1.13. *Consider the positive operator $T \colon C[0,1] \to C[0,1]$ defined by*

$$Tf(t) = \tfrac{1}{t} \int_0^t f(s)\, ds\,, \qquad (\star)$$

for all $0 < t \leq 1$ and $Tf(0) = f(0)$. Show that:

(a) *T defines indeed a positive operator on $C[0,1]$.*

(b) *T is one-to-one (and hence strictly positive).*

(c) *T is a Markov operator.*

(d) *T is not band irreducible—and hence not ideal irreducible either.*

(e) *The point spectrum of T is the set*

$$\sigma_p(T) = \{\lambda \in \mathbb{C} \colon \ |\lambda - \tfrac{1}{2}| < \tfrac{1}{2}\} \cup \{1\}.$$

(f) *T is not a compact operator.*

Solution: (a) We need to show that Tf is continuous at zero for each $f \in C[0,1]$. To see this, fix $f \in C[0,1]$, let $\epsilon > 0$, and then choose some $\delta > 0$ such that $u, v \in [0,1]$ with $|u - v| < \delta$ imply $|f(u) - f(v)| < \epsilon$. So, if $s \in [0,1]$ satisfies $0 < s < \delta$, then $|f(s) - f(0)| < \epsilon$. In particular, for each $t \in [0, \delta]$ we have

$$\big|Tf(t) - Tf(0)\big| = \tfrac{1}{t}\Big|\int_0^t [f(s) - f(0)]\, ds\Big| \leq \tfrac{1}{t} \int_0^t \big|f(s) - f(0)\big|\, ds < \tfrac{1}{t} \cdot \epsilon t = \epsilon\,.$$

This shows that Tf is continuous at zero. Now it should be clear that T as defined by $(\star)$ is a positive operator on $C[0,1]$.

(b) Assume that some $f \in C[0,1]$ satisfies $Tf = 0$. This implies $\int_0^t f(s)\, ds = 0$ for each $0 < t \leq 1$. The Fundamental Theorem of Calculus shows that we have $f(t) = \frac{d}{dt}\int_0^t f(s)\, ds = 0$ for all $0 \leq t \leq 1$. So, T is one-to-one—and consequently T is a strictly positive operator.

(c) Notice that $\mathbf{1}$ (the constant function one) is the unit of $C[0,1]$. It follows that for each $0 < t \leq 1$ we have $T\mathbf{1}(t) = \frac{1}{t}\int_0^t 1\, ds = 1$. Therefore, $T\mathbf{1} = \mathbf{1}$, and so T is a Markov operator. In particular, $\lambda = 1$ is an eigenvalue of T.

(d) Consider the band $B = \{f \in C[0,1]\colon\ f(s) = 0 \text{ for all } s \in [\frac{1}{2}, 1]\}$. It is easy to see that T does not leave B invariant. Thus, T is not band irreducible.

(e) To find the eigenvalues of T, we must find all complex numbers $\lambda = \alpha + \beta\imath$ for which there exists a non-zero continuous function $f\colon [0,1] \to \mathbb{C}$ such that

$$Tf(t) = \tfrac{1}{t}\int_0^t f(s)\,ds = \lambda f(t) \qquad (\boldsymbol{\alpha})$$

for each $0 < t \leq 1$. Since T is one-to-one, we see that $\lambda = 0$ cannot be an eigenvalue of T. So, we can assume from the outset that $\lambda \neq 0$. In view of (c), we can also assume that $\lambda \neq 1$.

Now a glance at $(\boldsymbol{\alpha})$ guarantees that for every solution f of $(\boldsymbol{\alpha})$ the derivative $f'(t)$ exists for each $0 < t \leq 1$. Rewriting $(\boldsymbol{\alpha})$ in the form $\int_0^t f(s)\,ds = \lambda t f(t)$ and differentiating, we get $f(t) = \lambda f(t) + \lambda t f'(t)$ or

$$\lambda t f'(t) + (\lambda - 1) f(t) = 0\,. \qquad (\boldsymbol{\beta})$$

We are looking for eigenvalues different than 0 and 1. In this case, aside from scalar multiples, the solution of $(\boldsymbol{\beta})$ is given by

$$f(t) = t^{\frac{1}{\lambda}-1} = e^{(\frac{1}{\lambda}-1)\ln t} \qquad (\boldsymbol{\gamma})$$

for each $0 < t \leq 1$. Letting $t \to 0^+$ in $(\boldsymbol{\alpha})$ yields $f(0) = \lambda f(0)$ or $f(0) = 0$. This shows that the function $f(t) = t^{\frac{1}{\lambda}-1}$ as defined in $(\boldsymbol{\gamma})$ is an eigenfunction of T if and only if $\lim_{t\to 0^+} t^{\frac{1}{\lambda}-1} = 0$. Taking into account the formula

$$t^{\frac{1}{\lambda}-1} = e^{(\frac{1}{\alpha+\imath\beta}-1)\ln t} = e^{\frac{\alpha-\alpha^2-\beta^2}{\alpha^2+\beta^2}\ln t}\, e^{-\imath\frac{\beta}{\alpha^2+\beta^2}\ln t}\,,$$

we see that $\lim_{t\to 0^+} t^{\frac{1}{\lambda}-1} = 0$ if and only if $\alpha - \alpha^2 - \beta^2 > 0$. Now note that

$$\begin{aligned} \alpha - \alpha^2 - \beta^2 > 0 \quad &\Longleftrightarrow \quad \alpha^2 + \beta^2 - \alpha < 0 \\ &\Longleftrightarrow \quad \left(\alpha - \tfrac{1}{2}\right)^2 + \beta^2 < \left(\tfrac{1}{2}\right)^2 \\ &\Longleftrightarrow \quad \left|\lambda - \tfrac{1}{2}\right| < \tfrac{1}{2}\,. \end{aligned}$$

Therefore, the point spectrum of T is $\sigma_p(T) = \{\lambda \in \mathbb{C}\colon\ |\lambda - \frac{1}{2}| < \frac{1}{2}\} \cup \{1\}$.

(f) Since the spectrum of T is uncountable, T cannot be a compact operator. A direct proof that T is not compact goes as follows. Consider the sequence $\{f_n\}$ in $C[0,1]$ defined by

$$f_n(s) = \begin{cases} ns & \text{if } 0 \leq s \leq \frac{1}{n} \\ 1 & \text{if } \frac{1}{n} < s \leq 1\,. \end{cases}$$

Clearly, $\|f_n\|_\infty = 1$ holds for each n. Moreover, a direct computation shows that

$$Tf_n(t) = \begin{cases} \frac{nt}{2} & \text{if } 0 \leq t \leq \frac{1}{n} \\ 1 - \frac{1}{2nt} & \text{if } \frac{1}{n} < t \leq 1\,. \end{cases}$$

Notice that $\{Tf_n\}$ converges pointwise to the discontinuous function $g\colon [0,1] \to \mathbb{R}$ given by $g(0) = 0$ and $g(t) = 1$ for $0 < t \leq 1$. This shows that no subsequence of $\{Tf_n\}$ can converge uniformly, and consequently T is not a compact operator. ∎

Problem 9.1.14. *Let $1 < p < \infty$. We consider the operator T of Problem 9.1.13 defined on $L_p[0,\infty)$ via the same formula*

$$Tf(t) = \tfrac{1}{t}\int_0^t f(s)\,ds\,, \qquad (\star\star)$$

for all $t > 0$ and $Tf(0) = f(0)$.

Show that the formula $(\star\star)$ defines a one-to-one strictly positive operator from $L_p[0,\infty)$ into $L_p[0,\infty)$ such that $\|T\|_p = \frac{p}{p-1}$.

Solution: Consider an arbitrary continuous function $f\colon (0,\infty) \to \mathbb{R}$ with compact support, i.e., $f \in C_c(0,\infty)$. Choose some $M > 0$ so that $0 \le f(s) \le M$ for all $s > 0$. If $K = \int_0^\infty f(t)\,dt$, then the function

$$g(s) = \begin{cases} M & \text{if } 0 < s \le 1 \\ \frac{K}{s} & \text{if } s > 1 \end{cases}$$

belongs to $L_p(0,\infty)$. Clearly, $0 \le Tf \le g$, and so Tf belongs to $L_p(0,\infty)$ for each $1 < p < \infty$. Also, in view of the inequalities

$$0 \le t\Big[\tfrac{1}{t}\int_0^t f(s)\,ds\Big]^p = t\big[Tf(t)\big]^p \le t\big[g(t)\big]^p\,,$$

it follows that

$$t\Big[\tfrac{1}{t}\int_0^t f(s)\,ds\Big]^p\Big|_0^\infty = 0\,.$$

Now integrating by parts and using Hölder's inequality, we get

$$\begin{aligned}
\|Tf\|_p^p &= \int_0^\infty \Big[\tfrac{1}{t}\int_0^t f(s)\,ds\Big]^p dt = \tfrac{1}{1-p}\int_0^\infty \Big[\int_0^t f(s)\,ds\Big]^p d\big(t^{1-p}\big)\\
&= \tfrac{1}{1-p}\Big[t\Big(\tfrac{1}{t}\int_0^t f(s)ds\Big)^p\Big|_0^\infty - p\int_0^\infty f(t)\Big(\tfrac{1}{t}\int_0^t f(s)ds\Big)^{p-1}dt\Big]\\
&= \tfrac{p}{p-1}\int_0^\infty f(t)\big[Tf(t)\big]^{p-1}\,dt\\
&\le \tfrac{p}{p-1}\,\|f\|_p\Big(\int_0^\infty \big[Tf(t)\big]^{q(p-1)}\,dt\Big)^{\frac{1}{q}}\\
&= \tfrac{p}{p-1}\,\|f\|_p\cdot\big(\|Tf\|_p\big)^{\frac{p}{q}}.
\end{aligned}$$

This easily implies $(\|Tf\|_p)^{p-\frac{p}{q}} \le \frac{p}{p-1}\|f\|_p$ or

$$\|Tf\|_p \le \tfrac{p}{p-1}\,\|f\|_p$$

for all $f \in C_c(0,\infty)$. In other words, the mapping

$$T\colon C_c(0,\infty) \longrightarrow L_p(0,\infty)$$

defines a continuous operator such that $\|T\| \le \frac{p}{p-1}$.

Since $C_c(0,\infty)$ is norm dense in $L_p(0,\infty)$, the operator T has a unique continuous (linear) extension $\hat{T}$ to all of $L_p(0,\infty)$ and $\|\hat{T}\| \le \frac{p}{p-1}$. Our next objective is to show that $\hat{T}f(t) = \frac{1}{t}\int_0^t f(s)\,ds$ holds for all $f \in L_p(0,\infty)$ and all $t > 0$.

To this end, let $0 \leq \phi$ be a step function. Choose some $C > 0$ satisfying $\phi(s) \leq C$ for all $s > 0$, and then pick a sequence $\{f_n\}$ in $C_c(0,\infty)$ with $\lim \int |f_n(s) - \phi(s)|^p \, ds = 0$. We can assume that $\lim f_n(s) = \phi(s)$ for almost all s; see Problem 1.2.13. In view of the inequality

$$\big| f_n \wedge C - \phi \big| = \big| f_n \wedge C - \phi \wedge C \big| \leq \big| f_n - C \big| ,$$

replacing $\{f_n\}$ by $\{f_n \wedge C\}$ if necessary, we can assume that $0 \leq f_n(s) \leq C$ for all $s > 0$ and all n. Since $\lim \|Tf_n - \hat{T}\phi\|_p = 0$, we can also assume (by passing to a subsequence) that $Tf_n(t) \to \hat{T}\phi(t)$ for almost all t. Next, observe that for each fixed $t > 0$ we have $\phi \in L_1(0,t)$ and so, by the Lebesgue Dominated Convergence Theorem, we see that

$$\hat{T}\phi(t) = \lim_{n\to\infty} Tf_n(t) = \lim_{n\to\infty} \tfrac{1}{t} \int_0^t f_n(s)\, ds = \tfrac{1}{t} \int_0^t \phi(s)\, ds .$$

Now let $0 \leq f \in L_p(0,\infty)$. Choose a sequence $\{\phi_n\}$ of step functions with $0 \leq \phi_n \uparrow f$. In view of

$$\lim \|\hat{T}\phi_n - \hat{T}f\|_p = 0 ,$$

we can assume that $\hat{T}\phi_n(t) \to \hat{T}f(t)$ for almost all t. Taking into account that for each fixed $t > 0$, we have $f \in L_p(0,t) \subseteq L_1(0,t)$, the Lebesgue Dominated Convergence Theorem yields

$$\hat{T}f(t) = \lim_{n\to\infty} \hat{T}\phi_n(t) = \lim_{n\to\infty} \tfrac{1}{t} \int_0^t \phi_n(s)\, ds = \tfrac{1}{t} \int_0^t f(s)\, ds$$

for each $t > 0$. Thus, $\hat{T} = T$.

Next, we shall show that $\|T\| = \frac{p}{p-1}$. We already know that $\|T\| \leq \frac{p}{p-1}$. So, it must be established that $\|T\| \geq \frac{p}{p-1}$. To this end, let

$$f_n(s) = \begin{cases} s^{(n^{-1}-1)p^{-1}} & \text{if } 0 < s < 1 \\ 0 & \text{if } s \geq 1 . \end{cases}$$

Then $\|f_n\|_p^p = \int_0^1 s^{n^{-1}-1}\, ds = n$, and also

$$Tf_n(t) = \tfrac{np}{1+n(p-1)} \begin{cases} t^{(n^{-1}-1)p^{-1}} & \text{if } 0 < t < 1 \\ t^{-1} & \text{if } t \geq 1 . \end{cases}$$

Consequently, we have

$$\begin{aligned} \big\|Tf_n\big\|_p^p &= \int_0^\infty \big|Tf_n(t)\big|^p\, dt \\ &= \Big[\tfrac{np}{1+n(p-1)}\Big]^p \Big[\int_0^1 t^{n^{-1}-1}\, dt + \int_1^\infty t^{-p}\, dt\Big] \\ &= \Big[\tfrac{np}{1+n(p-1)}\Big]^p \big(n + \tfrac{1}{p-1}\big) , \end{aligned}$$

and so

$$\tfrac{np}{1+n(p-1)} \big[n + \tfrac{1}{p-1}\big]^{\frac{1}{p}} = \big\|Tf_n\big\|_p \leq \big\|T\big\| \cdot \big\|f_n\big\|_p = \big\|T\big\| \cdot n^{\frac{1}{p}} .$$

This implies

$$\big\|T\big\| \geq \tfrac{p}{p-1+\frac{1}{n}} \cdot \big[1 + \tfrac{1}{n(p-1)}\big]^{\frac{1}{p}} \longrightarrow \tfrac{p}{p-1} ,$$

from which it follows immediately that $\|T\| \geq \frac{p}{p-1}$.

Finally, we establish that T is one-to-one. Assume that $Tf = 0$ for some function $f \in L_p(0,\infty)$. Then $\int_0^t f(s)\,ds = 0$ for all $t > 0$. This implies $f = 0$ a.e., and so the operator T is one-to-one. (See also the solution to Problem 7.1.4.) ∎

Problem 9.1.15. *Prove the following converse of Corollary* 9.14. *If for a positive operator* $T\colon E \to E$ *on a Banach lattice there exists some* $\lambda > r(T)$ *such that* $R(\lambda, T)$ *is strongly expanding, then* T *is ideal irreducible.*

Solution: Let $T\colon E \to E$ be a positive operator on a Banach lattice. Also, assume that for some $\lambda > r(T)$ the operator $R(\lambda, T)$ is strongly expanding. By Theorem 9.6 (or by Problem 9.1.1), the operator $R(\lambda, T)$ is ideal irreducible. Suppose that $T(J) \subseteq J$ for some closed ideal J in E.

Now fix some vector $x \in J$, and note that $T^n x \in J$ for each $n = 0, 1, \ldots$. This implies $R(\lambda, T)x = \sum_{n=0}^{\infty} \lambda^{-(n+1)} T^n x \in J$, and therefore the closed ideal J is $R(\lambda, T)$-invariant. Since $R(\lambda, T)$ is ideal irreducible, we get that either $J = \{0\}$ or $J = E$. This shows that T is ideal irreducible. ∎

Problem 9.1.16. *For a positive operator* $T\colon E \to E$ *on a Banach lattice show the following.*

(a) *The operator* T *is ideal irreducible if and only if* $\{0\}$ *and* E^* *are the only* w^**-closed bands in* E^* *that are invariant under the adjoint operator* $T^*\colon E^* \to E^*$.

(b) *If the adjoint operator* T^* *is band irreducible, then* T *is ideal irreducible. Give an example of an ideal irreducible operator whose adjoint is not band irreducible.*

(c) *If* E *has order continuous norm, then* T *is ideal irreducible if and only if* T^* *is band irreducible.*

Solution: We use the notation suggested in the hint of the exercise. For each ideal J in E and each band B in E^* we define the polars

$$\begin{aligned} J^{\text{o}} &= \{x^* \in E^*\colon\ \langle x, x^* \rangle = 0 \ \text{ for all } \ x \in J\} \ \text{ and} \\ B^{\text{o}} &= \{x \in E\colon\ \langle x, x^* \rangle = 0 \ \text{ for all } \ x^* \in E^*\}. \end{aligned}$$

Next, we shall present several properties of the polars. For the discussion below J denotes an ideal in E and B a band in E^*.

1. J^{o} *is a* w^**-closed band in* E^* *and* B^{o} *is a norm closed ideal in* E.

Clearly, J^{o} and B^{o} are both vector subspaces. We verify that J^{o} is a w^*-closed band. To see that J^{o} is an ideal, assume that $x^*, y^* \in E^*$ satisfy $|y^*| \leq |x^*|$ with $x^* \in J^{\text{o}}$. Then for each $0 \leq x \in J$, we have

$$\begin{aligned} |y^*(x)| &\leq |y^*|(|x|) \leq |x^*|(x) = \sup\{x^*(z)\colon\ z \in E \ \text{ and } \ |z| \leq x\} \\ &= \sup\{x^*(z)\colon\ z \in J \ \text{ and } \ |z| \leq x\} = 0\,. \end{aligned}$$

So, $y^*(x) = 0$ for all $x \in J$. That is, $y^* \in J^{\text{o}}$, and hence J^{o} is an ideal in E^*. Next, notice that if $\{x^*_\alpha\} \subseteq J^{\text{o}}$ satisfies $0 \leq x^*_\alpha \uparrow x^*$ in E^*, then for each $0 \leq x \in J$, we have $x^*(x) = \sup_\alpha x^*_\alpha(x) = 0$. This implies $x^* \in J^\circ$, and so J° is a band in E^*. The corresponding properties for the polars of B can be proven in a similar manner.

2. *If J is a closed ideal in E, then $J^{\circ\circ} = (J^\circ)^\circ = J$. Similarly, if B is a w^*-closed band in E^*, then $B^{\circ\circ} = B$.*

Assume that B is a w^*-closed band in E^*. It should be obvious that $B \subseteq B^{\circ\circ}$. Suppose by way of contradiction that $B^{\circ\circ} \neq B$. This means that there exists some $x^* \in B^{\circ\circ}$ such that $x^* \notin B$. Since B is w^*-closed, there exists (by the Separation Theorem) some $x \in X$ such that $x^*(x) \neq 0$ and $y^*(x) = 0$ for all $y^* \in B^{\circ\circ}$. In particular, $y^*(x) = 0$ for all $y^* \in B$. This implies $x \in B^\circ$, and therefore $x^*(x) = 0$, which is impossible. Hence, $B^{\circ\circ} = B$. The corresponding result for J can be proven in a similar manner.

We now return to our problem. Let $T\colon E \to E$ be a positive operator on a Banach lattice.

(a) Assume that T is ideal irreducible, and let B be a T^*-invariant w^*-closed band in E^*. If $x \in B^\circ$, then for each $x^* \in B$ we have $\langle Tx, x^*\rangle = \langle x, T^*x^*\rangle = 0$. This shows that $Tx \in B^\circ$, and so B° is T-invariant. Since T is ideal irreducible, it follows that either $B^\circ = \{0\}$ or $B^\circ = E$. This implies either $B = B^{\circ\circ} = \{0\}$ or $B = E^*$.

For the converse, suppose that the only w^*-closed T^*-invariant bands in E^* are $\{0\}$ and E^*. Let J be a closed T-invariant ideal in E. If $x^* \in J^\circ$ and $x \in J$, then $\langle x, T^*x^*\rangle = \langle Tx, x^*\rangle = 0$. This implies $T^*x^* \in J^\circ$, and so the w^*-closed band J° is T^*-invariant. From our hypothesis, $J^\circ = \{0\}$ or $J^\circ = E^*$. This means $J = J^{\circ\circ} = E$ or $J = J^{\circ\circ} = \{0\}$. Consequently, T is ideal irreducible.

(b) If T^* is band irreducible, then T^* does not leave invariant any non-trivial w^*-closed band in E^*. By part (a), T must be ideal irreducible.

The adjoint of any positive operator on $C[0,1]$ cannot be ideal irreducible since the norm dual of $C[0,1]$ does not have any quasi-interior points; see Problem 4.2.5 and Theorem 9.3. Therefore, any ideal irreducible positive operator T on $C[0,1]$ provides the desired example.

(c) This follows from (a) and the fact that a Banach lattice has order continuous norm if and only if every band of E^* is w^*-closed; see [**5**, Theorem 9.1, p. 60]. ∎

Problem 9.1.17. *Give examples of positive operators S and T on atomic and non-atomic Banach lattices such that T is ideal irreducible, $ST \leq S$, and S is a non-expanding strictly positive operator. (This shows that the inequality $TS \leq \lambda S$ in Lemma* 9.13 *cannot be replaced by $ST \leq \lambda S$.)*

Solution: For the atomic case let $E = \mathbb{R}^2$, and consider the pair of positive operators $S, T\colon E \to E$ given by the matrices

$$T = \begin{bmatrix} 0 & 1 \\ 1 & 0 \end{bmatrix} \quad \text{and} \quad S = \begin{bmatrix} 1 & 1 \\ 0 & 0 \end{bmatrix}.$$

Clearly, T is ideal irreducible and S is strictly positive. Since $Se_1 = e_1$, the operator S is not expanding. Finally, note that for each $(\alpha, \beta) \in E$ we have

$$ST\begin{bmatrix}\alpha\\ \beta\end{bmatrix} = S\begin{bmatrix}\beta\\ \alpha\end{bmatrix} = (\beta+\alpha, 0) = S\begin{bmatrix}\alpha\\ \beta\end{bmatrix}.$$

This shows that $ST = S$, and certainly $ST \leq S$ is also true.

For the non-atomic case, let $E = L_p[0,1]$, where $1 \leq p < \infty$. Fix any strictly positive functional $\phi \in E^*$ such that $\phi(\mathbf{1}) = 1$, for instance let $\phi(x) = \int_0^1 x(t)dt$, and consider the positive rank-one operator $T = \phi \otimes \mathbf{1}$. Clearly, the operator T is an ideal irreducible positive operator on E. Next, let $A = \chi_{[0,\frac{1}{2}]}$ and consider the operator $S = \chi_A T$ on E. Clearly S is strictly positive but fails to be expanding. In fact $R(S)$, the range of S, is the one-dimensional vector space $R(S) = \{\lambda\chi_A \colon \lambda \in \mathbb{R}\}$. Now note that if $x \in E^+$, then

$$Sx = \chi_A Tx = \chi_A \phi(x)\mathbf{1} = \phi(x)\chi_A\,,$$

and from $\phi(\mathbf{1}) = 1$ it follows that

$$STx = S(\phi(x)\mathbf{1}) = \phi(x)S(\mathbf{1}) = \phi(x)\phi(\mathbf{1})\chi_A = \phi(x)\chi_A\,.$$

Therefore, $ST = S$, and certainly $ST \leq S$ is also true. ∎

9.2. Ideal Irreducibility and the Spectral Radius

Problem 9.2.1. *Let E be a Banach lattice and let $T\colon E \to E$ be a positive operator having a non-trivial invariant closed ideal. Show that there exists some $0 < u \in E$ satisfying $\overline{E}_u \neq E$ and $T(\overline{E}_u) \subseteq \overline{E}_u$.*

Solution: Let J be a non-trivial T-invariant closed ideal and fix some $0 < x \in J$. Now consider the element $u = \sum_{n=0}^{\infty} \frac{1}{2^n} \cdot \frac{T^n x}{\|T\|^n}$. Since J is T-invariant, we see that $u \in J$. If $y \in E$ satisfies $|y| \leq \lambda u$, then

$$|Ty| \leq T|y| \leq \lambda Tu = \lambda \sum_{n=0}^{\infty} \tfrac{1}{2^n} \cdot \tfrac{T^{n+1}x}{\|T\|^n} = 2\lambda\|T\| \sum_{n=0}^{\infty} \tfrac{1}{2^{n+1}} \cdot \tfrac{T^{n+1}x}{\|T\|^{n+1}} \leq 2\lambda\|T\|u\,.$$

This implies $Ty \in E_u$, i.e., E_u is T-invariant. Consequently, $T(\overline{E}_u) \subseteq \overline{E}_u \subseteq J$. Thus, the vector $u > 0$ satisfies the desired properties. ∎

Problem 9.2.2. *Let $T\colon X \to X$ be a bounded operator on a Banach space. If λ is an eigenvalue of T, then show that the corresponding eigenspace $N_T(\lambda) = \{x \in X\colon\ Tx = \lambda x\}$ is a T-hyperinvariant closed subspace.*

Solution: Assume that a bounded operator $S\colon X \to X$ commutes with T, i.e., $ST = TS$. If $x \in N_T(\lambda)$, then

$$T(Sx) = S(Tx) = S(\lambda x) = \lambda Sx\,,$$

and so $Sx \in N_T(\lambda)$. In other words, $N_T(\lambda)$ is S-invariant, and thus the closed subspace $N_T(\lambda)$ is T-hyperinvariant. ∎

Problem 9.2.3 (Ercan–Onal [**29**]). *Let $T\colon X \to X$ be a bounded operator on a Banach space and let x be a vector in X. Show that T is locally quasinilpotent at x if and only if $\big|x^*(T^n x)\big|^{\frac{1}{n}} \longrightarrow 0$ for each $x^* \in X^*$.*

If X is also a Banach lattice, then show that T is locally quasinilpotent at x if and only if $\big|x^*(T^n x)\big|^{\frac{1}{n}} \longrightarrow 0$ for each $0 \le x^* \in X^*$.

Solution: If $\big\|T^n x\big\|^{\frac{1}{n}} \to 0$ and $x^* \in X^*$, then clearly

$$0 \le \big|x^*(T^n x)\big|^{\frac{1}{n}} \le \|x^*\|^{\frac{1}{n}} \cdot \big\|T^n x\big\|^{\frac{1}{n}} \to 0\,,$$

whence $\big|x^*(T^n x)\big|^{\frac{1}{n}} \longrightarrow 0$.

For the converse assume that $\big|x^*(T^n x)\big|^{\frac{1}{n}} \longrightarrow 0$ for each $x^* \in X^*$. We must establish that $\big\|T^n x\big\|^{\frac{1}{n}} \longrightarrow 0$. We shall present two solutions below. The first one is due to V. G. Troitsky. Let $\epsilon > 0$ and consider the sequence $\big\{\frac{T^n x}{\epsilon^n}\big\}$. We claim that the sequence $\big\{\frac{T^n x}{\epsilon^n}\big\}$ is weakly bounded. To see this, fix $x^* \in X^*$. From $\big|x^*(T^n x)\big|^{\frac{1}{n}} \longrightarrow 0$, it follows that there exists some n_0 such that $\big|x^*(T^n x)\big|^{\frac{1}{n}} < \epsilon$ for all $n \ge n_0$. The latter is equivalent to $\big|x^*\big(\frac{T^n x}{\epsilon^n}\big)\big| < 1$ for all $n \ge n_0$. This implies that the sequence of real numbers $\big\{x^*\big(\frac{T^n x}{\epsilon^n}\big)\colon\ n \in \mathbb{N}\big\}$ is bounded for each $x^* \in X^*$, i.e., the sequence $\big\{\frac{T^n x}{\epsilon^n}\big\}$ is weakly bounded. By the Principle of Uniform Boundedness, the sequence $\big\{\frac{T^n x}{\epsilon^n}\big\}$ is also norm bounded. Pick some $M > 0$ satisfying $\big\|\frac{T^n x}{\epsilon^n}\big\| \le M$ for each $n \in \mathbb{N}$ or, equivalently, $\|T^n x\|^{\frac{1}{n}} \le \epsilon M^{\frac{1}{n}}$ for each n. Consequently, we have

$$0 \le \liminf_{n\to\infty} \|T^n x\|^{\frac{1}{n}} \le \limsup_{n\to\infty} \|T^n x\|^{\frac{1}{n}} \le \limsup_{n\to\infty} \epsilon M^{\frac{1}{n}} = \epsilon\,.$$

Since $\epsilon > 0$ is arbitrary, we get $\liminf_{n\to\infty} \|T^n x\|^{\frac{1}{n}} = \limsup_{n\to\infty} \|T^n x\|^{\frac{1}{n}} = 0$. This shows that $\lim_{n\to\infty} \|T^n x\|^{\frac{1}{n}} = 0$.

The second solution of the converse is due to Z. Ercan and S. Onal [**29**]. Assume by way of contradiction that $\big\|T^n x\big\|^{\frac{1}{n}} \not\to 0$. So, there exist some $\epsilon > 0$ and a strictly increasing sequence $\{k_n\}$ of natural numbers such that $\epsilon < \big\|T^{k_n} x\big\|^{\frac{1}{k_n}}$ for each n. Now consider the sequence $\{x_n\}$ of X defined by

$$x_n = \tfrac{1}{\epsilon^{k_n}} T^{k_n} x\,.$$

Clearly, $\|x_n\| = \frac{1}{\epsilon^{k_n}} \|T^{k_n} x\| \ge \frac{1}{\epsilon^{k_n}} \cdot \epsilon^{k_n} = 1$ holds for each n. Moreover, we claim that $\{x_n\}$ converges weakly to zero. To establish this, fix $x^* \in X^*$. From $\big|x^*(T^{k_n} x)\big|^{\frac{1}{k_n}} \to 0$, it follows that there exists some ℓ such that $\big|x^*(T^{k_n} x)\big|^{\frac{1}{k_n}} < \frac{\epsilon}{2}$ for all $n \ge \ell$. Therefore, for $n \ge \ell$ we have

$$\big|x^*(x_n)\big| = \tfrac{1}{\epsilon^{k_n}} \big|x^*(T^{k_n} x)\big| < \tfrac{1}{\epsilon^{k_n}} \cdot \big(\tfrac{\epsilon}{2}\big)^{k_n} = \big(\tfrac{1}{2}\big)^{k_n}\,.$$

This implies $x^*(x_n) \to 0$, and thus $x_n \xrightarrow{\ w\ } 0$.

A glance at Problem 1.3.5 reveals that the sequence $\{x_n\}$ has a basic subsequence. By passing to a subsequence and relabelling, we can assume without loss of generality that $\{x_n\}$ is itself a basic sequence. If $\{x_n^*\}$ denotes the sequence of biorthogal functionals of the basic sequence $\{x_n\}$ defined on the vector subspace Y generated by $\{x_n\}$, then we can extend each x_n^* to X with preservation of its norm. From part (c) of Theorem 1.37 we know that $\|x_n^*\| \le \frac{\kappa}{\|x_n\|} \le \kappa < \infty$ for each n, where κ is the basis constant of the basic sequence $\{x_n\}$. That is, the sequence $\{x_n^*\}$ is norm bounded and satisfies $x_n^*(x_m) = \delta_{nm}$. Now consider the continuous linear functional $g = \sum_{n=1}^{\infty} \frac{1}{2^{k_n}} x_n^*$ and note that

$$\big|g(T^{k_n}x)\big|^{\frac{1}{k_n}} = \big|\epsilon^{k_n} g(x_n)\big|^{\frac{1}{k_n}} = \big|\epsilon^{k_n} \cdot \tfrac{1}{2^{k_n}}\big|^{\frac{1}{k_n}} = \tfrac{\epsilon}{2} \not\to 0\,.$$

However, the last conclusion contradicts our assumption. This contradiction establishes that $\|T^n x\|^{\frac{1}{n}} \to 0$.

Finally, let us assume that X is a Banach lattice and that $\big|x^*(T^n x)\big|^{\frac{1}{n}} \to 0$ for each $0 \le x^* \in X^*$. If $f \in X^*$ and $\epsilon > 0$ are given, then choose some $n_0 > 1$ such that $\big|f^+(T^n x)\big|^{\frac{1}{n}} < \frac{\epsilon}{2}$ and $\big|f^-(T^n x)\big|^{\frac{1}{n}} < \frac{\epsilon}{2}$ hold for each $n \ge n_0$. So, if $n \ge n_0$, then we have

$$\big|f(T^n x)\big| \le \big|f^+(T^n x)\big| + \big|f^-(T^n x)\big| < \tfrac{\epsilon^n}{2^n} + \tfrac{\epsilon^n}{2^n} = \tfrac{\epsilon^n}{2^{n-1}} < \epsilon^n\,.$$

Therefore, $\big|f(T^n x)\big|^{\frac{1}{n}} < \epsilon$ for each $n \ge n_0$. This shows that $\big|f(T^n x)\big|^{\frac{1}{n}} \to 0$ for each $f \in X^*$. By the previous case, we get $\|T^n x\|^{\frac{1}{n}} \to 0$. ∎

Problem 9.2.4. *This problem presents a one-to-one positive operator which is locally quasinilpotent at some positive vector but fails to be locally quasinilpotent. Consider the positive operator $T\colon \ell_2 \to \ell_2$ defined by*

$$T(x_1, x_2, x_3, \ldots) = \big(x_1, x_1, \tfrac{x_2}{2}, \tfrac{x_3}{3}, \ldots\big)\,.$$

Show that T is one-to-one, is locally quasinilpotent at e_2, and fails to be quasinilpotent.

Solution: Notice that $Te_n = \frac{1}{n} e_{n+1}$ for all $n \ge 2$. Now using an inductive argument, we see that $T^n e_2 = \frac{1}{(n+1)!} e_{n+2}$. Hence $\|T^n e_2\| = \frac{1}{(n+1)!}$ for each n. This implies

$$\lim_{n\to\infty} \|T^n e_2\|^{\frac{1}{n}} = \lim_{n\to\infty} \big[\tfrac{1}{(n+1)!}\big]^{\frac{1}{n}} = 0\,.$$

That is, T is locally quasinilpotent at e_2.

On the other hand, $\|T^n\| \ge \|T^n e_1\| \ge 1$ implies $\|T^n\|^{\frac{1}{n}} \ge 1$ for each n. Hence, T is locally quasinilpotent at $e_2 > 0$ but it is not a quasinilpotent operator. ∎

Problem 9.2.5. *Let $T\colon X \to X$ be a bounded operator on a Banach space and denote by $\mathcal{Q}_T$ the set of all points where T is locally quasinilpotent, i.e.,*

$$\mathcal{Q}_T = \big\{x \in X\colon \lim_{n\to\infty} \|T^n x\|^{\frac{1}{n}} = 0\big\}\,.$$

Show that $\mathcal{Q}_T$ is a T-hyperinvariant vector subspace.

Solution: We shall show first that $\mathcal{Q}_T$ is a vector space. To this end, let $x, y \in \mathcal{Q}_T$ and fix $\epsilon > 0$. Pick some n_0 such that $\|T^n x\| < \epsilon^n$ and $\|T^n y\| < \epsilon^n$ hold for all $n \ge n_0$. It follows that $\|T^n(x+y)\|^{\frac{1}{n}} \le (\|T^n x\| + \|T^n y\|)^{\frac{1}{n}} < (2\epsilon^n)^{\frac{1}{n}} < 2\epsilon$ for all

$n \geq n_0$. Therefore, $\lim_{n\to\infty} \|T^n(x+y)\|^{\frac{1}{n}} = 0$, and so $x+y \in \mathcal{Q}_T$. Also note that if λ is a scalar, then

$$\lim_{n\to\infty} \|T^n(\lambda x)\|^{\frac{1}{n}} = \lim_{n\to\infty} \|\lambda T^n(x)\|^{\frac{1}{n}} = \lim_{n\to\infty} |\lambda|^{\frac{1}{n}} \cdot \lim_{n\to\infty} \|T^n x\|^{\frac{1}{n}} = 0\,,$$

and so $\lambda x \in \mathcal{Q}_T$. Consequently, $\mathcal{Q}_T$ is a vector subspace of X.

Finally, let us show that $\mathcal{Q}_T$ is a T-hyperinvariant subspace. To see this, assume that an operator $S \in \mathcal{L}(X)$ satisfies $TS = ST$ and let $x_0 \in \mathcal{Q}_T$. Then we have

$$\|T^n(Sx_0)\|^{\frac{1}{n}} = \|S(T^n x_0)\|^{\frac{1}{n}} \leq \|S\|^{\frac{1}{n}} \cdot \|T^n x_0\|^{\frac{1}{n}} \xrightarrow[n\to\infty]{} 0\,.$$

This implies $Sx_0 \in \mathcal{Q}_T$, and so $\mathcal{Q}_T$ is a T-hyperinvariant subspace. ∎

Problem 9.2.6. *For the notation used here see Problem 9.2.5. For a bounded operator $T\colon X \to X$ on a Banach space establish the following properties.*

(a) *$\mathcal{Q}_T = \{0\}$ is possible.*

(b) *$\mathcal{Q}_T$ can be dense without being equal to X.*

Solution: (a) Let $T\colon X \to X$ be an isometry. Then for each non-zero vector $x \in X$, we have $\|T^n x\| = \|x\|$, and so $\lim_{n\to\infty} \|T^n x\|^{\frac{1}{n}} = \lim_{n\to\infty} \|x\|^{\frac{1}{n}} = 1$. This shows that T cannot be locally quasinilpotent at any non-zero vector, and hence $\mathcal{Q}_T = \{0\}$.

(b) Consider the "backward shift" operator $S\colon \ell_1 \to \ell_1$ defined by the formula $S(x_1, x_2, \ldots) = (x_2, x_3, \ldots)$. Notice that if a sequence $x \in \ell_1$ is eventually zero, then $S^n x = 0$ for all sufficiently large n. This implies $\lim_{n\to\infty} \|S^n x\|^{\frac{1}{n}} = 0$ for all sequences $x \in \ell_1$ that are eventually zero. In other words, $\mathcal{Q}_T$ contains all sequences in ℓ_1 that are eventually zero, and hence $\mathcal{Q}_T$ is dense in ℓ_1.

Now consider the vector $u = \big(1, \frac{1}{2}, \frac{1}{2^2}, \frac{1}{2^3}, \ldots\big) \in \ell_1$, and note that

$$T^n u = \big(\tfrac{1}{2^n}, \tfrac{1}{2^{n+1}}, \tfrac{1}{2^{n+2}}, \ldots\big) = \tfrac{1}{2^n}\big(1, \tfrac{1}{2}, \tfrac{1}{2^2}, \tfrac{1}{2^3}, \ldots\big) = \tfrac{1}{2^n} u\,.$$

This implies $\|T^n u\| = \frac{1}{2^{n-1}}$ or, equivalently, $\|T^n u\|^{\frac{1}{n}} = \frac{\sqrt[n]{2}}{2}$ for each n. Consequently, $\lim_{n\to\infty} \|T^n u\|^{\frac{1}{n}} = \frac{1}{2} \neq 0$, and so $u \notin \mathcal{Q}_T$. Hence, $\mathcal{Q}_T \neq \ell_1$. ∎

Problem 9.2.7. *This problem presents a pair of positive operators that satisfy the hypotheses of Corollary 9.21. Consider the positive operator $T\colon \ell_\infty \to \ell_\infty$ defined by*

$$T(x_1, x_2, x_3, \ldots) = \big(\tfrac{x_1+x_2}{2}, \tfrac{x_1+x_3}{2}, \tfrac{x_1+x_4}{2}, \ldots\big)\,.$$

Establish the following:

(a) *The operator T is a Krein (and hence an ideal irreducible) non-compact positive operator.*

(b) *The spectral radius of T equals one, i.e., $r(T) = 1$.*

(c) *There is some $0 \leq \phi \in \ell_\infty^*$ satisfying $\|\phi\| = \phi(e) = 1$ and $T^*\phi = \phi$, where e denotes the constant one sequence, i.e., $e = (1, 1, 1, \ldots)$.*

(d) *The rank-one positive operator $S = \phi \otimes e$ on ℓ_∞ satisfies*

$$TS = ST = \phi \otimes e\,.$$

Solution: (a) Clearly, T is a positive operator. Now, let $x = (x_1, x_2, \ldots) > 0$ be a bounded sequence, and let $k = \min\{i \in \mathbb{N}\colon\ x_i > 0\}$. Then it is easy to see that $T^k x \geq \frac{x_k}{2^k}(1, 1, 1, \ldots)$. This shows that $T^k x$ is a strong unit of ℓ_∞, and so T is a Krein operator.

To see that T is not a compact operator, consider the norm bounded sequence $\{x_n\} \subseteq \ell_\infty$ defined by $x_n = (1, 1, \ldots, 1, 0, 0, \ldots)$, where the 1's occupy the first n coordinates. It follows that

$$Tx_n = \left(1, 1, \ldots, 1, \tfrac{1}{2}, \tfrac{1}{2}, \tfrac{1}{2}, \ldots\right),$$

where the 1's occupy the first $n-1$ coordinates. This implies $\|Tx_n - Tx_m\|_\infty = \frac{1}{2}$ for all $n \neq m$, and consequently $\{Tx_n\}$ has no convergent subsequence in ℓ_∞. Therefore, T is not a compact operator.

(b) From $T(1, 1, 1, \ldots) = (1, 1, 1, \ldots)$, we see that T is a Markov operator. This implies $\|T^n\| = \|T^n(1, 1, \ldots)\|_\infty = \|(1, 1, 1, \ldots)\|_\infty = 1$ for each n, and so $r(T) = \lim_{n\to\infty} \|T^n\|^{\frac{1}{n}} = 1$.

(c) This follows immediately from part (b) of Problem 9.1.12.

(d) This follows again from part (b) of Problem 9.1.12. ∎

Problem 9.2.8. *Consider the operator $T\colon C[0,1] \to C[0,1]$ defined by the formula*

$$Tx(t) = \int_0^1 \sin(st)x(s)\,ds\,.$$

Show the following:

(a) *T is one-to-one and strictly positive.*

(b) *T is a compact operator.*

(c) *If $J = \{x \in C[0,1]\colon\ x(0) = 0\}$, then J is a non-trivial closed ideal and $T(J) \subseteq J$ holds—and so T is not ideal irreducible.*

(d) *For each $x > 0$ the function Tx is a quasi-interior point of J. In particular, the operator $T\colon J \to J$ (the restriction of T to J) is strongly expanding—and hence ideal irreducible.*

(e) *The spectral radius of the operator $T\colon C[0,1] \to C[0,1]$ is positive (that is, $r(T) > 0$) and coincides with the spectral radius of the operator $T\colon J \to J$ (the restriction of T to J).*

Solution: (a) Assume that $Tx = 0$, i.e., $\int_0^1 \sin(st)x(s)\,ds = 0$ for each $t \in [0,1]$. This implies

$$\begin{aligned}\frac{\partial^{2n}}{\partial t^{2n}} \int_0^1 \sin(st)x(s)\,ds &= \int_0^1 \frac{\partial^{2n}}{\partial t^{2n}}[\sin(st)x(s)]\,ds \\ &= (-1)^n \int_0^1 s^{2n}\sin(st)x(s)\,ds = 0\end{aligned}$$

for each $t \in [0,1]$ and all $n = 0, 1, \ldots$. Consequently, $\int_0^1 p(s^2)\sin(st)x(s)\,ds = 0$ for each polynomial p. Now observe that the algebra generated by $\{\mathbf{1}, s^2\}$ is (by the Stone–Weierstrass theorem) norm dense in $C[0,1]$, and so $\int_0^1 f(s)\sin(st)x(s)\,ds = 0$ for each continuous function $f \in C[0,1]$ and each $t \in [0,1]$. Letting $f(s) = x(s)$, we get $\int_0^1 \sin(st)[x(s)]^2\,ds = 0$ for each $t \in [0,1]$. Since $\sin(st) > 0$ for all $0 < s, t \leq 1$, we see that $x(s) = 0$ for all $0 < s \leq 1$, and so $x = 0$. This proves that T is one-to-one.

It should be clear that $0 < x \in C[0,1]$ implies $Tx > 0$, and thus T is a strictly positive operator.

(b) If $f \in C[0,1]$ satisfies $\|f\|_\infty \leq 1$, then note that

$$\begin{aligned} |Tf(t_1) - Tf(t_2)| &= \Big| \int_0^1 [\sin(st_1) - \sin(st_2)] f(s)\,ds \Big| \\ &\leq \int_0^1 \big|\sin(st_1) - \sin(st_2)\big|\,ds \\ &\leq \int_0^1 \big|st_1 - st_2\big|\,ds = |t_1 - t_2| \int_0^1 s\,ds = \tfrac{1}{2}|t_1 - t_2| \end{aligned}$$

holds for all $t_1, t_2 \in [0,1]$. This shows that $T([-\mathbf{1}, \mathbf{1}])$ is an equicontinuous subset of $C[0,1]$. By the classical Ascoli–Arzelà Compactness Theorem, $T([-\mathbf{1}, \mathbf{1}])$ is a norm totally bounded subset of $C[0,1]$, and so T is a compact operator.

(c) Clearly, J is a non-trivial closed ideal of $C[0,1]$. Since $Tf(0) = 0$ for each function $f \in C[0,1]$, it follows that J is T-invariant. This shows that T is not ideal irreducible.

(d) We claim that if some $u \in J$ satisfies $u(t) > 0$ for all $0 < t \leq 1$, then u is a quasi-interior point of J. Indeed, if $0 < y \in J$, then $(y \wedge nu)(t) \uparrow y(t)$ for each $t \in [0,1]$, and so (by Dini's theorem) $\|y \wedge nu - y\|_\infty \to 0$. Thus, u is a quasi-interior point of J.

Now let $0 < x \in C[0,1]$. Since $\sin(st) > 0$ for all $0 < s, t \leq 1$, it follows that $Tx(t) = \int_0^1 \sin(st)x(s)\,ds > 0$ for each $0 < t \leq 1$. Therefore, Tx is a quasi-interior point of J. This shows that $T\colon J \to J$ is strongly expanding, and consequently it is also ideal irreducible.

(e) The operator $T\colon J \to J$ is a compact ideal irreducible positive operator. By Corollary 9.20, the spectral radius r of $T|_J$ is positive, i.e., $r = r(T|_J) > 0$. Hence, by the Krein–Rutman Theorem 7.10, there exists some $0 < x \in J$ such that $Tx = rx$. This shows that $r(T) \geq r > 0$.

Since $T\colon C[0,1] \to C[0,1]$ is a compact positive operator, again the Krein–Rutman Theorem 7.10 guarantees that $r(T)$ is an eigenvalue of T. So, there exists some positive function $y \in C[0,1]$ satisfying $Ty(t) = \int_0^1 \sin(st)y(s)\,ds = r(T)y(t)$ for each $t \in [0,1]$. This implies $y(0) = 0$, and so $y \in J$. Consequently, $r(T) \leq r$ is also true, and therefore $r = r(T|_J) = r(T)$. ∎

Problem 9.2.9. *If an ideal irreducible positive operator $T\colon E \to E$ commutes with a compact positive operator $S\colon E \to E$, then neither T nor S is locally quasinilpotent at any positive vector.*

Solution: Assume that $S, T\colon E \to E$ are two commuting positive operators on a Banach lattice such that T is ideal irreducible and S is compact. Suppose by way of contradiction that either T or S is locally quasinilpotent at a vector $u > 0$. Then from

$$\begin{aligned} \|(ST)^n u\|^{\frac{1}{n}} &= \|S^n T^n u\|^{\frac{1}{n}} \le \|S\| \cdot \|T^n u\|^{\frac{1}{n}}, \text{ and} \\ \|(ST)^n u\|^{\frac{1}{n}} &= \|T^n S^n u\|^{\frac{1}{n}} \le \|T\| \cdot \|S^n u\|^{\frac{1}{n}}, \end{aligned}$$

it follows that the compact positive operator ST is quasinilpotent at the positive $u > 0$. So, by de Pagter's Theorem 9.19, there exists a non-trivial ℓ-hyperinvariant closed ideal J in E. In particular, since T commutes with ST, we have $T(J) \subseteq J$, which contradicts the ideal irreducibility of T. Hence, neither T nor S is locally quasinilpotent at any positive vector. ∎

Problem 9.2.10. *Corollary 9.22 asserts that if T is a compactly dominated ideal irreducible positive operator on a Banach lattice, then every positive operator S in the commutant of T has positive spectral radius, i.e., $r(S) > 0$.*

Generalize this result as follows: If $T\colon E \to E$ is a compactly dominated ideal irreducible positive operator on a Banach lattice, then every positive operator on E that commutes with T (in particular, T itself) cannot be locally quasinilpotent at any positive vector of E.

Solution: Let $T\colon E \to E$ be an ideal irreducible positive operator on a Banach lattice that is dominated by a compact operator. By Corollary 2.35, $T^3 > 0$ is a compact operator. Now let $S\colon E \to E$ be a positive operator that commutes with T, and assume by way of contradiction that S is locally quasinilpotent at some vector $u > 0$. Then the positive operator $T^3 S$ is compact, commutes with T, and is locally quasinilpotent at u (why?). By Theorem 9.19, there exists a non-trivial closed ℓ-hyperinvariant ideal J. In particular, we have $T(J) \subseteq J$, which is impossible. So, S cannot be locally quasinilpotent at any positive vector of E. ∎

Problem 9.2.11. *Show that in Corollary 9.22 the hypothesis "S commutes with T" cannot be replaced by "S commutes with some power of T" even when T is compact.*

Solution: Consider the pair of positive operators $S, T\colon \mathbb{R}^2 \to \mathbb{R}^2$ defined by $S(x, y) = (y, 0)$ and $T(x, y) = (y, x)$. It should be clear that $T^2 = I$ and so $ST^2 = T^2 S$. Therefore, S commutes with a power of the compact operator T. Now note that $S^2 = 0$ and this implies $r(S) = 0$. ∎

Problem 9.2.12. *If an ideal irreducible positive operator $T\colon E \to E$ on a Banach lattice has a compact power (in particular, if T is dominated by a compact operator), then show that $r(T) > 0$.*

Solution: Let $T\colon E \to E$ be an ideal irreducible positive operator on a Banach lattice such that the operator $S = T^k$ is compact for some k. Clearly, S commutes with T. Now Corollary 9.21 implies that $r(T) > 0$. ∎

Problem 9.2.13 (de Pagter [**61**]). *Show that if a power of an ideal irreducible positive operator T dominates a compactly dominated positive operator, then $r(T) > 0$.*

Solution: Let $T\colon E \to E$ be an ideal irreducible positive operator such that for some k the operator $S = T^k$ dominates a compactly dominated positive operator. Now use Corollary 9.29 to conclude that $r(T) > 0$. ∎

9.3. Band Irreducibility and the Spectral Radius

Problem 9.3.1. *Show that the conclusion of Lemma 9.30 is false if the "compactness" of T is replaced by "ideal irreducibility" of T. That is, give an example of two commuting positive operators $T, S\colon E \to E$ on a Banach lattice such that:*

(a) *T is strictly positive, order continuous, and ideal irreducible.*

(b) *S is not order continuous.*

Solution: Let $E = C(\Gamma)$, where Γ is the unit circle. We denote by $\mathbf{1}$ the constant one function. Fix a non-zero rational angle θ and let T be the operator defined by means of the rotation by the angle θ. That is, $Tx(\gamma) = x(\gamma e^{\imath\theta})$ for each $x \in C(\Gamma)$ and $\gamma \in \Gamma$. Clearly, T is a surjective lattice isometry. This easily implies that T is strictly positive and order continuous.

To continue the solution, we need the following result:

- *If θ is non-zero and rational, then the set $\{e^{\imath n\theta}\colon\ n \in \mathbb{N}\}$ is dense in Γ.*

To see this, fix a natural number k and let $\epsilon > 0$. By Kronecker's theorem (see Problem 7.3.3) the subgroup $\{e^{\imath n\theta}\colon\ n \in \mathbb{Z}\}$ of Γ is dense. This implies that the set $\{e^{\imath n\theta}\colon\ n \in \mathbb{Z}$ and $|n| > k\}$ is dense in Γ. So, there exists some $m \in \mathbb{Z}$ with $|m| > k$ such that $|e^{\imath m\theta} - 1| < \epsilon$. From $|e^{\imath m\theta} - 1| = |e^{\imath m\theta}(1 - e^{-\imath m\theta})| = |1 - e^{-\imath m\theta}|$, we can assume that $m \in \mathbb{N}$ (and so $m > k$). Now if we let $n = m - k \in \mathbb{N}$, then we have

$$\left|e^{\imath n\theta} - e^{-\imath k\theta}\right| = \left|e^{-\imath k\theta}(e^{\imath m\theta} - 1)\right| = \left|e^{\imath m\theta} - 1\right| < \epsilon\,.$$

The latter, coupled with Kronecker's theorem, shows that the set $\{e^{\imath n\theta}\colon\ n \in \mathbb{N}\}$ is dense in Γ.

Next, we claim that the rationality of θ implies that the operator T is ideal irreducible. To see this, let J be a T-invariant closed ideal in $C(\Gamma)$ such that $J \neq \{0\}$. Pick a function $x \in J^+$ and some $\gamma_0 \in \Gamma$ such that $x(\gamma_0) \neq 0$. Now let $\gamma \in \Gamma$ be an arbitrary point. Since θ is a rational angle, it follows from (•) that the set $\{\gamma e^{\imath n\theta}\colon\ n \in \mathbb{N}\}$ is dense in Γ. From $T^n x \in J^+$ and $T^n x(\gamma) = x(\gamma e^{\imath n\theta})$ for each $n \in \mathbb{N}$ (and $\gamma_0 \in \overline{\{\gamma e^{\imath n\theta}\colon\ n \in \mathbb{N}\}}$), it follows that for some $n \in \mathbb{N}$ we must have

$T^n x(\gamma) > 0$. Therefore, for each $\gamma \in \Gamma$ there exists a function $z_\gamma \in J^+$ such that $z_\gamma(\gamma) > 0$. In particular, for each $\gamma \in \Gamma$ we can find a neighborhood V_γ of γ such that $z_\gamma(\eta) > 0$ for all $\eta \in V_\gamma$. The compactness of Γ guarantees the existence of a finite subset $\{\gamma_1, \ldots, \gamma_k\}$ of Γ such that $\Gamma = \bigcup_{i=1}^k V_{\gamma_i}$. Now let $z = \sum_{i=1}^k z_{\gamma_i} \in J^+$, and note that $z(\gamma) > 0$ holds for each $\gamma \in \Gamma$. This implies that z is an order unit of E, and consequently $J = E$.

Now, in view of Problem 9.1.12 (or in view of Corollary 9.46) there exists some $0 < \phi \in E^*$ satisfying $T^*\phi = \phi$ and $\|\phi\| = \phi(\mathbf{1}) = 1$. It is well known that $C(\Gamma)$ does not admit any non-trivial order continuous linear functionals; see for instance [**6**, Example 4.5, p. 45]. Hence, ϕ is not order continuous and consequently the positive operator $S = \phi \otimes \mathbf{1}$ on E is not order continuous either. The reader should also observe the following two properties:

(i) Both operators T and S are Markov operators—and so $r(T) = r(S) = 1$.

(ii) The pair of closed subspaces $(\operatorname{Ker}\phi, \mathbb{R}\mathbf{1})$, where $\mathbb{R}\mathbf{1} = \{\lambda \cdot \mathbf{1}\colon\ \lambda \in \mathbb{R}\}$, is a reducing pair for the operator S.

Finally, we shall verify that T and S commute. Indeed, for each $x \in E$ we have $T(Sx) = T(\phi(x)\mathbf{1}) = \phi(x)\mathbf{1}$ and $S(Tx) = \phi(Tx)\mathbf{1} = [T^*\phi](x)\mathbf{1} = \phi(x)\mathbf{1}$. ∎

Problem 9.3.2. *Let J be an ideal in a Riesz space E. Show that J is a band if and only if the quotient map $x \mapsto [x]$, from E onto E/J, is order continuous.*

Solution: We know that the quotient map $x \mapsto [x]$ is a surjective lattice homomorphism; see Problem 3.4.2.

Assume first that J is a band and let $x_\alpha \downarrow 0$ in E. To see that $[x_\alpha] \downarrow 0$ in E/J, suppose that $[x_\alpha] \geq [x]$ in E/J for all α and some $x \in E^+$. For each α put $y_\alpha = (x - x_\alpha)^+$ and note that $0 \leq y_\alpha \uparrow x$. Now using that the quotient map is a lattice homomorphism, we see that $[y_\alpha] = [(x - x_\alpha)^+] = ([x] - [x_\alpha])^+ = 0$ for each α. This implies $\{y_\alpha\} \subseteq J$, and since J is a band, we get $x \in J$ or $[x] = 0$. Therefore, $[x_\alpha] \downarrow 0$ holds in E/J, and so the quotient map is order continuous.

For the converse, assume that the quotient map is order continuous, and let a net $\{x_\alpha\} \subseteq J$ satisfy $0 \leq x_\alpha \uparrow x$ in E. Then $0 = [x_\alpha] \uparrow [x]$ holds in E/J. This implies $[x] = 0$ or $x \in J$. This shows that J is a band. ∎

Problem 9.3.3. *Let $T\colon E \to E$ be a σ-order continuous positive operator on a Banach lattice. If the commutant of T contains a σ-order continuous band irreducible positive operator and a σ-order continuous compact positive operator, then show that $r(T) > 0$.*

Solution: Let $TS = ST$ and $TK = KT$, where $S > 0$ is a σ-order continuous band irreducible positive operator and $K > 0$ is a σ-order continuous compact positive operator. Fix some $\lambda > r(S)$ and observe that $T\big[R(\lambda, S)K\big] = \big[R(\lambda, S)K\big]T > 0$. Corollary 9.14 guarantees that $R(\lambda, S)$ is a σ-order continuous strongly expanding positive operator. Since TK is a σ-order continuous positive operator, it follows from Theorem 9.35 that $r\big(R(\lambda, S)TK\big) > 0$. Finally, from

$$0 < r\big(R(\lambda, S)TK\big) = r\big(T[R(\lambda, S)K]\big) \leq r(T)r\big(R(\lambda, S)K\big),$$

we can conclude that $r(T) > 0$. ∎

Problem 9.3.4. *Let $S, T\colon E \to E$ be two commuting positive operators on a Banach lattice. If T is a σ-order continuous band irreducible operator and some power of S dominates a σ-order continuous compactly dominated positive operator, then show that both operators have positive spectral radius, i.e., $r(S) > 0$ and $r(T) > 0$.*

Solution: Assume that S^m dominates some σ-order continuous compactly dominated positive operator. Fix some $\lambda > r(T)$ and let $Q = R(\lambda, T)T$. By Corollary 9.14, the operator Q is strongly expanding. Now from Theorem 9.35 it follows that

$$0 < r\big(QS^m\big) = r\big(R(\lambda, T)TS^m\big) \le r\big(R(\lambda, T)\big)r(T)\big[r(S)\big]^m,$$

and so $r(S) > 0$ and $r(T) > 0$ both hold. ∎

Problem 9.3.5. *Let $Q, S\colon E \to E$ be two positive operators such that:*

(1) *Q is σ-order continuous and expanding.*

(2) *S dominates a compactly dominated σ-order continuous positive operator; in particular this is so if S is itself σ-order continuous and dominated by a compact operator.*

Show that $r(QS) > 0$.

Solution: Let Q and S satisfy the properties stated in the problem. Pick a σ-order continuous positive operator L and a compact operator K satisfying $0 < L \le S$ and $L \le K$.

Consider the operator $C = QL$ and note that $C > 0$ and that if $C^k x > 0$ for some $x > 0$ and some k, then $C^k x$ is a weak unit. (This follows from the fact that $C^k = QR$ for some positive operator R.) In particular, we have $C^k > 0$ for each k. Since $0 < C = QL \le QK$, Corollary 2.35 implies that $C^3 > 0$ is a compact operator. Notice also that $(LQ)^3 = L(QL)(QL)Q$ is a compact strictly positive operator. Therefore, since C^3 is also strictly positive, it follows that E has the countable sup property (see Lemma 9.15). This guarantees that the operators Q and L are order continuous.

Assume by way of contradiction $r(C) = 0$. This implies $r(C^3) = [r(C)]^3 = 0$. Now taking into account $(C^3)^2 = C^6 > 0$ and the order continuity of C, Theorem 9.31 guarantees the existence of a non-trivial ℓ-hyperinvariant band B for C^3 such that $C^6(B) \ne \{0\}$ (and hence $C^3(B) \ne \{0\}$). This implies that there exists some $0 < x \in B$ such that $C^3 x > 0$. The preceding discussion guarantees that $C^3 x$ is a weak unit. Since the weak unit $C^3 x$ belongs to the band B, we see that $B = E$, a contradiction. So, $r(QS) \ge r(C) > 0$ holds. ∎

Problem 9.3.6 (Schaefer [**71**]; Grobler [**34**]). *Let $T\colon E \to E$ be a σ-order continuous band irreducible quasinilpotent positive operator on a Banach lattice. If $S\colon E \to E$ is a compact positive operator, then for each $k \ge 1$ show that $[0, S] \cap [0, T^k] = \{0\}$.*

Solution: Assume that the operators T and S satisfy the properties stated in the problem. Fix some $\lambda > r(T)$ and let $Q = R(\lambda, T)$. Then Q is a σ-order continuous positive operator, it commutes with T, and (in view of Corollary 9.14) it is also expanding. Now if $0 < A \leq S$ and $0 < A \leq T^k$, then from Theorem 9.35 it follows that $0 < r(QT^k) \leq r(Q)[r(T)]^k$ and so $r(T) > 0$, a contradiction. Hence, $[0, S] \cap [0, T^k] = \{0\}$ for each $k \geq 1$. ∎

Problem 9.3.7 (Caselles [**21**]). *A **Harris operator** is a positive operator $T\colon E \to E$ on a Dedekind complete Banach lattice such that some power of T is not disjoint from the band of integral operators.*

Prove that every σ-order continuous band irreducible Harris operator has a positive spectral radius.

Solution: Assume that $T\colon E \to E$ is a σ-order continuous band irreducible Harris operator. Then there exists an integral positive operator $K\colon E \to E$ such that $T \wedge K > 0$.

Fix some $\lambda > r(T)$. Then T commutes with $R(\lambda, T)$, and by Corollary 9.14 the operator $R(\lambda, T)$ is a σ-order continuous strongly expanding positive operator. Now observe that the operator T dominates the compactly dominated σ-order continuous positive operator $T \wedge K$. Therefore, according to Theorem 9.35, we have $0 < r(TR(\lambda, T)) \leq r(T)r(R(\lambda, T))$. This implies $r(T) > 0$, as desired. ∎

9.4. Krein Operators and $C(\Omega)$-spaces

Problem 9.4.1. *Let $T\colon C(\Omega) \to C(\Omega)$ be a positive operator. Show that for each unit $u \in C(\Omega)$ we have*

$$\|T\|_u = \|Tu\|_u \quad \textit{and} \quad r(T) = \lim_{n\to\infty} \left(\|T^n u\|_u\right)^{\frac{1}{n}}.$$

Solution: Assume that $u > 0$ is a unit. By Problem 3.1.2 we know that the formula

$$\|x\|_u = \inf\{\lambda > 0\colon\ |x| \leq \lambda u\}$$

defines a lattice norm that is equivalent to the sup norm and

$$\{x \in C(\Omega)\colon\ \|x\|_u \leq 1\} = [-u, u].$$

Since $T\colon C(\Omega) \to C(\Omega)$ is a positive operator, the norm of T^n with respect to the lattice norm $\|\cdot\|_u$ is given by $\|T^n\|_u = \|T^n u\|_u$. Therefore,

$$r(T) = \lim_{n\to\infty} \left(\|T^n\|_u\right)^{\frac{1}{n}} = \lim_{n\to\infty} \left(\|T^n u\|_u\right)^{\frac{1}{n}},$$

and the solution is finished. ∎

Problem 9.4.2. *Consider a positive operator $T\colon C(\Omega) \to C(\Omega)$ and put*

$$m = \min_{\omega\in\Omega} T\mathbf{1}(\omega) \quad \textit{and} \quad M = \max_{\omega\in\Omega} T\mathbf{1}(\omega).$$

Show that $m \leq r(T) \leq M$ and conclude from this that if $T\mathbf{1} \gg 0$, then the operator T has a positive spectral radius.

Solution: From the definition of the real numbers m and M, it follows that $m\mathbf{1} \leq T\mathbf{1} \leq M\mathbf{1}$. This implies $mT\mathbf{1} \leq T^2\mathbf{1} \leq MT\mathbf{1}$. Consequently, we have $m^2\mathbf{1} \leq T^2\mathbf{1} \leq M^2\mathbf{1}$. By induction, we get $m^n\mathbf{1} \leq T^n\mathbf{1} \leq M^n\mathbf{1}$ for each n.

Since the sup norm of the positive operator T^n satisfies $\|T^n\| = \|T^n\mathbf{1}\|_\infty$, it follows that $m^n \leq \|T^n\| \leq M^n$ or $m \leq \|T^n\|^{\frac{1}{n}} \leq M$ for each n. Therefore,

$$m \leq r(T) = \lim_{n\to\infty} \|T^n\|^{\frac{1}{n}} \leq M . \tag{$\star$}$$

Now if $T\mathbf{1} \gg 0$ holds, then $m > 0$, and so from ($\star$) it follows that $r(T) > 0$. ∎

Problem 9.4.3. *Let μ be a non-zero regular Borel measure on Ω with full support, and assume that $T\colon C(\Omega) \to C(\Omega)$ is a positive integral operator generated by a continuous strictly positive kernel. That is, there exists a continuous real-valued function K on $\Omega \times \Omega$ satisfying $K(t,s) > 0$ for all $(t,s) \in \Omega \times \Omega$ and*

$$Tf(t) = \int_\Omega K(t,s)f(s)\,d\mu(s)$$

for each $f \in C(\Omega)$ and all $t \in \Omega$. Show that T is a compact strong Krein operator.

Solution: The compactness of T follows immediately from Problem 3.1.13. We must verify that T is a strong Krein operator.

To this end, let $0 < f \in C(\Omega)$. Then the open set $V = \{\omega \in \Omega\colon\ f(\omega) > 0\}$ is non-empty. Since μ has full support, we have $\mu(V) > 0$. In particular, there exists some $\epsilon > 0$ such that the Borel set $B = \{\omega \in \Omega\colon\ f(\omega) \geq \epsilon\}$ satisfies $\mu(B) > 0$.

Now if we choose some $\delta > 0$ such that $K(t,s) \geq \delta$ for all $s, t \in \Omega$, then

$$Tf(t) = \int_\Omega K(t,s)f(s)\,d\mu(s) \geq \int_B K(t,s)f(s)\,d\mu(s) \geq \delta\epsilon\mu(B) > 0$$

for each $t \in \Omega$. This shows that Tf is a unit of $C(\Omega)$, and so T is a strong Krein operator. ∎

Problem 9.4.4. *Let $K\colon [a,b] \times [a,b] \to [0,\infty)$ be a continuous function such that for some $\epsilon > 0$ we have $K(t,s) > 0$ for all $|s-t| < \epsilon$. Show that the integral operator with kernel $K(t,s)$ is a compact Krein operator.*

Solution: We claim first that T^2 is also an integral operator. Indeed, for each $f \in C[a,b]$ we have

$$\begin{aligned}(T^2f)(t) &= [T(Tf)](t) = \int_a^b K(t,u)\Big[\int_a^b K(u,s)f(s)\,ds\Big]\,du \\ &= \int_a^b \Big[\int_a^b K(t,u)K(u,s)\,du\Big]f(s)\,ds .\end{aligned}$$

This shows that T^2 is a integral operator with kernel given by the formula

$$K_2(t,s) = \int_a^b K(t,u)K(u,s)\,du\,.$$

(For this conclusion see also Problem 5.2.15.) By induction, the operator T^n is an integral operator. We shall denote the kernel of the operator T^n by $K_n(t,s)$. Notice that $K_n(t,s) = \int_a^b K(t,u)K_{n-1}(u,s)\,du$ holds for $n = 2,3,4,\ldots$, where, of course, $K_1(s,t) = K(s,t)$.

Now observe that $K_2(t,s) > 0$ for all $|s-t| < 2\epsilon$. To see this, let $s < t$ and $t - s = |s-t| < 2\epsilon$. Let $u_0 = \frac{s+t}{2}$, the midpoint of the interval $[s,t]$, and put $\delta = \epsilon - \frac{|s-t|}{2}$. Clearly, $0 < \delta < \epsilon$, and if $|u_0 - u| < \delta$, then

$$|s-u| \le |s-u_0| + |u-u_0| < \tfrac{|s-t|}{2} + \delta = \epsilon\,,$$

and similarly $|t-u| < \epsilon$. Consequently, $K(t,u) > 0$ and $K(u,s) > 0$ for all $u \in (u_0-\delta, u_0+\delta)$. This implies

$$K_2(t,s) = \int_a^b K(t,u)K(u,s)\,du \ge \int_{u_0-\delta}^{u_0+\delta} K(t,u)K(u,s)\,du > 0\,.$$

Repeating the above process, we see that there exists some positive integer n such that the kernel $K_n(t,s)$ of the operator T^n satisfies $K_n(t,s) > 0$ for all (t,s) in $[a,b]\times[a,b]$. Now a glance at Problem 9.4.3 shows that T^n is a strong Krein operator, and consequently T is a Krein operator. The compactness of T follows from Problem 3.1.13. ∎

Problem 9.4.5. *By Theorem* 9.42 *we know that a Krein operator can have only one (up to a scalar multiple) positive eigenvector corresponding to a positive eigenvalue. Give an example of a Krein operator whose eigenvalues are all positive.*

Solution: Let α and β be two arbitrary positive numbers. Consider the operator $T = \begin{bmatrix} 1 & \beta \\ \alpha & 1 \end{bmatrix}$ on $\mathbb{R}^2$. Clearly, T is a strong Krein operator. A straightforward computation shows that $\lambda_1 = 1 + \sqrt{\alpha\beta}$ and $\lambda_2 = 1 - \sqrt{\alpha\beta}$ are eigenvalues of T. These eigenvalues are both positive provided $\alpha\beta < 1$. Two corresponding eigenvectors are $(\sqrt{\beta}, \sqrt{\alpha})$ and $(\sqrt{\beta}, -\sqrt{\alpha})$. ∎

Problem 9.4.6. *Show that the product of two commuting Krein operators is also a Krein operator. Give an example of two non-commuting Krein operators whose product is not a Krein operator.*

Solution: Let $S, T\colon C(\Omega) \to C(\Omega)$ be two arbitrary commuting Krein operators, and let $0 < x \in C(\Omega)$. Pick some $k \in \mathbb{N}$ such that $T^k x$ is a unit. Since T and S commute, we have $(ST)^k x = S^k(T^k x)$. By Lemma 9.40 each Krein operator carries units to units, and hence $S^k(T^k x)$ is a unit.

To see that the preceding conclusion is not valid for non-commuting Krein operators, consider the positive matrices $A = \begin{bmatrix} 0 & 1 \\ 1 & 1 \end{bmatrix}$ and $B = \begin{bmatrix} 1 & 1 \\ 1 & 0 \end{bmatrix}$. From

$A^2 = \begin{bmatrix} 1 & 1 \\ 1 & 2 \end{bmatrix}$ and $B^2 = \begin{bmatrix} 2 & 1 \\ 1 & 1 \end{bmatrix}$, we see that both A and B define Krein operators on $\mathbb{R}^2$. Now notice that $AB = \begin{bmatrix} 1 & 0 \\ 2 & 1 \end{bmatrix} \neq BA = \begin{bmatrix} 1 & 2 \\ 0 & 1 \end{bmatrix}$ and that AB is not a Krein operator. ∎

Problem 9.4.7. *Theorem 9.47 was originally proved by M. G. Krein under the extra hypothesis $T_\alpha(\mathcal{U}) \subseteq \mathcal{U}$ for each α, where $\mathcal{U} = \operatorname{Int}[C(\Omega)^+]$. Theorem 9.47 implies, of course, that version. Show that the original Krein theorem implies Theorem 9.47.*

Solution: For each $0 < \epsilon < 1$ let $T_{\alpha,\epsilon} = T_\alpha + \epsilon I$, where I is the identity operator on $C(\Omega)$. Clearly, $T_{\alpha,\epsilon}(\mathcal{U}) \subseteq \mathcal{U}$, and so for each $\epsilon > 0$ there exists some $0 < \phi_\epsilon \in \mathcal{C}$, where

$$\mathcal{C} = \{\psi \in C(\Omega)^*_+ :\ \psi(\mathbf{1}) = 1\},$$

and a family of non-negative scalars $\{\lambda_{\alpha,\epsilon}\}_{\alpha\in A}$ such that

$$T^*_\alpha \phi_\epsilon + \epsilon\phi_\epsilon = \lambda_{\alpha,\epsilon}\phi_\epsilon\,. \qquad (\star)$$

We consider the open interval $(0,1)$ directed by the relation $\succeq$ defined by $\epsilon_1 \succeq \epsilon_2$ whenever $\epsilon_1 \leq \epsilon_2$. Notice that the net $\{x_\epsilon\}_{\epsilon\in(0,1)}$ of real numbers, defined by $x_\epsilon = \epsilon$, satisfies $x_\epsilon \to 0$.

Next, observe that the net $\{\phi_\epsilon\}_{0<\epsilon<1}$ of $\mathcal{C}$ has a w^*-convergent subnet. We can assume that $\phi_\epsilon \xrightarrow{w^*} \phi$ holds in $\mathcal{C}$, and so $T^*_\alpha\phi_\epsilon \xrightarrow[\epsilon]{w^*} T^*_\alpha\phi$ for each $a \in A$. Now fix $\alpha \in A$ and then pick some $t > 0$ with $T_\alpha\mathbf{1} \leq t\mathbf{1}$. From $(\star)$ it follows that

$$0 \leq \lambda_{\alpha,\epsilon} = \lambda_{\alpha,\epsilon}\phi_\epsilon(\mathbf{1}) = T^*_\alpha\phi_\epsilon(\mathbf{1}) + \epsilon\phi_\epsilon(\mathbf{1}) \leq t + 1\,.$$

This implies that for each fixed α the net $\{\lambda_{\alpha,\epsilon}\}_{0<\epsilon<1}$ has an accumulation point $\lambda_\alpha \geq 0$. Without loss of generality we can assume that $\lambda_{\alpha,\epsilon} \xrightarrow[\epsilon]{} \lambda_\alpha$. Finally, taking w^*-limits in $(\star)$, we get

$$T^*_\alpha\phi = \lim_\epsilon [T^*_\alpha\phi_\epsilon + \epsilon\phi_\epsilon] = \lim_\epsilon \lambda_{\alpha,\epsilon}\phi_\epsilon = \lambda_\alpha\phi$$

for each $a \in A$, and we are finished. ∎

Problem 9.4.8. *Let $\{T_\alpha\}_{\alpha\in A}$ be a family of pairwise commuting positive operators on a $C(\Omega)$-space. If there exist some $\lambda > 0$ and some unit u such that $T_\alpha u = \lambda u$ for each α, then show that there is some $0 < \phi \in C(\Omega)^*$ such that $T^*_\alpha\phi = \lambda\phi$ for all α.*

Solution: For each $\alpha \in A$, let $S_\alpha = \frac{1}{\lambda}T_\alpha$. Then each S_α is positive and $S_\alpha u = u$ for each α. Moreover, $S_\alpha S_\beta = S_\beta S_\alpha$ for all $\alpha, \beta \in A$.

According to Theorem 9.48 there exists some function $0 < \phi \in C(\Omega)^*$ satisfying $\frac{1}{\lambda}T^*_\alpha\phi = S^*_\alpha\phi = \phi$ for each $\alpha \in A$. So, $T^*_\alpha\phi = \lambda\phi$ for each $\alpha \in A$. ∎

Problem 9.4.9. *For a pairwise commuting family $\{T_\alpha\}_{\alpha\in A}$ of positive operators on a $C(\Omega)$-space consider the following statements.*

(a) *If for some non-zero vector $v \in C(\Omega)$ we have $T_\alpha v = v$ for each α, then there exists some $0 < \phi \in C(\Omega)^*$ such that $T^*_\alpha\phi = \phi$ for all α.*

(b) *If for some unit $u \in C(\Omega)$ we have $T_\alpha u = u$ for all α, then there exists some $0 < \phi \in C(\Omega)^*$ such that $T_\alpha^* \phi = \phi$ for all α.*

Show that (a) $\Longrightarrow$ (b) *but* (b) *does not imply* (a).

Solution: The implication (a) $\Longrightarrow$ (b) is trivial. To see that (b) does not imply (a), let $C(\Omega) = \mathbb{R}^2$ and consider the positive projection $T\colon C(\Omega) \to C(\Omega)$ defined by the matrix $\begin{bmatrix} 1 & 0 \\ 0 & 0 \end{bmatrix}$. A direct computation shows that $\begin{bmatrix} 1 & 0 \\ 0 & 0 \end{bmatrix} \begin{bmatrix} 1 \\ 0 \end{bmatrix} = \begin{bmatrix} 1 \\ 0 \end{bmatrix}$. However, it is easy to see that T does not have any unit as an eigenvector. ∎

Problem 9.4.10. *Show that the commutativity assumption regarding the family of operators in Theorems* 9.48 *and* 9.50 *cannot be dropped.*

Solution: Let $\mathcal{P}$ denote the family of all (Borel) probability measures on $[0,1]$. For each $\mu \in \mathcal{P}$ consider the positive operator $T_\mu\colon C[0,1] \to C[0,1]$ defined by $T_\mu f(t) = \int_0^1 f(s)\,d\mu(s)$, i.e., $T_\mu = \mu \otimes \mathbf{1}$. A direct computation shows that $T_\mu T_\nu = T_\nu$ for all $\mu, \nu \in \mathcal{P}$. In particular, it follows that $\{T_\mu\colon\ \mu \in \mathcal{P}\}$ consists of non-commuting positive projections on $C[0,1]$. In addition, notice that $T_\mu \mathbf{1} = \mathbf{1}$ for each $\mu \in \mathcal{P}$, i.e., each T_μ is a Markov operator.

Assume by way of contradiction that there is some $0 < \phi \in \mathcal{M}[0,1] = [C(\Omega)]^*$ such that $T_\mu^* \phi = \phi$ for each μ. Scaling appropriately, we can suppose $\phi([0,1]) = 1$. Hence, $\phi = T_\mu^* \phi = (\mathbf{1} \otimes \mu)\phi = \langle \mathbf{1}, \phi \rangle \mu = \mu$ for each $\mu \in \mathcal{P}$, which is impossible. ∎

Chapter 10

Invariant Subspaces

10.1. A Smorgasbord of Invariant Subspaces

Problem 10.1.1 (Schaefer [**69**]). *Show that every linear operator on an infinite dimensional vector space has a non-trivial invariant subspace.*

Solution: Let $T\colon X \to X$ be a linear operator on an infinite dimensional vector space. Fix some $x \neq 0$ and consider the T-orbit space $\mathcal{O}_T$ of x. If $\mathcal{O}_T \neq X$, then $\mathcal{O}_T$ is a non-trivial T-invariant subspace. So, assume that $\mathcal{O}_T = X$. In this case, we claim that the set of vectors $\{x, Tx, T^2x, \ldots\}$ is linearly independent. Indeed, if the set of vectors $\{x, Tx, T^2x, \ldots\}$ is linearly dependent, then there exists some $k \in \mathbb{N}$ and scalars $\alpha_0, \alpha_1, \ldots, \alpha_{k-1}$ such that $T^kx = \sum_{j=0}^{k-1} \alpha_j T^j x$, and so $\mathcal{O}_T$ coincides with the vector subspace generated by the finite set of vectors $\{x, Tx, T^2x, \ldots, T^{k-1}x\}$. But then, $X = \mathcal{O}_T$ is finite dimensional, which is a contradiction. Hence, the set $\{x, Tx, T^2x, \ldots\}$ is linearly independent.

The above properties show that every vector $y \in X$ has a unique representation of the form $y = \sum_{i=0}^{\infty} \lambda_i T^i x$ with the λ_i equal to zero for all but finitely many i. Now define the non-zero linear functional $\phi\colon X \to \mathbb{R}$ by $\phi(y) = \sum_{i=0}^{\infty} \lambda_i$. Clearly, $V = \operatorname{Ker}\phi$ is a proper vector subspace of X. To finish the solution, we shall show that V is T-invariant.

To see this, let $v \in V$. So, if $v = \sum_{i=0}^{\infty} \lambda_i T^i x$, then $\phi(x) = \sum_{i=0}^{\infty} \lambda_i = 0$. Now notice that $Tv = \sum_{i=0}^{\infty} \lambda_i T^{i+1} x$, and so $\phi(Tv) = \sum_{i=0}^{\infty} \lambda_i = 0$. This shows that $Tv \in V$, and so V is T-invariant, as claimed. ∎

Problem 10.1.2. *Give an example of a non-zero operator on* $\mathbb{R}^4$ *without non-trivial hyperinvariant subspaces.*

Solution: For each $\beta \in [0, 2\pi)$ let $R_\beta\colon \mathbb{R}^2 \to \mathbb{R}^2$ be the rotation operator through the angle β. That is, $R_\beta = \begin{bmatrix} \cos\beta & -\sin\beta \\ \sin\beta & \cos\beta \end{bmatrix}$.

Fix a non-zero angle θ that is an irrational multiple of 2π and consider the operator $T\colon \mathbb{R}^4 \to \mathbb{R}^4$ defined by

$$T(x_1, x_2, x_3, x_4) = \big(R_\theta(x_1, x_2)\,, R_\theta(x_3, x_4)\big)\,.$$

We shall show that this operator T does not have any non-trivial hyperinvariant subspaces. To establish this, we need to describe first the non-trivial invariant subspaces of T. Two obvious non-trivial invariant subspaces are:

$$V_\ell = \big\{(x_1, x_2, 0, 0)\colon\ x_1, x_2 \in \mathbb{R}\big\} \quad \text{and} \quad V_r = \big\{(0, 0, x_3, x_4)\colon\ x_3, x_4 \in \mathbb{R}\big\}\,.$$

These invariant subspaces are not T-hyperinvariant. To verify this, consider the operator $S\colon \mathbb{R}^4 \to \mathbb{R}^4$ defined by

$$S(x_1, x_2, x_3, x_4) = (x_3, x_4, x_1, x_2)\,,$$

and note that S commutes with T. However, it is easy to see that S leaves neither V_ℓ nor V_r invariant.

Now for each $\gamma > 0$ and each angle $\beta \in [0, 2\pi)$ let

$$V_{\beta,\gamma} = \big\{(x_1, x_2, \gamma R_\beta(x_1, x_2))\colon\ x_1, x_2 \in \mathbb{R}\big\}\,. \tag{$\star$}$$

Clearly, each $V_{\beta,\gamma}$ is a subspace of $\mathbb{R}^4$. Moreover, from

$$\begin{aligned} T\big(x_1, x_2, \gamma R_\beta(x_1, x_2)\big) &= \big(R_\theta(x_1, x_2), \gamma R_\theta R_\beta(x_1, x_2)\big) \\ &= \big(R_\theta(x_1, x_2), \gamma R_\beta R_\theta(x_1, x_2)\big)\,, \end{aligned}$$

it follows that $T(V_{\beta,\gamma}) \subseteq V_{\beta,\gamma}$. That is, $V_{\beta,\gamma}$ is T-invariant.

Our next goal is to prove that the T-invariant subspaces described in $(\star)$ are all invariant subspaces of T that are different from V_r and V_ℓ. To this end, assume that V is a non-trivial T-invariant subspace that is different from V_r and V_ℓ. We will need the following fact.

($\bullet$) *If $V \cap V_\ell \neq \{0\}$, then $V = V_\ell$. Similarly, if $V \cap V_r \neq \{0\}$, then $V = V_r$.*

Let us verify the first claim. To see this, assume that there exists a non-zero vector $x = (x_1, x_2, 0, 0) \in V$. Consequently, the vector $Tx = \big(R_\theta(x_1, x_2), 0, 0\big)$ also belongs to V and, since $R_\theta(x_1, x_2)$ and (x_1, x_2) are linearly independent, it follows immediately that $V_\ell \subseteq V$. For the reverse inclusion, assume that there is some non-zero vector $v = (v_1, v_2, v_3, v_4) \in V$ that does not belong to V_ℓ. Then the vector $(0, 0, v_3, v_4) = (v_1, v_2, v_3, v_4) - (v_1, v_2, 0, 0)$ is non-zero and belongs to V. Hence, a similar to the above argument applies and yields that $V_r \subseteq V$. But then $X = V_\ell \oplus V_r \subseteq V$, contrary to $V \neq X$. The second statement can be proven in a similar manner.

We now return to our proof that V should coincide with one of the T-invariant subspaces described in $(\star)$. Since V is different from V_ℓ and V_r, there exists some vector $v = (v_1, v_2, v_3, v_4) \in V$ such that $|v_1| + |v_2| > 0$ and $|v_3| + |v_4| > 0$. Without loss of generality we can assume that $v_1^2 + v_2^2 = 1$. Let $\gamma = \sqrt{v_3^2 + v_4^2} > 0$ and let $\beta \in [0, 2\pi)$ be the angle satisfying the equality $(v_3, v_4) = \gamma R_\beta(v_1, v_2)$. We shall establish that $V = V_{\beta,\gamma}$.

Assume that $x = (x_1, x_2, x_3, x_4) \in V$. Clearly, $|x_1| + |x_2| > 0$ since otherwise $x \in V_r$, and ($\bullet$) would imply that $V = V_r$, a contradiction. Therefore, we can assume that $x_1^2 + x_2^2 = 1$. By Kronecker's classical theorem (see Problem 7.3.3), there exists a strictly increasing sequence of natural numbers $\{k_n\}$ such that $R_\theta^{k_n}(v_1, v_2) \to (x_1, x_2)$. By passing to a subsequence of $\{k_n\}$ if necessary, we can also assume that $R_\theta^{k_n}(v_3, v_4) \to (w_3, w_4)$. Since for each n we have

$$\big(R_\theta^{k_n}(v_1, v_2), R_\theta^{k_n}(v_3, v_4)\big) = T^{k_n}(v_1, v_2, v_3, v_4) \in V\,,$$

it follows that the vector $w = (x_1, x_2, w_3, w_4)$ belongs to V. This implies that $x - w = (0, 0, x_3 - w_3, x_4 - w_4) \in V$, and from this (in view of ($\bullet$)), we see that $x - w = 0$, i.e., $x_3 = w_3$ and $x_4 = w_4$. Now from

$$\begin{aligned}(x_3, x_4) = (w_3, w_4) &= \lim_{n\to\infty} R_\theta^{k_n}(v_3, v_4) = \lim_{n\to\infty} R_\theta^{k_n}\big[\gamma R_\beta(v_1, v_2)\big] \\ &= \gamma R_\beta\big[\lim_{n\to\infty} R_\theta^{k_n}(v_1, v_2)\big] = \gamma R_\beta(x_1, x_2)\,,\end{aligned}$$

we see that $(x_1, x_2, x_3, x_4) \in V_{\beta,\gamma}$. Consequently, we have shown that $V \subseteq V_{\beta,\gamma}$.

The reverse inclusion can be proven in a similar fashion. To this end, assume that $x = \big(x_1, x_2, \gamma R_\beta(x_1, x_2)\big)$ is a non-zero vector in $V_{\beta,\gamma}$. It should be clear that $(x_1, x_2) \neq 0$; otherwise $x = 0$, which is a contradiction. As before, we can assume that $x_1^2 + x_2^2 = 1$, and so there exists a strictly increasing sequence of natural numbers $\{k_n\}$ such that $R_\theta^{k_n}(v_1, v_2) \to (x_1, x_2)$. This implies

$$\gamma R_\beta(x_1, x_2) = \gamma R_\beta \lim_{n\to\infty}\Big[R_\theta^{k_n}(v_1, v_2)\Big] = \lim_{n\to\infty} R_\theta^{k_n}(v_3, v_4)\,.$$

Since $\big(R_\theta^{k_n}(v_1, v_2), R_\theta^{k_n}(v_3, v_4)\big) \in V$, it follows that $\big(x_1, x_2, \gamma R_\beta(x_1, x_2)\big) \in V$. Therefore, $V_{\beta,\gamma} \subseteq V$, and consequently $V = V_{\beta,\gamma}$.

Finally, we shall prove that no subspace of the form $V_{\beta,\gamma}$ is T-hyperinvariant. To see this, fix any angle $\alpha \in (0, 2\pi)$ and consider the operator $S\colon \mathbb{R}^4 \to \mathbb{R}^4$ defined by

$$S(x_1, x_2, x_3, x_4) = \big(x_1, x_2, R_\alpha(x_3, x_4)\big)\,.$$

For each $x = (x_1, x_2, x_3, x_4) \in \mathbb{R}^4$, we have

$$\begin{aligned}TSx &= T\big(x_1, x_2, R_\alpha(x_3, x_4)\big) = \big(R_\theta(x_1, x_2), R_\theta R_\alpha(x_3, x_4)\big) \\ &= \big(R_\theta(x_1, x_2), R_{\theta+\alpha}(x_3, x_4)\big)\,,\end{aligned}$$

and similarly

$$\begin{aligned}STx &= S\big(R_\theta(x_1, x_2), R_\theta(x_3, x_4)\big) = \big(R_\theta(x_1, x_2), R_\alpha R_\theta(x_3, x_4)\big) \\ &= \big(R_\theta(x_1, x_2), R_{\theta+\alpha}(x_3, x_4)\big)\,.\end{aligned}$$

Hence, $TSx = STx$ for each $x \in \mathbb{R}^4$, i.e., S commutes with T. Now notice that for each $\big(x_1, x_2, \gamma R_\beta(x_1, x_2)\big) \in V_{\beta,\gamma}$, we have

$$S\big(x_1, x_2, \gamma R_\beta(x_1, x_2)\big) = \big(x_1, x_2, \gamma R_{\alpha+\beta}(x_1, x_2)\big) \notin V_{\beta,\gamma}\,.$$

This implies that no $V_{\beta,\gamma}$ is invariant under S, and the solution is finished. ∎

Problem 10.1.3. *Let $T\colon X \to X$ be a bounded operator on a Banach space. Denote by* Lat (T) *the set of all closed T-invariant subspaces of X. Show that, under the standard set inclusion,* Lat (T) *is an order complete lattice with smallest and largest elements.*

Solution: We consider Lat (T) ordered by the inclusion relation $\subseteq$. Clearly, $\{0\}$ and X belong to Lat (T). So, Lat (T) has a smallest ($\{0\}$) and a largest (X) element with respect to the introduced ordering $\subseteq$.

Let $\mathcal{F}$ be a non-empty subset of Lat (T). Put $V = \bigcap_{F\in\mathcal{F}} F$ and note that V is a closed subspace of X. Now if $v \in V$, then $v \in F$ for each $F \in \mathcal{F}$, and so $Tv \in T(F) \subseteq F$. This implies $Tv \in V$, and hence V is T-invariant, i.e., $V \in$ Lat (T). It is easy to see that V is the greatest lower bound of the subset $\mathcal{F}$ in Lat (T). That is, $V = \inf \mathcal{F}$ in Lat (T).

Next, let us denote by W the vector subspace generated by $\bigcup_{F\in\mathcal{F}} F$. We claim that W is T-invariant. To see this, let $x \in W$. This means that there exist subspaces $F_1, \dots, F_n \in \mathcal{F}$ and vectors $x_i \in F_i$ $(i = 1, \dots, n)$ such that $x = \sum_{i=1}^n x_i$. Since $T(F_i) \subseteq F_i$ for each i, it follows that $Tx = \sum_{i=1}^n Tx_i \in F_1 + F_2 + \cdots + F_n \subseteq W$, i.e., the vector subspace W is T-invariant. This implies that its norm closure $\overline{W}$ is also T-invariant, i.e., $\overline{W} \in$ Lat (T). We claim that $\overline{W}$ is the least upper bound of the family $\mathcal{F}$ in Lat (T). To see this, let U be a closed subspace satisfying $F \subseteq U$ for each $F \in \mathcal{F}$. This implies $W \subseteq U$, and thus $\overline{W} \subseteq U$. Therefore, $\overline{W} = \sup \mathcal{F}$ holds in Lat (T).

The above show that, under the operations

$$Y \wedge Z = Y \cap Z \quad \text{and} \quad Y \vee Z = \overline{Y + Z}\,,$$

the partially ordered set (Lat $(T), \subseteq$) is an order complete lattice having $\{0\}$ as its smallest element and X as its largest. ∎

Problem 10.1.4. *For a bounded operator $T\colon X \to X$ on an infinite dimensional real Banach space establish the following.*

(a) *The operator T has a non-zero finite dimensional invariant subspace if and only if the operator $T_{\mathbf{c}}\colon X_{\mathbf{c}} \to X_{\mathbf{c}}$ likewise has a non-zero finite dimensional invariant subspace.*

(b) *The operator T has a non-zero finite dimensional hyperinvariant subspace if and only if the operator $T_{\mathbf{c}}\colon X_{\mathbf{c}} \to X_{\mathbf{c}}$ likewise has a non-zero finite dimensional hyperinvariant subspace.*

Solution: It should be clear that if V is a non-zero finite dimensional T-invariant (resp. T-hyperinvariant) subspace of X, then $V_{\mathbf{c}} = V \oplus \imath V_{\mathbf{c}}$ is a non-zero finite dimensional $T_{\mathbf{c}}$-invariant (resp. $T_{\mathbf{c}}$-hyperinvariant) subspace of $X_{\mathbf{c}}$.

(a) Assume that $\mathcal{V}$ is a non-zero finite dimensional $T_{\mathbf{c}}$-invariant subspace. Pick a basis $\{z_1, z_2, \dots, z_k\}$ of $\mathcal{V}$ and for each $1 \le n \le k$ let $z_n = x_n + \imath y_n \in X_{\mathbf{c}}$. Now let V be the non-zero vector subspace of X generated by the set of vectors $\{x_1, \dots, x_k, y_1, \dots, y_k\}$. Clearly, V is finite dimensional, and so (since X is infinite dimensional) V is a non-trivial closed subspace of X. Since $\mathcal{V}$ is $T_{\mathbf{c}}$-invariant, for each $1 \le n \le k$ there exist scalars $\lambda_{1,n}, \lambda_{2,n}, \dots, \lambda_{k,n}$ such that $T_{\mathbf{c}} z_n = \sum_{j=1}^k \lambda_{j,n} z_j$. So, if $\lambda_{j,n} = \alpha_{j,n} + \imath\beta_{j,n} \in \mathbb{C}$, then it follows from

$$Tx_n + \imath Ty_n = T_{\mathbf{c}} z_n = \sum_{j=1}^k \lambda_{j,n} z_j = \sum_{j=1}^k (\alpha_{j,n} x_j - \beta_{j,n} y_j) + \imath \sum_{j=1}^k (\beta_{j,n} x_j + \alpha_{j,n} y_j)$$

that $Tx_n = \sum_{j=1}^k(\alpha_{j,n}x_j - \beta_{j,n}y_j) \in V$ and $Ty_n = \sum_{j=1}^k(\beta_{j,n}x_j + \alpha_{j,n}y_j) \in V$. These show that V is T-invariant.

(b) Now assume that $\mathcal{V}$ is $T_{\mathbf{c}}$-hyperinvariant. As in the preceding part, let $\{z_1, z_2, \ldots, z_k\}$ be a basis of $\mathcal{V}$, where $z_n = x_n + \imath y_n \in X_{\mathbf{c}}$, and let V be the non-zero finite dimensional vector subspace of X generated by the set of vectors $\{x_1, \ldots, x_k, y_1, \ldots, y_k\}$. We shall show that V is T-hyperinvariant.

To this end, assume that an operator $S \in \mathcal{L}(X)$ satisfies $ST = TS$. This implies $S_{\mathbf{c}}T_{\mathbf{c}} = T_{\mathbf{c}}S_{\mathbf{c}}$, and so (since $\mathcal{V}$ is $T_{\mathbf{c}}$-hyperinvariant) $\mathcal{V}$ is $S_{\mathbf{c}}$-invariant. In particular, for each $n = 1, 2, \ldots, k$ there exist scalars $\mu_{j,n} = \gamma_{j,n} + \imath\delta_{j,n} \in \mathbb{C}$ $(j = 1, 2, \ldots, k)$ such that

$$Sx_n + \imath Sy_n = S_{\mathbf{c}}z_n = \sum_{j=1}^k \mu_{j,n}z_j = \sum_{j=1}^k(\gamma_{j,n}x_j - \delta_{j,n}y_j) + \imath\sum_{j=1}^k(\gamma_{j,n}y_j + \delta_{j,n}x_j)\,.$$

Therefore, $Sx_n = \sum_{j=1}^k(\gamma_{j,n}x_j - \delta_{j,n}y_j) \in V$ and $Sy_n = \sum_{j=1}^k(\gamma_{j,n}y_j + \delta_{j,n}x_j) \in V$ for each $n = 1, 2, \ldots, k$. This implies that V is S-invariant, and consequently V is T-hyperinvariant. ∎

Problem 10.1.5. *Let $T\colon X \to X$ be a bounded operator on a Banach space. If X is not reflexive, then show that the double adjoint $T^{**}\colon X^{**} \to X^{**}$ has a non-trivial norm closed T^{**}-invariant subspace.*

Solution: If X is not a reflexive Banach space, then X is a non-trivial closed subspace of X^{**} and $T^{**}(X) = T(X) \subseteq X$. That is, X is T^{**}-invariant, and so T^{**} has a non-trivial closed invariant subspace. ∎

Problem 10.1.6. *Let Ω be a compact Hausdorff space. Fix some function $\phi \in C(\Omega)$ and consider the multiplication operator M_ϕ on $C(\Omega)$ defined by $M_\phi(f) = \phi f$. If Ω_0 is a non-empty subset of Ω and*

$$V = \{f \in C(\Omega)\colon\ f(\omega) = 0 \text{ for all } \omega \in \Omega_0\}\,,$$

then show that V is a closed M_ϕ-invariant subspace of $C(\Omega)$.

Solution: Clearly, V is a closed subspace of $C(\Omega)$. Now if $f \in V$ and $\omega \in \Omega_0$, then we have $[M_\phi(f)](\omega) = \phi(\omega)f(\omega) = \phi(\omega)\cdot 0 = 0$. This means $M_\phi(f) \in V$, and so V is M_ϕ-invariant. ∎

Problem 10.1.7. *Let Ω be a compact Hausdorff space. Fix a continuous mapping $\tau\colon \Omega \to \Omega$ and consider the composition operator C_τ on $C(\Omega)$ defined, as usual, by $C_\tau(f) = f\circ\tau$. If Ω_0 is a τ-invariant non-empty subset of Ω (that is, $\tau(\Omega_0) \subseteq \Omega_0$), then show that the vector subspace*

$$V = \big\{f \in C(\Omega)\colon\ f(\omega) = 0 \text{ for all } \omega \in \Omega_0\big\}$$

is a closed C_τ-invariant subspace of $C(\Omega)$.

Solution: Clearly V is a closed subspace of $C(\Omega)$. Now if $f \in V$ and $\omega \in \Omega_0$, then we have $\tau(\omega) \in \Omega_0$, and so $[C_\tau f](\omega) = f(\tau(\omega)) = 0$. This implies $C_\tau(f) \in V$, and hence V is C_τ-invariant. ∎

Problem 10.1.8. *If $T\colon X \to X$ is a bounded operator on a Banach space, then show that:*

(a) *Its commutant $\{T\}' = \{S \in \mathcal{L}(X)\colon\ ST = TS\}$ is a closed unital subalgebra of $\mathcal{L}(X)$.*

(b) *T has a non-trivial closed hyperinvariant subspace if and only if $\{T\}'$ is a non-transitive algebra.*

Solution: (a) Let $S, R \in \{T\}'$ and let λ be a scalar. Then, we have

$$\begin{aligned} (S+R)T &= ST + RT = TS + TR = T(R+S)\,,\\ (\lambda S)T &= \lambda(ST) = \lambda(TS) = T(\lambda S)\,,\\ (SR)T &= S(RT) = S(TR) = (ST)R = (TS)R = T(SR)\,,\\ IT &= TI = T\,. \end{aligned}$$

Therefore, $S+T$, λS, SR, and I all belong to $\{T\}'$, and so $\{T\}'$ is a unital subalgebra of the Banach algebra $\mathcal{L}(X)$.

To see that $\{T\}'$ is closed in $\mathcal{L}(X)$, assume that a sequence $\{S_n\}$ in $\{T\}'$ satisfies $S_n \to S$ in $\mathcal{L}(X)$. Then $TS_n = S_nT$ for each n, and by letting $n \to \infty$ and using the joint continuity of the map $(A, B) \mapsto AB$ (see Problem 1.1.14), we get $TS = ST$, i.e., $S \in \{T\}'$. Therefore, $\{T\}'$ is a closed subalgebra of $\mathcal{L}(X)$.

(b) By definition, a vector subspace V is T-hyperinvariant if and only if it is invariant under every bounded operator that commutes with T. It remains to notice that the latter is equivalent to saying that V is a $\{T\}'$-invariant subspace. ∎

Problem 10.1.9. *If X is a Banach space, then show that the only I-hyperinvariant subspaces are $\{0\}$ and X.*

Solution: Assume that V is a non-zero I-hyperinvariant subspace of X. Since $\{I\}' = \mathcal{L}(X)$, it follows that V is invariant under every bounded operator on X. Fix some no-zero vector $v \in V$, and let $x \in X$. Also, choose some $x^* \in X^*$ with $x^*(v) = 1$. Now if we let $T = x^* \otimes x$, then T is a bounded operator, and so it leaves V invariant. In particular, we have $Tv = x \in V$. Since $x \in X$ is arbitrary, it follows that $V = X$. ∎

Problem 10.1.10. *Let $\mathcal{A}$ be an algebra of operators on a vector space. Show that $\mathcal{A}_1 = \{\alpha I + A\colon \alpha$ scalar and $A \in \mathcal{A}\}$ is the unital algebra generated by $\mathcal{A}$. (This implies that any algebra of operators $\mathcal{A}$ and the unital algebra generated by $\mathcal{A}$ have the same invariant subspaces.)*

Solution: Let $\mathcal{A}$ be an algebra of operators on a Banach space X. It should be clear that $\mathcal{A}_1 = \{\alpha I + A\colon \alpha \text{ scalar and } A \in \mathcal{A}\}$ is the vector subspace of $\mathcal{L}(X)$ generated by $\mathcal{A}$ and the identity operator I. It remains to verify that $\mathcal{A}_1$ is a subalgebra of $\mathcal{L}(X)$. To see this, take any two operators $\alpha_1 I + A_1$ and $\alpha_2 I + A_2$ in $\mathcal{A}_1$ and note that

$$(\alpha_1 I + A_1)(\alpha_2 I + A_2) = \alpha_1\alpha_2 I + (\alpha_1 A_2 + \alpha_2 A_1 + A_1 A_2)\,.$$

Since $\alpha_1 A_2 + \alpha_2 A_1 + A_1 A_2 \in \mathcal{A}$, the latter shows that $(\alpha_1 I + A_1)(\alpha_2 I + A_2) \in \mathcal{A}_1$.

The obtained representation of $\mathcal{A}_1$ shows at once that $\mathcal{A}$ and $\mathcal{A}_1$ have the same invariant subspaces. ∎

Problem 10.1.11. *Show that an algebra of operators $\mathcal{A} \subseteq \mathcal{L}(X)$ has no non-trivial closed invariant subspaces if and only if for any pair $x, y \in X$ with $x \neq 0$ and every $\epsilon > 0$ there exists some operator $A \in \mathcal{A}$ such that $\|Ax - y\| < \epsilon$.*

Solution: Assume first that $\mathcal{A}$ has no non-trivial closed $\mathcal{A}$-invariant subspaces. This implies that $\overline{A}_x = X$ for each $x \neq 0$, and the validity of the desired condition follows.

For the converse, suppose that for any pair $x, y \in X$ with $x \neq 0$ and every $\epsilon > 0$ there exists some operator $A \in \mathcal{A}$ such that $\|Ax - y\| < \epsilon$. Also, let V be a non-zero closed $\mathcal{A}$-invariant subspace. Fix a non-zero vector $x_0 \in V$. Now if $y \in X$ and $\epsilon > 0$ are given, choose some $A \in \mathcal{A}$ such that $\|Ax - y\| < \epsilon$. Since V is A-invariant, it follows that $Ax \in V$, and from this we infer that $y \in \overline{V} = V$. Therefore, $X \subseteq V$ so that $V = X$. This shows that $\mathcal{A}$ has no non-trivial closed invariant subspaces. ∎

Problem 10.1.12. *Let $\mathcal{A}$ be a unital algebra of continuous operators on a Banach space X and let $\mathcal{A}x = \{Ax\colon\ A \in \mathcal{A}\}$. Show that $\mathcal{A}$ is non-transitive if and only if $\mathcal{A}$ has a non-trivial closed subspace of the form $\overline{\mathcal{A}x}$.*

Solution: Assume first that for some $x \in X$ the closed vector space $\overline{\mathcal{A}x}$ is non-trivial. Note that if $A \in \mathcal{A}$, then for each $T \in \mathcal{A}$, we have $TA \in \mathcal{A}$, and so $T(Ax) = TA(x) \in \mathcal{A}x$. This shows that $\mathcal{A}x$ is $\mathcal{A}$-invariant. By continuity, we obtain that $\overline{\mathcal{A}x}$ is also $\mathcal{A}$-invariant.

Now suppose that there is a non-trivial closed $\mathcal{A}$-invariant subspace V. Fix some non-zero vector $v \in V$, and note that for each $A \in \mathcal{A}$ we have $Av \in A(V) \subseteq V$. This implies that $\overline{\mathcal{A}v} \subseteq \overline{V} = V \neq X$. Since $v = Iv \in \mathcal{A}v$, it follows that $\overline{\mathcal{A}v}$ is a non-trivial closed subspace of X. ∎

Problem 10.1.13. *Let $T\colon X \to X$ be a bounded operator on a Banach space and consider the set*

$$V = \left\{x \in X\colon\ \lim_{n\to\infty} \|T^n x\| = 0\right\}.$$

Establish the following.

(a) *V is a T-hyperinvariant subspace.*

(b) *If T is power bounded, then V is a closed T-hyperinvariant subspace.*

Solution: (a) We show first that V is a vector subspace. To this end, let $x, y \in V$ and let λ be a scalar. Then, from $\|T^n(\lambda x)\| = |\lambda| \cdot \|T^n x\| \to 0$, and

$$0 \leq \|T^n(x+y)\| = \|T^n x + T^n y\| \leq \|T^n x\| + \|T^n y\| \longrightarrow 0,$$

it easily follows that $x + y \in V$ and $\lambda x \in V$. Thus, V is a vector subspace.

To see that V is T-hyperinvariant, let $S \in \mathcal{L}(X)$ satisfy $ST = TS$. Now if $x \in V$, then

$$\|T^n(Sx)\| = \|S(T^n x)\| \leq \|S\| \cdot \|T^n x\| \to 0.$$

This implies $Sx \in V$, and so V is S-invariant. Therefore, the vector subspace V is T-hyperinvariant.

(b) Now assume that there exists some $M > 0$ such that $\|T^n\| \le M$ for each n, and let $v \in \overline{V}$. Fix $\epsilon > 0$, and then pick some $w \in V$ such that $\|v - w\| < \epsilon$. Next, choose some n_0 such that $\|T^n w\| < \epsilon$ for all $n \ge n_0$. Consequently, for each $n \ge n_0$ we have

$$\|T^n v\| \le \|T^n(v-w)\| + \|T^n w\| \le M\|v-w\| + \|T^n w\| \le M\epsilon + \epsilon = (M+1)\epsilon\,.$$

This shows that $\|T^n v\| \to 0$, and so $v \in V$. Thus, $V = \overline{V}$ and hence V is a closed T-hyperinvariant subspace. ∎

Problem 10.1.14. *Let $S, T\colon X \to X$ be a pair of bounded operators on a Banach space. As usual, the* **direct sum** *of S and T is the operator $S \oplus T$ on $X \oplus X$ defined by $(S \oplus T)(x \oplus y) = Sx \oplus Ty$.*

If a bounded operator $R\colon X \to X$ satisfies $RS = TR$, then show that its graph $G_R = \{x \oplus Rx\colon\ x \in X\}$ is a closed $S \oplus T$-invariant subspace of $X \oplus X$.

Solution: It should be clear that G_R is a vector subspace of $X \oplus X$. For each $x \oplus Rx \in G_R$ we have

$$(S \oplus T)(x \oplus Rx) = Sx \oplus T(Rx) = Sx \oplus R(Sx) \in G_R\,.$$

That is, G_R is $S \oplus T$-invariant.

To see that G_R is a closed subspace, assume that a sequence $\{x_n \oplus Rx_n\} \subseteq G_R$ satisfies $x_n \oplus Rx_n \to x \oplus y$ in $X \oplus X$. This is equivalent to having $x_n \to x$ and $Rx_n \to y$ in X. Since R is continuous, it follows that $Rx_n \to Rx$, and so $y = Rx$. Therefore, $x \oplus y = x \oplus Rx \in G_R$, and thus G_R is a closed subspace of $X \oplus X$. ∎

Problem 10.1.15. *Show that the shift operator on any ℓ_p-space has a non-trivial closed hyperinvariant subspace.*

Solution: Recall that the shift operator $S\colon \ell_p \to \ell_p$ is defined by

$$S(x_1, x_2, x_3, \ldots) = (0, x_1, x_2, x_3, \ldots)\,.$$

Let $V = \{(x_1, x_2, \ldots) \in \ell_p\colon\ x_1 = 0\}$. Clearly, V is a closed subspace of ℓ_p, and we claim that V is S-hyperinvariant. To see this, notice that V is the range of S (that is, $S(\ell_p) = V$), and then invoke Lemma 10.2. ∎

Problem 10.1.16. *Show that the converse of Corollary* 10.13 *is not true. That is, show that the dual algebra $\mathcal{A}^*$ of a transitive algebra of operators $\mathcal{A}$ may be non-transitive.*

Solution: According to C. J. Read [**66**] there is a bounded operator $T\colon \ell_1 \to \ell_1$ without non-trivial closed invariant subspaces. Let $\mathcal{A}$ be the unital algebra generated by T. That is,

$$\mathcal{A} = \{p(T)\colon\ \ p \text{ is a polynomial}\}\,.$$

Clearly, $\mathcal{A}$ is transitive.

Next, notice that the adjoint $T^*\colon \ell_\infty \to \ell_\infty$ has non-trivial closed invariant subspaces. For instance, for any arbitrary non-zero vector $x \in \ell_\infty$ the closed subspace V generated by $\{x, T^*x, (T^*)^2x, \ldots\}$ is a non-trivial closed T^*-invariant subspace. This is a simple consequence of the fact that ℓ_∞ is not separable. Now it should be clear that V is a non-trivial closed $\mathcal{A}^*$-invariant subspace. ∎

Problem 10.1.17. *Let $\phi\colon \Omega \to \mathbb{R}$ be a continuous function separating the points of a compact Hausdorff space Ω and let $M_\phi\colon C(\Omega) \to C(\Omega)$ denote the usual multiplication operator defined by $M_\phi(f) = \phi f$. Establish that $\{M_\phi\}' = \{M_g\colon\ g \in C(\Omega)\}$.*

Solution: Clearly $M_\phi M_g = M_g M_\phi$ for each $g \in C(\Omega)$, i.e.,

$$\big\{M_g\colon\ g \in C(\Omega)\big\} \subseteq \{M_\phi\}'\,.$$

Let $T \in \{M_\phi\}'$ and put $h = T(\mathbf{1})$. Since $TM_\phi = M_\phi T$ implies $T(\phi f) = \phi T(f)$ for each $f \in C(\Omega)$, it follows that $T(\phi) = h\phi$, and by induction $T(\phi^n) = h\phi^n$ for each $n \geq 0$. The latter implies

$$T(p(\phi)) = hp(\phi) \tag{$\star$}$$

for each polynomial p. Since the function ϕ separates the points of Ω, it follows from the Stone–Weierstrass Approximation Theorem that the algebra

$$\mathcal{A} = \big\{p(\phi)\colon\ p \text{ a polynomial}\big\}$$

is norm dense in $C(\Omega)$. Hence, in view of $(\star)$, we get $Tf = hf$ for each $f \in C(\Omega)$. This shows that $T = M_h$, and consequently $\{M_\phi\}' \subseteq \{M_g\colon\ g \in C(\Omega)\}$ is also true. We now conclude that $\{M_\phi\}' = \{M_g\colon\ g \in C(\Omega)\}$. ∎

10.2. The Lomonosov Invariant Subspace Theorem

Problem 10.2.1. *Present an example of a polynomially compact operator which is not compact.*

Solution: Any nilpotent operator which is not compact provides a desired example. We will produce such an operator. Let X be an infinite dimensional Banach space and consider the operator $T\colon X \oplus X \to X \oplus X$ defined by $T(x \oplus y) = 0 \oplus x$. Obviously, T is a bounded operator which is not compact. It remains to notice that $T^2 = 0$. ∎

Problem 10.2.2. *If Ω is a compact Hausdorff space without isolated points and a function $\phi \in C(\Omega)$ is one-to-one, then show that the multiplication operator M_ϕ does not commute with any non-zero weakly compact operator.*

Solution: Let $M_\phi T = TM_\phi$ for some weakly compact operator $T\colon C(\Omega) \to C(\Omega)$. By Lemma 4.19(3), we have $T = M_f$ for some function $f \in C(\Omega)$. Now a glance at Lemma 4.18 reveals that f must vanish at every accumulation point of Ω. Since Ω has no isolated points, this implies that $f = 0$. Consequently, $T = M_f = 0$. ∎

Problem 10.2.3. *Show that every multiplication operator on $C(\Omega)$ is a Lomonosov operator.*

Solution: If Ω is finite, then $C(\Omega) = \mathbb{R}^n$ or $C(\Omega) = \mathbb{C}^n$, and so every operator on $C(\Omega)$ is compact—and hence, a Lomonosov operator. Consequently, we can assume that Ω is an infinite set.

Now let $\phi \in C(\Omega)$. Fix $\omega_0 \in \Omega$ and then choose two open sets V and W such that $\omega_0 \in V \subset \overline{V} \subseteq W$ and $W^c \neq \emptyset$. Next, select a function $f \in C(\Omega)$ with $f = 1$ on $\overline{V}$ and $f = 0$ on W^c. Then the operator M_f is non-zero, is not a multiple of the identity operator and commutes with M_ϕ. By Theorem 4.20 the operator M_f commutes with a rank-one operator. Therefore, M_ϕ is a Lomonosov operator. ∎

Problem 10.2.4. *Let (Ω, Σ, μ) be a σ-finite measure space such that there exist two disjoint sets $A, B \in \Sigma$ with $0 < \mu(A) < \infty$ and $0 < \mu(B) < \infty$. Show that every multiplication operator on $L_p(\Omega, \Sigma, \mu)$, where $1 \leq p \leq \infty$, is a Lomonosov operator.*

Solution: We know that the multiplication operators on $L_p(\mu)$ are determined by the functions of $L_\infty(\mu)$; see Problems 3.3.5 and 5.2.7. So, fix any $\phi \in L_\infty(\mu)$, and let $f = \chi_A$. Clearly, the multiplication operator M_f is not a multiple of the identity operator and commutes with the multiplication operator M_ϕ. Now a glance at Problem 4.2.9 shows that M_f commutes with a rank-one positive operator. Consequently, M_ϕ is a Lomonosov operator. ∎

Problem 10.2.5. *Show that every Markov operator T on an AM-space with unit commutes with a rank-one positive operator—and hence T is a Lomonosov operator.*

Solution: Let E be an AM-space with unit e and suppose that $T\colon E \to E$ is a Markov operator, that is, T is positive and $Te = e$. Consider the non-empty w^*-compact convex set

$$\mathcal{C} = \{\psi \in E^*_+ : \ \psi(e) = 1\}.$$

Notice that if $\psi \in \mathcal{C}$, then $T^*\psi(e) = \psi(Te) = \psi(e) = 1$, that is, $\mathcal{C}$ is T^*-invariant. Moreover, it is easy to see that the mapping $T^*\colon (\mathcal{C}, w^*) \to (\mathcal{C}, w^*)$ is continuous. Therefore, by Tychonoff's Fixed Point Theorem (see, for instance, [**4**, Corollary 14.51, p. 483]) there exists some $\phi \in \mathcal{C}$ such that $T^*\phi = \phi$.

To finish the solution, we shall show that the rank-one positive operator on E given by $K = \phi \otimes e$ commutes with T. Indeed, for each $x \in E$ we have

$$\begin{aligned} KTx &= \phi(Tx)e = [T^*\phi](x)e = \phi(x)e = Kx\,, \ \text{and} \\ TKx &= T(\phi(x)e) = \phi(x)Te = \phi(x)e = Kx\,. \end{aligned}$$

Consequently, $KT = TK = K$. ∎

Problem 10.2.6. *Consider the operator $T\colon C[-1,1] \to C[-1,1]$ defined by $Tx(t) = x(-t)$ for each $x \in C[-1,1]$ and all $t \in [-1,1]$.*

(a) *Show that T is a non-scalar surjective lattice isometry which is also a Markov operator—and so, by Problem 10.2.6, T commutes with a rank-one positive operator and is a Lomonosov operator.*

(b) *Exhibit a rank-one operator on $C[-1,1]$ that commutes with T.*

(c) *Show that the compact operator $S\colon C[-1,1] \to C[-1,1]$, defined by*

$$Sx(t) = \int_{-1}^{1} x(s)\sin(st)\,ds\,,$$

commutes with the operator T.

(d) *Show that 1 and -1 are the only eigenvalues of T.*

(e) *Exhibit non-trivial invariant subspaces different from the eigenspaces N_1 and N_{-1}.*

Solution: (a) Clearly, T is non-scalar, one-to-one and surjective—if $y \in C[-1,1]$ consider the function $x \in C[-1,1]$ defined by $x(t) = y(-t)$ and note that $Tx = y$. From $Tx \geq 0$ if and only if $x \geq 0$, it follows that T is a lattice isomorphism, and $T\mathbf{1} = \mathbf{1}$ guarantees that T is a Markov operator. It should be obvious that

$$\|Tx\|_\infty = \sup_{-1\leq t\leq 1} |Tx(t)| = \sup_{-1\leq t\leq 1} |x(-t)| = \sup_{-1\leq t\leq 1} |x(t)| = \|x\|_\infty\,,$$

and so T is an isometry.

(b) Assume that $\phi \in C[-1,1]$ is an even non-zero function, i.e., $\phi(-t) = \phi(t)$ for each $t \in [-1,1]$, and let $K\colon C[-1,1] \to C[-1,1]$ be the rank-one operator defined by $Kx = x(0)\phi$. Now note that for each $x \in C[-1,1]$ and each $t \in [-1,1]$ we have

$$\begin{aligned} TKx(t) &= T\big(x(0)\phi\big)(t) = x(0)\phi(-t) = x(0)\phi(t) = Kx(t)\,, \text{ and} \\ KTx(t) &= x(0)\phi(t) = Kx(t)\,. \end{aligned}$$

Thus, $TK = KT = K$, and so T commutes with the rank-one operator K.

(c) If $x \in C[-1,1]$ and $t \in [-1,1]$, then we have

$$\begin{aligned} TSx(t) &= \int_{-1}^{1} x(s)\sin(-st)\,ds = -\int_{-1}^{1} x(s)\sin(st)\,ds\,, \text{ and} \\ STx(t) &= \int_{-1}^{1} x(-s)\sin(st)\,ds = \int_{1}^{-1} x(u)\sin(ut)\,du = -\int_{-1}^{1} x(s)\sin(st)\,ds\,. \end{aligned}$$

This shows that the compact operator S commutes with T.

(d) Assume that $Tx = \lambda x$ for some non-zero $x \in C[-1,1]$. This implies that $x(-t) = \lambda x(t)$ for each $t \in [-1,1]$. If $\lambda = 0$, then it follows that $x = 0$, and so $\lambda = 0$ cannot be an eigenvalue of T. Thus, we can assume that $\lambda \neq 0$. Fix some $t_0 \in [-1,1]$ with $x(t_0) \neq 0$ and note that $x(-t_0) = \lambda x(t_0) \neq 0$. From

$$x(-t_0)x(t_0) = x(-t_0)x\big(-(-t_0)\big) = \lambda x(t_0)\lambda x(-t_0) = \lambda^2 x(-t_0)x(t_0)\,,$$

it follows that $\lambda^2 = 1$. Therefore, the only possible eigenvalues of T are 1 and -1. To see that these are indeed the eigenvalues of T observe that $T\mathbf{1} = \mathbf{1}$ and that if $y(t) = t$, then $Ty(t) = y(-t) = -t = -y(t)$, i.e., $Ty = -y$.

(e) To exhibit non-trivial closed invariant subspaces different from the eigenspaces N_1 and N_{-1}, for each $0 \leq \delta < 1$ let

$$V_\delta = \big\{x \in C[-1,1]\colon\ x = 0 \ \text{ on } \ [-\delta,\delta]\big\}\,.$$

It is easy to check that each V_δ is a non-trivial closed T-invariant subspace that is different from the eigenspaces of T. ∎

Problem 10.2.7. *This exercise is a combination of results obtained by W. B. Arveson and J. Feldman* [**15**], *C. Pearcy and N. Salinas* [**62**], *and P. Meyer-Nieberg* [**56**]. *Let X be a complex Banach space and for an operator $T \in \mathcal{L}(X)$ let $\mathcal{R}(T)$ denote the norm closure in $\mathcal{L}(X)$ of the collection of all operators $q(T)$, where $q(\lambda)$ is a rational function (i.e., a quotient of two polynomials with complex coefficients) that is analytic in a neighborhood of the spectrum $\sigma(T)$. Clearly, $\mathcal{R}(T)$ is a commutative unital algebra of operators.*

Show that if an operator $T \in \mathcal{L}(X)$ is non-scalar and $\mathcal{R}(T)$ contains a non-zero compact operator, then $\mathcal{R}(T)$ is non-transitive.

Solution: Let X be a complex Banach space and fix an operator $T \in \mathcal{L}(X)$. Clearly, $\mathcal{R}(T) \subseteq \{T\}'$ and $\mathcal{R}(T)$ is a commutative unital algebra. If K is a non-zero compact operator in $\mathcal{R}(T)$, then it follows from Lomonosov's Theorem 10.19 that T has a non-trivial closed hyperinvariant subspace V. Thus, V is $\mathcal{R}(T)$-invariant, and so $\mathcal{R}(T)$ is a non-transitive algebra. ∎

10.3. Invariant Ideals for Positive Operators

Problem 10.3.1. *Let $S\colon E \to E$ be a positive operator on a Banach lattice such that $Sx_0 \geq \gamma x_0$ holds for some vector $x_0 > 0$ and some scalar $\gamma > 0$. Show that S is not locally quasinilpotent at x_0.*

Solution: It follows that for each $n \in \mathbb{N}$ we have $S^n x_0 \geq \gamma^n x_0$. Therefore, $\|S^n x_0\| \geq \gamma^n \|x_0\|$ for each n, and so $\liminf \|S^n x_0\|^{\frac{1}{n}} \geq \gamma \liminf \|x_0\|^{\frac{1}{n}} = \gamma > 0$. This proves that S is not quasinilpotent at x_0. ∎

Problem 10.3.2. *Let J be an ideal in a Banach lattice E. If J is invariant under some positive operator $B\colon E \to E$, then J is also invariant under every operator dominated by B.*[1]

Solution: Let J be an ideal in a Banach lattice E that is invariant under a positive operator $B\colon E \to E$. Assume that an operator $T\colon E \to E$ is dominated by B. Then for any $x \in J$ we have $|Tx| \leq B|x| \in J$, and so $Tx \in J$. This shows that J is T-invariant. ∎

Problem 10.3.3. *Give an example of a pair of non-commuting positive operators A and B such that $A \in [B\rangle$, that is, $AB - BA > 0$.*

Solution: Consider the positive operators $A, B\colon \ell_2 \to \ell_2$ defined by

$$A(x_1, x_2, x_3, \ldots) = (x_2, x_3, \ldots) \quad \text{and} \quad B(x_1, x_2, x_3, \ldots) = (0, x_1, x_2, x_3, \ldots)\,.$$

[1] Recall that an operator $T\colon E \to E$ is **dominated** by B if $|Tx| \leq B|x|$ for each $x \in E$.

That is, A is the backward shift and B is the forward shift. Then for each vector $x = (x_1, x_2, x_3, \ldots)$ in ℓ_2 we have

$$ABx = x = (x_1, x_2, x_3, \ldots) \quad \text{and} \quad BAx = (0, x_2, x_3, \ldots).$$

This shows that $AB \geq BA$, that is, $A \in [B\rangle$ and $AB \neq BA$, i.e., A and B do not commute.

For another example, take for T the operator B from above and let $S\colon \ell_2 \to \ell_2$ be the (projection) operator defined by $S(x_1, x_2, \ldots) = (x_1, x_2, \ldots, x_n, 0, 0, \ldots)$. Then for each $x = (x_1, x_2, \ldots)$, we have

$$TSx = (0, x_1, x_2, \ldots, x_{n-1}, x_n, 0, 0, \ldots) \text{ and } STx = (0, x_1, x_2, \ldots, x_{n-1}, 0, 0, \ldots).$$

This implies $TS \geq ST$ and $TS \neq ST$. ∎

Problem 10.3.4. *For a positive operator $B\colon E \to E$ on a Banach lattice establish the following.*

(a) *The **range ideal** R_B of B (i.e., the ideal generated by the range of the operator B) is given by*
$$R_B = \{y \in E\colon\ \exists\ x \in E^+ \ \text{ such that } \ |y| \leq Bx\}.$$

(b) *The range ideal R_B is ℓ-hyperinvariant.*

(c) *For each $A \in [B\rangle$ the range ideal R_A is B-invariant.*

(d) *The null ideal N_B is A-invariant for each $A \in [B\rangle$. In particular, N_B is ℓ–hyperinvariant.*

(e) *Present an example of an operator $A \in \langle B]$ for which the ideal N_B fails to be A-invariant.*

(f) *If B is non-zero and $[B\rangle$ contains an ideal irreducible operator, then B is strictly positive.*

Solution: (a) It should be clear that

$$\{y \in E\colon\ \exists\ x \in E^+ \ \text{ such that } \ |y| \leq Bx\} \subseteq R_B.$$

Now let $y \in R_B$. This means that there exist positive scalars $\alpha_1, \ldots, \alpha_n$ and vectors $x_1, \ldots, x_n$ such that $|y| \leq \sum_{i=1}^n \alpha_i |Bx_i| \leq B\big(\sum_{i=1}^n \alpha_i |x_i|\big)$. Therefore, if we put $x = \sum_{i=1}^n \alpha_i |x_i| \in E^+$, then $|y| \leq Bx$. This shows that

$$R_B \subseteq \{y \in E\colon\ \exists\ x \in E^+ \ \text{ such that } \ |y| \leq Bx\}.$$

Consequently, $R_B = \{y \in E\colon\ \exists\ x \in E^+ \ \text{ such that } \ |y| \leq Bx\}$.

(b) Take any positive operator A commuting with B and let $y \in R_B$. Pick some $x \in E^+$ such that $|y| \leq Bx$. Therefore, $|Ay| \leq A|y| \leq ABx = B(Ax)$. This shows that $Ay \in R_B$, and thus R_B is A-invariant.

(c) Let $y \in R_A$. By part (a) there exists some $x \in E^+$ such that $|y| \leq Ax$. This implies $|By| \leq B|y| \leq BAx \leq A(Bx)$, and so $By \in R_A$. Therefore, R_A is B-invariant.

(d) Fix $A \in [B\rangle$ and let $x \in N_B$. The inequalities $B|Ax| \leq BA|x| \leq AB|x| = 0$ imply $B|Ax| = 0$, and so $Ax \in N_B$. That is, N_B is A-invariant.

(e) For an example, let B be the backward shift and A be the forward shift on ℓ_2. By Problem 10.3.3 we know that $A \in \langle B]$, and an easy argument shows that

$$N_B = \{x = (x_1, 0, 0, 0, \ldots) \colon\ x_1 \in \mathbb{R}\}.$$

Clearly, N_B is not invariant under A.

(f) Assume that an operator $A \in [B\rangle$ is ideal irreducible. Clearly, $A \neq 0$. According to the conclusion of part (a), the closed ideal N_B is A-invariant. So, by the ideal irreducibility of A, we have either $N_B = \{0\}$ or $N_B = E$. Since $B \neq 0$, it follows that $N_B = \{0\}$, and so B is strictly positive. ∎

Problem 10.3.5. *Assume that $K \colon E \to E$ is a compact operator on a Banach lattice satisfying $Kx_0 \neq 0$ for some $x_0 \in E$. Show that there exists a compact operator C such that $Cx_0 > 0$ and $|Cx| \leq |Kx|$ for each $x \in E$.*

Solution: Let $J = R_K$ be the range ideal of K, i.e., J is the ideal generated by the range $K(E)$ of the operator K. We claim that $\overline{J}$ has a quasi-interior point.

To see this, start by observing that the compactness of K implies that the range of K is separable; see Problem 2.4.6. So, there exists a countable set $\{u_n\}$ of non-zero vectors in $K(E)$ that is norm dense in $K(E)$. Let $u = \sum_{n=1}^{\infty} 2^{-n} \frac{|u_n|}{\|u_n\|}$. It should be clear that $u \in \overline{J}$ and so $E_u \subseteq \overline{J}$, where E_u is the principal ideal generated by u in E. Therefore $\overline{E_u} \subseteq \overline{J}$. Conversely, from the inequalities $|u_n| \leq 2^n \|u_n\| u$ and the fact that $\{u_n\}$ is dense in $K(E)$ it follows that $J \subseteq \overline{E_u}$. This proves that $\overline{J} = \overline{E_u}$ and hence u is a quasi-interior point in $\overline{J}$.

Since the vector Kx_0 belongs to J and is non-zero, Lemma 4.16 guarantees that there is an operator $V \colon \overline{J} \to \overline{J}$ such that $V(Kx_0) > 0$ and $|Vy| \leq |y|$ for each $y \in \overline{J}$. Finally, consider the operator $C = VK$. Clearly C is a compact operator and $Cx_0 = VKx_0 > 0$. To complete the solution, it remains to notice that $|Cx| = |VKx| \leq K|x|$ for each $x \in E$. ∎

Problem 10.3.6. *For a positive operator T on a Banach lattice E let*

$$\mathfrak{A}_T = \{A \in \mathcal{L}(E) \colon\ \exists\ B \in \{T\}'_+ \text{ such that } |Ax| \leq B|x| \text{ for all } x \in E\},$$

where $\{T\}'_+$ denotes as usual the set of all positive operators in the commutant of T. Show that

(a) *The center of E is included in $\mathfrak{A}_T$, i.e., $\mathcal{Z}(E) \subseteq \mathfrak{A}_T$.*

(b) *$\mathfrak{A}_T$ is a unital subalgebra of $\mathcal{L}(E)$.*

(c) *An ideal J is ℓ-hyperinvariant for T if and only if J is $\mathfrak{A}_T$-invariant.*

Solution: (a) Since every operator $S \in \mathcal{Z}(E)$ satisfies $|Sx| \leq \lambda |x| = \lambda I|x|$ for some $\lambda > 0$ and the identity operator I commutes with T, it follows that $S \in \mathfrak{A}_T$, i.e., $\mathcal{Z}(E) \subseteq \mathfrak{A}_T$.

(b) A straightforward verification shows $\mathfrak{A}_T$ is a vector subspace of $\mathcal{L}(E)$. Now assume that $S_1, S_2 \in \mathfrak{A}_T$. Pick two positive operators $B_1, B_2 \in \{T\}'$ such that B_i dominates S_i for $i = 1, 2$. Now note that for each $x \in E$ we have

$$|S_1 S_2 x| \leq B_1 |S_2 x| \leq B_1 B_2 |x|,$$

i.e., S_1S_2 is dominated by B_1B_2. Since $B_1B_2 \in \{T\}'$, we see that $S_1S_2 \in \mathfrak{A}_T$. The above show that $\mathfrak{A}_T$ is a unital subalgebra of $\mathcal{L}(E)$.

(c) Assume that an ideal J in a Banach lattice E is ℓ-hyperinvariant for T. Also, let $S \in \mathfrak{A}_T$. This means that there exists some positive operator $B\colon E \to E$ that commutes with T and dominates S. Since J is ℓ-hyperinvariant for T, it follows that J is also B-invariant. But then, by Problem 10.3.2, J is automatically S-invariant. Therefore, J is $\mathfrak{A}_T$-invariant.

For the converse, assume that an ideal J in E is $\mathfrak{A}_T$-invariant. Next notice that if $B\colon E \to E$ is a positive operator that commutes with T, then $B \in \mathfrak{A}_T$, and so J must be B-invariant. This shows that the ideal J is ℓ-hyperinvariant for T. ∎

Problem 10.3.7. *Recall that an operator $T\colon E \to F$ between two Banach lattices is said to be AM-**compact**, provided that T maps order bounded sets to norm totally bounded sets, i.e., if $T[a,b]$ is a norm totally bounded set for each order interval $[a,b]$.*

Give an example of an AM-compact operator which is not compact.

Solution: We shall use the following fact:

- *If $1 \le p < \infty$, then every order interval in ℓ_p is a norm compact set.*

To see this, let $a = (a_1, a_2, \ldots)$ be a positive vector in some ℓ_p-space, where $1 \le p < \infty$, and consider the order interval $[0,a]$. Clearly $[0,a]$ is a norm closed subset of ℓ_p. Pick some k such that $\big(\sum_{n=k}^{\infty} a_n^p\big)^{\frac{1}{p}} < \epsilon$ for all $n > k$. If we let $b = (a_1, a_2, \ldots, a_k, 0, 0 \ldots)$, then the interval $[0,b]$ is obviously compact, and

$$[0,a] \subseteq [0,b] + \epsilon U\,,$$

where U is the closed unit ball of ℓ_p. This implies that $[0,a]$ is a norm totally bounded subset of ℓ_p, and thus $[0,a]$ is a norm compact subset of ℓ_p.

The preceding statement shows that every positive operator from any Banach lattice to any ℓ_p-space with $1 \le p < \infty$ is an AM-compact operator. However, not every positive operator on an ℓ_p-space is a compact operator. For instance, the identity operator and the shift operator are positive and non-compact. Hence, every non-compact positive operator on an ℓ_p-space (with $1 \le p < \infty$) provides an example of an AM-compact positive operator which is not a compact operator. ∎

Problem 10.3.8. *Let $B\colon E \to E$ be a positive operator on a Banach lattice and assume that there exists a positive operator $S\colon E \to E$ such that*

(1) *$S \in \langle B]$, that is, $SB \le BS$.*

(2) *There exists some $x_0 > 0$ satisfying $\liminf\limits_{n\to\infty} \|S^n x_0\|^{\frac{1}{n}} = 0$.*

(3) *S dominates a non-zero AM-compact operator.*

Show that the operator B has a non-trivial closed invariant ideal.

Solution: The proof is similar to that of Theorem 10.24. Let B, S, and x_0 satisfy the properties stated in the theorem, and let K be a non-zero AM-compact operator dominated by the operator S, i.e., $|Kx| \le S(|x|)$ for each $x \in E$. We can assume

$\|B\| < 1$. Then the series $A = \sum_{n=0}^{\infty} B^n$ defines a positive operator on E and $A \in [S\rangle$, i.e., $AS \geq SA$. To see this, use first the hypothesis that $S \in \langle B]$ to show that $SB^n \leq B^n S$ for each $n \geq 0$, and then note that

$$SA = S\Big(\sum_{n=0}^{\infty} B^n\Big) = \sum_{n=0}^{\infty} SB^n \leq \sum_{n=0}^{\infty} B^n S = \Big(\sum_{n=0}^{\infty} B^n\Big)S = AS\,.$$

Thus, $A \in [S\rangle$. This also implies that $SA^k \leq A^k S$ for all $k \geq 0$.

For each $x > 0$ we denote by J_x the principal ideal generated by Ax. From $0 < x \leq Ax$ we see that $x \in J_x$, and so J_x is non-zero. Moreover, we claim that J_x is B-invariant. Indeed, if $y \in J_x$, then $|y| \leq \lambda Ax$ for some $\lambda > 0$, and so

$$|By| \leq B|y| \leq \lambda B(Ax) = \lambda \sum_{n=1}^{\infty} B^n x \leq \lambda Ax\,,$$

which implies $By \in J_x$. So, $\overline{J}_x$ is a non-zero closed B-invariant ideal.

The proof will be finished if we show that $\overline{J}_x \neq E$ for some $x > 0$. To establish this, assume by way of contradiction that:

$$\overline{J}_x = E \quad \text{for each} \quad x > 0\,. \tag{$\star$}$$

We claim that without loss of generality we can suppose $Kx_0 \neq 0$. To see this, consider the ideal J_{x_0}. If $Ky = 0$ for each $0 < y \in J_{x_0}$, then $K = 0$ on J_{x_0}, and so by ($\star$) we get $K = 0$, a contradiction. Therefore, $Ky_0 \neq 0$ for some $0 < y_0 \in J_{x_0}$. We claim that $\liminf_{n\to\infty} \|S^n y_0\|^{\frac{1}{n}} = 0$. To prove this, pick some scalar $\lambda > 0$ such that $0 < y_0 \leq \lambda Ax_0$ holds and then notice that $0 \leq S^n y_0 \leq \lambda S^n Ax_0 \leq \lambda AS^n x_0$ also holds for each n. This implies $\|S^n y_0\|^{\frac{1}{n}} \leq \lambda^{\frac{1}{n}} \|A\|^{\frac{1}{n}} \cdot \|S^n x_0\|^{\frac{1}{n}}$, from which it follows that $\liminf_{n\to\infty} \|S^n y_0\|^{\frac{1}{n}} = 0$. So, replacing (if necessary) x_0 by y_0, we can assume $Kx_0 \neq 0$.

By scaling, we can assume $\|K\| = 1$. Also, replacing x_0 by αx_0 for an appropriate scalar $\alpha > 1$, we can suppose that $\|x_0\| > 1$ and $\|Kx_0\| > 1$. Now denote by U_0 the closed unit ball centered at $x_0 > 0$. By our choice of x_0, we have

$$0 \notin U_0 \qquad \text{and} \qquad 0 \notin \overline{K(U_0)}\,. \tag{$\star\star$}$$

From ($\star$), we know that $\overline{J_{|x|}} = E$ for each $x \neq 0$. Hence, for each $y \in E_+$ the sequence $\{y \wedge nA(|x|)\}$ is norm convergent and moreover, in view of Lemma 4.15, $\lim_{n\to\infty} y \wedge nA(|x|) = y$. In particular, for each $x \neq 0$ there exists some n such that $\|x_0 - x_0 \wedge nA(|x|)\| < 1$. Since the function $z \mapsto x_0 \wedge nA(|z|)$ is continuous, the set $\{z \in E\colon\ \|x_0 - x_0 \wedge nA(|z|)\| < 1\}$ is open for each n. In view of the fact that $0 \notin \overline{K(U_0)}$, the previous arguments guarantee that

$$\overline{K(U_0 \cap [0, x_0])} \subseteq \bigcup_{n=1}^{\infty} \{z \in E\colon\ \|x_0 - x_0 \wedge nA(|z|)\| < 1\}\,.$$

Since the AM-compactness of K implies that $\overline{K(U_0 \cap [0, x_0])}$ is compact and since $\{z \in E\colon\ \|x_0 - x_0 \wedge nA(|z|)\| < 1\} \subseteq \{z \in E\colon\ \|x_0 - x_0 \wedge (n+1)A(|z|)\| < 1\}$ for each n, it follows that

$$\overline{K(U_0 \cap [0, x_0])} \subseteq \{z \in E\colon\ \|x_0 - x_0 \wedge mA(|z|)\| < 1\}$$

for some m. In other words, there exists some fixed natural number m such that $x \in \overline{K(U_0 \cap [0,x_0])}$ implies $x_0 \wedge mA(|x|) \in U_0$.

In particular, $x_1 = x_0 \wedge mA(|Kx_0|) \in U_0 \cap [0,x_0]$. Since $Kx_1 \in K(U_0 \cap [0,x_0])$, it follows that $x_2 = x_0 \wedge mA(|Kx_1|) \in U_0 \cap [0,x_0]$. Proceeding inductively, we obtain a sequence $\{x_n\}$ of positive vectors in $U_0 \cap [0,x_0]$ defined recursively by the formula $x_{n+1} = x_0 \wedge mA(|Kx_n|)$. Now we claim that

$$0 \le x_n \le m^n A^n S^n x_0$$

for each n. The proof is by induction. First recall that $SA^k \le A^k S$ holds for all $k \ge 0$. For $n = 1$, the desired inequality $x_1 = x_0 \wedge mA\big(|Kx_0|\big) \le mA(Sx_0)$ is trivially true. For the induction step, note that if $0 \le x_n \le m^n A^n S^n x_0$ for some n, then the inequality $|Kx_n| \le Sx_n$ implies

$$\begin{aligned} 0 \le x_{n+1} &= x_0 \wedge mA(|Kx_n|) \le mA(|Kx_n|) \le mASx_n \\ &\le m^{n+1}AS(A^nS^nx_0) = m^{n+1}A(SA^n)S^nx_0 \\ &\le m^{n+1}A(A^nS)S^nx_0 \le m^{n+1}A^{n+1}S^{n+1}x_0. \end{aligned}$$

Thus, $\|x_n\| \le m^n\|A\|^n\|S^nx_0\|$, and so $\|x_n\|^{\frac{1}{n}} \le m\|A\| \cdot \|S^nx_0\|^{\frac{1}{n}}$ for each n. Since $\liminf_{n\to\infty} \|S^nx_0\|^{\frac{1}{n}} = 0$, it follows that $\liminf_{n\to\infty} \|x_n\|^{\frac{1}{n}} = 0$. Thus, there exists a subsequence $\{x_{k_n}\}$ of $\{x_n\}$ such that $\|x_{k_n}\|^{\frac{1}{k_n}} \to 0$. In particular, $x_{k_n} \to 0$. Since $\{x_{k_n}\} \subseteq U_0$, this implies that $0 \in \overline{U_0} = U_0$, contrary to $(\star\star)$, and the solution is finished. ∎

Problem 10.3.9. *Let $B\colon E \to E$ be a positive operator on a Banach lattice and assume that there exists a positive operator $S\colon E \to E$ such that:*

1. *$S \in [B\rangle$, that is, $SB \ge BS$.*
2. *S is quasinilpotent.*
3. *S dominates a non-zero AM-compact operator.*

Show that the operator B has a non-trivial closed invariant ideal.

Solution: Let B, S satisfy the properties stated in the theorem, and let K be a non-zero AM-compact operator dominated by S, i.e., $|Kx| \le S(|x|)$ for each $x \in E$.

We can assume that $\|B\| < 1$. Then, the series $A = \sum_{n=0}^{\infty} B^n$ defines a positive operator on E. Using the condition $S \in [B\rangle$, we can easily verify that the inequalities $SB^k \ge B^kS$ and $S^kA \ge AS^k$ hold for each k.

For each $x > 0$ we denote by J_x the principal ideal in E generated by Ax. Since $x \le Ax \in J_x$, the ideal J_x is non-zero. Moreover, we claim that J_x is B-invariant. Indeed, if $y \in J_x$, then $|y| \le \lambda Ax$ for some $\lambda > 0$, and so

$$|By| \le B|y| \le \lambda B(Ax) = \lambda \sum_{n=1}^{\infty} B^n x \le \lambda Ax\,,$$

which implies $By \in J_x$. So, $\overline{J_x}$ is a non-zero closed B-invariant ideal.

The proof will be finished, if we show that $\overline{J_x} \ne E$ for some $x > 0$. To establish this, assume by way of contradiction that

$$\overline{J_x} = E \text{ for each } x > 0\,. \tag{$\star$}$$

Since $K \neq 0$, we can find some $x_0 > 0$ for which $Kx_0 \neq 0$. By scaling, we can assume $\|K\| = 1$. Also, replacing x_0 by ax_0 for an appropriate scalar $a > 1$, we can suppose that $\|x_0\| > 1$ and $\|Kx_0\| > 1$. Now let $U_0 = \{z \in E\colon\ \|x_0 - z\| \leq 1\}$ be the closed unit ball centered at $x_0 > 0$. By our choice of x_0, we have

$$0 \notin U_0 \qquad \text{and} \qquad 0 \notin \overline{K(U_0)}\,. \tag{$\star\star$}$$

By ($\star$), we know that $\overline{J_{|x|}} = E$ for each $x \neq 0$. Hence, for each $y \geq 0$ the sequence $\{y \wedge nA(|x|)\}$ is norm convergent and $\lim_{n\to\infty} y \wedge nA(|x|) = y$. In particular, for each $x \neq 0$ there exists some n such that $\|x_0 - x_0 \wedge nA(|x|)\| < 1$. Since the function $z \mapsto x_0 \wedge nA(|z|)$ is continuous, we see that the set $\{z\colon\ \|x_0 - x_0 \wedge nA(|z|)\| < 1\}$ is open for each n. In view of $0 \notin \overline{K(U_0)}$, the above arguments guarantee that

$$\overline{K(U_0 \cap [0,x_0])} \subseteq \bigcup_{n=1}^{\infty} \{z \in E\colon\ \|x_0 - x_0 \wedge nA(|z|)\| < 1\}\,.$$

The AM-compactness of the operator K guarantees that the set $\overline{K(U_0 \cap [0,x_0])}$ is compact. This and the fact that the sets $\{z \in E\colon\ \|x_0 - x_0 \wedge nA(|z|)\| < 1\}$ are increasing as n increases imply that

$$\overline{K(U_0)} \subseteq \{z \in E\colon\ \|x_0 - x_0 \wedge mA(|z|)\| < 1\}$$

for some m. In other words, there exists some fixed m such that $x \in \overline{K(U_0 \cap [0,x_0])}$ implies that $x_0 \wedge mA(|x|) \in U_0$.

In particular, we have $x_1 = x_0 \wedge mA(|Kx_0|) \in U_0 \cap [0,x_0]$. Consequently $Kx_1 \in K(U_0 \cap [0,x_0])$ and so $x_2 = x_0 \wedge mA(|Kx_1|) \in U_0 \cap [0,x_0]$. Proceeding inductively, we obtain a sequence $\{x_n\}$ of positive vectors in $U_0 \cap [0,x_0]$ defined by $x_{n+1} = x_0 \wedge mA(|Kx_n|)$. We claim that

$$0 \leq x_n \leq m^n S^n A^n x_0$$

holds for every n. The proof is by induction in n. For $n = 1$ the inequality $x_1 = x_0 \wedge mA\big(|Kx_0|\big) \leq mA(Sx_0) \leq mSAx_0$ is trivially true. For the induction step, recall that $AS^k \leq S^k A$ and note that if $0 \leq x_n \leq m^n S^n A^n x_0$ for some n, then

$$\begin{aligned} 0 \leq x_{n+1} &= x_0 \wedge mA(|Kx_n|) \leq mA(|Kx_n|) \leq mA(Sx_n) \\ &\leq m^{n+1} A(SS^n A^n x_0) = m^{n+1} A(S^{n+1} A^n x_0) \\ &\leq m^{n+1} S^{n+1} A^{n+1} x_0. \end{aligned}$$

Thus $\|x_n\| \leq m^n \|S^n\|^{\frac{1}{n}} \cdot \|A\|^n \cdot \|x_0\|$, and so $\|x_n\|^{\frac{1}{n}} \leq m\|S^n\| \cdot \|A\| \cdot \|x_0\|^{\frac{1}{n}}$ for each n. Since $\lim\limits_{n\to\infty} \|S^n\|^{\frac{1}{n}} = 0$, it follows that $\lim\limits_{n\to\infty} \|x_n\|^{\frac{1}{n}} = 0$, and hence $\lim\limits_{n\to\infty} \|x_n\| = 0$. However, since $\{x_n\} \subseteq U_0$, we can conclude that $0 \in \overline{U_0} = U_0$, contrary to ($\star\star$), and the solution of the problem is finished. ∎

Problem 10.3.10. *Recall that a bounded operator $T\colon X \to Y$ between Banach spaces is called a* ***Dunford–Pettis operator*** *if $x_n \xrightarrow{w} 0$ in X implies that $\|Tx_n\| \to 0$ in Y.*

Show that every positive Dunford–Pettis operator on a Banach lattice which is quasinilpotent at a non-zero positive vector has a non-trivial closed invariant ideal.

Solution: Let $B\colon E \to E$ be a positive Dunford–Pettis operator on a Banach lattice and let B be quasinilpotent at a non-zero positive vector. We can assume that B is strictly positive; otherwise N_B is a non-trivial closed invariant ideal.

Since B is a Dunford–Pettis operator, it follows that B carries order intervals onto relatively weakly compact sets [**6**, Theorem 19.12, p. 339]. This guarantees that the strictly positive operator B^2 carries order intervals to norm totally bounded sets. That is, B^2 is a non-zero AM-compact operator. Since B commutes with B^2, the desired conclusion follows immediately from Corollary 10.28. ∎

10.4. Invariant Subspaces of Families of Positive Operators

Problem 10.4.1. *Let $\mathcal{S}$ be a multiplicative semigroup of operators on a Banach space and let $\mathcal{S}_1 = \mathcal{S} \cup \{I\}$, where I is the identity operator. Show that $\mathcal{S}_1$ is again a multiplicative semigroup.*

Solution: Take two arbitrary operators T_1, T_2 in $\mathcal{S}_1$. If neither of them is I, then both of them belong to $\mathcal{S}$ and so $T_1T_2 \in \mathcal{S}$ since $\mathcal{S}$ is a multiplicative semigroup. If one of them, say T_1, equals I, then $T_1T_2 = T_2$ and therefore the product again belongs to $\mathcal{S}_1$. ∎

Problem 10.4.2. *Let $\mathcal{S}$ be an additive semigroup of operators on a Banach space and let $\mathcal{S}_0 = \mathcal{S} \cup \{0\}$, where 0 is the zero operator. Show that $\mathcal{S}_0$ is again an additive semigroup.*

Solution: Take two arbitrary operators T_1, T_2 in $\mathcal{S}_0$. If neither of them is the zero operator, then both of them belong to $\mathcal{S}$ and so $T_1 + T_2 \in \mathcal{S}$ since $\mathcal{S}$ is an additive semigroup. If one of them, say T_1, equals to the zero operator, then $T_1 + T_2 = T_2$ and therefore the sum again belongs to $\mathcal{S}_0$. ∎

Problem 10.4.3. *Let $\mathcal{C}$ be a collection of bounded operators on a Banach space X and let $Q^f_{\mathcal{C}}$ be the set of all vectors in X at which $\mathcal{C}$ is finitely (locally) quasinilpotent. Show that $Q^f_{\mathcal{C}}$ is a vector subspace of X which is invariant under $\mathcal{C}$ and $\mathcal{C}'$, the commutant of $\mathcal{C}$.*

Solution: It is obvious that $Q^f_{\mathcal{C}}$ is closed with respect to scalar multiplication. To show that $Q^f_{\mathcal{C}}$ is closed with respect to addition, choose two arbitrary vectors $x, y \in Q^f_{\mathcal{C}}$ and let $\mathcal{F}$ be any finite subset of $\mathcal{C}$. We have

$$\|\mathcal{F}^n(x+y)\|^{\frac{1}{n}} \leq \big[\|\mathcal{F}^n x\| + \|\mathcal{F}^n y\|\big]^{\frac{1}{n}} \leq 2^{\frac{1}{n}} \max\{\|\mathcal{F}^n x\|^{\frac{1}{n}}, \|\mathcal{F}^n y\|^{\frac{1}{n}}\} \to 0,$$

since both $\|\mathcal{F}^n x\|^{\frac{1}{n}}$ and $\|\mathcal{F}^n y\|^{\frac{1}{n}}$ converge to zero by the definition of $Q^f_{\mathcal{C}}$. Therefore, $\|\mathcal{F}^n(x+y)\|^{\frac{1}{n}} \to 0$, and so $x + y \in Q^f_{\mathcal{C}}$.

To prove that $Q^f_{\mathcal{C}}$ is invariant under $\mathcal{C}$, pick any $C \in \mathcal{C}$ and let us verify that $Cx \in Q^f_{\mathcal{C}}$. Notice that for each finite subset $\mathcal{F}$ of $\mathcal{C}$ and for each $n \in \mathbb{N}$ we have

$\mathcal{F}^n C \subseteq \mathcal{G}^{n+1}$, where $\mathcal{G} = \mathcal{F} \cup \{C\}$. Consequently, from

$$\|\mathcal{F}^n Cx\|^{\frac{1}{n}} \le \|\mathcal{G}^{n+1}x\|^{\frac{1}{n}} = \left[\|\mathcal{G}^{n+1}x\|^{\frac{1}{n+1}}\right]^{\frac{n+1}{n}} \to 0\,,$$

we see that $\|\mathcal{F}^n Cx\|^{\frac{1}{n}} \to 0$. This shows that $Cx \in Q^f_{\mathcal{C}}$. The fact that $Tx \in Q^f_{\mathcal{C}}$ for each $T \in \mathcal{C}'$ can be proved in a similar manner. Moreover, in Lemma 10.40 we proved the stronger statement that $Q^f_{\mathcal{C}}$ is invariant under $[\mathcal{C}\rangle$. ∎

Problem 10.4.4. *Let $\mathcal{C}$ be a collection of positive operators on a Banach lattice E and let $N_{\mathcal{C}} = \bigcap_{C\in\mathcal{C}} N_C$. Show that $N_{\mathcal{C}}$ is invariant under $\mathcal{C}$ and $[\mathcal{C}\rangle$.*

Solution: Take any $x \in N_{\mathcal{C}}$ and two arbitrary operators $C_0 \in \mathcal{C}$ and $T_0 \in [\mathcal{C}\rangle$. We need to show that $C_0x \in N_{\mathcal{C}}$ and $T_0x \in N_{\mathcal{C}}$, that is, $C|C_0x| = 0$ and $C|T_0x| = 0$ for each $C \in \mathcal{C}$. The first equality follows from $C|C_0x| \le CC_0|x| = C0 = 0$. For the second, note that $CT_0 \le T_0C$ implies $C|T_0x| \le CT_0|x| \le T_0C|x| = T_00 = 0$, and so $C|T_0x| = 0$, as desired. ∎

Problem 10.4.5. *Let $\mathcal{C}$ be a collection of positive operators on a Banach lattice and let $\mathcal{S}_{\mathcal{C}}$ be the multiplicative semigroup generated by $\mathcal{C}$. Establish the following identities: $[\mathcal{C}\rangle = [\mathcal{S}_{\mathcal{C}}\rangle$, $\langle\mathcal{C}] = \langle\mathcal{S}_{\mathcal{C}}]$, and $\mathcal{C}' = \mathcal{S}'_{\mathcal{C}}$.*

Solution: We will verify the first equality only. The other two can be verified similarly. We begin by verifying the inclusion $[\mathcal{C}\rangle \subseteq [\mathcal{S}_{\mathcal{C}}\rangle$. Take an operator $T \in [\mathcal{C}\rangle$, and an arbitrary operator $S \in \mathcal{S}_{\mathcal{C}}$. We want to show that $TS \ge ST$. From the definition of $\mathcal{S}_{\mathcal{C}}$ we know that $S = C_1C_2\cdots C_n$ for some operators $C_1, C_2, \ldots, C_n$ in $\mathcal{C}$. We are also given that $TC \ge CT$ for each $C \in \mathcal{C}$. Using this, we obtain

$$TS = TC_1C_2\cdots C_n \ge C_1TC_2\cdots C_n \ge \ldots \ge C_1C_2\cdots C_nT = ST\,,$$

proving that $T \in [\mathcal{C}\rangle$.

To verify the reverse inclusion $[\mathcal{S}_{\mathcal{C}}\rangle \subseteq [\mathcal{C}\rangle$, note that if $T \in [\mathcal{S}_{\mathcal{C}}\rangle$, then T automatically belongs to $[\mathcal{C}\rangle$ because $\mathcal{C} \subseteq \mathcal{S}_{\mathcal{C}}$. ∎

Problem 10.4.6. *Let $\mathcal{C}$ be a commutative collection of positive operators on a Banach lattice. If $\mathcal{C}$ is quasinilpotent at some vector $x_0 > 0$, then $\mathcal{C}$ is finitely quasinilpotent at x_0.*

Solution: Let $\mathcal{F} = \{C_1, \ldots, C_k\}$ be an arbitrary finite subset of $\mathcal{C}$. We claim that $\|\mathcal{F}^n x_0\|^{\frac{1}{n}} = 0$. To prove this, fix $0 < \epsilon < 1$ and let $M = 1 + \max_{1\le i\le k}\|C_i\|$. Since $\lim_{n\to\infty}\|C_i^n x_0\|^{\frac{1}{n}} = 0$ for each $i = 1, \ldots, k$, we can find some n_0 such that $\|C_i^n x_0\| < (\epsilon^k)^n$ for each $n \ge n_0$ and each $i = 1, \ldots, k$.

Recall that the set $\mathcal{F}^n$ consists of all the products of arbitrary n operators from $\mathcal{F}$. Therefore, in view of the commutativity of the operators in $\mathcal{C}$, an arbitrary operator $S \in \mathcal{F}^n$ has a representation of the form $S = C_1^{\alpha_1}C_2^{\alpha_2}\cdots C_k^{\alpha_k}$ for some integers $\alpha_i \ge 0$ satisfying $\alpha_1 + \alpha_2 + \cdots + \alpha_k = n$. The latter condition implies that for each n there exists some $1 \le i_n \le k$ such that $\alpha_{i_n} \ge \frac{n}{k}$. We will assume that $n \ge kn_0$ and hence $\alpha_{i_n} \ge n_0$. Using the commutativity of the operators once again, we see that

$$C_1^{\alpha_1}C_2^{\alpha_2}\cdots C_k^{\alpha_k}x_0 = C_1^{\alpha_1}\cdots C_{i_n-1}^{\alpha_{i_n-1}}C_{i_n+1}^{\alpha_{i_n+1}}\cdots C_k^{\alpha_k}C_{i_n}^{\alpha_{i_n}}x_0\,,$$

and so without loss of generality we can assume that $i(n) = k$. Then we have

$$\begin{aligned} \|Sx_0\|^{\frac{1}{n}} &= \|C_1^{\alpha_1} C_2^{\alpha_2} \cdots C_k^{\alpha_k} x_0\|^{\frac{1}{n}} \le \|C_1^{\alpha_1} \cdots C_{k-1}^{\alpha_{k-1}}\|^{\frac{1}{n}} \cdot \|C_k^{\alpha_k} x_0\|^{\frac{1}{n}} \\ &\le \|C_1\|^{\frac{\alpha_1}{n}} \cdots \|C_{k-1}\|^{\frac{\alpha_{k-1}}{n}} \cdot (\epsilon^k)^{\frac{\alpha_k}{n}} \\ &\le M^{k-1} \cdot (\epsilon^k)^{\frac{1}{k}} = M^{k-1}\epsilon\,. \end{aligned}$$

This shows that $\|\mathcal{F}^n x_0\|^{\frac{1}{n}} \le M^{k-1}\epsilon$ for all $n \ge kn_0$, and so $\lim_{n\to\infty}\|\mathcal{F}^n x_0\|^{\frac{1}{n}} = 0$. Therefore, the collection $\mathcal{C}$ is indeed finitely quasinilpotent. ∎

Problem 10.4.7. *Let Δ be a locally compact Hausdorff space and let $\mathcal{S}$ be a non-zero multiplicative semigroup of positive operators on $C_0(\Delta)$ which are quasinilpotent at some $x_0 > 0$. If $\mathcal{S}$ is also an additive semigroup, then show that $\mathcal{S}$ has a common non-trivial closed invariant ideal.*

Solution: If $Sx_0 = 0$ for each $S \in \mathcal{S}$, then there is nothing to prove, since the intersection of the null ideals $J = \bigcap_{S\in\mathcal{S}}\{x \in C_0(\Delta)\colon S|x| = 0\}$ is a non-trivial closed $\mathcal{S}$-invariant ideal. (The ideal J is non-trivial since S is non-zero and $x_0 \in J$.)

So, we assume that $S_0x_0 > 0$ for some $S_0 \in \mathcal{S}$. Without loss of generality we can assume that the support set $F = \overline{\{t \in \Delta\colon x_0(t) > 0\}}$ of the function x_0 is compact. Indeed, each $S \in \mathcal{S}$ is certainly quasinilpotent at each element from the order interval $[0, x_0]$. Furthermore, the functions with a compact support in $[0, x_0]$ are norm dense in $[0, x_0]$ and this proves our claim.

We claim next that there exists a point $t_0 \in \Delta$ such that

$$[Sx_0](t_0) = 0\,, \tag{$\star$}$$

for each $S \in \mathcal{S}$. To see this, assume to the contrary that $(\star)$ is not true. Then for each $t \in \Delta$ there exists an $S_t \in \mathcal{S}$ such that $[S_tx_0](t) \ne 0$, and so there exists an open neighborhood $G(t)$ of t such that

$$[S_tx_0](s) \ge \epsilon_t > 0 \ \text{ for each } s \in G(t)\,.$$

Take a finite cover $G(t_1), \ldots, G(t_m)$ of the compact set F, and consider the operator

$$S = S_{t_1} + S_{t_2} + \cdots + S_{t_m}\,.$$

By our hypothesis $S \in \mathcal{S}$, and so S is quasinilpotent at x_0. On the other hand, we clearly have

$$[Sx_0](t) = \sum_{i=1}^{m}[S_{t_i}x_0](t) \ge \min\{\epsilon_{t_1}, \ldots, \epsilon_{t_m}\} > 0$$

for each $t \in F$. This implies that there exists some $\gamma > 0$ such that

$$Sx_0 \ge \gamma x_0\,.$$

But, as we know from Problem 10.3.1, the last inequality implies that S is not quasinilpotent at x_0, a contradiction. Therefore, a point t_0 with the required properties exists.

Finally, consider the ideal of $C_0(\Delta)$:

$$J = \Big\{x \in C_0(\Delta)\colon\ |x| \le \lambda\Big[\sum_{i=1}^{m} S_ix_0\Big],\ \text{ where } S_i \in \mathcal{S} \text{ and } \lambda \in \mathbb{R}^+\Big\}.$$

Obviously $J \neq \{0\}$ as $0 \neq S_0x_0 \in J$. It is also obvious that J is $\mathcal{S}$-invariant. Furthermore, notice that J cannot be dense in $C_0(\Delta)$ since $[Sx_0](t_0) = 0$ for each operator $S \in \mathcal{S}$. Consequently, the norm closure $\overline{J}$ of the ideal J is a non-trivial closed $\mathcal{S}$-invariant ideal. ∎

Problem 10.4.8. *Let $\mathcal{C}$ be a norm bounded collection of operators in $\mathcal{L}(X)$. The **joint** (or the **Rota–Strang**) **spectral radius** of $\mathcal{C}$ is defined by*

$$r(\mathcal{C}) = \inf_{n\in\mathbb{N}} \sqrt[n]{\|\mathcal{C}^n\|}\,,$$

where we let $\mathcal{C}^n = \{C_1C_2\cdots C_n\colon\ C_i \in \mathcal{C}\}$ and $\|\mathcal{C}^n\| = \sup\{\|S\|\colon\ S \in \mathcal{C}^n\}$. Show that $r(\mathcal{C}) = \lim_{n\to\infty} \sqrt[n]{\|\mathcal{C}^n\|}$.

Solution: The solution is absolutely similar to that of the case of a single operator considered in Problem 6.1.6. One should observe only that for a collection of operators $\mathcal{C}$ the inequality $\|\mathcal{C}^{n+m}\| \le \|\mathcal{C}^n\| \cdot \|\mathcal{C}^m\|$ still holds for all $n, m \in \mathbb{N}$. ∎

Problem 10.4.9. *Assume that a positive matrix $[a_{ij}]$ defines a continuous operator on some ℓ_p, where $1 \le p < \infty$, and this operator is quasinilpotent at some non-zero positive vector. Consider also the collection W of all double sequences $\boldsymbol{w} = \{w_{ij}\colon\ i,j = 1,2,\ldots\}$ of scalars such that the operator $T_{\boldsymbol{w}}$ defined by the weighted matrix $[w_{ij}a_{ij}]$ is a regular operator on ℓ_p. (Clearly, W contains all bounded double sequences.)*

Show that the family of operators $\{T_{\boldsymbol{w}}\colon\ \boldsymbol{w} \in W\}$ has a common non-trivial closed invariant ideal.

Solution: We denote by A the operator defined by the matrix $[a_{ij}]$. Let J be any non-trivial closed A-invariant ideal. Such an ideal always exists by virtue of Corollary 10.31 since A is quasinilpotent at a non-zero positive vector. Since the norm on ℓ_p is order continuous, the ideal J is a band.

Fix an arbitrary $\mathbf{w} = \{w_{ij}\colon\ i,j = 1,2,\ldots\} \in W$. It is should be clear that if $\mathbf{w}$ is bounded, that is, $|w_{ij}| \le c$ for some constant $c > 0$, then $|T_{\mathbf{w}}| \le cA$, and consequently the band J is also $|T_{\mathbf{w}}|$-invariant.

To consider the case of a general $\mathbf{w} \in W$, assume first that $\mathbf{w}$ consists of positive numbers and let $\mathbf{w}_n = \{w_{ij} \wedge n\}$. Obviously $\mathbf{w}_n \in W$ and $\mathbf{w}_n \uparrow \mathbf{w}$. Therefore, $T_{\mathbf{w}_n} \uparrow T_{\mathbf{w}}$, and consequently for each $x \in J_+$ we have $T_{\mathbf{w}_n}x \uparrow T_{\mathbf{w}}x$. Since $T_{\mathbf{w}_n}x \in J$ and J is a band, it follows that $T_{\mathbf{w}}x \in J$. That is, J is $T_{\mathbf{w}}$-invariant. The general case follows immediately from the previous one by observing that for each $\mathbf{w} \in W$ we have $|T_{\mathbf{w}}| \le T_{|\mathbf{w}|}$. ∎

10.5. Compact-friendly Operators

Problem 10.5.1. *Prove that the following classes of operators consist of compact-friendly operators:*

(a) *compact positive operators,*

(b) *positive operators commuting with non-zero positive compact operators,*

(c) *positive operators that dominate non-zero compact positive operators,*

(d) *positive E^*-integral operators (in particular, abstract integral operators), where E is a Dedekind complete Banach lattice,*

(e) *positive integral operators.*

Solution: Recall that a positive operator $B\colon E \to E$ on a Banach lattice is said to be **compact-friendly** if there exist two non-zero positive operators $R, K\colon E \to E$ with K compact and a non-zero operator $C\colon E \to E$ such that $RB = BR$ and

$$\big|Cx\big| \leq R|x| \quad \text{and} \quad \big|Cx\big| \leq K|x|$$

for each $x \in E$.

(a) Assume that $B\colon E \to E$ is a compact positive operator. To see that B is compact-friendly, we simply let $R = C = K = B$.

(b) Assume that $B\colon E \to E$ is a positive operator that commutes with a non-zero compact positive operator $T\colon E \to E$. If we let $R = C = K = T$, then it is easy to see that B is a compact-friendly operator.

(c) Assume that a positive operator $B\colon E \to E$ dominates a non-zero compact positive operator $T\colon E \to E$, i.e., $0 < T \leq B$. If we let $R = B$ and $C = K = T$, then we see that B is a compact-friendly operator.

(d) Let $T\colon E \to E$ be a positive E^*-integral operator on a Dedekind complete Banach lattice. This means that there exists a net $\{T_\alpha\}$ of positive operators such that $0 \leq T_\alpha \uparrow T$ in $\mathcal{L}_r(E)$ and each T_α is dominated by a finite-rank operator.

Now pick any index β such that $T_\beta > 0$, and then choose a finite-rank operator S satisfying $T_\beta \leq S$. Clearly, S is a compact positive operator. Now if we let $R = T$, $C = T_\beta$, and $K = S$, then we easily infer that T is a compact-friendly operator.

(e) According to Theorem 5.28 each integral operator on a Banach function space is an abstract integral operator. The conclusion now follows immediately from (d). ∎

Problem 10.5.2. *Let $B\colon E \to E$ be a positive operator on a Banach lattice. If there exists a non-trivial B-invariant projection band Δ in E such that $Bx = 0$ for each $x \in \Delta^{\mathrm{d}}$, then show that B is compact-friendly.*

Solution: Pick any vector $0 < v \in \Delta^{\mathrm{d}}$ and any non-zero positive functional ϕ such that $\phi(x) = 0$ for each $x \in \Delta$. (Such v and ϕ clearly exist because Δ is a non-trivial projection band.) Now consider the positive (non-zero) rank-one operator $T = \phi \otimes v$. A straightforward verification shows that $BT = TB = 0$. By letting $R = C = K = T$, we obtain the three operators R, C, and K needed to guarantee that B is a compact-friendly operator. ∎

Problem 10.5.3 (Sirotkin). *For a non-zero positive operator $T\colon E \to E$ on a Banach lattice establish the following.*

(a) *If $K \in \mathcal{L}(E)$ is a non-zero compact positive operator, then $T+K$ is compact-friendly.*

(b) *There exists a compact-friendly operator $B \in \mathcal{L}(E)$ such that $T \leq B$.*

(c) *If E is Dedekind complete, then there exists a compact-friendly operator S such that $0 < S \leq T$.*

Solution: (a) Let $K \in \mathcal{L}(E)$ be a non-zero compact positive operator and let $B = T + K$. Letting $R = B$ and $C = K$, we see that R commutes with B, and R dominates the non-zero operator C which is dominated by the compact positive operator K. Hence, B is compact-friendly.

(b) Pick any non-zero compact positive operator K on E. For instance, let K be a rank-one positive operator. By part (a), the operator $B = T + K$ is compact-friendly and satisfies $T \leq B$.

(c) Let E be Dedekind complete. We assume that $\dim E = \infty$, since otherwise there is nothing to prove. Choose three pairwise disjoint non-zero bands Δ_1, Δ_2 and Δ_3 such that $\Delta_1 \oplus \Delta_2 \oplus \Delta_3 = E$. This is always possible since E is infinite dimensional (how?). Let P_i denote the band-projection on the band Δ_i. Since $P_1 + P_2 + P_3 = I$, it should be clear that we can find $i, j \in \{1, 2, 3\}$ such that the operator $S = P_i T P_j$ is non-zero. Obviously, $0 < S \leq T$, and we claim that S is compact-friendly.

To verify that S is compact-friendly, we will establish that S satisfies the conditions of Problem 10.5.2. Let $\Delta = \Delta_i \oplus \Delta_j$. Clearly, Δ is a band in E which is also S-invariant. It remains to notice that $Sx = 0$ for each $x \in \Delta^{\mathrm{d}} = \Delta_k$, where $k \neq i, j$. Therefore, Problem 10.5.2 guarantees that S is a compact-friendly operator.

The assumption of Dedekind completeness can be considerably relaxed. The same proof goes through whenever there exist three pairwise disjoint non-zero bands Δ_1, Δ_2, and Δ_3 such that $\Delta_1 \oplus \Delta_2 \oplus \Delta_3 = E$. ∎

Problem 10.5.4. *Let ϕ be a positive one-to-one function in $C(\Omega)$, where Ω is a compact Hausdorff space without isolated points. Show directly, i.e., without using Theorem* 10.65, *that the multiplication operator M_ϕ is not compact-friendly.*

Solution: Suppose that Ω and ϕ satisfy the properties stated in the problem. Assume by way of contradiction that M_ϕ is compact-friendly. Then, there exist non-zero operators $R, K, A\colon C(\Omega) \to C(\Omega)$ with R and K positive and K compact such that

$$RM_\phi = M_\phi R, \quad R \text{ dominates } A \quad \text{and} \quad K \text{ dominates } A\,.$$

By Lemma 4.19(3), there exists some $0 \leq h \in C(\Omega)$ such that $R = M_h$. From the domination of A by R we have $|Ax| \leq hx$ for each $0 \leq x \in C(\Omega)$. Therefore, $A \in \mathcal{Z}\big(C(\Omega)\big)$ and so, by Theorem 3.33, there exists a non-zero function $\psi \in C(\Omega)$ such that $Ax = \psi x$ and $|\psi| \leq h$. Since A is dominated by a compact operator K, it follows from Theorem 2.34 that the operator A^3 is compact. Clearly $A^3x = \psi^3 x$ and the function ψ^3 is non-zero since ψ is non-zero. However, according to Lemma 4.18, a multiplication operator with a non-zero multiplier cannot be compact if Ω has no isolated points. This contradiction shows that M_ϕ cannot be compact-friendly. ∎

Problem 10.5.5. *A positive operator $B\colon E \to E$ on a Dedekind complete Banach lattice is called a **generalized Harris operator** if some power of B is not disjoint from the band $(E^* \otimes E)^{\mathrm{dd}}$ of E^*-integral operators on E, i.e., if some power of B dominates a non-zero positive E^*-integral operator. Recall that if some power of B is not disjoint from the band $(E^*_{\mathrm{n}} \otimes E)^{\mathrm{dd}}$ of abstract integral operators, then B is called a Harris operator. Of course, each Harris operator is a generalized Harris operator.*

For a generalized Harris operator $B\colon E \to E$ establish the following.

(a) *B is compact-friendly.*

(b) *If B is quasinilpotent at a non-zero positive vector, then B has a non-trivial closed ℓ-hyperinvariant ideal.*

Solution: (a) Let $B\colon E \to E$ be a generalized Harris operator on a Dedekind complete Banach lattice. This means that there exist some natural number m and a positive E^*-integral operator $0 < S \in (E^* \otimes E)^{\mathrm{dd}}$ such that $S \le B^m$. Therefore, there exist a positive operator $A\colon E \to E$ and a finite-rank operator $T\colon E \to E$ such that $0 < A \le S$ and $A \le T$. If we let $R = B^m$, $C = A$, and $K = T$, then we see that B is a compact-friendly operator.

(b) Use part (a) and invoke Theorem 10.55. ∎

Problem 10.5.6. *Show that Theorems* 10.55 *and* 10.60 *remain true if we replace the assumption* $\lim_{n\to\infty} \|T^n x_0\|^{\frac{1}{n}} = 0$ *with* $\liminf_{n\to\infty} \|T^n x_0\|^{\frac{1}{n}} = 0$.

Solution: The validity of the proofs of Theorems 10.55 and 10.60 under this new assumption can be established exactly as in the solution of Problem 10.3.8. ∎

Problem 10.5.7. *Let $T\colon \ell_p \to \ell_p$ $(1 \le p < \infty)$ be a continuous operator with modulus. Assume that there is a non-zero positive operator $S\colon \ell_p \to \ell_p$ such that:*

(a) *S commutes with the modulus of T.*

(b) *S is quasinilpotent at a non-zero positive vector.*

Show directly, i.e., without using Theorem 10.24, *that T has a non-trivial invariant closed ideal.*

Solution: It is enough to prove that the modulus $|T|$ has a non-trivial closed invariant ideal. Clearly, this ideal is T-invariant as well. Therefore, we can assume without loss of generality that $T \ge 0$. So, suppose that $S > 0$ satisfies $TS = ST$ and $\lim_{n\to\infty} \big\|S^n x_0\big\|^{\frac{1}{n}} = 0$ for some $x_0 > 0$. We distinguish two cases: $Sx_0 > 0$ and $Sx_0 = 0$.

We begin by assuming that $Sx_0 > 0$. Scaling x_0 if necessary and using the continuity of S, we obtain some k such that $x_0 \geq e_k > 0$ and $Se_k > 0$.

Now let $P\colon \ell_p \to \ell_p$ denote the natural projection onto the vector subspace generated by e_k. Clearly, $0 \leq Px \leq x$ for each $0 \leq x \in \ell_p$. We claim that

$$PT^m Se_k = 0 \tag{$\star$}$$

for each $m \geq 0$. To this end, fix $m \geq 0$ and let $PT^m Se_k = \alpha e_k$ for some $\alpha \geq 0$. Clearly $T^m S = ST^m$, and hence

$$0 \leq \alpha^n e_k = \big(PT^m S\big)^n e_k \leq \big(T^m S\big)^n e_k = T^{mn} S^n e_k \leq T^{mn} S^n x_0\,.$$

Consequently,

$$0 \leq \alpha \leq \big\|T\big\|^m \cdot \big\|S^n x_0\big\|^{\frac{1}{n}} \xrightarrow[n\to\infty]{} 0\,,$$

from which it follows that $\alpha = 0$.

Now consider the order ideal generated by the set $\{T^m Se_k\colon\ m = 0, 1, \ldots\}$, that is,

$$J = \big\{x \in \ell_p\colon\ \exists\, \lambda \geq 0 \text{ and } r \geq 0 \text{ such that } |x| \leq \lambda \textstyle\sum_{i=0}^r T^i Se_k\big\}\,.$$

Since $0 < Se_k \in J$, we see that $J \neq \{0\}$. Also, from ($\star$), it follows that

$$\langle e_k, x\rangle = \langle e_k, Px\rangle = 0$$

for all $x \in J$, and consequently $\langle e_k, x\rangle = 0$ for all $x \in \overline{J}$. The latter shows that $\overline{J}$ is a non-trivial closed ideal of ℓ_p. We claim that $\overline{J}$ is T-invariant. Indeed, if $x \in J$, then there exists some positive scalar $\lambda > 0$ and some $r \geq 0$ such that $|x| \leq \lambda \sum_{i=0}^r T^i Se_k$. Thus, $|Tx| \leq T(|x|) \leq \lambda \sum_{i=0}^{r+1} T^i Se_k$, and hence $Tx \in J$. That is, J (and hence $\overline{J}$) is T-invariant.

Finally, consider the case $Sx_0 = 0$. Let $\mathcal{A}$ denote the unital algebra generated by T in $\mathcal{L}(\ell_p)$. Also, let

$$J = \big\{x \in \ell_p\colon\ |x| \leq |A|x_0 \text{ for some } A \in \mathcal{A}\big\}\,.$$

Clearly, J is a T-invariant ideal which is non-zero since $x_0 \in J$. To finish the proof, it suffices to show that $\overline{J} \neq \ell_p$.

To see this, let $x \in J$. Pick some $A \in \mathcal{A}$ with $|x| \leq |A|x_0$ and then select $r \geq 0$ and a constant $c > 0$ such that $|A| \leq c \sum_{i=0}^r T^i$. But then, from

$$|Sx| \leq S|x| \leq S|A|x_0 \leq S\Big(c\sum_{i=0}^r T^i\Big)x_0 = c\sum_{i=0}^r T^i(Sx_0) = 0\,,$$

it follows that $Sx = 0$ for all $x \in J$. If $\overline{J} = \ell_p$, then $S = 0$, which is a contradiction. Therefore, $\overline{J}$ is a non-trivial closed T-invariant ideal. ∎

Problem 10.5.8 (Sirotkin). *Let $B\colon E \to E$ be a compact-friendly operator on a Banach lattice such that no positive operator commuting with B dominates a non-zero compact operator. Show that for each at most countable family of positive operators commuting with B there exists a non-trivial closed ideal that is B-invariant and $\mathcal{F}$-invariant.*

Solution: Let $B\colon E \to E$ be a non-zero compact-friendly operator on a Banach lattice such that no positive operator commuting with B dominates a non-zero compact operator. Also, let $\mathcal{F} = \{T_n\}$ be a sequence of positive operators such that $T_nB = BT_n$ holds for each n.

The solution is an adaptation of the proof of Theorem 10.55. Fix three non-zero operators $R, K, C\colon E \to E$ with K, R positive, K compact, and satisfying

$$RB = BR, \quad |Cx| \le R(|x|) \quad \text{and} \quad |Cx| \le K(|x|) \quad \text{for each} \quad x \in E\,.$$

Without loss of generality we can assume that $\|B\| < 1$.

For each n pick some $\alpha_n > 0$ such that the series $T = \sum_{n=1}^{\infty} \alpha_n T_n$ is norm convergent—and so it defines a positive operator—such that $\|B + T\| < 1$. Consequently, the series $A = \sum_{n=0}^{\infty}(B+T)^n$ also defines a positive operator. Clearly, T commutes with B, and A commutes with B and T. For each $x > 0$ we denote by J_x the principal ideal generated by Ax, that is,

$$J_x = \big\{y \in E\colon\ |y| \le \lambda Ax \text{ for some } \lambda > 0\big\}\,.$$

This ideal is non-zero because $x \in J_x$ in view of the obvious inequality $x \le Ax$. If the norm closure $\overline{J}_x \ne E$ for some vector $x > 0$, then $\overline{J}_x$ is a non-trivial closed $(B+T)$-invariant ideal. From $0 \le B, T \le B + T$, it follows that $\overline{J}_x$ is invariant under both B and T. Clearly, this ideal is also T_n-invariant for each n. We claim now that there exists some $x > 0$ such that $\overline{J}_x \ne E$. To prove this, assume to the contrary that

$$\overline{J}_x = E \tag{$\star$}$$

for each $x > 0$, that is, Ax is a quasi-interior point in E for each $x > 0$.

Since $C \ne 0$, there exists some $x_1 > 0$ such that $Cx_1 \ne 0$. Then, $A|Cx_1|$ is a quasi-interior point satisfying $A|Cx_1| \ge |Cx_1|$, and it follows from Lemma 4.16(1) that there exists an operator $M_1\colon E \to E$ dominated by the identity operator such that $x_2 = M_1Cx_1 > 0$. Put $\Pi_1 = M_1C$, and note that the operator Π_1 is dominated by the compact positive operator K and by the operator R.

By $(\star)$ we have $\overline{J}_{x_2} = E$ and so, since $C \ne 0$, there exists $0 < y \le Ax_2$ such that $Cy \ne 0$. Hence (again by $(\star)$) the element $A|Cy|$ is a quasi-interior point. Since $|Cy| \le A|Cy|$, there exists, by Lemma 4.16(1) again, an operator $M_2\colon E \to E$ dominated by the identity operator such that $x_3 = M_2Cy > 0$. Since $|y| \le Ax_2$ and Ax_2 is a quasi-interior point, it follows from Lemma 4.16(2) that there exists an operator $M\colon E \to E$ dominated by the identity operator such that $MAx_2 = y$. So, $x_3 = M_2Cy = M_2CMAx_2$. Let $\Pi_2 = M_2CMA$ and note that Π_2 is dominated by the compact positive operator KA and by the operator RA.

If we repeat the preceding arguments with the vector x_2 replaced by x_3, then we obtain one more operator $\Pi_3\colon E \to E$ that satisfies $\Pi_3x_3 > 0$ and that is dominated by the compact positive operator KA and by the operator RA.

Consider the operator $\Pi_3\Pi_2\Pi_1$. From $\Pi_3\Pi_2\Pi_1x_1 = \Pi_3x_3 > 0$, we see that the operator $\Pi_3\Pi_2\Pi_1$ is non-zero. Moreover, as shown above, each Π_i is dominated by a compact operator and, therefore, Theorem 2.34 guarantees that the operator $\Pi_3\Pi_2\Pi_1$ is necessarily compact. A straightforward verification shows also that

$$\big|\Pi_3\Pi_2\Pi_1x\big| \le RARAR|x|$$

for each $x \in E$. But this contradicts our assumption, since the operator $RARAR$ clearly commutes with B, and the solution of the problem is complete. ∎

Problem 10.5.9 (Troitsky [**79**]). *Show that every essentially quasinilpotent compact-friendly operator on a Banach lattice of dimension greater than one has a non-trivial closed invariant subspace.*

Solution: Let $B\colon E \to E$ be an essentially quasinilpotent compact-friendly operator on a Banach lattice of dimension greater than two. If B is quasinilpotent, then by Theorem 10.55 there exists a non-trivial closed B-invariant ideal. So, we can assume that $r(B) > 0$. Since the operator B is positive, we know (in view of Theorem 7.9) that $r(B) \in \sigma(B)$ and so, by Corollary 7.49, the spectral radius $r(B)$ is an eigenvalue of B. This guarantees (see Problem 10.1.4) that B has a non-trivial closed invariant subspace in $E_{\mathbf{C}}$ and also in E. ∎

Problem 10.5.10 (Troitsky [**79**]). *Let $S\colon E \to E$ be a quasinilpotent integral operator on a Banach function space such that S^- is compact. Then each of the operators $|S|$ and S^+ has a non-trivial closed invariant subspace.*

Solution: Let $B = |S|$. This is a positive integral operator and so it is compact-friendly. If B is quasinilpotent, then it has a non-trivial closed invariant ideal. So, we can assume that B is not quasinilpotent, i.e., $r(B) > 0$.

Since $B = S + 2S^-$, it follows that $r_{\mathrm{ess}}(B) = r(S) = 0$. That is, B is essentially quasinilpotent. But then, according to Problem 10.5.9, the operator $B = |S|$ has a non-trivial closed invariant subspace. The proof for S^+ is similar. ∎

Problem 10.5.11. *Show that the set of all compact-friendly operators on a Banach lattice E need not be closed or open in $\mathcal{L}(E)$.*

Solution: Let $E = C[0,1]$ and let $\phi(t) = t$. By Theorem 10.65 (or by Problem 10.5.4) the multiplication operator M_ϕ is not compact-friendly. It should be clear that there exists a sequence of positive functions $\phi_n \in C[0,1]$ such that $\|\phi_n - \phi\|_\infty \to 0$ and each ϕ_n has a flat. This implies that each multiplication operator M_{ϕ_n} is compact-friendly and that $M_{\phi_n} \to M_\phi$ in $\mathcal{L}\big(C[0,1]\big)$. This proves that the set of compact-friendly operators is not a closed subset of $\mathcal{L}\big(C[0,1]\big)$.

Now, for each n consider the function $\psi_n \in C[0,1]$ defined by $\psi_n(t) = 1 + \frac{1}{n}t$, and note that $\|\psi_n - \mathbf{1}\|_\infty \to 0$. This implies that the sequence of multiplication operators $\{M_{\psi_n}\}$ converges to the operator $M_{\mathbf{1}} = I$ in $\mathcal{L}\big(C[0,1]\big)$, which is compact-friendly. However, since no function ψ_n has a flat, it follows that each operator M_{ψ_n} is not compact-friendly. This proves that the set of compact-friendly operators is not an open subset of $\mathcal{L}\big(C[0,1]\big)$ either. ∎

Problem 10.5.12. *This problem presents an example of a Hilbert–Schmidt operator on $L_2[0,1]$ which is one-to-one, positive, locally quasinilpotent at a positive vector, but fails to be quasinilpotent. The construction of such an operator will be done in three steps as follows.*

(a) *Let* $0 \le a < b \le 1$ *and* $0 \le c < d \le 1$ *and consider the positive integral operator* $T\colon L_2[a,b] \to L_2[c,d]$ *defined by*

$$Tx(t) = \int_a^b \sin(st)x(s)\,ds\,, \quad c \le t \le d\,.$$

Then the operator T *is one-to-one and compact.*

(b) *Assume that* $0 \le a < b \le 1$. *Then the positive integral operator* $T\colon L_2[a,b] \to L_2[a,b]$, *defined by*

$$Tx(t) = \int_a^b \sin(st)x(s)\,ds, \quad a \le t \le b\,,$$

has a positive spectral radius—and so T *is a one-to-one compact positive integral operator which is not quasinilpotent.*

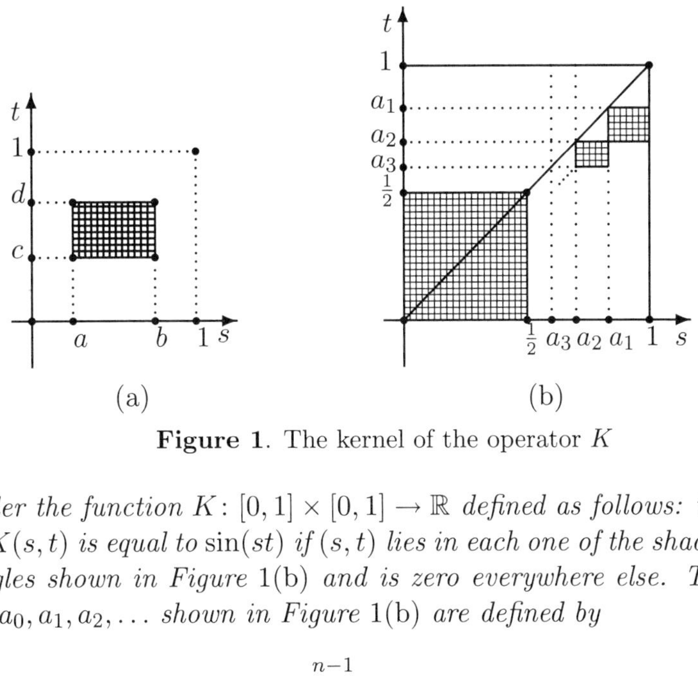

Figure 1. The kernel of the operator K

(c) *Consider the function* $K\colon [0,1] \times [0,1] \to \mathbb{R}$ *defined as follows: the value* $K(s,t)$ *is equal to* $\sin(st)$ *if* (s,t) *lies in each one of the shaded rectangles shown in Figure* 1(b) *and is zero everywhere else. The points* $a_0, a_1, a_2, \ldots$ *shown in Figure* 1(b) *are defined by*

$$a_0 = 1 \quad \textit{and} \quad a_n = 1 - \tfrac{1}{2e}\sum_{k=0}^{n-1} \tfrac{1}{k!} \quad \textit{for} \quad n \ge 1\,.$$

The length of the interval (a_{n+1}, a_n) *is equal to* $a_n - a_{n+1} = \frac{1}{n!2e}$. *As usual, we shall denote by* K *the operator on* $L_2[0,1]$ *defined be the kernel* $K(s,t)$. *That is, the operator* $K\colon L_2[0,1] \to L_2[0,1]$ *is defined by*

$$Kx(t) = \int_0^1 K(s,t)x(s)\,ds, \quad 0 \le t \le 1\,.$$

Then the integral operator $K\colon L_2[0,1] \to L_2[0,1]$ is a one-to-one compact positive operator which is quasinilpotent at some positive vectors but fails to be quasinilpotent.

Solution: (a) The compactness of T follows easily from the fact that T is a Hilbert–Schmidt operator; see also Theorem 7.7. We shall prove that T is one-to-one. The geometric situation is shown in Figure 1(a), where the kernel function $K(s,t) = \sin(st)$ is defined over the rectangle $Q = [a,b] \times [c,d]$. It is easy to see that Tx is a continuous function for each $x \in L_2[a,b]$. As a matter of fact (see [**7**, Theorem 20.4]) the function $Tx \in C^\infty[c,d]$ for each $x \in L_2[a,b]$ and

$$\big[Tx\big]^{(n)}(t) = \int_a^b \big[\tfrac{\partial^n}{\partial t^n} \sin(st)\big] x(s)\, ds\,. \qquad (\star)$$

Assume that $Tx = 0$ for some function $x \in L_2[a,b]$. Then for each n, it follows from $(\star)$ that $\int_a^b s^{2n} \sin(st) x(s)\, ds = 0$ for almost all $t \in [c,d]$.

The subalgebra $\mathcal{A}$ of $C[c,d]$ consisting of all even polynomials contains the constant function one and separates the points of $[a,b]$. (Notice that the polynomial $p(s) = s^2$ is one-to-one on $[a,b]$.) So, by the Stone–Weierstrass theorem, $\mathcal{A}$ is uniformly dense in $C[a,b]$. Consequently, $\mathcal{A}$ is $\|\cdot\|_2$-dense in $L_2[a,b]$. From $(\star)$ we see immediately that $\int_a^b p(s)\sin(st)x(s)\, ds = 0$ for each $p \in \mathcal{A}$ and almost all $t \in [c,d]$. Consequently, there exists a sequence $\{p_n\}$ in $\mathcal{A}$ satisfying $p_n \to x$ a.e. and $|p_n| \le y$ a.e. for all n and some $y \in L_2[a,b]$; see [**7**, p. 207]. So, for almost all $s \in [a,b]$ we have

$$p_n(s)\sin(st)x(s) \longrightarrow \sin(st)[x(s)]^2$$

and $\big|p_n(s)\sin(st)x(s)\big| \le y(s)|x(s)|$. Since $y|x| \in L_1[a,b]$, the Lebesgue Dominated Convergence Theorem implies

$$\int_a^b \sin(st)[x(s)]^2\, ds = \lim_{n\to\infty} \int_a^b p_n(s)\sin(st)x(s)\, ds = 0$$

for almost all t. Therefore, $x = 0$ a.e.

(b) Since the kernel of T is symmetric, it follows that T is a symmetric operator. Consequently, $r(T) = \|T\|_2 > 0$; see also Problem 11.1.10.

If for $1 \le p \le \infty$ we consider the operator $T\colon L_p[0,1] \to L_p[0,1]$ defined by

$$Tx(t) = \int_0^1 \sin(st)x(s)\, ds\,, \quad 0 \le t \le 1\,,$$

then we have the following norm estimates:

$$\begin{aligned}
\|T\|_\infty &= \|T\mathbf{1}\|_\infty = \sup_{t\in[0,1]} \int_0^1 \sin(st)\, ds \\
&= \sup_{t\in[0,1]} \tfrac{1-\cos t}{t} = 1 - \cos 1 \approx 0.4597 \\
\|T\|_2 &\le \Big[\int_0^1\!\!\int_0^1 \big[\sin(st)\big]^2\, dsdt\Big]^{\frac{1}{2}} \\
&= \Big[\tfrac{1}{2} - \tfrac{1}{4}\int_0^1 \tfrac{\sin 2t}{t}\, dt\Big]^{\frac{1}{2}} \approx 0.7009\,.
\end{aligned}$$

(c) The fact that K is one-to-one follows easily from part (a). Since the kernel $K(s,t)$ is positive and belongs to $L_2([0,1]\times[0,1])$, it follows that the operator K is Hilbert–Schmidt. In particular, K is a compact positive operator.

To see that K is not quasinilpotent, notice that K leaves $L_2\big[0,\frac{1}{2}\big]$ invariant. (As usual, here we identify $L_2\big[0,\frac{1}{2}\big]$ with the closed subspace of $L_2[0,1]$ given by $\big\{x\in L_2[0,1]\colon\ x=0 \text{ a.e. on } \big[\frac{1}{2},1\big]\big\}$.) Moreover, observe that the restriction of K to $L_2\big[0,\frac{1}{2}\big]$ is the integral operator with kernel $\sin(st)$. By part (b), the restriction of K to $L_2\big[0,\frac{1}{2}\big]$ has a positive spectral radius. That is, K is not quasinilpotent when restricted to $L_2\big[0,\frac{1}{2}\big]$. Therefore, K cannot be quasinilpotent.

Finally, consider the positive function $x_0=\chi_{[a_1,1]}>0$. An easy verification shows that $0\le K^n x_0\le \chi_{[a_{n+2},a_{n+1}]}$. Consequently,

$$\big[\|K^n x_0\|_2\big]^{\frac{1}{n}}\le\sqrt{\big(\tfrac{1}{n!2e}\big)^{\frac{1}{n}}}\,,$$

from which it follows that $\lim_{n\to\infty}\big[\|K^n x_0\|_2\big]^{\frac{1}{n}}=0$. This shows that K is quasinilpotent at the non-zero positive function x_0. ∎

10.6. Positive Operators on Banach Spaces with Bases

Problem 10.6.1. *Let $\{x_n\}$ be a basis in a Banach space X and let*

$$C=\Big\{x=\sum_{n=1}^{\infty}\alpha_n x_n\colon\ \alpha_n\ge 0 \text{ for each } n=1,2,\dots\Big\}.$$

*Show that C is a closed cone of X (called the **cone generated by** $\{x_n\}$).*

Solution: Let $x=\sum_{n=1}^{\infty}\alpha_n x_n$ and $y=\sum_{n=1}^{\infty}\beta_n x_n$. If $x,y\in C$, then clearly $x+y\in C$ and $\alpha x\in C$ for each $\alpha\ge 0$. That is, $C+C\subseteq C$ and $\alpha C\subseteq C$ for each $\alpha\ge 0$. Now assume that $x\in C\cap(-C)$. From the definition of the basis, we see that $\alpha_n=-\alpha_n$ and hence $\alpha_n=0$ for each n. This shows that $C\cap(-C)=\{0\}$. Therefore, C is a cone in X.

To show that C is a closed cone, assume that $x_n=\sum_{i=1}^{\infty}\alpha_{n,i}x_i\in C$ and $x_n\to x\in X$. From Theorem 1.37 we know that each coordinate functional c_k^* is continuous. Consequently, for each $k\in\mathbb{N}$ we have $0\le\alpha_{n,k}=c_k^*(x_n)\to c_k^*(x)=\alpha_k$. This implies $\alpha_k\ge 0$ and hence $x\in C$. Therefore, C is a closed cone in X. ∎

Problem 10.6.2. *If a Banach space X has an unconditional basis $\{u_n\}$, then show that:*

(a) *The cone C generated by the basis $\{u_n\}$ is a lattice cone.*
(b) *The Riesz space (X,C) has weak units.*
(c) *After an equivulent rcnorming the ordered Banach space (X,C) becomes a Banach lattice with order continuous norm.*

Solution: (a) According to Problem 10.6.1 we know that the set

$$C = \Big\{ x = \sum_{n=1}^{\infty} \alpha_n u_n \colon \ \alpha_n \geq 0 \ \text{for each} \ n = 1, 2, \dots \Big\}$$

is a closed cone. In the usual way this cone induces the coordinatewise ordering $\geq$ on X. That is, for $x = \sum_{n=1}^{\infty} \alpha_n u_n$ and $y = \sum_{n=1}^{\infty} \beta_n u_n$ in X we have $x \geq y$ if and only if $\alpha_n \geq \beta_n$ for each n.

We shall show next that for each $x = \sum_{n=1}^{\infty} \alpha_n u_n$ and $y = \sum_{n=1}^{\infty} \beta_n u_n$ in X the series $\sum_{n=1}^{\infty} \max\{\alpha_n, \beta_n\} u_n$ and $\sum_{n=1}^{\infty} \min\{\alpha_n, \beta_n\} u_n$ are norm convergent in X. If this is done, then clearly (X, C) is a Riesz space and

$$x \vee y = \sum_{n=1}^{\infty} \max\{\alpha_n, \beta_n\} u_n \quad \text{and} \quad x \wedge y = \sum_{n=1}^{\infty} \min\{\alpha_n, \beta_n\} u_n \,.$$

We shall consider the first series only. To this end, let $A = \{n \in \mathbb{N} \colon \ \alpha_n \geq \beta_n\}$ and $B = \{n \in \mathbb{N} \colon \ \alpha_n < \beta_n\}$. Since the basis $\{u_n\}$ is unconditional, both series $\sum_{n\in A} \max\{\alpha_n, \beta_n\} u_n = \sum_{n\in A} \alpha_n u_n$ and $\sum_{n\in B} \max\{\alpha_n, \beta_n\} u_n = \sum_{n\in B} \beta_n u_n$ are norm convergent. This implies that the series $\sum_{n=1}^{\infty} \max\{\alpha_n, \beta_n\} u_n$ is norm convergent.

(b) Pick a sequence $\{\lambda_n\}$ of strictly positive real numbers such that the series $w = \sum_{n=1}^{\infty} \lambda_n u_n$ is norm convergent in X. We claim that w is a weak unit.

To see this, assume that a vector $x = \sum_{n=1}^{\infty} \alpha_n u_n$ satisfies $w \wedge x = 0$. From

$$w \wedge x = \sum_{n=1}^{\infty} \min\{\lambda_n, \alpha_n\} u_n = 0 \,,$$

it follows that $\min\{\lambda_n, \alpha_n\} = 0$ for each n, and from this we infer that $\alpha_n = 0$ for each n. Therefore, $x = 0$ and so w is a weak unit in (X, C). (This conclusion can also be obtained from Problem 4.2.3 by observing that every Banach space with a basis is necessarily separable.)

(c) The equivalent norm $|||\cdot|||$ on X that turns (X, C) into a Banach lattice is defined for $x = \sum_{n=1}^{\infty} \alpha_n x_n \in X$ by

$$|||x||| = \sup \Big\{ \Big\| \sum_{n=1}^{\infty} \beta_n x_n \Big\| \colon \ (\beta_1, \beta_2, \dots) \in \mathbb{R}^{\mathbb{N}} \ \text{and} \ |\beta_n| \leq |\alpha_n| \ \text{for each} \ n \Big\} .$$

The formula defining $|||\cdot|||$ makes sense by virtue of Theorem 1.51. The verification that $|||\cdot|||$ is a lattice norm is straightforward. From Lemma 1.49, it follows that for each $x = \sum_{n=1}^{\infty} \alpha_n x_n$ we have

$$\|x\| \leq |||x||| \leq \kappa_u \|x\| \,,$$

where κ_u is the unconditional constant of the basis $\{u_n\}$. Therefore, the ordered vector space (X, C) equipped with the norm $|||\cdot|||$ is a Banach lattice.

To verify that the norm is order continuous, let $x_\gamma \downarrow 0$. Fix any $\epsilon > 0$ and γ_0. Then find some $n \in \mathbb{N}$ such that $|||(I - P_n)x_{\gamma_0}||| < \epsilon$, where P_n is the natural projection of X onto the vector subspace generated by $\{x_1, x_2, \dots, x_n\}$. Applying Lemma 1.49 again, we see that $|||(I - P_n)x_\gamma||| < \epsilon$ for each $\gamma \geq \gamma_0$. It remains to note that $|||P_n x_\gamma||| \downarrow 0$ since $x_\gamma \downarrow 0$ and since we are dealing here with a finite number of coordinates only. Thus, $|||x_\gamma||| < \epsilon$ for all γ that are large enough. ∎

Problem 10.6.3. *Let X be a Banach space with an unconditional basis $\{u_n\}$. If a linear operator $T: X \to X$ is positive with respect to the cone generated by the basis $\{u_n\}$, then show that T is a bounded operator.*

Solution: Let C be the cone generated by the unconditional basis $\{u_n\}$. By Problem 10.6.2 we know that C is a lattice cone and that we can renorm X so that (X, C) becomes a Banach lattice. Now a glance at Problem 1.2.14 guarantees that the positive operator T is continuous. ∎

Problem 10.6.4 (Spalsbury [**76**]). *Consider the Volterra operator defined in Example 7.8. Show that there is no basis in $L_2[0,1]$ with respect to which the Volterra operator is positive.*

Solution: Recall that the Volterra operator $V: L_2[0,1] \to L_2[0,1]$ is defined via the formula

$$Vf(t) = \int_0^t f(s)\,ds\,.$$

From Problems 7.1.3 and 7.1.4 we know that V is a quasinilpotent one-to-one compact positive operator.

Assume by way of contradiction that there exists a basis $\{x_n\}$ in $L_2[0,1]$ such that the Volterra operator V is positive with respect to the cone generated by this basis $\{x_n\}$. If P_1 is the projection of $L_2[0,1]$ onto the vector subspace generated by x_1, then, as shown in the proof of Theorem 10.66, we have

$$P_1 V^m x_1 = 0 \quad \text{for} \quad m = 1, 2, \ldots\,. \tag{†}$$

So, if M denotes the closed vector subspace generated by the set $\{V^m x_1 : m \geq 1\}$, then M is V-invariant, and in view of (†) we have

$$P_1 x = 0 \quad \text{for each} \quad x \in M\,.$$

Since $P_1 x_1 = x_1 \neq 0$, it follows that $x_1 \notin M$, and thus M is a non-trivial closed V-invariant subspace. By the well-known characterization of the non-trivial closed invariant subspaces of V (see [**44**]), there exists some $0 < \alpha < 1$ such that

$$M = \big\{f \in L_2[0,1] : \ f = 0 \ \text{a.e. on} \ [0, \alpha]\big\}\,.$$

Because $Vx_1 \in M$, we must have $Vx_1 = 0$ a.e. on $[0, \alpha]$. Since the Volterra operator is one-to-one on $L_2[0,\alpha]$ (see Problem 7.1.4), it follows that $x_1 = 0$ a.e. on $[0,\alpha]$, and consequently $x_1 \in M$, a contradiction. This contradiction shows that the Volterra operator cannot be positive with respect to any basis for $L_2[0,1]$. ∎

10.7. Non-transitive Algebras

Problem 10.7.1. *If Ω is a compact topological space and X is a Banach space, then show that the space $C(\Omega, X)$ of all X-valued continuous functions* *on Ω equipped with the norm*

$$\|f\| = \max_{\omega \in \Omega} \|f(\omega)\|$$

is a Banach space—which is a Banach lattice if X is a Banach lattice.

Solution: The verification of the norm axioms of the indicated function $f \mapsto \|f\|$ is straightforward. We show the completeness of the norm. Notice that convergence with respect to the norm in $C(\Omega, X)$ is equivalent to uniform convergence. So, if $\{f_n\}$ is a Cauchy sequence in $C(\Omega, X)$, then $\{f_n\}$ is convergent uniformly to a function $f\colon \Omega \to X$. It follows that f is continuous and $\|f_n - f\| \to 0$. Hence, $C(\Omega, X)$ is a Banach space.

Now assume that X is a Banach lattice. It is not difficult to see that $C(\Omega, X)$ is a Riesz space under the pointwise lattice operations

$$(f \vee g)(\omega) = f(\omega) \vee g(\omega) \quad \text{and} \quad (f \wedge g)(\omega) = f(\omega) \wedge g(\omega)\,.$$

Moreover, if $|f| \le |g|$ in $C(\Omega, X)$, then $|f(\omega)| \le |g(\omega)|$ for each $\omega \in \Omega$, and so $\|f(\omega)\| \le \|g(\omega)\|$, from which it follows that $\|f\| \le \|g\|$. That is, the norm on $C(\Omega, X)$ is a lattice norm, and so $C(\Omega, X)$ is a Banach lattice. ∎

Problem 10.7.2. *Let X be a Banach space, and let the closed unit ball U^* of X^* be equipped with the w^*-topology. Establish the following.*

(a) *If $T\colon X \to X$ is a compact operator, then $T^*\colon U^* \to X^*$ (the restriction of T^* to U^*) is a completely continuous function.*

(b) *The collection of all completely continuous functions $\mathcal{K}(U^*, X^*)$ is a norm closed subspace of $C(U^*, X^*)$, and so it is a Banach space in its own right.*

Solution: (a) Recall that a function $f\colon U^* \to X^*$ is said to be **completely continuous** if it is continuous for the weak* topology on U^* and the norm topology on X^*. Now assume that $T\colon X \to X$ is a compact operator, and let $x^*_\alpha \xrightarrow{w^*} x^*$ in U^*. If $x \in X$ is fixed, then from

$$\langle x, T^*x^*_\alpha - T^*x^*\rangle = \langle x, T^*(x^*_\alpha - x^*)\rangle = \langle Tx, x^*_\alpha - x^*\rangle \longrightarrow 0\,,$$

it follows that

$$T^*x^*_\alpha \xrightarrow{w^*} T^*x^* \qquad (\star)$$

in X^*. Since T is compact, the operator T^* is also compact, and hence $T^*(U^*)$ is a norm totally bounded subset of X^*. It follows that every subnet of the net $\{Tx^*_\alpha\}$ has a norm convergent subnet. In view of $(\star)$, every subnet of the net $\{Tx^*_\alpha\}$ has a norm convergent subnet to T^*x^*. This implies that $\|T^*x^*_\alpha - T^*x^*\| \to 0$, and so $T^*\colon U^* \to X^*$ is completely continuous.

(b) Let $\{f_n\} \subseteq \mathcal{K}(U^*, X^*)$ satisfy $\|f_n - f\| \to 0$ for some $f \in C(U^*, X^*)$, and let $\epsilon > 0$. Pick some k such that $\|f_n - f\| < \epsilon$ for all $n \ge k$.

Now assume that a net $\{x^*_\alpha\} \subseteq U^*$ satisfies $x^*_\alpha \xrightarrow{w^*} x^*$ in U^*. Since f_k belongs to $\mathcal{K}(U^*, X^*)$, there exists some index α_0 such that $\big\|f_k(x^*_\alpha) - f_k(x^*)\big\| < \epsilon$ for all

$\alpha \geq \alpha_0$. Consequently, for $\alpha \geq \alpha_0$ we have

$$\begin{aligned}\|f(x^*_\alpha) - f(x^*)\| &\leq \|f(x^*_\alpha) - f_k(x^*_\alpha)\| + \|f_k(x^*_\alpha) - f_k(x^*)\| + \|f_k(x^*) - f(x^*)\| \\ &< \epsilon + \epsilon + \epsilon = 3\epsilon\,.\end{aligned}$$

This shows that $\|f(x^*_\alpha) - f(x^*)\| \to 0$, and so $f \in \mathcal{K}(U^*, X^*)$. Therefore, $\mathcal{K}(U^*, X^*)$ is a closed subspace of $C(U^*, X^*)$. ∎

Problem 10.7.3. *Assume that for a subalgebra $\mathcal{A}$ of $\mathcal{L}(X)$ there exists a non-zero operator $T_0 \in \mathcal{L}(X)$ such that:*

(a) *T_0 commutes with each operator in $\mathcal{A}$.*

(b) *For some $B \in \mathcal{L}(X)$ its adjoint operator B^* is not in the τ_s-closure of the vector subspace spanned by $\{\alpha T^* T_0^*\colon\ \alpha \in C(U^*)$ and $T \in \mathcal{A}\}$ in $C(U^*, X^*)$.*

Show that $\mathcal{A}^*$ is non-transitive.

Solution: By Lemma 10.73, there exist vectors $x^{**} \in X^{**}$ and $x^* \in U^*$ satisfying

$$\langle x^{**} \otimes x^*, B^* \rangle = x^{**}(B^* x^*) = 1$$

and

$$\langle x^{**} \otimes x^*, \alpha T^* T_0^* \rangle = \langle \alpha(x^*) T^* T_0^* x^*, x^{**} \rangle = \alpha(x^*) \langle T^* T_0^* x^*, x^{**} \rangle = 0$$

for all $T \in \mathcal{A}$ and all $\alpha \in C(U^*)$. Therefore, $\langle T^* T_0^* x^*, x^{**} \rangle = 0$ for each $T \in \mathcal{A}$. If $y^* = T_0^* x^* \neq 0$, then the relation $\langle T^* y^*, x^{**} \rangle = 0$ for each $T \in \mathcal{A}$ coupled with Theorem 10.12 shows that $\mathcal{A}^*$ admits a non-trivial $\mathcal{A}^*$-invariant closed subspace.

Now assume that $T_0^* x^* = 0$, and let N denote the null space of T_0^*. Since T_0 commutes with $\mathcal{A}$, it follows that N is a closed $\mathcal{A}^*$-invariant subspace. From $x^{**}(B^* x^*) = 1$, we infer that $x^* \neq 0$. Since $x^* \in N$, we see that N is non-zero. On the other hand, if $N = X^*$, then $T_0^* = 0$, and so $T_0 = 0$, a contradiction. Therefore, in this case, N must be a non-trivial closed $\mathcal{A}^*$-invariant subspace. ∎

Problem 10.7.4. *Recall that a Banach space X is said to have the **Schur property** if $x_n \xrightarrow{w} 0$ implies $\|x_n\| \to 0$.*

If X is a Banach space with the Schur property and $\mathcal{A}$ is a commutative subalgebra of $\mathcal{L}(X)$, then show that the dual algebra $\mathcal{A}^*$ of $\mathcal{L}(X^*)$ is non-transitive.

Solution: If $\mathcal{A}$ consists entirely of scalar operators, then the conclusion is trivial. So, we can assume that there exists some non-scalar operator $T \in \mathcal{A}$. Fix some λ_0 in the approximate point spectrum of T and consider the operator $T_0 = T - \lambda_0 I$. Also, select a sequence $\{x_n\}$ of unit vectors of X satisfying

$$\|T_0 x_n\| = \|T x_n - \lambda_0 x_n\| \to 0\,.$$

Clearly, T_0 commutes with $\mathcal{A}$. Next, consider the subspace $\mathcal{M}$ of $C(U^*, X^*)$ generated by the collection $\{\alpha T^* T_0^*\colon\ \alpha \in C(U^*)$ and $T \in \mathcal{A}\}$. We claim that the restriction to U^* of the identity operator $I^*\colon X^* \to X^*$ does not belong to the τ_s-closure of $\mathcal{M}$.

To see this, assume by way of contradiction that I^* belongs to the τ_s-closure of $\mathcal{M}$. Therefore, if $\epsilon > 0$ and $x^* \in U^*$ are fixed, then there exist $\alpha_i \in C(U^*)$ and $T_i \in \mathcal{A}$ $(i = 1, \ldots, m)$ satisfying $\big\|\big[\, I^* - \sum_{i=1}^{m} \alpha_i T_i^* T_0^* \,\big](x^*)\big\| < \epsilon$. This implies

$$\Big|\langle x, x^* \rangle - \sum_{i=1}^{m} \alpha_i(x^*)\big\langle x, T_i^* T_0^*(x^*)\big\rangle\Big| < \epsilon \tag{$\star$}$$

for all $x \in X$ with $\|x\| \leq 1$. Taking into account that $\mathcal{A}$ is a commutative algebra, it follows from ($\star$) that $\big|\langle x, x^* \rangle - \sum_{i=1}^{m} \alpha_i(x^*)\langle T_0 x, T_i^* x^* \rangle\big| < \epsilon$ for each $x \in X$ with $\|x\| \leq 1$. In particular, we have $\big|\langle x_n, x^* \rangle - \sum_{i=1}^{m} \alpha_i(x^*)\langle T_0 x_n, T_i^* x^* \rangle\big| < \epsilon$ for each n, which, in view of ($\star\star$), yields $|\langle x_n, x^* \rangle| < \epsilon$ for all sufficiently large n. In other words, the sequence $\{x_n\}$ converges weakly to zero in X. Since X has the Schur property, it follows that $\|x_n\| \to 0$, contrary to the assumption that $\|x_n\| = 1$ for each n. Therefore, the identity operator I^* does not belong to the τ_s-closure of $\mathcal{M}$ and the conclusion follows from Problem 10.7.3. ∎

Problem 10.7.5 (Honor [**38**]). *If $\mathcal{A}$ is a commutative algebra of bounded operators on ℓ_1, then show that the dual algebra $\mathcal{A}^*$ of operators on ℓ_∞ is non-transitive.*

Solution: This follows from Problem 10.7.4 by recalling that ℓ_1 has the Schur property; see [**6**, Theorem 13.1, p. 200]. ∎

Chapter 11

The Daugavet Equation

11.1. The Daugavet Equation and Uniform Convexity

Problem 11.1.1. *Show that if the vectors $x_1, \dots, x_n$ in a normed space satisfy*

$$\Big\|\sum_{k=1}^{n} x_k\Big\| = \sum_{k=1}^{n} \|x_k\|\,,$$

then $\big\|\sum_{k=1}^{n} \alpha_k x_k\big\| = \sum_{k=1}^{n} \alpha_k \|x_k\|$ holds for each choice of non-negative scalars $\alpha_1, \dots, \alpha_n$.

Solution: The proof is by induction. For $n = 1$, the claim is trivially true. So, assume the claim is true for some n and let the vectors $x_1, \dots, x_n, x_{n+1}$ satisfy $\|\sum_{k=1}^{n+1} x_k\| = \sum_{k=1}^{n+1} \|x_k\|$. Also, let $\alpha_1, \dots, \alpha_n, \alpha_{n+1}$ be non-negative real numbers; we can suppose $\alpha_1 \geq \alpha_k$ for each k. From

$$\sum_{k=1}^{n+1} \|x_k\| = \Big\|\sum_{k=1}^{n+1} x_k\Big\| \leq \|x_1\| + \Big\|\sum_{k=2}^{n+1} x_k\Big\| \leq \sum_{k=1}^{n+1} \|x_k\|\,,$$

it follows that $\|\sum_{k=2}^{n+1} x_k\| = \sum_{k=2}^{n+1} \|x_k\|$. Now notice that the identity

$$\sum_{k=1}^{n+1} \alpha_k x_k = \alpha_1\Big(\sum_{k=1}^{n+1} x_k\Big) - \sum_{k=2}^{n+1} (\alpha_1 - \alpha_k) x_k\,,$$

in conjunction with the induction hypothesis, yields

$$\begin{aligned}\Big\|\sum_{k=1}^{n+1}\alpha_k x_k\Big\| &\geq \alpha_1\Big\|\sum_{k=1}^{n+1}x_k\Big\| - \Big\|\sum_{k=2}^{n+1}(\alpha_1-\alpha_k)x_k\Big\| \\ &= \alpha_1\Big(\sum_{k=1}^{n+1}\|x_k\|\Big) - \sum_{k=2}^{n+1}(\alpha_1-\alpha_k)\|x_k\| \\ &= \sum_{k=1}^{n+1}\alpha_k\|x_k\| \geq \Big\|\sum_{k=1}^{n+1}\alpha_k x_k\Big\|.\end{aligned}$$

Therefore, $\big\|\sum_{k=1}^{n+1}\alpha_k x_k\big\| = \sum_{k=1}^{n+1}\alpha_k\|x_k\|$, and the solution is finished. ∎

Problem 11.1.2. *This problem presents another version of Lemma 11.4. Let X be a normed space and assume that a vector $u \in X$ and some unit vector v satisfy $\|u+v\| \geq 2-\epsilon$ for some $\epsilon > 0$. Show that for all $\alpha, \beta \in \mathbb{R}_+$ we have*

$$\|\alpha u + \beta v\| \geq (1-\epsilon)(\alpha+\beta).$$

Solution: Assume first that $\alpha+\beta = 1$. Without loss of generality we can suppose that $\alpha \geq \beta$. Then we have

$$\begin{aligned}\|\alpha u + \beta v\| &\geq \|\alpha(u+v)\| - \|(\alpha-\beta)v\| \geq \alpha(2-\epsilon) - (\alpha-\beta) \\ &= \alpha+\beta-\alpha\epsilon \geq 1-\epsilon.\end{aligned}$$

Therefore, using the preceding case, we see that for all $\alpha, \beta \in \mathbb{R}_+$ we have

$$\|\alpha u + \beta v\| = (\alpha+\beta)\big\|\tfrac{\alpha}{\alpha+\beta}u + \tfrac{\beta}{\alpha+\beta}v\big\| \geq (\alpha+\beta)(1-\epsilon),$$

as claimed. ∎

Problem 11.1.3. *Let V be a vector subspace of a normed space X, and assume that for a vector $u \in X$ and some $\epsilon > 0$ we have $\|u+v\| \geq 2-\epsilon$ for each $v \in V$ with $\|v\| = 1$. Show that for each $v \in V$ and all scalars λ we also have $\|\lambda u + v\| \geq (1-\epsilon)\big[|\lambda| + \|v\|\big]$.*

Solution: Let $v \in V$ be a non-zero vector. If $\lambda \geq 0$, then Problem 11.1.2 implies

$$\|\lambda u + v\| = \big\|\lambda u + \|v\|\tfrac{v}{\|v\|}\big\| \geq (1-\epsilon)\big[|\lambda| + \|v\|\big].$$

If $\lambda < 0$, then

$$\|\lambda u + v\| = \big\||\lambda|u - v\big\| \geq (1-\epsilon)\big[|\lambda| + \|v\|\big],$$

and the solution is finished. ∎

Problem 11.1.4 (C.-S. Lin [**50**]). *Assume that two sequences $\{u_n\}$ and $\{v_n\}$ in a normed space satisfy $\lim_{n\to\infty}[\|u_n+v_n\| - (\|u_n\| + \|v_n\|)] = 0$. Show that $\lim_{n\to\infty}\big[\,\|\alpha u_n + \beta v_n\| - (\alpha\|u_n\| + \beta\|v_n\|)\,\big] = 0$ for all $\alpha \geq 0$ and $\beta \geq 0$.*

Solution: Without loss of generality we can assume that $\alpha \geq \beta$. Then note that

$$\begin{aligned} 0 &\geq \|\alpha u_n + \beta v_n\| - (\alpha\|u_n\| + \beta\|v_n\|) \\ &= \|\alpha u_n + \alpha v_n - (\alpha-\beta)v_n\| - (\alpha\|u_n\| + \beta\|v_n\|) \\ &\geq \alpha\|u_n + v_n\| - (\alpha-\beta)\|v_n\| - (\alpha\|u_n\| + \beta\|v_n\|) \\ &= \alpha\|u_n + v_n\| - \alpha(\|u_n\| + \|v_n\|)\,. \end{aligned}$$

That is, we have established that

$$\alpha\big[\,\|u_n + v_n\| - (\|u_n\| + \|v_n\|)\,\big] \leq \|\alpha u_n + \beta v_n\| - (\alpha\|u_n\| + \beta\|v_n\|) \leq 0$$

holds for each n. This implies $\lim_{n\to\infty}\big[\,\|\alpha u_n + \beta v_n\| - (\alpha\|u_n\| + \beta\|v_n\|)\,\big] = 0$. ∎

Problem 11.1.5 (C.-S. Lin [**50**]). *Generalize Problem* 11.1.4 *as follows. Assume that m sequences $\{u_n^k\}$ $(k = 1, 2, \ldots, m)$ in a normed space satisfy*

$$\lim_{n\to\infty}\Big[\Big\|\sum_{k=1}^m u_n^k\Big\| - \sum_{k=1}^m \|u_n^k\|\Big] = 0\,.$$

Then show that $\lim_{n\to\infty}\big[\,\|\sum_{k=1}^m \alpha_k u_n^k\| - \sum_{k=1}^m \alpha_k\|u_n^k\|\,\big] = 0$ *for any choice of non-negative scalars* $\alpha_1, \ldots, \alpha_m$.

Solution: Assume that m bounded sequences $\{u_n^k\}$ $(k = 1, 2, \ldots, m)$ in a normed space satisfy

$$\lim_{n\to\infty}\Big[\Big\|\sum_{k=1}^m u_n^k\Big\| - \sum_{k=1}^m \|u_n^k\|\Big] = 0\,,$$

and let $\alpha_1, \alpha_2, \ldots, \alpha_m$ be m non-negative scalars. Without loss of generality, we can suppose that $\alpha_1 \geq \alpha_k$ for each k. Now note that

$$\begin{aligned} 0 &\geq \Big\|\sum_{k=1}^m \alpha_k u_n^k\Big\| - \sum_{k=1}^m \alpha_k\|u_n^k\| \\ &= \Big\|\alpha_1\Big(\sum_{k=1}^m u_n^k\Big) - \sum_{k=1}^m(\alpha_1 - \alpha_k)u_n^k\Big\| - \sum_{k=1}^m \alpha_k\|u_n^k\| \\ &\geq \alpha_1\Big\|\sum_{k=1}^m u_n^k\Big\| - \sum_{k=1}^m(\alpha_1 - \alpha_k)\|u_n^k\| - \sum_{k=1}^m \alpha_k\|u_n^k\| \\ &= \alpha_1\Big[\Big\|\sum_{k=1}^m u_n^k\Big\| - \sum_{k=1}^m \|u_n^k\|\Big]\,. \end{aligned}$$

Therefore, we have shown that

$$\alpha_1\Big[\Big\|\sum_{k=1}^m u_n^k\Big\| - \sum_{k=1}^m \|u_n^k\|\Big] \leq \Big\|\sum_{k=1}^m \alpha_k u_n^k\Big\| - \sum_{k=1}^m \alpha_k\|u_n^k\| \leq 0$$

holds for each n. This implies $\lim_{n\to\infty}\big[\,\|\sum_{k=1}^m \alpha_k u_n^k\| - \sum_{k=1}^m \alpha_k\|u_n^k\|\,\big] = 0$. ∎

Problem 11.1.6 (C.-S. Lin [**50**]). *Generalize Theorem* 11.10 *by showing that for a uniformly convex Banach space X and two operators $S, T \in \mathcal{L}(X)$ the following statements are equivalent.*

(a) $\|I+S+T\| = 1+\|S\|+\|T\|$.

(b) *There exists a sequence* $\{x_n\}$ *of unit vectors such that*

$$\lim_{n\to\infty} \|(I+S+T)x_n\| = 1+\|S\|+\|T\|.$$

Solution: (a) $\Longrightarrow$ (b) Assume that $\|I+S+T\| = 1+\|S\|+\|T\|$. From the inequalities

$$1+\|S\|+\|T\| = \|I+S+T\| \le 1+\|S+T\| \le 1+\|S\|+\|T\|,$$

it follows that $\|I+S+T\| = 1+\|S+T\|$. This implies that $S+T$ satisfies the Daugavet equation. So, by Theorem 11.10, $\|S+T\| \in \sigma_a(S+T)$. This means that there exists a sequence $\{x_n\}$ of unit vectors such that $\big\|(S+T)x_n - \|S+T\|x_n\big\| \to 0$. Now note that

$$\begin{aligned}\|I+S+T\| &\ge \|(I+T+S)x_n\| \\ &\ge \big\|x_n+\|S+T\|x_n\big\| - \big\|(S+T)x_n - \|S+T\|x_n\big\| \\ &= 1+\|S+T\| - \big\|(S+T)x_n - \|S+T\|x_n\big\|.\end{aligned}$$

Since $1+\|S+T\| - \big\|(S+T)x_n - \|S+T\|x_n\big\| \to 1+\|S+T\|$, it easily follows from the above inequalities that $\lim_{n\to\infty}\|(I+S+T)x_n\| = 1+\|S+T\| = \|I+S+T\|$.

(b) $\Longrightarrow$ (a) Assume that a sequence $\{x_n\}$ of unit vectors satisfies

$$\lim_{n\to\infty} \|(I+S+T)x_n\| = 1+\|S\|+\|T\|.$$

Letting $n\to\infty$ and using the inequalities

$$1+\|S\|+\|T\| \ge 1+\|S+T\| \ge \|I+S+T\| \ge \|(I+S+T)x_n\|,$$

we get $1+\|S\|+\|T\| \ge 1+\|S+T\| \ge \|I+S+T\| \ge 1+\|S\|+\|T\|$. This implies $\|I+S+T\| = 1+\|S\|+\|T\|$. ∎

Problem 11.1.7 (C.-S. Lin [**50**]). *If* $T\colon X\to X$ *is an invertible operator on a uniformly convex or uniformly smooth Banach space, then show that* $\|I-2T\|=1$ *implies* $\|I-T\|<1$.

Solution: Assume that $\|I-2T\|=1$. We claim that $1=\|I-2T\| \notin \sigma(I-2T)$. Indeed, if we had $1\in\sigma(I-2T)$, then the operator $I-(I-2T)=2T$ would not be invertible, contradicting our assumption that T is invertible.

Since the Banach space X is either uniformly convex or uniformly smooth, it follows from Theorem 11.10 that the operator $I-2T$ does not satisfy the Daugavet equation. Therefore,

$$2\|I-T\| = \|2I-2T\| = \|I+(I-2T)\| < 1+\|I-2T\| = 2.$$

This implies $\|I-T\|<1$, as desired. ∎

Problem 11.1.8. *Let* $T\colon X\to X$ *be a non-zero bounded operator on a uniformly convex or uniformly smooth Banach space. Show that* $\|T\|I-T$ *is non-invertible if and only if* T *satisfies the Daugavet equation.*

Solution: Notice that $\|T\|I - T$ is non-invertible if and only if $\|T\| \in \sigma(T)$. Since X is either a uniformly convex or a uniformly smooth Banach space, it follows from Theorem 11.10 that the latter is equivalent to saying that the operator T satisfies the Daugavet equation. ∎

Problem 11.1.9. *If a continuous operator $T\colon X \to X$ on a uniformly convex or uniformly smooth Banach space satisfies the Daugavet equation, then show that for each $\alpha \geq 0$ the continuous operator $e^{\alpha T}$ satisfies the Daugavet equation and that $\|e^{\alpha T}\| = e^{\alpha\|T\|}$ holds.*

Solution: Fix $\alpha > 0$ and consider the formal series $f(\lambda) = e^{\lambda} = \sum_{n=0}^{\infty} \frac{1}{n!}\lambda^n$. Clearly, this series has non-negative coefficients and $f(\|\alpha T\|) = e^{\alpha\|T\|}$. The desired conclusion now follows from Theorem 11.14. ∎

Problem 11.1.10. *Let (Ω, Σ, μ) be an arbitrary σ-finite measure space. A linear operator $T\colon L_2(\mu) \to L_2(\mu)$ is called* ***symmetric*** *if*

$$\langle Tf, g\rangle = \langle f, Tg\rangle$$

for all $f, g \in L_2(\mu)$, where $\langle\cdot,\cdot\rangle$ denotes the usual inner product in $L_2(\mu)$ defined by $\langle f, g\rangle = \int_\Omega f(\omega)g(\omega)\,d\mu(\omega)$. Establish the following.

(a) *Symmetric operators are bounded.*

(b) *An integral operator T on $L_2(\mu)$ with a kernel $K \in L_2(\mu \times \mu)$ is symmetric if and only if $K(s,t) = K(t,s)$ holds for $\mu \times \mu$-almost all $(s,t) \in \Omega \times \Omega$.*

(c) *The integral operator $T\colon L_2[0,1] \to L_2[0,1]$, defined by*

$$Tx(t) = \int_0^1 \sin(st)x(s)\,ds\,,$$

is symmetric.

(d) *Every symmetric operator $T\colon L_2(\mu) \to L_2(\mu)$ satisfies $r(T) = \|T\|$.*

(e) *Every symmetric positive operator $T\colon L_2(\mu) \to L_2(\mu)$ satisfies the Daugavet equation, i.e., $\|I + T\| = 1 + \|T\|$.*

Solution: (a) Let $T\colon L_2(\mu) \to L_2(\mu)$ be an arbitrary symmetric operator. If $f \in L_2(\mu) = L_2^*(\mu)$, then for each $g \in U$ (the closed unit ball of $L_2(\mu)$) we have

$$\bigl|\langle Tg, f\rangle\bigr| = \bigl|\langle g, Tf\rangle\bigr| \leq \|g\| \cdot \|Tf\| \leq \|Tf\|\,.$$

This shows that the set $T(U)$ is a weakly bounded subset of $L_2(\mu)$, and hence $T(U)$ is also a norm bounded subset of $L_2(\mu)$. Therefore, T is a bounded operator.

(b) Assume first that $K(s,t) = K(t,s)$ for $\mu \times \mu$-almost all $(s,t) \in \Omega \times \Omega$. If $f, g \in L_2(\mu)$, then using Fubini's theorem, we see that

$$\begin{aligned} \langle Tf, g\rangle &= \int_\Omega \Big[\int_\Omega K(s,t)f(t)\,d\mu(t)\Big] g(s)\,d\mu(s) \\ &= \iint_{\Omega\times\Omega} K(s,t)f(t)g(s)\,d(\mu\times\mu)(s,t) \\ &= \int_\Omega \Big[\int_\Omega K(t,s)g(s)\,d\mu(s)\Big] f(t)\,d\mu(t) \\ &= \int_\Omega f(t)\Big[\int_\Omega K(t,s)g(s)\,d\mu(s)\Big]\,d\mu(t) = \langle f, Tg\rangle\,. \end{aligned}$$

For the converse, assume that $\langle Tf, g\rangle = \langle f, Tg\rangle$ holds for all $f, g \in L_2(\mu)$. Define the function $F \in L_2(\mu\times\mu)$ by $F(s,t) = K(s,t) - K(t,s)$. If $A, B \in \Sigma$ have finite measure, then the functions χ_A and χ_B belong to $L_2(\mu)$ and satisfy

$$\begin{aligned} &\iint_{\Omega\times\Omega} F(s,t)\chi_A(s)\chi_B(t)\,d(\mu\times\mu)(s,t) \\ &\quad= \iint_{\Omega\times\Omega} K(s,t)\chi_A(s)\chi_B(t)\,d(\mu\times\mu)(s,t) \\ &\qquad - \iint_{\Omega\times\Omega} K(t,s)\chi_A(s)\chi_B(t)\,d(\mu\times\mu)(s,t) \\ &\quad= \int_\Omega \Big[\int_\Omega K(s,t)\chi_B(t)\,d\mu(t)\Big]\chi_A(s)\,d\mu(s) \\ &\qquad - \int_\Omega \Big[\int_\Omega K(t,s)\chi_A(s)\,d\mu(s)\Big]\chi_B(t)\,d\mu(t) \\ &\quad= \langle T\chi_B, \chi_A\rangle - \langle \chi_B, T\chi_A\rangle = 0\,. \end{aligned}$$

This implies that for each $\mu\times\mu$-step function $\phi = \sum_{i=1}^n a_i\chi_{A_i\times B_i}$ we have

$$\iint_{\Omega\times\Omega} F(s,t)\phi(s,t)\,d(\mu\times\mu)(s,t) = 0\,. \qquad (\star)$$

Since F^+ belongs to $L_2(\mu)$, there exists a sequence of $\mu\times\mu$-step functions $\{\phi_n\}$ such that $0 \le \phi_n \uparrow F^+$. This implies $|\phi_n F| \le F^+F = (F^+)^2 \in L_1(\mu)$ and $\phi_n F \to (F^+)^2$ $\mu\times\mu$-a.e. So, using $(\star)$ and the Lebesgue Dominated Convergence Theorem, we get

$$0 = \iint_{\Omega\times\Omega} F(s,t)\phi_n(s,t)\,d(\mu\times\mu)(s,t) \longrightarrow \iint_{\Omega\times\Omega} (F^+)^2\,d(\mu\times\mu)\,.$$

Hence, $\iint_{\Omega\times\Omega}(F^+)^2\,d(\mu\times\mu) = 0$, and so $F^+ = 0$ $\mu\times\mu$-a.e. Similarly, $F^- = 0$ $\mu\times\mu$-a.e. Therefore, $F = F^+ - F^- = 0$. This shows that $K(s,t) = K(t,s)$ holds for $\mu\times\mu$-almost all $(s,t) \in \Omega\times\Omega$.

(c) Use (a) and the obvious equality $\sin(st) = \sin(ts)$ valid for all $s, t \in [0,1]$.

(d) Let $T\colon L_2(\mu) \to L_2(\mu)$ be a symmetric operator. Then for each $f \in L_2(\mu)$ we have $\|Tf\|^2 = \langle Tf, Tf\rangle = \langle T^2f, f\rangle \le \|T^2\|\cdot\|f\|^2$. So, $\|T\|^2 \le \|T^2\| \le \|T\|^2$, whence $\|T^2\| = \|T\|^2$.

Since T^k is also a symmetric operator for each k, an inductive argument guarantees that $\|T^{2^n}\| = \|T\|^{2^n}$ for each n. Hence, $r(T) = \lim_{n\to\infty} \|T^{2^n}\|^{\frac{1}{2^n}} = \|T\|$.

(e) To see that every symmetric positive operator satisfies the Daugavet equation, we need to observe that the spectral radius $r(T) = \|T\|$ belongs to the spectrum of T (see Theorem 7.9) and that $L_2(\mu)$ is uniformly convex. Now invoke Theorem 11.10. ∎

Problem 11.1.11. *Let $\{x_n\}$ and $\{y_n\}$ be two sequences of unit vectors in a uniformly convex Banach space X, and let $\{\phi_n\}$ be a sequence of unit vectors in X^*. If $\phi_n(x_n) \to 1$ and $\phi_n(y_n) \to 1$, then show that $\|x_n - y_n\| \to 0$.*

Solution: It follows from the hypotheses that $\phi_n(x_n + y_n) \to 2$. Hence, from $\phi_n(x_n+y_n) \le \|x_n+y_n\| \le 2$ we get $\|x_n+y_n\| \to 2$. Now the uniform convexity of X guarantees that $\|x_n - y_n\| \to 0$. ∎

Problem 11.1.12 (Milman [**57**]). *Show that uniformly convex Banach spaces are reflexive.*

Solution: The arguments below are due to B. J. Pettis [**64**]. Let X be a uniformly convex Banach space. Fix some $x^{**} \in X^{**}$ such that $\|x^{**}\| = 1$. Since the closed unit ball U of X is $\sigma(X^{**}, X^*)$-dense in the closed unit ball U^{**} of X^{**}, it follows that there exists a net $\{x_\alpha\}_{\alpha\in A}$ in U such that $x_\alpha \xrightarrow{\sigma(X^{**},X^*)} x^{**}$. Consider the net $\{x_\alpha + x_\beta\}_{(\alpha,\beta)\in A\times A}$ and notice that $x_\alpha + x_\beta \xrightarrow[\alpha,\beta]{\sigma(X^{**},X^*)} 2x^{**}$ in X^{**}. Moreover, from $\|x_\alpha + x_\beta\| \le 2$ for all $(\alpha,\beta) \in A\times A$, we see that $\limsup_{\alpha,\beta} \|x_\alpha + x_\beta\| \le 2$. On the other hand, by Problem 1.1.17, the norm function $y^{**} \mapsto \|y^{**}\|$ is $\sigma(X^{**}, X^*)$-norm lower semicontinuous. This implies $2 = \|2x^{**}\| \le \liminf_{\alpha,\beta} \|x_\alpha + x_\beta\|$. The above properties show that $\lim_{\alpha,\beta} \|x_\alpha + x_\beta\| = 2$.

Now use the fact that X is a uniformly convex Banach space to infer that $\|x_\alpha - x_\beta\| \xrightarrow[\alpha,\beta]{} 0$. (See Lemma 11.6.) This shows that the net $\{x_\alpha\}$ is a norm Cauchy net. If $x \in X$ is its norm limit, then it should be immediate that $x^{**} = x \in X$. Thus, $X^{**} = X$ holds true, and so X is a reflexive Banach space.

There is another way of proving this result using Problem 11.1.11 and the following famous theorem of R. C. James [**39**]:

- *A non-empty weakly closed bounded subset of a Banach space is weakly compact if and only if every continuous linear functional attains a maximum on the set.*

So, according to James' theorem it suffices to show that every continuous linear functional ϕ on X attains a maximum on the closed unit ball of X, i.e., it attains its norm. To see this, assume that $\|\phi\| = 1$. Pick a sequence $\{x_n\}$ of unit vectors such that $\phi(x_n) \to 1$. We claim that $\{x_n\}$ is a Cauchy sequence. Indeed, if $\{x_n\}$ is not a norm Cauchy sequence, then there exist a subsequence $\{y_n\}$ of $\{x_n\}$ and some $\epsilon > 0$ such that $\|y_{n+1} - y_n\| > \epsilon$ holds for each n. Now notice that $\phi(y_{n+1}) \to 1$ and $\phi(y_n) \to 1$. But then Problem 11.1.11 guarantees that $\|y_{n+1} - y_n\| \to 0$, which is impossible. This contradiction shows that $\{x_n\}$ is a Cauchy sequence. If $x_n \to x$, then $\|x\| = 1$ and $\phi(x) = 1 = \|\phi\|$, as desired. ∎

Problem 11.1.13. *Recall that a Banach space X is said to be* ***uniformly smooth*** *if for each $\epsilon > 0$ there exists some $\delta = \delta(\epsilon) > 0$ such that if $\|x\| \geq 1$, $\|y\| \geq 1$ and $\|x - y\| < \delta$, then*

$$\|x + y\| \geq \|x\| + \|y\| - \epsilon\|x - y\| . \tag{$\star$}$$

Show that a Banach space X is uniformly smooth if and only if for each $\epsilon > 0$ there exists some $\gamma > 0$ such that for each unit vector u and for each $v \in X$ with $\|v\| \leq \gamma$ we have

$$\|u + v\| + \|u - v\| \leq 2 + \epsilon\|v\| . \tag{$\star\star$}$$

Solution: Assume first that X is uniformly smooth, and let $\epsilon > 0$. Fix some $\delta > 0$ so that $\|x\| \geq 1$, $\|y\| \geq 1$ and $\|x - y\| < \delta$ imply

$$\|x + y\| \geq \|x\| + \|y\| - \tfrac{\epsilon}{2}\|x - y\| . \tag{$\dagger$}$$

Next, choose some $0 < \gamma < 1$ such that $\frac{2\gamma}{1-\gamma} < \delta$. Now take any unit vector $u \in X$ and any v such that $\|v\| < \gamma$. Let $a = \frac{u+v}{1-\gamma}$ and $b = \frac{u-v}{1-\gamma}$. Clearly, $\|a\| \geq 1$, $\|b\| \geq 1$ and $\|a - b\| = \frac{2}{1-\gamma}\|v\| < \frac{2\gamma}{1-\gamma} < \delta$, and so ($\dagger$) yields

$$\|a + b\| \geq \|a\| + \|b\| - \tfrac{\epsilon}{2}\|a - b\| . \tag{$\dagger\dagger$}$$

Substituting the values of a and b in ($\dagger\dagger$), we get:

$$\tfrac{1}{1-\gamma}\|2u\| \geq \tfrac{1}{1-\gamma}\|u + v\| + \tfrac{1}{1-\gamma}\|u - v\| - \tfrac{\epsilon}{2(1-\gamma)}\|2v\| .$$

This implies $2\|u\| \geq \|u + v\| + \|u - v\| - \epsilon\|v\|$ or

$$\|u + v\| + \|u - v\| \leq 2 + \epsilon\|v\| .$$

That is, the validity ($\star\star$) has been established.

For the converse, suppose that X satisfies ($\star\star$). Fix $\epsilon > 0$ and then choose some $0 < \delta < 1$ such that $\frac{\delta}{2-\delta} \leq \gamma$. Choose any $x, y \in X$ such that $\|x\| \geq 1$, $\|y\| \geq 1$ and $\|x - y\| < \delta$. Clearly, $\|x + y\| = \|2x - (x - y)\| \geq 2 - \delta > 0$. Now consider the vectors $u = \frac{x+y}{\|x+y\|}$ and $v = \frac{x-y}{\|x+y\|}$. Then $\|u\| = 1$ and $\|v\| = \frac{\|x-y\|}{\|x+y\|} < \frac{\delta}{2-\delta} \leq \gamma$. Therefore, ($\star\star$) applies and yields

$$\|u + v\| + \|u - v\| \leq 2 + \epsilon\|v\| .$$

Taking into consideration that $u + v = \frac{2x}{\|x+y\|}$ and $u - v = \frac{2y}{\|x+y\|}$, we get

$$\tfrac{2}{\|x+y\|}\|x\| + \tfrac{2}{\|x+y\|}\|y\| \leq 2 + \epsilon\tfrac{1}{\|x+y\|}\|x - y\| .$$

Consequently,

$$\|x + y\| \geq \|x\| + \|y\| - \tfrac{\epsilon}{2}\|x - y\| \geq \|x\| + \|y\| - \epsilon\|x - y\| ,$$

which is ($\star$), and the solution is finished. ∎

Problem 11.1.14 (Šmulian [**74**]). *For a Banach space X establish the following duality properties:*

(a) *X is uniformly convex if and only if its dual X^* is uniformly smooth.*

(b) *X is uniformly smooth if and only if its dual X^* is uniformly convex.*

Solution: We shall establish the following two properties:

(α) *If a Banach space X is uniformly convex, then X^* is uniformly smooth.*

(β) *If a Banach space X is uniformly smooth, then X^* is uniformly convex.*

Notice that if (α) and (β) have been established, then the desired conclusions follow. For instance, if X^* is uniformly smooth, then X^{**} is (by (β)) uniformly convex. Since X is a closed subspace of X^{**}, the latter implies that X is likewise uniformly convex. Similarly, if X^* is uniformly convex, then X must be uniformly smooth. We now proceed to establish (α) and (β).

(α) Let X be a uniformly convex Banach space and assume by way of contradiction that X^* is not uniformly smooth. Then, according to Problem 11.1.13, there exist some $\epsilon > 0$ and two sequences $\{\phi_n\}$ and $\{\psi_n\}$ in X^* such that $\|\psi_n\| \to 0$, $\|\phi_n\| = 1$, and

$$\|\phi_n + \psi_n\| + \|\phi_n - \psi_n\| > 2 + \epsilon\|\psi_n\| \qquad (\star)$$

for each n. Since X is reflexive (see Problem 11.1.12), we can find unit vectors x_n and y_n in X such that

$$[\phi_n + \psi_n](x_n) = \|\phi_n + \psi_n\| \quad \text{and} \quad [\phi_n - \psi_n](y_n) = \|\phi_n - \psi_n\|\,.$$

It is easy to check that $\big|\|\phi_n + \psi_n\| - 1\big| \le \|\psi_n\|$, i.e., $\big|[\phi_n + \psi_n](x_n) - 1\big| \le \|\psi_n\|$. Therefore, $\big|\phi_n(x_n) - 1\big| \le 2\|\psi_n\|$. Similarly, $\big|\phi_n(y_n) - 1\big| \le 2\|\psi_n\|$. Since $\|\psi_n\| \to 0$, we get $\phi_n(x_n) \to 1$ and $\phi_n(y_n) \to 1$. But then the uniform convexity of X guarantees (see Problem 11.1.11) that $\|x_n - y_n\| < \epsilon$ for all $n \ge n_0$. Now for any $n \ge n_0$ we have

$$\begin{aligned}\|\phi_n + \psi_n\| + \|\phi_n - \psi_n\| &= [\phi_n + \psi_n](x_n) + [\phi_n - \psi_n](y_n)\\ &= [\phi_n(x_n) + \phi_n(y_n)] + \psi_n(x_n - y_n)\\ &\le 2 + \|\psi_n\| \cdot \|x_n - y_n\|\\ &\le 2 + \epsilon\|\psi_n\|\,,\end{aligned}$$

which contradicts ($\star$). Hence, X^* must be uniformly smooth.

(β) Assume that X is uniformly smooth and let $\epsilon > 0$. According to Problem 11.1.13 there exists some $0 < \gamma < 1$ such that for each unit vector u and for all $v \in X$ with $\|v\| \le \gamma$ we have

$$\|u + v\| + \|u - v\| \le 2 + \tfrac{\epsilon}{4}\|v\|\,. \qquad (\star\star)$$

Now suppose that two linear functionals $x^*, y^* \in U_{X^*}$ satisfy $\|x^* - y^*\| > \epsilon$. Pick some unit vector $y \in X$ such that $(x^* - y^*)(y) > \frac{\epsilon}{2}$ and let $z = \gamma y$. Clearly,

$\|z\| = \gamma$ and $(x^* - y^*)(z) > \frac{\epsilon\gamma}{2}$. Finally, using $(\star\star)$, we obtain that

$$\begin{aligned} \|x^* + y^*\| &= \sup_{\|x\|=1} \big[x^* + y^*\big](x) \\ &= \sup_{\|x\|=1} \big[x^*(x+z) + y^*(x-z) - \big(x^*(z) - y^*(z)\big)\big] \\ &\le \sup_{\|x\|=1} \big[\,\|x+z\| + \|x-z\| - \tfrac{\epsilon\gamma}{2}\big] \\ &\le 2 + \tfrac{\epsilon}{4}\|z\| - \tfrac{\epsilon\gamma}{2} = 2 + \tfrac{\epsilon\gamma}{4} - \tfrac{\epsilon\gamma}{2} = 2 - \tfrac{\epsilon\gamma}{4}\,. \end{aligned}$$

That is, we have shown that if $x^*, y^* \in U_{X^*}$ and $\|x^* - y^*\| > \epsilon$, then $\big\|\frac{x^*+y^*}{2}\big\| \le 1-\delta$, where $\delta = \frac{\epsilon\gamma}{4}$. In other words, we have shown that X^* is uniformly convex. ∎

Problem 11.1.15. *Show that a locally uniformly convex Banach space X satisfies the* ***Kadets–Klee property****. That is, show that in a locally uniformly convex Banach space a sequence $\{x_n\}$ is norm convergent to some vector x if and only if $x_n \xrightarrow{w} x$ and $\|x_n\| \to \|x\|$.*

Solution: Let a sequence $\{x_n\}$ in a locally uniformly convex Banach space satisfy $x_n \xrightarrow{w} x$ and $\|x_n\| \to \|x\|$. If $x = 0$, then $\|x_n - x\| = \|x_n\| \to 0$ is trivially true. So, we can assume that $x \neq 0$ and $x_n \neq 0$ for each n.

Put $y_n = \frac{x_n}{\|x_n\|}$ for each n and $y = \frac{x}{\|x\|}$. Then $\|y_n\| = 1$ for each n and $\|y\| = 1$. Moreover, it should be clear (from the linearity of the weak topology) that $y_n \xrightarrow{w} y$. From the identity

$$x_n - x = \|x_n\|(y_n - y) + \big[\tfrac{\|x_n\|}{\|x\|} - 1\big]x\,,$$

it follows that $\|x_n - x\| \to 0$ if and only if $\|y_n - y\| \to 0$. Therefore, in order to complete the solution, we must show that $\|y_n - y\| \to 0$.

To this end, assume by way of contradiction that $\|y_n - y\| \not\to 0$. This means that there exist a subsequence of $\{y_n\}$ (which we shall denote by $\{y_n\}$ again) and some $0 < \epsilon \le 2$ such that $\|y_n - y\| \ge \epsilon$ holds for all n. Since X is locally uniformly convex, it follows that $\|y_n + y\| \not\to 2$ (see Lemma 11.15). Since $\|y_n + y\| \le 2$ for each n, this means that there exists another subsequence of $\{y_n\}$ (which we shall denote by $\{y_n\}$ once more!) and some $0 < \delta < 2$ such that $\|y_n + y\| < 2 - \delta$ for each n. Now pick some $x^* \in X^*$ satisfying $\|x^*\| = 1$ and $x^*(y) = 1$ and note that $|x^*(y_n + y)| \to |x^*(y) + x^*(y)| = 2$. On the other hand, we obviously have $|x^*(y_n+y)| \le \|x^*\|\cdot\|y_n+y\| \le 2-\delta$ for each n, a contradiction. Hence, $\|y_n - y\| \to 0$ and consequently $\|x_n - x\| \to 0$ is true.

The converse is true in any Banach space. That is, if $\|x_n - x\| \to 0$ in any Banach space, then clearly $x_n \xrightarrow{w} x$ and $\|x_n\| \to \|x\|$ are automatically true. ∎

Problem 11.1.16. *Show that each L_p-space with $1 < p < \infty$ is uniformly convex and uniformly smooth. Also show that the infinite dimensional L_1- and L_∞-spaces are neither uniformly convex nor uniformly smooth.*

Solution: We shall establish first that the $L_p(\mu)$-spaces with $1 < p < \infty$ are uniformly convex. So, let (Ω, Σ, μ) be a measure space, and fix some $1 < p < \infty$. We shall verify that for arbitrary functions $f, g \in L_p(\mu)$ the following equalities and inequalities hold.

(1) (**The Parallelogram Law**) *If* $p = 2$, *then*

$$\|f+g\|_p^2 + \|f-g\|_p^2 = 2\big(\,\|f\|_p^2 + \|g\|_p^2\,\big)\,.$$

(2) (**Clarkson's inequality # 1**; [**22**]) *If* $p > 2$, *then*

$$\|f+g\|_p^p + \|f-g\|_p^p \le 2^{p-1}\big(\,\|f\|_p^p + \|g\|_p^p\,\big)\,.$$

(3) (**Clarkson's inequality # 2**; [**22**]) *If* $1 < p < 2$ *and* $1 < q < \infty$ *satisfy* $\frac{1}{p} + \frac{1}{q} = 1$, *then*

$$\|f+g\|_p^q + \|f-g\|_p^q \le 2\big(\,\|f\|_p^p + \|g\|_p^p\,\big)^{\frac{1}{p-1}}\,.$$

If these properties are established, then $\|f_n + g_n\|_p \to 2$ with $\|f_n\|_p \le 1$ and $\|g_n\|_p \le 1$ imply $\|f_n - g_n\|_p \to 0$. That is, $L_p(\mu)$ is a uniformly convex Banach space for each $1 < p < \infty$.

First of all, we will show that the Parallelogram Law is true in any inner product space X. Indeed, if X is any inner product space and $x, y \in X$, then

$$\begin{aligned}
&\|x+y\|^2 + \|x-y\|^2 \\
&\qquad = (x+y, x+y) + (x-y, x-y) \\
&\qquad = \|x\|^2 + (x,y) + (y,x) + \|y\|^2 + \|x\|^2 - (x,y) - (y,x) + \|y\|^2 \\
&\qquad = 2\|x\|^2 + 2\|y\|^2\,.
\end{aligned}$$

Next, we shall verify Clarkson's inequality #1. To this end, assume that $p > 2$. In this case, we claim that for all $\alpha, \beta \in \mathbb{R}$ we have

$$\big|\alpha+\beta\big|^p + \big|\alpha-\beta\big|^p \le 2^{p-1}\big(|\alpha|^p + |\beta|^p\big)\,. \tag{$\star$}$$

If $(\star)$ is established and $f, g \in L_p(\mu)$, then

$$\big|f(\omega)+g(\omega)\big|^p + \big|f(\omega)-g(\omega)\big|^p \le 2^{p-1}\big(|f(\omega)|^p + |g(\omega)|^p\big)$$

for μ-almost all $\omega \in \Omega$, and the validity of Clarkson's inequality #1 follows by integrating.

Now let us prove $(\star)$. Observe first that if $(\star)$ is true for all $\alpha, \beta \in \mathbb{R}_+$, then $(\star)$ is also true for all $\alpha, \beta \in \mathbb{R}$. If $\alpha = 0$ or $\beta = 0$, then $(\star)$ is trivially true. So, we can suppose that $\alpha \ge \beta > 0$. Letting $t = \frac{\alpha}{\beta} \ge 1$, we can rewrite $(\star)$ as

$$(t+1)^p + (t-1)^p \le 2^{p-1}\big(1 + t^p\big)\,. \tag{$\star\star$}$$

To see that $(\star\star)$ is true, consider the function $f(t) = 2^{p-1}\big(1+t^p\big) - (t+1)^p - (t-1)^p$ for $t \ge 1$. Clearly, $(\star\star)$ is true if and only if $f(t) \ge 0$ holds for each $t \ge 1$. Taking derivatives, we see that

$$\begin{aligned}
f'(t) &= p\big[\,2^{p-1}t^{p-1} - (t+1)^{p-1} - (t-1)^{p-1}\,\big]\,, \text{ and} \\
f''(t) &= p(p-1)\big[\,2^{p-1}t^{p-2} - (t+1)^{p-2} - (t-1)^{p-2}\,\big] \\
&= p(p-1)\big[\,2(2t)^{p-2} - (t+1)^{p-2} - (t-1)^{p-2}\,\big] \\
&= p(p-1)\big(\,[(2t)^{p-2} - (t+1)^{p-2}] + [(2t)^{p-2} - (t-1)^{p-2}]\,\big) > 0\,.
\end{aligned}$$

The latter implies that the derivative f' is a strictly increasing function, and consequently $f'(t) \ge f'(1) = 0$ for all $t \ge 1$. In turn, this implies that f is an increasing function. In particular, we have $f(t) \ge f(1) = 0$, and this establishes the validity of

$(\star\star)$. (It is interesting to note that $(\star)$ is also true for any pair of complex numbers α and β.)

Next, we shall prove Clarkson's inequality #2. To this end, let $1 < p < 2$. Recall that $q = \frac{p}{p-1} > 1$. In this case, we can prove with an argument similar to the preceding one that for each $\alpha, \beta \in \mathbb{R}$ we have

$$|\alpha + \beta|^q + |\alpha - \beta|^q \le 2\big(|\alpha|^p + |\beta|^p\big)^{\frac{1}{p-1}}. \tag{$\dagger$}$$

Now let $f, g \in L_p(\mu)$. Then from $(\dagger)$ we get

$$\big|f(\omega) + g(\omega)\big|^q + \big|f(\omega) - g(\omega)\big|^q \le 2\big(|f(\omega)|^p + |g(\omega)|^p\big)^{\frac{1}{p-1}}$$

for μ-almost all $\omega \in \Omega$. This implies

$$\begin{aligned} \big\||f+g|^q + |f-g|^q\big\|_{p-1} &= \Big[\int_\Omega \big(|f+g|^q + |f-g|^q\big)^{p-1}\, d\mu\Big]^{\frac{1}{p-1}} \\ &\le \Big[2^{p-1}\int_\Omega \big(\,|f(\omega)|^p + |g(\omega)|^p\,\big)\, d\mu(\omega)\Big]^{\frac{1}{p-1}} \\ &= 2\big(\,\|f\|_p^p + \|g\|_p^p\,\big)^{\frac{1}{p-1}}. \end{aligned}$$

From $q(p-1) = p$, we see that $|f+g|^q, |f-g|^q \in L_{p-1}(\mu)$. Now taking into account that $0 < p-1 < 1$, it follows from the "reverse" Minkowski inequality[1] that

$$\big\||f+g|^q\big\|_{p-1} + \big\||f-g|^q\big\|_{p-1} \le \big\||f+g|^q + |f-g|^q\big\|_{p-1}.$$

Since $\big\||f+g|^q\big\|_{p-1} = \|f+g\|_p^q$ and $\big\||f-g|^q\big\|_{p-1} = \|f-g\|_p^q$, the preceding inequality can be written as

$$\|f+g\|_p^q + \|f-g\|_p^q \le \big\||f+g|^p + |f-g|^p\big\|_{p-1},$$

and so

$$\|f+g\|_p^q + \|f-g\|_p^q \le 2\big(\|f\|_p^p + \|g\|_p^p\big)^{\frac{1}{p-1}},$$

which is Clarkson's inequality # 2.

If $L_1(\mu)$ and $L_\infty(\mu)$ are not finite dimensional, then they are not reflexive. So, in this case, according to Problems 11.1.12 and 11.1.14 neither $L_1(\mu)$ nor $L_\infty(\mu)$ is uniformly convex or uniformly smooth. ∎

Problem 11.1.17. *This problem discusses various differentiability properties of the norm. Let x be an arbitrary non-zero vector in a Banach space X. Establish the following.*

(a) *If $0 < t_1 < t_2$, then for each $v \in X$ we have*

$$-\|v\| \le \frac{\|x + t_1 v\| - \|x\|}{t_1} \le \frac{\|x + t_2 v\| - \|x\|}{t_2} \le \|v\|.$$

This shows that the limit

$$\phi_x(v) = \lim_{t \to 0^+} \frac{\|x + tv\| - \|x\|}{t}$$

[1] It is well known that if $0 < r < 1$, then for every pair of functions $x, y \in L_r(\mu)$ we have $\|x + y\|_r \ge \|x\|_r + \|y\|_r$, where $\|z\|_r = \big(\int_\Omega |z|^r\, d\mu\big)^{\frac{1}{r}}$ for each $z \in L_r(\mu)$.

exists and $|\phi_x(v)| \le \|v\|$. *Also,* $\phi_x(x) = \|x\|$ *and* $\phi_x(-x) = -\|x\|$.

(b) *The function* $v \mapsto \phi_x(v)$ *from* X *to* $\mathbb{R}$ *is sublinear.*

(c) *A linear functional* $x^* \in X^*$ *is said to be a* ***supporting functional*** *at* x *if* $\|x^*\| = 1$ *and* $x^*(x) = \|x\|$. *We denote by* J_x *the non-empty collection of all supporting functionals at* x. *Show that* J_x *consists of all linear functionals on* X *that are dominated by* $\phi_x(\cdot)$, *i.e.,*

$$J_x = \big\{x^* \in X^* \colon\ x^*(v) \le \phi_x(v) \text{ for all } v \in X\big\}. \tag{$\star$}$$

(d) *For the function* $v \mapsto \phi_x(v)$ *the following statements are equivalent.*

  (i) $\phi_x(\cdot)$ *is linear.*

  (ii) $\phi_x(-v) = -\phi_x(v)$ *for each* $v \in X$.

  (iii) *The norm of the Banach space* X *is* ***Gateaux differentiable*** *at* x, *i.e.,* $\lim_{t\to 0} \frac{\|x+tv\|-\|x\|}{t}$ *exists in* $\mathbb{R}$ *for each* $v \in X$.

  (iv) *There exists a unique supporting functional at* x, *that is, the set* J_x *is a singleton.*

  (v) $\phi_x(\cdot)$ *is linear and is the only supporting functional at* x.

(e) *The norm of a Banach space* X *is said to be* ***Fréchet differentiable*** *at* x *if the limit* $\lim_{t\to 0} \frac{\|x+tv\|-\|x\|}{t}$ *exists uniformly in* v *in the unit sphere* $S_X = \{v \in X : \|v\| = 1\}$. *Establish the following.*

  α. *If the norm is Fréchet differentiable at* x, *then it is also Gateaux differentiable at* x.

  β. *The norm is Fréchet differentiable at* x *if and only if there exists a (necessarily unique) linear functional* $x^* \in X^*$ *such that: for every* $\epsilon > 0$ *there is some* $\delta > 0$ *so that for each* $u \in X$ *with* $\|u\| < \delta$ *we have*

$$\big|\, \|x+u\| - \|x\| - x^*(u) \,\big| \le \epsilon \|u\|. \tag{$\star\star$}$$

Solution: (a) Fix a vector $v \in X$ and consider the function $\psi \colon \mathbb{R} \to \mathbb{R}$ defined by $\psi(t) = \|x+tv\| - \|x\|$ for $t \in \mathbb{R}$. Clearly, $\psi(0) = 0$. The triangle inequality implies that ψ is a convex function satisfying $-|t|\|v\| \le \psi(t) \le |t|\|v\|$. The well-known properties of the convex functions guarantee that their difference ratio is monotone and so for $0 < t_1 < t_2$ we have

$$-\|v\| \le \frac{\psi(t_1) - \psi(0)}{t_1} = \frac{\psi(t_1)}{t_1} \le \frac{\psi(t_2)}{t_2} = \frac{\psi(t_2) - \psi(0)}{t_2} \le \|v\|,$$

which is (a). This certainly implies that the limit

$$\phi_x(v) = \lim_{t\to 0^+} \frac{\psi(t)}{t} = \lim_{t\to 0^+} \frac{\|x+tv\| - \|x\|}{t} = \lim_{t\downarrow 0} \frac{\|x+tv\| - \|x\|}{t}$$

exists and satisfies $|\phi_x(v)| \le \|v\|$.

Letting $v = x$, we see that $\psi(t) = (|1+t|-1)\|x\|$ and so $\psi(t) = t\|x\|$ for any $t \geq -1$. Thus,

$$\phi_x(x) = \lim_{t\to 0^+} \frac{\|x+tx\| - \|x\|}{t} = \|x\|\,,$$

and similarly

$$\phi_x(-x) = \lim_{t\to 0^+} \frac{\|x-tx\| - \|x\|}{t} = -\|x\|\,.$$

(b) If $\lambda > 0$, then we have $\frac{\|x+t(\lambda v)\|-\|x\|}{t} = \lambda\big[\frac{\|x+(\lambda t)v\|-\|x\|}{\lambda t}\big]$ for each $v \in X$. Letting $t \to 0^+$, we get $\phi_x(\lambda v) = \lambda\phi_x(v)$. That is, $\phi_x(\cdot)$ is positively homogeneous.

To verify the subadditivity of $\phi_x(\cdot)$ pick any $v, w \in X$ and notice that for each $t > 0$ we have

$$\begin{aligned}\frac{\|x+tv\|-\|x\|}{t} + \frac{\|x+tw\|-\|x\|}{t} &= \frac{\|x+tv\|+\|x+tw\|-2\|x\|}{t}\\ &\geq \frac{\|2x+t(v+w)\|-2\|x\|}{t}\\ &= \frac{\|x+\frac{t}{2}(v+w)\|-\|x\|}{\frac{t}{2}}\,.\end{aligned}$$

So, letting $t \to 0^+$, we get $\phi_x(v) + \phi_x(w) \geq \phi_x(v+w)$. This establishes that $\phi_x(\cdot)$ is a sublinear function.

(c) Let $J_x = \{x^* \in X^*\colon\ \|x^*\| = 1 \text{ and } x^*(x) = \|x\|\}$ be the set of all supporting functionals at x. The Hahn–Banach theorem guarantees that J_x is not empty. To verify ($\star$) assume first that $x^* \in J_x$. Then for every scalar t and every $v \in X$, we have

$$\|x\| + tx^*(v) = x^*(x) + tx^*(v) = x^*(x+tv) \leq \|x+tv\|\,,$$

and consequently $x^*(v) \leq \frac{\|x+tv\|-\|x\|}{t}$ for all $t > 0$ and $v \in X$. This implies that $x^*(v) \leq \phi_x(v)$ for each $v \in X$, i.e., $J_x \subseteq \big\{x^* \in X^*\colon\ x^*(v) \leq \phi_x(v) \ \text{ for all } v \in X\big\}$.

For the converse suppose that some $x^* \in X^*$ satisfies $x^*(v) \leq \phi_x(v)$ for all $v \in X$. Then $\pm x^*(v) \leq \max\{\phi_x(v), \phi_x(-v)\} \leq \|v\|$ or equivalently $|x^*(v)| \leq \|v\|$ for all $v \in X$, that is, $\|x^*\| \leq 1$. However, $-x^*(x) = x^*(-x) \leq \phi_x(-x) = -\|x\|$ and so $x^*(x) \geq \|x\|$. This proves that $x^*(x) = \|x\|$ and $\|x^*\| = 1$, that is, $x^* \in J_x$. Hence $\big\{x^* \in X^*\colon\ x^*(v) \leq \phi_x(v) \text{ for all } v \in X\big\} \subseteq J_x$, and so

$$J_x = \big\{x^* \in X^*\colon\ x^*(v) \leq \phi_x(v) \quad \text{for all } v \in X\big\}\,.$$

(d) The implication (i) $\Longrightarrow$ (ii) is trivial since if $\phi_x(\cdot)$ is linear, then of course $\phi_x(-v) = -\phi_x(v)$ for each $v \in X$. Notice also that whenever $\phi_x(\cdot)$ is linear, it follows from (a) that $\phi_x(\cdot) \in J_x$.

(ii) $\Longrightarrow$ (iii) We have

$$\lim_{t\to 0^-} \frac{\|x+tv\|-\|x\|}{t} = -\lim_{t\to 0^+} \frac{\|x+t(-v)\|-\|x\|}{t} = -\phi_x(-v) = \phi_x(v)\,.$$

This proves that $\lim_{t\to 0} \frac{\|x+tv\|-\|x\|}{t}$ exists and is equal to $\phi_x(v)$.

(iii) $\Longrightarrow$ (iv) Assume contrary to our claim that the set J_x is not a singleton. That is, suppose there are two distinct unit linear functionals $x^*, y^* \in X^*$ satisfying

$x^*(x) = y^*(x) = \|x\|$. Since $x^* \neq y^*$, the linear functionals x^* and y^* must be linearly independent. Otherwise, $x^* = \lambda y^*$ implies $\|x\| = x^*(x) = \lambda y^*(x) = \lambda\|x\|$. This yields $\lambda = 1$, and so $x^* = y^*$, a contradiction.

Since x^* and y^* are linearly independent, there is some $u \in X$ with $x^*(u) = 0$ and $y^*(u) > 0$. Similarly, there exists some $v \in X$ with $y^*(v) = 0$ and $x^*(v) > 0$. Now consider the vector $z = v - u$ and note that $x^*(z) > 0$ and $y^*(z) < 0$.

From $\|x\| + tx^*(z) = x^*(x) + tx^*(z) = x^*(x + tz) \leq \|x + tz\|$, it follows that $\frac{\|x+tz\|-\|x\|}{t} \geq x^*(z)$ for each $t > 0$, and so

$$\lim_{t\to 0} \frac{\|x + tz\| - \|x\|}{t} = \lim_{t\to 0^+} \frac{\|x + tz\| - \|x\|}{t} \geq x^*(z) > 0\,.$$

Similarly, for each $t < 0$ we have

$$\|x\| + ty^*(z) = y^*(x) + ty^*(z) = y^*(x + tz) \leq \|x + tz\|\,,$$

and so $\frac{\|x+tz\|-\|x\|}{t} \leq y^*(z)$ for each $t < 0$. This implies

$$\lim_{t\to 0} \frac{\|x + tz\| - \|x\|}{t} = \lim_{t\to 0^-} \frac{\|x + tz\| - \|x\|}{t} \leq y^*(z) < 0\,.$$

The above inequalities show that $\lim_{t\to 0} \frac{\|x+tz\|-\|x\|}{t}$ does not exist, which is a contradiction. Hence, J_x is a singleton.

(iv) $\Longrightarrow$ (v) Assume that the set J_x is a singleton, that is, $J_x = \{x^*\}$ for some $x^* \in X^*$. We want to prove that $x^* = \phi_x(\cdot)$. For a singleton J_x the identity ($\star$) established in (b) means that there exists exactly one linear functional which is dominated by the sublinear functional $\phi_x(\cdot)$. We claim then that ϕ_x itself must be linear. To see this, assume by way of contradiction that $\phi_x(\cdot)$ is not linear. Then there exists some $v \in X$ such that $-\phi_x(-v) < \phi_x(v)$. (Here we use the facts that a sublinear functional p on X always satisfies the inequality $-p(-z) \leq p(z)$ for each $z \in X$ and that p is linear if and only if $p(-z) = -p(z)$ for each $z \in X$.) Consider the linear subspace $M = \{\lambda v\colon \lambda \in \mathbb{R}\}$ generated by v, and define the linear functionals $g, h\colon M \to \mathbb{R}$ by

$$g(\lambda v) = -\lambda\phi_x(-v) \quad \text{and} \quad h(\lambda v) = \lambda\phi_x(v)$$

for $\lambda v \in M$. A moment's thought reveals that both g and h are dominated by the restriction of the sublinear functional $\phi_x(\cdot)$ to M. So, by the Hahn–Banach Extension Theorem, g and h have linear extensions to all of X that are dominated by ϕ_x. We denote these extensions again by g and h. It remains to notice that $g(v) = -\phi_x(-v) < \phi_x(v) = h(v)$ implies that $g \neq h$, and this contradicts the preceding conclusion. Hence, $\phi_x(\cdot)$ is a linear functional. Therefore $\phi_x(\cdot) \in J_x$, and so $\phi_x(\cdot) = x^*$, as desired.

(v) $\Longrightarrow$ (i) is trivial.

(e) We first prove (α). So, assume that the norm is Fréchet differentiable at x. That is, the limit $\phi_x(v) = \lim_{t\to 0} \frac{\|x+tv\|-\|x\|}{t}$ exists uniformly over $v \in S_X$. Now notice that if $v \in X$ is an arbitrary non-zero vector, then we have

$$\tfrac{\|x+tv\|-\|x\|}{t} = \|v\|\, \frac{\left\|x+(t\|v\|)\frac{v}{\|v\|}\right\|-\|x\|}{t\|v\|} \xrightarrow[t\to 0]{} \|v\|\phi_x\big(\tfrac{v}{\|v\|}\big)\,.$$

In particular, the limit $\lim_{t\to 0}\frac{\|x+tv\|-\|x\|}{t}$ exists in $\mathbb{R}$ for each $v\in X$. This shows that the norm is Gateaux differentiable at x.

Next, we shall prove (β). Assume first that the norm is Fréchet differentiable at x. By part (α) we know that the norm is Gateaux differentiable at x, and so $x^*=\phi_x(\cdot)$ is a linear functional. In other words, we have

$$\lim_{t\to 0}\tfrac{\|x+tv\|-\|x\|}{t}=x^*(v)$$

uniformly for $v\in S_X$. This means that for each $\epsilon>0$ there exists $\delta>0$ such that

$$\Big|\tfrac{\|x+tv\|-\|x\|}{t}-x^*(v)\Big|<\epsilon$$

for each $v\in S_X$ and each non-zero $t\in(-\delta,\delta)$. It follows that

$$\Big|\,\|x+tv\|-\|x\|-tx^*(v)\Big|\le\epsilon|t|$$

for all $|t|<\delta$. Now if $u\in X$ satisfies $0<\|u\|<\delta$, then we can write $u=tv$ where $v=\frac{u}{\|u\|}\in S_X$ and $t=\|u\|<\delta$. This implies that

$$\Big|\,\|x+u\|-\|x\|-x^*(u)\Big|\le\epsilon\|u\|\,,$$

holds for all $u\in X$ with $\|u\|<\delta$, and the validity of ($\star\star$) has been established.

For the converse, assume that ($\star\star$) is true. Fix $\epsilon>0$ and then choose some $\delta>0$ that satisfies ($\star\star$). Next pick any $v\in S_X$, any non-zero $t\in(-\delta,\delta)$, and let $u=tv$. By ($\star\star$) we have

$$\Big|\,\|x+tv\|-\|x\|-x^*(tv)\Big|\le\epsilon\|tv\|=\epsilon|t|$$

or, equivalently,

$$\Big|\tfrac{\|x+tv\|-\|x\|}{t}-x^*(v)\Big|\le\epsilon\,.$$

This proves that the limit $\lim_{t\to 0}\frac{\|x+tv\|-\|x\|}{t}=x^*(v)$ exists uniformly over $v\in S_X$. Therefore, the norm is Fréchet differentiable at x, and the solution is finished. ∎

Problem 11.1.18. *Show that a Banach space X is uniformly smooth if and only if its norm is uniformly Fréchet differentiable over the unit sphere of X, that is, the limit*

$$\phi_x(y)=\lim_{t\to 0}\frac{\|x+ty\|-\|x\|}{t}$$

exists in $\mathbb{R}$ uniformly for all unit vectors $x,y\in X$.

Solution: As usual we denote the unit sphere of the Banach space X by S_X, i.e., $S_X=\big\{x\in X\colon\ \|x\|=1\big\}$. Assume first that the limit

$$\phi_x(y)=\lim_{t\to 0}\frac{\|x+ty\|-\|x\|}{t}$$

exists uniformly over $x,y\in S_X$. In particular, this implies that for each unit vector x the sublinear functional $\phi_x(\cdot)$ is linear; see part (d) of Problem 11.1.17. A consequence of this is that the limit

$$\begin{aligned}\lim_{t\to 0}\Big[\tfrac{\|x+ty\|-\|x\|}{t}+\tfrac{\|x-ty\|-\|x\|}{t}\Big] &= \lim_{t\to 0}\tfrac{\|x+ty\|-\|x\|}{t}+\lim_{t\to 0}\tfrac{\|x-ty\|-\|x\|}{t}\\ &= \phi_x(y)+\phi_x(-y)=\phi_x(y)-\phi_x(y)=0\end{aligned}$$

exists uniformly over $x, y \in S_X$. So, if $\epsilon > 0$ is given, then there exists some $0 < \gamma < 1$ such that

$$\Big| \tfrac{\|x+ty\|-\|x\|}{t} + \tfrac{\|x-ty\|-\|x\|}{t} \Big| < \epsilon$$

for all $x, y \in S_X$ and all non-zero $t \in [-\gamma, \gamma]$. This implies that for all unit vectors x and y and for all $0 < |t| \le \gamma$ we have

$$\|x + ty\| + \|x - ty\| < 2 + \epsilon|t|\,.$$

In particular, if u is any unit vector and v is any vector with $0 < \|v\| \le \gamma$, then by letting $x = u$, $t = \|v\|$ and $y = \frac{v}{\|v\|}$, we get

$$\|u + v\| + \|u - v\| \le 2 + \epsilon\|v\|\,.$$

Clearly, this inequality is also true for $v = 0$. By Problem 11.1.13 the above condition is equivalent to the uniform smoothness of X.

For the converse, assume that X is uniformly smooth. We must show that there exists a function $\psi\colon S_X \times S_X \to \mathbb{R}$ such that for each $\epsilon > 0$ there exists some $\gamma > 0$ such that

$$\Big| \tfrac{\|x+ty\|-\|x\|}{t} - \psi(x, y) \Big| < \epsilon \tag{$\star$}$$

for all $x, y \in S_X$ and all t with $0 < |t| < \gamma$. The arguments below establishing this claim are due to J. Diestel.

We start by observing that according to Problem 11.1.14 the norm dual X^* of X is uniformly convex. So, given $\epsilon > 0$ there exists some $0 < \delta < 1$ such that:

$$f, g \in S_{X^*} \ \text{ and } \ \|f + g\| > 2 - \delta \implies \|f - g\| < \epsilon\,. \tag{$\star\star$}$$

Next, for each $x \in S_X$ select a functional $f_x \in X^*$ supporting at x, i.e., $\|f_x\| = 1$ and $f_x(x) = \|x\| = 1$. We claim that:

$$x, y \in S_X \ \text{ and } \ \|x - y\| < \delta \implies \|f_x - f_y\| < \epsilon\,. \tag{$\star\star\star$}$$

To see this, assume that two vectors $x, y \in S_X$ satisfy $\|x - y\| < \delta$. Then we have

$$\begin{aligned} \|f_x + f_y\| + \|x - y\| &\ge (f_x + f_y)(y) + f_x(x - y) \\ &= f_x(y) + f_y(y) + f_x(x) - f_x(y) = 2\,, \end{aligned}$$

and so $\|f_x + f_y\| > 2 - \delta$. Now a glance at ($\star\star$) guarantees that $\|f_x - f_y\| < \epsilon$.

We can extend the mapping $x \mapsto f_x$ from S_X to all of X by letting $f_z = \|z\| f_{\frac{z}{\|z\|}}$ if $z \ne 0$ and $f_0 = 0$. Clearly, $\frac{f_z}{\|z\|} = f_{\frac{z}{\|z\|}}$ if $z \ne 0$ and $f_z(z) = \|z\|^2$ for all $z \in X$. In addition, note that $|f_u(v)| \le \|u\| \cdot \|v\|$ for all $u, v \in X$.

Now fix $x, y \in S_X$, $t \in (0, 1)$, and let $z_t = \frac{x+ty}{\|x+ty\|}$. Then we have

$$\begin{aligned}
f_x(y) &= \frac{f_x(ty)}{t} = \frac{f_x(x)+f_x(ty)-1}{t} = \frac{f_x(x+ty)-1}{t} \le \frac{|f_x(x+ty)|-1}{t}\\
&\le \frac{\|f_x\|\cdot\|x+ty\|-1}{t} = \frac{\|x+ty\|-1}{t} = \frac{\|x+ty\|-\|x\|}{t}\\
&= \frac{\|x+ty\|^2-\|x\|\cdot\|x+ty\|}{t\|x+ty\|} \le \frac{\|x+ty\|^2-|f_{x+ty}(x)|}{t\|x+ty\|}\\
&= \frac{f_{x+ty}(x+ty)-|f_{x+ty}(x)|}{t\|x+ty\|} = \frac{f_{x+ty}(x)+tf_{x+ty}(y)-|f_{x+ty}(x)|}{t\|x+ty\|}\\
&\le \frac{tf_{x+ty}(y)}{t\|x+ty\|} = \frac{f_{x+ty}(y)}{\|x+ty\|} = f_{z_t}(y)\,.
\end{aligned}$$

In particular $f_x(y) \le \frac{\|x+ty\|-\|x\|}{t} \le f_{z_t}(y)$ for all $t \in (0,1)$. Similarly, for each $t \in (-1,0)$ we have $f_{z_t}(y) \le \frac{\|x+ty\|-\|x\|}{t} \le f_x(y)$. Therefore, for each non-zero $t \in (-1,1)$ we have

$$\Big|\tfrac{\|x+ty\|-\|x\|}{t} - f_x(y)\Big| \le \big|f_{z_t}(y)-f_x(y)\big| \le \big\|f_{z_t}-f_x\big\|\,. \tag{\dagger}$$

Next, put $\gamma = \min\{\frac{\delta}{4},\frac{1}{2}\}$ and choose any t satisfying $0<|t|<\gamma$. Clearly, $\|x+ty\| \ge \|x\|-|t|\cdot\|y\| = 1-|t| > \frac{1}{2}$, and an easy computation shows that the vectors $x, z_t \in S_X$ satisfy

$$\begin{aligned}
\|z_t - x\| &= \tfrac{\|ty+(1-\|x+ty\|)x\|}{\|x+ty\|} \le 2\big(\big|1-\|x+ty\|\,\big|+|t|\big)\\
&= 2\big(\big|\|x\|-\|x+ty\|\,\big|+|t|\big) \le 2\big(\|x-(x+ty)\|+|t|\big)\\
&= 4|t| < \delta\,.
\end{aligned}$$

Consequently, it follows from $(\star\star\star)$ that $\big\|f_{z_t}-f_x\big\| < \epsilon$. Finally, a glance at $(\dagger)$ shows that for all $x,y \in S_X$ and all $0<|t|<\gamma$ we have

$$\big|\tfrac{\|x+ty\|-\|x\|}{t} - f_x(y)\big| < \epsilon\,,$$

which is $(\star)$. This completes the solution. ∎

11.2. The Daugavet Property in AL- and AM-spaces

Problem 11.2.1. *Show that a compact Hausdorff space Ω has an isolated point if and only if the Banach lattice $C(\Omega)$ has an atom.*

Solution: Assume first that ω_0 is an isolated point of Ω, i.e., assume that the set $\{\omega_0\}$ is a clopen subset of Ω. We claim that the function $f=\chi_{\{\omega_0\}} \in C(\Omega)$ is an atom. To see this, assume that $0 \le g,h \le f$ satisfy $g \wedge h = 0$. Since $f(\omega)=0$ holds for all $\omega \ne \omega_0$, it follows that $g(\omega)=h(\omega)=0$ for all $\omega\ne\omega_0$. Assume $g \ne 0$, i.e., $g(\omega_0)>0$. Then, from $\min\{g(\omega_0),h(\omega_0)\} = (g\wedge h)(\omega_0) = 0$, it follows that $h(\omega_0)=0$ or $h=0$. That is, either $g=0$ or $h=0$ is true, proving that f is an atom of $C(\Omega)$.

For the converse, suppose that $0 < u \in C(\Omega)$ is an atom. Fix some $\omega_0 \in \Omega$ such that $u(\omega_0) > 0$. We claim that $u = \chi_{\{\omega_0\}}$. If this is established, then $\{\omega_0\}$ is a clopen subset of Ω, and so ω_0 must be an isolated point of Ω.

To see that $u = \chi_{\{\omega_0\}}$, pick any $\omega \neq \omega_0$. Since Ω is a compact Hausdorff space, there exist functions $f, g\colon \Omega \to [0,1]$ satisfying $f \wedge g = 0$ and $f(\omega_0) = g(\omega) = 1$. Now consider the continuous functions $u_1 = fu$ and $u_2 = gu$. Clearly, $0 \le u_1, u_2 \le u$, $u_1 \wedge u_2 = 0$, and $u_1(\omega_0) > 0$. Since, u is an atom, it follows that $u_2 = gu = 0$. In particular, we have $u(\omega) = g(\omega)u(\omega) = u_2(\omega) = 0$ for all $\omega \neq \omega_0$. This shows that $u = \chi_{\{\omega_0\}}$, as desired. ∎

Problem 11.2.2. *This problem shows that a compact operator on an AM- or an AL-space with atoms need not satisfy the Daugavet property. Let $E = C(\Omega)$, where Ω is a compact Hausdorff space having an isolated point, say ω_0. Consider the multiplication operator $T\colon C(\Omega) \to C(\Omega)$, defined by $Tf = -\chi_{\{\omega_0\}}f$. Establish that:*

(a) *T is compact.*

(b) *$\|T\| = 1$.*

(c) *$\|I + T\| \le 1 < 1 + \|T\| = 2$.*

Solution: (a) Let $\{f_n\}$ be a sequence in $C(\Omega)$ satisfying $\|f_n\|_\infty \le 1$ for each n. Then $\{f_n(\omega_0)\}$ is a bounded sequence of real numbers, and so it has a convergent subsequence, say $f_{k_n}(\omega_0) \to \alpha$. If $f = -\alpha\chi_{\{\omega_0\}}$, then note that $f \in C(\Omega)$ and

$$\|Tf_{k_n} - f\|_\infty = |\alpha - f_{k_n}(\omega_0)| \longrightarrow 0\,.$$

This shows that $T(U)$, where U is the closed unit ball of $C(\Omega)$, is a norm totally bounded subset of $C(\Omega)$. Consequently, T is a compact operator.

(b) If $f \in U$, then note that

$$\|Tf\|_\infty = \max_{\omega\in\Omega} |Tf(\omega)| = \max_{\omega\in\Omega} |(-\chi_{\{\omega_0\}}f)(\omega)| = |f(\omega_0)| \le 1\,.$$

So, $\|T\| = \sup_{f\in U} \|Tf\|_\infty \le 1$. Since $\|T\mathbf{1}\|_\infty = 1$ and $\mathbf{1} \in U$, we see that $\|T\| = 1$.

(c) For the norm of $I + T$ observe that if $f \in U$, then $f(\omega) + Tf(\omega) = f(\omega)$ if $\omega \neq \omega_0$ and $f(\omega) + Tf(\omega) = 0$ if $\omega = \omega_0$. This implies $|(f + Tf)(\omega)| \le 1$ for each $\omega \in \Omega$ and each $f \in U$, and so $\|I + T\| = \sup_{f\in U} \|f + Tf\|_\infty \le 1$.

Therefore, $\|I + T\| \le 1 < 2 = 1 + \|T\|$. This shows that the compact operator T does not satisfy the Daugavet equation. ∎

Problem 11.2.3. *If $J\colon E \to E$ is an isometry on an AL-space, then show that $\|J + T\| = 1 + \|T\|$ for each operator $T \in \mathcal{L}(E)$ that is disjoint from J.*

Solution: Recall that every bounded operator on an *AL*-space has a modulus and that its norm coincides with its r-norm; see Theorem 3.9. Now assume that an operator $T \in \mathcal{L}(E)$ is disjoint from an isometry $J \subset \mathcal{L}(E)$, i.e., $|J| \wedge |T| = 0$. This implies $|J + T| = |J| + |T|$. So, if $x \in E$ satisfies $\|x\| = 1$, then using that E is an

AL-space, we obtain

$$\begin{aligned}\|J+T\| &= \big\|\,|J+T|\,\big\| = \big\|\,|J|+|T|\,\big\| \\ &\geq \big\|\,|J||x|+|T||x|\,\big\| \\ &= \big\|\,|J||x|\,\big\| + \big\|\,|T||x|\,\big\| \\ &\geq \|Jx\|+\|Tx\| \\ &= 1+\|Tx\|\,.\end{aligned}$$

This implies $1+\|T\| = 1+\sup_{\|x\|=1}\|Tx\| \leq \|J+T\|$, and so $\|J+T\| = 1+\|T\|$. ∎

Problem 11.2.4. *Let $f, g\colon [0,1] \to \mathbb{R}$ be two non-zero bounded real measurable functions, and consider the rank-one operator $T = f \otimes g$ acting on each $L_p[0,1]$-space. That is, the operator $T\colon L_p[0,1] \to L_p[0,1]$ is defined via the formula*

$$Th = (f\otimes g)h = \langle h, f\rangle g = \Big[\int_0^1 h(s)f(s)\,ds\Big]g$$

for each $h \in L_p[0,1]$.

(a) *Verify directly (i.e., without using Theorem 11.20) that the operator $T\colon L_\infty[0,1] \to L_\infty[0,1]$ satisfies the Daugavet equation.*

(b) *Show that $T\colon L_1[0,1] \to L_1[0,1]$ satisfies the Daugavet equation.*

(c) *If $\langle g, f\rangle = 0$, then show that $T\colon L_p[0,1] \to L_p[0,1]$ does not satisfy the Daugavet equation for any $1 < p < \infty$.*

Solution: (a) We shall denote the Lebesgue measure on $[0,1]$ by λ. Note that $\|T\| = \|f\|_1 \cdot \|g\|_\infty$. Thus, according to Corollary 11.5, we can assume without loss of generality that $\|f\|_1 = \|g\|_\infty = 1$. In particular, we have $\|T\| = 1$.

Now fix $0 < \epsilon < 1$. From $\|g\|_\infty = 1$, it follows that there exists a measurable subset A of $[0,1]$ of positive Lebesgue measure such that $|g(t)| \geq 1-\epsilon$ holds for each $t \in A$. Let $A_1 = \{t \in A\colon\ g(t) > 1-\epsilon\}$ and $A_2 = \{t \in A\colon\ -g(t) > 1-\epsilon\}$. If $\lambda(A_1) = 0$, then $\lambda(A_2) > 0$. So, in this case, if we replace g by $-g$, f by $-f$ and A by A_2, we can assume that $g(t) \geq 1-\epsilon$ for each $t \in A$.

Next, let $x = \operatorname{Sgn} f$. That is, $x(t) = 1$ if $f(t) \geq 0$ and $x(t) = -1$ if $f(t) < 0$. Clearly, $x \in L_\infty[0,1]$ and $|x(t)| = 1$ for each t. By the absolute continuity of the integral, there exists some $\delta > 0$ such that for each Lebesgue measurable subset B of $[0,1]$ with $\lambda(B) < \delta$ we have $\int_B |f(t)|\,dt < \frac{\epsilon}{2}$. Pick any Lebesgue measurable subset B of A with $0 < \lambda(B) < \delta$ and consider the Lebesgue measurable function $y = x\chi_{B^c} + \chi_B$. Clearly, $|y(t)| = 1$ for each $t \in [0,1]$, and so $\|y\|_\infty = 1$. Now note that

$$\begin{aligned}\int_0^1 f(t)y(t)\,dt &= \int_{B^c}|f(t)|\,dt + \int_B f(t)\,dt \\ &= \int_0^1 |f(t)|\,dt + \int_B f(t)\,dt - \int_B |f(t)|\,dt \\ &\geq 1 - 2\int_B |f(t)|\,dt > 1-\epsilon\,.\end{aligned}$$

Consequently, we have

$$\begin{aligned}\|I+T\| &\geq \|y+Ty\|_\infty = \left\|y+\Big(\int_0^1 f(t)y(t)\,dt\Big)g\right\|_\infty\\ &\geq \left\|\Big[y+\Big(\int_0^1 f(t)y(t)\,dt\Big)g\Big]\chi_B\right\|_\infty.\end{aligned}$$

Since $y(t)=1$ and $g(t)>1-\epsilon$ for each $t\in B$, it follows that

$$\|I+T\|\geq 1+(1-\epsilon)^2$$

for each $0<\epsilon<1$. Letting $\epsilon\to 0^+$, we get $\|I+T\|\geq 2$. Therefore, $\|I+T\|=2$, and so T satisfies the Daugavet equation.

(b) From part (a) we know that the operator $S=g\otimes f\colon L_\infty[0,1]\to L_\infty[0,1]$ satisfies the Daugavet equation. It remains to note that $S=T^*$, and so T also satisfies the Daugavet equation.

(c) It is easy to see that $T^2=0$. This implies $r(T)=0$ or $\sigma(T)=\{0\}$. Since T is non-zero, we have $\|T\|>0$. Since for $1<p<\infty$ the Banach space $L_p(\mu)$ is uniformly convex (see Problem 11.1.16), it follows from Theorem 11.10 that T satisfies the Daugavet equation if and only if $0<\|T\|\in\sigma(T)=\{0\}$, which is impossible. Hence, for each $1<p<\infty$ the operator $T\colon L_p[0,1]\to L_p[0,1]$ does not satisfy the Daugavet equation. ∎

Problem 11.2.5. *If μ is a σ-finite non-atomic measure, then show that every finite-rank operator on $L_p(\mu)$, where $1\leq p<\infty$, is disjoint from the identity operator—and so in this case, by Lemma 11.21, every finite-rank operator on $L_1(\mu)$ satisfies the Daugavet equation.*

Solution: Assume first that (Ω,Σ,μ) is a non-atomic finite measure space. Fix $f\in L_q^+(\mu)$ and $u\in L_p^+(\mu)$, where $1\leq p<\infty$ and $1<q\leq\infty$ satisfy $\frac{1}{p}+\frac{1}{q}=1$. To prove $I\wedge(f\otimes u)=0$, it suffices to establish that $I\wedge(f\otimes u)(\mathbf{1})=0$.

To verify this, assume that some $g\in L_p^+(\mu)$ satisfies $g\leq I\wedge(f\otimes u)(\mathbf{1})$. Fix $\epsilon>0$. The absolute continuity of the integral guarantees the existence of some $\delta>0$ such that $\int_A f(t)\,d\mu(t)<\epsilon$ for each measurable set A with $\mu(A)<\delta$. Next, choose some n with $\frac{\mu(\Omega)}{n}<\epsilon$ and then select pairwise disjoint measurable sets $A_1,\ldots,A_n$ with $\mu(A_i)=\frac{\mu(\Omega)}{n}$ for each i; this is possible since μ is non-atomic. Since $\chi_{A_i^c}+\chi_{A_i}=\mathbf{1}$, it follows that

$$\begin{aligned}g &\leq I\wedge(f\otimes u)(\mathbf{1})\leq I(\chi_{A_i^c})+(f\otimes u)(\chi_{A_i})\\ &= \chi_{A_i^c}+\Big[\int_{A_i} f(t)\,d\mu(t)\Big]u\leq \chi_{A_i^c}+\epsilon u\end{aligned}$$

for each i. This implies $g\leq\epsilon u$. Since $\epsilon>0$ is arbitrary, the latter shows that $g=0$, and from this it follows that $I\wedge(f\otimes u)=0$.

Now assume that (Ω,Σ,μ) is a σ-finite measure space. Pick a sequence $\{\Omega_n\}$ in Σ such that $\mu(\Omega_n)<\infty$ for each n and $\Omega_n\uparrow\Omega$. Also, fix $f\in L_q^+(\mu)$ and $u\in L_p^+(\mu)$, where $1\leq p<\infty$ and $1<q\leq\infty$ satisfy $\frac{1}{p}+\frac{1}{q}=1$. By the preceding case, we have $I\wedge(f\otimes u)(\chi_{\Omega_n})=0$ for each n. So, if $0\leq g\in L_p(\mu)$, then $I\wedge(f\otimes u)(g\wedge\chi_{\Omega_n})=0$ for each n. Since $g\wedge\chi_{\Omega_n}\uparrow g$, it follows from the order continuity of the operator

$I \wedge (f \otimes u)$ that $I \wedge (f \otimes u)(g) = 0$. This implies that $I \wedge (f \otimes u) = 0$ holds true in this case too.

Now let $S = \sum_{i=1}^{k} f_i \otimes u_i$ be a finite rank-operator. Then

$$\begin{aligned} 0 \leq I \wedge |S| &= I \wedge \Big| \sum_{i=1}^{k} f_i \otimes u_i \Big| \leq I \wedge \Big(\sum_{i=1}^{k} |f_i \otimes u_i| \Big) \\ &= I \wedge \Big(\sum_{i=1}^{k} |f_i| \otimes |u_i| \Big) \\ &\leq \sum_{i=1}^{k} I \wedge (|f_i| \otimes |u_i|) = 0\,. \end{aligned}$$

(For the validity of the last inequality see the solution to Problem 3.3.1; see also Lemma 4.13.) So, $I \wedge |S| = 0$, and thus the identity operator I is disjoint from the vector space of all finite-rank operators. ∎

Problem 11.2.6. *Present an example of a bounded operator on some L_p-space with $1 < p < \infty$ that is disjoint from the identity operator and fails to satisfy the Daugavet equation.*

Solution: Let $f = \chi_{[0,\frac{1}{2}]}$, $u = \chi_{[\frac{1}{2},1]}$, and consider the rank-one positive operator $T = f \otimes u\colon L_p[0,1] \to L_p[0,1]$. By Problem 11.2.5, T is disjoint from the identity operator, and by Problem 11.2.4(c) T fails to satisfy the Daugavet equation. ∎

11.3. The Daugavet Property in Banach Spaces

Problem 11.3.1. *If $T\colon X \to Z$ and $S\colon Y \to Z$ are bounded operators between normed spaces, then show that*

$$\sup\big\{\|Tx + Sy\|\colon\ \|x\| \leq 1\,,\ \|y\| \leq 1\big\} = \sup\big\{\|Tx + Sy\|\colon\ \|x\| = \|y\| = 1\big\}\,.$$

Solution: Assume that $T\colon X \to Z$ and $S\colon Y \to Z$ are two bounded operators between normed spaces. Let

$$\begin{aligned} \alpha &= \sup\big\{\|Tx + Sy\|\colon\ \|x\| \leq 1\,,\ \|y\| \leq 1\big\}\,, \text{ and} \\ \beta &= \sup\big\{\|Tx + Sy\|\colon\ \|x\| = \|y\| = 1\big\}\,. \end{aligned}$$

Clearly, $\beta \leq \alpha$.

Pick any $x \in X$ and $y \in Y$ satisfying $\|x\| \leq 1$ and $\|y\| \leq 1$. If $x = 0$ and $y = 0$, then $\|Tx + Sy\| = 0 \leq \beta$ is trivially true. Now consider the case where $x \neq 0$ and $y = 0$. Put $u = \frac{x}{\|x\|}$, and then pick a unit vector $v \in Y$. Then, using Lemma 11.24, we see that

$$\begin{aligned} \|Tx + Sy\| &= \|x\| \cdot \|Tu\| \leq \|Tu\| = \|1Tu + 0Sv\| \\ &\leq \max\big\{\|Tu + Sv\|, \|Tu + S(-v)\|\big\} \leq \beta\,. \end{aligned}$$

Finally, consider the case $x \neq 0$ and $y \neq 0$. Again using Lemma 11.24, we get

$$\begin{aligned} \|Tx + Sy\| &= \big\| \|x\| T\big(\tfrac{x}{\|x\|}\big) + \|y\| S\big(\tfrac{y}{\|y\|}\big) \big\| \\ &\leq \max\big\{ \big\| T\big(\tfrac{x}{\|x\|}\big) + S\big(\tfrac{y}{\|y\|}\big) \big\|, \big\| T\big(\tfrac{x}{\|x\|}\big) + S\big(-\tfrac{y}{\|y\|}\big) \big\| \big\} \leq \beta . \end{aligned}$$

So, we have shown that if $x \in X$ and $y \in Y$ satisfy $\|x\| \leq 1$ and $\|y\| \leq 1$, then $\|Tx + Ty\| \leq \beta$. This implies $\alpha \leq \beta$, and so $\alpha = \beta$, as desired.

It should be clear that if $T \oplus S \colon X \oplus Y \to Z$ denotes the direct sum operator, defined by $(T \oplus S)(x \oplus y) = Tx + Sy$, and $X \oplus Y$ is equipped with the norm $\|x \oplus y\| = \max\{\|x\|, \|y\|\}$, then $\|T \oplus S\| = \alpha = \beta$. ∎

Problem 11.3.2 (Abramovich [**1**]). *Let E be a σ-Dedekind complete Banach lattice and let U_E be its closed unit ball. Consider the set D of all vectors x for which there exists some collection $\{x_1, \ldots, x_n\}$ of pairwise disjoint positive vectors in U_E such that $|x| = x_1 \vee x_2 \vee \cdots \vee x_n$, and let*

$$M(E) = \sup\big\{ \|x\| \colon \ x \in D \big\} .$$

Show that E is lattice isomorphic to an AM-space if and only if $M(E) < \infty$.

Solution: Assume first that E is lattice isomorphic to an AM-space. That is, assume that there exists an M-norm $|||\cdot|||$ on E such that

$$\alpha\|x\| \leq |||x||| \leq \beta\|x\| \tag{$\star$}$$

for each $x \in E$ and some constants $0 < \alpha \leq \beta$. Pick any collection $\{x_1, x_2, \ldots, x_n\}$ of pairwise disjoint positive vectors in U_E and let $x = x_1 \vee x_2 \vee \cdots \vee x_n$. Then we have

$$\|x\| \leq \tfrac{1}{\alpha}\, |||x||| = \tfrac{1}{\alpha} \max_{1 \leq i \leq n} |||x_i||| \leq \tfrac{\beta}{\alpha} \max_{1 \leq i \leq n} \|x_i\| \leq \tfrac{\beta}{\alpha} ,$$

whence $M(E) \leq \frac{\beta}{\alpha} < \infty$.

For the converse assume that $M(E) < \infty$. Recall that in any σ-Dedekind complete vector lattice every principal band is a projection band. For $z \in E^+$ the band projection on the principal band B_z will be denoted by P_z; as is well known, $P_z(x) = \sup_m x \wedge mz$ for each $x \in E^+$.

A direct verification shows that D is a solid subset of E satisfying the inclusions $U_E \subseteq D \subseteq M(E)U_E$. We claim that D is also a convex set. To verify this, pick any $x, y \in D$ and $\alpha \in [0,1]$. We must establish that $\alpha x + (1-\alpha)y \in D$. Since D is solid, we can assume without loss of generality that x and y are in E_+. By the definition of D there are pairwise disjoint positive vectors $x_1, \ldots, x_n$ in U_E such that $x = x_1 \vee \cdots \vee x_n$ and pairwise disjoint positive vectors $y_1, \ldots, y_k$ in U_E such that $y = y_1 \vee \cdots \vee y_k$.

For each $1 \leq i \leq n$ and each $1 \leq j \leq k$ let $u_{ij} = P_{y_j}(x_i)$ and $v_{ij} = P_{x_i}(y_j)$. From the identity

$$u_{ij} \wedge v_{\mu\nu} = \sup_m \big[x_i \wedge m y_j\big] \wedge \sup_\ell \big[y_\nu \wedge \ell x_\mu\big] = \sup_{m,\ell} \big[x_i \wedge m y_j \wedge y_\nu \wedge \ell x_\mu\big] ,$$

it follows that $u_{ij} \wedge v_{\mu\nu} = 0$ for $(i,j) \neq (\mu,\nu)$. Similarly, $(i,j) \neq (\mu,\nu)$ implies $u_{ij} \wedge u_{\mu\nu} = 0$ and $v_{ij} \wedge v_{\mu\nu} = 0$.

Now let $u_i = x_i - \sum_{j=1}^k u_{ij}$ and $v_j = v_j - \sum_{i=1}^n v_{ij}$. Since $\sum_{j=1}^k u_{ij} = P_y(x_i)$ and $\sum_{i=1}^n v_{ij} = P_x(y_j)$, we see that $0 \le u_i \le x_i$ and $0 \le v_j \le y_j$ and so $\|u_i\| \le 1$, $\|v_j\| \le 1$, $u_i \perp y$ and $v_j \perp x$ for all i and j, and hence $u_i \perp y_j$ and $v_j \perp x_i$ for all i and j. In particular, we have $u_i \perp v_j$ for all i and j. Also, from the definitions of the u_i and v_j it is easy to see that $u_i \perp u_\mu$ if $i \ne \mu$ and $v_j \perp v_\nu$ if $j \ne \nu$. Now notice that

$$x = \sum_{i=1}^n \Big(u_i + \sum_{j=1}^k u_{ij}\Big) \quad \text{and} \quad y = \sum_{j=1}^k \Big(v_j + \sum_{i=1}^n v_{ij}\Big).$$

Letting $w_{ij} = \alpha u_{ij} + (1-\alpha)v_{ij}$, we have $\|w_{ij}\| \le 1$ for all i and j and

$$\alpha x + (1-\alpha)y = \sum_{i=1}^n \sum_{j=1}^k w_{ij} + \sum_{i=1}^n \alpha u_i + \sum_{j=1}^n (1-\alpha) v_j\,,$$

where, as shown above, the vectors in this representation are pairwise disjoint and belong to U_E. Hence, $\alpha x + (1-\alpha)y \in D$, and thus D is a convex set.

Now we consider the Minkowski functional of D, i.e., the functional $p_{{}_D} \colon E \to \mathbb{R}$, defined by

$$p_{{}_D}(x) = \inf\big\{\lambda > 0\colon\; x \in \lambda D\big\}\,.$$

From $U_E \subseteq D \subseteq M(E)U_E$, it follows that $p_{{}_D}(x) \le \|x\| \le M(E)p_{{}_D}(x)$ for all $x \in E$. That is, $p_{{}_D}$ is a lattice norm on E that is equivalent to the original norm $\|\cdot\|$ on E. Finally, we claim that $p_{{}_D}$ is an M-norm. To see this, assume that two vectors $x, y \in E^+$ satisfy $x \wedge y = 0$. We must show that

$$p_{{}_D}(x \vee y) \le \max\big\{p_{{}_D}(x), p_{{}_D}(y)\big\}\,.$$

Let $\lambda > \max\big\{p_{{}_D}(x), p_{{}_D}(y)\big\}$. Clearly, $x \in \lambda D$ and $y \in \lambda D$ and from $x \wedge y = 0$, it easily follows that $x \vee y \in \lambda D$. This implies $p_{{}_D}(x \vee y) \le \lambda$, and from this we obtain

$$p_{{}_D}(x \vee y) \le \max\big\{p_{{}_D}(x), p_{{}_D}(y)\big\}\,.$$

Hence, $p_{{}_D}(x \vee y) = \max\big\{p_{{}_D}(x), p_{{}_D}(y)\big\}$ and so $p_{{}_D}$ is an M-norm, as desired. ∎

Problem 11.3.3. *Let $E_n = L_1[0,1]$ and $F_n = L_\infty[0,1]$ for each n. For the Banach lattices $E = (\sum_{n=1}^\infty \oplus E_n)_\infty$ and $F = (\sum_{n=1}^\infty \oplus F_n)_1$ establish the following.*

(a) *Neither E nor F is lattice isomorphic to an AL- or an AM-space.*

(b) *Neither $(E \oplus F)_1$ nor $(E \oplus F)_\infty$ is lattice isomorphic to an AL- or an AM-space.*

(c) *The Banach spaces $(E \oplus F)_1$ and $(E \oplus F)_\infty$ satisfy the Daugavet property for weakly compact operators.*

Solution: (a) We use the notation introduced in Problem 11.3.2. We claim that $M(E) = M(F) = \infty$, and so neither E nor F is lattice isomorphic to an AM-space.

To see that $M(E) = \infty$, start by letting $x_n = n(n+1)\chi_{(\frac{1}{n+1}, \frac{1}{n}]} \in L_1[0,1]$. Clearly, $\|x_n\|_1 = 1$ for each n and if $n \ne m$, then $x_n \wedge x_m = 0$ holds in $L_1[0,1]$. Next, for each n let

$$\mathbf{x}_n = (x_n, 0, 0, 0, \ldots)\,.$$

Note that $\|\mathbf{x}_n\|_\infty = 1$ for each n and $\mathbf{x}_n \wedge \mathbf{x}_m = 0$ in E for all $n \neq m$. In particular, we have

$$\begin{aligned}\big\|\mathbf{x}_1 \vee \mathbf{x}_2 \vee \cdots \vee \mathbf{x}_n\big\|_\infty &= \big\|(x_1 \vee x_2 \vee \cdots \vee x_n, 0, 0, 0, \ldots)\big\|_\infty \\ &= \big\|(x_1 + x_2 + \cdots + x_n, 0, 0, 0, \ldots)\big\|_\infty \\ &= \big\|x_1 + x_2 + \cdots + x_n\big\|_1 = \sum_{i=1}^{n} \|x_i\|_1 = n\,.\end{aligned}$$

This implies $M(E) \geq n$ for each n, and so $M(E) = \infty$.

For the Banach lattice F, for each n let

$$\mathbf{y}_n = (0, 0, \ldots, 0, 0, \mathbf{1}, 0, 0, \ldots)\,,$$

where the function $\mathbf{1} = \chi_{[0,1]} \in L_\infty[0,1]$ occupies the nth coordinate. Clearly, $\mathbf{y}_n \wedge \mathbf{y}_m = 0$ holds in F for all $n \neq m$. Now notice that

$$\big\|\mathbf{y}_1 \vee \mathbf{y}_2 \vee \cdots \vee \mathbf{y}_n\big\|_1 = \big\|(\underbrace{\mathbf{1}, \mathbf{1}, \ldots, \mathbf{1}}_{n\ positions}, 0, 0, 0, \ldots)\big\|_1 = \sum_{i=1}^{n} \|\mathbf{1}\|_\infty = n\,.$$

This implies $M(F) \geq n$ for each n, and so $M(F) = \infty$.

From $F^* = E$, it follows that $M(F^*) = \infty$. An easy argument also shows that $M(E^*) = \infty$. Therefore, neither E^* nor F^* is lattice isomorphic to an AM-space, and hence neither E nor F is lattice isomorphic to an AL-space.

(b) The above arguments applied to $(E \oplus F)_1$ and $(E \oplus F)_\infty$ show that none of these Banach lattices is lattice isomorphic to an AL- or an AM-space.

(c) By Theorems 11.20 and 11.30 both Banach lattices E and F satisfy the Daugavet property with respect to weakly compact operators. Now, applying Theorem 11.30 once again, we see that $(E \oplus F)_1$ and $(E \oplus F)_\infty$ also satisfy the Daugavet property with respect to weakly compact operators. ∎

11.4. The Daugavet Property in $C(\Omega)$-spaces

Problem 11.4.1. *Give an example of a continuous operator on $C[0,1]$ that factors through c_0 but that is not weakly compact.*

Solution: As the hint to the problem suggests, we can use here the following well-known theorem of A. Sobczyk [**75**]: *Every closed subspace of $C[0,1]$ isomorphic to c_0 is complemented.*

We begin by constructing a closed subspace X of $C[0,1]$ which is isomorphic to c_0. To do so, for each n pick a non-negative function $f_n \in C[0,1]$ of norm one and with support in $[\frac{1}{n+1}, \frac{1}{n}]$. Now consider the positive operator $T\colon c_0 \to C[0,1]$ defined by $T(\alpha_1, \alpha_1, \ldots) = \sum_{n=1}^{\infty} \alpha_n f_n$, where the convergence of the series is uniform, i.e., in the norm of $C[0,1]$. It is not difficult to see that T is a lattice isometry. So, if $X = T(c_0)$, then X is a closed subspace of $C[0,1]$ and $T\colon c_0 \to X$ is a surjective lattice isometry. Let $J\colon X \to C[0,1]$ denote the natural embedding of X into $C[0,1]$.

By Sobczyk's theorem, X is complemented in $C[0,1]$. Let $P\colon C[0,1] \to C[0,1]$ be a continuous projection with range X. Now notice that if we consider the scheme of bounded operators

$$C[0,1] \xrightarrow{P} X \xrightarrow{T^{-1}} c_0 \xrightarrow{T} X \xrightarrow{J} C[0,1],$$

then it is easy to see that $P = (JT)(T^{-1}P)$. This shows that $P\colon C[0,1] \to C[0,1]$ factors through c_0.

Finally, we claim that $P\colon C[0,1] \to C[0,1]$ is not a weakly compact operator. To see this, assume by way of contradiction that P is a weakly compact operator. Since the restriction of P to X is the identity operator on X, it follows that I_X (= the identity operator on X) is weakly compact. In particular, the closed unit ball of X is a weakly compact set. This implies that X (and hence c_0) must be a reflexive Banach space, a contradiction. So, P is an example of a non-weakly compact operator on $C[0,1]$ that factors through c_0. ∎

Problem 11.4.2. *For a series $\sum_{n=1}^{\infty} x_n$ in a (real or complex) Banach space X the following statements are equivalent.*

(a) *The series $\sum_{n=1}^{\infty} x_n$ is weakly unconditionally Cauchy.*

(b) *The supremum $\sup_{|\epsilon_n|\le 1}\sup_{m\in\mathbb{N}}\left\|\sum_{n=1}^{m}\epsilon_n x_n\right\|$ is finite.*

(c) *The supremum $\sup_{|\epsilon_n|= 1}\sup_{m\in\mathbb{N}}\left\|\sum_{n=1}^{m}\epsilon_n x_n\right\|$ is finite.*

Solution: (a) $\Longrightarrow$ (b) Assume that the series $\sum_{n=1}^{\infty} x_n$ is weakly unconditionally Cauchy and fix $x^* \in X^*$. This implies that $s = \sum_{n=1}^{\infty}|x^*(x_n)| < \infty$. Now for each $m \in \mathbb{N}$ and each choice of scalars $\epsilon_1, \ldots, \epsilon_m$ satisfying $|\epsilon_n| \le 1$ for each n we have

$$\left|x^*\Big(\sum_{n=1}^{m}\epsilon_n x_n\Big)\right| = \left|\sum_{n=1}^{\infty}\epsilon_n x^*(x_n)\right| \le \sum_{n=1}^{m}|\epsilon_n|\cdot|x^*(x_n)| = \sum_{n=1}^{m}|x^*(x_n)| \le s < \infty.$$

Consequently, the set $\left\{\sum_{n=1}^{m}\epsilon_n x_n\colon\ m \in \mathbb{N} \text{ and } |\epsilon_n| \le 1 \text{ for each } n\right\}$ is weakly bounded. By the Uniform Boundedness Principle this set is also norm bounded, and the validity of (b) is established.

(b) $\Longrightarrow$ (c) This is obvious.

(c) $\Longrightarrow$ (a) Assume that $M = \sup_{|\epsilon_n|=1}\sup_{m\in\mathbb{N}}\left\|\sum_{n=1}^{m}\epsilon_n x_n\right\| < \infty$ and let $x^* \in X^*$. For each n put $\epsilon_n = \frac{|x^*(x_n)|}{x^*(x_n)}$ if $x^*(x_n) \neq 0$ and $\epsilon_n = 1$ if $x^*(x_n) = 0$. Clearly, $|\epsilon_n| = 1$ for each n and for any $m \in \mathbb{N}$ we have

$$\sum_{n=1}^{m}|x^*(x_n)| = \sum_{n=1}^{m}\epsilon_n x^*(x_n) = x^*\Big(\sum_{n=1}^{m}\epsilon_n x_n\Big) \le \|x^*\|\cdot\left\|\sum_{n=1}^{m}\epsilon_n x_n\right\| \le M\|x^*\| < \infty.$$

This implies that $\sum_{n=1}^{\infty}|x^*(x_n)| \le M\|x^*\| < \infty$ for each $x^* \in X^*$, and so the series $\sum_{n=1}^{\infty} x_n$ is weakly unconditionally Cauchy. ∎

Problem 11.4.3. *Prove that a series $\sum_{n=1}^{\infty} x_n$ in a Banach space X is weakly unconditionally Cauchy if and only if for each $\lambda = (\lambda_1, \lambda_2, \ldots) \in c_0$*

the series $\sum_{n=1}^{\infty} \lambda_n x_n$ is unconditionally norm convergent. Furthermore, show that the mapping $T\colon c_0 \to X$, defined by

$$T(\lambda_1, \lambda_2, \ldots) = \sum_{n=1}^{\infty} \lambda_n x_n \,,$$

is a bounded operator that need not be weakly compact.

Solution: If $\sum_{n=1}^{\infty} x_n$ is weakly unconditionally Cauchy, then according to Problem 11.4.2 there exists a constant $M > 0$ such that for each finite subset σ of $\mathbb{N}$ and for each collection of scalars $\{\epsilon_n\colon\ n \in \sigma\}$ satisfying $|\epsilon_n| \leq 1$, we have

$$\Big\| \sum_{n \in \sigma} \epsilon_n x_n \Big\| \leq M\,.$$

Next, fix $\lambda = (\lambda_1, \lambda_2, \ldots) \in c_0$ and for each k let $b_k = \frac{1}{k} + \sup\{|\lambda_n|\colon\ n \geq k\} > 0$. Clearly, $b_k \to 0$. Now note that for any finite subset A of $\{k, k+1, \ldots\}$ we have

$$\Big\| \sum_{n \in A} \lambda_n x_n \Big\| = \Big\| b_k \sum_{n \in A} \tfrac{\lambda_n}{b_k} x_n \Big\| \leq M b_k \xrightarrow[k\to\infty]{} 0\,.$$

Since A is an arbitrary subset of $\{k, k+1, \ldots\}$, it follows that the series $\sum_{n=1}^{\infty} \lambda_n x_n$ is unconditionally norm convergent. (See Problem 1.3.8.)

For the converse, assume that the series $\sum_{n=1}^{\infty} \lambda_n x_n$ is unconditionally norm convergent for each $\lambda = (\lambda_1, \lambda_2, \ldots) \in c_0$. This implies that the series of scalars $\sum_{n=1}^{\infty} x^*(\lambda_n x_n) = \sum_{n=1}^{\infty} \lambda_n x^*(x_n)$ converges for each $x^* \in X^*$. Since $\lambda \in c_0$ is arbitrary and $c_0^* = \ell_1$, it follows that $\{x^*(x_n)\} \in \ell_1$, i.e., $\sum_{n=1}^{\infty} |x^*(x_n)| < \infty$ holds for each $x^* \in X$. This shows that the series $\sum_{n=1}^{\infty} x_n$ is weakly unconditionally Cauchy. (See also the footnote in the proof of Theorem 4.9.)

The fact that T as defined above is a bounded operator follows immediately from statement (b) of Problem 11.4.2. To see that T might fail to be weakly compact, let $X = c_0$ and $\{x_n\}$ be the standard basis in c_0. A direct verification shows that the series $\sum_{n=1}^{\infty} x_n$ is weakly unconditionally Cauchy. To finish the solution notice that in this case $T = I$, the identity operator on c_0. ∎

Problem 11.4.4. *Let S be a topological space without isolated points and let G be an arbitrary open subset of S. Show that neither G nor its closure $\overline{G}$ has isolated points—where, of course, G and $\overline{G}$ are equipped with the induced topology.*

Solution: Assume that S has no isolated points and let G be an open subset of S. Consider first an arbitrary point $s \in G$. If s is an isolated point in G, then $\{s\}$ is an open subset of G. This means that there exists an open subset V of S such that $\{s\} = G \cap V$. Since G is an open subset of S, it follows that $\{s\}$ is also an open subset of S, i.e., s is an isolated point of S, which is impossible. Thus, G has no isolated points.

Now pick an arbitrary point $p \in \overline{G}$. If p is an isolated point of $\overline{G}$, then there exists an open subset W of S such that $\{p\} = W \cap \overline{G}$. This implies that W is an open neighborhood of p. As $p \in \overline{G}$, it follows that $W \cap G \neq \emptyset$. Since $W \cap G \subseteq W \cap \overline{G} = \{p\}$, we infer that $\{p\} = W \cap G$. That is, $\{p\}$ is an open subset of S, i.e., p is an isolated point of S, which is a contradiction. This shows that $\overline{G}$ has no isolated points, and the solution is finished. ∎

Problem 11.4.5. *For a compact Hausdorff space Ω show that.*

(a) *If μ is a regular Borel measure with full support (i.e.,* Supp $\mu=\Omega$*), then the natural embedding $J_\mu\colon C(\Omega) \to L_\infty(\mu)$ is a lattice isometry.*

(b) *If Ω is also metrizable and has no isolated points, then there exists a non-atomic regular Borel probability measure on Ω with full support.*

Solution: (a) Assume that a regular Borel measure μ has full support. It should be clear that the natural embedding $J_\mu\colon C(\Omega) \to L_\infty(\mu)$ is a lattice homomorphism. Now let $f \in C(\Omega)$. We must show that $\|f\|_\infty = \|f\|_{\mu,\infty}$. Since $\|f\|_\infty \geq \|f\|_{\mu,\infty}$ is obviously true, it suffices to establish that $\|f\|_{\mu,\infty} \geq \|f\|_\infty$.

To this end, let $\epsilon > 0$. Notice that the set $V = \{\omega \in \Omega\colon\ |f(\omega)| > \|f\|_\infty - \epsilon\}$ is a non-empty open set. Therefore, from $\operatorname{Supp}\mu = \Omega$, it follows that $\mu(V) > 0$. This implies $\|f\|_{\mu,\infty} \geq \|f\|_\infty - \epsilon$. Since $\epsilon > 0$ is arbitrary, the latter inequality yields $\|f\|_{\mu,\infty} \geq \|f\|_\infty$, as desired.

(b) Assume that Ω is a metrizable compact space without isolated points. In this case there exists a countable collection $\{O_1, O_2, \ldots\}$ of open sets that form a base of the topology on Ω. Since Ω has no isolated points, it follows from Problem 11.4.4 that each $\overline{O}_n$ has no isolated points either. But then, according to Theorem 11.34, for each n there exists a non-atomic regular Borel probability measure μ_n on $\overline{O}_n$.

Now consider the measure $\mu = \sum_{n=1}^\infty 2^{-n}\mu_n$. Then μ is a non-atomic regular Borel probability measure on Ω. Moreover, we claim that it has full support. To see this, assume by way of contradiction that the open set $\Omega\setminus\operatorname{Supp}\mu$ is non-empty. This implies that there exists some k such that $\overline{O}_k \subseteq \Omega\setminus\operatorname{Supp}\mu$. Consequently,

$$0 = \mu(\overline{O}_k) = \sum_{n=1}^{\infty} 2^{-n}\mu_n(\overline{O}_n) > 2^{-k}\mu_k(\overline{O}_k) = 2^{-k} > 0\,,$$

which is impossible. Therefore, μ has full support. ∎

Problem 11.4.6. *Establish the following separability properties.*

(a) *If the adjoint of a bounded operator $T\colon X \to Y$ between Banach spaces has a separable range, then T likewise has a a separable range.*

(b) *If a Banach space X has a separable dual, then X itself is separable. Also, give an example of a separable Banach space whose norm dual is not separable.*

Solution: (a) Let $T\colon X \to Y$ be a bounded operator between Banach spaces such that the range $R(T^*) = \{T^*y^*\colon\ y^* \in Y^*\}$ of its adjoint operator T^* is separable. Pick a countable collection $\{y_1^*, y_2^*, \ldots\}$ of linear functionals in Y^* such that the set $\{T^*y_1^*, T^*y_2^*, \ldots\}$ lies in the unit sphere of $R(T^*)$ and is norm dense there. That is, $\|T^*y_n^*\| = 1$ for each n and if $z^* \in Y^*$ satisfies $\|T^*z^*\| = 1$, then for each $\epsilon > 0$ there exists some m satisfying $\|T^*z^* - T^*y_m^*\| < \epsilon$.

Next, for each n pick a unit vector $x_n \in X$ such that $|(T^*y_n^*)(x_n)| > \frac{1}{2}$, and consider the closed linear subspace Y_0 generated in Y by the countable set $\{Tx_1, Tx_2, \ldots\}$. Obviously Y_0 is separable, and so it suffices to verify that the range of T satisfies $R(T) \subseteq Y_0$. If this is not the case, then there exists some $x \in X$ such that $Tx \notin Y_0$. But then, by the Hahn–Banach theorem, there exists some non-zero $y^* \in Y^*$ such that $y^*(Tx) = T^*y^*(x) \neq 0$ and $y^*(y) = 0$ for all $y \in Y_0$. Scaling appropriately, we can assume that $\|T^*y^*\| = 1$. Now note that for each n we have

$$\begin{aligned} \|T^*y^* - T^*y_n^*\| &\geq \big|(T^*y^* - T^*y_n^*)(x_n)\big| = \big|y^*(Tx_n) - (T^*y_n^*)(x_n)\big| \\ &= \big|(T^*y_n^*)(x_n)\big| > \tfrac{1}{2}\,, \end{aligned}$$

which is impossible. This contradiction establishes that T has a separable range.

(b) Assume that the norm dual X^* of a Banach space X is separable. Consider the identity operator $I\colon X \to X$, and note that the adjoint operator $I^*\colon X^* \to X^*$ coincides with the identity operator on X^*. Since, by our hypothesis, $R(I^*) = X^*$ is separable, it follows from part (a) that $X = R(I)$, the range of I, is also separable.

For the last part notice that if we let $X = C[0,1]$, then X is separable while $C[0,1]^* = \mathrm{ca}[0,1]$ (the Banach lattice of all finite Borel signed measures on $[0,1]$) is non-separable. ∎

Problem 11.4.7 (Weis–Werner [**82**]). *If Ω is a metrizable compact space without isolated points, then show that every weakly compact operator on $C(\Omega)$ satisfies the Daugavet equation.*

Solution: Assume that Ω is a metrizable compact space without isolated points and let $T \in \mathcal{L}\big(C(\Omega)\big)$ be a weakly compact operator. Since Ω is metrizable, it follows that $C(\Omega)$ is separable. So, by Gantmacher's theorem (see Problem 2.4.4), T^{**} maps $C(\Omega)^{**}$ into $C(\Omega)$. This shows that T^{**} has a separable range. Therefore, by Problem 11.4.6, the adjoint operator T^* (and hence also the operator T) has a separable range. Now a glance at Theorem 11.43 guarantees that T satisfies the Daugavet equation. ∎

Problem 11.4.8 (Weis–Werner [**82**]). *Modify the arguments in the proof of Theorem* 11.43 *using the Bartle–Dunford–Schwartz theorem* [**28**, p. 306] *to show that the assumption of metrizability made in Problem* 11.4.7 *is not necessary. That is, show that if Ω is a compact Hausdorff space without isolated points, then every weakly compact operator on $C(\Omega)$ satisfies the Daugavet equation.*

Solution: The Bartle–Dunford–Schwartz theorem is the following result.

- *A non-empty subset A of $C(\Omega)^* = \mathrm{ca}(\Omega)$ is weakly compact if and only if it is norm bounded and there exists a measure $\mu \in \mathrm{ca}(\Omega)$ such that $\nu(E) \to 0$ uniformly in $\nu \in A$ when $\mu(E) \to 0$.*

We shall use this theorem for the solution of our problem. Let $T \in \mathcal{L}\big(C(\Omega)\big)$ be a weakly compact operator, where Ω is a compact Hausdorff space. It follows that the representing kernel $\{\mu_\omega\colon\ \omega \in \Omega\} = \{T^*\delta_\omega\colon\ \omega \in \Omega\}$ of T is a relatively weakly compact set. So, by the Bartle–Dunford–Schwartz theorem there exists a regular Borel measure $\mu \in \mathrm{ca}(\Omega)$ such that each μ_ω is absolutely continuous with respect to μ. Since μ is a finite measure, it follows that the set

$$B = \{\omega \in \Omega\colon\ \mu(\{\omega\}) \neq 0\}$$

is at most countable. The absolute continuity of each μ_ω with respect to μ implies that the set

$$A = \{\omega \in \Omega\colon\ \mu_\omega(\{\omega\}) \neq 0\}$$

is a subset of B. Consequently, A is at most countable.

Now for each $a \in A$ let $\Omega_a = \{\omega \in \Omega\colon\ \omega \neq a\} = \{a\}^c$. Since Ω has no isolated points, it follows that each Ω_a is an open dense subset of Ω. From the Baire Category Theorem (see, for instance [**59**, p. 294]), we conclude that the set

$$\bigcap_{a\in A} \Omega_a = A^c = \{\omega \in \Omega\colon\ \mu_\omega(\{\omega\}) = 0\}$$

is also dense in Ω. This implies that condition (†) of Lemma 11.42 is satisfied, and so T satisfies the Daugavet equation. ∎

Problem 11.4.9 (Ansari [**12**]). *Fix $q \in [0,1]$ and consider the rank-one positive operator $T = \delta_q \otimes \mathbf{1}$ on $C[0,1]$. That is, $Tx = x(q)\mathbf{1}$ for each $x \in C[0,1]$. Establish the following properties of T.*

(a) *T factors through c_0 (and hence T satisfies the Daugavet equation).*

(b) *The operator $jT\colon C[0,1] \to C[0,1]^{**}$ is not disjoint to jI, where $j\colon C[0,1] \to C[0,1]^{**}$ is the canonical embedding and I is the identity operator on $C[0,1]$.*

(c) *The adjoint operator $T^*\colon C[0,1]^* \to C[0,1]^*$ is not disjoint from I^*, the identity operator on $C[0,1]^*$.*

Solution: (a) Since T is of finite-rank, it easily follows that T factors through c_0. Consequently, T must satisfy the Daugavet equation.

(b) To see that jI and jT are not disjoint in $\mathcal{L}(C[0,1], C[0,1]^{**})$, note that

$$\begin{aligned}[jI \wedge jT](\mathbf{1}) &= \inf\{u + jTv\colon\ 0 \le u, v \in C[0,1] \text{ and } u + v = \mathbf{1}\} \\ &= \inf\{u + v(q)\mathbf{1}\colon\ 0 \le u, v \in C[0,1] \text{ and } u + v = \mathbf{1}\},\end{aligned}$$

where the infima are taken in $C[0,1]^{**}$. Consider on $[0,1]$ the characteristic function $w = \chi_{\{q\}}$. This is a bounded Borel function, and so $w \in C[0,1]^{**}$. Obviously $\langle w, \delta_q\rangle = 1$. We shall show that $u + v(q)\mathbf{1} \ge w$ for all $0 \le u, v \in C[0,1]$ with $u + v = \mathbf{1}$. Indeed, for each $0 \le u, v \in C[0,1]$ satisfying $u + v = \mathbf{1}$, we necessarily have that $[u + v(q)\mathbf{1}](q) \ge 1$, i.e., $u + v(q)\mathbf{1} \ge w$. Therefore, $[jI \wedge jT](\mathbf{1}) \ge w > 0$.

(c) To show that I^* and T^* are not disjoint, we will verify that $[I^* \wedge T^*](\delta_q) > 0$. Note that $T^* = \mathbf{1} \otimes \delta_q$ and so $T^*(\delta_q) = \delta_q$ and that if $0 \le \mu, \nu \in C[0,1]^*$ and

$\mu + \nu = \delta_q$, then $\mu = \alpha\delta_q$, $\nu = (1-\alpha)\delta_q$ for some scalar $\alpha \in [0,1]$. Now we have

$$\begin{aligned}[I^* \wedge T^*](\delta_q) &= \inf\{\mu + T^*(\nu)\colon\ 0 \le \mu, \nu \in C[0,1]^* \text{ and } \mu + \nu = \delta_q\} \\ &= \inf\{\alpha\delta_q + T^*((1-\alpha)\delta_q)\colon\ \alpha \in [0,1]\} \\ &= \inf\{\alpha\delta_q + (1-\alpha)\delta_q\colon\ \alpha \in [0,1]\} = \delta_q\,.\end{aligned}$$

This shows that $I^* \wedge T^* > 0$. ∎

11.5. Slices and the Daugavet Property

Problem 11.5.1. *Let X be a linear subspace of a Banach space Y and let $J\colon X \to Y$ denote the inclusion operator. Show that the following statements are equivalent.*

(1) *Every rank-one operator $T\colon X \to Y$ satisfies the Daugavet equation, i.e., $\|J + T\| = 1 + \|T\|$.*

(2) *If $y_0 \in Y$ and $x_0^* \in X^*$ are unit vectors, then for each $0 < \epsilon < 1$ there exist a unit vector $y \in Y$ and some $0 < \delta < 1$ such that:*
   (a) *$Slice(U_{Y^*}, y, \delta) \subseteq Slice(U_{Y^*}, y_0, \epsilon)$.*
   (b) *$y^* \in Slice(U_{Y^*}, y, \delta)$ implies $\|x_0^* + J^*y^*\| \ge 2 - \epsilon$.*

Solution: (1) $\Longrightarrow$ (2) Assume that $y_0 \in Y$ and $x_0^* \in X^*$ are unit vectors, and let $0 < \epsilon < 1$. Start by observing that the rank-one operator $T = x_0^* \otimes y_0 \in \mathcal{L}(X,Y)$ satisfies $\|T\| = 1$. So, by our hypothesis, $\|J + T\| = 1 + \|T\| = 2$. Fix some unit vector $x \in X$ satisfying $\|x + Tx\| > 2 - \epsilon > 0$ and $x_0^*(x) \ge 0$, and then let

$$y = \tfrac{x+Tx}{\|x+Tx\|} \quad\text{and}\quad \delta = 1 - \tfrac{2-\epsilon}{\|x+Tx\|}\,.$$

Clearly, $\|y\| = 1$ and $0 < \delta < 1$.

Next, fix $y^* \in Slice(U_{Y^*}, y, \delta)$, and note that

$$\begin{aligned}y^*(x) + x_0^*(x)y^*(y_0) &= \langle x + Tx, y^*\rangle = \|x+Tx\|\langle y, y^*\rangle \\ &\ge \|x+Tx\|(1-\delta) = 2 - \epsilon\,.\end{aligned}$$

From this, $y^*(x) \le 1$ and $0 \le x_0^*(x) \le 1$, we get

$$x_0^*(x)y^*(y_0) \ge 2 - \epsilon - y^*(x) \ge 1 - \epsilon > 0\,.$$

This implies $0 < x_0^*(x) \le 1$ and $0 < y^*(y_0) \le 1$, and therefore,

$$y^*(y_0) \ge x_0^*(x)y^*(y_0) \ge 1 - \epsilon > 0\,.$$

That is, $y^* \in Slice(U_{Y^*}, y_0, \epsilon)$, and so $Slice(U_{Y^*}, y, \delta) \subseteq Slice(U_{Y^*}, y_0, \epsilon)$.

Finally, taking into account that $y^* \in Slice(U_{Y^*}, y, \delta)$ implies $0 < y^*(y_0) \le 1$, we get

$$\begin{aligned}\|x_0^* + J^*y^*\| &\ge \langle x, x_0^* + J^*y^*\rangle = J^*y^*(x) + x_0^*(x) \\ &\ge y^*(x) + y^*(y_0)x_0^*(x) \ge 2 - \epsilon\,.\end{aligned}$$

(2) $\Longrightarrow$ (1) Let $T = x_0^* \otimes y_0 \in \mathcal{L}(X,Y)$ be a rank-one operator. We must verify that $\|J+T\| \geq 1 + \|T\|$. Without loss in generality we can assume that $\|x_0^*\| = \|y_0\| = 1$ (and so $\|T\| = \|x_0^*\| \cdot \|y_0\| = 1$).

Fix any $0 < \epsilon < 1$. According to our assumption, there exist a unit vector $y \in Y$ and some $0 < \delta < 1$ such that $y^* \in Slice(U_{Y^*}, y, \delta)$ implies $y^* \in Slice(U_{Y^*}, y_0, \epsilon)$ and $\|x_0^* + J^*y^*\| \geq 2 - \epsilon$. Clearly, $Slice(U_{Y^*}, y, \delta) \neq \emptyset$, and thus if we fix any $y^* \in Slice(U_{Y^*}, y, \delta)$, then we have $y^*(y_0) \geq 1 - \epsilon$ and $0 \leq 1 - y^*(y_0) < \epsilon$. So,

$$\begin{aligned}
\|J+T\| &= \|J^* + T^*\| \geq \big\|(J^* + y_0 \otimes x_0^*)y^*\big\| = \big\|J^*y^* + y^*(y_0)x_0^*\big\| \\
&= \big\|y^*(y_0)\big[x_0^* - J^*y^*\big] - \big[y^*(y_0) - 1\big]J^*y^*\big\| \\
&\geq y^*(y_0)\big\|x_0^* + J^*y^*\big\| - \big[1 - y^*(y_0)\big]\big\|J^*y^*\big\| \\
&\geq (1-\epsilon)(2-\epsilon) - \epsilon\,.
\end{aligned}$$

Letting $\epsilon \to 0^+$, we get $\|J+T\| \geq 2$, and the solution is finished. ∎

Problem 11.5.2 ([43]). *Let X be a linear subspace of a Banach space Y and let $J\colon X \to Y$ denote the inclusion operator. Show that the following statements are equivalent.*

(a) *Every rank-one operator $T\colon X \to Y$ satisfies the Daugavet equation, i.e., $\|J+T\| = 1 + \|T\|$.*

(b) *For any pair of unit vectors $y_0 \in Y$ and $x_0^* \in X^*$ and each $0 < \epsilon < 1$ there is $x \in X$ satisfying $\|x\| = 1$, $x_0^*(x) \geq 1 - \epsilon$, and $\|x + y_0\| \geq 2 - \epsilon$.*

(c) *For each pair of unit vectors $y_0 \in Y$ and $x_0^* \in X^*$ and each $0 < \epsilon < 1$ there is some unit linear functional $y^* \in Y^*$ satisfying $y^*(y_0) \geq 1 - \epsilon$ and $\|x_0^* + J^*y^*\| \geq 2 - \epsilon$.*

Solution: (a) $\Longleftrightarrow$ (b) Assume that (a) is true. Let $y_0 \in Y$ and $x_0^* \in X^*$ be unit vectors, and let $0 < \epsilon < 1$. According to statement (2) of Lemma 11.46 there exist a linear functional $x^* \in X^*$ of norm one and some $0 < \delta < 1$ such that:

(1) $Slice(U_X, x^*, \delta) \subseteq Slice(U_X, x_0^*, \epsilon)$.

(2) For each $x \in Slice(U_X, x^*, \delta)$ we have $\|x + y_0\| \geq 2 - \epsilon$.

Clearly, any $x \in Slice(U_X, x^*, \delta)$ satisfies the desired properties.

Now suppose that (b) is true. Let $T = x_0^* \otimes y_0 \in \mathcal{L}(X,Y)$ be a rank-one operator. We must verify that $\|J+T\| \geq 1 + \|T\|$. Without loss of generality we can assume that $\|x_0^*\| = \|y_0\| = 1$ (and so $\|T\| = 1$). Fix $0 < \epsilon < 1$. By our assumption there exists a unit vector $x \in X$ satisfying $x_0^*(x) \geq 1 - \epsilon$ and $\|x + y_0\| \geq 2 - \epsilon$. The latter implies $0 \leq 1 - x_0^*(x) \leq \epsilon$, and so

$$\begin{aligned}
\|J+T\| &\geq \|x + Tx\| = \big\|x + x_0^*(x)y_0\big\| \\
&= \big\|[x_0^*(x)x + x_0^*(x)y_0] + [1 - x_0^*(x)]x\big\| \\
&\geq \big\|x_0^*(x)(x + y_0)\big\| - \big\|[1 - x_0^*(x)]x\big\| \\
&\geq (1-\epsilon)(2-\epsilon) - \epsilon
\end{aligned}$$

for each $0 < \epsilon < 1$. Letting $\epsilon \downarrow 0$ yields $\|J+T\| \geq 2$. Therefore, $\|J+T\| = 2$, and so T satisfies the Daugavet equation.

(a) $\Longleftrightarrow$ (c) Suppose that (a) is true. Then statement (1) of Lemma 11.46 is true and the validity of (c) follows immediately from statement (3) of this lemma.

For the converse assume that (c) is true. Let $T = x_0^* \otimes y_0 \in \mathcal{L}(X,Y)$ be a rank-one operator. We must verify that $\|J+T\| \geq 1 + \|T\|$. Without loss in generality we can assume that $\|x_0^*\| = \|y_0\| = 1$ (and so $\|T\| = \|x_0^*\| \cdot \|y_0\| = 1$).

Now fix any $0 < \epsilon < 1$. According to our assumption, there exists a unit linear functional $y^* \in Y^*$ satisfying $\|x_0^* + J^*y^*\| \geq 2 - \epsilon$ and $y^*(y_0) \geq 1 - \epsilon$. The latter implies $0 \leq 1 - y^*(y_0) < \epsilon$, and so

$$\begin{aligned} \|J+T\| &= \|J^* + T^*\| \geq \|(J^* + y_0 \otimes x_0^*)y^*\| = \|J^*y^* + y^*(y_0)x_0^*\| \\ &= \|y^*(y_0)[x_0^* - J^*y^*] - [y^*(y_0) - 1]J^*y^*\| \\ &\geq y^*(y_0)\|x_0^* + J^*y^*\| - [1 - y^*(y_0)]\|J^*y^*\| \\ &\geq (1-\epsilon)(2-\epsilon) - \epsilon\,. \end{aligned}$$

Letting $\epsilon \to 0^+$, we get $\|J+T\| \geq 2$, and the solution is finished. ∎

Problem 11.5.3. *If X is a reflexive Banach space, then X does not satisfy the Daugavet property for rank-one operators.*

Solution: If X has the Daugavet property for rank-one operators, then Theorem 11.50 implies that X satisfies the Daugavet property for weakly compact operators. Since X is reflexive, the operator $T = -I$ is weakly compact and hence T satisfies the Daugavet equation. However, this is impossible since $\|I+T\| = 0$. ∎

Problem 11.5.4. *Let $C = \ell_2^+$, the positive cone of ℓ_2. Show that 0 is an exposed point of C that fails to be strongly exposed.*

Solution: Consider the Banach space $X = \ell_2$ and let $C = \ell_2^+$. Clearly, C is a norm closed convex subset of ℓ_2. We claim that the point $0 \in C$ is an exposed point of C that fails to be strongly exposed. To prove this, let $u = \big(1, \frac{1}{2}, \frac{1}{3}, \frac{1}{4}, \ldots\big) \in \ell_2^+$.

If we consider the linear functional $z^* = -u \in \ell_2 = \ell_2^*$, then it is not difficult to see that $z^*(0) = 0 > z^*(x)$ for each $x = (x_1, x_2, \ldots) > 0$. That is, the continuous linear functional z^* exposes 0, and so 0 is an exposed point. Now for each n let

$$y_n = \big(0, 0, \ldots, 0, 1, \tfrac{1}{2}, \tfrac{1}{3}, \tfrac{1}{4}, \ldots\big)\,,$$

where the zeros occupy the first n coordinates, and note that $\|y_n\|_2 = \|u\|_2 = \frac{\pi}{\sqrt{6}}$ for each n. Moreover, for any $v = (v_1, v_2, \ldots) \in \ell_2 = \ell_2^*$ we have

$$\begin{aligned} |\langle y_n, v\rangle| &= \Big|\sum_{i=n+1}^{\infty} \tfrac{v_i}{i-n}\Big| \leq \Big[\sum_{i=n+1}^{\infty} \tfrac{1}{(i-n)^2}\Big]^{\frac{1}{2}} \cdot \Big[\sum_{i=n+1}^{\infty} v_i^2\Big]^{\frac{1}{2}} \\ &= \tfrac{\pi}{\sqrt{6}}\Big[\sum_{i=n+1}^{\infty} v_i^2\Big]^{\frac{1}{2}} \xrightarrow[n\to\infty]{} 0\,. \end{aligned}$$

That is, $\langle y_n, v\rangle \to \langle 0, v\rangle$ for each $v \in \ell_2 = \ell_2^*$. Since $\|y_n\| = \frac{\pi}{\sqrt{6}} \not\to 0$, this implies that 0 is not a strongly exposed point of C. ∎

Problem 11.5.5. *Use Theorem 11.50 to present an alternative proof of the second part of Theorem 11.20:* If E is an atomless AL-space, then E satisfies the Daugavet property for weakly compact operators.

Solution: Let E be an atomless AL-space. Without loss of generality we can assume that $E = L_1(\Omega, \Sigma, \mu)$, where (Ω, Σ, μ) is a non-atomic measure space. To establish that E satisfies the Daugavet property for weakly compact operators, we must show (according to Theorem 11.50) that E satisfies the Daugavet property for rank-one operators.

To this end, let $\phi \in L_1^*(\mu)$ and $u \in L_1(\mu)$. Here it should be pointed out that we do not make any assumptions regarding the measure space (Ω, Σ, μ), and therefore we cannot claim that $L_1^*(\mu)$ coincides with $L_\infty(\mu)$. It is exactly this circumstance that makes the proof more complicated than that of Problem 11.2.5. To establish that $\phi \otimes u$ satisfies the Daugavet equation, it suffices to show (according to Lemma 11.21) that $\phi \otimes u$ is disjoint from the identity operator, i.e., $I \wedge |\phi \otimes u| = 0$ holds in $\mathcal{L}_r\big(L_1(\mu), L_1(\mu)\big)$. Since $|\phi \otimes u| = |\phi| \otimes |u|$, we can assume without loss of generality that $\phi \geq 0$ and $u \geq 0$. Next, let $0 \leq x \in L_1(\mu)$, and put

$$0 \leq w = [I \wedge (\phi \otimes u)](x) = \inf\big\{z + \phi(x - z)u \colon\ 0 \leq z \leq x\big\}. \qquad (\star)$$

Now fix $\epsilon > 0$. Since ϕ is a continuous linear functional, there exists some $\delta > 0$ such that $\|u\|_1 < \delta$ implies $|\phi(u)| < \epsilon$. The absolute continuity of the norm of the vector x guarantees the existence of some $\eta > 0$ such that if $A \in \Sigma$ satisfies $\mu(A) < \eta$, then $\|x\chi_A\|_1 < \delta$ (and so $\phi(x\chi_A) < \epsilon$ also holds).

Since the function x is integrable, its support is σ-finite. That is, we can write $\operatorname{Supp} x = \big\{\omega \in \Omega \colon\ x(\omega) > 0\big\} = \bigcup_{n=1}^\infty \Omega_n$, where $\Omega_n \in \Sigma$ and $\mu(\Omega_n) < \infty$ for each n. Next, fix some n and consider the measurable set Ω_n. Pick some k such that $\frac{\mu(\Omega_n)}{k} < \eta$, and then select k pairwise disjoint measurable subsets $\mathcal{S}_1, \ldots, \mathcal{S}_k$ of Ω such that $\mu(\mathcal{S}_i) = \frac{\mu(\Omega_n)}{k}$ for each $1 \leq i \leq k$ and $\bigcup_{i=1}^k \mathcal{S}_i = \Omega_n$. (This selection is possible since the measure μ is non-atomic.) If for each $1 \leq i \leq k$ we let $f_i = x\chi_{\mathcal{S}_i} \in L_1(\mu)$ and $g_i = x - f_i = \sum_{j \neq i} f_j$, then we have $0 \leq g_i \leq x$, and so from $(\star)$, we see that $w \leq g_i + \phi(f_i)u \leq f_i + \epsilon u$ for each $0 \leq i \leq k$. This implies $w(\omega) \leq \epsilon u(\omega)$ for μ-almost all $\omega \in \mathcal{S}_i$ and for each $0 \leq i \leq k$. Therefore, $w(\omega) \leq \epsilon u(\omega)$ for μ-almost all $\omega \in \Omega_n$. In turn, this shows that $w(\omega) \leq \epsilon u(\omega)$ for μ-almost all $\omega \in \operatorname{Supp} x$. Since clearly w vanishes outside of the support of x, it follows that $0 \leq w \leq \epsilon u$ for each $\epsilon > 0$. This shows that $w = 0$, and so $[I \wedge (\phi \otimes u)](x) = 0$ for each $x \in L_1^+(\mu)$. Therefore, $I \wedge (\phi \otimes u) = 0$ holds in $\mathcal{L}\big(L_1(\mu), L_1(\mu)\big)$, as desired. (See also Problem 5.1.8.) ∎

Problem 11.5.6 ([**43, 47**]). *Let (Ω, Σ, μ) be a non-atomic measure space and let Z be a Banach space. Show that the Banach space $X = L_1(\mu, Z)$ of Z-valued Bochner integrable functions satisfies the Daugavet property for weakly compact operators.*

Solution: It is enough to show that condition (b) of Problem 11.5.2 is fulfilled. To this end, let $y \in X$ and $x^* \in X^*$ be two unit vectors and fix some $0 < \epsilon < 1$. It is well known that the linear functional x^* can be identified with a weak* measurable function taking values in Z^*; see [**24**].

Choose some $B \in \Sigma$ such that $\|\chi_B y\|_X \leq \frac{\epsilon}{2}$ and $\|\chi_B x^*\|_{X^*} \geq 1 - \frac{\epsilon}{2}$. Next, choose any unit vector $x \in X$ with support in B and such that $x^*(x) \geq 1 - \epsilon$. This easily implies $\|x + y\|_X \geq 2 - \epsilon$, and hence condition (b) of Problem 11.5.2 is fulfilled, as desired. ∎

Problem 11.5.7 (Kadets [**40**]; Khalil [**47**]). *Let Ω be a compact Hausdorff space without isolated points and let Z be an arbitrary Banach space. Show that the Banach space $X = C(\Omega, Z)$ of Z-valued continuous functions satisfies the Daugavet property for weakly compact operators.*

Solution: It suffices to establish that condition (b) of Problem 11.5.2 is fulfilled. So, fix $0 < \epsilon < 1$ and consider two unit vectors $y \in X$ and $x^* \in X^*$. Also pick a unit vector $u \in X$ such that $x^*(u) > 1 - \epsilon$. It is well known that the linear functional x^* can be identified with a Z^*-valued regular Borel measure of bounded variation; see [**24**]. In particular, x^* has at most countably many atoms. Therefore (since Ω has no isolated points and $\|y\| = 1$), we can find a point $\omega_0 \in \Omega$ and an open neighborhood V of ω_0 such that:

(a) $\|y(\omega_0)\|_Z > 1 - \epsilon$.

(b) $|x^*|(V) < \epsilon$, that is, the total variation of x^* on V is less than ϵ.

Now the Urysohn–Tietze extension theorem guarantees the existence of a function $v \in X$ such that $v(\omega) = u(\omega)$ for each $\omega \in \Omega \setminus V$, $v(\omega_0) = y(\omega_0)$ and $\|v(\omega)\|_Z \le \|y(\omega)\|_Z$ for each $\omega \in V$. In particular, $1 - \epsilon < \|v\|_X \le 1$. Let $x = \frac{1}{\|v\|_X} v$. Clearly, $\|x\|_X = 1$ and

$$\begin{aligned} \|x + y\|_X &\ge \|x(\omega_0) + y(\omega_0)\|_Z = \left\| \tfrac{1}{\|v\|_X} v(\omega_0) + y(\omega_0) \right\|_Z \\ &= \left\| \tfrac{1}{\|v\|_X} y(\omega_0) + y(\omega_0) \right\|_Z = \left(\tfrac{1}{\|v\|_X} + 1 \right) \|y(\omega_0)\|_Z \\ &\ge 2(1 - \epsilon)\,. \end{aligned}$$

A straightforward calculation also shows that

$$x^*(x) = x^*(x\chi_{\Omega \setminus V}) + x^*(x\chi_V) > 1 - 4\epsilon\,.$$

Thus, the unit vector x satisfies condition (b) of Problem 11.5.2, and so the Banach space X has the Daugavet property for weakly compact operators. ∎

11.6. Narrow Operators

Problem 11.6.1. *Give an example of a weakly compact operator that is not narrow.*

Solution: Let $E = L_p[0, 1]$, where $1 < p < \infty$. Since E is reflexive, the identity operator I on E is weakly compact. However, I is not a narrow operator. Indeed, if we let $A = [0, 1]$, then for any measurable subset B of A with $\lambda(B) \ge \frac{1}{2}$ (where λ denotes the Lebesgue measure on $[0, 1]$) and any function $x \in L_p[0, 1]$ with $|x| = \chi_B$ we have

$$\|Ix\|_p = \|x\|_p \ge 2^{-\frac{1}{p}}\,.$$

This shows that the identity I is not a narrow operator. ∎

Problem 11.6.2. *Show that each weakly compact operator on $L_1(\mu)$ is a narrow operator.*

Solution: It is well known that each weakly compact operator $T\colon L_1(\mu) \to Y$ is a Dunford–Pettis operator; see for instance [**6**, pp. 335–336]. Therefore, by Theorem 11.57, the operator T is narrow. ∎

Problem 11.6.3. *If in a scheme of bounded operators $E \xrightarrow{S} Y \xrightarrow{T} Z$ the operator S is narrow, then show that the operator $TS\colon E \to Z$ is narrow.*

Solution: If $A \in \Sigma$ and $\epsilon > 0$ are given, then there exists some $x \in E$ such that $|x| = \chi_A$ and $\|Sx\| < \epsilon$. But then we have $\|(TS)x\| = \|T(Sx)\| < \epsilon\|T\|$, and this shows that TS is a narrow operator. ∎

Problem 11.6.4. *Give an example of a narrow operator that is not regular.*

Solution: It is well known that there exists a compact non-regular operator $T\colon L_2[0,1] \to L_2[0,1]$. (See for instance [**6**, Example 16.6].) By Theorem 11.57 the operator T is automatically narrow. ∎

Problem 11.6.5. *Let $T\colon E \to Y$ be a narrow operator. Show that for each $A \in \Sigma$ there exists a sequence of Rademacher functions $\{r_n\}$ supported by A and such that $\|Tr_n\| \to 0$.*

Solution: The proof is by induction. Fix $A \in \Sigma$. By Theorem 11.53 there exists a measurable subset B_1 of A such that $\mu(B_1) = \frac{1}{2}\mu(A)$ and $\|Tr_1\| < 2^{-1}$, where $r_1 = \chi_{B_1} - \chi_{A \setminus B_1}$.

For the induction step assume that the Rademacher function r_n satisfying

$$r_n = \sum_{k=1}^{2^n} (-1)^{k+1} \chi_{B_k^n}$$

has been constructed, where $B_i^n \cap B_j^n = \emptyset$ for $i \neq j$, $\mu(B_k^n) = 2^{-n}\mu(A)$, and $\|Tr_n\| < 2^{-n}$. Applying Theorem 11.53 again, we can write each B_k^n as a disjoint union $B_k^n = C_k^n \cup D_k^n$ with $\mu(C_k^n) = \mu(D_k^n) = \frac{1}{2}\mu(B_k^n)$ such that the function $x_k = \chi_{C_k^n} - \chi_{D_k^n}$ satisfies $\|Tx_k\| < 2^{-(2n+1)}$. Now let $r_{n+1} = \sum_{k=1}^{2^n} x_k$.

It is easy to verify that the above inductive argument produces a sequence $\{r_n\}$ of Rademacher functions with the desired properties. ∎

Problem 11.6.6 (Plichko–Popov [**65**]). *Show that the sum of a narrow and a compact operator is narrow.*

Solution: Let $T\colon E \to Y$ be a narrow operator and let $K\colon E \to Y$ be a compact operator. We want to prove that $T + K$ is a narrow operator.

Fix $A \in \Sigma$. By Problem 11.6.5 there exists a sequence $\{r_n\}$ of Rademacher functions satisfying $|r_n| = \chi_A$ for each n and $\|Tr_n\| \to 0$. By Lemma 11.56, we know that $r_n \xrightarrow{w} 0$. Since the operator K is compact, it follows that $\|Kr_n\| \to 0$ and hence $\|(T+K)r_n\| \to 0$. This, coupled with the condition $|r_n| = \chi_A$ for each n, implies that $T + K$ satisfies the definition of a narrow operator. (Note that, in fact, this solution shows that the sum of a narrow operator and a Dunford–Pettis operator is narrow.) ∎

Problem 11.6.7 (Plichko–Popov [**65**]). *Let $E = L_p[0,1]$, where $1<p<\infty$. Show that the identity operator I is the sum of two narrow projection operators—and hence the sum of two narrow operators need not be narrow.*

Solution: Let $\{h_1, h_2, \ldots\}$ denote the standard Haar system in E; see Problem 1.3.3 for a precise definition of the Haar functions. In that problem we showed that the Haar system is a basis in E, and we also mentioned a stronger result that will be used here: *The Haar system is an unconditional basis in E.* (It is not that easy to establish this fact. For a proof see [**72**, p. 407].)

From the definition of the Haar functions we know that the support of each Haar function is an interval of the form $\left[\frac{k}{2^n}, \frac{k+1}{2^n}\right]$ for some $k, n \in \mathbb{N}$. Next, let $\mathbb{N} = \mathbb{N}_1 \cup \mathbb{N}_2$ be a splitting of $\mathbb{N}$ into disjoint infinite subsets. For $j = 1, 2$ we let E_j be the closed linear span of the Haar functions h_i such that $2^{m-1} \le i < 2^m$ holds for some $m \in \mathbb{N}_j$.

Since the sequence $\{h_1, h_2, \ldots\}$ is an unconditional basis, the closed subspaces E_1 and E_2 are complemented and $E = E_1 \oplus E_2$. We will show now that the natural projection P_1 of E onto E_1 along the subspace E_2 is a narrow operator.

To this end, fix $A \in \Sigma$ and $\epsilon > 0$, and then pick some $\delta > 0$ such that $\|\chi_B\| < \epsilon$ whenever $B \in \Sigma$ satisfies $\mu(B) < \delta$; see Lemma 11.51. Let us also choose a finite collection $\{A_1, A_2, \ldots, A_n\}$ of pairwise disjoint intervals that are support sets of the Haar functions such that $\mu(A \Delta \widehat{A}) < \delta$, where $\widehat{A} = \bigcup_{k=1}^n A_k$. We will also assume that δ is small enough so that $\mu(A \cap \widehat{A}) \ge \frac{1}{2}\mu(A)$.

Next, choose $m_1 \in \mathbb{N}_2$ large enough so that for each h_i with $2^{m_1-1} \le i < 2^{m_1}$ its support $\operatorname{Supp} h_i$ has measure at most $\mu(A_1)$ and some of the supports of these functions cover A_1. That is,

$$A_1 = \bigcup_{i \in \Delta_{m_1}} \operatorname{Supp} h_i\,,$$

for an appropriate subset Δ_{m_1} of $\left[2^{m_1-1}, 2^{m_1}\right) \cap \mathbb{N}$.

At the next step, let us choose $m_2 \in \mathbb{N}_2$ and a subset Δ_{m_2} of $\left[2^{m_2-1}, 2^{m_2}\right) \cap \mathbb{N}$ such that $m_2 > m_1$ and

$$A_2 = \bigcup_{i \in \Delta_{m_2}} \operatorname{Supp} h_i\,.$$

We repeat the same procedure inductively for the sets $A_3, \ldots, A_n$ and obtain the sets of integers Δ_{m_k} for $k = 3, \ldots, n$. Let $\widehat{x}$ be the sum of the Haar functions corresponding to the indices from the above sets Δ_{m_k}, i.e., $\widehat{x} = \sum_{k=1}^n \sum_{i \in \Delta_{m_k}} h_i$. By construction the function $\widehat{x} \in E_2$ and $|\widehat{x}| = \chi_{\widehat{A}}$.

Finally, let $B = \widehat{A} \cap A$ and $x = \widehat{x}\chi_B$. As mentioned above, $\mu(B) > \frac{1}{2}\mu(A)$ and clearly $|x| = \chi_B$. Also note that $|x - \widehat{x}| = \chi_{\widehat{A}\Delta A}$, and hence $\|x - \widehat{x}\| < \epsilon$ since $\mu(\widehat{A}\Delta A) < \delta$. Clearly, $\widehat{x} \in E_2$, and therefore $P_1\widehat{x} = 0$. This implies that $\|P_1 x\| = \|P_1(x - \widehat{x})\| < \epsilon\|P_1\|$, and consequently the projection P_1 is a narrow operator.

By symmetry, the natural projection P_2 of E onto E_2 along the subspace E_1 is also a narrow operator, and thus the identity operator I is the sum of two narrow operators, $I = P_1 + P_2$. (It should be noted that the above proof shows that our previous conclusion is also true for each rearrangement invariant Banach function space on $[0,1]$ with an unconditional basis.) ∎

Problem 11.6.8 (Plichko–Popov [**65**]). *Show that each operator on $L_p[0,1]$, where $1 < p < \infty$, is the sum of two narrow operators.*

Solution: Let $E = L_p[0,1]$. According to Problem 11.6.7 there exist two narrow operators P_1 and P_2 such that $I = P_1 + P_2$. Therefore, for each operator T on E we have $T = TI = TP_1 + TP_2$. Now a glance at Problem 11.6.3 shows that TP_1 and TP_2 are both narrow operators, and the solution is finished. ∎

Problem 11.6.9. *Let (Ω, Σ, μ) be a non-atomic probability space. Show that every generalized sequence of Rademacher functions $\{r_n(A)\}$ satisfies the conclusion of the Riemann–Lebesgue lemma, i.e., $\int_\Omega x r_n \, d\mu \to 0$ for each $x \in L_1(\mu)$.*

Solution: We begin by noting that the desired conclusion does not follow directly from Lemma 11.56 since the norm on $L_\infty(\mu)$ is not order continuous. However, a simple approximation does the job. Given $x \in L_1(\mu)$, fix any $\epsilon > 0$ and pick any function $y \in L_\infty(\mu)$ such that $\|x - y\|_1 < \epsilon$. By Lemma 11.56 there is $n_0 \in \mathbb{N}$ such that $|\int_\Omega y r_n \, d\mu| < \epsilon$ for all $n \geq n_0$. Also we have $|\int_\Omega (x-y) r_n \, d\mu| \leq \|x - y\|_1 < \epsilon$. This yields that $|\int_\Omega x r_n \, d\mu| < 2\epsilon$ for each $n \geq n_0$. ∎

Problem 11.6.10. *Let (Ω, Σ, μ) be a non-atomic probability space, and assume that E is a Banach function space with order continuous norm satisfying $L_\infty(\mu) \subseteq E \subseteq L_1(\mu)$. Also, let $\{r_n\}$ be a sequence of Rademacher functions supported by a measurable subset A of Ω. Establish the following.*

(a) *For each $x \in E$ we have $x r_n \xrightarrow{w} 0$ in E.*

(b) *For each $x \in L_1(\mu)$ we have $\|x + x r_n\|_1 \to \|x\|_1$.*

Solution: (a) Fix $x \in E$ and let $f \in E^*$. Since E has an order continuous norm, we know that $E^* = E_n^*$, that f can be identified with a measurable function, and that $fx \in L_1(\mu)$. We must show that $f(x r_n) = \int_\Omega f \cdot x r_n \, d\mu \to 0$. Since $fx \in L_1(\mu)$, this follows immediately from Problem 11.6.9 in view of the obvious identity

$$\int_\Omega f \cdot x r_n \, d\mu = \int_\Omega f x \cdot r_n \, d\mu \, .$$

(b) Letting $E = L_1(\mu)$, it follows from part (a) that $\int_\Omega |x| r_n \, d\mu \to 0$ for each $x \in L_1(\mu)$. Hence $\int_\Omega (|x| + |x| r_n) \, d\mu \to \int_\Omega |x| \, d\mu = \|x\|_1$. It remains to note that $\|x + x r_n\|_1 = \int_\Omega |(1 + r_n) x| \, d\mu = \int_\Omega (|x| + |x| r_n) \, d\mu$ since $\mathbf{1} + r_n \geq 0$. Thus indeed $\|x + x r_n\|_1 \to \|x\|_1$. ∎

11.7. Some Applications of the Daugavet Equation

Problem 11.7.1. *Let X be a Banach space and assume that there exists an increasing continuous function $\psi \colon \mathbb{R}_+ \to \mathbb{R}$ and a scalar $\gamma > 0$ such that* *for every finite-rank projection P we have*

$$\psi(\|I-P\|) \geq \gamma + \psi(\|P\|)\,. \tag{$\star$}$$

Then show that X does not have an unconditional basis.

Solution: Assume by way of contradiction that the Banach space X has an unconditional basis $\{e_n\}$.

Let $\mathcal{A}$ be the collection of all finite subsets of $\mathbb{N}$. For each $A \in \mathcal{A}$ we denote by P_A the natural projection on the closed subspace generated by $\{e_k\colon\ k \in A\}$ and by Q_A the complementary projection, i.e., $Q_A = I - P_A$. Since the basis is unconditional, Lemma 1.53 guarantees that

$$\alpha = \sup\bigl\{\|P_A\|\colon\ A \in \mathcal{A}\bigr\} < \infty\,.$$

Consequently

$$\beta = \sup\bigl\{\|Q_A\|\colon\ A \in \mathcal{A}\bigr\} < \infty\,.$$

By the hypothesis $\psi(\|Q_A\|) \geq \gamma + \psi(\|P_A\|)$ for each $A \in \mathcal{A}$. Since ψ is increasing, we get $\psi(\beta) \geq \gamma + \psi\bigl(\|P_A\|\bigr)$ for each $A \in \mathcal{A}$. Taking the supremum over $A \in \mathcal{A}$, we obtain

$$\psi(\beta) \geq \gamma + \psi(\alpha)\,. \tag{$\star\star$}$$

On the other hand, for each $A \in \mathcal{A}$ and for each vector $x \in X$ we obviously have $\|P_{A_n}x - Q_Ax\| \to 0$, where $A_n = \{1,\ldots,n\} \cap A^{\mathrm{c}}$, and this implies that $\|Q_A\| \leq \sup_{A\in\mathcal{A}}\|P_A\| = \alpha$. Taking the supremum over $\mathcal{A}$ on the left side of the last inequality, we get $\beta \leq \alpha$, whence $\psi(\beta) \leq \psi(\alpha)$ in contradiction to ($\star\star$).

If X satisfies the Daugavet property for rank-one operators, then the function $\psi(t) = t$ and the constant $\gamma = 1$ satisfy ($\star$). Therefore, Theorem 11.60 is a special case of this problem. ∎

Problem 11.7.2. *Show that every Banach space can be renormed so that in the new norm it fails the Daugavet property for rank-one operators.*

Solution: Let X be a Banach space. It suffices to establish that X can be renormed in such a way that its closed unit ball has strongly exposed points; see Theorem 11.63.

To see that this is possible, take any x_0 that does not belong to the original closed unit ball U and then renorm X by taking for the new closed unit ball V the closure of the convex circled hull of $U \cup \{x_0\}$. Note that

$$V = \bigl\{\alpha u + \beta x_0\colon\ \alpha \geq 0 \ \text{and}\ \alpha + |\beta| \leq 1\bigr\}\,.$$

Clearly, $U \subseteq V \subseteq \bigl(1 + \|x_0\|\bigr)U$. This implies that the Minkowski functional of V defines an equivalent norm with closed unit ball V. To finish the proof, we shall establish that x_0 is a strongly exposed point of V.

Since $x_0 \notin U$, it follows that there exists some $\phi \in X^*$ with $\|\phi\| = 1$ such that $\phi(x_0) > 1$. Now assume that a sequence $\{\alpha_n u_n + \beta_n x_0\} \subseteq V$ satisfies

$$\phi(\alpha_n u_n + \beta_n x_0) \to \phi(x_0)\,. \tag{$\dagger$}$$

We must show that $\|\alpha_n u_n + \beta_n x_0 - x_0\| \to 0$. For this, it suffices to establish that $\alpha_n \to 0$ and $\beta_n \to 1$. In view of $\alpha_n \geq 0$, the inequality $\alpha_n + |\beta_n| \leq 1$ and ($\dagger$), it is enough to prove that $|\beta_n| \to 1$.

If $|\beta_n| \not\to 1$, then by passing to an appropriate subsequence, we can assume that $|\beta_n| \to \beta < 1$. Now note that for each n we have

$$\big|\phi(\alpha_n u_n + \beta_n x_0)\big| \le \alpha_n \big|\phi(u_n)\big| + |\beta_n|\phi(x_0) \le 1 - |\beta_n| + |\beta_n|\phi(x_0)\,.$$

Taking limits and using (†) once more, we see that

$$\phi(x_0) \le 1 - \beta + \beta\phi(x_0) < (1-\beta)\phi(x_0) + \beta\phi(x_0) = \phi(x_0)\,,$$

which is impossible. This contradiction completes the solution. ∎

Bibliography

1. Y. A. Abramovich, Some theorems on normed lattices, *Vestnik Leningr. Univ. Mat. Meh. Astronom.* no. 13 (1971), 5–11. (English translation: *Vestnik Leningr. Univ. Math.* **4** (1977), 153–159.)
2. Y. A. Abramovich, C. D. Aliprantis, and O. Burkinshaw, Local quasinilpotence, cycles and invariant subspaces, in: *Interaction between Functional Analysis, Harmonic Analysis, and Probability* (N. Kalton, E. Saab, and S. Montgomery-Smith, eds.), Lecture Notes in Pure and Applied Mathematics, **175**, Marcel Dekker, New York, 1995, pp. 1–12.
3. Y. A. Abramovich, C. D. Aliprantis, and W. R. Zame, A representation theorem for Riesz spaces and its applications to economics, *Economic Theory* **5** (1995), 527–535.
4. C. D. Aliprantis and K. C. Border, *Infinite Dimensional Analysis: A Hitchhikers Guide*, 2nd Edition, Springer–Verlag, Heidelberg and New York, 1999.
5. C. D. Aliprantis and O. Burkinshaw, *Locally Solid Riesz Spaces*, Academic Press, New York and London, 1978.
6. C. D. Aliprantis and O. Burkinshaw, *Positive Operators*, Academic Press, New York and London, 1985.
7. C. D. Aliprantis and O. Burkinshaw, *Principles of Real Analysis*, 3rd Edition, Academic Press, New York and London, 1998.
8. C. D. Aliprantis and O. Burkinshaw, *Problems in Real Analysis*, 2nd Edition, Academic Press, New York and London, 1998.
9. C. D. Aliprantis, O. Burkinshaw, and P. Kranz, On lattice properties of the composition operator, *Manuscripta Math.* **36** (1981), 19–31.
10. T. Andô, Contractive projections in L_p-spaces, *Pacific J. Math.* **17** (1966), 391–405.
11. T. Andô, Banachverbände und positiver Projektionen, *Math. Z.* **109** (1969), 121–130.
12. S. I. Ansari, Essential disjointness and the Daugavet equation, *Houston J. Math.* **19** (1993), 587–601.
13. W. Arendt, Über das spektrum regularer operatoren, Ph.D. Dissertation, University of Tübingen, 1979.
14. N. Aronszajn and K. T. Smith, Invariant subspaces of completely continuous operators, *Ann. of Math.* **60** (1954), 345–350.

15. W. B. Arveson and J. Feldman, A note on invariant subspaces, *Michigan Math. J.* **15** (1968), 61–64.

16. A. Benedek and R. Panzone, The spaces with mixed norm, *Duke Math. J.* **28** (1961), 301–324.

17. A. R. Bernstein and A. Robinson, Solution of an invariant subspace problem of K. T. Smith and P. R. Halmos, *Pacific J. Math.* **16** (1966), 421–431.

18. C. Bessaga and A. Pelczynski, On bases and unconditional convergence of series in Banach spaces, *Studia Math.* **17** (1958), 151–164.

19. E. Bishop and R. R. Phelps, The support functionals of a convex set, in: V. Klee ed., *Convexity* (Proceedings of Symposia in Pure Mathematics, **7**, Amer. Math. Soc., Providence, RI, 1963), 27–35.

20. A. V. Bukhvalov, V. B. Korotkov, A. G. Kusraev, S. S. Kutateladze, and B. M. Makarov, *Vector Lattices and Integral Operators*, Nauka, Novosibirsk, 1992. (English translation edited by S. S. Kutateladze was published by Kluwer Academic Publishers, Dordrecht and Boston, 1996.)

21. V. Caselles, On irreducible operators on Banach lattices, *Indag. Math.* **48** (1986), 11–16.

22. J. A. Clarkson, Uniformly convex spaces, *Trans. Amer. Math. Soc.* **40** (1936), 396–414.

23. J. Diestel, H. Jarchow, and A. Tong, *Absolutely Summing Operators*, Cambridge Studies in Advanced Mathematics, **43**, Cambridge University Press, 1995.

24. J. Diestel and J. J. Uhl, Jr., *Vector Measures*, Math. Surveys, **15**, Amer. Math. Soc., Providence, Rhode Island, 1977.

25. P. G. Dodds and D. H. Fremlin, Compact operators in Banach lattices, *Israel J. Math.* **34** (1979), 287–320.

26. P. G. Dodds, C. B. Huijsmans, and B. de Pagter, Characterizations of conditional expectation-type operators, *Pacific J. Math.* **141** (1990), 55–77.

27. R. G. Douglas, *Banach Algebra Techniques in Operator Theory*, Academic Press, New York and London, 1972.

28. N. Dunford and J. T. Schwartz, *Linear Operators I*, Wiley (Interscience), New York, 1958.

29. Z. Ercan and S. Onal, Invariant subspaces for positive operators acting on a Banach space with Markushevich basis, *Positivity*, forthcoming.

30. V. Gantmacher, Über schwache totalstetige operatoren, *Mat. Sb. (N.S.)* **7** (49) (1940), 301–308.

31. S. A. Gershgorin, Über die Albrenzung Eigenwerte einer Matrix, *Izv. Akad. Nauk SSSR Ser. Fiz.-Mat.* **6** (1931), 749–754.

32. W. T. Gowers and B. Maurey, The unconditional sequence problem, *J. Amer. Math. Soc.* **6** (1993), 851–874.

33. W. T. Gowers and B. Maurey, Banach spaces with small spaces of operators, *Math. Ann.* **307** (1997), 541–568.

34. J. J. Grobler, Band irreducible operators, *Indag. Math.* **48** (1986), 405–409.

35. A. Grothendieck, Une characterisation vectorielle metrique des espaces L^1, *Canad. J. Math.* **7** (1955), 552–561.

36. A. Grothendieck, Produits tensoriels toplogiques et espaces nuclaires, *Mem. Amer. Math. Soc.*, **16**, 1965.

37. F. L. Hernández and B. Rodriguez-Salinas, On ℓ^p-complimented copies in Orlicz spaces II, *Israel J. Math.* **68** (1989), 27–55.

38. R. B. Honor, Density and transitivity results on ℓ^∞ and ℓ^1, *J. London Math. Soc.* **32** (1985), 521–527.

39. R. C. James, Weakly compact sets, *Trans. Amer. Math. Soc.* **113** (1964), 129–140.

40. V. M. Kadets, Some remarks concerning the Daugavet equation, *Quaestiones Math.* **19** (1996), 225–235.

41. V. M. Kadets and M. M. Popov, On the Liapunov convexity theorem with applications to sign-embeddings, *Ukrainian Math. J.* **44** (1992), 1192–1200.

42. V. M. Kadets and M. M. Popov, The Daugavet property for narrow operators in rich subspaces of $C[0,1]$ and $L_1[0,1]$, *St. Petersburg Math. J.* **8** (1997), 571–584.

43. V. M. Kadets, R. V. Shvidkoy, G. G. Sirotkin, and D. Werner, Banach spaces with the Daugavet property, *Trans. Amer. Math. Soc.* **352** (2000), 855–873.

44. G. K. Kalisch, On similarity, reducing manifolds, and unitary equivalences of certain Volterra operators, *Ann. of Math.* **66** (1957), 481–494.

45. L. V. Kantorovich and G. P. Akilov, *Functional Analysis*, Pergamon Press, Oxford and New York, 1982.

46. L. V. Kantorovich, B. Z. Vulikh, and A. G. Pinsker, *Functional Analysis in Partially Ordered Spaces*, Gostekhizdat, Moscow and Leningrad, 1950.

47. R. Khalil, The Daugavet equation in vector-valued function spaces, *Panamer. Math. J.* **6** (1996), 51–53.

48. J. Kim, The characterization of a lattice homomorphism, *Canadian J. Math.* **27** (1975), 172–175.

49. V. B. Korotkov, *Integral Operators*, Nauka, Novosibirsk, 1983.

50. C.-S. Lin, Generalized Daugavet equations and invertible operators on uniformly convex Banach spaces, *J. Math. Anal. Appl.* **197** (1996), 518–528.

51. J. Lindenstrauss and L. Tzafriri, On the complemented subspaces problem, *Israel J. Math.* **9** (1971), 263–269.

52. J. Lindenstrauss and L. Tzafriri, *Classical Banach Spaces I*, Springer–Verlag, Berlin and New York, 1977.

53. V. I. Lomonosov, A counterexample to the Bishop–Phelps theorem in complex spaces, *Israel J. Math.* **115** (2000), 25–28.

54. G. Ya. Lozanovsky, On a theorem of N. Dunford, *Izv. Vyssh. Uchebn. Zaved. Mat.* **8** (1974), 58–59. (Russian)

55. H. P. Lotz, Minimal and reflexive Banach lattices, *Trans. Amer. Math. Soc.* **211** (1975), 85–100.

56. P. Meyer-Nieberg, Quasitriangulierbare Operatoren und invariante Untervekrorräume stetiger linearer Operatoren, *Arch. Math. (Basel)* **22** (1971), 186–199.

57. D. P. Milman, On some criteria for the regularity of spaces of the type (B), *Dokl. Akad. Nauk SSSR, N. S.,* **20** (1938), 243–246. (In Russian.)

58. S-T. C. Moy, Characterization of conditional expectation as a transformation on function spaces, *Pacific J. Math.* **4** (1954), 47–63.

59. J. R. Munkres, *Topology*, Prentice-Hall, Englewood Cliffs, NJ, 1975.

60. H. Nakano, *Modern Spectral Theory*, Maruzen Co., Tokyo, 1950.

61. B. de Pagter, Irreducible compact operators, *Math. Z.* **192** (1986), 149–153.

62. C. Pearcy and N. Salinas, An invariant-subspace theorem, *Michigan Math. J.* **20** (1973), 21–31.

63. A. Pelczynski, On the impossibility of embedding of the space L in certain Banach spaces, *Coll. Math.* **8** (1961), 199–203.

64. B. J. Pettis, A proof that every uniformly convex space is reflexive, *Duke Math. J.* **5** (1939), 369–374.

65. A. M. Plichko and M. M. Popov, *Symmetric function spaces on atomless probability spaces*, Dissertationes Mathematicae, **306**, 1990.

66. C. J. Read, A solution to the invariant subspace problem on the space ℓ_1, *Bulletin London Math. Soc.* **17** (1985), 305–317.

67. C. J. Read, A short proof concerning the invariant subspace problem, *J. London Math. Soc.* (2) **34** (1986), 335–348.

68. C. J. Read, Quasinilpotent operators and the invariant subspace problem, *J. London Math. Soc.* (2) **56** (1997), 595–606.

69. H. H. Schaefer, Spekraleigenschaften positwer linearer Operatoren, *Math. Z.* **82** (1963), 303–313.

70. H. H. Schaefer, *Banach Lattices and Positive Operators*, Springer–Verlag, Berlin and New York, 1974.

71. H. H. Schaefer, On theorems of de Pagter and Andô–Krieger, *Math. Z.* **192** (1986), 155–157.

72. I. Singer, *Bases in Banach Spaces*, Vol. 1, Springer–Verlag, Berlin and New York, 1970.

73. G. G. Sirotkin, Compact-friendly multiplication operators on Banach function spaces, *J. Funct. Analysis*, forthcoming.

74. V. L. Šmulian, Sur la structure de la sphere unitaire dans l'espace de Banach, *Mat. Sb. (N.S.)* **9** (1941), 545–561.

75. A. Sobczyk, Projection of the space m on its subspace c_0, *Bull. Amer. Math. Soc.* **47** (1941), 938–947.

76. A. Spalsbury, Operators not positive with respect to any basis, *Quaestiones Math.* **23** (2000), 489–494.

77. A. Taylor and D. C. Lay, *Introduction to Functional Analysis*, R. E. Krieger, Malabar, Florida, 1986.

78. T. Terzioglu, A characterization of compact linear mappings, *Arch. Math. (Basel)* **22** (1971), 76–78.

79. V. G. Troitsky, On the modulus of C. J. Read's operator, *Positivity* **3** (1998), 257–264.

80. L. Tzafriri, Remarks on contractive projections in L_p-spaces, *Israel J. Math.* **7** (1969), 9–15.

81. L. Tzafriri, An isomorphic characterization of L_p- and c_0-spaces, II, *Michigan Math. J.* **18** (1971), 21–31.

82. L. Weis and D. Werner, The Daugavet equation for operators not fixing a copy of $C(S)$, *J. Operator Theory* **39** (1998), 89–98.

83. A. C. Zaanen, *Riesz Spaces II*, North-Holland, Amsterdam, 1983.

Index

AL-space, 87, 90
AM-compact operator, 313
AM-space, 87, 90
  represented as $L_\infty(\pi)$, 152
  with unit, 87
abstract integral operator, 154
additive semigroup of operators, 319
adjoint of resolvent, 190
adjoint of strictly singular operator, 142
adjoint operator, 66
algebra of operators, 304
  non-transitive, 304, 305, 331
  transitive, 306
  unital, 304
algebraic complement of subspace, 136
algebraic homomorphism, 129
analytic function, 50, 205
annihilator, 19
  of a set, 257
  of a vector subspace, 19, 257
antisymmetry property, 20
approximate eigenvalue, 198, 222
approximate point spectrum, 198, 232
approximation of vectors in E_u, 104
approximation property, 122
Archimedean Riesz space, 22, 25, 72, 108, 212
ascent of operator, 68, 137, 210
atom, 72, 74, 352
  in $C(\Omega)$-space, 352
atomic Banach lattice, 282
atomic measure space, 154
atomless Riesz space, 73
automatic continuity, 132

backward shift, 69, 286, 311
Banach function space, 154
Banach lattice, 332
  atomic, 282
  complex, 96
  with mixed norm, 163
  with order continuous norm, 71, 76, 90, 126, 329
Banach space
  direct sum, 7, 195
  quotient, 18
  reflexive, 11, 40, 341, 367
  renormed, 373
  separable, 362
  uniformly convex, 11, 338, 341, 342, 344
  uniformly smooth, 338, 342–344
band, 24, 72, 75, 76, 106, 159, 265, 291
  generated by a set, 24
  generated by a vector, 24
  generated by an ideal, 24
  principal, 110
  projection, 29, 321
band irreducibility, 290
band irreducible operator, 273, 276, 281
  quasinilpotent, 292
band projection, 108
Bartle–Dunford–Schwartz theorem, 363
basic sequence, 39
basis, 31, 329
  in $C[0,1]$, 37
  in $L_p[0,1]$, 34
  in a B-space, 31, 33
  orthogonal, 244
  orthonornal, 244
  unconditional, 36, 40, 329
biorthogonal sequence, 40

Bochner integrable function, 368
Borel measure, 94, 126, 231, 362
bounded below operator, 63

Calkin algebra, 241
Carleman operator, 147
carrier, 31
  of functional, 156
  of ideal, 31
Cayley–Hamilton theorem, 250, 258
center, 105
  of $L_p(\mu)$, 107, 157
  of a Banach function space, 157
  of Banach lattice, 105, 106
central operator, 106, 274
Cesaro operator, 195
chain of open sets, 51
characteristic polynomial, 245, 250, 252, 255, 256, 268
Clarkson's inequality, 345
clopen set, 26
closed ideal, 112
closest point to a set, 11
cofinal set, 51
collection of operators, 317
  finitely quasinilpotent, 318
  locally quasinilpotent, 317
commutant of an operator, 304
commutator of two operators, 209
commuting Krein operators, 295
commuting operators, 191, 296
compact operator, 78, 82
compact-friendly operator, 321
compactly dominated operator, 289, 292
complement
  algebraic of subspace, 136
  of closed subspace, 121, 359
complemented subspace, 121, 359
complete measure, 52
completely continuous function, 332
completeness of normed Riesz spaces, 28
completion
  Dedekind, 30, 93
  Maeda–Ogasawara–Vulikh, 108
  norm, 112
complex Banach lattice, 96, 243
complex ideal, 99
complex lattice homomorphism, 130
complexification, 3, 96, 97
  of $\mathbb{R}^n$, 97
  of $C_{\mathbb{R}}(\Omega)$, 97
  of a B-lattice, 96
  of a B-space, 227
  of a real normed space, 4
  of a real vector space, 3, 5
component, 26, 77
composition of analytic functions, 206
composition operator, 83, 128, 303
conditional expectation operator, 169, 170
cone, 15, 20, 21, 329
  generated by a Schauder basis, 329
  generating, 21, 183
  in a B-space, 329
  in a normed Riesz space, 27
conjugate, 98
  of complex number, 98
  of matrix, 98
continuity of positive operators, 27, 331
contour, 205
  Jordan, 205
contraction, 64, 218
contractive projection, 170, 175
convergence
  $\star$, 55
  in measure, 57
  notions in L_0, 158
  order, 22, 53
  ultrafilter, 45
convex hull of components, 77
convolution of kernels, 162
countable spectrum, 197
countable sup property, 54
curve in a topological space, 50
cyclic vector, 260

Daugavet equation, 335, 363, 365
Daugavet property, 352, 353, 356, 359, 367
Dedekind complete Riesz space, 26, 30, 54, 105, 108, 155, 233
Dedekind completion, 30, 93
derivatives of the resolvent function, 189
descent of operator, 68, 137, 210
  finite, 138
diagonal matrix, 109
diagonal projection, 109
diagonalizable matrix, 246
diagonalizing matrix, 246
direct sum Banach space, 7, 195
direct sum of operators, 306
direct sum operator, 195
Dirichlet's theorem, 186
discrete vector, 72
disjoint linear functionals, 74
disjoint sequence, 212
disjoint vectors, 74, 125
disjointly strictly singular operator, 144
disjointness of I from integral operators, 151
domination of operators, 100, 290, 310
double adjoint, 303
double power bounded operator, 195
dual of ℓ_p-sum of B-spaces, 224
duality of AM- and AL-spaces, 88
Dunford's theorem, 160
Dunford–Pettis operator, 316

dyadic point, 38

$E_n^\sim$, order continuous dual of E, 31
eigenspace, 283
eigenvalue, 201, 206, 240, 262
  approximate, 198, 222
eigenvector of matrix, 262, 267
equivalent norms, 97
essential singularity, 210
essential spectrum, 237
  of forward shift, 238
essentially nilpotent operator, 239
expanding operator, 273
exposed point of a set, 367
extension of additive function, 21
extension of linear functional, 10

factorization of compact operators, 82
filter of sets, 44
Finite Dimensional Separation Theorem, 16
finite-rank operator, 79, 119, 123, 127, 215, 355
finitely quasinilpotent collection, 318
forward shift, 69, 238, 311
Fréchet differentiability of the norm, 347
Fredholm alternative, 138
Fredholm operator, 134
Frobenius' theorem, 262
function
  linear, 184
  analytic, 50, 205
  Bochner integrable, 368
  completely continuous, 332
  harmonic, 186
  matrix-valued, 267
  piecewise linear, 184
  Rademacher, 91, 370
  retraction, 13
functional
  multiplicative, 221
  supporting set at a point, 347
functional calculus, 205

Gantmacher's theorem, 80
Gateaux differentiability of the norm, 347
generalized Harris operator, 323
generating cone, 21, 183
Gershgorin's theorem, 263
Gribanov's theorem, 149

Haar system, 34
harmonic function, 186
Harris operator, 293, 323
hereditarily indecomposable B-space, 143
Hermitian matrix, 243, 246, 249
Hilbert space, 102
Hilbert–Schmidt operator, 153

homomorphism
  algebraic, 129
  lattice, 74, 105, 117, 129, 227, 264
hyperinvariant closed subspace, 283

$\mathcal{I}(X,Y)$, set of isomorphisms from X to Y, 67
ideal, 99, 265
  complex, 99
  invariant, 310
  null, 116, 311
  null of an ℓ-seminorm, 115
  null of functional, 31
  order dense, 58
  order dense in $L_0(\pi)$, 58
  principal, 90
  range of operator, 311
ideal irreducibility, 283
ideal irreducible operator, 100, 273, 276
identity operator, 160, 356
implication scheme, 273
independence
  of eigenvectors, 197
  of linear functionals, 119
index of operator, 134, 137
Index Theorem, 137
inequality
  Clarkson's, 345
  Jensen's, 173
  Khintchine's, 143
  triangle, 96
inner product preserving matrix, 244
integral operator, 94, 145, 215, 294
internal point of a set, 87
interval preserving operator, 46, 132, 231, 264
invariant ideal, 310
invariant measure, 276
invariant subspace, 5, 203, 299
inverse of operator, 1, 7, 193, 232, 235
invertible elements in a B-algebra, 237
invertible matrix, 263
invertible operator, 7, 193, 232, 235, 251
irreducibility
  band, 290
  ideal, 283
irreducible matrix, 265, 266
irreducible operator, 273
isolated point, 352, 361
isolated point of $\sigma(T)$, 197, 210, 239
isometry, 8, 353
  lattice, 225
  linear, 8, 142, 225
isomorphism
  lattice, 117

James' theorem, 341

Jensen's inequality, 173
joint continuity of composition, 10
joint spectral radius, 320
Jordan contour, 205

KB-space, 75
Kadets–Klee property, 344
Khintchine's inequality, 143
Krein operator, 266, 267, 271, 273, 293
  compact, 294
  with positive eigenvalues, 295
Krein–Rutman theorem, 219
Kronecker's theorem, 231

L-space, 114
$L_0(\mu)$, space of measurable functions, 54
$L_{p,1}$, B-lattice of mixed norms, 163
$L_{p,\infty}$, B-lattice of mixed norms, 163
ℓ_p-sum of B-spaces, 223
$\mathcal{L}(X,Y)$, the B-space of bounded operators, 2, 10
$\mathcal{L}_r(E,F)$, the B-lattice of regular operators, 28
Lat (T), algebra of T-invariant closed subspaces, 301
Laplace's equation, 186
lattice homomorphism, 46, 74, 105, 117, 129, 132, 227, 229
  complex, 130
  order continuous, 133
  spectrum, 230
lattice isometry, 225
lattice isomorphism, 117
lattice operations
  in $\mathcal{L}_r(E,F)$, 26
  weakly σ-continuous, 91
lattice seminorm, 115
lattice-subspace, 183
  finite dimensional, 183
law
  Parallelogram, 345
limit functional, 134
linear function, 184
linear functional
  limit, 134
  multiplicative, 131, 142
linear isometry, 8, 114, 142, 225
linearly independent functionals, 119
locally quasinilpotent collection, 317
locally quasinilpotent operator, 285
Lomonosov operator, 308
Lomonosov's Invariant Subspace Theorem, 307

M-space, 113
Maeda–Ogasawara–Vulikh Representation Theorem, 108, 110
majorizing vector subspace, 183
Markov matrix, 95, 263, 269
Markov operator, 95, 130, 276, 308
Markov projection, 177
matrix
  diagonal, 109
  diagonalizable, 246
  diagonalizing, 246
  Hermitian, 243, 246, 249
  inner product preserving, 244
  invertible, 263
  irreducible, 265, 266
  Markov, 95, 263, 269
  nilpotent, 249, 265
  non-negative, 262
  norm preserving, 244
  positive, 262
  positive semidefinite, 248
  primitive, 271
  quasinilpotent, 249
  stochastic, 95
  strongly positive, 262
  unitary, 244, 246
  upper triangularizable, 251
  upper triangularizing, 251
matrix representing an operator, 6, 255
matrix-valued function, 267
maximum modulus principle, 187
measure, 107, 126
  atomic, 154
  Borel, 94, 126, 362
  complete, 52
  invariant, 276
  of bounded variation, 126
  on a semiring, 59
  separable, 148
measure convergence, 57
measure of non-compactness, 240
  of operator, 241
  of set, 240
minimal extension, 155
minimal polynomial, 253
mixed norm, 163
Miyajima projection, 183
  continuous, 185
modulus, 123, 262
  of finite-rank operator, 123
  of integral operator, 94
  of matrix, 262
  of regular operator, 74
Morera's theorem, 48
multiplication operator, 125, 127, 303, 307
  weakly compact, 220
multiplicative linear functional, 131, 142, 221
multiplicative semigroup, 319
multiplicity of eigenvalues, 252

Nakano's theorem, 156
narrow operator, 369
nearest point mapping, 11
nearest point to a set, 11
Neumann series, 191
nilpotent matrix, 249, 265
nilpotent operator, 70, 195
non-negative matrix, 262
non-transitive algebra, 304, 305, 331
norm, 1
  Fréchet differentiable, 347
  Gateaux differentiable, 347
  mixed, 163
  of integral operator, 94
  of operator, 1
  quotient, 18, 112
  uniformly convex, 11
  uniqueness in a B-lattice, 28
norm completion, 112, 114
  of AL-space, 114
  of AM-space, 113
norm extension, 233
norm of integral operator, 94
norm preserving matrix, 244
norm totally bounded set, 81
norming subspace, 49
norms on $\mathbb{C}^n$, 98
nuclear operator, 124
null ideal, 31, 65, 311
  of an ℓ-seminorm, 115
  of functional, 116

one-to-one operator, 7, 251
operator
  AM-compact, 313
  abstract integral, 154
  adjoint, 66
  band irreducible, 273, 276, 281
  bounded below, 63
  Carleman, 147
  central, 106, 274
  Cesaro, 195
  compact, 78, 82
  compact-friendly, 320, 321
  compactly dominated, 289, 292
  composition, 83, 128, 303
  conditional expectation, 169, 170
  contraction, 64
  direct sum, 195
  disjointly strictly singular, 144
  double power bounded, 195
  Dunford–Pettis, 316
  essentially nilpotent, 239
  expanding, 273
  finite-rank, 79, 119, 123, 127, 215, 355
  Fredholm, 134
  generalized Harris, 323
  Harris, 293, 323
  Hilbert–Schmidt, 153
  ideal irreducible, 100, 273, 276
  identity, 151, 160, 356
  integral, 94, 145, 215, 294
  interval preserving, 46, 132, 231, 264
  invertible, 7, 193, 232, 235, 251
  irreducible, 273
  isometry, 8
  Krein, 266, 267, 271, 273, 293
  locally quasinilpotent, 285
  locally quasinilpotent but not quasinilpotent, 285
  Lomonosov, 308
  Markov, 95, 130, 276, 308
  multiplication, 125, 127, 303, 307
  narrow, 369, 372
  nilpotent, 70, 195
  non-strictly singular, 143
  nuclear, 124
  one-to-one, 7, 251
  order continuous, 25, 202
  polynomially compact, 307
  power compact, 219, 239
  principal ideal preserving, 106
  projection, 176
  quasinilpotent, 192, 195
  rank-one, 127, 354, 365
  regular, 93
  shift, 219, 306
  strictly positive, 127, 311
  strictly singular, 139, 220, 240
  strong Krein, 273
  strongly expanding, 274, 275, 281
  surjective, 7, 63, 226, 251
  symmetric, 328, 339
  Volterra, 216, 219, 331
  weakly compact, 78, 80, 84, 220, 367
  weakly expanding, 274
  with closed range, 65–67, 139
  with countable spectrum, 197
  with separable range, 81, 362
operator domination, 100, 310
operator norm, 1
orbit of vector, 260
order closed set, 23
order complete lattice of invariant subspaces, 301
order complete Riesz space, 22
order completeness of $L_0(\mu)$, 54
order continuity of the norm
  in an AL-space, 90
order continuous dual, 31
  of L_∞, 157
order continuous operator, 25, 202
order convergence, 22, 53
order dense ideal, 58

order dense Riesz subspace, 54
order interval, 92
  weakly compact, 92
order spectrum, 232
order unit, 87, 125
ordered vector space
  Archimedean, 21
orthogonal basis, 244
orthogonal complement, 247
orthomorphism, 105
orthonornal basis, 244

Parallelogram Law, 345
partition of set, 171
Perron–Frobenius theorem, 268, 270
piecewise linear function, 184
point
  closest to a set, 11
  dyadic, 38
  exposed, 367
  internal of set, 87
  isolated, 352, 361
  isolated of $\sigma(T)$, 239
  nearest to a set, 11
  quasi-interior, 125, 126
  strongly exposed, 367
  support of a set, 14
Poisson's formula, 186
pole of the resolvent, 210
polynomial
  characteristic, 245, 250, 252, 255, 256, 268
  minimal, 253
polynomially compact operator, 307
positive matrix, 262
positive operator
  on a B-space, 329
positive projection, 169, 180
positive semidefinite matrix, 248
power compact operator, 219, 239
primitive matrix, 271
principal band, 110, 159
principal ideal, 90
principal ideal preserving operator, 106
product rule for derivatives, 48
projection, 29, 121, 169, 176, 179, 199
  band, 108, 110
  contractive, 170, 175
  diagonal, 109
  finite-rank operator, 121
  Markov, 177
  orthogonal, 179
  positive, 180
  spectral, 211, 239
projection band, 321
property
  antisymmetry, 20
  Daugavet, 352
  Kadets–Klee, 344
  reflexivity, 20
  Schur, 84, 333
  transitivity, 20
pseudoinverse of operator, 139

quasi-interior point, 125, 126
quasinilpotent matrix, 249
quasinilpotent operator, 192, 195
quotient Banach space, 18
quotient map, 18, 111
quotient norm, 18, 112

Rademacher functions, 91, 370
range ideal of operator, 311
range of operator, 66, 139
rank of a matrix, 258
rank-one operator, 127, 354, 365
reducing subspace, 207
reflexive B-space, 11, 40, 341, 367
reflexivity property, 20
regular operator, 93
representation
  Kakutani–Bohnenblust–Krein, 130
  Maeda–Ogasawara–Vulikh, 108
  of AM-space as $L_\infty(\pi)$, 152
  of finite-rank operator, 120
  of operator by matrix, 255
resolvent
  of positive operator, 201
resolvent identity
  second, 189
retract of a space, 13
retraction, 13
Riemann–Lebesgue lemma, 91
Riesz space, 133, 155
  Archimedean, 22, 24, 52, 72, 108, 212
  atomless, 73
  Dedekind complete, 26, 30, 54, 105, 108, 155, 233
  order complete, 22
Riesz subspace, 110, 112
  order dense, 54
Riesz's lemma, 135
Riesz–Kantorovich formula, 26, 125

$\star$-convergence, 55
S-invariant subspace, 318
Schauder system, 37
Schur property, 84, 333
second resolvent identity, 189
semigroup of operators, 319
  additive, 319
  multiplicative, 319
semiring of sets, 59
separable B-space, 362
separable measure, 148

separable range, 81, 362
separation of points, 157, 183
separation of sets, 16
Separation Theorem
  Finite Dimensional, 16
sequence
  basic, 39
  biorthogonal, 40
  disjoint, 212
  weakly unconditionally Cauchy, 360
series
  Neumann, 191
  unconditionally convergent, 41, 361
  weakly unconditionally Cauchy, 360
set
  clopen, 26
  cofinal, 51
  solid, 23
  spectral, 209, 211
  totally bounded, 169
  weakly compact, 84
shift, 69
shift operator, 219, 306
  backward, 69, 286, 311
  forward, 69, 238, 311
similar matrices, 244, 255
similar operators, 199
similarity invariance, 121
singularity
  essential, 210
slice of a set, 365
solid set, 23
space
  L, 114
  M, 113
  Hilbert, 102
special points of the spectrum, 197
Spectral Mapping Theorem
  for polynomials, 205
spectral projection, 211, 239
spectral radius, 190, 192, 283, 289, 290, 292, 294
  joint, 320
  of sum of operators, 202
  Rota–Strang, 320
spectral set, 209, 211
spectrum, 189, 220
  approximate, 198, 232
  essential, 237
  of a direct sum operator, 195
  of compact operator, 215
  of composition of operators, 196, 201
  of lattice homomorphism, 230
  of multiplication operator, 194
  of operator, 189
  of order bounded operator, 236
  of projection, 199
  of similar operators, 200
  order, 232
spectrum of order bounded operator, 236
stochastic matrix, 95
strictly positive operator, 127, 311
strictly singular operator, 139, 220, 240
strong Krein operator, 273
strong unit, 87
strongly expanding operator, 274, 275, 281
strongly exposed point, 367
strongly positive matrix, 262
subgroup of the unit circle, 230, 231
  dense, 231
  finite, 230
subspace
  complemented, 121
  invariant, 5, 203
  norming, 49
  reducing, 207
subspace invariant under a collection, 318
subspace invariant under an operator, 5
subspace of an ultrapower, 46
sum of B-spaces, 223
sum of narrow operators, 370, 372
sum of unconditionally convergent series, 42
sum of vector spaces closed, 111
support point of a set, 14
supporting functional, 347
surjective operator, 7, 63, 226, 251
symmetric operator, 328, 339
system
  Haar, 34
  Schauder, 37

totally bounded set, 76, 81, 169
trace, 120
  of finite-rank operator, 120
  of matrix, 244
transformer, 132
transitive algebra, 306
transitivity property, 20
transpose of matrix, 259
triangle inequality, 96

$\mathcal{U}$-limit of sequence, 45
ultrafilter, 44, 229
  free, 229
ultrafilter convergence, 45
ultrapower of a B-space, 44
unbounded component of $\rho(T)$, 202
unconditional basis, 36, 40, 329
unconditionally convergent series, 41, 361
uniformly convex B-space, 11, 341, 342, 344
uniformly convex norm, 11
uniformly smooth B-space, 342–344
uniqueness of kernel, 145
unit, 87, 330

order, 87, 125
strong, 87
weak, 25, 125, 126, 156, 169, 275, 330
unital algebra, 127
unital algebra of operators, 304
unitary matrix, 244, 246
upper triangularizable matrix, 251
upper triangularizing matrix, 251

vector
cyclic, 260
vector sublattice, 178
closed, 178
generated by a vector space, 182
Volterra operator, 216, 219, 331

weak unit, 25, 125, 126, 169, 275, 330
in $E_{\rm n}^*$, 156
weakly compact interval, 92
weakly compact operator, 78, 80, 84, 220, 367
weakly compact set, 84
weakly expanding operator, 274
weakly sequentially continuous lattice operations, 91
weakly unconditionally Cauchy sequence, 360
weakly unconditionally Cauchy series, 360

Titles in This Series

51 **Y. A. Abramovich and C. D. Aliprantis,** Problems in operator theory, 2002
50 **Y. A. Abramovich and C. D. Aliprantis,** An invitation to operator theory, 2002
49 **John R. Harper,** Secondary cohomology operations, 2002
48 **Y. Eliashberg and N. Mishachev,** Introduction to the h-principle, 2002
47 **A. Yu. Kitaev, A. H. Shen, and M. N. Vyalyi,** Classical and quantum computation, 2002
46 **Joseph L. Taylor,** Several complex variables with connections to algebraic geometry and Lie groups, 2002
45 **Inder K. Rana,** An introduction to measure and integration, second edition, 2002
44 **Jim Agler and John E. McCarthy,** Pick interpolation and Hilbert function spaces, 2002
43 **N. V. Krylov,** Introduction to the theory of random processes, 2002
42 **Jin Hong and Seok-Jin Kang,** Introduction to quantum groups and crystal bases, 2002
41 **Georgi V. Smirnov,** Introduction to the theory of differential inclusions, 2002
40 **Robert E. Greene and Steven G. Krantz,** Function theory of one complex variable, 2002
39 **Larry C. Grove,** Classical groups and geometric algebra, 2002
38 **Elton P. Hsu,** Stochastic analysis on manifolds, 2002
37 **Hershel M. Farkas and Irwin Kra,** Theta constants, Riemann surfaces and the modular group, 2001
36 **Martin Schechter,** Principles of functional analysis, second edition, 2002
35 **James F. Davis and Paul Kirk,** Lecture notes in algebraic topology, 2001
34 **Sigurdur Helgason,** Differential geometry, Lie groups, and symmetric spaces, 2001
33 **Dmitri Burago, Yuri Burago, and Sergei Ivanov,** A course in metric geometry, 2001
32 **Robert G. Bartle,** A modern theory of integration, 2001
31 **Ralf Korn and Elke Korn,** Option pricing and portfolio optimization: Modern methods of financial mathematics, 2001
30 **J. C. McConnell and J. C. Robson,** Noncommutative Noetherian rings, 2001
29 **Javier Duoandikoetxea,** Fourier analysis, 2001
28 **Liviu I. Nicolaescu,** Notes on Seiberg-Witten theory, 2000
27 **Thierry Aubin,** A course in differential geometry, 2001
26 **Rolf Berndt,** An introduction to symplectic geometry, 2001
25 **Thomas Friedrich,** Dirac operators in Riemannian geometry, 2000
24 **Helmut Koch,** Number theory: Algebraic numbers and functions, 2000
23 **Alberto Candel and Lawrence Conlon,** Foliations I, 2000
22 **Günter R. Krause and Thomas H. Lenagan,** Growth of algebras and Gelfand-Kirillov dimension, 2000
21 **John B. Conway,** A course in operator theory, 2000
20 **Robert E. Gompf and András I. Stipsicz,** 4-manifolds and Kirby calculus, 1999
19 **Lawrence C. Evans,** Partial differential equations, 1998
18 **Winfried Just and Martin Weese,** Discovering modern set theory. II: Set-theoretic tools for every mathematician, 1997
17 **Henryk Iwaniec,** Topics in classical automorphic forms, 1997
16 **Richard V. Kadison and John R. Ringrose,** Fundamentals of the theory of operator algebras. Volume II: Advanced theory, 1997
15 **Richard V. Kadison and John R. Ringrose,** Fundamentals of the theory of operator algebras. Volume I: Elementary theory, 1997
14 **Elliott H. Lieb and Michael Loss,** Analysis, 1997
13 **Paul C. Shields,** The ergodic theory of discrete sample paths, 1996

12 **N. V. Krylov,** Lectures on elliptic and parabolic equations in Hölder spaces, 1996
11 **Jacques Dixmier,** Enveloping algebras, 1996 Printing
10 **Barry Simon,** Representations of finite and compact groups, 1996
9 **Dino Lorenzini,** An invitation to arithmetic geometry, 1996
8 **Winfried Just and Martin Weese,** Discovering modern set theory. I: The basics, 1996
7 **Gerald J. Janusz,** Algebraic number fields, second edition, 1996
6 **Jens Carsten Jantzen,** Lectures on quantum groups, 1996
5 **Rick Miranda,** Algebraic curves and Riemann surfaces, 1995
4 **Russell A. Gordon,** The integrals of Lebesgue, Denjoy, Perron, and Henstock, 1994
3 **William W. Adams and Philippe Loustaunau,** An introduction to Gröbner bases, 1994
2 **Jack Graver, Brigitte Servatius, and Herman Servatius,** Combinatorial rigidity, 1993
1 **Ethan Akin,** The general topology of dynamical systems, 1993